T0183153

Lecture Notes in Computer Science 10485

More information about this series at http://www.springer.com/series/7412

Sebastiano Battiato · Giovanni Gallo
Raimondo Schettini · Filippo Stanco (Eds.)

Image Analysis and Processing - ICIAP 2017

19th International Conference
Catania, Italy, September 11–15, 2017
Proceedings, Part II

Springer

Editors
Sebastiano Battiato
University of Catania
Catania
Italy

Giovanni Gallo
University of Catania
Catania
Italy

Raimondo Schettini
University of Milano-Bicocca
Milan
Italy

Filippo Stanco
University of Catania
Catania
Italy

ISSN 0302-9743 ISSN 1611-3349 (electronic)
Lecture Notes in Computer Science
ISBN 978-3-319-68547-2 ISBN 978-3-319-68548-9 (eBook)
https://doi.org/10.1007/978-3-319-68548-9

Library of Congress Control Number: 2017956081

LNCS Sublibrary: SL6 – Image Processing, Computer Vision, Pattern Recognition, and Graphics

Printed on acid-free paper

This Springer imprint is published by Springer Nature
The registered company is Springer International Publishing AG
The registered company address is: Gewerbestrasse 11, 6330 Cham, Switzerland

Preface

The 2017 International Conference on Image Analysis and Processing, ICIAP 2017, was the 19th edition of a series of conferences promoted biennaly by the Italian Member Society (GIRPR) of the International Association for Pattern Recognition (IAPR). The conference traditionally covers both classic and the most recent trends in image processing, computer vision, and pattern recognition, addressing both theoretical and applicative aspects.

ICIAP 2017 (http://www.iciap2017.com) was held in Catania, during September 11–15, 2017, in the Benedictine Monastery of San Nicolò l'Arena. The monastery is a UNESCO World Heritage Site and today it hosts the Department of Humanities (DISUM) of the University of Catania. The conference was organized by Image Processing Laboratory, Department of Mathematics and Computer Science (DMI) of the University of Catania. Moreover, ICIAP 2017 was endorsed by the International Association for Pattern Recognition (IAPR), the Italian Member Society of IAPR (GIRPR), and received the institutional support of the University of Catania. Notable sponsorship came from several industrial partners such as STMicroelectronics, Micron, and iCTLab.

ICIAP is traditionally a venue for discussing image processing and analysis, pattern recognition, computer vision, and machine learning, from both theoretical and applicative perspectives, promoting connections and synergies among senior scholars and students, universities, research institutes, and companies. ICIAP 2017 followed this trend, and the program was subdivided into eight main topics, covering a broad range of scientific areas, which were managed by two area chairs per each topic. They were: Biomedical and Assistive Technology; Image Analysis, Detection and Recognition; Information Forensics and Security; Imaging for Cultural Heritage and Archaeology; Multimedia; Multiview Geometry and 3D Computer Vision; Pattern Recognition and Machine Learning; Video Analysis and Understanding.

Moreover, we hosted several prominent companies as well as start-ups to show their activities while assessing them with respect to the cutting-edge research in the respective areas.

ICIAP 2017 received 229 paper submissions coming from all over the world, including Australia, Austria, Brazil, Canada, China, Colombia, Cuba, France, Germany, Hungary, Iran, Ireland, Italy, Israel, Japan, Korea, Kuwait, Malaysia, Mexico, Poland, Portugal, Romania, Russia, Saudi Arabia, Serbia, Spain, South Africa, The Netherlands, Tunisia, Turkey, UK, USA. The paper review process was managed by the program chairs with the invaluable support of 15 area chairs, together with the Program Committee and a number of additional reviewers. The peer-review selection process was carried out by three distinct reviewers in most of the cases. This ultimately led to the selection of 138 high-quality manuscripts, 23 oral presentations, and 115 interactive papers/posters, with an overall acceptance rate of about 60%

(about 10% for oral presentations). The ICIAP 2017 proceedings are published as volumes of the *Lecture Notes in Computer Science* (LNCS) series by Springer.

The program also included five invited talks by distinguished scientists in computer vision pattern recognition and image analysis. We enjoyed the plenary lectures of Daniel Cremers, Technische Universität München, Irfan Essa, Georgia Institute of Technology, Fernando Peréz-Gonzalez, University of Vigo, Nicu Sebe, University of Trento, Roberto Scopigno, ISTI-CNR, and Alain Tremeau, University Jean Monnet, who addressed very interesting and recent research approaches and paradigms such as deep learning and semantic scene understanding in computer vision, multimedia forensics, and applications in the field of color retrieval and management and cultural heritage.

While the main conference was held during September 13–15, 2017, ICIAP 2017 also included five tutorials and seven workshops, held on Monday, September 11, and Tuesday, September 12, 2017, on a variety of topics.

The organized tutorials were: "Virtual Cell Imaging (Methods and Principles)" by David Svoboda; "Image Tag Assignment, Refinement, and Retrieval" by Xirong Li, Tiberio Uricchio, Lamberto Ballan, Marco Bertini, Cees Snoek, Alberto Del Bimbo; "Active Vision and Human Robot Collaboration" by Dimitri Ognibene, Fiora Pirri, Guido De Croon, Lucas Paletta, Mario Ceresa, Manuela Chessa, Fabio Solari; "Humans Through the Eyes of a Robot: How Human Social Cognition Could Shape Computer Vision" by Nicoletta Noceti and Alessandra Sciutti.

There was a special session, "Imaging Solutions for Improving the Quality of Life (I-LIFE'17)," organized by Dan Popescu and Loretta Ichim with eight interesting works selected by the organizers.

ICIAP 2017 also hosted seven half- or full-day satellite workshops: the "First International Workshop on Brain-Inspired Computer Vision (WBICV 2017)" organized by George Azzopardi, Laura Fernández-Robles, Antonio Rodríguez-Sánchez; "Third International Workshop on Multimedia Assisted Dietary Management (MADiMa 2017)" organized by Stavroula Mougiakakou, Giovanni Maria Farinella, Keiji Yanai; "Social Signal Processing and Beyond (SSPandBE 2017)" organized by Mariella Dimiccoli, Petia Ivanova Radeva, Marco Cristani; "Natural Human–Computer Interaction and Ecological Perception in Immersive Virtual and Augmented Reality (NIVAR 2017)" organized by Manuela Chessa, Fabio Solari, Jean-Pierre Bresciani; "Automatic Affect Analysis and Synthesis" organized by Nadia Berthouze, Simone Bianco, Giuseppe Boccignone, Paolo Napoletano; "International Workshop on Biometrics As-a-Service: Cloud-Based Technology, Systems, and Applications" organized by Silvio Barra, Arcangelo Castiglione, Kim-Kwang Raymond Choo, Fabio Narducci; "Background Learning for Detection and Tracking from RGBD Videos" organized by Massimo Camplani, Lucia Maddalena, Luis Salgado. The workshop papers were all collected in a separate volume of the LNCS series by Springer.

We thank all the workshop organizers and tutorial speakers who made possible such an interesting pre-conference program.

Several awards were conferred during ICIAP 2017. The "Eduardo Caianiello" Award was attributed to the best paper authored or co-authored by at least one young researcher (PhD student, postdoc, or similar); a Best Paper Award was also assigned after a careful selection made by an ad hoc appointed committee provided by Springer and IAPR.

The organization and the success of ICIAP 2017 were made possible thanks to the cooperation of many people. First of all, special thanks should be given to the area chairs, who made a big effort for the selection of the papers, together with all the members of the Program Committee. Second, we would also like to thank the industrial, special session, publicity, publication, and Asia and US liaison chairs, who, operating in their respective fields, made this event a successful forum of science.

Special thanks go to the workshop and tutorial chairs as well as all workshop organizers and tutorial lecturers for making richer the conference program with notable satellite events. Last but not least, we are indebted to the local Organizing Committee, mainly colleagues from IPLAB, who dealt with almost every aspects of the conference.

Thanks very much indeed to all the aforementioned people, since without their support we would have not made it.

We hope that ICIAP 2017 met its aim to serve as a basis and inspiration for future ICIAP editions.

September 2017

Sebastiano Battiato
Giovanni Gallo
Raimondo Schettini
Filippo Stanco

Organization

General Chairs

Sebastiano Battiato	University of Catania, Italy
Giovanni Gallo	University of Catania, Italy

Program Chairs

Raimondo Schettini	University of Milano-Bicocca, Italy
Filippo Stanco	University of Catania, Italy

Workshop Chairs

Giovanni Maria Farinella	University of Catania, Italy
Marco Leo	ISASI- CNR Lecce, Italy

Tutorial Chairs

Gian Luca Marcialis	University of Cagliari, Italy
Giovanni Puglisi	University of Cagliari, Italy

Special Session Chairs

Carlo Sansone	University of Naples Federico II, Italy
Cesare Valenti	University of Palermo, Italy

Industrial and Demo Chairs

Cosimo Distante	ISASI – CNR Lecce, Italy
Michele Nappi	University of Salerno, Italy

Publicity Chairs

Antonino Furnari	University of Catania, Italy
Orazio Gambino	University of Palermo, Italy

Video Proceedings Chair

Concetto Spampinato	University of Catania, Italy

US Liaison Chair

Francisco Imai — Canon US Inc., USA

Asia Liaison Chair

Lei Zhang — The Polytechnic University, Hong Kong, SAR China

Steering Committee

Virginio Cantoni	University of Pavia, Italy
Luigi Pietro Cordella	University of Naples Federico II, Italy
Rita Cucchiara	University of Modena and Reggio Emilia, Italy
Alberto Del Bimbo	University of Florence, Italy
Marco Ferretti	University of Pavia, Italy
Fabio Roli	University of Cagliari, Italy
Gabriella Sanniti di Baja	ICAR-CNR, Italy

Area Chairs

Biomedical and Assistive Technology

Domenico Tegolo	University of Palermo, Italy
Sotirios Tsaftaris	University of Edinburgh, UK

Image Analysis, Detection and Recognition

Edoardo Ardizzone	University of Palermo, Italy
M. Emre Celebi	University of Central Arkansas, USA

Imaging for Cultural Heritage and Archaeology

Matteo Dellepiane	ISTI-CNR, Italy
Herbert Maschner	University of South Florida, USA

Information Forensics and Security

Stefano Tubaro	Polytechnic University of Milan, Italy
Zeno Geradts	University of Amsterdam, The Netherlands

Multimedia

Costantino Grana — University of Modena and Reggio Emilia, Italy

Multiview Geometry and 3D Computer Vision

Andrea Fusiello	Università degli Studi di Udine, Italy
David Fofi	University of Burgundy, France

Pattern Recognition and Machine Learning

Dima Damen	University of Bristol, UK
Vittorio Murino	Italian Institute of Technology (IIT), Italy

Video Analysis and Understanding

François Brémond	Inria, France
Andrea Cavallaro	Queen Mary University of London, UK

Invited Speakers

Daniel Cremers	Technische Universität München, Germany
Irfan Essa	Georgia Institute of Technology, USA
Fernando Peréz-Gonzalez	University of Vigo, Vigo, Spain
Nicu Sebe	University of Trento, Italy
Roberto Scopigno	ISTI-CNR, Italy
Alain Tremeau	Jean Monnet University, France

Program Committee

Lourdes Agapito	University College London, UK
Jake Aggarwal	University of Texas at Austin, USA
Irene Amerini	University of Florence, Italy
Djamila Aouada	University of Luxemburg, Luxemburg
Federica Arrigoni	University of Udine, Italy
Lamberto Ballan	University of Padova, Italy
Fabio Bellavia	University of Florence, Italy
Simone Bianco	University of Milan-Bicocca, Italy
Silvia Biasotti	CNR-IMATI, Italy
Manuele Bicego	University of Verona, Italy
Giulia Boato	University of Trento, Italy
Giuseppe Boccignone	University of Milan, Italy
Alex Bronstein	Israel Institute of Technology, Israel
Alfred Bruckstein	Israel Institute of Technology, Israel
Joachim Buhmann	ETH Zurich, Switzerland
Francesco Camastra	University of Naples Parthenope, Italy
Barbara Caputo	University of Rome La Sapienza, Italy
Modesto Castrillon-Santana	University of Las Palmas de Gran Canaria, Spain
Rama Chellappa	University of Maryland, USA
Aladine Chetouani	University of Orleans, France

Paolo Cignoni	CNR-ISTI, Italy
Gianluigi Ciocca	University of Milan-Bicocca, Italy
Carlo Colombo	University of Florence, Italy
Antonio Criminisi	University of Oxford, UK
Marco Cristani	University of Verona, Italy
Jin Dakai	National Institutes of Health, USA
Tanasi Davide	University of South Florida, USA
Maria De Marsico	University of Rome La Sapienza, Italy
Alessio Del Bue	Italian Institute of Technology, Italy
Cedric Demonceaux	Univ. Bourgogne Franche-Comte, France
Adrien Depeursinge	University of Applied Sciences Western Switzerland, Switzerland
Luigi Di Stefano	University of Bologna, Italy
Naveed Ejaz	Fraunhofer Institute of Integrated Circuits, Germany
Sabu Emmanuel	Kuwait University, Kuwait
Francisco Escolano	University of Alicante, Spain
Gianluca Foresti	University of Udine, Italy
Ana Fred	Technical University of Lisbon, Portugal
Carlo Gatta	Computer Vision Center, Spain
Andrea Giachetti	University of Verona, Italy
Giorgio Giacinto	University of Cagliari, Italy
Mehmet Gonen	School of Medicine, Koç University, Turkey
Marco Gori	University of Siena, Italy
Giorgio Grasso	University of Messina, Italy
Adlane Habed	University of Strasbourg, France
Edwin Hancock	University of York, UK
Anders Hast	Uppsala University, Sweden
Loretta Ichim	Polytechnic University of Bucharest, Romania
Sebastiano Impedovo	University of Bari, Italy
Ignazio Infantino	National Research Council of Italy, Italy
Federico Iuricich	University of Maryland, USA
Richard Jiang	Northumbria University, UK
Michal Kawulok	Silesian University of Technology, Poland
Marco La Cascia	University of Palermo, Italy
Michela Lecca	Fondazione Bruno Kessler, Italy
Ales Leonardis	University of Birmingham, UK
Salvatore Livatino	University of Hertfordshire, UK
Giosuè Lo Bosco	University of Palermo, Italy
Marco Loog	Delft University of Technology, The Netherlands
Carmen Alina Lupascu	Italian National Research Council, Italy
Lucia Maddalena	ICAR-CNR, Italy
Luca Magri	University of Verona, Italy
Simone Marinai	University of Florence, Italy
Eleonora Maset	University of Udine, Italy
Pier Luigi Mazzeo	Italian Research Council, Italy
Christian Micheloni	University of Udine, Italy

Paolo Napoletano	University of Milan-Bicocca, Italy
Ram Nevatia	University of Southern California, USA
Francesca Odone	University of Genova, Italy
Pietro Pala	University of Florence, Italy
Alfredo Petrosino	University of Naples Parthenope, Italy
Wilfried Philips	University of Gent, Belgium
Massimo Piccardi	University of Technology Sydney, Australia
Ruggero Pintus	Center for Advanced Studies, Research, and Development in Sardinia, Italy
Alessandro Piva	University of Florence, Italy
Dan Popescu	Universitatea Lucian Blaga Sibiu, Romania
Andrea Prati	University of Parma, Italy
Maria Giulia Preti	Université de Genève, Switzerland
Daniele Ravì	University College London, UK
Carlo Regazzoni	University of Genova, Italy
Haoyu Ren	Simon Fraser University, Italy
Elena Ricci	Technologies of Vision, Italy
Daniel Riccio	University of Naples, Italy
Karina Rodriguez-Echavarra	University of Brighton, UK
Bodo Rosenhahn	Leibniz-University of Hannover, Germany
Beatrice Rossi	STMicroelectronics, Italy
Albert Ali Salah	Bogazici University, Turkey
Paul Scheunders	University of Antwerp, Belgium
Roberto Scopigno	CNR-ISTI, Italy
Giuseppe Serra	University of Udine, Italy
Désiré Sidibé	Univ. Bourgogne Franche-Comte, France
Patricio Simari	Catholic University of America, USA
Bogdan Smolka	Silesian University of Technology, Poland
Michela Spagnuolo	CNR-IMATI, Italy
Davide Tanasi	University of South Florida, USA
Le Thi Lan	International Research Institute MICA, Vietnam
Massimo Tistarelli	University of Sassari, Italy
Andrea Torsello	University of Venice, Italy
Francesco Tortorella	University of Cassino, Italy
Alain Tremeau	University Jean Monnet, France
Mario Vento	University of Salerno, Italy
Luisa Verdoliva	University of Naples, Italy
Alessandro Verri	University of Genova, Italy
Salvatore Vitabile	University of Palermo, Italy
Domenico Vitulano	National Research Council of Italy, Italy
Marcel Worring	University of Amsterdam, The Netherlands
Tony Xiang	Queen Mary, University of London, UK
Hao Zhang	Carnegie Mellon University, USA
Huiyu Zhou	Queen's University Belfast, UK

Additional Reviewers

Dario Allegra
Lorenzo Baraldi
Catarina Barata
Federico Bolelli
Rodu Nicolae Dobrescu
Amr Elkhouli
Recep Erol
Fausto Galvan
Messina Giuseppe
Francesco Gugliuzza
Sen Jia
Corneliu Lazar
Dario Lo Castro
Liliana Lo Presti
Giuseppe Mazzola
Filippo Luigi Maria Milotta
Marco Moltisanti
Vito Monteleone
Oliver Moolan-Feroze
Pietro Morerio
Alessandro Ortis
Toby Perrett
Roberto Pirrone
Giuseppa Sciortino
Diego Sona
Valeria Tomaselli
Roberto Vezzani

Endorsing Institutions

International Association for Pattern Recognition (IAPR)
Italian Group of Researchers in Pattern Recognition (GIRPR)
Springer

Institutional Patronage

University of Catania
Image Processing Laboratory IPLab

Sponsoring and Supporting Institutions

iCTLab
Micron
STMicroelectronics

Contents – Part II

Image Analysis, Detection and Recognition

Multimedia

Biomedical and Assistive Technology

Imaging for Cultural Heritage and Archaeology

Imaging Solutions for Improving the Quality of Life

Contents – Part I

Video Analysis and Understanding

Pattern Recognition and Machine Learning

Multiview Geometry and 3D Computer Vision

Image Analysis, Detection and Recognition

Image Analysis, Detection and Recognition

3D Face Recognition in Continuous Spaces

Francisco José Silva Mata[1], Elaine Grenot Castellanos[1], Alfredo Muñoz-Briseño[1], Isneri Talavera-Bustamante[1], and Stefano Berretti[2(✉)]

[1] CENATAV, Havana, Cuba
[2] University of Florence, Florence, Italy
stefano.berretti@unifi.it

Abstract. This work introduces a new approach for face recognition based on 3D scans. The main idea of the proposed method is that of converting the 3D face scans into a functional representation, performing all the subsequent processing in the continuous space. To this end, a model alignment problem is first solved by combining graph matching and clustering. Fiducial points of the face are initially detected by analysis of continuous functions computed on the surface. Then, the alignment is performed by transforming the geometric graphs whose nodes are the critical points of the representative function of the surface in previously determined subspaces. A clustering step is finally applied to correct small displacement in the models. The 3D face representation is then obtained on the aligned models by functions carefully selected according to mathematical and computational criteria. In particular, the face is divided into regions, which are treated as independent domains where a set of functions is determined by fitting the surface data using the least squares method. Experimental results demonstrate the feasibility of the method.

Keywords: 3D face recognition · Functional representation

1 Introduction

Models of the face acquired by 3D devices consist of dense point clouds, where points correspond to coordinates of the face surface discretely sampled by the capture device. For high resolution 3D scans, a very large number of points is typically used to represent the face, and triangular mesh representations are then derived to connect points in a structured way. However, this low level representation cannot be used directly to compare faces in recognition tasks, but appropriate descriptors that reduce the high dimensionality of points keeping, at the same time, salient features of the face should be derived.

Face recognition using either high resolution or low-resolution 3D scans has received an increasing interest in the last few years (for a thorough discussion of existing methods we refer to the survey in [7] and the literature review in [3,18]). In general, 3D face recognition approaches proposed in the literature can be grouped as *global* (or *holistic*), and *local* (or *region-based*). Hybrid approaches

S. Battiato et al. (Eds.): ICIAP 2017, Part II, LNCS 10485, pp. 3–13, 2017.
https://doi.org/10.1007/978-3-319-68548-9_1

that combine solutions in these two categories are also possible as well as multimodal approaches that combine together 2D and 3D methods. Among the aspects that still are critical for most state of the art methods, we can count recognition across scans with different resolutions (high- or low-resolution as for consumer cameras like Kinect [5]), and recognition of scans with large/extreme pose variations or occlusions, which requires partial face matching. This is also reflected by the few face datasets that include face scans with different resolutions [5] or partial acquisitions [1,2,17]. Global 3D face representations for partial face matching have been proposed in a limited number of works [8,14]. More successful and scalable solutions used local representations of the face. In fact, one possible way to solve the problem of missing data in 3D faces is to detect locally the absence of regions of the face and use the existing data to reconstruct the missing parts (for example, exploiting the hypothesis of face symmetry to recover missing data in the case of scans with large pose variations [15]). The reconstructed scans can then be used as input to conventional 3D face recognition methods [10]. Tackling the problem from an opposite perspective, some methods divided the face into regions and tried to restrict the match to uncorrupted parts of the face [11,12]. Most of these methods used landmarks of the face to identify the regions to be matched; however, facial landmarks are difficult to detect when the pose significantly deviates from the frontal one. In addition, since parts of the regions can be missing or occluded, the extraction of effective descriptors is hindered so that regions comparison is mostly performed using rigid (ICP) or elastic registration (deformable models). Approaches that use keypoints of the face solve some of these limitations. Rather than relying on the detection of specific regions of the face that can fail in the presence of occlusions and missing parts, they detect keypoints on the face surface and describe the face locally at the keypoints. Matching keypoints can thus naturally account for occlusions and missing parts of the face [4,13].

In this work, we propose an original solution to 3D face recognition that, on the one hand, exploits keypoints for face alignment, on the other, accurately represents locally the face surface. Our solution is robust to the presence of scans acquired with large pose variations (and thus with missing parts), and is based on two main original contributions: a graph-based solution to align 3D face scans with missing parts; a *functional representation* that provides a locally continuous approximation of the face surface. The idea of approximating the face surface with continuous functions is a well known and used techniques in Computer Graphics. In that case, recovering the exact form of the surface is important for visualization. Differently, in the case of recognition tasks, the necessary optimization of the functional model must be able to obtain more discriminative representations of the face with the least number of coefficients. Indeed, the process of optimizing the functional model as well as the selection of the set of base functions are crucial for this method [16]. Functional representations are attractive for the recognition scenario because they show some interesting aspects. First, they demonstrate great power in compacting the data thanks to the small-dimensional vectors of used coefficients. In addition, they allow recovering the original continuous nature of biometric objects or their parts. This representation also allows capturing the correlation between the different values of 2D pixels or 3D vertices. The ability to use the existing theory of continuous functions often simplifies calculations and analysis. The representation of dynamic aspects of the original data and the possibility of extracting some important features through the analysis of the properties of functions, such as monotonicity, derivability and smoothness, makes attractive the use of functions to represent data that naturally vary in space continuously. However, an essential element to make functional representations comparable is that the origin of coordinates and the directions of the axes coincide across different objects. To achieve this, a process of prior alignment of the 3D faces is necessary. To this end, in this paper, we also propose a solution for aligning face scans with missing parts. This relies on three steps: first, the face is divided in rectangular domains and fiducial points of the face are detected as critical points of a local functional representation of the face surface based on *Local Thin Plate Bivariate Splines* (LTPBVS) [6]; then, a graph-like structure is constructed from the fiducial points connections; finally, matching these graphs permits face alignment.

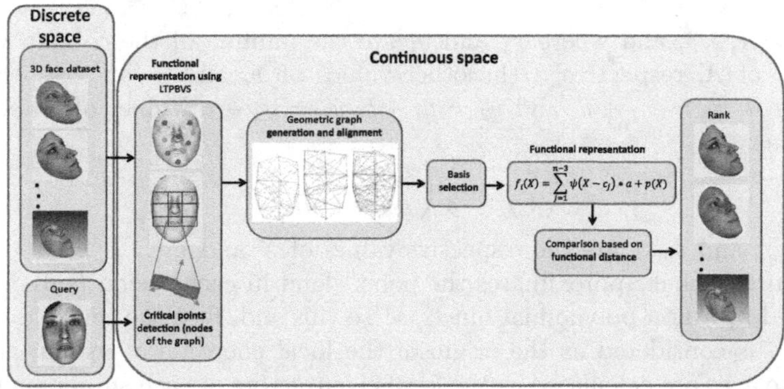

Fig. 1. The proposed 3D face recognition approach in continuous space

The processing steps of the proposed solution are summarized in Fig. 1: first, face scans are subdivided, approximated with LTPBVS and aligned using a graph of critical points; then, a LTPBVS basis is selected to approximate the face surface; finally, coefficients of the functional representation are used in the match. The rest of the paper is organized as follows: in Sect. 2, the method used for the detection of critical points of 3D faces is presented; the construction of a graph based on these points and its use for face alignment are illustrated in Sect. 3; the functional representation of the face is discussed in Sect. 4; experiments performed to evaluate our proposed method are reported in Sect. 5; discussion and conclusions in Sect. 6 close the paper.

2 Detection of Characteristic Points in 3D Faces

For the detection of characteristic points in 3D faces, Principal Component Analysis (PCA) was first performed to normalize the cloud of points of the whole face in such a way that their coordinate axes coincide with the principal components. To this end, given a set of points represented in the matrix A of size $N \times 3$, where each row is a point in space, it was necessary to calculate the covariance matrix. The eigenvectors of this matrix were used as the new coordinate axes: the z-axis captures the direction of the data with the smallest variance, i.e., the eigenvector corresponding to the lowest eigenvalue (this axis is also an estimate of the actual normal vector of the face surface corresponding to the points cloud); the y-axis corresponds to the vector of greatest variance and, finally, the x-axis corresponds to the vector associated with the eigenvalue of intermediate value.

The surface of the face is divided in rectangular domains, and a non-polynomial function is fitted to the surface of each domain. These rectangles have the same size determined according to the mesh size, and are represented as:

$$\mathrm{D}_{ij} = \{(x, y, z) : (x, y) \in [x_i, x_i + d] \times [y_j, y_j + t]\}, \tag{1}$$

with $i, j = 1, 2, 3$, and where x_1 and y_1 are the minima of the column vectors X and Y of A, respectively; the other values of x_i and y_i are, respectively, $x_2 = x_1 + d$, $x_3 = x_2 + d$, and $y_2 = y_1 + t$, $y_3 = y_2 + t$. Values of d and t are obtained as:

$$\begin{pmatrix} d \\ t \end{pmatrix} = \frac{1}{3} \begin{pmatrix} x_{max} - x_1 \\ y_{max} - y_1 \end{pmatrix}, \tag{2}$$

where x_{max} and y_{max} are the respective values of X and Y.

The surface that approximates the point cloud in each region (sub-domain) is obtained by a non-polynomial function. To this end, first the centroid of each region D_{ij} is considered as the origin of the local coordinate system, and the coordinate axes are calculated as the local eigenvectors of each sub-domain. The smallest of the three eigenvectors corresponds to the normal direction of each sub-domain. This ensures that the local z-axis is perpendicular to the surface. Then, the function that approximates the region of the points cloud has the form of a scattered translates, namely, a Bivariate Thin-Plate Spline. It uses arbitrary or scattered translates $\psi(. - c_j)$ of one fixed function ψ, in addition to some polynomial terms. Explicitly, such a form describes a function:

$$f(X) = \sum_{j=1}^{n-3} \psi(X - c_j)a_j + p(X), where\ X = (x, y), \tag{3}$$

where the basis function is $\psi(X) = \varphi(\|X\|^2)$, with $\|\cdot\|$ the Euclidean norm, and $\varphi(t) = t \log t$; c_j, a sequence of sites called *centers*, and a_j a corresponding sequence of n coefficients with the final three coefficients involved in the polynomial part:

$$P(X) = a_{n-2} \cdot x + a_{n-1} \cdot y + a_n. \tag{4}$$

```
Procedure classifyPoints()
for j = 1 to 8 do
   if (x_j, y_j) isReal then
      if det(h_j) > 0 then
         if h_j[0,0] > 0 then
            (x_i, y_i) → minimum
         else if h_j[0,0] < 0 then
            (x_i, y_i) → maximum
         end if
      else if det(h_j) < 0 then
         (x_i, y_i) → saddle
      end if
   end if
end for
return
```

Fig. 2. Procedure *classifyPoints*()

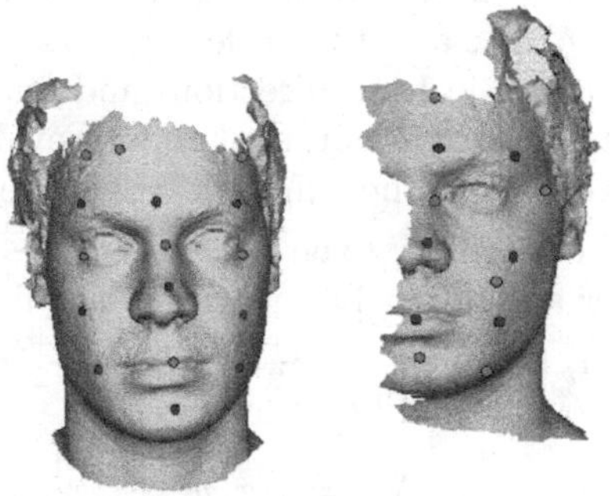

(a) Front capture (b) Side capture

Fig. 3. Results of the detection process.

The critical points (*maxima*, *minima*, and *saddles*) of the polynomial P correspond to the characteristic points of the face. These points are found with a subsequent inverse transformation to reach the points of the original face. To this end, the *gradient* G of the polynomial P is computed:

$$G(P) = \left(\frac{\partial P}{\partial x}(x, y), \frac{\partial P}{\partial y}(x, y) \right). \tag{5}$$

solving the following system:

$$\begin{cases} \frac{\partial P}{\partial x}(x, y) = 0 \\ \frac{\partial P}{\partial y}(x, y) = 0 \end{cases}. \tag{6}$$

As result, the eight possible solutions $\{(x_j, y_j)\}_{j=1}^{8}$ for the system are found. An evaluation of every real solution is performed on the *Hessian* matrix H of P. This evaluation is denoted as $h_j = H(P)(x_j, y_j)$. In this way, each real solution is classified according to its type (*minimum*, *maximum* or *saddle*) by following the procedure described in Fig. 2. As can be seen in this procedure, the classification is performed by computing the determinant of h_i and evaluating its first element (Fig. 3 shows some detected critical points).

On the other hand, when the determinant of h turns out to be zero, the point (x_i, y_i) in the polynomial function is evaluated, and its behavior is analyzed in such a way that: if $P(x_i, y_i) < P(x, y)$ it is a maximum; if $P(x_i, y_i) > P(x, y)$ it is a minimum; and if $P(x, y)_{(x,y)<(x_i,y_i)} < P(x_i, y_i) < P(x, y)_{(x,y)>(x_i,y_i)}$ it is a saddle point.

Some automatic adjustments of the position of windows or sub-domains were made to achieve greater efficacy of the method. These adjustments were executed starting by placing the first sub-domain in the approximate area of the nose (usually located in the center of the face for D_{22}), where in almost 100% of the cases there is a detectable maximum. Given the windows:

$$\mathrm{V}_{i2} = \{(x, y, z) : (x, y) \in [x_i, x_i + d] \times [y_2, y_2 + t]\}, \tag{7}$$

with $i = 2, 3, \ldots$, making shifts of five units to the right $x_2 = x_1 + 5$, $x_3 = x_2 + 5, \ldots$, and to the left $x_i = x_1 - 5$, $x_{i+1} = x_2 - 5, \ldots$. Until to lose the maximum in both directions and find points of minima and saddles at the end of the nose; it is obtained an intermediate sub-domain that is used like reference for the rest of the windows of the face. The length d of this intermediate window in the x-axis is given by $\frac{x_1^d + x_2^d}{2} - \frac{x_1^i + x_2^i}{2}$, where $x_{1,2}^d$ and $x_{1,2}^i$ are the respective lower and upper boundaries of the final windows given the right and left shifts. Then, being $x_1 = \frac{x_1^i + x_2^i}{2}$, $x_2 = \frac{x_1^i + x_2^i}{2}$ and $x_3 = \frac{x_1^d + x_2^d}{2} - d$, and the remaining windows would be as follows:

$$\mathrm{V}_{ij} = \{(x, y, z) : (x, y) \in [x_i, x_i + d] \times [y_j, y_j + t]\}. \tag{8}$$

3 Alignment of Two Faces

Before performing the recognition step between two faces, an alignment must be performed. Let $P_1 = \{p_1, p_2, \ldots, p_n\}$ and $P_2 = \{p_1, p_2, \ldots, p_n\}$ be the sets of fiducial points extracted from the representations of two 3D faces. Each point of these sets can be represented by the tuple $p_i = (x_i, y_i, z_i, l_i)$, where x_i, y_i and z_i are the coordinates of the described point in $\mathbb{R}^3$, and l_i is a label that can take three values depending on the kind of fiducial point detected (i.e., maximum, minimum or saddle).

The proposed alignment is based on finding a labeled geometric graph for each set of points. This is performed by computing Delaunay triangulation in 3D of the sets P_1 and P_2, denoted by $DT_3(P_1)$ and $DT_3(P_2)$. This triangulation is a generalization of the classic Delaunay triangulation in which no point in P_i is inside the circum-hypersphere of any simplex (tetrahedron) in $DT_3(P_i)$. It is known that $DT_3(P_i)$ is unique if P_i is a set of points in general position. This means that the affine hull of P_i is 3-dimensional and no set of 5 points in P_i lie on the boundary of a ball whose interior does not intersect P_i [9]. In this way, $DT_3(P_i)$ can be decomposed in simplexes, each one conformed by four facets. The main objective of computing $DT_3(P_1)$ and $DT_3(P_2)$ is to find a tolerant to distortions and unique geometrical structure for each P_i.

On the other hand, a labeled geometric graph can be defined as follows:

Definition 1 (Geometric graph). *A geometric graph is a 4-tuple,* $G = (V, E, I, K)$*, where* V *is a set of vertexes,* $E \subseteq \{\{u, v\} \mid u, v \in V, u \neq v\}$ *is a set of edges (the edge* $\{u, v\}$ *connects the vertexes* u *and* v*),* $I : V \rightarrow L_V$ *is a function that assigns labels to vertexes where* L *is the domain of labels and, finally,* $K : V \rightarrow \mathbb{R}^3$ *is a function that assigns coordinates to vertexes,* $\mathbb{R}$ *represents the set of real numbers, and* $K(u) \neq K(v)$ *for each* $u \neq v$.

Using the previous definition and the triangulations $DT_3(P_1)$ and $DT_3(P_2)$, the labeled geometric graphs $G_1 = (V_1, E_1, I_1, K_1)$ and $G_2 = (V_2, E_2, I_2, K_2)$ are obtained, respectively, from P_1 and P_2. It results that:

- V_i represents the points of P_i;
- E_i contains all the edges generated by $DT_3(P_i)$;

1: **Procedure** $alignPoints(P_1, P_2)$
2: $G_1(V_1, E_1, I_1, K_1) \leftarrow buildGraph(P_1)$
3: $G_2(V_2, E_2, I_2, K_2) \leftarrow buildGraph(P_2)$
4: $H \leftarrow \{\emptyset\}$
5: **for all** $face\ v_i, v_j, v_k \in G_1$ **do**
6:     **for all** $face\ v_l, v_m, v_n \in G_2$ **do**
7:         $sortVertexes(v_i, v_j, v_k)$
8:         $sortVertexes(v_l, v_m, v_n)$
9:         **if** $I_1(v_i) == I_2(v_l)$ **and** $I_1(v_j) == I_2(v_m)$ **and** $I_1(v_k) == I_2(v_n)$ **then**
10:             $addTrans(H, E_1(v_i, v_j), E_2(v_l, v_m))$
11:             $addTrans(H, E_1(v_j, v_k), E_2(v_m, v_n))$
12:             $addTrans(H, E_1(v_k, v_i), E_2(v_n, v_l))$
13:         **end if**
14:     **end for**
15: **end for**
16: $T_M \leftarrow getMaxRepTrans(H)$
17: $applyTransformation(P_2, T_M)$
18: **return**

Fig. 4. Procedure $alignPoints(P_1, P_2)$

- I_i is a function that assigns labels from $L_V = \{1, 2, 3\}$ depending on the type of the point represented (i.e., maximum, minimum or saddle);
- K assigns coordinates to the vertexes.

In Fig. 6 a geometric graph generated with this procedure is shown. After this step, a graph matching technique between G_1 and G_2 is done. With this technique, the geometric transformation T that best aligns G_1 with G_2 is found.

In Figs. 4 and 5 the procedures used for aligning two sets of points P_1 and P_2 are shown. In lines 2–3 of the procedure $alignPoints()$, the graphs are created. Then, a map structure H is initialized. In lines 5–15, the faces of each simplex of G_1 are compared with those contained in G_2. For this, the vertexes of the faces are sorted according to the lengths of its segments. After this, if the analyzed pair of faces have the same labels, the procedure $addTrans()$ is called. In this latter procedure, the transformation matrix T used to convert the second segment into the first one is computed. Then, if T or a similar transformation is contained in H, the respective counter is augmented; otherwise a new entry is added with the counter set to 0. Finally, the transformation T_M with higher counter in H is used to rotate and translate P_2 with respect to P_1.

The main idea of this algorithm is based on finding the geometric transformation T_M that aligns the highest number of edges belonging to G_1 and G_2. This algorithm assumes that the fiducial points extracted from all the 3D faces have a similar geometric disposition and labeling. As an example, Fig. 6 shows the representation as geometric graphs and the alignment of two faces.

In order to refine the geometric graphs alignment, a posterior clustering procedure is performed. First of all, the PCA algorithm is applied to the whole model to determine the z-axis, as the direction of lower variance. Then, a

```
Procedure addTrans(H, E1(vi, vj), E2(vl, vm))
s1 ← sizeofSegment(E1(vi, vj))
s2 ← sizeofSegment(E2(vl, vm))
if abs(s1 − s2) then
    T ← getTransMatrix(E1(vi, vj), E2(vl, vm))
    if isContained(H, T) then
        H(T).counter + +
    else
        H ← {T, 1}
    end if
end if
return
```

Fig. 5. Procedure $addTrans(H, E_1(v_i, v_j), E_2(v_l, v_m))$

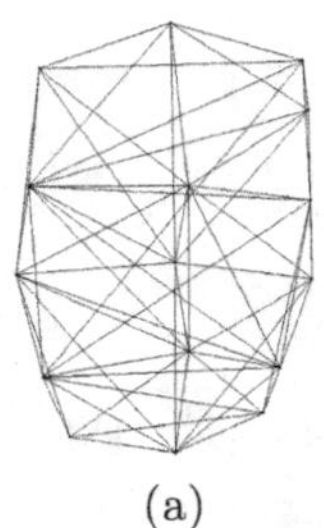

(a)

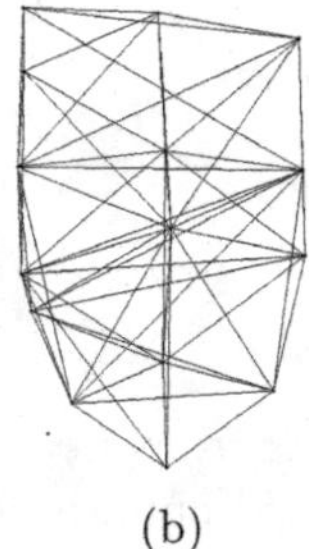

(b)

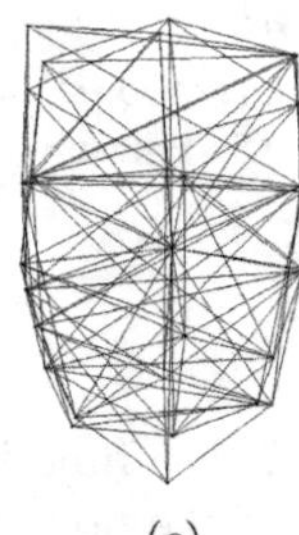

(c)

Fig. 6. (a)–(b) Two graphs of faces; (c) alignment of the graphs in (a) and (b)

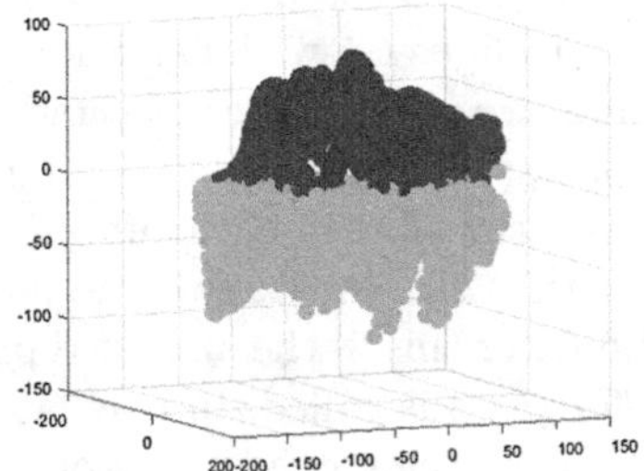

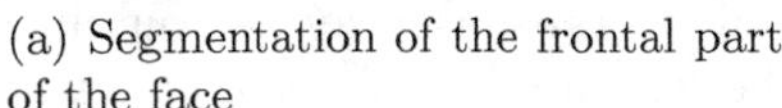

(a) Segmentation of the frontal part of the face

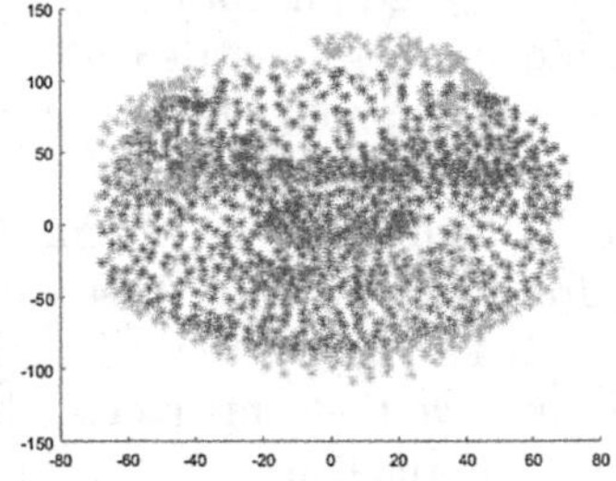

(b) Alignment of two meshes

Fig. 7. Refinement of the alignment process

k-means clustering is applied to the z values, in order to segment the frontal part of the face (see Fig. 7a). Finally, PCA is applied again, using as origin the maximum value of the z coordinates found. In this way, the y-axis is given by the direction of lower variance. In Fig. 7 the results of this process are shown.

The main advantages of the proposed method over other state of the art approaches [16] are the following:

- The localization of specific fiducial points in the faces, like pronasal points, are not needed.
- The use of a point set registration algorithm, like ICP is avoided. These algorithms are computationally expensive.
- Our proposal performs the alignment process with high precision, between lateral and frontal views of the faces. This is not possible in previous works.

4 Functional Representation

Once aligned, the next step is to obtain the representation of the points cloud as a surface corresponding to a function $z = f(x, y)$ over a spatial domain. The appropriate domain in terms of its dimensions and geometry must correspond to the completion of the functional representation. As base functions, we use the LTPBVS (see (3)), but adjusted to the surface of the new regions obtained after the alignment of the faces. In this way, the representation is constructed by the same procedure used to detect the points for alignment, which simplifies the implementation of the process. The decision to use LTPBVS is supported by the well known advantages of these functions. Among them, we can mention that LTPBVS produces smooth surfaces, which are infinitely differentiable. Also, they do not have free parameters that need manual tuning.

The matching step is performed by comparing the coefficients of the corresponding representative functions of the faces, in a way similar to [16]. However, in this work we obtain one functional representation for each one of the m regions in which the face is divided. Given two faces F and G, their distance can be computed as in (9), where f_i and g_i are the corresponding functions of the i-th region, defined on a common domain $[a, b] \times [c, d]$ for the norm L_n:

$$d(F, G) = \sum_{i=1}^{m} \sqrt[n]{\int_a^b \int_c^d |f_i(x, y) - g_i(x, y)|^n dxdy}. \tag{9}$$

5 Experimental Results

The proposed 3D face recognition approach has been evaluated on the 2D/3D Florence dataset [2]. This dataset includes 3D faces acquired with different devices and challenges (i.e., non-frontal pose, presence of hair, neck, shoulders). For the whole dataset, the representations were constructed based on local thin plate bivariate splines (LTPBVS) defined, respectively, on twelve disjoint regions oriented by the normals to the origin and over the correspondent control grid. Thus, in the case of side faces, they contain six disjoint rectangular regions. The face recognition problem was modeled as a classification task, using a k-NN classifier with Euclidean distance. Results are reported in Table 1. For each one of the disjoint regions found, 39 coefficients were computed.

Table 1. Rank-1 recognition accuracy on the 2D/3D Florence face dataset

Method	# coef.	Frontal	Left	Right
Our proposal	**468**	98%	**97.1%**	**97.2%**
GBVS [16]	4096	100%	96.2%	96.4%
LBVS [16]	4608	100%	96.3%	96.7%

It can be noted, the results do not outperform those obtained in a previous approach on the frontal case, but it reports better results on lateral cases. This feature makes the proposal of this work more suitable for environments in which occlusions are common. Also, the number of coefficients used on this approach is lesser than previous works, which reduces the dimension of the data and improves the efficiency of the method.

6 Discussion and Conclusions

Recognizing faces from 3D scans is becoming a problem of increasing interest, with applications in several practical contexts. Though effective solutions exist for the cooperative case, where faces are acquired in frontal pose, the recognition is much more difficult when acquisitions include facial expressions or pose variations (missing parts).

In this paper, we have presented an original 3D face recognition solution, which is capable of recognizing faces also in the case of expressions and missing parts. The proposed method relies on the idea of constructing a functional representation of the face locally. First, keypoints of the face are detected using surface analysis, and they are used to partition the face into local rectangular domains, which are subsequently aligned. Then, the surface is approximated locally to each domain using Local Thin Plate Bivariate Splines (LTPBVS). The LTPBVS provide a descriptive and compact representation of the face, where coefficients of the functions are used for effective and efficient face matching. On the other hand, the proposed alignment method is very robust in presence of position variation or omission of fiducial points. This occurs because the alignment can be performed by using only a small subset of fiducial points, which allows a higher degree of tolerance. The proposed method has good performance even when a certain amount of spurious fiducial points are located. Recognition results obtained on the UF-3D [2] database show performances, which are comparable or superior to state of the art solutions.

References

1. University of Notre Dame biometrics database (2008). http://www.nd.edu/@cvrl/UNDBiometricsDatabase.html
2. Bagdanov, A.D., Del Bimbo, A., Masi, I.: The Florence 2D/3D hybrid face dataset. In: Proceedings of Joint ACM Workshop on Human Gesture and Behavior Understanding, pp. 79–80 (2011)

3. Berretti, S., Del Bimbo, A., Pala, P.: 3D face recognition using iso-geodesic stripes. IEEE Trans. Pattern Anal. Mach. Intell. **32**(12), 2162–2177 (2010)
4. Berretti, S., Del Bimbo, A., Pala, P.: Sparse matching of salient facial curves for recognition of 3-D faces with missing parts. IEEE Trans. Inf. Forensics Secur. **8**(2), 374–389 (2013)
5. Bondi, E., Pala, P., Berretti, S., Del Bimbo, A.: Reconstructing high-resolution face models from Kinect depth sequences. IEEE Trans. Inf. Forensics Secur. **11**(12), 2843–2853 (2016)
6. Bookstein, F.L.: Principal warps: thin-plate splines and the decomposition of deformations. IEEE Trans. Pattern Anal. Mach. Intell. **11**(6), 567–585 (1989)
7. Bowyer, K.W., Chang, K.I., Flynn, P.J.: A survey of approaches and challenges in 3D and multi-modal 3D+2D face recognition. Comput. Vis. Image Underst. **101**(1), 1–15 (2006)
8. Bronstein, A.M., Bronstein, M.M., Kimmel, R.: Robust expression-invariant face recognition from partially missing data. In: Leonardis, A., Bischof, H., Pinz, A. (eds.) ECCV 2006. LNCS, vol. 3953, pp. 396–408. Springer, Heidelberg (2006). doi:10.1007/11744078_31
9. Cohen-Steiner, D., de Verdière, E.C., Yvinec, M.: Conforming Delaunay triangulations in 3D. Comput. Geom. **28**(2–3), 217–233 (2004)
10. Colombo, A., Cusano, C., Schettini, R.: Gappy PCA classification for occlusion tolerant 3D face detection. J. Math. Imaging Vis. **35**(3), 193–207 (2009)
11. Drira, H., Ben Amor, B., Srivastava, A., Daoudi, M., Slama, R.: 3D face recognition under expressions, occlusions, and pose variations. IEEE Trans. Pattern Anal. Mach. Intell. **35**(9), 2270–2283 (2013)
12. Faltemier, T.C., Bowyer, K.W., Flynn, P.J.: A region ensemble for 3D face recognition. IEEE Trans. Inf. Forensics Secur. **3**(1), 62–73 (2008)
13. Huang, D., Ardabilian, M., Wang, Y., Chen, L.: 3D face recognition using eLBP-based facial representation and local feature hybrid matching. IEEE Trans. Inf. Forensics Secur. **7**(5), 1551–1564 (2012)
14. Lu, X., Jain, A.K., Colbry, D.: Matching 2.5D face scans to 3D models. IEEE Trans. Pattern Anal. Mach. Intell. **28**(1), 31–43 (2006)
15. Passalis, G., Perakis, P., Theoharis, T., Kakadiaris, I.A.: Using facial symmetry to handle pose variations in real-world 3D face recognition. IEEE Trans. Pattern Anal. Mach. Intell. **33**(10), 1938–1951 (2011)
16. Porro-Muñoz, D., José Silva-Mata, F., Revilla-Eng, A., Talavera-Bustamante, I., Berretti, S.: 3D face recognition by functional data analysis. In: Bayro-Corrochano, E., Hancock, E. (eds.) CIARP 2014. LNCS, vol. 8827, pp. 818–826. Springer, Cham (2014). doi:10.1007/978-3-319-12568-8_99
17. Savran, A., Alyüz, N., Dibeklioğlu, H., Çeliktutan, O., Gökberk, B., Sankur, B., Akarun, L.: Bosphorus database for 3D face analysis. In: Schouten, B., Juul, N.C., Drygajlo, A., Tistarelli, M. (eds.) BioID 2008. LNCS, vol. 5372, pp. 47–56. Springer, Heidelberg (2008). doi:10.1007/978-3-540-89991-4_6
18. Wang, Y., Liu, J., Tang, X.: Robust 3D face recognition by local shape difference boosting. IEEE Trans. Pattern Anal. Mach. Intell. **32**(12), 1858–1870 (2010)

Object Detection for Crime Scene Evidence Analysis Using Deep Learning

Surajit Saikia[1,2(✉)], E. Fidalgo[1,2], Enrique Alegre[1,2], and Laura Fernández-Robles[2,3]

[1] Department of Electrical, Systems and Automation, University of León, León, Spain
{ssai,efidf,ealeg}@unileon.es
[2] INCIBE (Spanish National Cybersecurity Institute), León, Spain
l.fernandez@unileon.es
[3] Department of Mechanical, Informatics and Aerospace Engineering, University of León, León, Spain

Abstract. Object detection is the key module in most visual-based surveillance applications and security systems. In crime scene analysis, the images and videos play a significant role in providing visual documentation of a scene. It allows police officers to recreate a scene for later analysis by detecting objects related to a specific crime. However, due to the presence of a large volume of data, the task of detecting objects of interest is very tedious for law enforcement agencies. In this work, we present a Faster R-CNN (Region-based Convolutional Neural Network) based real-time system, which automatically detects objects which might be found in an indoor environment. To test the effectiveness of the proposed system, we applied it to a subset of ImageNet containing 12 object classes and Karina dataset. We achieved an average accuracy of 74.33%, and the mean time taken to detect objects per image was 0.12 s in Nvidia-TitanX GPU.

Keywords: Object detection · Convolutional neural network · Deep learning · Video surveillance · Crime scenes · Cyber-security

1 Introduction

The problem of detecting objects of interest in videos and images plays a key role in most security and surveillance systems. In the domain of forensic science, the digital images and videos play a significant role in determining fingerprints, identifying criminals and understanding crime scenes. For instance, given an object detected in an image, further analysis can be done to extract crucial information (i.e. relating objects of various crime scenes). However, due to the presence of a large amount of visual data, the creation of tools to manage or categorize this data have an exceptional importance. If we now imagine the task of a single person, trying to extract some type of intelligence from thousands of

S. Battiato et al. (Eds.): ICIAP 2017, Part II, LNCS 10485, pp. 14–24, 2017.
https://doi.org/10.1007/978-3-319-68548-9_2

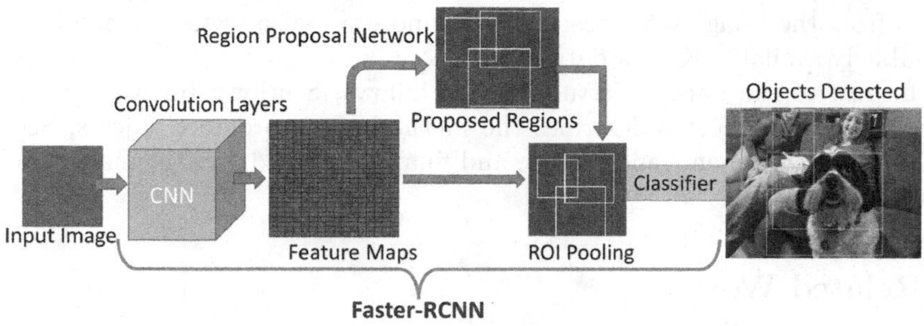

Fig. 1. Object detection pipeline using Faster R-CNN.

images or hours of video, i.e. a police officer reviewing digital evidence from a crime scene, the following question raises: *How is it possible to ease the task of detecting an object, that can be the evidence for a crime, inside this amount of visual information?* The answer could be an object detection system working in real-time.

Object detection is the task of detecting instances of objects belonging to a specific class. These systems are exploited in a wide range of applications in the field of AI (Artificial Intelligence), medical diagnosis [1,2], military [3] and crime prevention [4]. Furthermore, such systems can also be joined with other techniques to extract useful information in different types of cyber-crimes. Some examples might be Face detection and recognition [5] to detect and identify criminals on Internet videos, Video surveillance [6] to identify videos which can be a threat to society and nations, and Image Understanding [7] to recognize crime scenes based on the contents of images on the Internet.

In this work, we address the problem of analyzing the data gathered in a crime scene through the use of object detection. By detecting the objects present in the evidences found in a crime scene, it is possible to extract some intelligence or relations that can help a police officer, for example, to relate different crime scenes. We introduce an object detection method that uses the pre-trained VGG-16 [8] architecture of Faster R-CNN (Region-Based Convolutional Neural Network) [9], which was trained on MS-COCO [10] dataset. We selected 12 objects from ImageNet [11], which are most commonly found in a bedroom. Such objects are representative of the ones that might be found in an indoor environments. In Fig. 1 we briefly illustrate our object detection method. Given an input image, a Region Proposal Network (RPN) suggests regions based on the features generated by the last convolution layer. The proposed regions are then classified according to their detected class labels. Furthermore, this method can also be generalized for other objects and outdoor scenes. Our proposal aims at smoothing and reducing the labour of police while dealing with visual data (i.e. *a police officer can use this method to detect objects from one scene, and automatically relate them to other similar scenes efficiently*). To evaluate our proposal, we created a test-set and named it ImageNet-RoomObjects, which is a subset of images from the ImageNet dataset. Furthermore, we also tested our method on the publicly available Karina dataset[1].

The rest of the paper is structured as follows. Section 2 briefly introduces the related works, Sect. 3 describes the method used to detect objects, Sect. 4 presents the experiments and results, and finally, Sect. 5 draws the main conclusions of this work.

2 Related Work

Recently in the literature, many works use deep convolutional neural networks (e.g. AlexNet [12], GoogleNet [13], VGG-Net [8]) for detecting and locating objects with class specific bounding boxes. Typically a CNN consist of multiple convolution layers, followed by ReLU (Rectified Linear Units), pooling layers and fully connected layers. The activations which are generated by the last layers of a CNN can be used as a descriptor for object detection and classification. Razavian *et al.* [14] employed the activations generated by the fully connected layers as region descriptors. Babenko *et al.* [15] demonstrated that such activations can be even used for image retrieval task, and they named such descriptors as neural codes. Later, they established that such descriptors performs competitively even if a CNN is trained for unrelated classification task i.e. a CNN trained with ImageNet [11] dataset can be generalized to detect objects in MS-COCO [10] dataset.

The deep learning algorithms have improved the image classification and object detection tasks in manifolds as compared to SIFT or other variants [16]. The algorithm proposed by Lee *et al.* [17] learns high-level features i.e. object parts from natural scenes and unlabeled images. Simonyan and Zisserman *et al.* [8] investigated the depth of such algorithms on its accuracy, and achieved state-of-the-art results.

Girshick *et al.* [18] presented R-CNN (Region-based Convolutional Neural Network), which proposes regions before feeding into a CNN for classification. The network is a version of AlexNet, which has been trained using Pascal VOC Detection data [19]. The network contains a three-stage pipeline, thus making the training process slow. Since then, in terms of accuracy and speed, great improvements have been achieved. He *et al.* [20] proposed SPP-net, which is based on Spatial Pyramid pooling. The network improves the detection and classification time by pooling region features instead of passing each region into the CNN. Later, Girshik [21] proposed Fast R-CNN, which is similar to SPP-net, but replaced SVM classifiers with neural networks. Ren *et al.* [9] introduced Faster R-CNN, a faster version of Fast R-CNN, which replaces the previous region proposal method with RPN (Region proposal Network), which simultaneously predicts object bounds and scores.

[1] http://pitia.unileon.es/varp/node/373.

3 Methodology

In this section, we introduce a system for object detection which is based on the Faster R-CNN algorithm. At first, we briefly describe the algorithm, and then we present the architecture details of VGG-16 network and the Region Proposal Network (RPN).

3.1 Background of Faster-RCNN

Region-based Convolutional Neural Network (RCNN) is an object detection method based on visual information of images. The network first computes the region proposal (i.e. possible locations of objects), and then it feeds the proposed regions into the CNN for classification. Nevertheless, this algorithm has important drawbacks due to its three-stage pipeline, which makes the training process expensive from space and computation point of view. For each object proposal, the network does a CNN pass without sharing the computations, thus making the network slow. As an illustration, if there are 1000 proposals, then we have to do 1000 CNN passes. In order to speed up the method, a faster version of R-CNN algorithm known as Fast R-CNN [21] was introduced. During CNN passes, this algorithm shares the computations when there are overlaps between the proposals, resulting in faster detection. Since the algorithm processes images by resizing them into a fixed shape, the detection time is approximately same for all the images. It takes approximately 2 s to detect objects including the time taken to propose regions.

Faster-RCNN. The Faster R-CNN [9] algorithm has brought the task of object detection to real-time, which takes approximately 12 ms for detecting objects in a RGB image, including the time cost in region proposal. The algorithm has replaced the previous region proposal method with Region Proposal Network (RPN), which is further merged with Fast R-CNN, making the network a single unified module.

3.2 VGG-16 Network

We use the 16 layered VGG [8] network, which comprises of 13 convolution layers and 3 fully connected layers. The network is unified with the RPN to serve the purpose of object detection. The Fig. 2 presents the architecture of the network excluding the RPN and RoI (Region of Interest) pooling layer, and the values represents the dimension of response maps in each convolution layer, i.e. in the first convolution layer $224 \times 224 \times 64$ represent 64 response maps of size 224×224.

Region Proposal Network. A RPN contains a sliding window of spatial size $n \times n$ (we use $n = 3$), which is applied on the feature maps generated by the last convolution layer to obtain an intermediate layer in 512 dimension. Then,

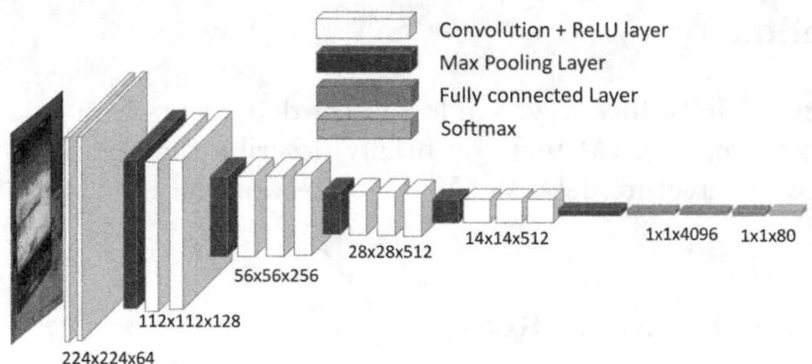

Fig. 2. Architecture of VGG net.

the intermediate layer feeds into a *box classification layer* and a *box regression layer*. There are k-anchor boxes with respect to each position of the sliding window, where k denotes the number of maximum possible proposals. The *box classification* layer determines whether the k anchor boxes contains object or not, and generates 2k scores (object/not object for each k). The *box regression layer* gives 4 coordinates with respect to each of the anchor boxes ($4k$ for k anchor boxes).

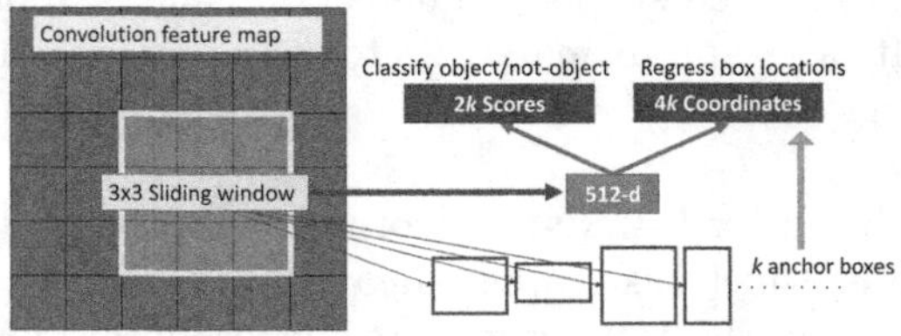

Fig. 3. Region proposal network (RPN).

The RPN is combined with a Convolutional Neural Network as a single module, which proposes regions within the network itself. It takes feature maps generated by the last convolution layer as an input and generates rectangular regions (object proposals) along with objectiveness scores. The RPN determines whether a region generated by a sliding window is an object or not, and if it is an object then the network does bounding box regression. Figure 3 shows the architecture of the RPN.

3.3 Object Detection

Once the complete system has been described, in Fig. 4 we present an example illustrating intermediate outputs of the system when we try to detect the objects. At first, an image (Fig. 4(a)) is given as input to the algorithm. The convolution layers generate activations, and the last layer activations are given as an input

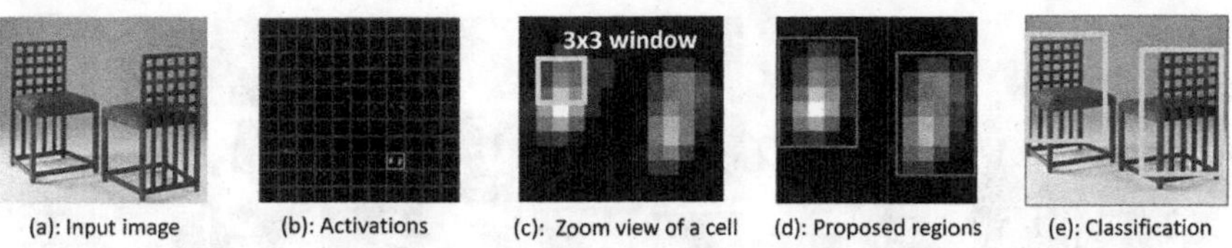

Fig. 4. Stages of object detection process.

to the RPN for region proposal. Figure 4(b) shows an example of the activation maps generated by the last convolution layer, and Fig. 4(c) presents how a cell in an activation maps looks like when it is zoomed. Next, the RPN slides a window of size 3×3 in each of the cell and fits the anchor boxes (Fig. 4(d)). Then, we classify the regions corresponding to each of the anchor boxes, and we obtain their class labels along with the bounding boxes (Fig. 4(e)). In this way, we can input images to the system to detect specific objects.

4 Experiments and Results

In this section, we describe the datasets, the experimental setup and the results achieved.

4.1 Datasets

Our network is pre-trained with the MS-COCO dataset. This is an image segmentation, recognition and captioning dataset by Microsoft corporation, which is a collection of more than 300,000 images and 80 object categories with multiple objects per image. Then, we tested the method in two different test-sets: a subset of images containing 12 indoor objects extracted from ImageNet, which we called it *ImageNet-RoomObjects*, and the *Karina dataset* [22]. We briefly describe the test-sets along with the datasets.

Test-Set: ImageNet-RoomObjects. This is a collection of 1345 images with 12 object categories, that are commonly found in an indoor environment, i.e. bedroom. We randomly selected images from the ImageNet, which is a huge dataset with a collection of 14,197,122 images with more than 1000 object classes, and each object class in this dataset contains thousands of images.

Karina Dataset [22]**.** This is a video dataset that was created to evaluate object recognition in environments which are similar to those that might appear in child pornography. The dataset contains 16 videos of 3 min which are filmed in 7 different rooms, and it contains 40 different categories, which represents to some of the objects that can be found most commonly in an indoor environment i.e. a bedroom.

Fig. 5. Examples of object detection results on ImageNet-IndoorObjects. Red squares overlaid on input images mark the bounding boxes of the detected objects. (Color figure online)

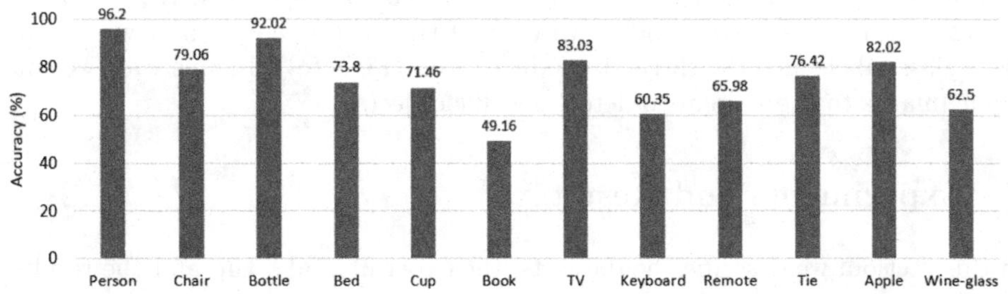

Fig. 6. Detection accuracy (in percentage) of each class.

4.2 Experimental Setup

We have used the pre-trained VGG-16 network architecture of Faster R-CNN to detect objects in our created test-set and the Karina dataset. The network was trained based on the following parameters: base learning rate: 0.001, learning policy: *step*, gamma: 0.1, momentum: 0.9, weight decay: 0.0005 and iterations 490000. All the experiments were carried out using the *Caffe* [23] deep learning framework in *Nvidia Titan X GPU* and in an *Intel Xeon* machine with 128 GB RAM.

4.3 Detection Accuracy

We present the detection accuracy (in percentage) of each class in the ImageNet-RoomObjects test-set, which is the percentage of true positives in each class of the test-set. Figure 6 shows the detection accuracies for all the 12 considered classes, and we obtained an average accuracy of 74.33%. We also present the mean of confidence scores (Fig. 7) for each class. Figure 5 shows the samples of object detection in each of the class category.

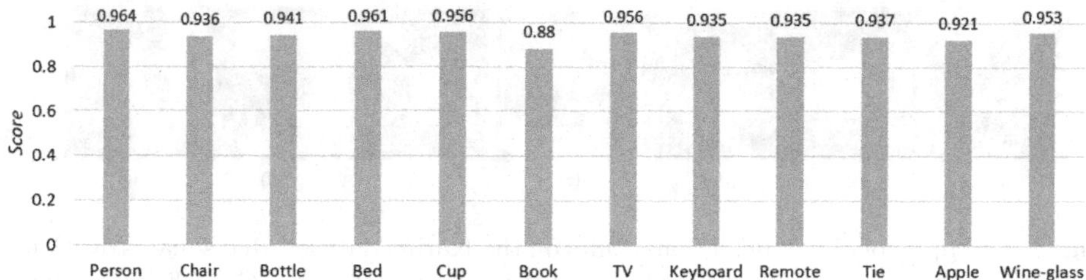

Fig. 7. Mean confidence scores per class.

Table 1. Detection time (in seconds) for each object class along with the number of images per class.

Class	Person	Chair	Bottle	Bed	Cup	Book	TV	Keyboard	Remote	Tie	Apple	Wine-glass
Time	31.35	26.8	37.02	35.6	21.65	21.65	27.22	31.62	24.74	28.27	27.94	34.53
Total images	104	100	110	129	121	120	130	110	105	114	106	105

4.4 Detection Time

This is the amount of time taken to detect specific objects in an image. Table 1 shows the time taken to detect objects specific to each of the class. For example, there are 104 images of person in the test-set, and the time needed to detect a person in those images was 31.35 s. The system takes 356.35 s in a GPU platform to detect objects in all the 1345 images, which also includes the time to propose regions. We also observe that, it takes an average of 2 s to detect objects in a single image in a CPU environment, and 0.12 s in a GPU environment, saving approximately 90% of the needed time.

4.5 Experiments on the Karina Dataset

To test the system, we have created a test-set from the Karina dataset by extracting image frames (size 640 × 480) from the videos. Out of 40 object categories we select 6 classes i.e. *bed, book, toy car, teddy, person and cup*. In Table 2 we present the detection accuracy for each class along with the total number of images present in each category. Figure 8 shows some samples, where we were able to detect some indoor objects like *(a) cup, (b) bed, (c) book, (d) toy car, (e) remote, (f) doll* and *(g) teddy bear*. Since the resolution of the images is low, we performed only preliminary test. In the future, we would like to apply *image super-resolution* techniques [24] to enhance the resolution before detection.

Table 2. Detection accuracy for each object class in the Karina dataset.

Class	Person	Book	Toy car	Teddy bear	Cup	Bed
Accuracy	81.3%	28.42%	5.4%	18.2%	45.61%	20.23%
Total images	100	120	105	80	109	91

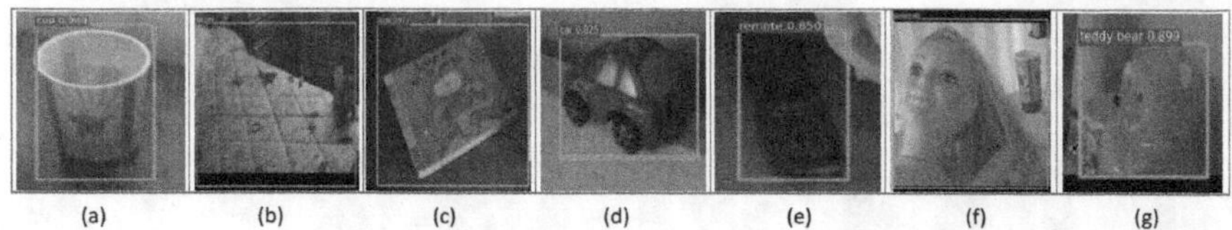

Fig. 8. Examples of object detection results on the Karina dataset. Red squares overlaid on input images mark the bounding boxes of the detected objects. (Color figure online)

4.6 Discussion

We have evaluated our object detection system in ImageNet-RoomObjects and the Karina Dataset, and we made several observations and conclusions based on our experiments.

First of all, our object detection system is pre-trained on the MS-COCO dataset. While testing on the ImageNet-RoomObjects, we observed that the system was able to detect objects with an average accuracy of 74.33%, where the highest accuracy was 96.2% obtained with the class *person*, and the lowest was 49.16% yielded with *books*. However, this performance can be further improved by fine tuning the architecture using datasets with respect to each class category, and we will address it in our future work.

We have also evaluated the performance of the system in the Karina dataset. But, due to the presence of low resolution images, the average accuracy obtained was 33.19%. In real world scenarios, we can not expect all the images to be of good resolution, and it is a difficult task for the forensic department to recreate crime scenes using such images. In the future, we will handle this issue by applying image-super resolution techniques to enhance the image quality, which will ease the task for police officers to detect objects effectively even in low quality images.

The average detection time per image was 0.12 s in Nvidia Titan X GPU, which makes the system suitable to be used as a real-time application. Furthermore, the system might be an application for a forensic department, which can help police officers to automatically detect objects of interest from large scale datasets in real-time.

5 Conclusion

In this work, we presented a real-time system which can detect objects related to indoor environments (i.e. bedroom). We have used the state-of-the-art VGG-16 network architecture of the Faster R-CNN algorithm, which can compute region proposal within the network itself. Due to this property, the algorithm can be widely used to develop real-time object detection applications. To evaluate the system, we have created a test-set "*ImageNet-RoomObjects*" comprising of images commonly found in an indoor environment, and we achieved state-of-the-art accuracy. The system has also been tested on the Karina dataset, but we have achieved poor accuracy due to the low quality of the images. In future works, we will address this issue by applying image super-resolution techniques, and we will train a new model containing a large number of categories based on the additional object types that the police might found interesting during their crime scene research. Finally, the method can be used as a surveillance application to detect objects of interest in videos and images in real time for analysing various crime scenes.

Acknowledgement. This research was funded by the framework agreement between the University of León and INCIBE (Spanish National Cybersecurity Institute) under addendum 22. We gratefully acknowledge the support of NVIDIA Corporation with the donation of the Titan X GPU used for this research.

References

1. Vaidehi, K., Subashini, T.: Automatic classification and retrieval of mammographic tissue density using texture features. In: 2015 IEEE 9th International Conference on Intelligent Systems and Control (ISCO), pp. 1–6. IEEE (2015)
2. Nosato, H., Sakanashi, H., Takahashi, E., Murakawa, M.: Method of retrieving multi-scale objects from optical colonoscopy images based on image-recognition techniques. In: 2015 IEEE Biomedical Circuits and Systems Conference (BioCAS), pp. 1–4. IEEE (2015)
3. Li, J., Ye, D.H., Chung, T., Kolsch, M., Wachs, J., Bouman, C.: Multi-target detection and tracking from a single camera in unmanned aerial vehicles (UAVs). In: 2016 IEEE/RSJ International Conference on Intelligent Robots and Systems (IROS), pp. 4992–4997. IEEE (2016)
4. Rao, R.S., Ali, S.T.: A computer vision technique to detect phishing attacks. In: 2015 Fifth International Conference on Communication Systems and Network Technologies (CSNT), pp. 596–601. IEEE (2015)
5. Herrmann, C., Beyerer, J.: Face retrieval on large-scale video data. In: 2015 12th Conference on Computer and Robot Vision (CRV), pp. 192–199. IEEE (2015)
6. Sidhu, R.S., Sharad, M.: Smart surveillance system for detecting interpersonal crime. In: 2016 International Conference on Communication and Signal Processing (ICCSP), pp. 2003–2007. IEEE (2016)
7. Vallet, A., Sakamoto, H.: Convolutional recurrent neural networks for better image understanding. In: 2016 International Conference on Digital Image Computing: Techniques and Applications (DICTA), pp. 1–7. IEEE (2016)
8. Simonyan, K., Zisserman, A.: Very deep convolutional networks for large-scale image recognition. arXiv preprint arXiv:1409.1556 (2014)
9. Ren, S., He, K., Girshick, R., Sun, J.: Faster R-CNN: towards real-time object detection with region proposal networks. In: Advances in Neural Information Processing Systems, pp. 91–99 (2015)
10. Lin, T.-Y., Maire, M., Belongie, S., Hays, J., Perona, P., Ramanan, D., Dollár, P., Zitnick, C.L.: Microsoft COCO: common objects in context. In: Fleet, D., Pajdla, T., Schiele, B., Tuytelaars, T. (eds.) ECCV 2014. LNCS, vol. 8693, pp. 740–755. Springer, Cham (2014). doi:10.1007/978-3-319-10602-1_48
11. Russakovsky, O., Deng, J., Su, H., Krause, J., Satheesh, S., Ma, S., Huang, Z., Karpathy, A., Khosla, A., Bernstein, M., et al.: Imagenet large scale visual recognition challenge. Int. J. Comput. Vis. **115**(3), 211–252 (2015)

12. Krizhevsky, A., Sutskever, I., Hinton, G.E.: Imagenet classification with deep convolutional neural networks. In: Advances in Neural Information Processing Systems, pp. 1097–1105 (2012)
13. Szegedy, C., Liu, W., Jia, Y., Sermanet, P., Reed, S., Anguelov, D., Erhan, D., Vanhoucke, V., Rabinovich, A.: Going deeper with convolutions. In: Proceedings of the IEEE Conference on Computer Vision and Pattern Recognition, pp. 1–9 (2015)
14. Razavian, A.S., Sullivan, J., Carlsson, S., Maki, A.: Visual instance retrieval with deep convolutional networks. arXiv preprint arXiv:1412.6574 (2014)
15. Babenko, A., Slesarev, A., Chigorin, A., Lempitsky, V.: Neural codes for image retrieval. In: Fleet, D., Pajdla, T., Schiele, B., Tuytelaars, T. (eds.) ECCV 2014. LNCS, vol. 8689, pp. 584–599. Springer, Cham (2014). doi:10.1007/978-3-319-10590-1_38
16. Fidalgo, E., Alegre, E., González-Castro, V., Fernández-Robles, L.: Compass radius estimation for improved image classification using edge-sift. Neurocomputing **197**, 119–135 (2016)
17. Lee, H., Grosse, R., Ranganath, R., Ng, A.Y.: Convolutional deep belief networks for scalable unsupervised learning of hierarchical representations. In: Proceedings of the 26th Annual International Conference on Machine Learning, pp. 609–616. ACM (2009)
18. Girshick, R., Donahue, J., Darrell, T., Malik, J.: Rich feature hierarchies for accurate object detection and semantic segmentation. In: Proceedings of the IEEE Conference on Computer Vision and Pattern Recognition, pp. 580–587 (2014)
19. Everingham, M., Van Gool, L., Williams, C.K., Winn, J., Zisserman, A.: The PASCAL visual object classes (VOC) challenge. Int. J. Comput. Vis. **88**(2), 303–338 (2010)
20. He, K., Zhang, X., Ren, S., Sun, J.: Spatial pyramid pooling in deep convolutional networks for visual recognition. In: Fleet, D., Pajdla, T., Schiele, B., Tuytelaars, T. (eds.) ECCV 2014. LNCS, vol. 8691, pp. 346–361. Springer, Cham (2014). doi:10.1007/978-3-319-10578-9_23
21. Girshick, R.: Fast R-CNN. In: Proceedings of the IEEE International Conference on Computer Vision, pp. 1440–1448 (2015)
22. Fernández-Robles, L., Castejón-Limas, M., Alfonso-Cendón, J., Alegre, E.: Evaluation of clustering configurations for object retrieval using SIFT features. In: Ayuso Muñoz, J.L., Yagüe Blanco, J.L., Capuz-Rizo, S.F. (eds.) Project Management and Engineering. LNMIE, pp. 279–291. Springer, Cham (2015). doi:10.1007/978-3-319-12754-5_21
23. Jia, Y., Shelhamer, E., Donahue, J., Karayev, S., Long, J., Girshick, R., Guadarrama, S., Darrell, T.: Caffe: convolutional architecture for fast feature embedding. In: Proceedings of the 22nd ACM International Conference on Multimedia, pp. 675–678. ACM (2014)
24. Dong, C., Loy, C.C., He, K., Tang, X.: Learning a deep convolutional network for image super-resolution. In: Fleet, D., Pajdla, T., Schiele, B., Tuytelaars, T. (eds.) ECCV 2014. LNCS, vol. 8692, pp. 184–199. Springer, Cham (2014). doi:10.1007/978-3-319-10593-2_13

Person Re-Identification Using Partial Least Squares Appearance Modelling

Gregory Watson(✉) and Abhir Bhalerao

Department of Computer Science, University of Warwick, Coventry, UK
{g.a.watson,abhir.bhalerao}@warwick.ac.uk

Abstract. Person Re-Identification is an important task in surveillance and security systems. Whilst most methods work by extracting features from the entire image, the best methods improve performance by prioritising features from foreground regions during the feature extraction stage. In this paper, we propose the use of a Partial Least Squares Regression model to predict the skeleton of a person, allowing us to prioritise features from a person's limbs rather than from the background. Once the foreground area has been identified, we use the LOMO [9] and Salient Colour Names [21] features. We then use the XQDA [9] Distance Metric Learning method to compute the distance between each of the feature vectors. Experiments on VIPeR [4], QMUL GRID [12–14] and CUHK03 [8] data sets demonstrate significant improvements against state-of-the-art.

1 Introduction

Person Re-Identification, or simply ReID, is the process of automatically identifying someone from a gallery of images that has the same identity as a person presented in an new image, and it has a number of important applications in surveillance, people-monitoring and biometrics. Re-Identification is challenging because often example images are taken with non-overlapping cameras, e.g. for a CCTV network, and consequently the images will exhibit large variations in person pose, illumination, and resolution. There are generally two main components in most ReID systems, feature extraction and distance metric learning. Feature extraction defines the process of obtaining a robust descriptor of the person, using features such as colour and texture, often chosen because of their robustness to varying illumination.

To overcome the problem of pose variation, many methods, such as [21], split person images into several bands of stripes, and extract colour histograms separately. This aims to maintain spatial information and enables matching to be done area-by-area rather than image-by-image, improving results. However, background information is also preserved, leading to some information irrelevant to the matching process. In [21], a Gaussian distribution is used to weight pixels according to their distance from the centre of the image where they are more likely to represent a persons body, but is less successful when a person is walking

S. Battiato et al. (Eds.): ICIAP 2017, Part II, LNCS 10485, pp. 25–36, 2017.
https://doi.org/10.1007/978-3-319-68548-9_3

perpendicular to the camera and has their legs spread widely apart. Symmetry-Driven Accumulation of Local Features is proposed in [2], where the person image is divided into three parts - the head, torso and legs, and a vertical axis of symmetry is found. This is achieved by finding vertical axes which separate regions with strongly different appearances. Then, a HSV histogram, Maximally Stable Colour Regions [3] and Recurrent Highly Structured Patches [2] are extracted. The use of three parts as well as a vertical axis of symmetry maintains some spatial information by allowing the system to partially know which areas represent a person's body. However, it still assumes that foreground information is always closer to the centre of the image.

Another method that has been used widely for foreground modelling is Stel Component Analysis (SCA) [5] and attempts to capture the structure of images of a given type, by splitting the image into small areas (stels) that have a common feature distribution [2,21]. However, if any single component of an image has a wider feature distribution, or if its feature distribution is similar to the background, both regions may be merged. This may lead to a significant portion of the image being misclassified. For feature extraction, Yang et al. [21] proposed Salient Colour Names, where pixels colours are quantised by their distance from sixteen named colours in the RGB space. The authors argue that whilst the foreground information is highly important, background information may also be important to give context, and extract features from both the foreground and the background, but placing priority on the foreground regions. We employ a similar approach to feature weighting, extracting features from the entire image but giving a higher weighting to those from the foreground. In LOMO [9], each person image is split into ten by ten pixel patches with an overlap of five pixels in each dimension. A HSV joint histogram and a Scale Invariant Local Ternary Pattern (SILTP) texture histogram [11] is then extracted from each patch. Afterwards, each row of patches is analysed, with the highest value in each bin taken to form the final histogram descriptor across that row. The histograms for each row are summed to form the descriptor. This helps achieve some invariance to viewpoint changes. The image is then downscaled by a factor of two and four and the process is repeated. The feature descriptors are then concatenated together. Our proposed foreground segmentation algorithm is used with the LOMO and Salient Colour Names [21] features, extracting from the entire image whilst weighting pixels more highly from the foreground.

In recent years, multi-layer convolutional neural networks (CNNs) have been shown to be effective for the ReID matching problem, e.g. [8], in some cases out-performing traditional feature extraction and matching learning methods. Because CNNs consists of many millions of weight parameters which have to be learned though training, they require many thousands of training samples, which restricts their use on some limited gallery ReID data, and where they are effective, the matching requires re-presentation of all the gallery data to the network during matching, and thus can be inefficient to use. One way to generalise training data is by data augmentation through warping, but the set of transformations required allow for small view point changes, but pose changes cannot easily be made without an appearance model of people.

In this paper, we propose a person Re-Identification method which predicts the skeleton and relative widths of the torso and limbs of a person prior to the feature extraction stage. We use supervised learning in order to calculate a regression between the appearance of an image and the skeleton landmarks and is achieved through Partial Least Squares. We show using this method that the Rank-1 rate can be significantly improved in a traditional ReID approach. We detail the method and present comparative experimental results using the VIPeR [4], QMUL GRID [12–14] and CUHK03 [8] data sets. In conclusion, we make proposals of how our appearance model might be used for video ReID and perhaps in conjunction with a CNN for data augmentation during network training.

2 Method

In this section, we describe our method (Fig. 1) which uses a novel skeleton fitting approach to model the image foreground, and then it is combined with robust feature extraction and a distance metric learning to perform matching.

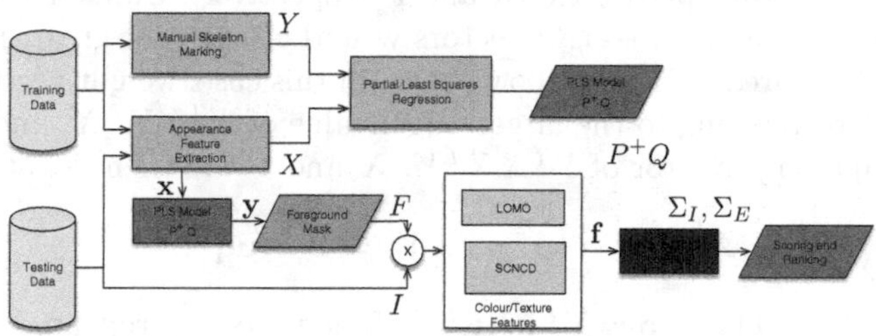

Fig. 1. System diagram of the proposed method showing PLS model building, feature extraction and matching.

2.1 Partial Least Squares Foreground Appearance Modelling

To predict the skeleton of each person, we learn a regression between the appearance of a person image and a set of landmarks which define a persons skeleton: head, torso and limbs. We use a Partial Least Squares Regression model to compute the regression.

Let $X = (\mathbf{x}_1, \mathbf{x}_2, \ldots, \mathbf{x}_n)$ be a matrix of image appearances, where each column is a feature vector extracted from a set of n images. Each vector consists of the concatenation of several local shape and texture features extracted from different patches of an image. Similarly, $Y = (\mathbf{y}_1, \mathbf{y}_2, \ldots, \mathbf{y}_n)$ is a matrix where each column consists of a series of co-ordinates representing skeleton keypoints, one for each corresponding image appearance. In constructing $\mathbf{y}$, each skeleton limb is specified by three points: $(\mathbf{m}, \mathbf{n}, \mathbf{o})$, representing the ends of the limbs along their centre line and the point located perpendicular to its axis, defining the limb width.

Fig. 2. Examples of PLS skeleton fitting. Each set of five images shows: original, input HOG appearance features, ground-truth skeleton data, PLS fitted skeleton result, foreground segmented mask.

Partial Least Squares [16,17] (PLS) is used to find a linear decomposition of X and Y such that:

$$X = TP^T + E, \qquad Y = UQ^T + F, \tag{1}$$

where T and U are the score matrices, P and Q are the loading matrices, and E and F represent residual matrices of X and Y respectively. Unlike PCA, the PLS algorithm initially computes weight vectors $\mathbf{w}$ and $\mathbf{c}$ such that greatest variation in X *and* Y is captured. It can be shown that, in this case, weight vector $\mathbf{w}$ is the eigenvector corresponding to the largest eigenvalue of X^TYY^TX, and similarly, $\mathbf{c}$ is the principal eigenvector of Y^TXX^TY. X and Y are then deflated by

$$X^{'} = X - \mathbf{t}\mathbf{p}^T, \qquad Y^{'} = Y - \mathbf{u}\mathbf{q}^T, \tag{2}$$

where $\mathbf{p} = X^T\mathbf{t}/(\mathbf{t}^T\mathbf{t})$ and $\mathbf{q} = Y^T\mathbf{u}/(\mathbf{u}^T\mathbf{u})$. The process is repeated for $X^{'}$ and $Y^{'}$ until the residuals are below a required threshold and all score vectors $\mathbf{t}$ and $\mathbf{u}$ have been extracted.

Having calculated the score matrices, T and U, to predict a skeleton given a test instance of an image appearance, $\mathbf{x}_i$, then

$$\hat{\mathbf{y}}_i = P^+Q\mathbf{x}_i \tag{3}$$

where $P = (T^TT)^{-1}T^TX'$ and $Q = (U^TU)^{-1}U^TY'$ and P^+ is the Moore-Penrose inverse of P.

To learn the PLS regression, we take a set of training images and extract HOG features on a regular grid across the image to form our appearances, and regress them to corresponding landmarks by solving for P^+Q. In our method, the appearance X is represented by these Histogram of Oriented Gradients features. Figure 2 shows examples of skeleton fitting on a trained PLS model. The PLS skeleton fitting result can be used to create a image mask, which is applied to the original image to produce a person silhouette. The method can be trained to work with both frontal and sideways views.

2.2 Foreground Feature Extraction and Feature Weighting

Having located the foreground regions (person) with the PLS skeleton fitting, we apply a feature extraction stage and use weighted Local Maximal Occurrence (LOMO) [9] and Salient Colour Names [21].

Weighted LOMO. We modify LOMO [9] such that foreground regions are prioritised over background by feature weighting. LOMO begins by applying a colour normalisation step using the Retinex algorithm [7] to make the images of the same person from different cameras with different illumination conditions appear more consistent. LOMO features are taken from overlapping image patches across the image. HSV histograms and SILTP histograms over three scales [11] are integrated in a combined feature. Then by taking the maximum value in each bin, some invariance to viewpoint variations is gained.

In [18,21], the authors discuss the benefits of background in providing context for the problem of People Re-Identification. However, both methods extract not only from the entire image, but also from the foreground areas again, concatenating both to form the final feature. In our work, because the skeleton of each person has been estimated, we use an image mask to weight the LOMO features as they are accumulated by the percentage of predicted foreground in each feature patch which overlaps with the foreground mask:

$$\mathbf{f}_w(B) = \frac{|F \cap B|}{|B|}\mathbf{f}(B), \tag{4}$$

where B is the set of pixels in the image patch and F is the set of pixels labelled foreground. Once all patches in a row have been weighted, the maximum value for each bin in that row can be taken towards the final descriptor. In the experiments presented below, we use the code given by [9] to extract the LOMO features, and alter them as described prior in order to prioritise the features from the foreground.

Salient Colour Names. Salient Colour Names [21] define sixteen coordinates in the RGB space of carefully chosen colours, e.g. fuchsia, blue, aqua, lime, etc., extending the Colour Names [19] method, which has only eleven. The RGB colour space is then first quantized into $32 \times 32 \times 32$ indexes, $\mathbf{d}$, with each index having 512 quite similar colours, $\mathbf{w}$. The set of colour names are defined as coordinates in the RGB space, $Z = \{\mathbf{z_1}, \mathbf{z_2}, \ldots, \mathbf{z_{16}}\}$, and a mapping (posterior probability) from a given index colour $\mathbf{d}$ and a colour name distribution over Z is calculated. The process is a form of vector quantisation, and similar to the concept of a Bag of Words, and has the advantage for being able to assign multiple similar colours to similar colour name distributions.

Then a mapping posterior probability distribution is factorised into two terms:

$$p(\mathbf{z}|\mathbf{d}) = \sum_{j=1}^{512} p(\mathbf{z}|\mathbf{w}_j)p(\mathbf{w}_j|\mathbf{d}), \tag{5}$$

where the first term is a distribution of probabilities $p(\mathbf{z}|\mathbf{w})$ calculated as normally distributed variates of the closest K colour name given a quantized colours $\mathbf{w}_j$ (i.e. one of those that fall within a discretization bin $\mathbf{d}_i$). Note that variance of this distribution is estimated over $K-1$ colours *not* in the nearest K nearest neighbour set, $\frac{1}{K-1}\sum_{\mathbf{z}_k \neq \mathbf{z}} p(\mathbf{z}_k|\mathbf{z}_j)$:

$$p(\mathbf{z}|\mathbf{w}_j) = \frac{\exp(-\|\mathbf{z}-\mathbf{w}_j\|^2 / \frac{1}{K-1}\sum_{\mathbf{z}_i \neq \mathbf{z}} \|\mathbf{z}_i - \mathbf{w}_j\|^2)}{\sum_k \exp(-\|\mathbf{z}-\mathbf{w}_k\|^2 / \frac{1}{K-1}\sum_{\mathbf{z}_i \neq \mathbf{z}} \|\mathbf{z}_i - \mathbf{w}_j\|^2)} \tag{6}$$

The second term of Eq. 5, $p(\mathbf{w}|\mathbf{d})$, models the variance $\mathbf{w}_j$ at sample $\mathbf{d}_i$, against its mean value, μ, capturing how likely many similar colours are being captured at this position. The more similar colours that are present, the larger this value.

$$p(\mathbf{w}_j|\mathbf{d}) = \frac{\exp(-\alpha|\mathbf{w}_j - \mu|^2)}{\sum_{k=1}^{512} \exp(-\alpha|\mathbf{w}_k - \mu|^2)}. \tag{7}$$

Together the two terms capturing how similar or salient colour names are, to the sample colour indexes $\mathbf{d}_i$. Multiple similar colours result in similar colour name distributions and thus providing greater illumination variation. The salient colour name colour distributions can be computed off-line, and so are computationally efficient when used.

Finally, similarly to the LOMO feature, a log transform is applied to the Salient Colour Names features, and each histogram is normalised to a unit length. This descriptor is concatenated with the weighted LOMO features to form our final feature descriptor.

2.3 Distance Metric Learning

A metric for measuring the distance between feature descriptors is used at the matching stage. KISSME [6] calculates the distance between two feature vectors as:

$$\tau_M^2(\mathbf{f_i}, \mathbf{f_j}) = (\mathbf{f}_i - \mathbf{f}_j)^T(\Sigma_I^{-1} - \Sigma_E^{-1})(\mathbf{f}_i - \mathbf{f}_j). \tag{8}$$

with the intra-personal, Σ_I, and extra-personal, Σ_E, scatter matrices. Here, we use Cross-view Quadratic Discriminant Analysis (XQDA) [9], which extends KISSME. With KISSME, it is possible to perform dimensionality reduction prior to estimating Σ_I and Σ_E by performing PCA on the input vectors. XQDA however considers the metric learning and the dimensionality reduction together. If D is the original dimensionality of the data and R the required reduced dimensionality, XQDA learns a subspace $W = (\mathbf{w}_1, \mathbf{w}_2, \ldots, \mathbf{w}_R) \in \mathbb{R}^R$, whilst simultaneously learning a distance function:

$$d_w(\mathbf{f}_i, \mathbf{f}_j) = (\mathbf{f}_i - \mathbf{f}_j)^T W(\Sigma_I^{'-1} - \Sigma_E^{'-1})W^T(\mathbf{f}_i - \mathbf{f}_j) \tag{9}$$

where $\Sigma_I^{'} = W^T \Sigma_I W$ and similarly $\Sigma_E^{'} = W^T \Sigma_E W$. Directly optimising d_w is not possible because of the presence of two inverse matrices. As the distribution of intra-personal and extra-personal distances have zero mean, a traditional LDA cannot be used to determine W. Instead, for any projection direction $\mathbf{w}$, which is a column of W, it is possible to maximise the ratio of variances $\sigma_E^2(\mathbf{w})/\sigma_I^2(\mathbf{w})$. Since however, $\sigma_E^2(\mathbf{w}) = \mathbf{w}^T \Sigma_E \mathbf{w}$ and similarly $\sigma_I^2(\mathbf{w}) = \mathbf{w}^T \Sigma_E \mathbf{w}$, the objective function is the Generalised Rayleigh Quotient:

$$J(\mathbf{w}) = \frac{\mathbf{w}^T \Sigma_E \mathbf{w}}{\mathbf{w}^T \Sigma_I \mathbf{w}}. \tag{10}$$

We can solve for $\mathbf{w}$ by a generalised eigenvalue decomposition in the same way as LDA is solved by maximising

$$\max_{\mathbf{w}} \mathbf{w}^T \Sigma_E \mathbf{w}, \; s.t. \; \mathbf{w}^T \Sigma_I \mathbf{w} = 1. \tag{11}$$

The columns of W are the R eigenvectors of $\Sigma_I^{-1} \Sigma_E$ taken in decreasing order of eigenvalue. Liao et al. [9], have advice on how to make the distance learning calculation robust and computationally efficient. Features extracted from training images using the ground-truth skeletons can be passed to XQDA in order to learn a distance metric.

3 Results and Discussion

In our experimentation, we define a skeleton of twenty-nine points representing fourteen limbs, where each limb consists of two end-points, and a third point to locate the limb edge. The bottom point of each limb is also the top point of the following limb. For our PLS models, we extract the top fifteen components for the Skeleton appearance models. We use three data sets for our experiments:

- VIPeR: VIPeR [4] contains 632 image pairs, each with a size of 128×48 pixels. Images in the VIPeR data set are captured using two cameras, and have large variations with pose and illumination, and also contain occlusion.
- QMUL GRID: QMUL GRID [12–14] consists of 250 person image pairs, taken from eight disjoint cameras in an underground public transport station. In addition, there are also 775 identities consisting of only one image. Images come in varying sizes. This data set suffers from severe occlusion, as well as variations in pose and illumination. Colour is not as vibrant as in the other data sets, and significant noise is present in the images.
- CUHK03: CUHK03 [8] is the largest widely-used data set in this area, consisting of 1360 identities across two cameras per identity. Each individual has an average of about five images per camera view. Images are obtained by taking stills from a video sequence over several months, and thus suffer from varying illumination conditions. This data set also suffers from pose variations and occlusion. Images are cropped using both manual cropping and a person detector.

From all data sets, we extract Histogram of Oriented Gradients (HOG) features from the standard, non-Retinex images, using a cell size of 6 pixels and a block size of 2 pixels. From the VIPeR data set, we extract the HOG features from the V channel of the HSV colour space, in order to build PLS regression models. The VIPeR data set provides person orientation information for each image, and thus we can split the images in to two partitions - perpendicular to the camera or otherwise. We build separate PLS models for each partition and learn a classifier for model selection. Examples of skeleton fits and the best and worst fitting results on VIPeR are shown in Fig. 3. The fitting fails on the few images where a person has their arms raised above their heads.

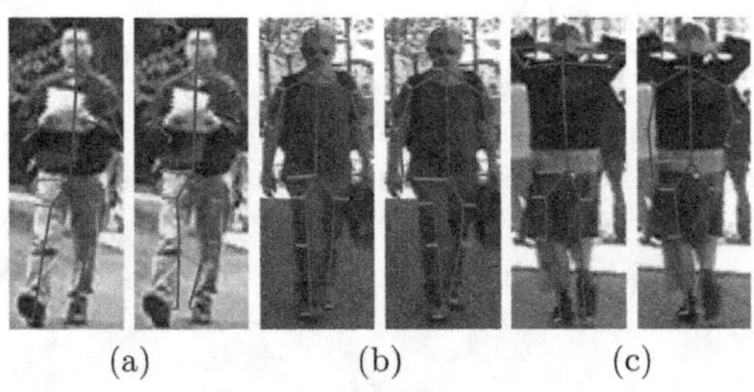

(a) (b) (c)

Fig. 3. Examples of the ground-truth and predicted skeletons from VIPeR: (a) A random image with a RMSE of 3.8 pixels; (b) The image with the minimum RMSE of 1.8 pixels; (c) The image with the maximum RMSE of 16.0 pixels. The average RMSE is 5.2 pixels.

We use the experimental procedures used in various literature, e.g. [2,9]. The training and testing sets are split randomly into even sized sets, with 316 identities used for each. We repeat our experiment ten times, averaging the scores to produce our final result. From Table 1, we can see that by concatenating the original LOMO features with features primarily from the foreground (PLSAM(v1)), we can achieve an increase in all measured Rank scores and demonstrates that our method produces a much more robust person descriptor. When concatenated with Salient Colour Names features (PLSAM(v2)), the results increase further. This is unsurprising, due to how distinct are the clothing colours in the VIPeR data set. Overall, we can see a 4.0% increase when comparing our method to the state-of-the-art. The CMC curve plots are given in Fig. 4.

For the QMUL GRID data set, we resize each image to 128 $\times$ 48 pixels. PLS models are built from HOG features from the V channel of the HSV colour model. Whilst we use two view PLS models for the VIPeR data set, only a single model is used for the QMUL GRID data set because most people in this data set are facing either towards or away from the camera. Examples of skeleton fits and the best and worst fitting results on QMUL GRID are shown in Fig. 5. Again, because of lack of sufficient training examples, the fitting fails on people with raised arms. We use the experimental protocols used in various literature [9, 15]. The training and testing sets are split evenly, with 125 identities used for

Table 1. A comparison of state-of-the-art methods: VIPeR [4] data set with 316 person identities were allocated for training, and 316 for testing; QMUL GRID [12–14] data set with 125 person identities were allocated for training, and 900 for testing, where the testing identities contained 125 image pairs and 775 single images; CUHK03 [8] data set with 1160 person identities were allocated for training, and 100 for testing. For the test set, one image of each identity was taken to form the gallery set. Every probe image in the test set was compared to every gallery image in the test set. PLSAM(v2) is with weighted LOMO and Salient Colour Names features and XQDA; PLSAM(v1) is with weighted LOMO and XQDA.

	VIPeR				QMUL GRID				CUHK03			
	r = 1	r = 5	r = 10	r = 20	r = 1	r = 5	r = 10	r = 20	r = 1	r = 5	r = 10	r = 20
PLSAM (v2)	**46.3**	**75.0**	**85.6**	**93.9**	**26.7**	**47.9**	**59.0**	**68.2**	**65.2**	**89.8**	**95.0**	97.9
PLSAM (v1)	42.8	71.9	82.0	91.9	23.9	41.8	51.0	61.4	64.6	89.2	94.9	**98.1**
Null Space [23]	42.3	71.5	82.9	92.1	-	-	-	-	58.9	85.6	92.5	96.3
MLAPG [10]	40.7	69.9	82.3	92.4	16.6	33.1	41.2	53.0	58.0	87.1	94.7	98.0
DeepList [20]	40.5	69.2	81.0	91.2	-	-	-	-	55.9	86.3	93.7	98.0
LOMO+XQDA [9]	40.3	68.3	80.9	91.1	17.3	36.3	44.8	55.4	54.9	85.3	92.6	97.1
FPNN [8]	-	-	-	-	-	-	-	-	20.7	50.9	67.0	83.0

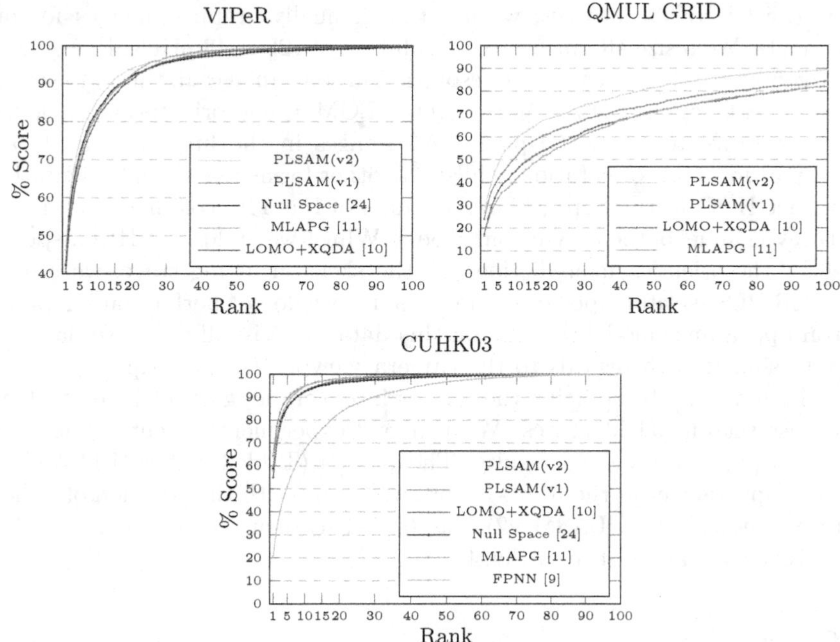

Fig. 4. CMC on the VIPeR data set [4], QMUL GRID data set [12–14] and, CUHK03 data sets [8]. All of our CMC curves are single-shot results. Results are reproduced from [8–10,22].

each. The 775 images which do not belong to an image pair are added to the gallery set and we run our experiment ten times, averaging the scores to produce the final result. Table 1 again shows that both PLSAM(v1) and PLSAM(v2)

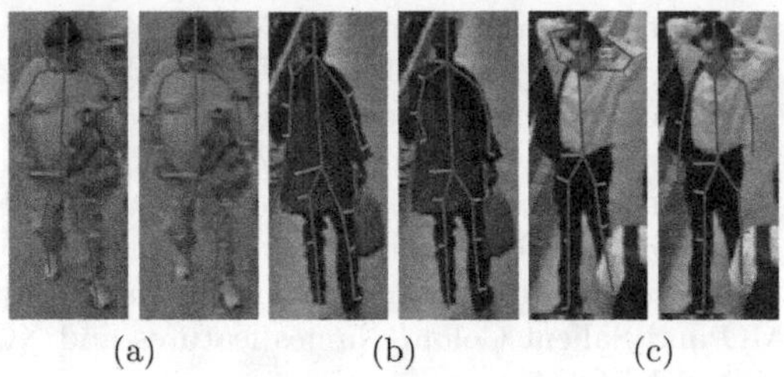

(a) (b) (c)

Fig. 5. Examples of the ground-truth and the predicted skeleton from the QMUL GRID data set: (a) A random image with a RMSE of 3.9 pixels; (b) The image with the minimum RMSE of 2.3 pixels; (c) The image with the maximum RMSE of 17.7 pixels. The average RMSE over the entire test is 5.3 pixels.

out-perform other methods. As the cameras are located in a busy station, this data has a high level of occlusion and overlapping people in the background and particularly benefits from foreground modelling, producing more representative person descriptors. The CMC curve is plotted in Fig. 4.

For the CUHK03 data set, we use the manually cropped images for our experiments. We resize all images to a resolution of 128 × 48 pixels. However, to take advantage of the higher source resolution in some images in this data set, for the feature extraction (LOMO and weighted LOMO) and orientation modelling stages, we resize to 160 × 60 pixels. All images in the first camera of each camera pair form one orientation, whilst the other forms the second orientation. We extract HOG features from the Y channel of the YIQ colour model, rather than the V channel of the HSV colour model. Whilst we build a CUHK03-specific model for the orientation prediction, for the skeleton fitting stage, we re-used the VIPeR PLS skeleton appearance models, i.e. we do not perform any separate skeleton appearance model training on this data set. Visually the two data sets are quite similar with regards to the camera viewpoints. Our experimentation protocol follow [8,9] by splitting the images into a training set of 1160 identities and a test set of 100 identities. We run our experiments twenty times, and average to produce the final results. The results (Table 1), show PLSAM(v1) gives an improvement in the Rank-1 score by 5.7%. With the addition of Salient Colour Names features, PLASM(v2), the Rank-1 score instead improves by 6.3%. The CMC curve can be seen in Fig. 4.

4 Conclusions

In this paper, we demonstrated the advantages of using a skeleton appearance model to identify locations of torso and limbs from a person image. Partial Least Squares regression is used on appearance features from training images, and corresponding hand-marked skeleton data to build a model for predicting the foreground region of an new image. We use the foreground to locate and prioritise feature extraction to create a robust feature descriptor less sensitive to background clutter and occlusion. Our comparative analysis, using state-of-art feature extraction (LOMO and Salient Colour Names) and XQDA distance metric learning, demonstrate a superior matching accuracy when feature extraction is weighted by our foreground estimation. Experiments on the VIPeR, QMUL GRID and CUHK03 data sets show that the proposed method achieves an improvement when vs. the LOMO feature of 6.0%, 9.4% and 10.3% respectively in the Rank-1 matching rate. An improvement of 4.0%, 9.4% and 6.3% respectively in the Rank-1 matching rate is observed vs. other state-of-the-art methods.

In the case of CUHK03, we show that the skeleton fitting foreground model learnt on one data set, in this case VIPeR, generalises to a different camera view without the need to retrain. For the orientation in VIPeR, we fitted two separate foreground models: one for frontal views and one for sideways views. The model selection process seems to work well in our experiments, and might be extended to multiple views from a network of cameras.

Our further work is focused on using the skeleton fitting in the context of a deep learning (CNN) architecture [1]. We are working on using the PLS model to train a deconvolution CNN to perform foreground modelling and will compare the performance. The non-linearities inherent in neural network regression models might preclude the need to use multiple linear PLS models for varying viewpoints. We also think the PLS models may prove useful in data augmentation for CNN training where the number of training examples is limited, which might be achieved by synthesising appearance model instances from a PCA of the learnt space of variation, e.g. [24].

References

1. Ahmed, E., Jones, M., Marks, T.K.: An improved deep learning architecture for person re-identification. In: Proceedings the IEEE Conference on CVPR, pp. 3908–3916 (2015)
2. Farenzena, M., Bazzani, L., Perina, A., Murino, V., Cristani, M.: Person re-identification by symmetry-driven accumulation of local features. In: Proceeding IEEE Conference on CVPR, pp. 2360–2367, June 2010
3. Forssen, P.E.: Maximally stable colour regions for recognition and matching. In: Proceedings of the IEEE Conference on CVPR, pp. 1–8, June 2007
4. Gray, D., Brennan, S., Tao, H.: Evaluating appearance models for recognition, reacquisition, and tracking. In: Proceedings IEEE International Workshop on Performance Evaluation for Tracking and Surveillance (PETS), vol. 3, no. 5, October 2007
5. Jojic, N., Perina, A., Cristani, M., Murino, V., Frey, B.: Stel component analysis: modeling spatial correlations in image class structure. In: IEEE Conference on CVPR, pp. 2044–2051, June 2009
6. Köstinger, M., Hirzer, M., Wohlhart, P., Roth, P.M., Bischof, H.: Large scale metric learning from equivalence constraints. In: IEEE Conference on CVPR, pp. 2288–2295, June 2012
7. Land, E.H., McCann, J.J.: Lightness and retinex theory. JOSA **61**(1), 1–11 (1971)
8. Li, W., Zhao, R., Xiao, T., Wang, X.: Deep filter pairing neural network for person re-identification. In: Proceedings of the IEEE Conference on CVPR, pp. 152–159 (2014)

9. Liao, S., Hu, Y., Zhu, X., Li, S.Z.: Person re-identification by local maximal occurrence representation and metric learning. In: Proceedings of the IEEE Conference on CVPR, pp. 2197–2206 (2015)
10. Liao, S., Li Stan, Z.: Efficient PSD constrained asymmetric metric learning for person re-identification. In: Proceedings of the IEEE Conference on ICCV, pp. 3685–3693, December 2015
11. Liao, S., Zhao, G., Kellokumpu, V., Pietikainen, M., Li, S.Z.: Modeling pixel process with scale invariant local patterns for background subtraction in complex scenes. In: IEEE Conference on CVPR, pp. 1301–1306, June 2010
12. Liu, C., Gong, S., Loy, C.C., Lin, X.: Person re-identification: what features are important? In: Fusiello, A., Murino, V., Cucchiara, R. (eds.) ECCV 2012. LNCS, vol. 7583, pp. 391–401. Springer, Heidelberg (2012). doi:10.1007/978-3-642-33863-2_39
13. Loy, C.C., Xiang, T., Gong, S.: Multi-camera activity correlation analysis. In: Proceedings of the IEEE Conference on CVPR, pp. 1988–1995, June 2009
14. Loy, C.C., Xiang, T., Gong, S.: Time-delayed correlation analysis for multi-camera activity understanding. IJCV **90**(1), 106–129 (2010)
15. Loy, C.C., Liu, C., Gong, S.: Person re-identification by manifold ranking. In: Proceedings of the IEEE Conference on ICIP, pp. 3567–3571, September 2013
16. Maitra, S., Yan, J.: Principle component analysis and partial least squares: two dimension reduction techniques for regression. Appl. Multivar. Stat. Models **79**, 79–90 (2008)
17. Rosipal, R., Krämer, N.: Overview and recent advances in partial least squares. In: Saunders, C., Grobelnik, M., Gunn, S., Shawe-Taylor, J. (eds.) SLSFS 2005. LNCS, vol. 3940, pp. 34–51. Springer, Heidelberg (2006). doi:10.1007/11752790_2
18. Russakovsky, O., Lin, Y., Yu, K., Fei-Fei, L.: Object-centric spatial pooling for image classification. In: Fitzgibbon, A., Lazebnik, S., Perona, P., Sato, Y., Schmid, C. (eds.) ECCV 2012. LNCS, pp. 1–15. Springer, Heidelberg (2012). doi:10.1007/978-3-642-33709-3_1
19. Van De Weijer, J., Schmid, C., Verbeek, J., Larlus, D.: Learning color names for real-world applications. IEEE Trans. Image Process. **18**(7), 1512–1523 (2009)
20. Wang, J., Wang, Z., Gao, C., Sang, N., Huang, R.: DeepList: learning deep features with adaptive listwise constraint for person re-identification. IEEE Trans. Circ. Syst. Video Technol. **27**, 513–524 (2017)
21. Yang, Y., Yang, J., Yan, J., Liao, S., Yi, D., Li, S.Z.: Salient color names for person re-identification. In: Fleet, D., Pajdla, T., Schiele, B., Tuytelaars, T. (eds.) ECCV 2014. LNCS, vol. 8689, pp. 536–551. Springer, Cham (2014). doi:10.1007/978-3-319-10590-1_35
22. Zhao, R., Ouyang, W., Wang, X.: Learning mid-level filters for person re-identification. In: Proceedings of the IEEE Conference on CVPR, pp. 144–151 (2014)
23. Zhang, L., Xiang, T., Gong, S.: Learning a discriminative null space for person re-identification. In: Proceedings of the IEEE Conference on CVPR, pp. 1239–1248 (2016)
24. Zhang, Q., Bhalerao, A., Helm, E., Hutchinson C.: Active shape model unleashed with multi-scale local appearance. In: Proceedings of the IEEE Conference on ICIP, pp. 4664–4668 (2015)

Gender and Expression Analysis Based on Semantic Face Segmentation

Khalil Khan[1], Massimo Mauro[2], Pierangelo Migliorati[1](✉), and Riccardo Leonardi[1]

[1] DII, University of Brescia, Via Branze, 38, Brescia, Italy
{khalil.khan,pierangelo.migliorati,riccardo.leonardi}@unibs.it
[2] Yonder S.r.l., Via Praga, 5, Trento, Italy
http://yonderlabs.com

Abstract. The automatic estimation of gender and face expression is an important task in many applications. In this context, we believe that an accurate segmentation of the human face could provide a good information about these mid-level features, due to the strong interaction existing between facial parts and these features. According to this idea, in this paper we present a gender and face expression estimator, based on a semantic segmentation of the human face into six parts. The proposed algorithm works in different steps. Firstly, a database consisting of face images was manually labeled for training a discriminative model. Then, three kinds of features, namely, location, shape and color have been extracted from uniformly sampled square patches. By using a trained model, facial images have then been segmented into six semantic classes: hair, skin, nose, eyes, mouth, and back-ground, using a Random Decision Forest classifier (RDF). In the final step, a linear Support Vector Machine (SVM) classifier was trained for each considered mid-level feature (i.e., gender and expression) by using the corresponding probability maps. The performance of the proposed algorithm was evaluated on different faces databases, namely FEI and FERET. The simulation results show that the proposed algorithm outperforms the state of the art competitively.

Keywords: Face segmentation · Gender classification · Face expression classification

1 Introduction

Human Face Segmentation (HFS) is an active research area in computer vision and multi-media signal processing communities until recently. In fact, face segmentation plays a crucial role in many face related applications such as, face tracking, human computer interaction, face recognition, face expression analysis, and head pose estimation. Moreover, face analysis could be useful also for the estimation of shot scale [1], whereas expression recognition could also be useful for the affective characterization of shots [2]. Performance of all these, and

S. Battiato et al. (Eds.): ICIAP 2017, Part II, LNCS 10485, pp. 37–47, 2017.
https://doi.org/10.1007/978-3-319-68548-9_4

of many other potential applications, can be boosted if a properly segmented face is provided in input. However, variations in illumination and visual angle, complex color background, different facial expressions and rotation of the head make it a challenging task.

Researches tried to solve many complicated problems of segmentation with semantic segmentation techniques. To investigate the problem of semantic segmentation, extensive research work has been carried out by PASCAL VOC challenge [3]. Haung et al. [4] performed joint work of face segmentation and pose estimation. They argued that features such as pose, gender and expression can be predicted easily if a well segmented face is provided in input. They used a database of 100 images with three simple poses: frontal, right and left profile images. Their algorithm segmented facial images into three semantic classes: skin, hair and background. Using the segmented image, the head pose was estimated. Psychology literature also confirmed their claim, as facial features are more illuminating for human visual system to recognize face identity [5,6]. Recently some attention have also been given to the approaches that uses the Convolutional Neural Networks (see, for example, [7]).

In some previous works, we already investigated the HFS challenge [8], and the problem of Head Pose Estimation (HPE) [9]. Unlike previous approaches which consider few classes, we extended the labeled set to six semantic classes. We performed also a preliminary experimental work on a small database of high resolution frontal images. We exploited position, color and shape features to build a discriminative model. The proposed classifier returned a probability value and a class label for each pixel of the testing image. A reasonable pixel-wise accuracy was obtained with the proposed algorithm. In this paper, we propose and test an algorithm which segments a human face image into six semantic classes (using RDF), and then, using the segmented image, address the problems of gender and expression classification (using SVM). In more detail, the proposed algorithm works in few steps. Initially, a database consisting of face images was manually labeled for training a discriminative model. Then, three kinds of features, namely, location, shape and color have been extracted from uniformly sampled square patches. Then, using a trained model, facial images have been segmented into six semantic classes: hair, skin, nose, eyes, mouth, and back-ground, using a Random Decision Forest classifier (RDF). In the end, a linear Support Vector Machine (SVM) classifier was trained for each considered mid-level feature (i.e., gender and expression) by using the corresponding probability maps. The reported results are encouraging, and confirm that extracted information are immensely helpful in the prediction of the considered mid-level features. The structure of the paper is as follow. In Sect. 2 we quickly review previous works on face segmentation and related applications. In Sect. 3, we describe the used HFS algorithm. In Sect. 4, we propose a framework which uses the segmentation results to try to solve GC, and EC problems. Section 5 is about experimental setup, and detail setting of the used databases. In Sect. 6 we discuss the obtained results. Final remarks are given in Sect. 7.

2 Related Works

Face labeling provides a robust representation by assigning each pixel a semantic label in the face image. Several authors proposed different algorithms to assign a semantic label to the different facial parts. Previous algorithms reported in HFS literature consider either three-classes or four-classes. In three-classes labeling methods, e.g.,[4,10–14], a semantic label is assigned to three prominent parts: skin, hair, and background. Similarly, four-classes labeling algorithms assign an additional label to clothes with skin, hair, and background [15–17].

Compared to state-of-the-art, in [8] we proposed a unified approach that generates a more complete labeling of facial parts. A database was also created consisting of manually labeled images. Manual labeling was performed with extreme care using Photo-shop software. An RDF classifier was trained by extracting location, color and shape features from square patches. By changing various parameters which tuned extracted features, the best possible configuration was explored. The output of the trained model is a segmented image with six classes labels; skin, nose, hair, mouth, eyes, and back.

Many methods have been proposed to try to solve the problems of GC and EC. Like HPE, GC and EC are classified into two categories: appearance-based and geometric-based approaches. Geometric-based methods are the same algorithms as model-based but with a different name in GC researchers community. Active Shape Modeling (ASM) [18] is used to locate facial landmarks both in GC and EC methods. A feature vector is created from these landmarks and a machine learning tool is used for training and prediction. However, ASM may fail to locate the landmark if lighting condition is changing, or if facial expressions are too much complicated. ASM also loose performance if the pose of the image is changed from frontal to extreme profile.

State-of-the-art shows that the three problems HPE, GC, and EC were handled separately. Haung et al. [4] is, at our best knowledge, the only work which combined HFS and HPE. The authors of the mentioned paper used a database where discrimination between consecutive poses was less, and three simple poses were used. We believe that if a good information is extracted from human face, all the three problems HPE, GC, and EC can be addressed in a single framework. We therefore extended our labeled data-set to six semantic classes. We used a challenging database Pointing'04, where difference between two consecutive poses was just 15°. We also integrated the framework to perform the task of GC and EC. The GC task was performed using both smiling and neutral images showing the fact that change in facial expression has little effect on performance of proposed algorithm. The obtained experimental results reflect that the proposed algorithm not only gives appreciable results for GC, but also outperforms EC on standard face databases FEI [19], and FERET [20].

3 The Face Segmentation Algorithm

The presentation of the adopted face segmentation algorithm is divided into two parts. Section 3.1 describes the features extracted from training images and how the label is assigned to each pixel. Section 3.2 describes how the classification is performed, and which classification strategy was used.

3.1 Feature Extraction

Most of the segmentation algorithms work at pixel or super-pixel level. We used pixel-based approach in our work. We considered square patches as processing primitives. We extracted patches from training and testing images while keeping a fixed step size. Every patch is classified by transferring the class label to the center pixel of each patch.

To capture information from spatial relationship between different classes, relative location of the pixel was used as a feature vector. Coordinates of the central pixel of each patch was extracted. Relative location of the central pixel at position (x, y) of a patch can be represented as:

$$f_{loc} = [x/W, y/H] \in R^2$$

where W is the width, and H is the height of the image.

Color features were extracted from patches by using HSV color space. A single feature vector was created for color by concatenating Hue, Saturation and Value histograms. We noted that the dimension of the patch and the number of bins used has an immense impact on the performance of the segmentation algorithm. We did a lot of experimentation previously, e.g., [8], to investigate the best possible setting for patch size and number of bins. Based on those experimental results, we concluded that patch size 16 × 16, and number of bins 32 is a good combination for our work. Using these values, feature vectors generated for each patch would be $f_{HSV}^{32} \in R^{96}$.

For extracting shape information we used Histogram of Oriented Gradient (HOG) [21] features. Each patch was transformed into HOG feature space. Like HSV color, we also investigated for proper patch dimension for HOG features. We found that results are best on a patch size 64 × 64. Using this patch size a feature vector created for each patch would be $f_{HOG}^{64 \times 64} \in R^{1764}$.

All the three features: spatial prior, HSV and HOG were then concatenated to form a single unique feature vector $f^I \in R^{1862}$, where f^I is the feature vector for patch $I(x, y)$.

3.2 Classification

A RDF classier was then trained using ALGLIB [22] implementation. The trained RDF classifier predicted probability value for each pixel. The class label was assigned to every pixel based on maximum probability:

$$\hat{c} = \arg\max_{c \in C} p(c|\mathbf{L}, \mathbf{C}, \mathbf{S})$$

where $C = \{skin, background, eye, nose, mouth, hair\}$ and $\mathbf{L}$, $\mathbf{C}$, and $\mathbf{S}$ are random variables for features f_{loc} (relative location), f_{HSV} (HSV color), and f_{HOG} (Shape information), respectively.

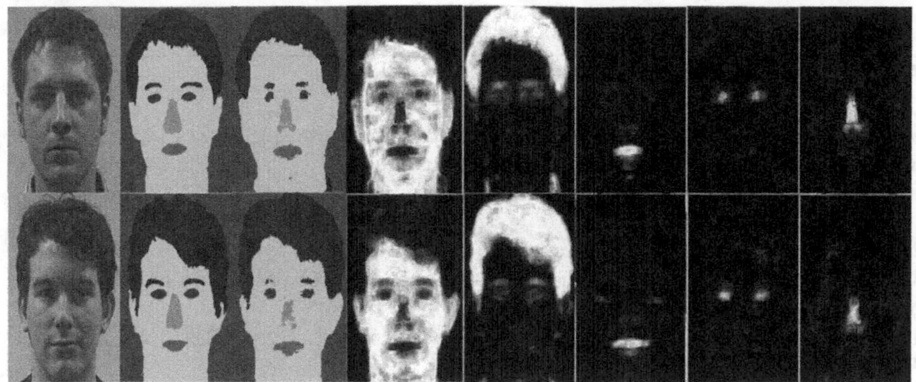

Fig. 1. Frontal images (Pointing'04 database): column 1: original images; column 2: ground truth images; column 3: images segmented by the proposed algorithm; column 4: probability map for skin; column 5: probability map for hair; column 6: probability map for mouth; column 7: probability map for eyes; column 8: probability map for nose.

4 The Proposed Gender and Expression Classification Algorithms

Rather than outputting the most likely class, we used probabilistic classification to get probability map for each class from HFS algorithm. Given a set C of classes $c \in C$, we created a probability map for each semantic class represented by $\boldsymbol{P}_{skin}$, $\boldsymbol{P}_{hair}$, $\boldsymbol{P}_{eyes}$, $\boldsymbol{P}_{nose}$, $\boldsymbol{P}_{mouth}$, and $\boldsymbol{P}_{back}$. The probability maps were computed by converting the probability of each pixel to a gray scale image. In probability maps, higher intensity represents higher value of probability for the most likely class on their respective position. Figure 1 shows two images of Pointing'04 database, and their respective probability maps for five classes. We investigated thoroughly which probability map is more helpful in GC and EC. A feature vector was created comprised of probability maps which is more helpful in GC and EC. We trained a linear SVM classifier using the feature vector. A binary SVM was trained for GC and EC. Our proposed algorithm flow is described in Table 1.

4.1 Gender Classification

Some features play a key role in gender discrimination in human faces. We categorize these features as: eyes, nose, mouth, hair, and forehead. Locating these features is easy for human, but it is a challenging task for machine. Our segmentation algorithm provides reasonable information about these main differences. The proposed HFS algorithm fully segment facial image into six parts: back, nose, mouth, eyes, skin, and hair. After a large pool of experiments, we found that four parts play a key role in male and female face differentiation. We list these as: skin, nose, eyes and hair. We are giving an overview below

Table 1. GC and EC algorithm flow.

Algorithm 1 GC and EC algorithm

Input: $\boldsymbol{T}_{train} = \{(I_n, L_n)\}_{n=1}^{m}$, $\boldsymbol{T}_{test}$.
where $\boldsymbol{T}_{train}$ is the training data and $\boldsymbol{T}_{test}$ is the testing image. I is the input training image, L ground truth corresponding labeling image with $L(i,j) \in \{1,2,3,4,5,6\}$, and n is the number of training pairs each for I and L.
1: Train an RDF classifier $\mathfrak{A}$ using $\boldsymbol{T}_{train}$.
2: Extract patches from $\boldsymbol{T}_{test}$ using a step size 1. Pass each feature vector to $\mathfrak{A}$.
3: Use the probabilistic classification strategy and get probability prediction for each pixel.
4: Create probability maps for five semantic classes: $\boldsymbol{P}_{skin}$, $\boldsymbol{P}_{hair}$, $\boldsymbol{P}_{nose}$, $\boldsymbol{P}_{mouth}$, and $\boldsymbol{P}_{eyes}$.
if: GC **Then**
$f^I = \boldsymbol{P}_{skin} + \boldsymbol{P}_{hair} + \boldsymbol{P}_{eyes} + \boldsymbol{P}_{nose}$.
else if: EC **Then**
$f^I = \boldsymbol{P}_{mouth}$.
where f^I is the feature vector.
5: Train a linear SVM classifier $\overline{\mathfrak{A}}$ by using feature vector f^I.
Output: Predicted gender/expression

how these parts are helpful in gender classification, if a fully segmented image is provided. Gender Classification literature reported that male forehead is larger than female. First reason for larger forehead; the hairline for male is back than a female. Even in some cases hairline is completely missing. Second reason; male neck is comparatively thicker, and on larger area. Forehead has a skin label in our HFS algorithm. As a result, skin pixels are bounded on larger area, and a more brighter probability maps is noted for male. Figure 2 compares probability maps of the two subjects (first line female, and second line male) from FERET database.

From our segmentation results we noted that male eyes are more efficiently segmented as compared to female. Usually female eyelashes are longer and curl outwards. Mostly, these eyelashes are mis-classified with hair by our HFS algorithm. As a result, location having eyes pixels are predicted with less probability. As compared to female, male eyelashes are smaller and hardly visible, so better segmentation results are produced (see Fig. 3, column 3). It was also noted that female nose is smaller, having smaller bridge and ridge [23]. Male nose is comparatively larger and its width is also more. The valid reason reported for this fact: male body mass is comparatively larger, which require larger lungs and sufficient passage for air supply to the lungs. As a result, male nostrils are usually bigger resulting in a comparatively big nose (e.g., Fig. 3, column 4). Hairstyle is another one of the effective features for gender discrimination. It provides discriminative clues which are extremely helpful in gender differentiation. In earlier work hair was not counted as a feature because of its complicated geometry which varies from person to person. Our proposed segmentation algorithm get 95.14% pixel-wise accuracy for hair. It is locating the geometry of hair in a good

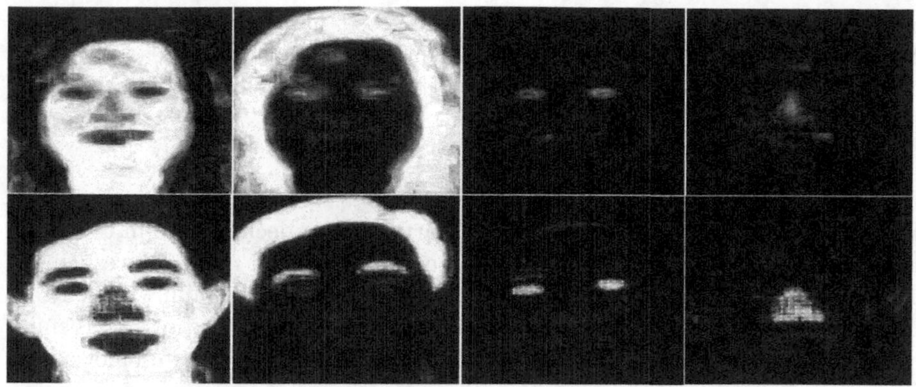

Fig. 2. Gender differentiation probability maps from FERET database (row 1 male, and row 2 female). Column 1: probability map for skin; column 2: probability map for hair; column 3: probability map for eyes; column 4: probability map for nose.

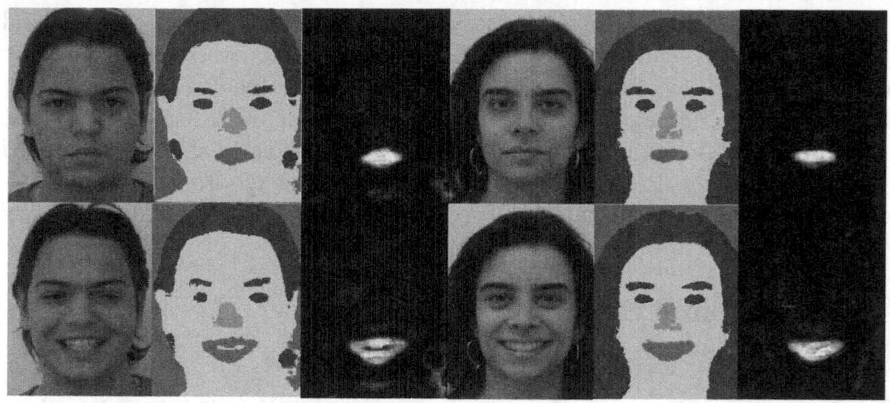

Fig. 3. Expression classification differentiation. Column 1: original database images; column 2: images segmented by proposed segmentation algorithm; column 3: probability maps for mouth; column 4: original database images; column 5: images segmented by proposed segmentation algorithm; column 6: probability maps for mouth.

way. Sometime eyebrows also help in discrimination. The class label for eyebrow is the same as hair in our HFS algorithm. In general, female eyebrows are thinner and longer. In contrary, male eyebrows are shorter and wide. The mouth can really help make a human face more masculine or feminine. Female lips are more clear as compared to male. Even in some cases upper lip is completely missing in male. But our segmentation algorithm does not help to utilize these differences, and we do not noted any increase in classification rate while using mouth. We trained an RDF classifier to create a feature vector for gender discrimination. We manually labeled 15 images for each gender, from each database for training the classifier. The probabilistic classification approach was used to get probability maps for each of the four facial features; skin, nose, hair, and eyes. These probability maps were concatenated to form a single feature vector for each training image. A binary linear SVM classifier was then trained to differentiate between the two classes.

4.2 Expression Classification

We believe that movement of some facial parts plays a decisive role in the identification of some facial expressions. In case of smiling and neutral faces, single movement of mouth pixels are sufficient to decide if a face is smiling or neutral. Movement of mouth pixels is more, and more area is covered if expression is smiling. In contrary, mouth pixels are bounded in a small area with limited movement if expression is neutral. However, this phenomenon is not implemented in complicated facial expression. Multiple movements of facial parts are necessary to be investigated in differentiating between complicated facial expressions. After some experimental work, we concluded to use probability map of mouth for expression discrimination.

We implemented our expression classification algorithm on two simple facial expressions, smiling and non-smiling faces. Experiments were carried out by using two standard face databases FEI and FERET. We labeled 15 images manually, each from smiling and neutral expression from both databases. These images were used to train an RDF classifier. Each testing image is then passed to an RDF classifier. Probability map was obtained for mouth from HFS algorithm. Mouth probability map was used as a feature vector to train a binary SVM classifier.

5 Experimental Setup

5.1 Gender and Expression Classification

To evaluate the performance of gender and expression classification algorithms we used two databases, FEI and FERET. The FEI database is composed of 200 subjects (100 male and 100 female). Each person has two frontal images (one with a smiling and the other neutral facial expression). We used both images for each person making a total number of 400 images. The FEI data-set is in two sessions; FEI A and FEI B. All neutral expression images are in FEI A and smiling expression images in FEI B. We used both FEI A and FEI B in gender and expression tests. In similar way, 400 images were used from FERET database for gender and expression tests. To measure the classification performance, 10-fold cross validation protocol was adopted for GC and EC as well. Finally, results are averaged and reported in the paper.

6 Discussion of Results

6.1 Gender Recognition

The FEI and FERET databases are large databases but we performed our experimental work only on frontal images. The criterion used to evaluate the system

Method	database	CR (%)
Proposed approach	FEI	98.50
Proposed approach	**FERET**	**96.50**
A priori-driven PCA [25]	**FEI**	**≈99.00**
A priori-driven PCA [25]	FERET	≈84.00
2D Gabor+2DPCA [24]	FEI	96.61

Fig. 4. Gender recognition experiments.

is Classification Rate (CR). The CR is a real life deployment criterion, and it finds how many images are correctly classified by the framework. Figure 4 shows reported results and its comparison with state-of-the-art. It should be noted that reported results and its comparison as in figure are on the same set of images. A large amount of literature exist specially on FERET database with different image settings. Preeti et al. [24] used the same set of images, but validation protcol was 2-fold. Figure 8 shows that classification results achieved with the proposed algorithm are better than those reported, except one case (FEI database), where [25] surpassed us. The proposed gender recognition algorithm is robust to illumination and expression variations. We used both neutral and smiling facial expression images in our experiments. However, further investigations are needed to know how much CR will drop with change of pose or with complicated facial expression.

6.2 Expression Recognition

The same FEI and FERET databases were used for training and testing. Ten-fold cross validation experiments were performed. CR is the evaluation criterion here as well. We obtained an average CR of 96.25 % on FEI and 83.25 % on FERET which are best till date on the same set of images and same databases. Figure 5 compares results obtained with previously reported results.

Method	database	CR (%)
Proposed approach	**FEI**	**96.25**
Proposed approach	**FERET**	**83.25**
A priori-driven PCA [25]	FEI	≈95.00
A priori-driven PCA [25]	FERET	≈74.00

Fig. 5. Expression classification experiments.

7 Conclusions

In this paper we have proposed a framework which try to solve the challenging problems of gender and expression classification. A face segmentation algorithm segments a facial image into six semantic classes. A probabilistic classification strategy is then used to generate probability maps for each semantic class. A series of experiments were carried out to identify which probability map is more supportive in gender and expression classification. The respective probability maps are used to generate a feature vector and train a linear SVM classifier for each of the possible cases. The proposed methods have been compared with state-of-the-art on standard face databases FEI, and FERET, and good results were obtained.

Several promising directions can be planned as future work, and we are considering two of these. Firstly, we believe that our segmentation results provide sufficient information for various hidden variables in face and provide a rout towards solving difficult recognition problems. Our planes also include adding the problems of age estimation and race classification into the architecture. Secondly, the current segmentation model can be improved to get nearly perfectly segmented results.

References

1. Benini, S., Svanera, M., Adami, N., Leonardi, R., Kovács, A.B.: Shot scale distribution in art films. Multimedia Tools Appl. **75**(23), 16499–16527 (2016)
2. Canini, L., Benini, S., Leonardi, R.: Affective analysis on patterns of shot types in movies. In: 2011 7th International Symposium on Image and Signal Processing and Analysis (ISPA), pp. 253–258. IEEE (2011)
3. Everingham, M., Van Gool, L., Williams, C.K., Winn, J., Zisserman, A.: The Pascal visual object classes (VOC) challenge. Int. J. Comput. Vis. **88**(2), 303–338 (2010)
4. Huang, G.B., Narayana, M., Learned-Miller, E.: Towards unconstrained face recognition. In: IEEE Computer Society Conference on Computer Vision and Pattern Recognition Workshops, CVPRW 2008, pp. 1–8. IEEE (2008)
5. Davies, G., Ellis, H., Shepherd, J.: Perceiving and Remembering Faces. Academic Press, Cambridge (1981)
6. Sinha, P., Balas, B., Ostrovsky, Y., Russell, R.: Face recognition by humans: nineteen results all computer vision researchers should know about. In: Proceedings of IEEE, vol. 94(11), pp. 1948–1962 (2006)
7. Levi, G., Hassner, T.: Age and gender classification using convolutional neural networks. In: Computer Vision and Pattern Recognition (CVPR) (2015)
8. Khan, K., Mauro, M., Leonardi, R.: Multi-class semantic segmentation of faces. In: 2015 International Conference on Image Processing, ICIP 2015, vol. 1, pp. 63–66. IEEE (2015)
9. Khan, K., Mauro, M., Migliorati, P., Leonardi, R.: Head pose estimation through multi-class face segmentation. In: Proceedings of ICME-2017 (2017, to appear)
10. Yacoob, Y., Davis, L.S.: Detection and analysis of hair. IEEE Trans. Pattern Anal. Mach. Intell. **28**(7), 1164–1169 (2006)

11. Lee, C., Schramm, M.T., Boutin, M., Allebach, J.P.: An algorithm for automatic skin smoothing in digital portraits. In: Proceedings of the 16th IEEE International Conference on Image processing, pp. 3113–3116. IEEE Press (2009)
12. Lafferty, J., McCallum, A., Pereira, F.C.: Conditional random fields: probabilistic models for segmenting and labeling sequence data (2001)
13. Kae, A., Sohn, K., Lee, H., Learned-Miller, E.: Augmenting CRFs with Boltzmann machine shape priors for image labeling. In: 2013 IEEE Conference on Computer Vision and Pattern Recognition (CVPR), pp. 2019–2026. IEEE (2013)
14. Eslami, S.A., Heess, N., Williams, C.K., Winn, J.: The shape Boltzmann machine: a strong model of object shape. Int. J. Comput. Vis. **107**(2), 155–176 (2014)
15. Li, Y., Wang, S., Ding, X.: Person-independent head pose estimation based on random forest regression. In: 2010 17th IEEE International Conference on Image Processing (ICIP), pp. 1521–1524. IEEE (2010)
16. Scheffler, C., Odobez, J.-M.: Joint adaptive colour modelling and skin, hair and clothing segmentation using coherent probabilistic index maps. In: British Machine Vision Association-British Machine Vision Conference (2011)
17. Ferrara, M., Franco, A., Maio, D.: A multi-classifier approach to face image segmentation for travel documents. Expert Syst. Appl. **39**(9), 8452–8466 (2012)
18. Cootes, T.F., Taylor, C.J., Cooper, D.H., Graham, J.: Active shape models-their training and application. Comput. Vis. Image Underst. **61**(1), 38–59 (1995)
19. da Fei, C.U.: Fei database. http://www.fei.edu.br/~cet/facedatabase.html
20. Phillips, P.J., Wechsler, H., Huang, J., Rauss, P.J.: The FERET database and evaluation procedure for face-recognition algorithms. Image Vis. Comput. **16**(5), 295–306 (1998)
21. Dalal, N., Triggs, B.: Histograms of oriented gradients for human detection. In: IEEE Computer Society Conference on Computer Vision and Pattern Recognition, CVPR 2005, vol. 1, pp. 886–893. IEEE (2005)
22. Bochkanov, S.: Alglib: http://www.alglib.net
23. Enlow, D.H., Moyers, R.E.: Handbook of Facial Growth. WB Saunders Company, Philadelphia (1982)
24. Rai, P., Khanna, P.: An illumination, expression, and noise invariant gender classifier using two-directional 2DPCA on real Gabor space. J. Vis. Lang. Comput. **26**, 15–28 (2015)
25. Thomaz, C., Giraldi, G., Costa, J., Gillies, D.: A Priori-Driven PCA. In: Park, J.-I., Kim, J. (eds.) ACCV 2012. LNCS, vol. 7729, pp. 236–247. Springer, Heidelberg (2013). doi:10.1007/978-3-642-37484-5_20

Two More Strategies to Speed Up Connected Components Labeling Algorithms

Federico Bolelli(✉), Michele Cancilla, and Costantino Grana

Dipartimento di Ingegneria "Enzo Ferrari",
Università Degli Studi di Modena e Reggio Emilia,
Via Vivarelli 10, 41125 Modena, MO, Italy
{federico.bolelli,michele.cancilla,costantino.grana}@unimore.it

Abstract. This paper presents two strategies that can be used to improve the speed of Connected Components Labeling algorithms. The first one operates on optimal decision trees considering image patterns occurrences, while the second one articulates how two scan algorithms can be parallelized using multi-threading. Experimental results demonstrate that the proposed methodologies reduce the total execution time of state-of-the-art two scan algorithms.

Keywords: Connected components labeling · Binary decision trees · Parallelization · Optimization

1 Introduction

Connected Components Labeling (CCL) of binary image is a fundamental task in several image processing and computer vision applications ranging from video surveillance to medical imaging. CCL transforms a binary image into a symbolic one in which all pixels belonging to the same connected component are given the same label. Thus, this transformation is required whenever a computer program needs to identify independent components. Moreover, given that labeling is the base step of most real time applications it is required to be as fast as possible. Since labeling is a well-defined problem and the exact solution for a given image should be provided as output, the proposals of the last twenty years have focused on performance optimization. A significant improvement was given by the introduction of the *Union-Find* [17] approach for label equivalences resolution and array-based data structures [11].

Most of the labeling algorithms employ a raster scan mask to look at neighborhood of a pixel and to determine its correct label. As shown in literature, the decision table associated to the mask which rules the scan step can be converted to an optimal binary decision tree by the use of a dynamic programming approach [8]. This approach leads to a reduction of total memory accesses and total execution time.

Another strategy to improve the performance of existing algorithms could be the parallelization. A simple process which divides the input image into horizontal stripes and computes labeling separately on each one is described in [3]. The

S. Battiato et al. (Eds.): ICIAP 2017, Part II, LNCS 10485, pp. 48–58, 2017.
https://doi.org/10.1007/978-3-319-68548-9_5

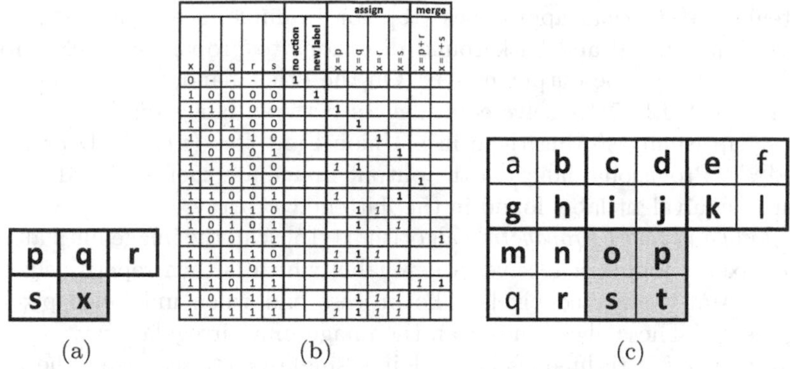

							assign				merge	
x	p	q	r	s	no action	new label	x=p	x=q	x=r	x=s	x=p+r	x=r+s
0	-	-	-	-	1							
1	0	0	0	0		1						
1	1	0	0	0			1					
1	0	1	0	0				1				
1	0	0	1	0					1			
1	0	0	0	1						1		
1	1	1	0	0			*1*	1				
1	1	0	1	0							1	
1	1	0	0	1			*1*			1		
1	0	1	1	0				1	*1*			
1	0	1	0	1				1		*1*		
1	0	0	1	1								1
1	1	1	1	0			*1*	1	*1*			
1	1	1	0	1			*1*	1		*1*		
1	1	0	1	1							*1*	1
1	0	1	1	1				1	*1*	*1*		
1	1	1	1	1			*1*	1	*1*	*1*		

Fig. 1. (a) Masks used to compute labels of pixel x by SAUF algorithm, (b) its associated *OR*-decision table. Finally, (c) is the mask used to compute label of pixels o, p, s, and t by the BBDT algorithm.

general problem of Connected Components Labeling on parallel architecture was exhaustively threated in a theoretical way by [1].

Recently, our research group released an open source C++ framework for performance evaluation of CCL algorithms. The benchmarking system called YACCLAB [7] (acronym for *Yet Another Connected Components Labeling Benchmark*) collects, runs, and tests the state-of-the-art labeling algorithms over an extremely variety of different datasets solving the problem of fair evaluation of different strategies. Results shown in this paper are produced using this tool.

With this work we experiment two strategies to improve the performance of Connected Components Labeling applying them to the SAUF (*Scan Array Union Find*) algorithm proposed by Wu *et al.* [18] and BBDT (*Block Based with Decision Tree*) strategy proposed by Grana *et al.* [8]. Firstly, we have transformed the decision trees used by both the algorithms considering the occurrence probability of each pattern in a mask and on a reference dataset. Secondly, we have parallelized both the algorithms, evaluating the benefits and limits of different approaches. Then we have tested all resulting algorithms on different datasets and environments to highlights their different behaviors.

The rest of this paper is organized as follows: in Sect. 2 we give an overview of existing Connected Components Labeling algorithms; Sects. 3 and 4 contain the description of the implemented strategies, which are then evaluated in Sect. 5. Finally, we draw the conclusions in Sect. 6.

2 Previous Works

Connected Components Labeling has a very long story full of different strategies that can be classified in three different main groups:

- *Raster Scan* algorithms which scan the image exploiting a mask (see for instance Fig. 1) and solve equivalences between labels using different

strategies. *Multiscans* approaches [10], for example, scan the image alternatively in forward and background directions to propagate labels until no changes occur in the output matrix. On the other hand, modern *Two Scan* algorithms [8,12,17,18] solve equivalences between labels on-line during the first scan, usually storing them in a Union-Find tree (*i.e.* a 1-D array also called P). Provisional labels in the output image are then replaced with the smallest equivalent label found in the flattened P array.
- *Searching and Label Propagation* algorithms [15] scan the image until an unlabeled pixel is found: it receives a new label which is then repetitively propagated to all connected pixels. The process end when unlabeled pixels no longer exist. These algorithms scan the image in an irregular way.
- *Contour Tracing* techniques [4] exploit a single raster scan over the image. During this process all pixels in both the contour and the immediately external background of an object are clockwise tagged in a single operation. Finally, the connected components have to be filled propagating contours' labels.

Two scan algorithms have revealed best performances [7], so our analysis focuses on them.

The SAUF algorithm [17,18] implements the Union-Find technique with path compression and exploits a decision tree for accessing only the minimum number of already labeled pixels.

In [8] it was proven that, when performing 8-connected labeling, the scanning process can be extended to block-based scanning, that scans the image in 2×2 blocks (BBDT). In that case, because of the large number of possible combinations, the final decision tree is generated automatically by another program [9].

He *et al.* [12] recently observed that BBDT checks many pixels repeatedly, because after labeling one pixel, the mask moves to the next one, but many pixels in the current mask are overlapped to the previous ones, which may have already been checked. They thus proposed a *Configuration Transition Based* (CTB) algorithm which introduces different configurations *states* in order to make further decisions. This procedure reduces the number of pixels checked and thus speeds up the labeling procedure.

Finally, Grana *et al.* [6] proposed an approach called *Optimized Connected Component Labeling with Pixel Prediction* (PRED) which employs a reproducible strategy able to avoid repeatedly checking the same pixels multiple times. The first scan phase of PRED is ruled by a forest of decision trees connected into a single graph. Each tree derives from a reduction of one complete optimal decision tree.

3 Pattern Analysis and Modeling of Decision Trees

In [17] Wu *et al.* have shown that when considering 8-connected components in a 2D image it is advantageous to exploit the dependencies between pixels in the scan mask to reduce memory accesses. This can be performed with the use of a decision tree which can be derived from the decision table (Fig. 1b) associated to the scan mask. In [8] Grana *et al.* proposed an automatic strategy to convert

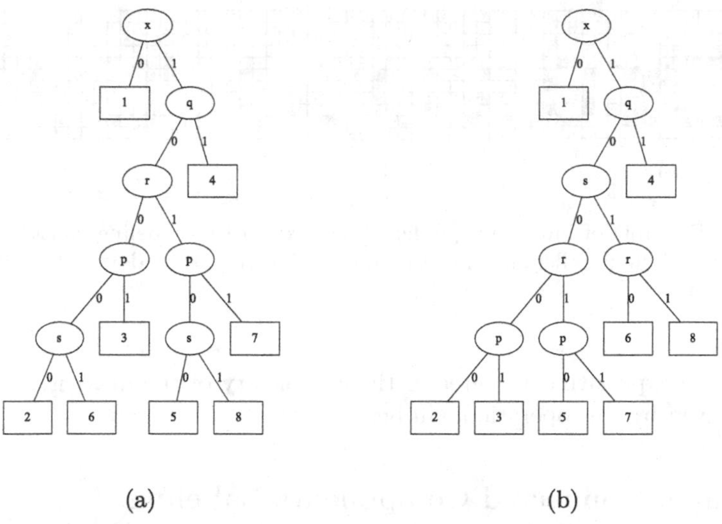

Fig. 2. Two of the optimal decision trees derivable from Rosenfeld's mask (Fig. 1a). Nodes (circle box) show the conditions to check and leaves (square box) contain the action to perform: (1) nothing to do, (2) new label, (3) $x = p$, (4) $x = q$, (5) $x = r$, (6) $x = s$, (7) $x = r + p$ and (8) $x = r + s$.

AND-decision tables to *OR*-decision tables and produce the associated optimal decision tree by the use of a dynamic programming technique originally described by Schumacher and Sevcik [16].

The basic concept behind the creation of a simplified tree from a decision table (with n conditions) is that if two branches lead to the same action the condition from which they originate may be removed. Thus, this conversion can be interpreted as the partitioning of an n-dimensional hypercube where the vertexes correspond to the 2^n possible rules. Associating to each condition removal a unitary gain we can select the tree which maximizes the total gain and minimizes the number of memory accesses. Moreover, in [9], an exhaustive search procedure is provided to select the most convenient action among the alternatives in the *OR*-decision table. This strategy is also proven to generate always one of the possible optimal decision trees. A more detailed description of the tree generation can be found in [8].

In Fig. 2 we have reported an example of two equivalent optimal decision tree associated to the SAUF algorithm. The Fig. 2a shows the tree presented by Wu *et al.* in [17], while Fig. 2b is another tree obtainable from the original *OR*-decision table.

The novelty of the proposed approach lies on the analysis of the occurrence frequencies of each pattern derived from a mask on a reference dataset. If we generate the trees considering pattern probabilities, we are able to reduce the average number of memory accesses for a specific kind of images. In this case the gain obtained by a condition removal is related to the frequency of associated

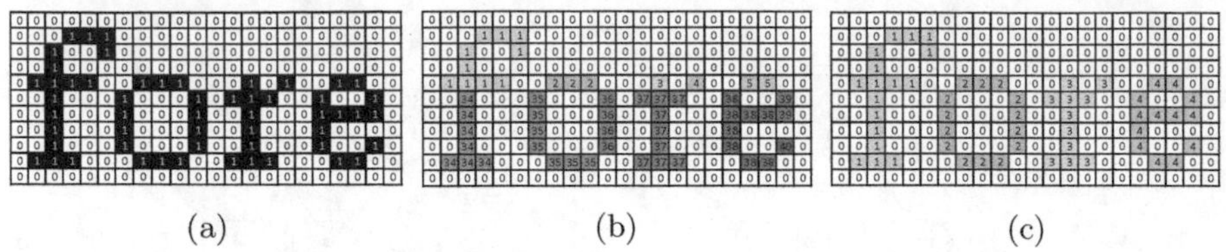

(a) (b) (c)

Fig. 3. (a) Example of binary image depicting text, (b) its labeling considering only the first scan of the parallel implementation of SAUF (two threads), and finally (c) its labeling result.

patterns. We expect that the greater the complexity of the mask is, the higher the gain given by this operation will be.

4 Parallel Connected Components Labeling

Modern computer architectures are multi-cores and support multi-threaded applications, so it is convenient to analyze how much the performance of CCL algorithms scale up when multi-threading is involved. The goal is to spread the computational cost among different threads without increasing the execution time with heavy synchronization mechanisms. A basic approach to parallelize two scan labeling algorithms divides images into stripes (chunks) and operates with the following three steps [3]:

1. Compute first scan on each stripe (in parallel);
2. Merge border labels and flatten equivalences array P;
3. Compute second scan on each stripe (in parallel).

Assuming to have n workers available to compute labeling, we can spread the computational cost for the first scan between them dividing the image in n parts. In order to be cache friendly we choose to cut the image horizontally. Each stripe is then labeled independently (first scan) by each worker using provisional labels.

To avoid getting overlapped labels among different stripes, each thread must use a different set of initial provisional labels. When using 8-connectivity, each 2×2 block of pixels can only contain a single label, so the starting label number for a given thread can be easily calculated as

$$\left\lfloor \frac{r_i + 1}{2} \right\rfloor \cdot \left\lfloor \frac{w + 1}{2} \right\rfloor + 1, \tag{1}$$

where r_i represents the index of the i-th chunk's first row and w is the image width. The background will always take label zero. During the first scan, threads have also to collect any possible equivalence between labels and store it in the P array. To avoid collisions each stripe deals with a disjoint set of labels and this implies non overlapped accesses to the array storing the equivalences. In

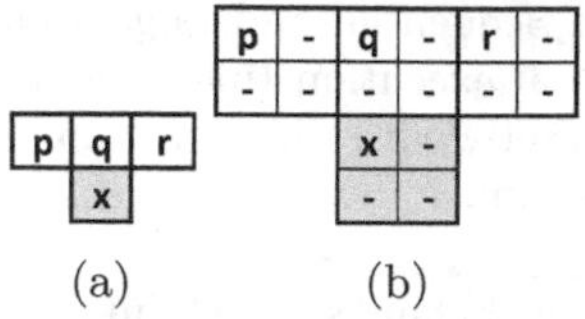

Fig. 4. (a) Merge mask used by SAUF parallel algorithm and (b) by BBDT parallel algorithm.

Fig. 3b an example of the resulting image from the first scan is reported when two threads are involved.

After the first scan we have to establish a bridge between two adjacent chunks with the operation of Merge (Union of the Union-Find approach). This is required since chunks are labeled with uncorrelated provisional labels. In order to compute Merge we use masks reported in Fig. 4.

There are basically two strategies to figure out this problem and they are listed below.

- *Sequential Merge* scans each chunk's border sequentially and does not employ multi-threading, avoiding collision management in label equivalences array.
- *Logarithmic Merge* already proposed in [3], employs multi-threading to speed up this operation. Indeed, we can operate concurrently on two or more chunks when they are not accessed at the same time by another thread.

For example, if we consider an image processed with eight threads hence divided into eight stripes the Sequential Merge requires 7 stages for merging chunks while the Logarithmic one only 3 ($\log_2 8$).

The second approach, compared to the Sequential one, leads, at least in theory, to better performance. However, the management of Logarithmic Merge is more complex when the number of threads is not a power of two, so a correct implementation of it must include additional conditional statements. Hence, as shown in Table 1 the Sequential Merge operation performs better if compared with Logarithmic. Due to this, all parallel results shown in Sect. 5 are obtained using Sequential Merge.

It is important to notice that the described approach leads to a fragmented array of equivalences and consequently the Union-Find tree needs to be flattened in a hop-by-hop way. This behavior is linked to the possibility of having a lot of unused provisional labels associated to each chunk in the P array.

To conclude the labeling process, the second scan is done in parallel by the same number of threads and using the same mask of the first scan. Second scan better suits the multi-threading approach because it contains few conditional statements.

5 Experimental Evaluation

In order to produce an overall view of the proposed methods performance, we ran each algorithm using the YACCLAB tool on different environments:

Table 1. Comparison between Sequential and Logarithmic Merge incidence (in percentage) with respect to the total execution time. The reported values are calculated considering all YACCLAB datasets, with different number of threads (hence chunks) and using BBDT parallel algorithm.

	2 chunks	4 chunks	8 chunks	16 chunks	24 chunks
Sequential	0.45 %	1.49 %	3.35 %	5.34 %	7.10 %
Logarithmic	0.84 %	3.64 %	6.11 %	7.63 %	8.73 %

a Windows server with two Intel Xeon X5650 *CPU* @ 2.67GHz and Microsoft Visual Studio 2013 and a Windows PC with an Intel i5-6600 *CPU* @ 3.30GHz running Microsoft Visual Studio 2013. All tests were repeated 10 times, and for each image the minimum execution time was considered in order to reduce the effects of background processes.

In the following of this section, we use acronyms to refer to the available algorithms: BBDT is the Block Based with Decision Trees algorithm by Grana *et al.* [8], BBDT_ALL and BBDT_ONE are the versions of the BBDT algorithm with optimal trees generated considering respectively patterns frequency of all datasets described in Sect. 5.1 and patterns frequency of the same dataset on which the algorithm is tested. BBDT_$\langle n \rangle$ is the parallel implementation of the BBDT algorithm run with n threads, SAUF is the Scan Array Union Find algorithm by Wu *et al.* [18], SAUF_ALL refers to the algorithm that uses the optimal tree generated considering patterns frequency on all YACCLAB datasets, and, finally, SAUF_$\langle n \rangle$ is the SAUF parallel version run using n threads. BBDT and SAUF are the algorithms currently included in the OpenCV's *connectedComponents* function.

5.1 Datasets

The datasets used for tests are currently included in YACCLAB. All images are provided in 1 bit per pixel PNG format, with 0 (black) being background and 1 (white) being foreground. The images can be grouped by their nature as follows:

- *MIRflickr* is composed by natural images, it contains 25,000 standard resolution images taken from Flickr, with an average resolution of 0.18 megapixels.
- *Hamlet* and *Tobacco* [13] are two set of document images. The first one contains 104 images scanned from a version of the Hamlet found on Project Gutenberg. The second one is composed of 1290 document images and it is a realistic database for document image analysis research.
- *Medical* dataset, provided by Dong *et al.* [5], is composed of histological images and allow to cover this fundamental field.
- *Fingerprints* [14] accommodates 960 fingerprint images collected by using low-cost optical sensors or synthetically generated.
- *3DPeS* [2] is a surveillance dataset mainly designed for people re-identification in multi-camera systems with non-overlapped fields of view. Images have an

average amount of 0.41 million of pixels to analyze and 320 components to label.
- *Random* contains black and white random noise images with nine different foreground densities (from 10% up to 90%).

A more detailed description of the YACCLAB's datasets can be found in [7].

5.2 Frequencies Results

Firstly, we have run the tree generator algorithm described in Sect. 3 on the SAUF mask considering the patterns frequency of all YACCLAB datasets. The resulting tree is the one reported in Fig. 2b, that is different from the one proposed by Wu *et al.* (Fig. 2a), but it is one of the optimal equivalent trees derivable from decision table in Fig. 1b without considering frequencies. Indeed, the small number of conditions for this case limits the number of possible operations on trees. However, Table 3b shows that the tree generated by our algorithm requires from 0.03% to 0.27% less memory accesses when applied on real datasets. Instead, as expected, on random dataset the number of accesses is almost the same due to its uniform distribution of patterns' probability. In Table 3a comparison between execution time is also reported.

Table 2. Results in ms on Windows PC with i5-6600 @ 3.30 GHz and Microsoft Visual Studio 2013 (lower is better).

	BBDT	BBDT_ALL	BBDT_ONE
3DPeS	0.678	**0.621**	0.645
Fingerprints	0.343	0.322	**0.320**
Hamlet	5.284	**5.048**	5.252
Medical	2.273	**2.154**	2.213
MIRflickr	0.450	**0.414**	0.424
Tobacco800	7.752	**7.283**	7.431

The BBDT algorithm, instead, requires much more complex trees because of the high number of conditions to check and thus leaves space for better optimization. In Table 2 the execution times of three BBDT algorithms which use different trees are compared: the first one is obtained with uniform frequencies (BBDT), the second one is generated considering all YACCLAB dataset frequencies (BBDT_ALL), and, finally, the third one employs the tree generated by the frequencies of dataset on which it is tested (BBDT_ONE). Table 2 demonstrates that the pattern analysis leads to better performance compared to the original BBDT algorithm. It is important to highlight that the best results are obtained with BBDT_ALL algorithm instead of BBDT_ONEs: this is an unexpected result considering how trees have been generated. From our knowledge the only reasonable explanation is related to the code optimizations applied by the compiler.

The improvements related to frequency analysis, albeit limited, are significant given the maturity of the problem and the performance achieved by the best existing algorithms.

Table 3. (a) Results in ms on Windows with Intel i5-6600 @ 3.30GHz and Microsoft Visual Studio 2013 (lower is better) with optimization disabled. (b) Analysis of memory accesses performed on YACCLAB dataset.

	SAUF	SAUF_ALL
3DPeS	2.059	2.037
Fingerprints	0.847	0.844
Hamlet	15.070	14.834
Medical	6.505	6.401
MIRflickr	1.053	1.059
Tobacco800	24.667	24.606
Random	28.731	29.065

(a)

	SAUF	SAUF_ALL
3DPeS	2 069 566	2 069 002
Fingerprints	809 288	807 136
Hamlet	14 325 113	14 310 510
Medical	7 035 408	7 028 268
MIRflickr	1 170 066	1 168 648
Tobacco800	23 873 949	23 857 166
Random	20 963 033	20 963 038

(b)

5.3 Parallel Results

The parallel implementation of both BBDT and SAUF is based on the OpenCV's build in *parallel_for_* function. This function runs the parallel loop of one of the available parallel frameworks, selecting it at compilation time. All parallel results presented in this paper are obtained using *Intel Threading Building Blocks* (TBB). Table 4 shows the average execution time of sequential and parallel version of both the algorithms on all dataset and with different number of threads. Experiments reveal that the overhead introduced by the *parallel_for_* with TBB is negligible *i.e.* the execution time of a sequential algorithm and its parallel version implemented with OpenCV parallel framework and tested with one thread is the same. As shown in Table 4, the speed up obtained with two threads on SAUF is ×1.5 in average and it increases up to ×4 on random dataset when 12 threads are involved. Starting from 16 threads (*i.e.* when hyper-threading is involved and threads share cache of both first and second level) the performance of SAUF decrease. This could be explained by the increment of instruction cache misses due to data overflow. For what concerns BBDT algorithm, experimental results demonstrate a greater speed up with low number of threads (*i.e.* ×1.7 with 2 threads, up to ×4.7 on random dataset with 8 threads). However, increasing the parallelism, as it happens with the SAUF algorithm, leads to worse performance. Once again this behavior could be related to the increase of instruction cache misses that, in this case, occur starting from 8 threads, because BBDT code footprint is much bigger than SAUF one.

Table 4. Average results in ms on a virtual Windows workstation with two Intel Xeon X5650 *CPU* @ 2.67GHz (6 physical cores and 12 logical processors per socket) and Microsoft Visual Studio 2013 (lower is better).

	SAUF	SAUF_2	SAUF_4	SAUF_8	SAUF_12	SAUF_16	SAUF_24
3DPeS	1.817	1.258	1.013	0.886	**0.845**	0.972	0.969
Fingerprints	0.793	0.548	0.431	0.361	**0.338**	0.402	0.426
Hamlet	13.449	8.682	6.726	5.651	**5.338**	5.615	5.446
Medical	6.316	4.414	3.104	2.626	**2.542**	2.742	2.745
MIRflickr	1.053	0.770	0.608	0.544	**0.543**	0.618	0.657
Tobacco800	22.075	14.362	11.083	9.445	**9.057**	9.457	9.166
Random	31.525	19.554	11.956	8.695	**7.795**	9.144	8.605

	BBDT	BBDT_2	BBDT_4	BBDT_8	BBDT_12	BBDT_16	BBDT_24
3DPeS	1.274	0.794	**0.609**	0.720	0.812	0.918	1.002
Fingerprints	0.606	0.386	0.298	**0.274**	0.290	0.347	0.388
Hamlet	10.528	5.989	**4.342**	4.528	5.186	5.423	6.146
Medical	4.945	3.051	**1.831**	2.060	2.407	2.621	2.998
MIRflickr	0.790	0.520	**0.393**	0.438	0.479	0.559	0.622
Tobacco800	16.587	9.590	**6.518**	7.207	8.206	9.215	10.375
Random	25.106	13.177	7.084	**5.375**	6.004	7.567	8.387

6 Conclusions

In this paper we presented two different approaches able to speed up two scan Connected Components Labeling algorithms. The first strategy operates on optimal decision trees in order to reduce memory accesses and improve performance. This transformation processes can be performed off-line and requires to know the occurrences of all possible patterns on a reference dataset or a group of them. As shown, the proposed approach leads to an improvement of the performance of CCL if applied to complex decision trees, such as the one implemented by BBDT. The second strategy employs multi-threading in order to spread computational cost among different processes. Differently from the first approach it is easily applicable to all two scan algorithms. The source code of the described algorithms is included in YACCLAB: we strongly believe that given the maturity of the problem and the subtlety involved in the implementation, it should be mandatory to allow the community to reproduce the results without forcing everyone to reimplement every proposal. To conclude, the parallel versions of BBDT and SAUF algorithms discussed in this paper have been submitted to OpenCV.

References

1. Alnuweiri, H.M., Prasanna, V.K.: Parallel architectures and algorithms for image component labeling. IEEE Trans. Pattern Anal. Mach. Intell. **14**(10), 1014–1034 (1992)

2. Baltieri, D., Vezzani, R., Cucchiara, R.: 3DPeS: 3D people dataset for surveillance and forensics. In: Proceedings of the 2011 Joint ACM Workshop on Human Gesture and Behavior Understanding, pp. 59–64. ACM (2011)
3. Cabaret, L., Lacassagne, L., Etiemble, D.: Parallel light speed labeling: an efficient connected component labeling algorithm for multi-core processors. In: 2015 IEEE International Conference on Image Processing (ICIP), pp. 3486–3489. IEEE (2015)
4. Chang, F., Chen, C.J., Lu, C.J.: A linear-time component-labeling algorithm using contour tracing technique. Comput. Vis. Image Underst. **93**(2), 206–220 (2004)
5. Dong, F., Irshad, H., Oh, E.Y., Lerwill, M.F., Brachtel, E.F., Jones, N.C., Knoblauch, N.W., Montaser-Kouhsari, L., Johnson, N.B., Rao, L.K., et al.: Computational pathology to discriminate benign from malignant intraductal proliferations of the breast. PLoS ONE **9**(12), e114885 (2014)
6. Grana, C., Baraldi, L., Bolelli, F.: Optimized connected components labeling with pixel prediction. In: Blanc-Talon, J., Distante, C., Philips, W., Popescu, D., Scheunders, P. (eds.) ACIVS 2016. LNCS, vol. 10016, pp. 431–440. Springer, Cham (2016). doi:10.1007/978-3-319-48680-2_38
7. Grana, C., Bolelli, F., Baraldi, L., Vezzani, R.: YACCLAB - yet another connected components labeling benchmark. In: 23rd International Conference on Pattern Recognition, ICPR (2016)
8. Grana, C., Borghesani, D., Cucchiara, R.: Optimized block-based connected components labeling with decision trees. IEEE Trans. Image Process. **19**(6), 1596–1609 (2010)
9. Grana, C., Montangero, M., Borghesani, D.: Optimal decision trees for local image processing algorithms. Pattern Recogn. Lett. **33**(16), 2302–2310 (2012)
10. Haralick, R.: Some neighborhood operators. In: Onoe, M., Preston, K., Rosenfeld, A. (eds.) Real-Time Parallel Computing, pp. 11–35. Springer, Heidelberg (1981)
11. He, L., Chao, Y., Suzuki, K.: A linear-time two-scan labeling algorithm. Int. Conf. Image Process. **5**, 241–244 (2007)
12. He, L., Zhao, X., Chao, Y., Suzuki, K.: Configuration-transition-based connected-component labeling. IEEE Trans. Image Process. **23**(2), 943–951 (2014)
13. Lewis, D., Agam, G., Argamon, S., Frieder, O., Grossman, D., Heard, J.: Building a test collection for complex document information processing. In: Proceedings of 29th Annual International ACM SIGIR Conference, pp. 665–666 (2006)
14. Maltoni, D., Maio, D., Jain, A., Prabhakar, S.: Handbook of Fingerprint Recognition. Springer, Heidelberg (2009)
15. Rosenfeld, A., Kak, A.: Digital Picture Processing. Computer Science and Applied Mathematics, vol. 1. Academic Press, Cambridge (1982)
16. Schumacher, H., Sevcik, K.C.: The synthetic approach to decision table conversion. Commun. ACM **19**(6), 343–351 (1976)
17. Wu, K., Otoo, E., Suzuki, K.: Two strategies to speed up connected component labeling algorithms. Technical Report LBNL-59102, Lawrence Berkeley National Laboratory (2005)
18. Wu, K., Otoo, E., Suzuki, K.: Optimizing two-pass connected-component labeling algorithms. Pattern Anal. Appl. **12**(2), 117–135 (2009)

Embedded Real-Time Visual Search with Visual Distance Estimation

Marco Paracchini[1](✉), Emanuele Plebani[2], Mehdi Ben Iche[2], Danilo Pietro Pau[2], and Marco Marcon[1]

[1] DEIB - Dipartimento di Elettronica, Informazione e Bioingegneria, Politecnico di Milano, Milan, Italy
{marcobrando.paracchini,marco.marcon}@polimi.it
[2] Advanced System Technology, STMicroelectronics, Agrate Brianza, Italy
{emanuele.plebani1,mehdi.beniche-ext,danilo.pau}@st.com

Abstract. Visual Search algorithms are a class of methods that retrieve images by their content. In particular, given a database of reference images and a query image the goal is to find an image among the database that depicts the same object as in the query, if any. Moreover, in many different real case applications more than one object of interest could be viewed in the query image. Furthermore, in this kind of situations, often, it is not sufficient to identify the object depicted on a query image but its precise localization inside the scene viewed by the camera is also requested. In this paper we propose to couple a Visual Search system, which can retrieve multiple objects from the same query image, with an additional Distance Estimation module that exploits the localization information already computed inside the Visual Search stage to estimate localization of the object in three dimensions. In this work we implement the complete image retrieval and spatial localization pipeline (including relative distance estimation) on two different embedded devices, exploiting also their GPU in order to get near real time performances on low-power devices. Lastly, the accuracy of the proposed distance estimation is evaluated on a dataset of annotated query-reference pairs ad-hoc created.

Keywords: Visual Search · CBIR · LoG · SIFT · Distance estimation

1 Introduction

Visual Search (VS), also known as Content Base Image Recognition (CBIR), is generally defined as a system, based on computer vision methods, that performs image retrieval based on the pixel content alone. In particular, given a query image depicting a (usually rigid and not reflective) object and a reference database of images containing the query object among many others, the system has to return the correct image. This task is generally different from image classification, since in VS the goal is to identify a single object instance in the image (e.g. a specific book or a building) instead of assigning a class to the object (e.g. a generic book or a generic building). VS systems can be modeled as a large Cyber

S. Battiato et al. (Eds.): ICIAP 2017, Part II, LNCS 10485, pp. 59–69, 2017.
https://doi.org/10.1007/978-3-319-68548-9_6

Physical System (CPS) [1] in which many mobile imaging nodes (i.e. the users' smartphones) acquire images and contact the central node to retrieve some information about the objects depicted in each image. Such systems can be used in a large amount of domains, such as mobile e-commerce applications, interactive museum guides and many others. In general, VS applications may potentially substitute QR codes in all the fields they are used in, essentially replacing a printed code attached to the object with the object itself and thus preserving the original object appearance.

In this paper we consider scenarios in which the query object must not only be identified but also precisely localized in the physical 3D space in front of the camera. An application of this kind could be implemented in many different scenarios, such as robots picking up different objects from an unordered shelf scaffold (e.g. in a kitchen or a warehouse) or a drone that has to reach a specific landing pad. In these particular scenarios the VS application could run entirely on the local device and on a small reference database, without the need of a central node. On the other hand, the application must be also able to identify and localize multiple objects depicted in a single query image and estimate the distance between them and the observing camera. The only assumptions are that the camera is already calibrated and the objects retrieved are rigid, non reflective, locally planar and of known size.

An application able to solve the aforementioned task is described in Sect. 3 and its implementation and testing on two different ARM devices is reported in Sect. 4. Finally in Sect. 5 the real time viability of this kind of implementation on embedded boards is discussed.

2 Related Work

Algorithms for VS/CBIR firstly arose in the mid '90s with the goal of organize photo collections and analyze the image content of a rapidly growing Internet. The first applications created to solve these kind of problems were Query By Image Content (QBIC) from IBM [2] and the Photobook system from MIT [3], among many others.

The rapid growth of the smartphone market in the late 2000 s and the introduction of high resolution cameras, coupled with increased CPU/GPU computational power, moved the first personal computer (PC) and web-based Visual Search experiments towards mobile Visual Search. The growing processing power of mobile CPUs showed that sending to the server an entire image, even compressed, is unnecessary and image analysis such as feature extraction could be performed directly on the smartphone, following the "analyze then compress" paradigm to minimize the loss of information. Furthermore, other factors such as unstable or limited bandwidth and network latency empathize the importance of features compression in mobile VS applications [4].

For this reason, starting from 2010, the Moving Picture Experts Group (MPEG) standardized content based access to multimedia data in the MPEG-7 standard by defining a set of descriptors for images, namely Compact Descriptor for Visual Search (CDVS) [4]. The standard proposal also includes a reference CDVS implementation in a *Test Model* and the formal definition of a bit stream which encodes in compressed form the information required to perform a server-side search. The information encoded consist of a global descriptor, which is a digest computed from SIFT [5] descriptors extracted from the image and compressed SIFT descriptors with the coordinates of the associated keypoints.

Regarding the distance estimation part, many different methods exist in literature, as, for example, Structure from Motion (SfM) [6] or Visual Simultaneous Localization And Mapping (VSLAM) [7] in which images acquired by a set of cameras (or a singular moving camera) depicting the same scene is used to 3d reconstruct the viewed scene via triangulation and bundle adjustment. The method used in this paper is instead based on homography decomposition [8].

The Visual Search application proposed in this paper is based on the work presented in [9], without the global descriptor stage, which is unnecessary on small databases such as the ones used to store the reference targets, and with the addition of the Distance Estimation module.

3 Proposed Pipeline

The Visual Search and Distance Estimation pipeline proposed in this paper can be split into four main blocks: three belongs the Visual Search part, the Image Analyzer (IA), the Retrieval Stage (RS) and the Multiple Results Loop (MRL), plus one additional module for the Distance Estimation (DE). The flowchart of the proposed method is depicted in Fig. 1.

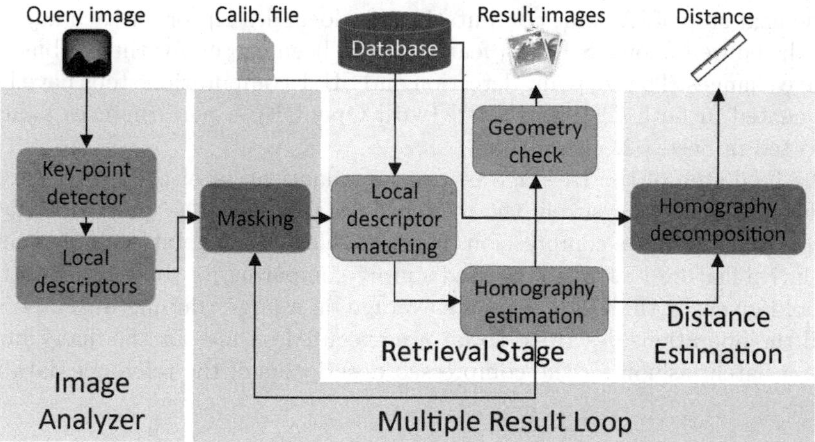

Fig. 1. The complete Visual Search and Distance Estimation pipeline proposed in this paper. The different blocks that make up the pipeline are highlighted in different colors. The query image, the reference database and the camera calibration file are the only inputs required. The outputs generated by the application are the list of objects found in the query image and their estimated distance from the observing camera. (Color figure online)

The first stage of the pipeline, IA (in green in Fig. 1), processes the image and extracts relevant visual descriptors from it whilst the second stage, RS (in blue in Fig. 1), compares the query and database image descriptors to determine if the query image shows the same object, or scene, of an image in the reference database. The MRL stage (in purple in Fig. 1) is implemented as a feedback module that is able to efficiently re-run the previous block (i.e. RS) if more than one object is depicted in the query image. The DE (the red part in Fig. 1) module is used to extract geometrical information for each of the results found by the MRL.

3.1 Image Analyzer

The Visual Search application proposed in this paper is based on local features, so the first step of the IA is to select a set of pixel coordinates $\{[u,v]^\intercal\}_k$ where relevant information about the image will be extracted. This step is implemented with a Laplacian of Gaussian (LoG) [10,11] image pyramid. The LoG filter acts as a corner detector, emphasizing regions of rapid intensity change thanks to its Laplacian component and de-noising the final result with its Gaussian component. As an alternative filter DoG (Difference of Gaussian) could be used, as in [5], which approximates the LoG response. LoG has been chosen mainly because DoG requires the computation of an additional filtered image, due to the fact that DoG is based on differences. The LoG filter is applied on a set of many down-sampled versions of the query image to detect interesting points (i.e. keypoints) on multiple scales. The set of obtained key-points is then ranked using a simple model that predict their importance and only a subset of the most relevant ones is kept.

The next step of IA is the computation of a local descriptor for each keypoint selected: the well-known SIFT [5] features have been chosen for their robustness to image changes. The keypoint detection and SIFT computation steps have been implemented in both CPU and GPU (with OpenGL); a performances analysis is reported in Sect. 4.2.

The final step of the IA stage is a scalar quantization of the SIFT features in order to reduce their size in the reference database and the query descriptor. In order to achieve the compression, for each selected keypoint each element of the relative feature vector is binarized simply comparing its value to a constant threshold given by the SIFT elements average on a large training database.

All the operations described so far are executed on-line for the query image and in a batch fashion for the compressed descriptors of the reference database images.

3.2 Retrieval Stage

The RS takes as inputs the descriptors generated by the IA and use them to determine if the query image depicts the same object as an image in the reference database. For each key-point selected and for each image in the database a local descriptor matching is performed. If I is a reference image in the database, the

k-th local descriptor of the query image matches with the j-th local feature of I if s_k^j is above a threshold, where the score s_k^j is defined as:

$$s_k^j = \cos\left(\frac{\pi}{2}\frac{h_k^j}{h_k^i}\right) \quad (1)$$

In this formula, h_k^j is the Hamming distance between the k-th query descriptor and the j-th reference descriptor, while h_k^i is the distance from the second nearest feature in the image I. This particular score function has been chosen in order to assign high values (i.e. 1) when the best match distance is much smaller in respect to the second best one and low values (i.e. 0) when the two distances are very similar. If a significant number of matched features is found, the image I is selected as a candidate match for the query image.

The described procedure generates many outlier matches between features, i.e. matches between visually similar area of the two images but geometrically inconsistent. For this reason, the matches are verified through a geometry check based on homography estimation, which assumes locally planar and rigid objects. The homography ($\boldsymbol{H}$) is rapidly computed using the widely used combination between the Direct Linear Transform (DLT) [8] algorithm and RANdom SAmple Consensus (RANSAC) [8] method in order to remove outlier matches. The list of candidate matches is then ranked using the geometry check score and the highest ranked image is chosen as the retrieval result.

3.3 Multiple Results Loop

The MRL retrieves multiple results from the database if a query picture depicts more than one object. As is shown on Fig. 1, the MRL is a feedback loop that performs cyclically the RS steps. After each RS iteration the top match is added to a multiple results list and the RS is performed again on a subset of the local features from the previous iteration. This subset is determined by mapping the bounding box of the previous result image I on the query one using the estimated homography $\boldsymbol{H}$ (adding also some padding) and computing an occupancy mask. This mask localizes accurately the region of the query image in which the retrieved object is located: then, all the features having coordinates inside the mask are removed for the next iteration of MRL. At each iteration the current mask is obtained from the union of all the previous masks and the new re-projected bounding box. This features removal operation are described in Fig. 2.

The MRL is performed until either the remaining non-masked surface in the query image is sufficiently small or the scores of the RS results drop below a selected threshold. Figure 1 also shows how the IA is not part of the MRL, so the query image is analyzed only once, independently from the number of objects depicted in it. This is particularly important if the Visual Search application is implemented on a distributed system, e.g. a Cyber-Physical System (CPS), with a central node and multiple imaging nodes, in which each leaf node can perform the IA stage independently.

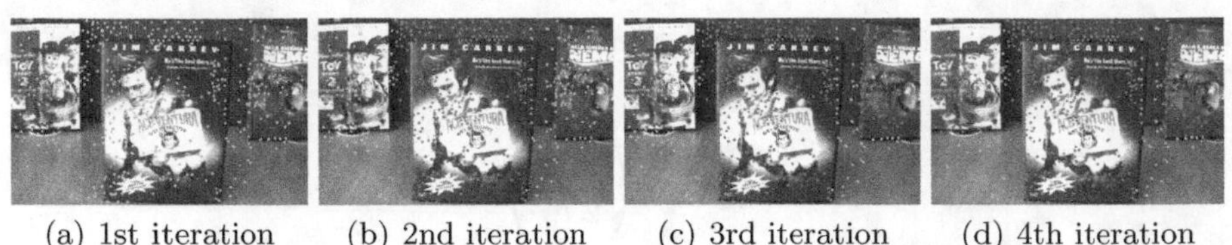

(a) 1st iteration (b) 2nd iteration (c) 3rd iteration (d) 4th iteration

Fig. 2. These figures show the MRL operation removal performed in each iteration of the loop when these method is applied on a query image depicting 3 different objects. The green circles represent the features considered in each iteration while the red ones are the removed features. (Color figure online)

3.4 Distance Estimation

The Distance Estimation module can estimate the relative position between the camera and the retrieved object, expressed as a rotation matrix $\boldsymbol{R}$ and a translation vector $\boldsymbol{t}$, by decomposing the homography $\boldsymbol{H}$ computed in the RS. The inputs needed by the module are the camera calibration parameters (for simplicity only the intrinsic parameters matrix $\boldsymbol{K}$ of a pinhole camera model) and the annotated object size, either the width w_o or the height h_o.

Having defined $\left[u\ v\ 1\right]$ as a homogeneous pixel coordinate on the query image, $\left[u^*\ v^*\ 1\right]^\intercal$ as the corresponding homogeneous pixel coordinate on the result image and $\boldsymbol{H}$ as the estimated homography, if λ is a positive scaling factor the formula:

$$\left[u\ v\ 1\right]^\intercal = \lambda \boldsymbol{H} \left[u^*\ v^*\ 1\right]^\intercal \tag{2}$$

follows from the homography definition. On the other hand, if $\left[x\ y\ 0\right]$ is a physical point lying on the observed planar object, where we assumed for simplicity a plane with equation $z = 0$, from the definition of the camera projection matrix the corresponding point on the query image $\left[u\ v\ 1\right]$ is expressed as:

$$\begin{aligned} \left[u\ v\ 1\right]^\intercal &= \gamma \boldsymbol{K} \left[\boldsymbol{R} \mid \boldsymbol{t}\right] \left[x\ y\ 0\ 1\right]^\intercal \\ &= \gamma \boldsymbol{K} \left[\boldsymbol{r_1}\ \boldsymbol{r_2}\ \boldsymbol{t}\right] \left[x\ y\ 1\right]^\intercal \\ &= \gamma \boldsymbol{K} \left[\boldsymbol{r_1}\ \boldsymbol{r_2}\ \boldsymbol{t}\right] \boldsymbol{S} \left[u^*\ v^*\ 1\right]^\intercal \end{aligned} \tag{3}$$

where γ is another positive scaling factor, $\boldsymbol{r_1}$ and $\boldsymbol{r_2}$ are the first two columns of $\boldsymbol{R}$. $\boldsymbol{S}$ is a scaling matrix transforming 3D points on the plane to pixel coordinates $[u^*, v^*, 1]^\intercal$ defined as:

$$\boldsymbol{S} = \begin{bmatrix} \frac{w_p}{w_o} & 0 & 0 \\ 0 & \frac{h_p}{h_o} & 0 \\ 0 & 0 & 1 \end{bmatrix} \tag{4}$$

having defined w_o and h_o as the width and height of the physical planar object and w_p and h_p as the pixel width and height of the associated image in the database. From Eqs. 2 and 3 follows that:

$$\left[\boldsymbol{r_1}\ \boldsymbol{r_2}\ \boldsymbol{t}\right] = \eta \boldsymbol{K}^{-1}\boldsymbol{H}\boldsymbol{S}^{-1} \tag{5}$$

where η is such that $||\boldsymbol{r_1}||_2 = ||\boldsymbol{r_2}||_2 = 1$ and $\boldsymbol{r_3} = \boldsymbol{r_1} \times \boldsymbol{r_2}$; $\boldsymbol{R}$ is then orthogonalized with SVD to make the estimation more robust. The estimated distance between the object and the camera is simply $d = ||\boldsymbol{t}||_2$ where $\boldsymbol{t}$ is the estimated translation vector. The factor η combined with the scaling matrix $\boldsymbol{S}$ ensures that the translation has the right value in metric units.

4 Experiments and Results

The Visual Search and Distance Estimation pipeline described in Sect. 3 has been implemented on two different systems based on two different ARM based devices: the Hardkernel Odroid XU3[1] and the STMicroelectronics B2260[2]. The Odroid, the most powerful of the two, is equipped with a Samsung Exynos5422 system-on-a-chip (SoC) including a Cortex-A15 2.0 GHz quad core, Cortex-A7 quad core CPU and an ARM Mali-T628 MP6 GPU. The B2260 uses a ST Cannes2-STiH410-EJB SoC including a dual ARMcortex-A9 at 1.5 GHz and an ARM Mali-400 MP2 GPU. Both systems are connected to an USB IDS Ueye UI-1221LE-M-GL[3] camera to acquire images. The setup of the Visual Search and Distance Estimation application running on the Odroid XU3 is shown in Fig. 3. For both systems the camera is mounted on the sliding horizontal headpiece of a stadiometer facing the ground; the object to be retrieved (e.g. an A4 print of

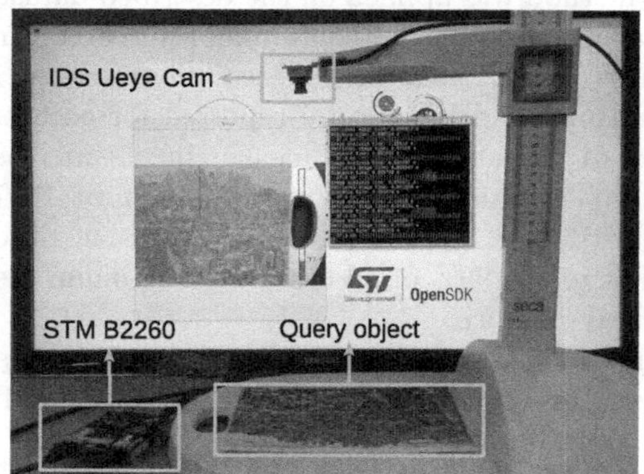

Fig. 3. Setup of the Visual Search and Distance Estimation application running on the STMicroelectronics B2260. The IDS Ueye camera is mounted on the sliding headpiece of the stadiometer, facing the ground, and it is connected with an USB cable to the B2260. The query object is a printed sheet depicting an aerial view of a city and its placed on the bottom platform of the stadiometer. The result of the application is shown on the monitor.

[1] http://www.hardkernel.com/main/products/prdt_info.php?g_code=g140448267127.
[2] http://www.96boards.org/documentation/ConsumerEdition/B2260/README.md.
[3] https://en.ids-imaging.com/store/ui-1221le.html.

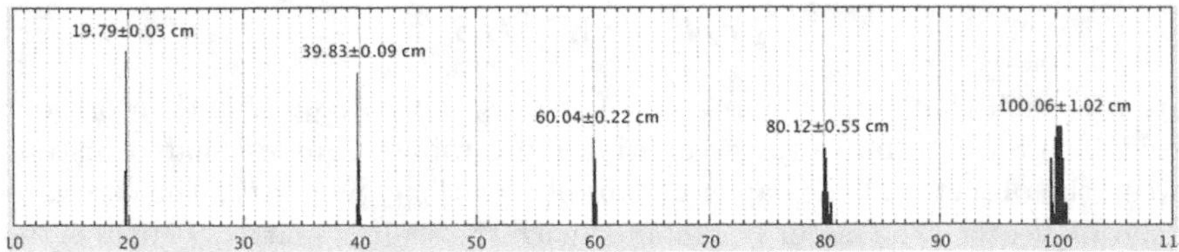

Fig. 4. Histograms of distance estimation results on 5 different set of images acquired at 5 difference distances (20, 40, 60, 80, 100 cm). The intervals represent $\pm 3\sigma$ and they are all below 1% of the estimated distances.

a building aerial view as in Fig. 3) has been placed on the bottom platform of the stadiometer in order to easily record the ground truth distance.

Regarding the quantitative evaluation of the the proposed method's retrieval accuracy, some results could be found in [9].

4.1 Distance Estimation Accuracy

To test the distance estimation accuracy of the proposed method, an ad-hoc test set of 250 images has been created. All the images were acquired by the same camera and all of them depicts the one object acquired at 5 different ground-truthed camera-object distances: 20, 40, 60, 80 and 100 cm. Each image contains a printed sheet depicting an aerial view of a building to mimic a drone application. The proposed method was applied on the described dataset while having a small reference database of 10 images containing the correct image. The object depicted in each query image was always successfully retrieved and in Fig. 4 the histograms of the estimated distances are reported for each true distance. As expected, the standard deviation of the measurements increase with the true distance: the standard deviation reaches up to 0.34 cm at 100 cm, producing a 99% confidence interval of ± 1.02 cm, assuming Gaussian observations; this is equivalent to say that in the 99% of all cases the maximum measurement error is below 1% of the real distance.

Figure 4 also shows that the average measured distance is slightly shifted from the ground-truth one (less than 30 mm in all cases), with a negative bias for small distances and positive bias for large distances. This bias is probably due to some small errors in the camera calibration, such as ignoring small lens distortions.

4.2 Performances

To test the performance of the Visual Search application proposed in this paper and comparing its computational speed on two different systems the same dataset described in Sect. 4.1 has been used. The average time for each module is reported in Table 1. As the queries show only one object, the MRL was not performed, but it would contribute linearly to RS and DE whereas the IA does not depend on

Table 1. Comparison of execution times for each module of the Visual Search and Distance Estimation application obtained on Odroid XU3 and B2260, either exploiting or not the on-board GPU for the IA stage. Only one object is depicted in the query image, so the MRL is never executed. The reference database used has size 10.

		IA	RS	DE	fps
CPU	Odroid XU3	416.9 ms	102.9 ms	0.3 ms	1.9
	B2260	1207.9 ms	159.0 ms	0.3 ms	0.7
GPU	Odroid XU3	105.9 ms	103.7 ms	0.3 ms	4.8
	B2260	157 ms	233 ms	0.3 ms	2.6

it, as described in Sect. 3.3; moreover, the reference database contains 10 images (not to be confused with the test dataset which, on the other hand, contains 250 images).

As it can be noticed from Table 1 the measured average speed of the CPU implementation is 1.9 fps on the Odroid XU3 and 0.7 fps on the B2260. This difference is mainly due to the inequality between the CPUs of the two devices. Moreover, as can be observed, the heavier computational stage of the VS is the IA that is about 4 times slower than the RS part (while the impact of the DE time is negligible). On the other hand, the RS scales linearly with the dimension of the reference database while the IA is independent from it. The RS time is also indirectly influenced by the use of the GPU in the IA because the descriptors are not computed in the exact same way as in the CPU implementation. For this reason the RS time measured could be slightly different between the GPU and CPU implementation of the VS, as it is in the case of the B2260. The Visual Search and Distance Estimation overall frame rate exploiting the system GPU is 4.8 fps on the Odroid XU3 and 2.6 fps on the B2260.

5 Conclusions

In this work we described and implemented a reliable Visual Search pipeline. In order to be able to work in a real case scenario the proposed application is able to retrieve multiple objects from the same query image. Moreover, due to the necessity in many real case scenarios to also spatially localize the retrieve object the proposed VS application is coupled with an accurate distance estimation module.

In particular, the Visual Search and Distance Estimation application proposed in this paper, takes as input a query image depicting one or more objects of interest, extracts some relevant features, searches inside a database of images the most similar one(s) and finally estimates the relative position between the camera and the pictured object(s). The distance estimation is efficiently performed exploiting the homography already estimated in the VS pipeline, in particular the Retrieval Stage, Sect. 3.2.

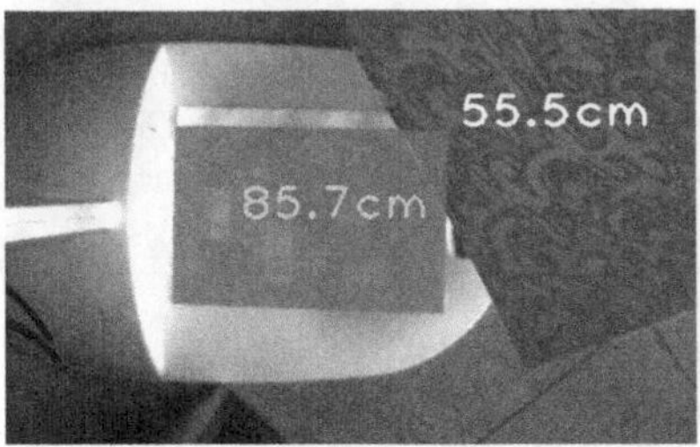

(a) Results on 2 objects

(b) Results on 3 objects

Fig. 5. The Visual Search and Distance Estimation output when the camera views more than one object. Objects highlighted in red are estimated to be closer to the camera while more distant ones are colored in green. (Color figure online)

As reported in Sect. 4.1, this method has proved to be very accurate when estimating object distances between 20 cm and 100 cm with a 99% confidence interval of just 1 cm. The complete pipeline has also been implemented on two different embedded devices (STMicroelectronics B2260 and Hardkernel Odroid XU3) with different computation capabilities, both connected to an USB camera. Furthermore, some qualitative results are also reported in Fig. 5. In particular the proposed method running on a B2260 is applied at two different query images. All the objects depicted in the two images are localized and their distance from the camera is displayed.

Two different implementations have been tested for each device, one exploiting the on-board GPU and the other only using the CPU, and, as shown in Sect. 4.2 multiple frames could be processed in each second achieving near real-time performances for the GPU implementations for both devices. Moreover, in the Visual Search and Distance Estimation method described in this paper, the object identification procedure is performed for every frame acquired by the camera but this is not necessary; in fact, this method could be easily coupled with a fast object tracking module performed only after retrieving the depicted object(s), as shown in [12], leading to an increment of the overall performances.

Finally, the proposed method could also be implemented on a distributed system, in a similar way as in [9], in which the embedded device(s), connected to a server, executes only the Image Analyzer part leaving the RS and the distance estimation to be executed on the central node. This would grant the application to run in real time (the IA spends 105.9 ms in the Odroid XU3 GPU implementation, 9.4 fps) also working on a big database (located in the server).

Acknowledgments. This work was supported in part by the H2020 European project COSSIM.

References

1. Gill, H.: Cyber-physical systems: beyond ES, SNs, SCADA. In: Presentation in the Trusted Computing in Embedded Systems (TCES) Workshop (2010)

2. Niblack, W.C., Barber, R., Equitz, W., Flickner, M.D., Glasman, E.H., Petkovic, D., Yanker, P., Faloutsos, C., Taubin, G.: The QBIC project: Querying images by content, using color, texture, and shape. In: Storage and Retrieval for Image and Video Databases (SPIE) (1994)
3. Pentland, A., Picard, R.W., Sclaroff, S.: Photobook: content-based manipulation of image databases. Int. J. Comput. Vis. **18**(3), 233–254 (1996)
4. The moving picture experts group website. http://mpeg.chiariglione.org/standards/mpeg-7/compact-escriptors-visual-search
5. Lowe, D.G.: Distinctive image features from scale-invariant keypoints. Int. J. Comput. Vis. **60**(2), 91–110 (2004)
6. Pagani, A., Stricker, D.: Structure from motion using full spherical panoramic cameras. In: ICCV Workshops, pp. 375–382. IEEE (2011)
7. Cadena, C., Carlone, L., Carrillo, H., Latif, Y., Scaramuzza, D., Neira, J., Reid, I.D., Leonard, J.J.: Simultaneous localization and mapping: present, future, and the robust-perception age. CoRR, vol. abs/1606.05830 (2016)
8. Hartley, R., Zisserman, A.: Multiple View Geometry in Computer Vision, 2nd edn. Cambridge University Press, New York (2003)
9. Paracchini, M., Marcon, M., Plebani, E., Pau, D.P.: Visual search of multiple objects from a single query. In: 2016 IEEE 6th International Conference on Consumer Electronics-Berlin (ICCE-Berlin), Berlin, Germany, pp. 41–45. IEEE (2016)
10. Haralick, R.M., Shapiro, L.G.: Computer and Robot Vision. Addison-Wesley Longman Publishing Co. Inc, Boston (1992)
11. Canny, J.: A computational approach to edge detection. IEEE Trans. Pattern Anal. Mach. Intell. **8**(6), 679–698 (1986)
12. Plebani, E., Buzzella, A., Pau, D.P., Marcon, M.: Mixing retrieval and tracking using compact visual descriptors. In: 2013 IEEE 3rd International Conference on Consumer Electronics-Berlin (ICCE-Berlin), Berlin, Germany. IEEE (2013)

Synchronization in the Symmetric Inverse Semigroup

Federica Arrigoni(✉), Eleonora Maset, and Andrea Fusiello

DPIA, Università di Udine, Via Delle Scienze, 208, Udine, Italy
arrigoni.federica@spes.uniud.it

Abstract. Matches between images are represented by partial permutations, which constitute the so-called *Symmetric Inverse Semigroup*. Synchronization of these matches is tantamount to joining them in multi-view correspondences while enforcing loop-closure constraints.

This paper proposes a novel solution for partial permutation synchronization based on a spectral decomposition. Experiments on both synthetic and real data shows that our technique returns accurate results in the presence of erroneous and missing matches, whereas a previous solution [12] gives accurate results only for total permutations.

Keywords: Permutation synchronization · Partial permutations · Spectral decomposition · Multi-view matching

1 Introduction

Consider a network of nodes where each node is characterized by an unknown state. Suppose that pairs of nodes can measure the ratio (or difference) between their states. The goal of *synchronization* [14] is to infer the unknown states from the pairwise measures. Typically, states are represented by elements of a group Σ. Solving a synchronization problem can be seen as upgrading from relative (pairwise) information, which involves two nodes at a time, onto absolute (global) information, which involves all the nodes simultaneously. In practice, a solution is found by minimizing a suitable cost function which evaluates the consistency between the unknown states and the pairwise measures.

Several instances of synchronization have been studied in the literature, which correspond to different instantiations of Σ [2–4,6–9,13,14,17]. In particular, $\Sigma = \mathcal{S}_d$ gives rise to *permutation synchronization* [12], where each state is an unknown permutation (i.e. reordering) of d objects. Permutation synchronization finds application in multi-image matching [5,18], where a set of matches between pairs of images is computed through standard techniques (e.g. SIFT [11]), and the goal is to combine them in a global way so as to reduce the number of false matches and complete the matches with new ones retrieved indirectly via loop closure.

The authors of [12] derive an approximate solution based on a spectral decomposition, which is then projected onto $\mathcal{S}_d$. This method can effectively handle

S. Battiato et al. (Eds.): ICIAP 2017, Part II, LNCS 10485, pp. 70–81, 2017.
https://doi.org/10.1007/978-3-319-68548-9_7

false matches, but it can not deal with missing matches (i.e. partial permutations), as confirmed by the experiments.

To overcome this drawback, we propose a novel method that extends [12] to the synchronization of *partial permutations*.

The paper is organized as follows. In Sect. 2 the concept of synchronization is formally defined, and it is shown that, if Σ admits a matrix representation, a solution can be obtained via spectral decomposition. In Sect. 3 the multi-view matching problem is introduced, and it is expressed as a synchronization problem. Section 4 reviews the spectral solution proposed in [12] for permutation synchronization, while Sect. 5 is devoted to explain our solution to partial permutation synchronization. Section 6 presents experiments on both synthetic and real data, and Sect. 7 draws the conclusion.

2 Synchronization

The goal of *synchronization* is to estimate elements of a group given a (redundant) set of measures of their ratios (or differences). Formally, let Σ be a group and let $*$ denote its operation. Suppose that a set of measures $z_{ij} \in \Sigma$ is known for some index pairs $(i,j) \subseteq \{1,\dots,n\} \times \{1,\dots,n\}$. The synchronization problem can be formulated as the problem of recovering $x_i \in \Sigma$ for $i = 1,\dots,n$ such that the following *consistency constraint* is satisfied

$$z_{ij} = x_i * x_j^{-1}. \tag{1}$$

It is understood that the solution is defined up to a global (right) product with any group element, i.e., if $x_i \in \Sigma$ satisfies (1) then also $x_i * y$ satisfies (1) for any (fixed) $y \in \Sigma$.

If the input measures are corrupted by noise, then the consistency constraint (1) will not be satisfied exactly, thus the goal is to recover the unknown elements $x_i \in \Sigma$ such that a *consistency error* is minimized, which measures the violation of the consistency constraint, as shown in Fig. 1. If we assume that Σ is equipped with a metric function $\delta : \Sigma \times \Sigma \to \mathbb{R}^+$, the consistency error can be defined as

$$\varepsilon(x_1,\dots,x_n) = \sum_{(i,j)} \delta\big(z_{ij}, x_i * x_j^{-1}\big). \tag{2}$$

Definition 1. *An* inverse semigroup $(\Sigma, *)$ *is a semigroup in which for all* $s \in \Sigma$ *there exists an element* $t \in \Sigma$ *such that* $s = s * t * s$ *and* $t = t * s * t$*. In this case, we write* $t = s^{-1}$ *and call* t *the* inverse *of* s*. If* Σ *has an identity element* 1_Σ *(i.e. it is a* monoid*), then it is called an* inverse monoid.

Inverses in an inverse semigroup have many of the same properties as inverses in a group, e.g., $(a * b)^{-1} = b^{-1} * a^{-1}$ for all $a, b \in \Sigma$.

We observe that the notion of synchronization can be extended to the case where Σ is an *inverse monoid*. In this case Eq. (1) still makes sense, with the provision that x_j^{-1} now denotes the inverse of x_j in the semigroup. Note that $x_j^{-1} * x_j$ and $x_j * x_j^{-1}$ are not necessarily equal to the identity. The solution to the synchronization problem in an inverse monoid is defined up to a global (right) product with any element $y \in \Sigma$ such that $y * y^{-1} = 1_\Sigma = y^{-1} * y$.

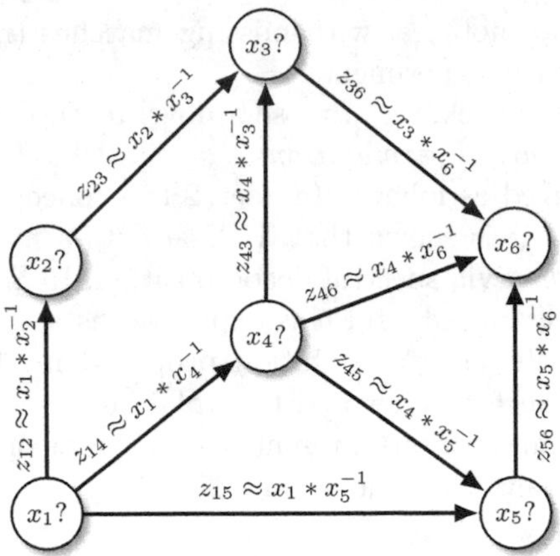

Fig. 1. The synchronization problem. Each node is characterized by an unknown state and measures on the edges are ratios of states. The goal is to compute the states that best agree with the measures.

2.1 Matrix Formulation

If Σ admits a matrix representation, i.e. Σ can be embedded in $\mathbb{R}^{d\times d}$, then the synchronization problem can be expressed as an eigenvalue decomposition, resulting in an efficient and simple solution. Specifically, the unknown states are derived from the leading eigenvectors of a matrix constructed from the pairwise measures.

Suppose, e.g., that Σ is the orthogonal group of dimension d, i.e. $\Sigma = O(d)$, which admits a matrix representation through orthogonal $d \times d$ real matrices, where the group operation reduces to matrix multiplication, the inverse becomes matrix transposition and $1_\Sigma = I_d$ (the $d \times d$ identity matrix). Let $X_i \in \mathbb{R}^{d\times d}$ and $Z_{ij} \in \mathbb{R}^{d\times d}$ denote the matrix representations of $x_i \in \Sigma$ and $z_{ij} \in \Sigma$, respectively. Using this notation, Eq. (1) rewrites $Z_{ij} = X_i X_j^\mathsf{T}$.

Let us collect the unknown group elements and all the measures in two matrices $X \in \mathbb{R}^{dn\times d}$ and $Z \in \mathbb{R}^{dn\times dn}$ respectively, which are composed of $d\times d$ blocks, namely

$$X = \begin{bmatrix} X_1 \\ X_2 \\ \dots \\ X_n \end{bmatrix}, \quad Z = \begin{bmatrix} I_d & Z_{12} & \dots & Z_{1n} \\ Z_{21} & I_d & \dots & Z_{2n} \\ \dots & & & \dots \\ Z_{n1} & Z_{n2} & \dots & I_d \end{bmatrix}. \tag{3}$$

Thus the consistency constraint can be expressed in a compact matrix form as

$$Z = XX^\mathsf{T}. \tag{4}$$

Proposition 1 [14]. *The columns of X are d (orthogonal) eigenvectors of Z corresponding to the eigenvalue n.*

Proof. Since $X^\intercal X = nI_d$ it follows that $ZX = XX^\intercal X = nX$, which means that the columns of X are d eigenvectors of Z corresponding to the eigenvalue n.

Note that, since $\text{rank}(Z) = d$ all the other eigenvalues are zero, so n is also the *largest* eigenvalue of Z. In the presence of noise, the eigenvectors of Z corresponding to the d largest eigenvalues can be seen as an estimate of X. Note that closure is not guaranteed, thus such solution must be projected onto $\Sigma = O(d)$, e.g., via singular value decomposition.

This procedure was introduced in [14] for $\Sigma = SO(2)$, extended in [1,15] to $\Sigma = SO(3)$, and further generalized in [2,4] to $\Sigma = SE(d)$. The same formulation appeared in [12] for $\Sigma = \mathcal{S}_d$, which is a subgroup of $O(d)$.

3 Problem Formulation

Consider a set of n *nodes.* A set of k_i *objects* out of d is attached to node i (we say that the node "sees" these k_i objects) in a random order, i.e., each node has its own local labeling of the objects with integers in the range $\{1, \dots, d\}$.

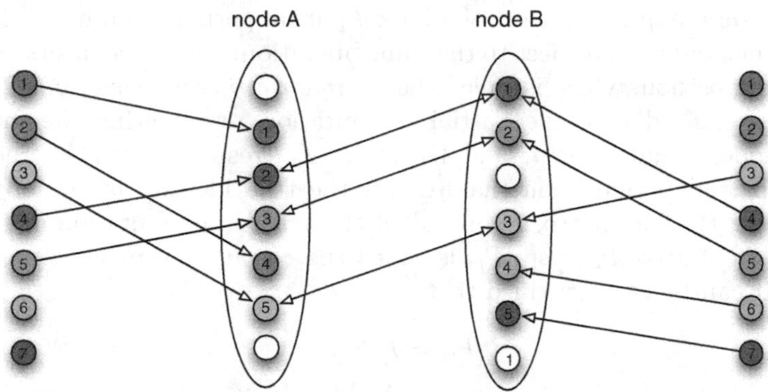

Fig. 2. In the center, two nodes with partial visibility match their three common objects. At the extrema the ground truth ordering of the objects. Each node sees some of the objects (white circles are missing objects) and puts them in a different order, i.e., it gives them different numeric labels.

For example, with reference to Fig. 2, the same object is referred to as n. 5, e.g., in node A and as n. 3 in node B. Pairs of nodes can *match* these objects, establishing which objects are the same in the two nodes, despite the different naming. For example, a match means that the two nodes agree that "my object n. 5 is your object n. 3". The goal is to infer a global labeling of the objects, such that the same object receives the same label in all the nodes.

A more concrete problem statement can be given in terms of feature matching, where nodes are *images* and objects are *features.* A set of matches between pairs of images is given, and the goal is to combine them in a multi-view matching, such that each feature has a unique label in all the images.

Each matching is a bijection between (different) subsets of objects, which is also known as *partial permutation* (if the subsets are improper then the permutation is *total*). Total and partial permutations admit a matrix representation through permutation and partial permutation matrices, respectively.

Definition 2. *A matrix P is said to be a* permutation matrix *if exactly one entry in each row and column is equal to 1 and all other entries are 0. A matrix P is said to be a* partial permutation matrix *if it has at most one nonzero entry in each row and column, and these nonzero entries are all 1.*

Specifically, the partial permutation matrix P representing the matching between node B and node A is constructed as follows: $[P]_{h,k} = 1$ if object k in node B is matched with object h in node A; $[P]_{h,k} = 0$ otherwise. If row $[P]_{h,\cdot}$ is a row of zeros, then object h in node A does not have a matching object in node B. If column $[P]_{\cdot,k}$ is a column of zeros, then object k in node B does not have a matching object in node A.

The set of all $d \times d$ permutation matrices forms a group with respect to matrix multiplication, where the inverse is matrix transposition, which is called the *symmetric group* $\mathcal{S}_d$. The set of all $d \times d$ partial permutation matrices forms an inverse monoid with respect to the same operation, where the inverse is again matrix transposition, which is called the *symmetric inverse semigroup* $\mathcal{I}_d$.

Let $P_{ij} \in \mathcal{I}_d$ denote the partial permutation representing the matching between node j and node i, and let $P_i \in \mathcal{I}_d$ (resp. $P_j \in \mathcal{I}_d$) denote the unknown partial permutation that reveals the true identity of the objects in node i (resp. j). The matrix P_{ij} is called the *relative* permutation of the pair (i, j), and the matrix P_i (resp. P_j) is called the *absolute* permutation of node i (resp. j). It can be easily verified that

$$P_{ij} = P_i P_j^{\mathsf{T}}. \tag{5}$$

Thus the problem of finding the global labeling can be modeled as finding n absolute permutations, assuming that a set of relative permutations is known, where the link between relative and absolute permutations is given by Eq. (5).

If permutations were total, Eq. 5 would be recognized as the consistency constraint of a synchronization problem over $\mathcal{S}_d$ [12], and this will be reviewed in Sect. 4. However, in all practical settings, permutations are partial, thus in Sect. 5 we address the synchronization problem over the inverse monoid $\mathcal{I}_d$, as the main contribution of this paper.

4 Permutation Synchronization

Let us describe the synchronization problem over $\Sigma = \mathcal{S}_d$ [12]. Since $\mathcal{S}_d$ is a subgroup of $O(d)$, permutation synchronization can be addressed as explained in Sect. 2.1. Specifically, as done in Eq. (3), all the absolute/relative permutations are collected in two matrices $X \in \mathbb{R}^{dn \times d}$ and $Z \in \mathbb{R}^{dn \times dn}$ respectively, namely

$$X = \begin{bmatrix} P_1 \\ P_2 \\ \dots \\ P_n \end{bmatrix}, \quad Z = \begin{bmatrix} P_{11} & P_{12} & \dots & P_{1n} \\ P_{21} & P_{22} & \dots & P_{2n} \\ \dots & & & \dots \\ P_{n1} & P_{n2} & \dots & P_{nn} \end{bmatrix} \tag{6}$$

where $P_{ii} = P_i P_i^\mathsf{T} = I_d$. Thus the consistency constraint becomes $Z = XX^\mathsf{T}$ with Z of rank d, and the columns of X are d (orthogonal) eigenvectors of Z corresponding to the eigenvalue n, as stated by Proposition 1. In the presence of noise, the eigenvectors of Z corresponding to the d largest eigenvalues are computed.

We now explain the link between this spectral solution and the consistency error of the synchronization problem. Let us consider the following metric $\delta(P,Q) = d - \operatorname{trace}(P^\mathsf{T} Q) = d - \sum_{r,s=1}^{d} [P]_{r,s}[Q]_{r,s}$, which simply counts the number of objects *differently* assigned by permutations P and Q. Thus Eq. (2) rewrites

$$\varepsilon(P_1, \dots, P_n) = \sum_{i,j=1}^{n} \left(d - \operatorname{trace}(P_{ij}^\mathsf{T} P_i P_j^\mathsf{T}) \right) \tag{7}$$

and hence the synchronization problem over $\mathcal{S}_d$ becomes equivalent to the following maximization problem

$$\max_{P_1,\dots,P_n \in \mathcal{S}_d} \sum_{i,j=1}^{n} \operatorname{trace}(P_{ij}^\mathsf{T} P_i P_j^\mathsf{T}) \iff \max_{X \in \mathcal{S}_d^n} \operatorname{trace}(X^\mathsf{T} Z X). \tag{8}$$

Solving (8) is computationally difficult since the feasible set is non-convex. A tractable approach consists in relaxing the constraints and considering the following optimization problem

$$\max_{U^T U = n I_d} \operatorname{trace}(U^\mathsf{T} Z U) \tag{9}$$

which is a generalized Rayleigh problem, whose solution is given by the d leading eigenvectors of Z.

Since Problem (9) is a *relaxed* version of Problem (8), the solution U formed by eigenvectors is not guaranteed to be composed of permutation matrices. Thus each $d \times d$ block in U is projected onto $\mathcal{S}_d$ via the Kuhn-Munkres algorithm [10]. As long as U is relatively close to the ground-truth X, this procedure works well, as confirmed by the experiments.

5 Partial Permutation Synchronization

Consider now the synchronization problem over $\Sigma = \mathcal{I}_d$. Despite the group structure is missing, we show that a spectral solution can be derived in an analogous way, which can be seen as the extension of [12] to the case of partial permutations.

We define two block-matrices X and Z containing absolute and relative permutations respectively – as done in (6) – so that the consistency constraint becomes $Z = XX^{\mathsf{T}}$ with Z of rank d. Note that here $P_i \in \mathcal{I}_d$, thus the $d \times d$ (diagonal) matrix $P_{ii} = P_i P_i^{\mathsf{T}}$ is not equal, in general, to the identity, unless $P_i \in \mathcal{S}_d$. Indeed, the (k,k)-entry in P_{ii} is equal to 1 if node i sees object k, and it is equal to 0 otherwise. When all the objects seen by node i are different from those seen by node j we have $P_i P_j^{\mathsf{T}} = 0$, resulting in a zero block in Z.

Proposition 2. *The columns of X are d (orthogonal) eigenvectors of Z and the corresponding eigenvalues are given by the diagonal of $V := \sum_{i=1}^{n} P_i^{\mathsf{T}} P_i$.*

Proof. Define the $d \times d$ diagonal matrix $V := X^{\mathsf{T}}X = \sum_{i=1}^{n} P_i^{\mathsf{T}} P_i$. Then $ZX = XX^{\mathsf{T}}X = XV$ which is a spectral decomposition, i.e. the columns of X are d eigenvectors of Z and the corresponding eigenvalues are on the diagonal of V.

Although $\mathcal{I}_d$ is not a group, we have obtained an eigenvalue decomposition problem where the non-zero (and hence *leading*) eigenvalues are d integers contained in the diagonal of V. Specifically, the k-th eigenvalue counts how many nodes see object k. In the case of total permutations all the nodes see all the objects, thus $V = nI_d$ and all the eigenvalues are equal, hence we get Proposition 1.

In the presence of noise, the eigenvectors of Z corresponding to the d largest eigenvalues, which are collected in a $nd \times d$ matrix U, solve Problem (9) as in the case of total permutations.

Note that the reverse of Proposition 2 is not true, i.e., the matrix U is not necessarily equal (up to scale) to X. Indeed, a set of eigenvectors is uniquely determined (up to scale) only if the corresponding eigenvalues are distinct. However, when eigenvalues are repeated, the corresponding eigenvectors span a linear subspace, and hence any (orthogonal) basis for such a space is a solution. Note that all the eigenvalues are integer numbers lower or equal to n. Thus, when the number of objects is larger than the number of nodes (i.e., $d > n$) – which is likely to happen in practice – the eigenvalues are indeed repeated, therefore U is not uniquely determined.

So we have to face the problem of how to select, among the infinitely many Us, the one that resembles X, a matrix composed of partial permutations. Projecting each $d \times d$ block of U onto $\mathcal{I}_d$ does not solve the problem, as confirmed by the experiments. A key observation is reported in the following proposition, suggesting that such a problem can be solved via clustering techniques.

Proposition 3. *Let U be the $nd \times d$ matrix composed by the d leading eigenvectors of Z; then U has $d+1$ different rows (in the absence of noise).*

Proof. Let $\lambda_1, \lambda_2, \ldots, \lambda_\ell$ denote all the *distinct* eigenvalues of Z (with $\ell \leq d$), and let $m_1, m_2, \ldots, m_\ell$ be their multiplicities such that $\sum_{k=1}^{\ell} m_k = d$. Let U_{λ_k} denote the m_k columns of U corresponding to the eigenvalue λ_k, and let X_{λ_k} be the corresponding columns of X. Up to a permutation of the columns, we have

$$U = [U_{\lambda_1}\ U_{\lambda_2}\ \ldots\ U_{\lambda_\ell}], \quad X = [X_{\lambda_1}\ X_{\lambda_2}\ \ldots\ X_{\lambda_\ell}]. \tag{10}$$

Since U_{λ_k} and X_{λ_k} are (orthogonal) eigenvectors corresponding to the same eigenvalue, there exists an orthogonal matrix $Q_k \in \mathbb{R}^{m_k \times m_k}$ representing a change of basis in the eigenspace of λ_k, such that $U_{\lambda_k} = X_{\lambda_k} Q_k$. In matrix form this rewrites

$$U = X \underbrace{\text{blkdiag}(Q_1, Q_2, \ldots, Q_\ell)}_{Q} \tag{11}$$

where blkdiag$(Q_1, Q_2, \ldots, Q_\ell)$ produces a $d \times d$ block-diagonal matrix with blocks $Q_1, Q_2, \ldots, Q_\ell$ along the diagonal. Note that the rows of X are the rows of I_d plus the zero row. Since Q is invertible (hence injective), $U = XQ$ has only $d+1$ different rows as well.

In the presence of noise, we can cluster the rows of U with k-means into $d+1$ clusters, and then assign the centroid which is closest to zero to the zero row, and arbitrarily assign each of the other d centroids to a row of I_d. Recall that the solution to partial permutation synchronization is defined up to a global total permutation.

Note that, if we assign each row of U to the centroid of the corresponding cluster, then we may not obtain partial permutation matrices. Indeed, since there are no constraints in the clustering phase, it may happen that different rows of a $d \times d$ block in U are assigned to the same cluster, resulting in more than one entry per column equal to 1. For this reason, for each $d \times d$ block in U we compute a partial permutation matrix that best maps such block into the set of centroids via the Kuhn-Munkres algorithm [10], and such permutation is output as the sought solution.

Remark 1. As observed (e.g.) in [5], the multi-view matching problem can be seen as a graph clustering problem, where each cluster corresponds to one object (feature) out of d. The underlying graph is constructed as follows: each vertex represents a feature in an image, and edges encode the matches. Note that the matrix Z defined in (6) coincides with the adjacency matrix of such a graph. Our procedure first constructs a $dn \times d$ matrix U from the d leading eigenvectors of Z, and then applies k-means to the rows of U. This coincides with solving the multi-view matching problem via *spectral clustering* applied to the adjacency matrix, rather than to the Laplacian matrix as customary.

6 Experiments

The proposed method – henceforth dubbed PartialSynch – was implemented in Matlab and compared to the solution of [12] (which will be referred to as TotalSynch[1]), considering both simulated and real data.

In the synthetic experiments, performances have been measured in terms of *precision* (number of correct matches returned divided by the number of matches returned) and *recall* (number of correct matches returned divided by the number

[1] The code is available at http://pages.cs.wisc.edu/~pachauri/perm-sync/.

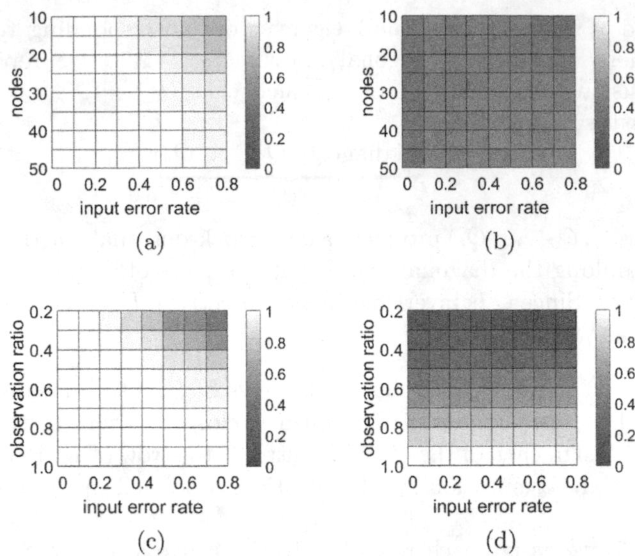

Fig. 3. F-score (the higher the better) of PARTIALSYNCH (a, c) and TOTALSYNCH (b, d). In (a, b) the number of nodes n and the input error rate are varying, while the observation ratio is constant and equal to 0.6. In (c, d) the observation ratio and input error rate vary with $n = 30$.

of correct matches that should have been returned). In order to provide a single figure of merit we computed the *F-score* (twice the product of precision and recall divided by their sum), which is a measure of accuracy and reaches its best value at 1. In the real experiments the number of matches that should have been returned is not known, hence only the precision could be computed.

For the synthetic case, a fixed number of $d = 20$ objects was chosen, while the number of nodes varied from $n = 10$ to $n = 50$. The *observation ratio*, i.e., the probability that an object is seen in a node, decreased from 1 (that corresponds to total permutations) to 0.2. After generating ground-truth absolute permutations, pairwise matches were computed from Eq. (5). Then random errors were added to relative permutations by switching two matches, removing true matches or adding false ones. The *input error rate*, i.e., the ratio of mismatches, varied from 0 to 0.8. For each configuration the test was run 20 times and the mean F-score was computed. In order to evaluate a solution, the total permutation that best aligns the estimated absolute permutations onto ground-truth ones was computed with the Kuhn-Munkres algorithm.

Results are reported in Fig. 3, which shows the F-score for the two methods as a function of number of nodes, observation ratio and input error rate. In the case of total permutations (observation ratio = 1) both techniques perform well. Our method (PARTIALSYNCH) correctly recovers the absolute permutations even when not all the objects are seen in every node, and in the presence of high contamination. On the contrary, TOTALSYNCH cannot deal with partial permutations, indeed its performances degrade quickly as the observation ratio decreases. In general, the accuracy increases with the number of nodes.

In the experiments with real data, the problem of feature matching across multiple images was considered. The Herz-Jesu-P8 and Fountain-P11 datasets [16] were chosen, which contain 8 and 11 images respectively. A set of features was detected with SIFT [11] in each image. Among them a subset was manually selected by looking at the tracks, with the aim of knowing the total number of objects (equal to $d = 30$). Subsequently, correspondences between each image pair (i, j) were established using nearest neighbor and ratio test as in [11] and refined using RANSAC. The resulting relative permutations P_{ij} were given as input to the considered methods.

Table 1. Precision [%] achieved by the two methods.

Data	PartialSynch	TotalSynch
Fountain-P11	95.4	43.6
Herz-Jesu-P8	93.6	50.7

When evaluating the output, matches were considered correct if they lie within a given distance threshold (set equal to 0.01 times the image diagonal) from the corresponding epipolar lines, computed from the available ground-truth camera matrices. The precision achieved by the two methods on these sets is reported in Table 1, which confirm the outcome of the experiments on simulated data. For illustration purposes, Fig. 4 shows the results for some sample images.

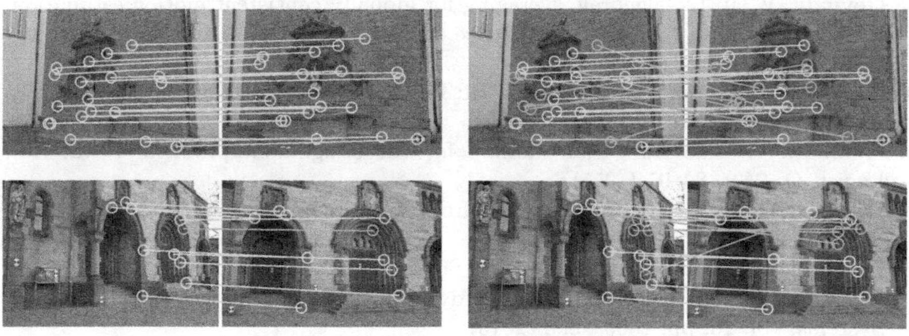

Fig. 4. Matches for a representative image pair from the Fountain-P11 dataset (top) and from the Herz-Jesu-P8 dataset (bottom). True matches and false matches are shown in light-blue and red, respectively, for PartialSynch (left) and TotalSynch (right). (Color figure online)

7 Conclusion

In this paper we proposed a novel solution for partial permutation synchronization based on a spectral decomposition, that can be applied to multi-image matching. The method was evaluated through synthetic and real experiments, showing that it handles both false and missing matches, whereas [12] gives accurate results only for total permutations.

References

1. Arie-Nachimson, M., Kovalsky, S.Z., Kemelmacher-Shlizerman, I., Singer, A., Basri, R.: Global motion estimation from point matches. In: Proceedings of the Joint 3DIM/3DPVT Conference: 3D Imaging, Modeling, Processing, Visualization and Transmission (2012)
2. Arrigoni, F., Rossi, B., Fusiello, A.: Spectral synchronization of multiple views in SE(3). SIAM J. Imaging Sci. **9**(4), 1963–1990 (2016)
3. Arrigoni, F., Fusiello, A., Rossi, B.: Camera motion from group synchronization. In: Proceedings of the International Conference on 3D Vision (3DV). IEEE (2016)
4. Bernard, F., Thunberg, J., Gemmar, P., Hertel, F., Husch, A., Goncalves, J.: A solution for multi-alignment by transformation synchronisation. In: Proceedings of the IEEE Conference on Computer Vision and Pattern Recognition (2015)
5. Chen, Y., Guibas, L., Huang, Q.: Near-optimal joint object matching via convex relaxation. In: Proceedings of the International Conference on Machine Learning, pp. 100–108 (2014)
6. Cucuringu, M.: Synchronization over Z_2 and community detection in signed multiplex networks with constraints. J. Complex Netw. **3**(3), 469–506 (2015)
7. Giridhar, A., Kumar, P.: Distributed clock synchronization over wireless networks: algorithms and analysis. In: Proceedings of the IEEE Conference on Decision and Control, pp. 4915–4920 (2006)
8. Govindu, V.M.: Lie-algebraic averaging for globally consistent motion estimation. In: Proceedings of the IEEE Conference on Computer Vision and Pattern Recognition, pp. 684–691 (2004)
9. Hartley, R., Trumpf, J., Dai, Y., Li, H.: Rotation averaging. Int. J. Comput. Vis. **103**(3), 267–305 (2013)
10. Kuhn, H.W.: The Hungarian method for the assignment problem. Naval Res. Logist. Q. **2**(2), 83–97 (1955)
11. Lowe, D.G.: Distinctive image features from scale-invariant keypoints. Int. J. Comput. Vis. **60**(2), 91–110 (2004)
12. Pachauri, D., Kondor, R., Singh, V.: Solving the multi-way matching problem by permutation synchronization. In: Advances in Neural Information Processing Systems, vol. 26, pp. 1860–1868 (2013)
13. Schroeder, P., Bartoli, A., Georgel, P., Navab, N.: Closed-form solutions to multiple-view homography estimation. In: 2011 IEEE Workshop on Applications of Computer Vision (WACV), pp. 650–657, January 2011
14. Singer, A.: Angular synchronization by eigenvectors and semidefinite programming. Appl. Comput. Harmon. Anal. **30**(1), 20–36 (2011)
15. Singer, A., Shkolnisky, Y.: Three-dimensional structure determination from common lines in cryo-EM by eigenvectors and semidefinite programming. SIAM J. Imaging Sci. **4**(2), 543–572 (2011)

16. Strecha, C., von Hansen, W., Gool, L.J.V., Fua, P., Thoennessen, U.: On benchmarking camera calibration and multi-view stereo for high resolution imagery. In: Proceedings of the IEEE Conference on Computer Vision and Pattern Recognition (2008)
17. Wang, L., Singer, A.: Exact and stable recovery of rotations for robust synchronization. Inf. Inference: J. IMA **2**(2), 145–193 (2013)
18. Zhou, X., Zhu, M., Daniilidis, K.: Multi-image matching via fast alternating minimization. In: Proceedings of the International Conference on Computer Vision, pp. 4032–4040 (2015)

A Fully Convolutional Network for Salient Object Detection

Simone Bianco, Marco Buzzelli(✉), and Raimondo Schettini

Dipartimento di Informatica, Sistemistica e Comunicazione,
Università degli Studi di Milano-Bicocca, Viale Sarca 336, 20126 Milan, Italy
{simone.bianco,marco.buzzelli,schettini}@disco.unimib.it

Abstract. In this paper we address the task of salient object detection without requiring an explicit object class recognition. To this end, we propose a solution that exploits intermediate activations of a Fully Convolutional Neural Network previously trained for the recognition of 1,000 object classes, in order to gather generic object information at different levels of resolution. This is done by using both convolution and convolution-transpose layers, and combining their activations to generate a pixel-level salient object segmentation. Experiments are conducted on a standard benchmark that involves seven heterogeneous datasets. On average our solution outperforms the state of the art according to multiple evaluation measures.

Keywords: Salient object detection · Fully convolutional neural network · Foreground/background segmentation

1 Introduction

Accurate visual saliency models are fundamental for multiple disciplines such as computer vision [5], neuroscience [12], and cognitive psychology [11]. In this paper we focus on salient object detection, which consists in segmenting the main foreground object from the background in a digital image. Salient object detection methods are commonly used in applications such as object-of-interest proposal, object recognition, adaptive image and video compression, content aware image editing, image retrieval, and object-level image manipulation [6].

In the literature there is no universal agreement for the definition of foreground and background. This will be evident later on in this paper, by comparing the annotation criteria adopted through the different benchmark datasets. During the training of our Fully Convolutional Network, we simultaneously exploit annotated data coming from different datasets. In this way we obtain a model that represents a good compromise among the different datasets on the definition of foreground/background segmentation.

Many different approaches and solutions have been proposed in the last years for salient object detection. The method proposed in Discriminative Regional Feature Integration (DRFI) [13] builds a multi-level representation of the input

S. Battiato et al. (Eds.): ICIAP 2017, Part II, LNCS 10485, pp. 82–92, 2017.
https://doi.org/10.1007/978-3-319-68548-9_8

image, and creates a regression model mapping the regional feature vector of each level to the corresponding saliency score. These scores are finally fused in order to determine the complete saliency map. In Quantum Cut (QCUT) [3] authors model salient object segmentation as an optimization problem. They then exploit the link between quantum mechanics and graph-cuts to develop an object segmentation method based on the ground state solution of a modified Hamiltonian. The authors of Minimum Barrier Distance (MBD) [26] present an approximation of the MBD transform, and combine it with an appearance-based backgroundness cue. The resulting method performs significantly better than other solutions having the same computational requirements. In Saliency Tree (ST) [18] authors simplify the image into primitive regions, with associated saliency based on multiple handcrafted measures. They generate a saliency tree using region merging, and perform a systematic analysis of such tree to derive the final saliency map. Robust Background Detection (RBD) [28] introduces boundary connectivity: a background measure based on an intuitive geometrical interpretation. This measure is then used along with multiple low level cues to produce saliency maps through a principled optimization framework.

In a very recent work Borji et al. [6] present an exhaustive review of state of the art methods for salient object detection. They compared more than forty methods on a benchmark composed of seven different datasets. In this paper we investigate the use of a Fully Convolutional Network (FCN) for salient object detection taking inspiration from the work of Long et al. [19], and evaluate it on the Borji et al. [6] benchmark. Differently from compared solutions, we propose a data-driven model that leverages semantic cues as the basis for saliency estimation. Other approaches using deep learning methods also exist [7,10,15], although they don't adhere to the data and methods in the reference benchmark we adopt here. The main contributions of this paper can be summarized as follows:

- we propose a semantically-aware FCN to address the problem of salient object detection that is able to produce a binary pixel-level saliency map;
- we systematically investigate the contribution of different kinds of synthetic data augmentation to train the FCN;
- we evaluate the effectiveness of our proposal on a standard benchmark for salient object detection composed of seven different datasets [6]. The proposed method on average outperforms the state of the art according to multiple evaluation measures.

2 Proposed Method

We propose a Fully Convolutional Network to address the problem of salient object detection, taking inspiration from a work originally developed for semantic segmentation [19], that uses layers previously trained for the recognition of 1,000 object classes (Visual Geometry Group, or VGG [22]). This allows our network to be semantically-aware, and therefore capable of exploiting high-order concepts for separating foreground from background. Furthermore, the fully convolutional architecture is specifically designed to produce a per-pixel prediction, which perfectly fits the task of generating an input-sized foreground/background mask.

The main difference with respect to the semantic segmentation proposed in [19] is that in our proposal the salient object could belong to any object category. Our network is in fact able to segment salient objects belonging to categories not restricted to the 20 classes defined in the original semantic segmentation task [19], or the 1,000 object classes used to train the VGG [22]. Finally, we adopt a different training procedure, as we find advantage in applying several kinds of data augmentation. The effects of such augmentation are analyzed and discussed in the experimental results section.

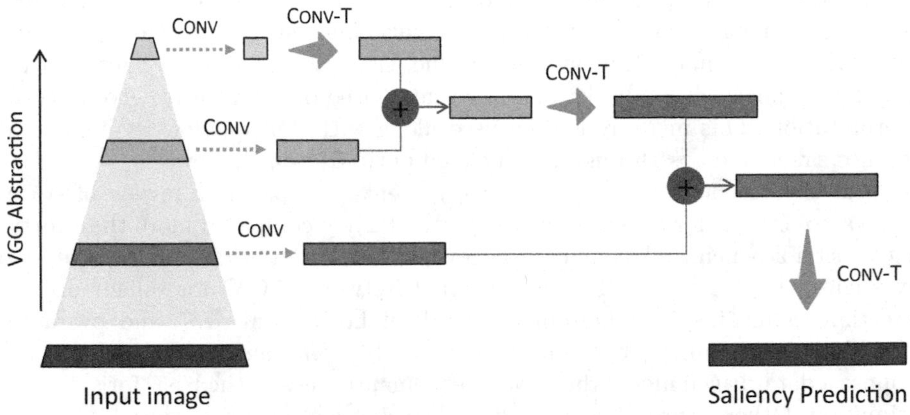

Fig. 1. Schematic view of the Fully Convolutional Network employed for salient object detection. Intermediate activations of a VGG-based processing are resized and combined in order to implement a multi-resolution analysis.

The network architecture is illustrated in Fig. 1, and adheres to the following logic:

1. Build abstractions of gradually decreasing spatial resolution, using [22].
2. Extract intermediate activations, and map their depth to the final problem size (2 classes for our task), using convolution layers.
3. Increase size of activations, using convolution-transpose layers.
4. Sum-up activations having now compatible size.
5. Produce as output a binary pixel-level saliency map.

Thanks to this strategy, the network can see both the whole picture and small details at the same time, thus producing a globally-aware yet precise output.

2.1 Training

Layers inherited from VGG (which supposedly only need fine-tuning) and new layers (trained from scratch), are all updated using the same learning-rate. The task of calibrating the gradients for the two strategies is implicitly left to the Adam optimizer [14].

Many methods in the state of the art generate a continuous-valued prediction [3,13,18,26,28] directly correlated to the saliency of pixels in the image. Most of the available datasets, though, are published with a binary ground truth [5,8,17,24,25]. For this reason we choose to approach the problem as a per-pixel binary classification task: all ground truth images are converted to binary data, setting to 1 all values greater than 0. The neural network is then trained with a softmax cross entropy loss, with the global loss of each batch computed by averaging all loss values from the single pixels.

All training examples are processed by an online data augmentation procedure in order to provide additional information to the learning process. The following perturbations are considered:

- Random crop. We select a square subwindow of random side between 256 pixels and the original image limits. The crop is then resized to the fixed training dimension, i.e. 256 $\times$ 256 pixels.
- Random horizontal flip.
- Random gamma between 0.3 and $\frac{1}{0.3}$.

Each perturbation category was individually tested on a small subset of the benchmark data, in order to assess its impact on performance. An analysis on such effects is reported in Sect. 3.2.

All models are trained with a learning rate of 5×10^{-5} and a batch size equal to 15 for a total of 20 epochs.

3 Experiments

3.1 Datasets

Experiments were performed according to the benchmark proposed in [6] concerning both the datasets and the evaluation protocol of the results. The benchmark is composed by seven different datasets that are presented in Table 1.

Table 1. Summary of tested datasets

Dataset	Images	Notes
PASCAL-S [16]	850	High background clutter
THUR15K [8]	6233	Only 6233/15000 annotated images
JuddDB [5]	900	Salient object typically very small
DUT-OMRON [25]	5166	-
MSRA10K [17]	10000	-
ECSSD [24]	1000	-
SED2 [2]	100	Two salient objects per picture

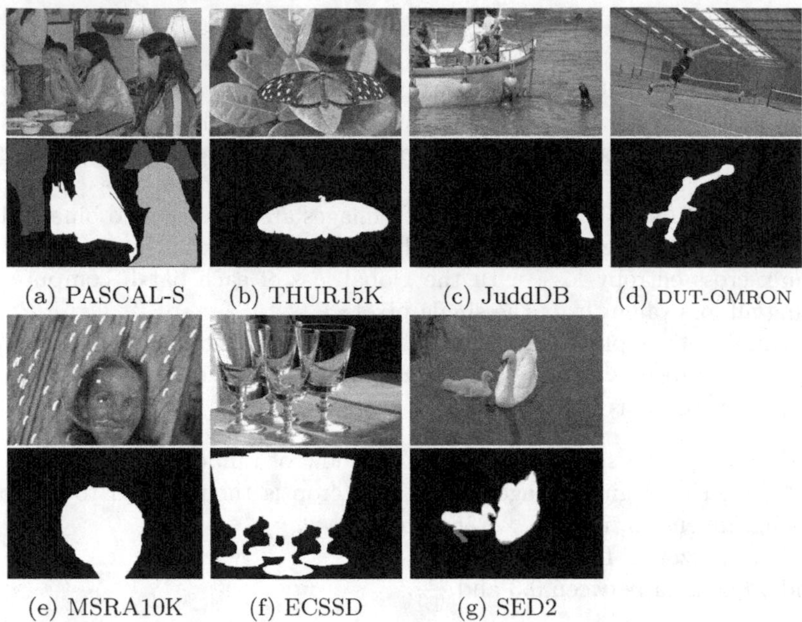

(a) PASCAL-S (b) THUR15K (c) JuddDB (d) DUT-OMRON

(e) MSRA10K (f) ECSSD (g) SED2

Fig. 2. Image-annotation examples for each of the seven datasets used in the benchmark [6].

Each dataset has different kinds of content and bias. Figure 2 shows an image-annotation pair for each dataset. The benchmark defines no official training/test split for the seven datasets, mainly because at the time of its original release few of the tested methods involved an explicit training phase. Our approach requires a significant amount of training data, so we adopted a Leave-One-Dataset-Out (LODO) solution. This allows us to have a fair comparison with the state of the art, as we test on the official datasets, and to avoid overfitting the model to the data. However, since in each LODO split we train the FCN on images collected and annotated with potentially different criteria than those used on the test set, our results could be lower than those we would obtain on homogeneous data (e.g. train/test split of the same dataset).

In order to ensure a totally fair evaluation procedure, we checked for any near-duplicates among dataset pairs. Following [4] we computed Structure Similarity measure (SSIM) [23] between all pairs of images, previously scaled to 64×64 pixels and converted to grayscale, and manually checked those having similarity higher than 0.9. Out of more than 200 million pairs, only five duplicates were found. Although this number of pairs is probably too small to have any overfitting effect, these images were excluded from the training set whenever the corresponding ones were present in the test set. Table 2 lists the found duplicate pairs.

Table 2. Duplicates found among the seven analyzed datasets.

Dataset/image		Dataset/Duplicate image
JuddDB/00854.jpg	$\approx$	DUT-OMRON/sun_acnpbyuckesqygsf.jpg
PASCAL-S/101.jpg	$\approx$	ECSSD/0046.jpg
PASCAL-S/180.jpg	$\approx$	ECSSD/0054.jpg
PASCAL-S/276.jpg	$\approx$	ECSSD/0062.jpg
PASCAL-S/277.jpg	$\approx$	ECSSD/0063.jpg

3.2 Data Augmentation

A preliminary investigation on the usefulness of the data augmentation as described in Sect. 2.1 was performed on the DUT-OMRON Leave-One-Dataset-Out (LODO) setting. Figure 3 shows the loss values on both the training and test sets for three different setups: no data augmentation, three separate perturbations, and the same three perturbations applied jointly. It can be seen that all the investigated perturbation strategies reduce the ability of fitting the training data, while at the same time enhancing the model predictive power on unseen data. Their joint application results in the best improvement, thanks to the little correlation among the single contributions. Thus, it is used for the training of the FCN on all the datasets.

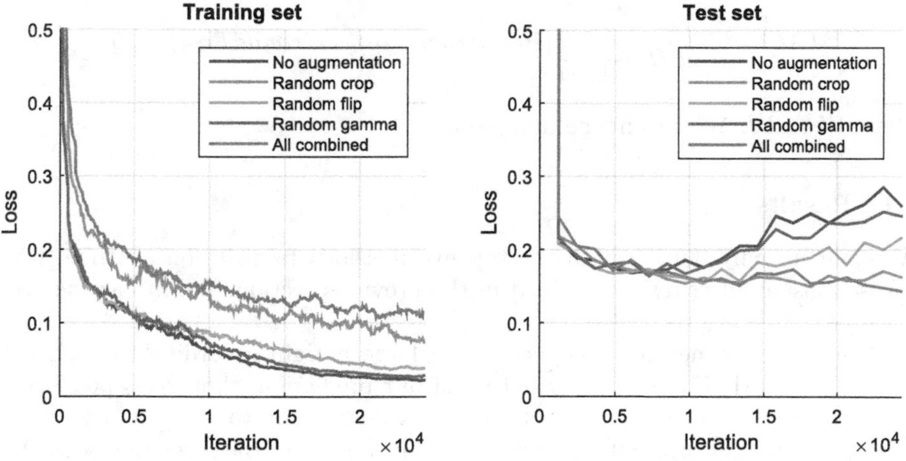

Fig. 3. Softmax cross entropy loss on the DUT-OMRON LODO setup under different kinds of data augmentation.

3.3 Evaluation Measures

Evaluation is performed under the following criteria, aimed at capturing different aspects of the quality of the predicted saliency region:

F-Measure (F_β) is the weighted harmonic mean between precision and recall:

$$F_\beta = \frac{(1+\beta^2)Precision \times Recall}{\beta^2 Precision + Recall} \tag{1}$$

According to [6] the weight β^2 is set to 0.3 in order to benefit precision, considered more important than recall for this specific task [1,17]. Since precision and recall require a binary input, the benchmark adopts three different alternatives for binarization of the methods that do not provide a binary prediction:

1. Varying fixed threshold: Precision and Recall are computed at all integer thresholds between 0 and 255, and then averaged.
2. Adaptive threshold [1]: The threshold for binarization is set to twice the mean value of the prediction map.
3. Saliency Cut [9]: The threshold is set to a low value, thus granting high recall rate. GrabCut [21] is then iteratively applied to the binarized prediction, typically producing a map with more precise edges.

Area Under Curve (AUC) is the area under the Receiver Operating Characteristic curve. The ROC curve is computed by varying the binarization threshold and plotting True Positive Rate (TPR) versus False Positive Rate (FPR) values.

Mean Absolute Error (MAE) is computed directly on the prediction, without any binarization step, as:

$$MAE = \frac{1}{W \times H} \sum_{x=1}^{W} \sum_{y=1}^{H} |Prediction(x,y) - GroundTruth(x,y)| \tag{2}$$

where W and H refer to image dimensions.

3.4 Results

We compare our solution with the top five methods from [6] on all the seven datasets using all criteria described in the previous section. Results are shown in Table 3.

The proposed method is superior by a large margin according to both F_β measures and MAE. The binary nature of our prediction, though, is penalized by AUC due to the particular benchmark evaluation protocol [6]. On average our method outperforms all compared solutions for five of the seven datasets. On JuddDB and MSRA10K, and to a lesser extent on THUR15K, we have lower performance compared to the state of the art. We may notice that images in the JuddDB dataset contain many different subjects, out of which only one is annotated as the main salient object, based on fixations gathered from different observers. This particular set of conditions, radically different from those of the other datasets used for training in our Leave-One-Dataset-Out setup, could be the root cause of sub-optimal performance of our method, and it is left to future work for further analysis. Figure 4 reports some example predictions from all

Table 3. Evaluation results for all measures on all datasets

Measure	Method	P [16]	T [8]	J [5]	D [25]	S [2]	M [17]	E [24]	Average
F_β Varying	Ours	**0.763**	0.666	0.406	**0.706**	**0.847**	0.850	**0.864**	**0.729**
	DRFI [13]	0.679	**0.670**	0.475	0.665	0.831	**0.881**	0.787	0.713
	QCUT [3]	0.695	0.651	**0.509**	0.683	0.810	0.874	0.779	0.714
	MBD [26]	N/A	0.622	0.472	0.624	0.799	0.849	0.739	0.684
	ST [18]	0.660	0.631	0.455	0.631	0.818	0.868	0.752	0.688
	RBD [28]	0.652	0.596	0.457	0.630	0.837	0.856	0.718	0.678
Measure	Method	P [16]	T [8]	J [5]	D [25]	S [2]	M [17]	E [24]	Average
F_β Adaptive	Ours	**0.688**	0.620	0.382	**0.678**	**0.857**	0.833	**0.783**	**0.692**
	DRFI [13]	0.615	0.607	0.419	0.605	0.839	0.838	0.733	0.665
	QCUT [3]	0.654	**0.625**	**0.454**	0.647	0.801	**0.843**	0.738	0.680
	MBD [26]	N/A	0.594	0.422	0.592	0.803	0.830	0.703	0.657
	ST [18]	0.601	0.580	0.394	0.577	0.805	0.825	0.690	0.639
	RBD [28]	0.607	0.566	0.403	0.580	0.825	0.821	0.680	0.640
Measure	Method	P [16]	T [8]	J [5]	D [25]	S [2]	M [17]	E [24]	Average
F_β Sal Cut	Ours	**0.778**	**0.702**	0.409	**0.712**	**0.791**	0.890	**0.888**	**0.739**
	DRFI [13]	0.690	0.674	0.447	0.669	0.702	**0.905**	0.801	0.698
	QCUT [3]	0.613	0.620	**0.480**	0.647	0.672	0.843	0.747	0.660
	MBD [26]	N/A	0.642	0.470	0.636	0.759	0.890	0.785	0.697
	ST [18]	0.671	0.648	0.459	0.635	0.768	0.896	0.777	0.693
	RBD [28]	0.667	0.618	0.461	0.647	0.750	0.884	0.757	0.683
Measure	Method	P [16]	T [8]	J [5]	D [25]	S [2]	M [17]	E [24]	Average
AUC	Ours	0.820	0.851	0.680	0.828	0.844	0.877	0.896	0.828
	DRFI [13]	**0.897**	**0.938**	**0.851**	**0.933**	**0.944**	**0.978**	**0.944**	**0.926**
	QCUT [3]	0.870	0.907	0.831	0.897	0.860	0.956	0.909	0.890
	MBD [26]	N/A	0.915	0.838	0.903	0.922	0.964	0.917	0.910
	ST [18]	0.868	0.911	0.806	0.895	0.922	0.961	0.914	0.897
	RBD [28]	0.867	0.887	0.826	0.894	0.899	0.955	0.894	0.889
Measure	Method	P [16]	T [8]	J [5]	D [25]	S [2]	M [17]	E [24]	Average
MAE	Ours	**0.122**	**0.106**	**0.210**	**0.079**	**0.080**	**0.073**	**0.065**	**0.105**
	DRFI [13]	0.221	0.150	0.213	0.155	0.130	0.118	0.166	0.165
	QCUT [3]	0.195	0.128	0.178	0.119	0.148	0.118	0.171	0.151
	MBD [26]	N/A	0.162	0.225	0.168	0.137	0.107	0.172	0.162
	ST [18]	0.224	0.179	0.240	0.182	0.145	0.122	0.193	0.184
	RBD [28]	0.199	0.150	0.212	0.144	0.130	0.108	0.173	0.159

datasets. False positives mostly correspond to actual objects that were not in the ground truth due to annotation guidelines (e.g. the flower in Fig. 4b and the fish in Fig. 4e), which could also be contributing to the lower performance on datasets MSRA10K and THUR15K. False negatives are often related to holes in our prediction (e.g. the window glasses in Fig. 4a), thus highlighting a current downside of the solution. Finally, we can also observe that the edges of our predictions are in general smoother and less precise than the reference annotations.

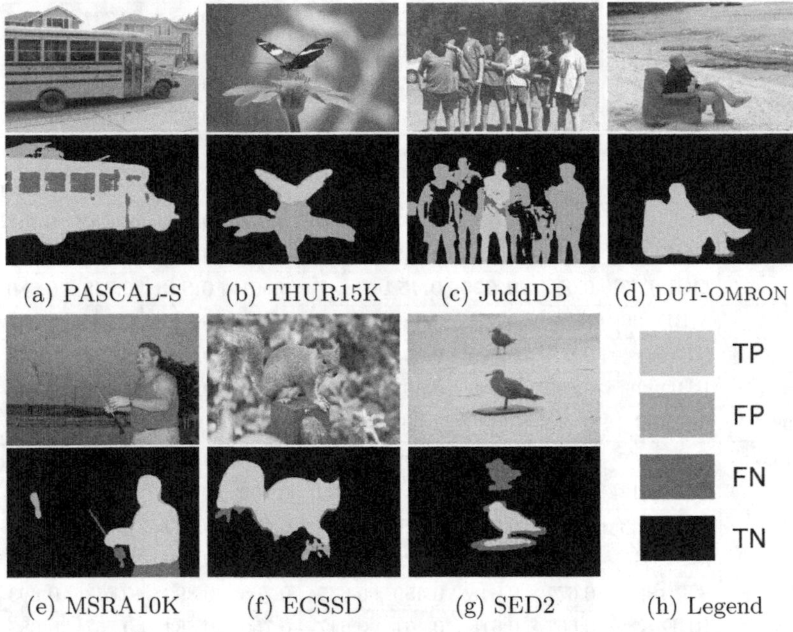

(a) PASCAL-S (b) THUR15K (c) JuddDB (d) DUT-OMRON (e) MSRA10K (f) ECSSD (g) SED2 (h) Legend

Fig. 4. Example predictions on different datasets.

A direct comparison with other methods in terms of computational complexity cannot be performed in a fair setup, as our solution is designed to run on GPU, unlike the compared methods. On a NVIDIA TITAN X GPU our prediction takes on average 0.09 s on each image of the MSRA10K dataset (typical image resolution 400×300). For reference, the fastest among compared methods (RBD [28]) takes 0.269 s using a desktop machine with Xeon E5645 2.4 GHz CPU [6].

4 Conclusions

In this work we exploited the semantic awareness of a Fully Convolutional Network to address the problem of salient object detection. We verified the effectiveness of this approach by comparing it on a standard benchmark, composed of seven datasets and more than forty methods (we reported here only the top five). Despite the challenging Leave-One-Dataset-Out setup, which naturally excludes the possibility of overfitting the model to the data, we outperformed the state of the art on most datasets.

In the future we might switch from a binary foreground/background prediction to a multiclass one, in order to also consider the different levels of saliency defined in some of the used datasets. Bringing this even further, we might directly treat the problem as a regression task, and study the effects of different training losses on the final performance.

Finally, we plan on extending evaluation and comparison to other datasets [15,20] and methods [15,27], which were currently left-out for not being contemplated in the adopted benchmark, as well as for space constraints.

References

1. Achanta, R., Hemami, S., Estrada, F., Susstrunk, S.: Frequency-tuned salient region detection. In: Computer vision and pattern recognition. In: IEEE Conference on VPR 2009, pp. 1597–1604. IEEE (2009)
2. Alpert, S., Galun, M., Brandt, A., Basri, R.: Image segmentation by probabilistic bottom-up aggregation and cue integration. IEEE Trans. Pattern Anal. Mach. Intell. **34**(2), 315–327 (2012)
3. Aytekin, C., Kiranyaz, S., Gabbouj, M.: Automatic object segmentation by quantum cuts. In: 2014 22nd International Conference on Pattern Recognition (ICPR), pp. 112–117. IEEE (2014)
4. Bianco, S., Buzzelli, M., Mazzini, D., Schettini, R.: Deep learning for logo recognition. Neurocomputing **245**, 23–30 (2017). http://dx.doi.org/10.1016/j.neucom.2017.03.051
5. Borji, A.: What is a salient object? a dataset and a baseline model for salient object detection. IEEE Trans. Image Process. **24**(2), 742–756 (2015)
6. Borji, A., Cheng, M.M., Jiang, H., Li, J.: Salient object detection: a benchmark. IEEE Trans. Image Process. **24**(12), 5706–5722 (2015)
7. Chen, T., Lin, L., Liu, L., Luo, X., Li, X.: Disc: deep image saliency computing via progressive representation learning. IEEE Trans. Neural Netw. Learn. Syst. **27**(6), 1135–1149 (2016)
8. Cheng, M.M., Mitra, N.J., Huang, X., Hu, S.M.: Salientshape: group saliency in image collections. Vis. Comput. **30**(4), 443–453 (2014)
9. Cheng, M.M., Mitra, N.J., Huang, X., Torr, P.H., Hu, S.M.: Global contrast based salient region detection. IEEE Trans. Pattern Anal. Mach. Intell. **37**(3), 569–582 (2015)
10. Han, J., Zhang, D., Hu, X., Guo, L., Ren, J., Wu, F.: Background prior-based salient object detection via deep reconstruction residual. IEEE Trans. Circ. Syst. Video Technol. **25**(8), 1309–1321 (2015)
11. Hayhoe, M., Ballard, D.: Eye movements in natural behavior. Trends Cogn. Sci. **9**(4), 188–194 (2005)
12. Itti, L., Koch, C.: Computational modelling of visual attention. Nat. Rev. Neurosci. **2**(3), 194–203 (2001)
13. Jiang, H., Wang, J., Yuan, Z., Wu, Y., Zheng, N., Li, S.: Salient object detection: a discriminative regional feature integration approach. In: Proceedings of the IEEE Conference on Computer Vision and Pattern Recognition, pp. 2083–2090 (2013)
14. Kingma, D., Ba, J.: Adam: a method for stochastic optimization. arXiv preprint (2014). arXiv:1412.6980
15. Li, G., Yu, Y.: Visual saliency based on multiscale deep features. In: Proceedings of the IEEE Conference on Computer Vision and Pattern Recognition, pp. 5455–5463 (2015)

16. Li, Y., Hou, X., Koch, C., Rehg, J.M., Yuille, A.L.: The secrets of salient object segmentation. In: Proceedings of the IEEE Conference on Computer Vision and Pattern Recognition, pp. 280–287 (2014)
17. Liu, T., Yuan, Z., Sun, J., Wang, J., Zheng, N., Tang, X., Shum, H.Y.: Learning to detect a salient object. IEEE Trans. Pattern Anal. Mach. Intell. **33**(2), 353–367 (2011)
18. Liu, Z., Zou, W., Le Meur, O.: Saliency tree: a novel saliency detection framework. IEEE Trans. Image Process. **23**(5), 1937–1952 (2014)
19. Long, J., Shelhamer, E., Darrell, T.: Fully convolutional networks for semantic segmentation. In: Proceedings of the IEEE Conference on Computer Vision and Pattern Recognition, pp. 3431–3440 (2015)
20. Movahedi, V., Elder, J.H.: Design and perceptual validation of performance measures for salient object segmentation. In: 2010 IEEE Computer Society Conference on Computer Vision and Pattern Recognition Workshops (CVPRW), pp. 49–56. IEEE (2010)
21. Rother, C., Kolmogorov, V., Blake, A.: Grabcut: interactive foreground extraction using iterated graph cuts. In: ACM Transactions on Graphics (TOG), vol. 23, pp. 309–314. ACM (2004)
22. Simonyan, K., Zisserman, A.: Very deep convolutional networks for large-scale image recognition. arXiv preprint (2014). arXiv:1409.1556
23. Wang, Z., Bovik, A.C., Sheikh, H.R., Simoncelli, E.P.: Image quality assessment: from error visibility to structural similarity. IEEE Trans. Image Process. **13**(4), 600–612 (2004)
24. Yan, Q., Xu, L., Shi, J., Jia, J.: Hierarchical saliency detection. In: Proceedings of the IEEE Conference on Computer Vision and Pattern Recognition, pp. 1155–1162 (2013)
25. Yang, C., Zhang, L., Lu, H., Ruan, X., Yang, M.H.: Saliency detection via graph-based manifold ranking. In: Proceedings of the IEEE Conference on Computer Vision and Pattern Recognition, pp. 3166–3173 (2013)
26. Zhang, J., Sclaroff, S., Lin, Z., Shen, X., Price, B., Mech, R.: Minimum barrier salient object detection at 80 fps. In: Proceedings of the IEEE International Conference on Computer Vision, pp. 1404–1412 (2015)
27. Zhao, R., Ouyang, W., Li, H., Wang, X.: Saliency detection by multi-context deep learning. In: Proceedings of the IEEE Conference on Computer Vision and Pattern Recognition, pp. 1265–1274 (2015)
28. Zhu, W., Liang, S., Wei, Y., Sun, J.: Saliency optimization from robust background detection. In: Proceedings of the IEEE Conference on Computer Vision and Pattern Recognition, pp. 2814–2821 (2014)

A Lightweight Mamdani Fuzzy Controller for Noise Removal on Iris Images

Andrea Francesco Abate[1], Silvio Barra[2], Gianni Fenu[2], Michele Nappi[1], and Fabio Narducci[1](✉)

[1] Department of Computer Science, University of Salerno, 84084 Salerno, Italy
{abate,mnappi,fnarducci}@unisa.it

[2] Department of Mathematics and Computer Science, University of Cagliari, 09124 Cagliari, Italy
barra.silvio@gmail.com, fenu@unica.it

Abstract. The iris segmentation step is usually the most time consuming stage of biometric systems when dealing with non ideal conditions, which produce diverse noise factors during the acquisition. On the other side, it also represents a crucial step since poor removal of noise leads to degradation of recognition performance. In this work, a lightweight fuzzy-based solution has been explored. The goal is to propose a fast but reliable segmentation approach which preserves the original resolution of the iris images. The preliminary results obtained on a subset of MICHE dataset, confirmed both acceptable performance in terms of time consumption and good quality of segmentation mask suitable for matching purposes.

Keywords: Iris segmentation · Noise removal · Fuzzy controller

1 Introduction

The recognition rate achieved by any biometric method has been widely accepted as the reference metric to evaluate the performance of the proposed system. However, the matching phase between biometric templates represents the final stage of all biometric processing pipeline, whatever its complexity. On the opposite side, the segmentation covers a significant initial set of operations useful to extract the biometric features and the related template. Segmentation is a very critical part in iris recognition systems, since errors in this initial stage are propagated to subsequent processing of the pipeline [17], thus affecting the overall performance of the proposed approach. In very controlled conditions, occlusions by eyelids/eyelashes and environmental noise like reflections and refractions on iris surface are quite limited. As a consequence, the constraints of the segmentation algorithm are negligible. When approaching at the problem of uncontrolled or non-cooperative users, the factors affecting the acquired trait can sensibly degrade the quality of the acquisition and, consequently, its suitability to the purpose of biometric identification [5]. Images may also suffer from motion blur,

S. Battiato et al. (Eds.): ICIAP 2017, Part II, LNCS 10485, pp. 93–103, 2017.
https://doi.org/10.1007/978-3-319-68548-9_9

camera diffusion, head rotation, gaze direction, camera angle, and problems due to contraction and dilation [33]. Removing the portions affected by these factors while preserving sufficient features for a reliable matching process is particularly challenging. Iris segmentation is a very active field of research with lots of approaches and techniques [7,19–21], even though less explored than iris recognition. One of the first categories of techniques of iris segmentation relied on the geometrical appearance of the iris, consisting of two quasi-concentric circles; such group of techniques exploited this information to approximate two circumferences [11,12,32], or two ellipses [14], in order to isolate the region of interest. Besides the geometrical information, other methods exploited the texture appearance and the local characteristics to segment the iris [28]. Another technique, often used in segmentation tasks, is based on the evolution of the active contours, or *snakes*: this class of algorithms iteratively adapts the segmented shape to the edges of the image [18,30,31]. Deep learning, which is a promising new algorithmic solution for several open problems, has been revealed to be effective for image segmentation [10,23] while the circular integro-differential operator by Daugman [13] was a pioneer of this kind of study. Among these two (chronologically opposite) approaches there is a wide and diverse collection of solutions aimed at preserving significant pixels from images containing a biometric trait. Most of them, however, quite often work on a strong down-sampled resolution of the original image to meet the requirements of (near) real-time processing. In this work, we explored a solution based on fuzzy logic to the problem of segmenting iris images acquired in uncontrolled conditions. The goal is that of achieving a fast but reliable segmentation without sacrificing the resolution (two requirements which are in contrast to each other). In the following sections, a clear description of the proposed method is provided, supported by preliminary results that highlight the potentials and the refinements still needed. In particular, Sect. 2 describes the tools and algorithm involved in the design of the solution. Particular attention is given to the localisation and to decomposition of the iris in the acquired image and at the implementation of the fuzzy controller used to automatically annotate "good" clusters of pixels. Section 3 presents the first results obtained and Sect. 4 draws conclusions and future improvements of the proposal.

2 The Proposed Algorithm

The process of segmenting the iris starts with the detection and extraction of the iris from the image of a whole human eye. The detection is a simple localisation of the circles in the images representing the iris and pupils. Several, and sometimes sophisticated, approaches to iris localisation in literature involve complex formulas but also require high processing power. The proposed method aims at reducing the computational demand so that it could be credibly implemented on limited processing platforms, including mobile devices. The key is that of using a lightweight processing for iris localisation to have a rough approximation of the significant area of the image where iris features can be found. Once such

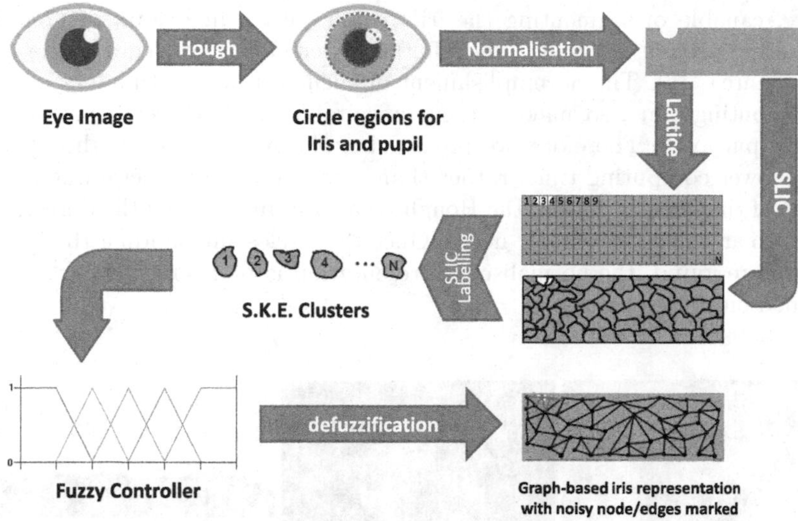

Fig. 1. The processing pipeline of the proposed method for iris location and segmentation. After locating the iris in the image by the Hough transformation, the normalised iris is processed by SLIC clustering to decompose the images in smaller uniform regions. On each regions the value of skewness, kurtosis and entropy are computed and used as input of a fuzzy controller. This one, according to the implemented membership functions, labels iris features separating significant areas from noisy ones. The final stage is a graph-based representation of the iris that discards areas of iris surface which might degrade the performance of biometric recognition methods.

an information is extracted (a normalised representation of iris pixels), a *lattice* structure is built above the pixels and adapted to clusters obtained by running SLIC clustering [4] on them. On each cluster, the triad SKE (Skewness, Kurtosis and Entropy) is computed considering all pixels in it. The selection of noisy areas is automatically performed by the fuzzy controller. The defuzzification applied to clusters of pixels allows to discard the regions in the normalised iris, which are affected by noise and which could impact on recognition performance. By removing disturbing factors like light spots, reflections of eyelashes, portion of sclera and so on, the final result is a representation of the iris where only significant pixels for a reliable iris matching are preserved. A graph-based representation of the clusters and their relationships is obtained. Vertices represent the centroids of the clusters and edges are preserved among all pairs of centroids if they are similar regions according to the SKE values. A schematic view of the processing pipeline is shown in Fig. 1. Details of the method are provided in the following subsections.

2.1 Iris Detection and Decomposition

The existing iris detection and segmentation algorithms need much time to extract the region of interest from the input image. Many of them are really effective, capable of segmenting the iris region with a high level of precision, so excluding eyelids, eyelashes and all those elements that degrade the quality of the area [16]. The accomplishment of such tasks, unfortunately, needs a high computing time, so making them unsuitable for both real-time and on-the-move purposes. Therefore, we preferred exploiting a more rough approach needing lower computing time, rather than using one more precise but heavier in terms of time consumption: the Hough transform method for the detection of circles into an image suited our needs. Once the circles surrounding the iris and the pupil are found, the normalised representation is built and SLIC clustering performed on it (see Fig. 2).

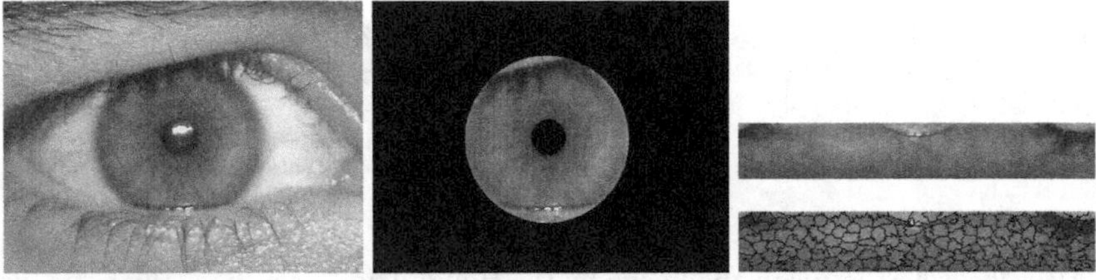

Fig. 2. From left to right, the iris detection process achieved by Hough. The first image shows the image to be processed. The method looks for circles in the image, in order to correctly detect the region of interest related to the iris (middle). The rightmost images show the polarised and normalised iris together with the SLIC decomposition.

SLIC is a simple and efficient method that aims at decomposing an image in a fixed number of region, according to their inner homogeneity. It starts by dividing the entire image into M tiles (hereafter called *superpixels*), each containing about the same number of pixels. For each superpixel, the centroid C is computed. SLIC adjusts iteratively both the size and the shape of the superpixels, by following three simple steps:

- K-means clustering is executed over C_i, the centroid of the superpixel S_i; this operation is done for $I = 1, \ldots, M$;
- For each of the regions generated, new centroids are computed;
- The superpixel is no longer expanded when the residual error of the Manhattan distance on new and previously-computed centroids is less than a threshold.

2.2 The SKE Values

Texture quality assessment of human irises has always been one of the most challenging practises in the biometric field [9,24] The assessment policies here adopted exploit three measurements, each one describing a different aspect of the human iris texture: *Skewness*, *Kurtosis* and *Entropy*, the SKE triad. The effectiveness of skewness and kurtosis in analysing the quality of an image, have been extensively demonstrated through the years, especially for segmenting purposes [1–3,15]. All three measurements get a float number that can vary in different range of values. The Skewness measures the asymmetry of the probability distribution and it is centred in zero (Fig. 3). It can be greater or smaller than its centre according to the following rules:

1. **Skewness > 0:** right skewed distribution - most values are concentrated on left of the mean, with extreme values to the right.
2. **Skewness < 0:** left skewed distribution - most values are concentrated on the right of the mean, with extreme values to the left.
3. **Skewness = 0:** mean = median, the distribution is symmetrical around the mean.

In a similar way to the concept of skewness, kurtosis is a descriptor of the shape of a probability distribution. More specifically, it represents the degree of *peakedness* of a distribution or, in alternative way, it is a measure of the *tailedness* of the probability distribution (see Fig. 5). The kurtosis is defined as the fourth standardized moment of a variable and it values 3 when the distribution is normal. To centre it in zero, often the *excess kurtosis* is used, which is the

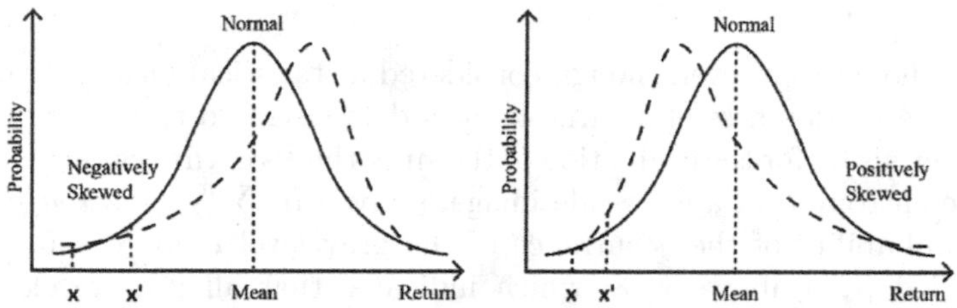

Fig. 3. The interpretation of skewness. The image on the left shows a negative skewed curve and how the values of x and x' changes with respect to the normal distribution. On the right the positively skewed distribution. Picture is from [34].

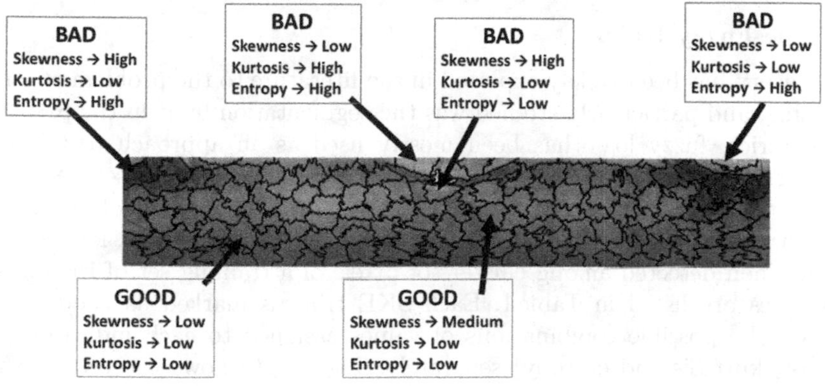

Fig. 4. A visual perception of how the superpixels are classified as *good* or *bad*.

kurtosis value minus 3 [22]. According to its value, the excess kurtosis of a curve defines tree different entities:

1. **mesokurtic:** zero kurtosis of the normal distribution,
2. **platykurtic:** kurtosis < 0, the levels of frequency in the distribution are close to be equal. It indicates a relatively flat curve.
3. **leptokurtic:** kurtosis > 0, high frequencies are concentrated in a small portion of the distribution. It indicates a relatively peaked curve.

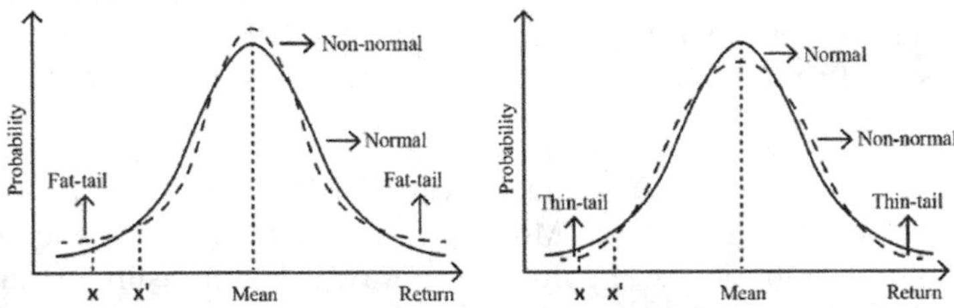

Fig. 5. The interpretation of kurtosis. The image on the left shows a leptokurtic curve (dotted line) and how the values of x and x' changes with respect to the normal distribution. On the right the platykurtic curve (positive kurtosis). Picture is from [34].

Finally, the entropy, even though considered a statistical measure capable of indicating the randomness of a variable, it is demonstrated to be very useful in the computer vision for the evaluation of the smoothness of the region of an image [6,29]. The entropy h of a grayscale image is given by $\sum_{i=1}^{n} p_i \times log_2 p_i$, where p_i is the probability of the occurrence of the gray level i (for $i = 0, \ldots, 255$). It is bounded from above by 8, which indicates that all gray levels has the same probability. This property has been revealed particularly useful to detect light spots on iris surface which are characterised by a high levels of entropy. Figure 4 shows some examples of combinations of SKE values that discriminate *good* superpixels from *bad* ones.

2.3 The Fuzzy Rules

Fuzzy theory has been widely explored in the literature to the problem of image processing, and particularly studied was the segmentation by it in the past [27]. In biometrics, fuzzy logic has been mostly used as an approach to template matching and encryption while only few research studies focus on segmentation. In this work, a simple Mamdani fuzzy controller [25] has been built. It consists of twelve fuzzy rules, defined by observing the behaviour of skewness, kurtosis and entropy when detected among clusters of pixels of a training set of iris images. These rules are listed in Table 1. Each SKE triad is marked as good or bad based on the possible combinations of values assigned to each indicator (i.e., skewness, kurtosis and entropy) separately. The Fig. 6 shows the relationships between pairs of SKE values which define the output of the controller on each cluster of pixels.

Table 1. The fuzzy rules used to discriminate among good and bad clusters of pixels.

#	1	2	3	4	5	6	7	8	9	10	11	12
Skewness	Low	High	High	Low	Medium	Medium	Medium	Low	Low	High	High	High
Kurtosis	Low	High	Low	Low	High	Low	Low	High	High	Low	High	Low
Entropy	Low	High	High	High	Low	High	Low	Low	High	Low	Low	Low
Output	Good	Bad	Bad	Bad	Bad	Bad	Good	Bad	Bad	Bad	Bad	Bad

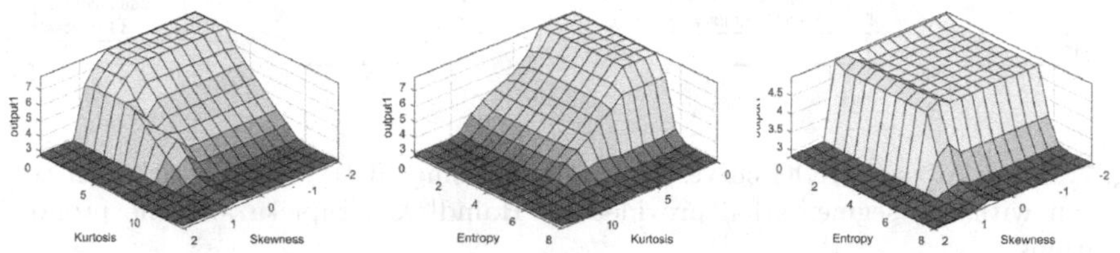

Fig. 6. The three-dimensional curves that represent the mapping from pairs of SKE values to the quality of the cluster

3 Preliminary Results

In order to evaluate the proposed method, a subset of images from MICHE dataset [26] was used; it can be downloaded from the BIPLab site biplab.unisa.it. It consists of 90 subjects whose images have been acquired under different light condition, in both outdoor and indoor environment and by using two different mobile devices, an iPhone5 and a Galaxy S4. The heterogeneous nature of such dataset helped to assess the overall quality of the method; on a side, the indoor acquired images allowed to evaluate its strength in controlled condition; on the other side, we were also able to test its reliability, by exploiting the images acquired in uncostrained conditions. We attempted a comparison against the method of Haindl and Krupička [16], whose segmentation algorithm ranked in first place in the MICHE-I challenge [26] and, at a later time, used as reference for the iris recognition challenge MICHE-II [8]. The comparison has been carried out by executing SKIPSOM [3] classifier over the segmentations produced by our method versus that of Haindl and Krupička. The experiments showed that the results between the two methods are quite similar in terms of Recognition Rate (RR), Area Under Curve (AUC), Equal Error Rate (EER) and Decidability (DEC), but they vary significantly in time consumption. On full resolution images of 800×600 the proposed method takes about 1/5th of the time required to the segmentation algorithm by Haindl. The curves are depicted in Fig. fig:roccurves, and detailed in Table 2.

Beside the very reduced processing time, the method offers the possibility to be executed over images represented in cartesian coordinates, and not only over those in polar coordinates. This leads to a twofold advantage: first of all, the polarisation process introduces sampled information in the iris image, in order to let the pixels on the outer bound of the iris (near the sclera) match with

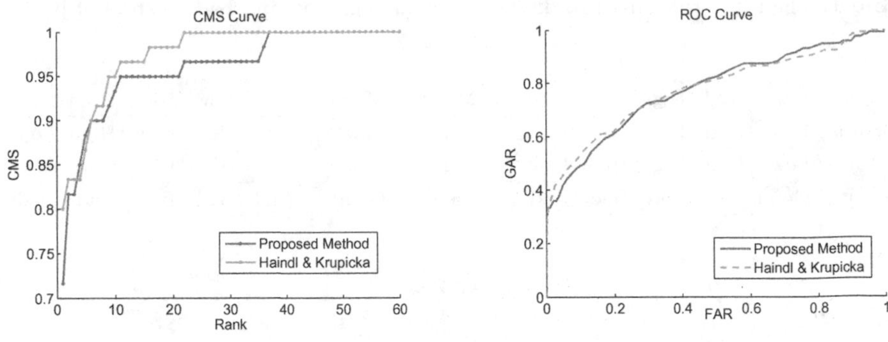

Fig. 7. The CMS and ROC curves obtained on using SKIPSOM algorithm in combination with the segmentation provided by Haindl & Krupička and the proposed algorithms.

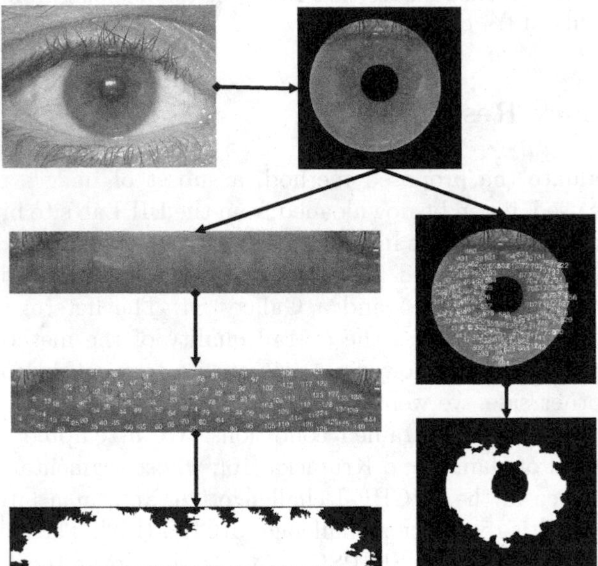

Fig. 8. The figure shows the output from each stage of the pipeline process: in the leftmost branch, the iris is transformed from the cartesian coordinates to the polar ones; in the rightmost, this transformation is not applied.

Table 2. The results obtained in terms of Recognition Rate (RR), Area Under Curve (AUC), Equal Error Rate (EER), Decidability (DEC) and processing time. Decidability formula can be found at http://nice2.di.ubi.pt/evaluation.htm

Method	RR	AUC	EER	DEC	Segmentation time + Noise removal
Proposed method	77%	**79%**	**0.27**	**1.14**	**∼11.3** s + **∼0.7** s = **∼12.0** s
Haindl and Krupička	**79%**	78%	0.28	1.15	∼1 min 6 s

those on the inner bound (near the pupil); this sampling process can affect the recognition stage. The second advantage regards the processing time: the normalisation process can take many seconds to be accomplished, especially whean dealing with high-resolution images. In the Fig. 8, the twofold application of the method is shown: on the leftmost branch of the figure, the segmentation and noise removal processes are executed over the polarised version of the iris, whereas in the rightmost one, the image is not further processed, and no polarisation phase takes place.

4 Conclusions

Segmentation is a very critical step in iris recognition since errors at this stage propagates to subsequent operations of the processing pipeline, thus potentially degrading the achievable performance rates. In iris recognition systems under uncontrolled conditions, the amount of disturbing factors that the segmentation algorithm should detect and remove is particularly wide and heterogeneous. When dealing with them, the execution flow of segmentation algorithm becomes progressively complex and, as direct consequence, more and more time demanding. In this paper a lightweight fuzzy controller has been proposed as a solution for iris segmentation. It explores the SLIC decomposition to provide a quick clustering of iris pixels which are then processed by computing the SKE values. The fuzzy controller processes the tree-values representation of the clusters to discard all those might not be contribute significantly to the matching phase.

The proposed algorithm has been compared to the segmentation algorithm by Haindl and Krupička, which won the MICHE-I competition. The quality of the segmentation has been tested by comparing the recognition performance of SKIPSOM algorithm when using the masks produced by both algorithms. The results obtained showed that the difference in terms of recognition performance achieved is negligible. Rather, a significant difference in time demand has been observed between the two segmentation algorithms.

The preliminary results looked encouraging and will be used to drive a deeper refinement and optimisation of the method with the aim of reducing the time demand while improving the accuracy of the detection and removal of noise in iris images.

References

1. Abate, A.F., Barra, S., D'Aniello, F., Narducci, F.: Two-tier image features clustering for iris recognition on mobile. In: Petrosino, A., Loia, V., Pedrycz, W. (eds.) WILF 2016. LNCS, vol. 10147, pp. 260–269. Springer, Cham (2017). doi:10.1007/978-3-319-52962-2_23
2. Abate, A., Barra, S., Gallo, L., Narducci, F.: SKIPSOM: Skewness & kurtosis of iris pixels in Self Organizing Maps for iris recognition on mobile devices. In: 2016 23rd International Conference on Pattern Recognition (ICPR), Cancun, pp. 155–159 (2016). doi:10.1109/ICPR.2016.7899625

3. Abate, A.F., Barra, S., Gallo, L., Narducci, F.: Kurtosis and skewness at pixel level as input for SOM networks to iris recognition on mobile devices. Pattern Recogn. Lett. (2017)
4. Achanta, R., Shaji, A., Smith, K., Lucchi, A., Fua, P., Süsstrunk, S.: Slic superpixels compared to state-of-the-art superpixel methods. IEEE Trans. Pattern Anal. Mach. Intell. **34**(11), 2274–2282 (2012)
5. Barra, S., Casanova, A., Narducci, F., Ricciardi, S.: Ubiquitous iris recognition by means of mobile devices. Pattern Recogn. Lett. **57**, 66–73 (2015). Mobile Iris CHallenge Evaluation part I (MICHE I)
6. Barra, S., De Marsico, M., Cantoni, V., Riccio, D.: Using mutual information for multi-anchor tracking of human beings. In: Cantoni, V., Dimov, D., Tistarelli, M. (eds.) BIOMET 2014. LNCS, vol. 8897, pp. 28–39. Springer, Cham (2014). doi:10.1007/978-3-319-13386-7_3
7. Bowyer, K.W., Hollingsworth, K.P., Flynn, P.J.: A survey of iris biometrics research: 2008–2010. In: Bowyer, K.W., Burge, M.J. (eds.) Handbook of Iris Recognition. ACVPR, pp. 23–61. Springer, London (2016). doi:10.1007/978-1-4471-6784-6_2
8. Castrillón-Santana, M., De Marsico, M., Nappi, M., Narducci, F., Proença, H.: Mobile iris challenge evaluation ii: results from the ICPR competition. In: International Conference on Pattern Recognition (2016)
9. Chaskar, U.M., Sutaone, M.S., Shah, N.S.: Iris image quality assessment for biometric application. IJCSI Int. J. Comput. Sci. Issues **9**(3(3)) (2012). ISSN (Online): 1694-0814. https://www.ijcsi.org/papers/IJCSI-9-3-3-474-478.pdf
10. Chen, L.C., Papandreou, G., Kokkinos, I., Murphy, K., Yuille, A.L.: Semantic image segmentation with deep convolutional nets and fully connected CRFs. arXiv preprint arXiv:1412.7062 (2014)
11. Daugman, J.: Statistical richness of visual phase information: update on recognizing persons by iris patterns. Int. J. Comput. Vis. **45**(1), 25–38 (2001)
12. Daugman, J.: How iris recognition works. IEEE Trans. Circ. Syst. Video Technol. **14**(1), 21–30 (2004). doi:10.1109/TCSVT.2003.818350
13. Daugman, J.G.: High confidence visual recognition of persons by a test of statistical independence. IEEE Trans. Pattern Anal. Mach. Intell. **15**(11), 1148–1161 (1993)
14. Du, Y., Arslanturk, E., Zhou, Z., Belcher, C.: Video-based noncooperative iris image segmentation. IEEE Trans. Syst. Man Cybern. Part B (Cybern.) **41**(1), 64–74 (2011)
15. El-Zaart, A.: Skin images segmentation. J. Comput. Sci. **6**(2), 217–223 (2010)
16. Haindl, M., Krupika, M.: Unsupervised detection of non-iris occlusions. Pattern Recogn. Lett. **57**, 60–65 (2015). Mobile Iris CHallenge Evaluation part I (MICHE I)
17. Hofbauer, H., Alonso-Fernandez, F., Wild, P., Bigun, J., Uhl, A.: A ground truth for iris segmentation. In: 2014 22nd International Conference on Pattern Recognition, pp. 527–532, August 2014
18. Jarjes, A.A., Wang, K., Mohammed, G.J.: Improved greedy snake model for detecting accurate pupil contour. In: 2011 3rd International Conference on Advanced Computer Control, pp. 515–519, January 2011
19. Jayalakshmi, S., Sundaresan, M.: A survey on iris segmentation methods. In: 2013 International Conference on Pattern Recognition, Informatics and Mobile Engineering, pp. 418–423, February 2013
20. Jeong, D.S., Hwang, J.W., Kang, B.J., Park, K.R., Won, C.S., Park, D.K., Kim, J.: A new iris segmentation method for non-ideal iris images. Image Vis. Comput. **28**(2), 254–260 (2010). http://www.sciencedirect.com/science/article/pii/S0262885609000638. Segmentation of Visible Wavelength Iris Images Captured At-a-distance and On-the-move

21. Labati, R., Genovese, A., Piuri, V., Scotti, F.: Iris segmentation: state of the art and innovative methods. Intell. Syst. Ref. Libr. **37**, 151–182 (2012)
22. Liang, Z., Wei, J., Zhao, J., Liu, H., Li, B., Shen, J., Zheng, C.: The statistical meaning of kurtosis and its new application to identification of persons based on seismic signals. Sensors **8**(8), 5106–5119 (2008)
23. Liu, Z., Li, X., Luo, P., Loy, C.C., Tang, X.: Semantic image segmentation via deep parsing network. In: Proceedings of the IEEE International Conference on Computer Vision, pp. 1377–1385 (2015)
24. Makinana, S., Malumedzha, T., Nelwamondo, F.V.: Iris image quality assessment based on quality parameters. In: Nguyen, N.T., Attachoo, B., Trawiński, B., Somboonviwat, K. (eds.) ACIIDS 2014. LNCS, vol. 8397, pp. 571–580. Springer, Cham (2014). doi:10.1007/978-3-319-05476-6_58
25. Mamdani, E.H., Assilian, S.: An experiment in linguistic synthesis with a fuzzy logic controller. Int. J. Man Mach. Stud. **7**(1), 1–13 (1975)
26. Marsico, M.D., Nappi, M., Riccio, D., Wechsler, H.: Mobile iris challenge evaluation (MICHE)-I, biometric iris dataset and protocols. Pattern Recogn. Lett. **57**, 17–23 (2015). http://www.sciencedirect.com/science/article/pii/S0167865515000574. Mobile Iris CHallenge Evaluation part I (MICHE I)
27. Pal, N.R., Pal, S.K.: A review on image segmentation techniques. Pattern Recogn. **26**(9), 1277–1294 (1993)
28. Proenca, H.: Iris recognition: on the segmentation of degraded images acquired in the visible wavelength. IEEE Trans. Pattern Anal. Mach. Intell. **32**(8), 1502–1516 (2010)
29. Rad, R.M., Attar, A., Atani, R.E.: A comprehensive layer based encryption method for visual data. Int. J. Sig. Process. Image Process. Pattern Recogn. **6**(1), 37–48 (2013)
30. Ross, A., Shah, S.: Segmenting non-ideal irises using geodesic active contours. In: 2006 Biometrics Symposium: Special Session on Research at the Biometric Consortium Conference, Baltimore, MD, pp. 1–6 (2006). doi:10.1109/BCC.2006.4341625
31. Shah, S., Ross, A.: Iris segmentation using geodesic active contours. IEEE Trans. Inf. Forensics Secur. **4**(4), 824–836 (2009)
32. Tian, Q.C., Pan, Q., Cheng, Y.M., Gao, Q.X.: Fast algorithm and application of hough transform in iris segmentation. In: vol. 7, pp. 3977–3980 (2004). https://www.scopus.com/inward/record.uri?eid=2-s2.0-6344285715partnerID=40 md5=d649c8c8884f96fecf730bf1bbad77e7, cited By 41
33. Vatsa, M., Singh, R., Noore, A.: Improving iris recognition performance using segmentation, quality enhancement, match score fusion, and indexing. IEEE Trans. Syst. Man Cybern. Part B (Cybern.) **38**(4), 1021–1035 (2008)
34. Wan, Y., Clutter, M.L., Mei, B., Siry, J.P.: Assessing the role of U.S. timberland assets in a mixed portfolio under the mean-conditional value at risk framework. Forest Policy Econ. **50**, 118–126 (2015)

Analysis of the Discriminative Generalized Hough Transform for Pedestrian Detection

Eric Gabriel[1(✉)], Hauke Schramm[1,2], and Carsten Meyer[1,2]

[1] Institute of Computer Science, Kiel University of Applied Sciences, Kiel, Germany
{eric.gabriel,hauke.schramm,carsten.meyer}@fh-kiel.de

[2] Department of Computer Science, Kiel University (CAU), Kiel, Germany

Abstract. Many approaches have been suggested for automatic pedestrian detection to cope with the large variability regarding size, occlusion, background variability etc., among them deformable part models, feature-based approaches (e.g. histograms of oriented gradients), and recently deep learning-based algorithms. Current deep learning-based frameworks rely either on a proposal generation mechanism (e.g. "Faster R-CNN") or on inspection of image quadrants/octants (e.g. "YOLO"), which are then further analyzed with deep convolutional neural networks (CNN). In this work, we analyze the Discriminative Generalized Hough Transform (DGHT), which operates on edge images, for pedestrian detection. The analysis motivates to use the DGHT as an efficient proposal generation mechanism, followed by accepting or rejecting the proposals (based on image data) using a deep CNN. Due to the low false negative rate of the DGHT and the high accuracy of the CNN we obtain competitive performance on several pedestrian detection databases.

Keywords: Pedestrian detection · Hough transform · Error analysis · Proposal generation · Patch classification · Convolutional neural network

1 Introduction

In the last decades, automatic pedestrian detection has been a very important and still challenging task [4] in computer vision exhibiting many sources of large variability, i.a. regarding the object size and pose, occlusion and background. A lot of detection approaches have been suggested, among them feature-based detectors such as Viola-Jones [33] and Two-layer histograms of oriented gradients (HOG) [36], deformable part models [12,18], Random Forest-based approaches [25] and recently deep learning algorithms. The latter mainly consist of architectures using region proposals and a subsequent patch analysis with convolutional neural networks (CNN) as e.g. in Faster R-CNN [29]. Alternatively, approaches with a constant trivial region generation scheme and a subsequent bounding box regression or those that directly operate on full images have been proposed, e.g. R-CNN minus R [24], YOLO [27,28] and SSD [35], respectively.

S. Battiato et al. (Eds.): ICIAP 2017, Part II, LNCS 10485, pp. 104–115, 2017.
https://doi.org/10.1007/978-3-319-68548-9_10

The Discriminative Generalized Hough Transform (DGHT) [30] is an efficient voting-based localization approach and has successfully been applied in single-object localization tasks with limited variability, such as joint [30] and epiphyses [15] localization in medical images or state-of-the-art iris localization [14].

In this work, we analyze in detail the performance of the DGHT in a pedestrian detection task with many sources of variability (background, object size, pose etc.). In particular, we suggest to use the DGHT as an efficient proposal generation mechanism, accepting or rejecting the generated candidates using a deep convolutional neural network. We compare our approach to state-of-the-art algorithms, obtaining competitive performance on three different databases.

2 Methods

2.1 Structured Edge Detection

We use the real-time edge detection approach of [10], which learns information on the object of interest. Here, a Random Forest [5] maps an input image patch to an output edge image patch using pixel-lookups and pairwise-difference features of 13 (3 color, 2 magnitude and 8 orientation) channels. The approach incorporates ideas of Structured Learning [22] for handling the large amount and variability of training patch combinations as well as for efficient training. While testing, densely sampled, overlapping image patches are fed into the trained detector. The edge patch outputs which refer to the same pixel are locally averaged. The resulting intensity value can be seen as a confidence measure for the current pixel belonging to an edge. Subsequently, a non-maximum suppression (NMS) can be applied in order to sharpen the edges and reduce diffusion. For an example see Fig. 1b; further details can be found in [10].

2.2 Discriminative Generalized Hough Transform

The Generalized Hough Transform (GHT) [2] is well-known as a general model-based approach for object localization. Each model point m_j of the shape model M (Fig. 1c) is represented by its coordinates with respect to the reference point. The Discriminative Generalized Hough Transform (DGHT) [30] extends the

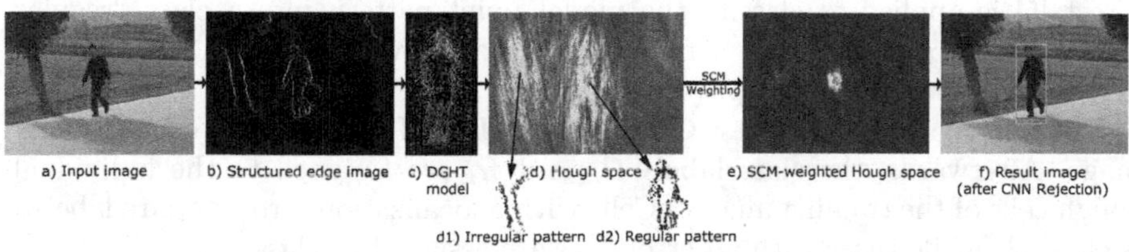

Fig. 1. Application of structured edge detection (Sect. 2.1), DGHT (Sect. 2.2) and SCM (Sect. 2.3) to an input image. Only a single image scale is shown.

GHT by individual model point weights λ_j for the J model points, which are optimized by a discriminative training algorithm that also accounts for automatic generation of M. Using this shape model M, the DGHT transforms a feature image X – in our work an edge image as outlined in Sect. 2.1 – into a parameter space H, called Hough space, by a simple voting procedure (see Fig. 1d):

$$H(c_i, X) = \sum_{\forall m_j \in M} \lambda_j f_j(c_i, X) \text{ with} \tag{1}$$

$$f_j(c_i, X) = \sum_{\forall e_l \in X} \begin{cases} 1, & \text{if } c_i = \lfloor (e_l - m_j)/\rho \rfloor \text{ and } |\varphi_{e_l} - \varphi_{m_j}| < \Delta\phi \\ 0, & \text{otherwise.} \end{cases} \tag{2}$$

The Hough space H (we set the quantization parameter $\rho = 2$) consists of discrete Hough cells c_i, which accumulate the weighted number of matching pairs of all model points m_j (with corresponding weight λ_j) and feature points e_l. $f_j(c_i, X)$ determines how often model point m_j votes for Hough cell c_i given a feature image X. A vote, however, is only counted, if the absolute orientation difference of the model and feature point, φ_{m_j} and φ_{e_l}, respectively, is below $\Delta\phi$. Each Hough cell c_i represents a target hypothesis with its image space coordinates given by $\lfloor (c_i + 0.5) \cdot \rho \rfloor$. The number of weighted votes for each c_i corresponds to the degree of matching between model M and feature image X.

For ensuring good localization quality, a high correlation with the feature image at target point locations and a small correlation at confusable objects is desired. The DGHT achieves this by an iterative training procedure starting with an initial model of superimposed annotated feature images at the reference point. In each iteration, the model point weights λ_j are optimized using a Minimum Classification Error (MCE) approach and, afterwards, the model is extended by target structures from training images which still have a high localization error. To reduce model size, those points with low absolute weights are eliminated. This procedure is repeated until all training images are used or have a low localization error. Further details on this technique can be found in [30].

2.3 Rejection of Proposals

Shape Consistency Measure (SCM). [14] suggested to analyze the model point pattern voting for a particular Hough cell c_i. More specifically, a Random Forest [5] is applied to classify the model point pattern into a class "regular shape" Ω_r (representing e.g. a frontal or a side view of a person) and a class "irregular shape" Ω_i (see Fig. 1(d1) and (d2)).

To train the Random Forest Classifier, the DGHT is applied to each training image. Afterwards, the class labels Ω_r and Ω_i are assigned to the individual Hough cells of the training images: Cells with a localization error $<\varepsilon_1$ are labeled as class Ω_r while those with an error $>\varepsilon_2$ are assigned to class Ω_i.

For a test image X, a DGHT model is applied to generate a Hough space H. For each local maximum c_i in H, the Random Forest Classifier is used to calculate the probability $p = p(\Omega_r, c_i)$ that the set of model points voting for c_i

has a regular shape. The obtained probability is used as an additional weighting factor for the Hough space votes, i.e. $S(c_i, X) = H(c_i, X) \cdot p(\Omega_r, c_i)$ (see Fig. 1e). The local maxima in H are now sorted according to decreasing $S(c_i, X)$ to provide an ordered list $C = \{c_i\}$ of most probable object positions c_i.

Deep Convolutional Neural Networks. Deep convolutional neural networks (CNN) have been successfully used for image classification tasks achieving state-of-the-art classification performance e.g. on the large-scale ImageNet classification challenge [23,31,38]. In combination with a separate region proposal generation step, deep CNNs have been successfully applied in object detection taks, e.g. R-CNN [19] or Fast R-CNN [20]. Furthermore, Faster R-CNN [29] combined these two components into one network sharing convolutional features.

In this work, we use a deep CNN to individually accept or reject each proposal c_i out of the list C generated by the DGHT+SCM (see Sects. 2.2 and 2.3). Specifically, each candidate position $c_i \in C$ (seen as a proposal) is transferred from Hough space to image space. Then, a bounding box corresponding to the mean object size (60×160 px) is centered around that position (Fig. 1f), and the image patch corresponding to the bounding box is rescaled to an input size of 64×64. The patch pixel intensities of all three color channels are normalized to $[0, 1]$, and then used as input to a deep CNN. We use the standard VGG16 classifier as described in [31], pre-trained on ImageNet and fine-tuned on the IAIR training corpus (see Sect. 3.1). The output of the CNN is a softmax layer with 2 classes, pedestrian and background. We use $p = p(\text{pedestrian}, c_i)$ for candidate rejection (see Sect. 3.2). With an appropriate rejection threshold θ, any candidate c_i is rejected if $p(\text{pedestrian}, c_i) < \theta$ (Fig. 1f).

3 Experimental Setup

3.1 Databases

IAIR-CarPed. We perform most experiments on the IAIR-CarPed [36] database, because it has a reasonable amount of independent 2D images and additionally offers difficulty labels (e.g. occlusion, low contrast) for each annotation. We use this additional information for our detailed error analysis. As suggested in [36], we train on a random 50%-split of the available pedestrian images, i.e. in total 1046 images containing 2341 pedestrians with an object height range from 45 to 383 px (mean height: 160 px). The remaining 1046 images (2367 pedestrians with a similar object height range and mean height) are used for evaluation. Training and test corpus each contain all types of difficulties present in the IAIR corpus.

INRIA Person. We also evaluate our approach on the well-known INRIA Person [6] database. The test set contains 288 images which contain 561 annotated persons with a height range from 100–788 px (mean height: 299 px).

TUD Pedestrians. Moreover, we apply our framework to the TUD Pedestrians [1] dataset. The test set consists of 250 images containing 311 annotated pedestrians with a height range from 71 to 366 px (mean height: 213 px).

3.2 Experimental Setup and System Parameters

Our system setup for the single-frame detection of pedestrians in non-consecutive 2D RGB images is organized as follows:

Feature Image Generation: As input images for training and testing, we use the output of the Structured Edge Detector (see Sect. 2.1). We train this edge detector specifically for pedestrians on the PennFudan database [34], because this is the only database we were aware of providing segmentation information needed to train the edge detector. The Structured Edge Detector suppresses most of the background edges and thus significantly reduces background variability [13].

DGHT Model and SCM Training: In order to generate a DGHT pedestrian model including a certain amount of size variability, we allow a size range of 144–176 px (mean object height ±10%). All training images with pedestrians not in this size range are scaled to a person size selected randomly from the allowed range (uniform distribution), separately for each pedestrian in an image. To train our DGHT shape model (see Sect. 2.2), we only use those full training images containing "simple" pedestrians (IAIR difficulty type "S", 1406 pedestrians/775 images). Having trained the DGHT shape model, we train the SCM on the full IAIR training set comprising all difficulty types and all pedestrians scaled to the range 144–176 px as described above (see also Sect. 2.3). We set ε_1 for class Ω_r to 5 and ε_2 for class Ω_i to 15 Hough cells.

Testing: To handle the large range of object sizes, we scale each test image by the following heuristic set of 10 scaling factors such that each pedestrian should roughly fit into the expected object range (mean object height ±10%):
50%, 62.5%, 75%, 100%, 150%, 200%, 225%, 250%, 275%, 300%.

The trained DGHT model is applied independently to each scaled image, i.e. independent Hough spaces are generated for each image scale and, afterwards, weighted by the SCM (Sect. 2.3). In each weighted Hough space, local maxima $C = \{c_i\}$ are identified using a NMS with a minimum distance of 1/3 of the model width, i.e. 20 px. To reduce the amount of candidates, we discard those candidates c_i with $S(c_i, X) < \max S(X) \cdot 0.2$. To reject possible false positive candidates e.g. due to inconsistent voting schemes [14], we investigate two rejection mechanisms:

SCM Rejection: For each image scale independently, a candidate is rejected if $p(\Omega_r, c_i) < \theta$, see Sect. 2.3. This is a purely Hough-based rejection method.

Our results and error analysis (Sect. 4) motivate an additional rejection step:

CNN Rejection: Here, any candidate position c_i is transferred to (scaled) image space, and a bounding box corresponding to the mean model size is centered around that position. A deep CNN is then used to reject c_i if $p(\text{pedestrian}, c_i) < \theta$. This is an image-based rejection, as opposed to the SCM rejection.

We use the standard Keras VGG16 model, which is initialized on ImageNet. We fine-tune this model on our IAIR training corpus, using the annotated pedestrian bounding boxes scaled to $(64 \times 64 \times 3)$ as positive samples and the same candidates as for class Ω_i in the SCM training as negative samples, i.e. high scoring peaks with a minimum error of 15 Hough cells. For fine-tuning we use the Adam optimizer [21] with categorical cross-entropy loss, a learning rate η of 0.001, which is reduced on plateaus, and an input dimension of $(64 \times 64 \times 3)$.

Combining Scales and Post-processing: Subsequent to the rejection step, the remaining candidate bounding boxes of all image scales are divided by the respective scaling factor to match the original image dimensions. Afterwards, the candidates are greedily grouped based on the mutual overlap[1] and finally a NMS is applied to each group using $S(c_i, X)$ (without CNN) or $p(\text{pedestrian}, c_i)$ as criterion, respectively, in order to avoid double detections.

Analysis: As suggested in [39], we conduct a detailed error analysis including oracle experiments: (a) localization oracle (all false positives (FP) that overlap with the ground truth are ignored) and (b) background vs. foreground oracle (all FP that do not overlap with the ground truth are ignored). In addition, we conduct another oracle experiment (c): perfect rejection oracle (DGHT as a proposal generator). For each ground truth annotation, the rejection oracle picks the best matching candidate out of the set C generated by the DGHT (including the SCM and post-processing over image scales) and rejects all other candidates. Thus, we quantify the minimal miss rate for the DGHT as proposal generator, assuming a perfect rejection mechanism.

3.3 Comparison to State-of-the-Art Approaches

We compare our approach against several state-of-the-art algorithms. For the IAIR-CarPed database, we compare to the results for the Two-layer HOG and the PASCAL deformable part model (DPM) published in [36]. We also downloaded the latest DPM release (DPMv5) [18] trained on PASCAL, the pre-trained YOLOv1 [27] full model as well as the pre-trained YOLOv2 [28] full model (both pre-trained on ImageNet and fine-tuned on PASCAL) and evaluated them on our IAIR-CarPed test corpus. Addtitionally, we used the pre-trained YOLOv1 full model and fine-tuned it on our IAIR-CarPed training set. For details on these state-of-the-art approaches see the respective references. For the other databases, we use the benchmark results from [16] and [37], respectively.

3.4 Evaluation Metrics

We evaluate our detections using the intersection over union (IoU) measure. According to [11], a detection of an object is correct if the IoU of the prediction and the ground truth exceeds 50%. As suggested in [9], for single frame evaluation we computed Detection Error Tradeoff (DET) curves plotting the miss rate against the false positives per image (FPPI) on a log-log scale by modifying the rejection threshold θ. For comparison, the miss rates at 0.5 and 1 FPPI are shown. For the TUD Pedestrians database, we use the recall at equal error rate (EER), as other groups have frequently used this measure. For measuring the candidate quality, we use the Average Best Overlap (ABO) score from [32].

[1] We set this parameter to 30%.

4 Results

4.1 SCM Rejection

Table 1 compares the DGHT + SCM (without CNN) detection performance to other approaches on the IAIR test corpus as described in Sect. 3.2.

As can be seen, the overall performance of the DGHT is comparable to the Two-Layer HOG and the published original DPM result, but worse than the current DPMv5, the fine-tuned YOLOv1 and the pre-trained YOLOv2.

Table 1. Comparison of detection results on the IAIR-CarPed test corpus[a] S: Simple, D1: Occlusion, D2: Low Contrast, D3: Infrequent Shape

Approach	Miss Rate at 0.5 FPPI					Miss Rate at 1 FPPI				
	S	D1	D2	D3	All	S	D1	D2	D3	All
DGHT + SCM	0.32	0.51	0.74	0.50	0.44	0.22	0.40	0.66	0.32	0.34
Two-Layer HOG [36]	N/A	N/A	N/A	N/A	N/A	0.25	0.47	0.44	0.50	0.35
DPM [36]	N/A	N/A	N/A	N/A	N/A	0.29	0.37	0.45	0.36	0.34
DPMv5 [18]	0.18	0.37	0.47	0.45	0.29	0.16	0.32	0.40	0.40	0.25
YOLOv1 Pre-trained [27]	0.37	0.47	0.87	0.45	0.49	0.32	0.41	0.81	0.42	0.44
YOLOv1 Fine-tuned [27]	0.06	0.18	0.23	0.18	0.13	0.06	0.17	0.22	0.17	0.13
YOLOv2 Pre-trained [28]	0.13	0.28	0.31	0.23	0.21	0.12	0.25	0.28	0.20	0.19

[a]Results for Two-Layer HOG and DPM taken from [36], i.e. potentially different training/test split of the IAIR corpus. Results for DGHT, DPMv5, YOLOv1 and YOLOv2 obtained on our test split. For details on pre-training and fine-tuning see Sect. 3.3

4.2 SCM Rejection: Error Analysis

In this section, we analyze in detail all errors of the DGHT with SCM rejection (without CNN), in comparison to the DPMv5. We specifically focus on the DPM since the code is publicly available (such that we can perform own experiments on the IAIR corpus) and since it is also a model-based approach operating on feature images. We use a similar error analysis as described in [39].

Generally, there are two basic types of errors: (I) false positives (FP), i.e. a false detection, e.g. because of confusable background structures, misaligned, larger or smaller predictions ($0\% < \text{IoU} < 50\%$) or double detections; (II) false negatives (FN), which are ground truth annotations that are not detected. FN mostly occur because of the "well known difficulty of detecting small and occluded objects" [39] as small persons are often over-/underexposed or blurry. Besides, there might be a dataset bias, e.g. side-views or cyclists are under-represented in the training set, which hampers the detection of such instances.

We manually evaluated all detection errors (FP and FN) of the DGHT and DPMv5 [18] on the IAIR test corpus (see Sect. 3.2) at 1 FPPI:

28% (DGHT)/31% (DPM) of all FP are due to localization errors, mostly because of body part (DGHT) and double detections (DPM).

For both approaches, the vast majority of FP are background detections (69% (DGHT)/66% (DPM)) due to confusing vertical structures or trees. The DPM often detects pedestrian groups as one detection. The remaining 3% are missing annotations, which actually are true instead of false positives.

For the DGHT, 29% of all FN are due to low contrast, i.e. insufficient or no feature representation at all. Another 20% are slightly below the rejection threshold at 1 FPPI indicating that the SCM as a rejection mechanism could still work more properly. The remaining 50% mainly consist of errors at small scales (16%), localization errors (12%) and occluded pedestrians (11%).

For the DPM, the main reasons for FN are small scales (28%), occlusion (26%) and low contrast (22%; mostly also at small scales). The remaining FN mainly consist of localization errors (7%), side views (6%), and cyclists (3%).

The DPM has more FN because of small scales or occlusion. Low contrast or missing edges are a problem for both approaches. Side views/cyclists are better detected with the DGHT pipeline. Moreover, the DGHT FN often are only slightly below θ (with a lower θ, too many FP would have been generated).

4.3 SCM Rejection: Oracle Experiments

The results of the oracles (a), (b) and (c) (see Sect. 3.2 "Analysis") are shown in Table 2. Since the localization oracle only reduces the miss rate at 1 FPPI by 0.03, it shows that the DGHT detections are usually quite accurate (except for a few outliers). On the contrary, there is still much room for improvement regarding background vs. foreground errors. If we would be able to reject FP in the background (IoU = 0%), the miss rate drops from 0.34 (current DGHT + SCM result at 1 FPPI) to 0.16 at only 0.3 FPPI. This again indicates that the DGHT candidates are very accurate, but the rejection using only the model point voting pattern on structured edge images is not sufficient to properly overcome the well-known problems of small-scale detections or those of confusable background structures (see Sect. 4.2). In case of perfect rejection of DGHT proposals, we would obtain a miss rate of only 0.04 with an ABO score of 78.2% (perfect rejection oracle). This clearly indicates that the main drawbacks of the DGHT are (a) FP in the background and (b) non-optimal selection of candidates based on $S(c_i, X)$. The low miss rate of the perfect rejection oracle motivates to use the DGHT as a proposal generator, and to improve the proposal rejection. To this end, we apply the CNN proposal rejection subsequently to the SCM rejection, as outlined in Sect. 3.2.

Table 2. DGHT oracle results (at highest or max. 1 FPPI)

Experiment	S	D1	D2	D3	All
DGHT + SCM at 1 FPPI	0.22	0.40	0.66	0.32	0.34
DGHT Localization oracle at 1 FPPI	0.20	0.35	0.49	0.30	0.31
DGHT BG vs. FG oracle at 0.3 FPPI	0.09	0.24	0.24	0.23	0.16
DGHT Perfect Rejection Oracle at 0 FPPI	0.01	0.01	0.20	0.00	0.04

4.4 CNN Rejection: Detection Results

Table 3 shows the detection results using additional CNN proposal rejection on top of SCM rejection ("DGHT + SCM + VGG16") as introduced in Sect. 3.2 and motivated in Sects. 4.2 and 4.3[2]. Additionally, we show the results of this setup on TUD Pedestrians and INRIA Person in Tables 4 and 5, respectively. Note that no component of our system has been retrained on TUD or INRIA. We obtain minimal miss rates of 0.04 (78.2% ABO) at 350 candidates per image, 0.01 (75.8% ABO) at 55 candidates per image and 0.01 (76.8% ABO) at 102 candidates per image on IAIR, TUD Pedestrians and INRIA Person, respectively.

Table 3. Comparison of detection results on the IAIR-CarPed test corpus S: Simple, D1: Occlusion, D2: Low Contrast, D3: Infrequent Shape. Results of other state-of-the-art algorithms are partly repeated from Table 1

Approach	Training data	Miss Rate at 0.5 FPPI				
		S	D1	D2	D3	All
DGHT + SCM	IAIR	0.32	0.51	0.74	0.50	0.44
DGHT + SCM + VGG16	ImageNet/IAIR	0.09	0.30	0.40	0.20	0.19
DPMv5 [18]	PASCAL	0.18	0.37	0.47	0.45	0.29
YOLOv1 Pre-trained [27]	ImageNet/PASCAL	0.37	0.47	0.87	0.45	0.49
YOLOv1 Fine-tuned [27]	Im.Net/PASC./IAIR	0.06	0.18	0.23	0.18	0.13
YOLOv2 Pre-trained [28]	ImageNet/PASCAL	0.13	0.28	0.31	0.23	0.21

Table 4. Recall at EER on TUD Pedestrians without retraining

Approach	DGHT + SCM + VGG16	PartISM [1]	HoughForests [17]	Yao et al. [37]
Training data	IAIR	TUD/INRIA	TUD/INRIA	TUD/INRIA
Recall at EER	0.88	0.84	0.87	0.92

Table 5. Miss Rate at 1 FPPI on INRIA Person without retraining. Ours: DGHT + SCM + VGG16

Approach	Ours	HOG	ICF [7]	Yao [37]	FPDW [8]	VeryFast [3]	Spat.Pool. [26]
Training data	IAIR	INRIA	INRIA	INRIA	INRIA	INRIA	INRIA/Caltech
Miss rate	0.14	0.23	0.14	0.12	0.09	0.07	0.04

5 Discussion

The experiments have shown that the DGHT in general is suitable for proposal generation due to the low false negative rate and the comparably small number

[2] Evaluation at 0.5 FPPI since this is the highest FPPI rate for DGHT+SCM+VGG16.

of candidates. The trained DGHT pedestrian model for proposal generation also generalizes well to other pedestrian databases without retraining any of the components. Our pedestrian detection pipeline achieves comparable results to other state-of-the-art approaches. In comparison to Selective Search [32] (2,000–10,000 candidates) or the region proposals of Faster R-CNN [29] (300+ candidates), the DGHT outputs a smaller number of candidates (see Sect. 4.4). Example detections are shown in Fig. 2.

Our approach still has some limitations: due to the edge feature images, we intrinsically miss those objects which are of low contrast, since they do not generate well pronounced edges or no edges at all. This limitation can be seen in Tables 2 and 3 at difficulty type D2. Additionally, we currently do not perform any bounding box refinement step which might further improve detection accuracy. However, the ABO scores of >75% for all three databases indicate that the candidates are already of good quality. Currently, our implementation does not aim at real-time performance. However, due to the independent voting of model points, the DGHT exhibits a high potential for parallelization.

Fig. 2. Example DGHT + SCM + VGG16 detections on IAIR. (green): ground truth, (yellow): correct detection, (blue): FP, (red): FN; best viewed in color

6 Conclusions

In this work, we applied the DGHT as a proposal generator - in combination with proposal rejection by a deep CNN - to a real-world multi-object detection task exhibiting many sources of variability, namely pedestrian detection. We obtained comparable performance to state-of-the-art approaches on the IAIR-CarPed, the TUD Pedestrians and the INRIA Person databases, demonstrating that our framework (trained on IAIR) generalizes well also to other datasets. The main advantages of the DGHT proposal generation are (a) the relatively low amount of training images needed for training of DGHT and SCM (on the order of 100 or less per variability class), (b) the low amount of resources needed at test time, (c) the relatively low amount of proposals generated per image. Thus, our framework could be useful especially for detecting specific object categories with limited available training material.

Acknowledgement. This work was funded by the Department of Social Affairs, Health, Science and Equality of Schleswig-Holstein, Germany. Thanks to Andrew J. Richardson for providing the VGG16 framework.

References

1. Andriluka, M., et al.: People-tracking-by-detection and people-detection-by-tracking. In: CVPR (2008)
2. Ballard, D.H.: Generalizing the Hough transform to detect arbitrary shapes. Pattern Recogn. **13**(2), 111–122 (1981)
3. Benenson, R., et al.: Pedestrian detection at 100 frames per second. In: CVPR (2012)
4. Benenson, R., et al.: Ten years of pedestrian detection, what have we learned? In: ECCV (2014)
5. Breiman, L.: Random forests. Mach. Learn. **45**(1), 5–32 (2001)
6. Dalal, N., Triggs, B.: Histograms of oriented gradients for human detection. In: CVPR (2005)
7. Dollar, P., et al.: Integral channel features. In: BMVC (2009)
8. Dollar, P., et al.: The fastest pedestrian detector in the west. In: BMVC (2010)
9. Dollar, P., et al.: Pedestrian detection: an evaluation of the state of the art. In: PAMI (2012)
10. Dollar, P., et al.: Fast edge detection using structured forests. In: PAMI (2015)
11. Everingham, M., et al.: The PASCAL VOC challenge. In: IJCV (2010)
12. Felzenszwalb, P., et al.: A discriminatively trained, multiscale, deformable part model. In: CVPR (2008)
13. Gabriel, E., et al.: Structured edge detection for improved object localization using the discriminative generalized Hough transform. In: Proceedings of VISAPP (2016)
14. Hahmann, F., et al.: A shape consistency measure for improving the generalized Hough transform. In: Proceedings of VISAPP (2015)
15. Hahmann, F., Böer, G., Deserno, T.M., Schramm, H.: Epiphyses localization for bone age assessment using the discriminative generalized Hough transform. In: Deserno, T.M., Handels, H., Meinzer, H.-P., Tolxdorff, T. (eds.) Bildverarbeitung für die Medizin 2014. I, pp. 66–71. Springer, Heidelberg (2014). doi:10.1007/978-3-642-54111-7_17
16. http://www.vision.caltech.edu/Image_Datasets/CaltechPedestrians/
17. Gall, J., Lempitsky, V.: Class-specific Hough forests for object detection. In: CVPR (2009)
18. Girshick, R.B., Felzenszwalb, P.F., McAllester, D.: Discriminatively Trained Deformable Part Models, Release 5. http://people.cs.uchicago.edu/~rbg/latent-release5/
19. Girshick, R.B., et al.: Rich feature hierarchies for accurate object detection and semantic segmentation. In: CVPR (2014)
20. Girshick, R.B.: Fast R-CNN. In: ICCV (2015)
21. Kingma, D.P., Ba, J.L.: Adam: a method for stochastic optimization. In: ICLR (2015)
22. Kontschieder, P., et al.: Structured class-labels in random forests for semantic image labelling. In: ICCV (2011)
23. Krizhevsky, A., et al.: ImageNet classification with deep CNNs. In: NIPS (2012)
24. Lenc, K., Vedaldi, A.: R-CNN minus R. In: BMVC (2015)
25. Marin, J., et al.: Random forests of local experts for pedestrian detection. In: ICCV (2013)
26. Paisitkriangkrai, S., et al.: Strengthening the effectiveness of pedestrian detection. In: ECCV (2014)

27. Redmon, J., Farhadi, A.: YOLO: unified, real-time object detection. In: CVPR (2016)
28. Redmon, J., Farhadi, A.: YOLO9000: better, faster, stronger. arXiv:1612.08242 (2016)
29. Ren, S., et al.: Faster R-CNN: towards real-time object detection with region proposal networks. In: NIPS (2015)
30. Ruppertshofen, H.: Automatic Modeling of Anatomical Variability for Object Localization in Medical Images. BoD-Books on Demand, Norderstedt (2013)
31. Simonyan, K., Zisserman, A.: Very deep convolutional networks for large-scale image recognition. In: ICLR (2015)
32. Uijlings, J.R.R., et al.: Selective search for object recognition. In: IJCV (2012)
33. Viola, P., et al.: Detecting pedestrians using patterns of motion and appearance. In: IJCV (2005)
34. Wang, L., et al.: Object detection combining recognition and segmentation. In: ACCV (2007)
35. Liu, W., Anguelov, D., Erhan, D., Szegedy, C., Reed, S., Fu, C.-Y., Berg, A.C.: SSD: single shot multibox detector. In: Leibe, B., Matas, J., Sebe, N., Welling, M. (eds.) ECCV 2016. LNCS, vol. 9905, pp. 21–37. Springer, Cham (2016). doi:10.1007/978-3-319-46448-0_2
36. Wu, Y., Liu, Y., Yuan, Z., Zheng, N.: IAIR-CarPed: a psychophysically annotated dataset with fine-grained and layered semantic labels for object recognition. Pattern Recogn. Lett. **33**(2), 218–226 (2012)
37. Yao, C., et al.: Human detection using learned part alphabet and pose dictionary. In: ECCV (2014)
38. Zeiler, M., Fergus, R.: Visualizing and understanding ConvNets. In: ECCV (2014)
39. Zhang, S., et al.: How far are we from solving pedestrian detection? In: CVPR (2016)

Bubble Shape Identification and Calculation in Gas-Liquid Slug Flow Using Semi-automatic Image Segmentation

Mauren Louise Sguario C. Andrade(✉), Lucia Valeria Ramos de Arruda, Eduardo Nunes dos Santos, and Daniel Rodrigues Pipa

Federal Technological University of Parana, Parana, Brazil
{mlsguario,vlarruda,ensantos,danielpipa}@utfpr.edu.br
http://www.utfpr.edu.br

Abstract. In this article, image segmentation techniques were used to identify the average size of gas bubbles in two-phase flow air-water. The technique of applying the results were compared to images reconstructed from the wire-mesh sensor. The method has the advantages of no intrusion and easy implementation. Preliminary results in the segmentation of gas bubbles images in horizontal flow proposed here help a better identification of the shapes of elongated bubbles. Those results also suggest that the methodology used in this case can be applied in the segmentation of different elongated bubble representations, for instance, in vertical flows.

Keywords: Priori shape · Level Set method · Two-phase flow · Wire-mesh sensor

1 Introduction

The multiphasic flows are involved in several industrial processes, and their dysfunctions have been associated to safety issues. In the oil industry, for instance, the involvement of oil, water, gas and even solid particles, such as sand, influence the flow stability. Due to the complexity of shaping, however, this situation is normally simplified by a gas-liquid two-phase flow. It is, though, required the correct rendering and identification of both the flow patterns and the relative properties of the fluids, such as void fractions, fluids velocity, among others. The studies about multiphasic flows are challenging the scientific community in its search for techniques and devices capable of describing the course of fluids in ducts and the establishment of its patterns.

Among the horizontal flow patterns, the slug pattern has been studied since 1969, being present in industrial applications, mainly at the oil industry, where this model is held, frequently, on the production lines. The slug pattern is a periodic flow whose technical features vary in space and time, and that is characterized by a series of liquid pistons followed by an elongated bubble, that could present scattered bubbles or not, depending on the phases velocity. Wallis introduced in 1969 [18] the concept of unit cell as illustrated on Fig. 1, which shows

S. Battiato et al. (Eds.): ICIAP 2017, Part II, LNCS 10485, pp. 116–126, 2017.
https://doi.org/10.1007/978-3-319-68548-9_11

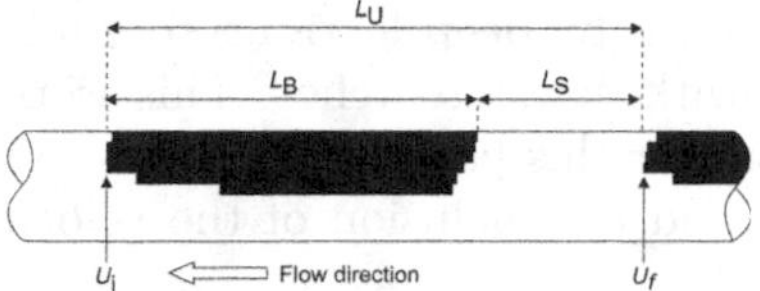

Fig. 1. Representation of a unit cell in horizontal gas-liquid slug flow as well as the liquid film, the elongated bubble and the liquid piston.

graphically the slug flow model. It can be observed by the illustration, the liquid piston, its adjacent elongated bubble and several bubbles scattered with the respective liquid film. With the description and analysis of an unit cell behaviour, all properties of the slug flow pattern through a duct can be predicted [3].

The study of this flow pattern has been done in experimental plants through mathematical shaping and the use of different kinds of sensors such as wire-mesh sensors, conductive and capacitive probes, ultrasound transducers, optical tomography, X-ray and gamma-ray tomography, and high-speed camera. In this context, image processing techniques have been used experimentally, because they allow identification and visual evaluation of important technical features through solutions with relatively simple projecting/implementing.

In relation to the horizontal air-water flow, aim of this article, several guidelines pull out regions of interest in images for later analysis based on basic image processing operations. [2] utilizes mathematical morphology and segmentation by Watershed transformation to extract information from the gas bubbles. In [10] images are used on the measuring of the area, length and volume of gas bubbles. In [14] the flow images are grouped in order to identify shape of the bubbles, but without bubble segmentation for later measuring of their properties.

In this context, this article investigates the typical gas bubbles shape, volume and length as a function of speed and property of the fluid. When studying the bubbles shape on the flow as a function of speed and length, will be possible to represent the fluids behaviour such as the generation of bubbles and disruption due to their dispersion and distribution. With this goal, a sequence of images taken by high definition camera and image processing techniques were used to obtain estimated results of the elongated bubbles shape in gas-liquid flow. For this, the application of an image segmentation method was used to correctly delimit the elongated bubble (or unit cell). The images were overlaid to define the average gas bubble shape, and then segmented by the Restrict Level Set method propose here and finally it was compared to the images generated by the wire-mesh sensor [7] and segmented by Level Set Method proposed in [5].

The segmentation by Level Set method has its application consolidated especially in the medical area [12,13,20] where the presence of standard shapes is observed, as well as in different problems associated with object segmentation [1,11,19]. Nevertheless, new approaches have been developed in order to contribute to the development of a more efficient solution. As well as in the work of [4] that develops a shape-driven approach for object segmentation. They propose a shape prior constraint term by deep learning to guide variational segmentation, which collaborates with new researches. This work is a continuity of the one published in [17], however this paper contributes to the validation of values by the wire-mesh sensor and the inclusion of the velocity term, that were not previously used.

2 Materials and Methods

2.1 Image Acquisition

The experimental data were collected in a horizontal air-water biphasic flow trial display located in the Thermal Science Laboratory (LACIT) at the Federal Technological University of Parana (UTFPR). The assay is described in [2,14]. The flow images were captured by a high speed camera Nano Sense MKIII (DantecDynamics) model in 232×500 pixels resolution. For this task, the images were taken with the gas superficial velocity varied between 0.5 and 1.5 m/s to a superficial liquid velocity of 0.5 and 0.7 m/s to observe the effect of the bubble shape. Thus, in this study, a total of 5 measuring points are evaluated with 150 air bubbles in each one of them.

2.2 Image Segmentation by Level Set Method

The image segmentation method consists in giving an image in a group of regions non-overlapped and homogenous that must match to significant objects for a certain application. In this article, one new approach of segmentation based on the Level Set method and a priori knowledge of the way it was used in order to separate the gas bubble from the liquid piston.

The main idea of the Level Set method, introduced by Osher and Sethian [16], consists on the representation of a surface as a zero level interface of a superior dimensional function (called Level Set Function). One of the main advantages of the method is its capacity of dealing with changes or topological discontinuities that can appear during the zero level evolution curve.

The Level Set method is based on two central points: firstly, the incorporation of the surface as the zero level (zero Level Set) of a dimensional superior function, and, secondly, the incorporation (or extension) of the surface velocity F to this more elevated dimension level of function. In this task, a new approach was applied to the image segmentation based on the Level Set method that unites active girth to a priori shape. For this, the shape model of the targeted object was trained and defined by the points distribution model, next, added as a function of extension speed to the zero level curve evolution. The goal of the new approach is to make an structure that consists in three energy terms and an extension velocity function $(\lambda L_g(\phi) + vA_g(\phi) + \mu P(0) + \phi_f)$.

The first three terms of this equation are the same terms introduced by [5] and the last ϕ_f (called global restricted shape) is based on the shape representation previously trained. The extension speed function, ϕ_f was introduced, in order to orientate the zero level evolution curve and consist on the shape model trained at first through the points distribution model. Thus, to each new interaction the curve goes toward its normal, respecting the limitation imposed by the initially defined shape, i.e., the function will guide the evolution of the level curve with the annexation of a seed that represents the object to be segmented, taking advantage when it gets closer to the previously trained shape.

2.3 A Priori Shape Definition

The use of the Level Set method to the image segmentation requires the selection of an inner marker (or outer) to the region of interest, which will represent the initial curve ϕ, what can be done with simple operations (such as threshold and mathematical morphology). In this work the priori knowledge is defined and used as an inner marker and will be hereafter described.

Firstly, it is made a selection and manual marking of the points that define a data display, or coordinates of what is considered by the expert as being the limits for the observed shape. In this case it is necessary the manual marking of several distinct images. In this work, 20 images for each one of the gas bubble parts were marked, in order to be trained to generate an average shape between the marked images. At the end of this stage, a considerable amount of intercepted shapes are obtained (the points manually marked are align and measured the distances between its points to generate a new mean shape) from the different images that make up the training set, that need to be converted into an only final pattern, able to embrace and recognise images of the same phenomenon, but different from those used during the series of training.

The points aligning stage is a prerequisite so that the statistic values that describe the standard shape can be obtained. By aligning and overlaying every marked point next to the average standard model, it is observed the overlapping of equivalent points as well as the clouds formation with diffuse points, as those reference markings are partially correlated, and do not move independently, the points distribution model (PDM) looks for a pattern of coordinates variation in this diffused points zone or clouds. (Details of the shape model definition can be found in [6]).

2.4 Level Set Priori Shape Restrict - LSR

After the average shape definition it is necessary to consider an image that contains only one object with a shape similar to a sample in the training set. Then the goal is to recognize the area of the image that corresponds to this object. Thus the shape model is defined as initial curve (marker) for the Level Set evolution - MLS [5].

In image segmentation, active contours are dynamic curves that move toward the boundaries of the object. To achieve this goal, it is explicitly set an external energy that can move the zero level curve toward the object boundaries. Consider I as a picture, and g is the boundary indicator function defined by:

$$g_I(x,y) = \frac{1}{1 + |\nabla(G_\sigma * I(x,y))|} \tag{1}$$

where $G_\sigma * I$ is the convolution of image I as a Gaussian Kernel with standard deviation σ. Term $\nabla(G_\sigma * I(x,y))$ is essentially zero except near significant variations of the gradient that typically corresponds to the object boundaries. Thus $g_I(x,y)$ goes to 1 (one) outside the boundaries and goes to 0 (zero) near them.

Consider a velocity function of the form $F = \pm 1 - \epsilon\kappa$, where ϵ is a constant that acts as an advection term and κ denotes the curvature at a point [16]. Note that, when multiplying F by g_I, the flow does not evolve through the edges.

Li et al. proposed the use of an external function energy $\varepsilon(\phi)$, to drive the movement of the zero level curve, as follows [5]:

$$\varepsilon_{g,\lambda,v}(\phi) = \lambda L_g(\phi) + v A_g(\phi) \tag{2}$$

where $\lambda > 0$ and v are constant. Term $L_g(\phi)$ calculates the length of the zero level curve of ϕ and term $A_g(\phi)$ is introduced to accelerates the movement of zere level contour during the Level Set evolution that is necessary when the initial contour is located far away from the boundaries of the desired object. The total functional energy defined by the authors is represented as:

$$\varepsilon(\phi) = \mu P(\phi) + \varepsilon_{g,\lambda,v}(\phi) \tag{3}$$

The energy $\varepsilon_{g,\lambda,v}(\phi)$ is called external energy and it drives the zero Level Set to the boundaries of the object, whereas the energy $\mu P(\phi)$ is called internal energy and it penalizes the deviation of ϕ from a signaled distance function during its evolution. (To complete understand the implementation of the Level Set curve, see [5]). According to the terms presented above, the proposed approach introduces a restriction to the initial curve imposed by the priori shape knowledge and incorporated to the external energy at each iteration as follows:

$$\phi(t+1) = \phi(t) + delt * (\lambda L_g(\phi) + v A_g(\phi) + \mu P(0)) + \phi_f \tag{4}$$

where *delt* is the time step and ϕ_f is the velocity function of the evolving form and it is obtained by the difference between the previously trained form ϕ_i, it is at the same time defined as a marker initial, and the time-evolving curve ϕ_t. The formulation of the curve evolution by the proposed approach is then defined by:

$$\phi_f = \lambda_2 * (\phi_i - \phi(t)) \tag{5}$$

where λ_2 is obtained empirically and reflects the weight that must be set to the shape energy. The additional inserted term is set such a way that it does not affect the important property of the signaled distance function developed by Li et al. [5]. That is necessary to ensure a good evolution of the initial curve.

With the inclusion priori shape knowledge the proposed approach evolves the initial curve (shape model) as a function of the average shape of the object and gradient of the image however it takes into account the characteristics of each object of different images. In this case its application facilitates the segmentation of overlapping objects with occlusions or missing parts.

3 Proposed Scheme to Two-Phase Flow Image Segmentation

The development of this new variation of the Level Set method aims to optimize the objects segmentation both in two-phase flow images and in any images that have shape and pose previously known, adding up to the other existing approaches that use priori knowledge and active contour.

To flow images it is executed the manual marking of a set of images that represent each of the three parts of the gas bubble (nose, body and tail, illustrated on Fig. 2(a, b and c)). The goal is to build a model that describes the typical shape, using examples of the Fig. 2 as a training group (60 images divided in: nose, body and tail). The key points are defined by the manual marking and are around the limit between the gas bubble and liquid. This must be done with each shape on the training group. With it, is extracted the representation of each example, as a group of labelled points measuring the points average positions and the main shapes in which the points tend to escape from the average.

In this case, the result will be the average shape that will be used as prior information conducting the segmentation. It is worth pointing out that this is a stage conducted only once. Then, the next step consists of inserting a new image of the two-phase flow set of images. At this point, the system classifies the new image according to one of the three pre-established shapes (nose, body and tail). After that, the algorithm segments part of the gas bubble, using the resultant shape of the ranking. As an effect, the images now segmented are saved into a new file having the complete flow flux. Finally, from the segmented images, the

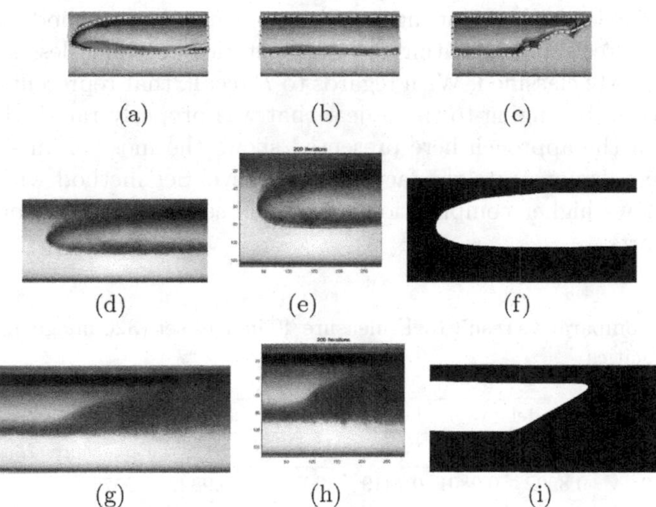

Fig. 2. Representation of the image marker (a–c) typicall gas bubble shape, (d–f) gas bubble noise results, (g–i) gas bubble tail results.

system does the gas bubble overlapping, in order to obtain the definition of an average gas bubble shape to each one of the speeds tested.

The proposed approach to the curve evolution is illustrated in Fig. 2. The Fig. 2(a, b and c) shows the marked points to compose the shape representation. In Fig. 2(d and g) the image overlapping result with the [14] technique. In Fig. 2(e and h) the end of the Level Set curve evolution by ϕ_f function. Finally the Fig. 2(f and i) illustrates the final segmentation by LSR.

4 Experimental Results

The image base is composed by 150 flow horizontal images air-water, to each of the superficial gas velocity that was analyzed (between 0.5 and 1.5 m/s). After the segmentation, the images were overlapped in order to identify the average gas bubble shape (the bubble aligning follows the method developed by [14]), so that they could be optically compared with the images generated by the wire-mesh sensor and by the flow images in gray levels used in [14]. Finally, from the segmentation result the algorithm was developed for the gas bubbles velocity measurement, which were compared to those obtained from the wire-mesh sensor and from the mathematical model developed by Bendiksen [9].

4.1 Validation

Table 1 shows the mean F-Measure index calculated regarding, to precision measuring calculation and recall, only the pixels that belong to the segmented region that better resemble to the aimed objects. Notice that Level Set method without priori knowledge leads the following results to the proposed method, mostly concerning the P precision, indicating one segmentation in which less significative pixels are correctly classified. With regards to R recall, that represents the entire pixels proportion belonging to the object that was properly rated, the segmentation through the approach here presented shows the most promising results. These numbers demonstrate the fact that the Level Set method without priori knowledge shows higher complication when applied on occlusion conditions or overlapped parts.

Table 1. The comparative result by F-measure 40 images set (320 images), to different superficial velocities

JL/JG	Active shape model			Level Set - MLS			Approach propose - LSR		
m/s	F-measure	R	P	F-measure	R	P	F-measure	R	P
0.5–0.5	0.7019	0.8992	0.9801	0.7419	0.6673	0.9822	0.9575	0.9288	0.9923
0.5–0.7	0.6867	0.9250	0.9298	0.7592	0.6860	0.9439	0.9733	0.9540	0.9938
0.5–1.0	0.6602	0.9037	0.8483	0.9301	0.8880	0.9349	0.9728	0.9534	0.9833
0.5–1.5	0.6624	0.86550	0.8130	0.9305	0.8834	0.8156	0.9750	0.9538	0.9186

4.2 Wire-Mesh Image Comparison and Grey Levels Images

In order to determine the applicability of the method developed on experimental plants, consider the results obtained by the proposed approach regarding the relation to the regenerated images by the wire-mesh sensor [15] and by the obtained mean shape with gray level images overlaid by the [14] essay. Figure 3 illustrates the obtained results by the three methods.

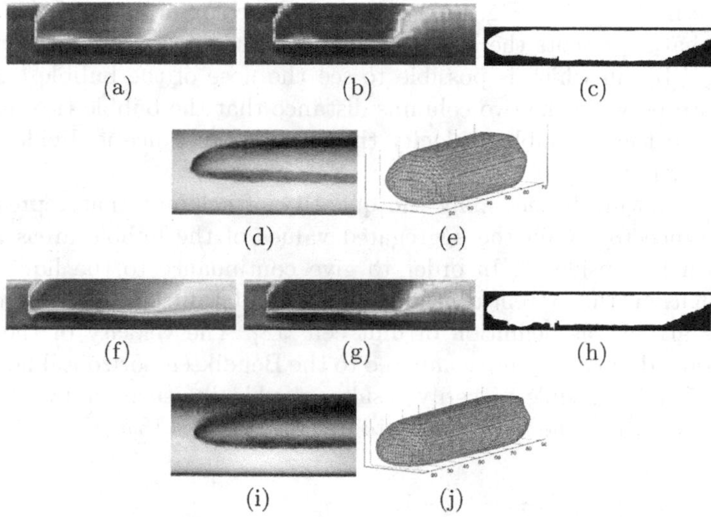

Fig. 3. $jl = 15\, jG = 10$: (a) Mean image - wire-mesh, (b) Median image - wire-mesh, (c) binary image segmentation, (d) Image gray level overlapping; (e) 3D superficial shape; $jl = 15\, jG = 15$: (f) Mean image - wire-mesh, (g) Median image - wire-mesh, (h) binary image segmentation, (i) Image gray level overlapping; (j) 3D superficial shape.

4.3 Gas Bubbles Velocity

Besides from the implementation of the new approach of the segmentation already discussed, a system capable of identify some unit cells properties was developed. With the goal of detecting automatically the beginning and the end of each gas bubble two points were defined, based on the lines and columns of the image representation. In order to detect the input and output of the bubble considered, respectively the last and the first column of the image, as shown on Fig. 4.

Each time that the nose of the bubble crosses the initial point, the system records the number of the frame in which this event occurs, that means, it is identified in which frame the analyzed bubble had its beginning. From there, the associated properties measuring start, including area, length gas bubble velocity, among others, until its tail hits the finishing point, determining the unit cell

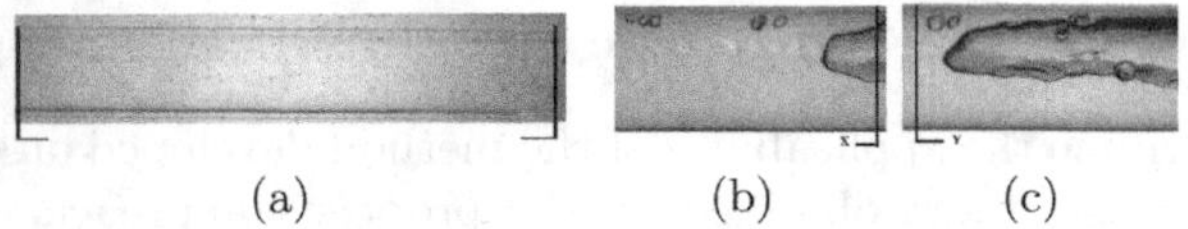

Fig. 4. Identification of the image column of the initial point (X) and end (Y).

finalization. Notice as example the Fig. 4. When the gas bubble is identified by crossing the initial point (Fig. 4(a)) is stored the column on which is held the first pixel that represents the bubble. This procedure is performed on the last frame (Fig. 4(b)) in which is possible to see the nose of the bubble and, from the difference between the two columns distance that the bubble travels can be calculated. To the J_G bubble velocity the obtained distance is divided by the frame number in this gap.

On the following frames, only the quantity of columns that represent the needed distance to obtain the aggregated values of the bubble areas and the liquid piston is considered. In order to give continuance to the liquid piston value calculation, the system continues to accumulate its values until the new gas bubble arrives, as definition of unit cell [18]. The velocity of the bubble results, obtained in trials, were compared to the Bendiksen horizontal flow model [9]. In Fig. 5 it is possible to verify results graphically, such values are shown satisfactory, in which the mean standard deviation was 0.13 m/s.

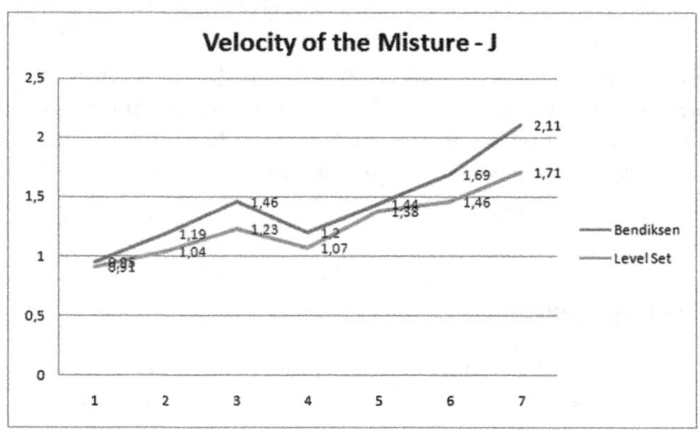

Fig. 5. Comparative graphic of the mixture velocity J.

Figure 6 exemplifies the obtained results about the properties associated to the gas length and the liquid piston. To evaluate the approach performance LSR, the values were compared to those generated by wire-mesh sensor [15].

It is possible to see that the achieved measures through the LSR segmentation are quite accurate regarding the values collected from the correlation of Bendiksen and to the wire-mesh sensor.

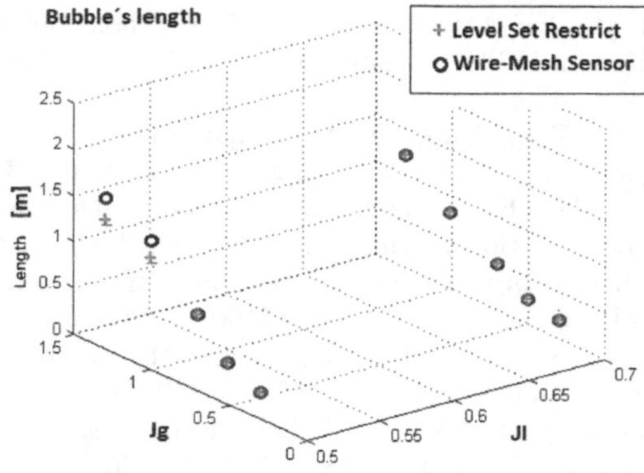

Fig. 6. Comparative graphic of the length of the bubble.

5 Conclusion

Data from the two-phase flow generated from the image segmentation contains high degree of detail and provide useful information on the flow phenomena. In this article, we investigated the use of two approaches to estimate the shape of elongated gas bubbles. The image segmentation techniques and image analysis by wire-mesh were applied to two sets of gas bubbles.

An image processing method, which can observe the interfacial information directly without disturb the flow while providing a high time-space resolution, has become an important measurement technique along with the development of the computer image processing technology.

We propose an approach for horizontal two-phase flow image segmentation based on Level Set methods and priori knowledge. The obtained results are very good, reaching precision and recall indexes higher than 98%.

We also compare these results against the ones obtained by the wellknown wire-mesh sensor. Preliminary results collaborate with the format proposition of the elongated bubble by [8] and suggest further research on other superficial velocities of gas bubbles.

References

1. Allaire, G., Dapogny, C., Frey, P.: Shape optimization with a level set based mesh evolution method. Comput. Methods Appl. Mech. Eng. **282**, 22–53 (2014)
2. do Amaral, C., Alves, R., da Silva, M., Arruda, L., Dorini, L., Morales, R., Pipa, D.: Image processing techniques for high-speed videometry in horizontal two-phase slug flows. Flow Measur. Instrum. **33**, 257–264 (2013)
3. Azzopardi, B.: Gas - Liquid Flows. Begell House, Danbury (2006)
4. Chen, F., Yu, H., Hu, R., Zeng, X.: Deep learning shape priors for object segmentation. In: The IEEE Conference on Computer Vision and Pattern Recognition (CVPR), June 2013

5. Li, C., Xu, C., Gui, C., Fox, M.: Level set evolution without re-initialization: a new variational formulation. IEEE Comput. Soc. Conf. Comput. Vis. Pattern Recogn. **1**, 430–436 (2005)
6. Cootes, T.F., Taylor, C.J., Cooper, D.H., Graham, J.: Active shape models-their training and application. Comput. Vis. Image Underst. **61**(1), 38–59 (1995)
7. Da Silva, M., Schleicher, E., Hampel, U.: Capacitance wire-mesh sensor for fast measurement of phase fraction distributions. Meas. Sci. Technol. **18**(7), 2245 (2007)
8. Fagundes Netto, J.R., Fabre, J., Paresson, L.: Shape of long bubbles in horizontal slug flow. Int. J. Multiphase Flow **25**, 1129–1160 (1999)
9. Bendiksen, K.H.: An experimental investigation of the motion of long bubbles in inclined tubes. In: Multiphase Flow, pp. 467–483 (1984)
10. Matamoros, L., Loureiro, J., Freire, A.S.: Length area volume of long bubbles in horizontal slug flow. Int. J. Multiph. Flow **65**, 24–30 (2014)
11. Mesadi, F., Cetin, M., Tasdizen, T.: Disjunctive normal parametric level set with application to image segmentation. IEEE Trans. Image Process. **26**(6), 2618–2631 (2017)
12. Ngo, T.A., Lu, Z., Carneiro, G.: Combining deep learning and level set for the automated segmentation of the left ventricle of the heart from cardiac cine magnetic resonance. Med. Image Anal. **35**, 159–171 (2017)
13. Peng, J., Zhang, Y., Liu, G., Jiang, J., Zhao, Y.: MRI segmentation of a prostate based on distance regularized level set evolution with a priori shape. Int. J. Hybrid Inf. Technol. **9**(11), 199–206 (2016)
14. Pipa, D.R., da Silva, M.J., Morales, R.E., Zibetti, M.V.: Typical bubble shape estimation in two-phase flow using inverse problem techniques. Flow Meas. Instrum. **40**, 64–73 (2014)
15. Santos, E.N., Pipa, D.R., Morales, R.E.M., da Silva, M.J.: Bubble shape estimation in gas-liquid slug flow using wire-mesh sensor and advanced data processing. In: 2014 IEEE International Conference on Imaging Systems and Techniques (IST) Proceedings, pp. 308–311, October 2014
16. Sethian, J.A.: Level Set Methods: Evolving Interfaces in Geometry, Fluid Mechanics, Computer Vision and Materials Sciences. [S.l.]: Cambridge University Press (1996)
17. Sguario, M., Arruda, L., Morales, R.: Segmentation of two-phase flow: a free representation for level set method with a priori knowledge. Comput. Methods Biomech. Biomed. Eng.: Imaging Vis. **4**(2), 98–111 (2016)
18. Wallis, G.B.: One-Dimensional Two-Phase Flow. McGraw-Hill, New York (1969)
19. Wang, B., Gao, X., Tao, D., Li, X., Balla-Arabé, S.: Salient object segmentation with a shape-constrained level set. In: Biomedical Image Segmentation: Advances and Trends, p. 371 (2016)
20. Yang, X., Zhan, S., Xie, D., Zhao, H., Kurihara, T.: Hierarchical prostate MRI segmentation via level set clustering with shape prior. Neurocomputing **257**(Suppl. C), 154–163 (2017). doi:10.1016/j.neucom.2016.12.071. http://www.sciencedirect.com/science/article/pii/S0925231217301467

Deep Face Model Compression Using Entropy-Based Filter Selection

Bingbing Han[1], Zhihong Zhang[1], Chuanyu Xu[1], Beizhan Wang[1(✉)], Guosheng Hu[2], Lu Bai[3], Qingqi Hong[1], and Edwin R. Hancock[4]

[1] Xiamen University, Xiamen, Fujian, China
wangbz@xmu.edu.cn
[2] Anyvision Group, Belfast, UK
[3] Central University of Finance and Economics, Beijing, China
[4] Department of Computer Science, University of York, York, UK

Abstract. The state-of-the-art face recognition systems are built on deep convolutional neural networks (CNNs). However, these CNNs contain millions of parameters, leading to the deployment difficulties on mobile and embedded devices. One solution is to reduce the size of the trained CNNs by model compression. In this work, we propose an entropy-based prune metric to reduce the size of intermediate activations so as to accelerate and compress CNN models both in training and inference stages. First the importance of each filter in each layer is evaluated by our entropy-based method. Then some unimportant filters are removed according to a predefined compressing rate. Finally, we fine-tune the pruned model to improve its discrimination ability. Experiments conducted on LFW face dataset shows the effectiveness of our entropy-based method. We achieve 1.92× compression and 1.88× speed-up on VGG-16 model, 2× compression and 1.74× speed-up on WebFace model, both with only about 1% accuracy decrease evaluated on LFW.

1 Introduction

In the past few years, as the emergence of big training data, Convolutional Neural Networks (CNNs) have attained great success in the field of computer vision, from image classification [1–4], to other applications such as image caption [5], super resolution [6] and others. To extract massive information from big training data, the researchers typically trained a large CNN or a CNN ensemble, which contains millions of parameters. Nevertheless, due to the need for mobile payment and security, it is very important to apply face recognition to mobile devices, computing large CNNs cost too much memory and running time, which prevents them from being widely used. Thus network compression has attracted great attention from researchers. In this paper, we focus on deep model compression in the field of face recognition.

In order to apply large CNNs to mobile devices, a large number of CNN methods have been put forward. The early ideas focused on how to compress

S. Battiato et al. (Eds.): ICIAP 2017, Part II, LNCS 10485, pp. 127–136, 2017.
https://doi.org/10.1007/978-3-319-68548-9_12

the parameters of fully-connected layers [7], and latter work which compressed convolutional layers [8,11] for the purpose of accelerating model speed becomes an important research field. However, very little research has been proposed to reduce the size of activation in each layer.

In this work, we apply model compression to face recognition. We propose a framework to remove redundant filters in order to accelerate and compress CNNs at the same time. Our main idea is that, in each layer, we should keep filters that are representative (i.e. these filters could extract as much information as original filters) and prune these less representative ones. As a result, we could reduce the number of channels directly without losing the information of feature map, converting a cumbersome network to a much smaller one without or with a little bit performance descend. We propose an entropy-based channel selection metric to evaluate the importance of each filter and prune 'weak' filters in order to keep some important filters that could capture nearly the same information as all the original filters. Then the pruned network is fine-tuned to regain its discrimination ability.

We evaluate our entropy-based channel selection metric for face recognition using two commonly used models: VGG-16 [2] and WebFace [19]. These two models are both trained and finetuned on the CASIA-WebFace dataset [16]. Our entropy-based prune metric achieve 1.92× compression and 1.88× speed-up on VGG-16 model, 2× compression and 1.74× speed-up on WebFace model, both with about 1% accuracy decrease.

Our strategy has the following advantages and contributions:

- We propose a simple yet efficient framework to compress CNN models in both training and inference stage. Our framework can reduce the number of filters so as to compress the size of activation in *each* layer, which catch almost no attention in previous work.
- We use an effective learning strategy to make a balance between training speed and classification accuracy in our pruning framework.
- Our framework does not rely on any specific libraries to gain the compression and acceleration performance, thus could be applied to any popular CNN library, such as caffe [15], tensorflow [17].

2 Related Works

Researchers have revealed that large CNNs suffered from over-parameterization, which causes not only the waste of memory and computation, but also serious over-fitting problem. Denil et al. [9] demonstrated that we can use only a small part of its original parameters to reconstruct a network almost without the accuracy dropping. However, very little work aims at optimizing the number of filters. Most previous works mainly focus on two fields: one focused on how to reconstruct the fully-connected layers, the other one focused on how to prune the weights of convolutional layers.

Focusing on the fully-connected layers, some researchers reconstructed layers or modules to substitute bottleneck components. GoogleNet [4] and ResNet [3] are famous examples which use the global average pooling to replace the dense fully-connected layers in order to reduce memory and computation consumption. Recently, SqueezeNet [10] uses a block called "Fire Module" and other strategies to achieve AlexNet [1] level accuracy with only 4.8 MB disk size, almost 50× fewer parameters than the original AlexNet size.

Focusing on the convolutional layers, different methods have been explored to reduce number of connections and weights in neural networks. Some researchers approximate the dense parameters matrix with several low-rank matrices in order to compute the matrix-vector with high-speed [7]. Other researchers focus on network pruning, which has widely been studied to reduce the number of connections and prevent over-fitting [12]. Li et al. [13] use the absolute weight to measure the importance of each filter and remove less useful filters, which has the similar idea with ours, but our method is more efficient.

Though the mentioned methods work well in compressing large CNNs, we look for a framework that has optimal number of filters in each layer for specified networks and given tasks.

3 Entropy-Based Model Compression

In this section, we detail the proposed entropy-based channel selection metric which performs important filter selection. Our main idea is to discard some unimportant filters, and to recover its performance via fine-tuning. Finally we describe our efficient learning strategy.

3.1 Framework

Figure 1 illustrates the framework of the proposed activation pruning method. For a specific layer that we want to prune (i.e. layer k), we just focus on the activation tensor, we use entropy-based channel selection metric to evaluate the importance of each filter, then some less important filters will be removed from the original model, which makes the architecture more compact. Clearly, the corresponding channels of filters in the next layer are removed too. This strategy not only makes the network with fewer parameters which lead to the decrease of running time and memory consumption, but also reduces the size of activations. In addition, some researchers have proved that each neuron is represented by many neurons, each neuron participates in the representation of many concepts [14], so whatever these filters are, they can extract some features from the feature maps of the former layer. Thus, generalization ability of the pruned model will be affected. We fine-tune the whole network after removing the less important filters in order to recovery the performance.

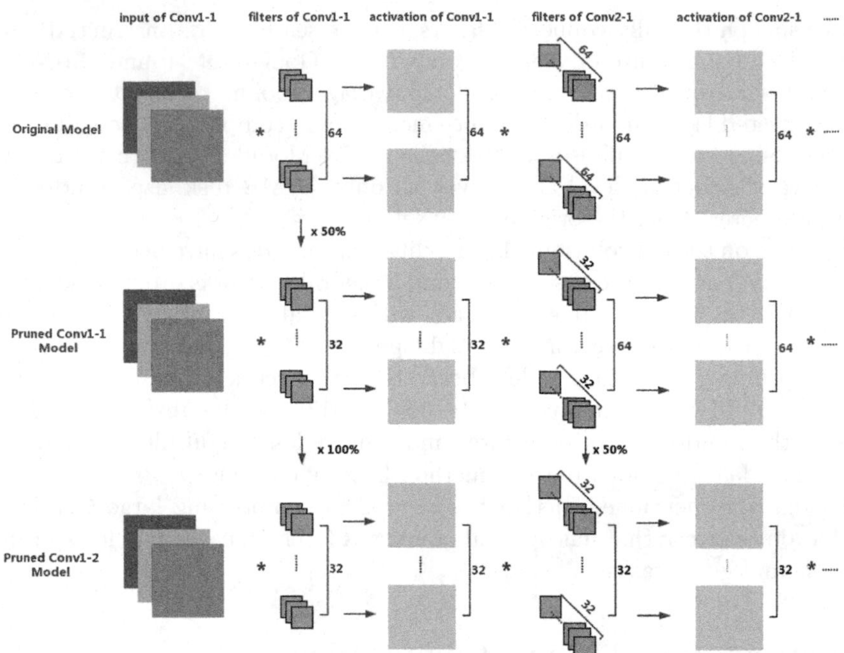

Fig. 1. Illustration of the framework of the iterative activation pruning approach on VGG-16 network. We predefine the compressing rate as 0.5, meaning that 50% filters in each layer will be removed. First, we compute the entropy of the activation of Conv1-1 and discard the unimportant filters. Then we fine-tune the pruned model with few iterations to recover the discrimination of pruned Conv1-1 model. Next we prune the Conv1-2 based on the former model in the same way. After the last layer being pruned, the final model will be fine-tuned carefully. $*$ is the convolution operator.

3.2 Entropy-Based Prune Metric

We use a triplet $(I_i, W_i, *)$ to denote the convolution in layer i, where $I_i \in \mathbb{R}^{(c \times h \times w)}$ means the input tensor, $W_i \in \mathbb{R}^{(d \times c \times k \times k)}$ means a set of filter weights, $*$ means the convolution operation, and d, c, $k \times k$ are the number of filters, the number of output channels on the upper layer and the size of the filters respectively. Our goal is to remove some unimportant filters, i.e. reduce d.

From the structure of networks, we know that each filter corresponds to a single channel of its activation tensor, so the ability of extracting features of each filter is closely related to its activation channel. In this paper, we propose an entropy-based metric to evaluate the importance of filters. Entropy is a commonly used metric to measure the disorder or uncertainty in information theory. A large entropy value means the system contains more information. In our filter pruning method, if a channel of activation tensor contains less information, its corresponding filter is less important, thus could be removed.

Specifically, we first calculate the mean value of each channel, converting a $c \times h \times k$ tensor into a $1 \times c$ vector. In this way, each channel of a $I_{(i+1)}$ (activation of layer i is also the input of layer $i+1$) has a corresponding value for one image. In order to calculate the entropy, more output values need to be collected, which can be obtained using an evaluation set. In practice, the evaluation set can simply be the original training set, or a subset of it. Finally, we get a matrix $M \in \mathbb{R}^{(n \times c)}$, where n is the number of images in the evaluation set, and c is the channel number. For each channel j, we compute its final score H_j according to the entropy of its output$M_{(:,j)}$:

$$H_j = E_p log \frac{1}{p(x)} = - \sum_{x \in M_{(:,j)}} p(x) log p(x)$$

Another important issue is to decide the pruning boundary. One possible method is to denote a specific value, all channels with score below this value are removed from the network. However the specific value is a hyperparameter, which is hard to be specified. Another more practical method is to predefine a compression rate, all the filters are sorted in the descending order corresponding to the entropy score, and only the top k filters are preserved. Of course, the corresponding channels are removed too.

3.3 Network Trimming

Our network trimmimg consists of three main steps. First the network is trained under conventional process and the number of filters in each layer is set empirically. Next, we run the network on a large validation dataset to obtain the entropy score of each filter, only the top k filters are preserved according to the compression rate. Of course, the connections to and from the removed filter are removed accordingly. After the filter pruning, the trimmed network is initialized using the weights before trimmimg. The trimmed network reduces some level of performance. So, at last we re-train the network to enhance the performance of the trimmed one.

4 Experiments

We use the standard caffe library [15] to train our deep models. The two CNN architectures we used are VGG-16 [2] model and WebFace model [19]. These two architectures achieve great succuss in the filed of face recognition. Our models are trained using the CASIA-WebFace dataset [16] consisting of 419922 face images of 10575 identities. The training images are horizontally flipped for data argumentation. The finetuning process during model compression is conducted on CASIA-WebFace dataset as well. The face recognition performance is evaluated on the famous LFW (Labeled Faces in the Wild) database [21]. LFW contains 5,749 subjects and 13,233 images. We follow the standard unrestricted protocol and conduct 10-fold cross validation. The face recognition rate is evaluated by mean accuracy. The sample images of LFW are shown in Fig. 2. All the experiments are conducted on a computer equipped with Nvidia Tesla K80 GPU.

Fig. 2. A column in the green box indicates the same person and the red box indicates different people. (Color figure online)

4.1 VGG-16 Network

The original VGG-16 model contains 13 convolution layers, 2 fully-connected layers and one softmax layer. As our training data CASIA-WebFace database is much smaller than ImageNet [20] which is used to train the original VGG-16 network, in this work, we remove the two fully-connected layers to avoid over-fitting. At the same time, the last convolutional layer is followed by a global average pooling layer whose kernel size is of 14×14. This type of global pooling can replace a fully connected layer but has much fewer parameters. It has widely been used to design an efficient network such as Google Inception network [4].

Before training, all the training images are aligned and then resized to 224×224. We achieve 97.55% face recognition rate on LFW using the full (not compressed) VGG-16 model. The next step is to prune the model based on our entropy-based prune metric. In this work, we set the compression rate as 0.5, meaning that half of the parameters of convolutional layers should be removed. We empirically find that the first few convolution layers have much more redundant information than the deeper layers which capture the semantic information. Therefore, we only prune the first 10 layers (Conv1-1 to Conv4-3 in Table 1). During finetuning/pruning, we set the original learning rate to 0.01 for each layer. The learning rate is reduced to 0.1 of the previous one when the loss stops decreasing. The number of iterations and the learning rate changing are detailed in Table 2.

In Table 1, the parameters before and after compression of VGG-16 network are detailed. The convolutional kernel weights are greatly reduced, leading to a more efficient inference. The computations measured by FLOPs are also dramatically reduced. Interestingly, our method halves the intermediate activation size which can greatly save the memory.

In Table 3, we report our compression results. We can see that our model gains $1.88\times$ speed up measured by the averaged running time of 1000 images during inference. Also, we achieve $1.92\times$ compression results in terms of the number of model parameters. Compared with original VGG-16 model, our method reduces the FLOPS computations by $3.29\times$. However, the face recognition rate only drops by around 1%.

Table 1. The performance of our method to reduce Parameters and FLOPs (FLoating-point OPerations) on the VGG-16 model. The activation size is the sum of convolutional, relu, pooling layers' output and the input data when batch size is set to 1.

Layer	Parameters		FLOPs		Intermediate Activation Size	
	Original	Pruned	Original	Pruned	Original	Pruned
Conv1-1	1.73 K	0.86 K	86.7 M	43.35 M	12.25 MB	6.125 MB
Conv1-2	36.86 K	9.22 K	1.85 B	462.42 M	12.25	6.125 MB
Conv2-1	73.73 K	18.43 K	0.92 B	231.21 M	6.13 MB	3.06 MB
Conv2-2	147.46 K	36.86 K	1.85 B	462.42 M	6.13 MB	3.06 MB
Conv3-1	294.91 K	73.73 K	0.92 B	231.21 M	3.06 MB	1.53 MB
Conv3-2	589.82 K	147.46 K	1.85 B	462.42 M	3.06 MB	1.53 MB
Conv3-3	589.82 K	147.46 K	1.85 B	462.42 M	3.06 MB	1.53 MB
Conv4-1	1.18 M	294.92 K	0.92 B	231.21 M	1.53 MB	0.77 MB
Conv4-2	2.36 M	589.82 K	1.85 B	462.42 M	1.53 MB	0.77 MB
Conv4-3	2.36 M	589.82vK	1.85 B	462.42 M	1.53 MB	0.77 MB
Conv5-1	2.36 M	1.18 M	462.42 M	231.21 M	392 KB	392 KB
Conv5-2	2.36 M	2.36 M	462.42 M	462.42 M	392 KB	392 KB
Conv5-3	2.36 M	2.36 M	462.42 M	462.42 M	392 KB	392 KB
Total	14.71 M	7.66 M	15.35 B	4.67B	109.89 MB	56.33 MB

Table 2. The number of iterations on finetuning/pruning each layer. For {10000, 20000} of Conv1-2, it means the learning rate changes to 0.1 of previous one at 10000th iteration, and the training stops at 20000th iteration.

Layer	Conv1-1	Conv1-2	Conv2-1	Conv2-2	Conv3-1
Iterations	-	10000	-	10000	-
	10000	20000	10000	20000	10000
Layer	Conv3-2	Conv3-3	Conv4-1	Conv4-2	Conv4-3
Iterations	-	10000	-	-	40000/80000
	10000	20000	10000	10000	12000

Table 3. The summarization of the performance of our compression method.

Model	Accuracy (LFW)	FLOPS	Compression	Speed-up
Original VGG-16	97.55%	1×	1×	1×
Pruned VGG-16	96.50%	3.29×	1.92×	1.88×

4.2 WebFace Network

The WebFace [19] is another popular face recognition CNN architecture. Compared with VGG-16, WebFace is more compact. We conduct our compression method on WebFace to verify the effectiveness of our method on smaller architecture. The architecture of WebFace is detailed in Fig. 3. Clearly, WebFace also stackes several Conv-Pool-Relu unites. Before fully-connected layers, a 7×7 global pooling is used. Following the original work [19], the input images are aligned and cropped to the size of 100×100. The training and finetuning strategies are similar to those shown in Sect. 4.1.

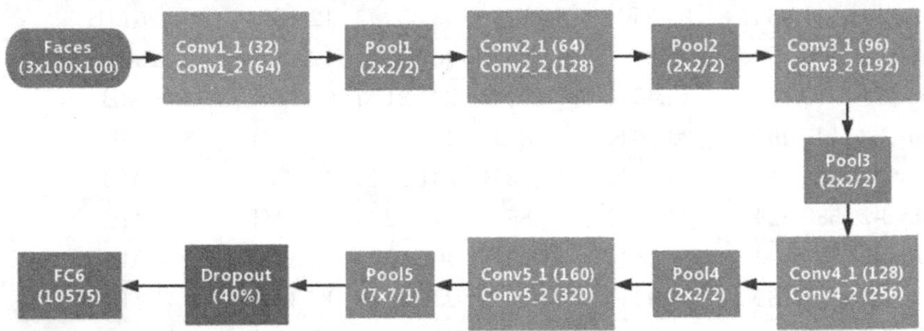

Fig. 3. The framework of Webface.

In Table 4, we report our compression results using WebFace network. Unlike VGG-16 network, we evaluate the compression results on different compression rates: 0.9 (Pruned-90 in Table 4), 0.75 and 0.5. The results in terms of face recognition rates (LFW), FLOPS, compression and running time are detailed. For the largest compression rate 0.5 (Pruned-50), the accuracy drops only 1.05%, but the FLOPS of Pruned-50 is only 28.07% of the original WebFace, the model size is only half, and the running time is only 57.60% compared with the original model. Another 2 models (Pruned-75, Pruned-90) also achieve great performance. In the real application, it is always a trade-off between accuracy and model size/computation. Therefore, Table 4 is an important reference for researchers and engineers to look for such a trade-off.

Table 4. WebFace compression results on different compression rates.

Model	Accuracy (LFW)	FLOPS		Compression		Speed-up	
		Nums	Percent	Nums	Percent	Nums	Percent
Original WebFace	96.92%	770 M	100%	1.75 M	100%	5.92 ms	100%
Pruned-90	97.28%	619 M	80.37%	1.53 M	87.53%	5.24 ms	88.51%
Pruned-75	96.62%	448 M	58.18%	1.26 M	71.71%	4.60 ms	77.70%
Pruned-50	95.87%	216 M	28.07%	0.88 M	50%	3.16 ms	57.60%

4.3 Effectiveness Analysis

To demonstrate the effectiveness of our entropy-based metric, we compare our method with random selection which prunes the CNN model parameters randomly. We prune 3 different layers of WebFace with these two methods. All the experimental setups are kept the same, except for channel selection method.

The results are summarized in Table 5. The 3 layers of WebFace we choose to prune are: Conv1-1, Conv3-1, and Conv4-2 (shown in Fig. 3). Each layer is pruned independently with 0.5 compression rate, and fine-tuned 10000 iterations. As can been seen, our entropy-based method consistently works better than random selection method on all the 3 selected layers, showing the effectiveness of our compression method. The advantage of our method is larger on deeper layers, e.g. 96.48% vs 96.01% (Conv1-1), 95.97% vs 95.02% (Conv4-2). In addition, the finetuing is very important for performance improvement for both our method and random selection.

Table 5. Comparison of entropy-based channel selection metric and random selection. 0 iteration means finetuning is not conducted after pruning.

Model	Conv1-1		Conv3-1		Conv4-2	
	0 Iteration	10 K Iterations	0 Iteration	10 K Iterations	0 Iteration	10 K Iterations
Entropy	94.48%	96.48%	93.00%	96.90%	92.30%	95.97%
Random	94.27%	96.01%	92.55%	96.32%	92.27%	95.02%

5 Conclusion

In this paper, we propose a network pruning strategy to trim redundant filters based on the entropy-based channel selection metric. Our method can remove unimportant filters that provide little contribution to the final performance without damaging results of our network. In addition, our strategy does not rely on any dedicated library, thus can widely be used in various applications with current deep learning libraries.

Acknowledgment. This work is supported by National Natural Science Foundation of China (Grant No.61402389, 61503422 and 61502402), the Fundamental Research Funds for the Central Universities in China (no. 20720160073) and the Natural Science Foundation of Fujian Province of China(Grant No. 2015J05129).

References

1. Krizhevsky, A., Sutskever, I., Hinton, G.E.: Imagenet classification with deep convolutional neural networks. In: Advances in Neural Information Processing Systems (NIPS), pp. 1097–1105 (2012)
2. Simonyan, K., Zisserman, A.: Very deep convolutional networks for large-scale image recognition. In: International Conference on Learning Representations (ICLR) (2015)
3. He, K., Zhang, X., Ren, S., Sun, J.: Deep residual learning for image recognition. In: IEEE Conference on Computer Vision and Pattern Recognition (CVPR), pp. 770–778 (2016)
4. Szegedy, C., Liu, W., Jia, Y., Sermanet, P., Reed, S., Anguelov, D., Erhan, D., Vanhoucke, V., Rabinovich, A.: Going deeper with convolutions. In: IEEE Conference on Computer Vision and Pattern Recognition (CVPR), pp. 1–9 (2015)

5. Fang, H., Gupta, S., Iandola, F., et al.: From captions to visual concepts and back. In: The IEEE Conference on Computer Vision and Pattern Recognition (CVPR), pp. 1473–1482 (2015)
6. Dong, C., Loy, C.C., He, K., Tang, X.: Learning a deep convolutional network for image super-resolution. In: Fleet, D., Pajdla, T., Schiele, B., Tuytelaars, T. (eds.) ECCV 2014. LNCS, vol. 8692, pp. 184–199. Springer, Cham (2014). doi:10.1007/978-3-319-10593-2_13
7. Sindhwani, V., Sainath, T., Kumar, S.: Structured transforms for small footprint deep learning. In: Advances in Neural Information Processing Systems (NIPS), pp. 3088–3096 (2015)
8. Wu, J., Leng, C., Wang, Y., Hu, Q., Cheng, J.: Quantized convolutional neural networks for mobile devices. In: IEEE Conference on Computer Vision and Pattern Recognition (CVPR), pp. 4820–4828 (2016)
9. Denil, M., Shakibi, B., Dinh, L., de Freitas, N.: Predicting parameters in deep learning. In: Advances in Neural Information Processing Systems (NIPS), pp. 2148–2156 (2013)
10. Iandola, F.N., Moskewicz, M.W., Ashraf, K., Han, S., Dally, W.J., Keutzer, K.: SqueezeNet: AlexNet-level accuracy with 50× fewer parameters and 0.5 MB model size. arXiv preprint arXiv:1602.07360 (2016)
11. Luo, J.-H., Wu, J.: An entropy-based pruning method for CNN compression. arXiv:1706.05791 (2017)
12. Han, S., Pool, J., Tran, J., Dally, W.: Learning both weights and connections for efficient neural network. In: Advances in Neural Information Processing Systems (NIPS), pp. 1135–1143 (2015)
13. Li, H., Kadav, A., Durdanovic, I., Samet, H., Graf, H.P.: Pruning filters for efficient ConvNets. arXiv preprint arXiv:1608.08710
14. Bengio, Y., Courville, A., Vincent, P.: Representation learning: a review and new perspectives. IEEE Trans. Pattern Anal. Mach. Intell. (TPAMI), **35**(8), 1798–1828 (2013)
15. Jia, Y., Shelhamer, E., Donahue, J., Karayev, S., Long, J., Girshick, R., Guadarrama, S., Darrell, T.: Learning distributed representations of concepts. In: ACM International Conference on Multimedia (ACM MM), pp. 675–678 (2014)
16. CASIA-WebFace dataset: http://www.cbsr.ia.ac.cn/english/CASIA-WebFace-Database.html
17. Abadi, M., Agarwal, A., Barham, P., et al.; Tensorflow: large-scale machine learning on heterogeneous distributed systems. arXiv preprint arXiv:1603.04467 (2016)
18. Lin, M., Chen, Q., Yan, S.: Network in Network. CoRR abs/1312.4400 (2013)
19. Yi, D., Lei, Z., Liao, S., Li, S.Z.: Learning face representation from scratch. arXiv:1411.7923
20. Deng, J., Dong, W., Socher, R., Li, L.-J., Li, K., Fei-Fei, L.: ImageNet: a large-scale hierarchical image database. In: CVPR 2009 (2009)
21. LFW: http://vis-www.cs.umass.edu/lfw/

Deep Passenger State Monitoring Using Viewpoint Warping

Ian Tu[1(✉)], Abhir Bhalerao[1], Nathan Griffiths[1], Mauricio Delgado[2], Thomas Popham[2], and Alex Mouzakitis[2]

[1] Department of Computer Science, University of Warwick, Coventry, UK
{i.tu,abhir.bhalerao,nathan.griffiths}@warwick.ac.uk
[2] Jaguar Land Rover, Engineering Centre, Coventry, UK
{mdelgado,tpopham,amouzak1}@jaguarlandrover.com

Abstract. The advent of autonomous and semi-autonomous vehicles has meant passengers now play a more significant role in the safety and comfort of vehicle journeys. In this paper, we propose a deep learning method to monitor and classify passenger state with camera data. The training of a convolutional neural network is supplemented by data captured from vehicle occupants in different seats and from different viewpoints. Existing driver data or data from one vehicle is augmented by viewpoint warping using planar homography, which does not require knowledge of the source camera parameters, and overcomes the need to re-train the model with large amounts of additional data. To analyse the performance of our approach, data is collected on occupants in two different vehicles, from different viewpoints inside the vehicle. We show that the inclusion of the additional training data and augmentation by homography increases the average passenger state classification rate by 11.1%. We conclude by proposing how occupant state may be used holistically for activity recognition and intention prediction for intelligent vehicle features.

Keywords: Passenger state monitoring · Camera homography · Convolutional neural networks · Classification · Deep learning

1 Introduction

With the advent of autonomous and semi-autonomous vehicles, and smart vehicles systems, all the occupants inside a vehicle have become relevant to safety and comfort. In recent years, the focus has been mainly on monitoring the driver, for example whether they are alert [16], or paying attention to the road [20]. In self-driving vehicles, the type of behaviour and actions being monitored are not limited to safe driving and it is important for the vehicle to sense if the driver is in a state to regain control if required, so called hand-over from autonomous driving to driver control. Knowing the state and behaviour of the occupants, including the driver, is also useful for optimising the in-vehicle experience, which involves monitoring the state of all occupants. Many car manufacturers, such as

S. Battiato et al. (Eds.): ICIAP 2017, Part II, LNCS 10485, pp. 137–148, 2017.
https://doi.org/10.1007/978-3-319-68548-9_13

Jaguar Land Rover, are developing and building new intelligent vehicle systems (e.g. ADAS [9]) for semi-autonomous and connected vehicles, and being able to monitor and predict occupant state is a vital parameter in designing, optimising and adapting a car's intelligent vehicle systems so as to maximise the safety and comfort of a journey [6].

A way to observe and analyse the actions and behaviours of vehicle occupants is to use inward facing cameras, which have the advantage that they are relatively cheap and general purpose sensors. Computer vision-based methods are proposed where image features are detected as proxies of driver state. The signals captured from imagery have been shown to be robust in identifying driver fatigue and distraction, for example see [2,10].

Deep learning using Convolutions Neural Networks (CNNs) classification methods have been demonstrated to work effectively for a variety of visual object classification problems, provided there is sufficient training data available, an appropriate deep architecture can be realised that generalises well to unseen data, e.g. [13]. CNNs have been shown to be more effective and have less overfitting when the size of the training data is increased [3,17]. One way to increase the amount of training data is to augment the data, data augmentation is the process of transforming the data without altering the data's labels - a common practice in visual-based problems is to apply geometric transformations such as rotations, scaling, flips, etc. [13].

Recent related works involving driver behaviour include a paper by Yan et al. [23], using the Southeast University Driving-posture Dataset (SEU dataset) by [25] designed a CNN model to identify 6 driver actions: calling, eating, braking, wheel use, phone use, and smoking. They used 2 inputs to the CNN, a primary input and a secondary input; the primary input was a bounding box around the whole driver in the image, and the secondary input was a set of skin regions. Their method was successful in determining the correct action, even when two actions were very similar, achieving a mean average precision of 97.8%. A CNN trained on 4 different driver postures, driving, answering phone call, eating and smoking, with an overall accuracy of 99.8% was used in [21]. For driver gaze detection, in the main step determines which of the 9 different gaze zones the driver was looking at. This method achieves an average of 95% accuracy [4]. Another method, [22] uses a CNN model to predict driver fatigue and distraction from the locating the eye, ears and mouth, and achieves overall accuracy of 95.6% in classifying the six states: eyes open/closed, mouth normal/eating, and ear normal/on phone.

Though actions are usually associated with movement and so video is required, it transpires that many simple actions can be identified from only still imagery. Deep learning utilising CNNs for action recognition has proved effective at image classification and object detection which rely heavily on image features [7,11,18]. An action can be determined from contextual cues, such as a person's pose or the presence of an object. For example, Gkioxari's et al. [7] action recognition method exploits this by marking bounding boxes around the subject and bounding boxes around relevant contextual cues and then using these as inputs to a CNN, in a manner similar to Yan's driver monitoring method [23].

Because passenger state monitoring and action detection is still in its infancy and focused on the driver, there is limited data available depicting passenger state, and much of this data relevant for driver behaviour, and action classification pertinent to vehicle safety. In this paper, we: (1) estimate projective transformations (planar homography) [8] to compute approximate viewpoint corrections to re-use driver monitoring data for passenger state classification; (2) widen the set of states detected to those of interest to semi-autonomous and autonomous vehicles; (3) augment the data set to generalise for small camera motions, including homography, and; (4) train and analyse the performance of a CNN for passenger state classification. We demonstrate that the approach has a number of benefits: it enables the large amounts of existing driver state monitoring data to be re-used for occupant monitoring in general so explicit changes in camera positions can be made without the need to recapture passenger data. It also gives flexibility to the transfer the learning model between vehicles; by augmenting the data set with randomised projective transformations, the learnt model also becomes more robust to small view-point changes and generalises better.

2 Method

The proposed method has two stages: image alignment and classification using a convolutional neural network. The image alignment stage chooses a single viewpoint to which all the other images are mapped and this mapping is calculated by marking corresponding points from two example images. The output of this stage is a view-transformed dataset where all images are approximately from the same viewpoint. The second stage consist of re-training a partially-trained CNN model, augmenting the training samples with small viewpoint variations, regressed to labelled output state, see Fig. 1.

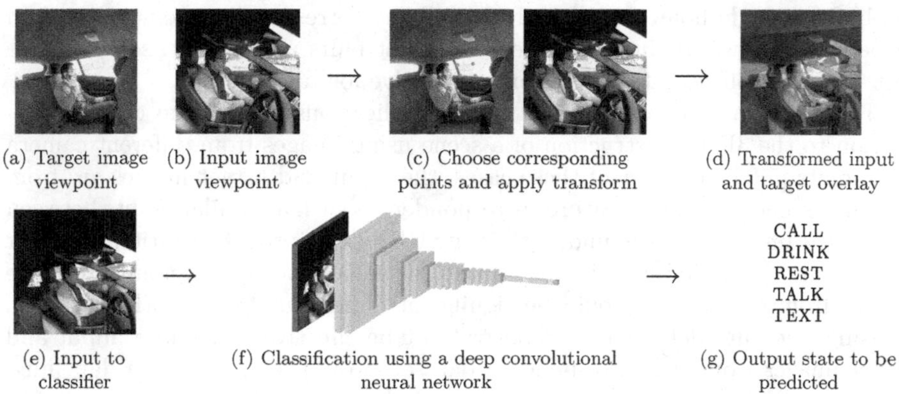

(a) Target image viewpoint (b) Input image viewpoint (c) Choose corresponding points and apply transform (d) Transformed input and target overlay

(e) Input to classifier (f) Classification using a deep convolutional neural network (g) Output state to be predicted

Fig. 1. The passenger state classification pipeline: (a), (b) A target viewpoint is chosen and all remaining image data will be aligned to that viewpoint using planar homography, (c); (d) Comparison of target and test viewpoints; (e)–(g) Deep state classification using a trained CNN.

2.1 Image Alignment Using Homography

A homography is a projective transformation from one plane to another and can be defined as the algebraic linear mapping $h : \mathbb{R}^2 \mapsto \mathbb{R}^2$ is a homography if and only if there exist a non-singular 3×3 matrix H such that for any point in $\mathbb{R}^2$, represented by a homogeneous coordinate x, $h(\mathrm{x}) = \mathrm{Hx}$ [8]. We can express the mapping of a 2D point (x, y) in homogeneous coordinates as a vector $\mathrm{x} = (x, y, 1)^T$, and likewise a target 2D point (u, v) as $\mathrm{u} = (u, v, 1)^T$, through a homography matrix $\mathbf{h} = (h_{ij})$:

$$\begin{pmatrix} u \\ v \\ 1 \end{pmatrix} \sim \begin{pmatrix} h_{11} & h_{12} & h_{13} \\ h_{21} & h_{22} & h_{23} \\ h_{31} & h_{32} & h_{33} \end{pmatrix} \begin{pmatrix} x \\ y \\ 1 \end{pmatrix} \tag{1}$$

This mapping can be solved using the Direct Linear Transform (DLT) algorithm. The homography matrix $\mathbf{h}$ is scale-invariant, meaning multiplying by a non-zero constant will not change the equations needed to be solved, so $\mathbf{h}$ is a homogeneous matrix representing only 8 degrees of freedom, therefore, there are only 8 unknowns to solve for. As a result, 4 pairs of (non-colinear) points are required, with each pair of source and target points providing two equations:

$$\begin{aligned} xh_{11} + yh_{12} + h_{13} - xuh_{31} - yuh_{32} - uh_{33} = 0 \\ yh_{21} + yh_{22} + h_{23} - xvh_{31} - yvh_{32} - vh_{33} = 0 \end{aligned} \tag{2}$$

The homographic matrix $\mathbf{h}$ can be found by solving, $A_i\mathbf{h} = \mathbf{0}$, with SVD, where

$$A_i = \begin{pmatrix} x_i & y_i & 1 & 0 & 0 & 0 & -u_ix_i & -u_iy_i & -u_i \\ 0 & 0 & 0 & x_i & y_i & 1 & -v_ix_i & -v_iy_i & -v_i \end{pmatrix}. \tag{3}$$

More accurate estimates of $\mathbf{h}$ are obtained with more that 4 point pairs, though to obtain a single homogeneous solution all the corresponding point pairs need to be exact. However, in most cases, the point pairs are inexact, so a suitable cost function will need to be minimised to solve for $\mathbf{h}$ [8].

The use of homography in real world applications range from camera calibration to the 3D reconstruction of a scene using images from different camera viewpoints [5,24]. In most of these cases, the input and target images are from the same scene, so similar feature correspondences such as similar points between images the images can be found, e.g. using SURF descriptors [1] to stitch together a panorama [12]. However, in the context of vehicle occupant monitoring, the images from one dataset could be significantly different from another dataset, so using a feature detection method will not be effective. Therefore, input and target images from the corresponding datasets would require manual labelling. Only a single input and target image needs be chosen, under the assumption that for each dataset the images are from a fixed or similar camera viewpoints, otherwise further homography matrices are needed from the additional images with different camera viewpoints. Using the homography matrix calculated from the input and target images, all the images from corresponding input dataset are transformed to be similar to the target dataset image's viewpoint. The resulting datasets are then be used for training the CNN model, Fig. 1a to e shows results of the image alignment process on two different datasets.

2.2 Synthesising Viewpoint Changes

For data augmentation, we apply image warping to our training data set by randomising using homography given knowledge of the intrinsic camera matrix of the target viewpoint. The camera projection (without lens distortion) is modelled as by the perfect pin-hole camera geometry such that 3D world coordinate points $\mathbf{X}$ project to the camera plane as $\mathbf{x}$ through the product of the intrinsic and extrinsic camera matrices, K and $(R|T)$,

$$\begin{pmatrix}\mathbf{x}\\1\end{pmatrix} \sim K \begin{pmatrix}R & T\\ \mathbf{0}^T & 1\end{pmatrix}\begin{pmatrix}\mathbf{X}\\1\end{pmatrix} \tag{4}$$

To synthesise small viewpoint changes around the principal axis of the camera i.e. $T = \mathbf{0}$, we induce random motions of the principal axis and rotations around this axis. Letting the quaternion, Q, represent the spatial rotation of θ about the principal axis, $(a_x, a_y, 1)^T$, the (z-axis) in camera coordinates:

$$Q = (a_x sin(\theta/2), a_y sin(\theta/2), sin(\theta/2), cos(\theta/2)) \tag{5}$$

then, random motions of the axis and random rotations can be created by drawing normal random samples $a_x, a_y \sim \mathcal{N}(0, \sigma_{xy})$ and $\theta \sim \mathcal{N}(0, \sigma_\theta)$. After normalisation, the rotations $Q(\sigma_{xy}, \sigma_\theta)$ are then converted to the matrices R and substituted into Eq. 4 to generate the 3×3 random homography matrix, $H = K(R|0)$.

2.3 Passenger State Classification

Convolutional neural networks (CNNs) are machine learning models which use multiple or deep neural network layer, combining a number of operational layers. They can be trained to learn regressions of pixel values from images and used as supervised classifiers to learn object classes [14]. Below, we detail the CNN network architecture and training regime.

Network Architecture. The CNN model architecture is based on the commonly used VGG19 architecture [19] as it has been shown generalise well on a wide range of datasets. The input to the network was a 224×224 RGB image, with the output being one the 5 states: calling on phone, drinking, resting, talking and texting. The network architecture used is outlined in Fig. 2. To prevent overfitting, dropout with $p = 0.5$ was applied to the fully connected layers.

Transfer Learning. The passenger state classification CNN was pre-trained on the ILSVRC 2012 dataset (ImageNet) is retrained to work on our occupant state datasets. In order to incorporate the model for our own use, the fully connected

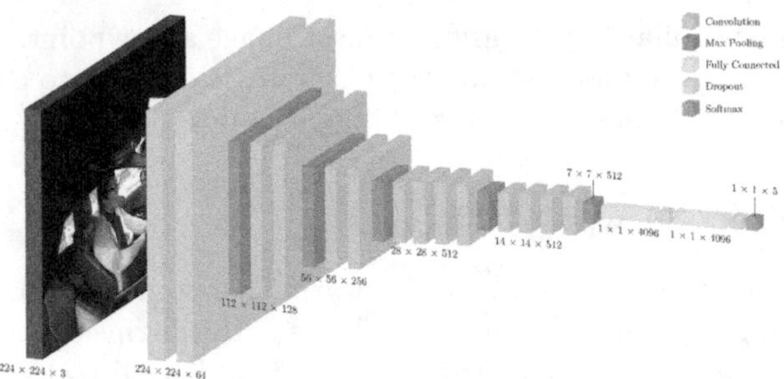

Fig. 2. Modified VGG19 network architecture.

layer weights were discarded in the pre-trained model, and randomly initialised weights were used for these layers. The number of epochs (iteration over entire dataset) ranged from 10–100. A smaller learning rate was used for the new model's weights thus effectively fine-tuning the results. The learning rate was set initially to 1e−3 and decreased accordingly after 5–10 epochs, weight decay was set to 1e−6, and was trained in batches of 16 using stochastic gradient descent (SGD) [15] with momentum 0.9.

3 Experiments and Results

3.1 Dataset

We conducted experiments to create two different datasets, one dataset captured in a Land Rover SUV and another BMW hatchback. Participants were asked to conduct typical in-vehicle actions whilst being filmed; these actions included mobile phone use, eating, drinking, sleeping, and talking. At the end of data collection, the video data was converted into 1920×1080 images, and each image was labelled with a state. States where there was not enough data were excluded from the final dataset. The states that are included are the following:

- **Call**: The occupant has their mobile phone up to their face for a phone call.
- **Drink**: The occupant is drinking or in the motion of drinking.
- **Rest**: The occupant is not engaging in any notable activity, this includes sleeping.
- **Talk**: The occupant is actively engaging in a conversation with another occupant.
- **Text**: The occupant is looking at their phone and actively using their phone.

The amount of data and the distribution of data is as following:

- **Land Rover SUV** dataset. This dataset contains 7 people enacting 5 states. There are 36151 images, from 3 different viewpoints, the approximate distribution of images to class is 30/5/20/15/30 respectively for call/drink/rest/talk/text.

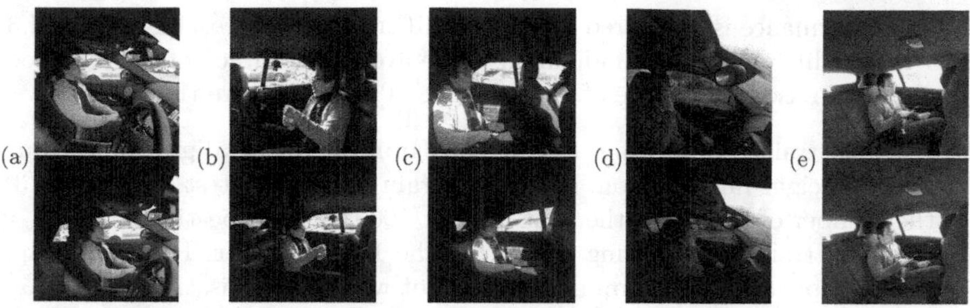

Fig. 3. The upper row shows the original images with the 4 different viewpoints from the camera situated at: (a) Front seat right in the Land Rover (rest state); (b) Back seat left in the Land Rover (drink state); (c) Back seat right in the Land Rover (talk state); (d) Front seat right in the BMW (call state); (e) Back seat right in the BMW (text state). The bottom row shows the transformed images, (e) is the reference viewpoint.

- **BMW Hatchback** dataset. This dataset contains 8 people enacting 5 states. There are 30340 images, from 2 different viewpoints, the approximate distribution of images to class is 12/8/30/20/30 respectively for call/drink/rest/talk/text.

Figure 3 shows example images from these datasets and their viewpoints.

3.2 Evaluation

For each dataset, for each viewpoint, we split the images into the following subsets: 80% of images for training and 20% of images for testing. The training data is further split into 80% for training and 20% for validation. The data is split according to individual, so for example, training using the Land Rover data will mean 4 individuals are used for training, 1 for validation, and 3 for testing. There is no overlap in individuals for the training, validation, and testing splits. There is no overlap in individuals for splits between vehicles. The images are randomly picked from each individual and each class for training, validation, and testing.

We combine the two datasets into one using the proposed homography alignment. We choose an image from the training set of the BMW back right seat camera viewpoint as the target image, then all other images not from that viewpoint are mapped to that specified target image using the proposed homography method. The RGB images are then cropped into a $1:1$ ratio, by equal clipping of the sides, and resized to a resolution of 224×224. The proposed CNN models are then trained, validated and tested with a different combination of viewpoints and datasets, and using different weight initialisations. No data augmentation is used apart from the horizontal flip required to map the back seat left viewpoint to back seat right viewpoint, and also the proposed homography alignment.

The performance is compared amongst 9 different models, each is repeated 3 times with a different split of individuals. The average accuracy of the repeated models will be used as a guide of performance. The models are the following:

(A) A model trained only using **BMW**. **No homography alignment**. **ImageNet** weight initialisation. Individual train/validation/test split is 4/1/3, the number of images in the split is 4000/1000/1250 respectively.
(B) A model trained only using data from the **Land Rover**. Horizontal flip. **No homography alignment**. **ImageNet** weight initialisation. Individual train/validation/test split is 3/1/3, the number of images in the split is 6000/1500/1875.
(C) A model trained only using data from the **Land Rover**. Horizontal flip. **No homography alignment**. **Model A** weight initialisation. Individual train/validation/test split is 3/1/3, the number of images in the split is 6000/1500/1875.
(D) A model trained only using data from the **BMW**. **Homography alignment**. **ImageNet** weight initialisation. Individual train/validation/test split is 4/1/3, the number of images in the split is 4000/1000/1250.
(E) A model trained only using data from the **Land Rover**. Horizontal flip. **Homography alignment**. **ImageNet** weight initialisation. Individual train/validation/test split is 3/1/3, the number of images in the split is 6000/1500/1875.
(F) A model trained only using data from the **Land Rover**. Horizontal flip. **Homography alignment**. **Model D** weight initialisation. Individual train/validation/test split is 3/1/3, the number of images in the split is 6000/1500/1875.
(G) A model trained using data from the **BMW** and **Land Rover**, but only using **front seat** viewpoint images. **Homography alignment**. **ImageNet** weight initialisation. Individual train/validation/test split is 10/2/3, the number of images in the split is 4000/1000/1250.
(H) A model trained using data from the **BMW** and **Land Rover**, but only using back seat viewpoint images. Horizontal flip. **Homography alignment**. **ImageNet** weight initialisation. Individual train/validation/test split is 10/2/3, the number of images in the split is 6000/1500/1875.
(I) A model trained using data from the **BMW** and **Land Rover**, but only using **back seat** viewpoint images. Horizontal flip. **Homography alignment**. **Model G** weight initialisation. Individual train/validation/test split is 10/2/3, the number of images in the split is 6000/1500/1875.

3.3 Results

Table 1 shows the accuracy results in confusion matrix form for every model. For the models with no homography alignment applied, the performance is poor; models A (Table 1a), B (Table 1b) and C (Table 1c) score under 60% for overall average accuracy. The models particularly struggle to classify the rest and talk states, often wrongly predicting between the two. This a persistent problem for

Table 1. Confusion matrices for evaluation models: (a)–(c) trained on non-aligned data; (d)–(f) trained with viewpoint-aligned data; (g)–(i) front seat and back seat state classification (also aligned data). See main text for details.

(a) Model A (**58.8%**)

Call	Drink	Rest	Talk	Text
0.49	0.32	0.00	0.19	0.00
0.01	0.73	0.03	0.10	0.13
0.00	0.06	0.32	0.54	0.08
0.02	0.07	0.32	0.51	0.09
0.00	0.06	0.00	0.04	0.89

(b) Model B (**52.0%**)

Call	Drink	Rest	Talk	Text
0.27	0.22	0.17	0.25	0.08
0.00	0.71	0.12	0.04	0.13
0.02	0.10	0.44	0.36	0.07
0.02	0.09	0.22	0.59	0.07
0.00	0.16	0.18	0.07	0.59

(c) Model C (**57.2%**)

Call	Drink	Rest	Talk	Text
0.36	0.15	0.09	0.32	0.08
0.01	0.72	0.10	0.12	0.05
0.02	0.11	0.55	0.28	0.04
0.01	0.09	0.12	0.75	0.03
0.00	0.19	0.16	0.16	0.49

(d) Model D (**64.1%**)

Call	Drink	Rest	Talk	Text
0.77	0.05	0.03	0.10	0.05
0.14	0.81	0.00	0.04	0.01
0.02	0.12	0.46	0.31	0.08
0.09	0.05	0.29	0.30	0.26
0.00	0.10	0.01	0.03	0.86

(e) Model E (**64.2%**)

Call	Drink	Rest	Talk	Text
0.65	0.10	0.19	0.05	0.01
0.01	0.83	0.08	0.05	0.03
0.04	0.04	0.76	0.11	0.05
0.01	0.07	0.49	0.42	0.02
0.00	0.20	0.19	0.05	0.56

(f) Model F (**75.3%**)

Call	Drink	Rest	Talk	Text
0.73	0.01	0.15	0.10	0.01
0.02	0.75	0.14	0.06	0.03
0.02	0.05	0.79	0.08	0.06
0.02	0.02	0.22	0.68	0.06
0.00	0.04	0.10	0.05	0.82

(g) Model G (**65.5%**)

Call	Drink	Rest	Talk	Text
0.59	0.32	0.00	0.06	0.03
0.03	0.85	0.08	0.02	0.03
0.02	0.05	0.55	0.29	0.10
0.00	0.01	0.39	0.41	0.19
0.00	0.08	0.02	0.03	0.88

(h) Model H (**68.6%**)

Call	Drink	Rest	Talk	Text
0.71	0.00	0.19	0.09	0.01
0.03	0.66	0.11	0.09	0.11
0.02	0.05	0.70	0.13	0.11
0.02	0.02	0.28	0.61	0.07
0.00	0.08	0.11	0.07	0.74

(i) Model I (**75.3%**)

Call	Drink	Rest	Talk	Text
0.89	0.01	0.07	0.03	0.01
0.07	0.73	0.08	0.04	0.08
0.02	0.03	0.71	0.13	0.11
0.03	0.02	0.22	0.65	0.08
0.00	0.08	0.11	0.04	0.77

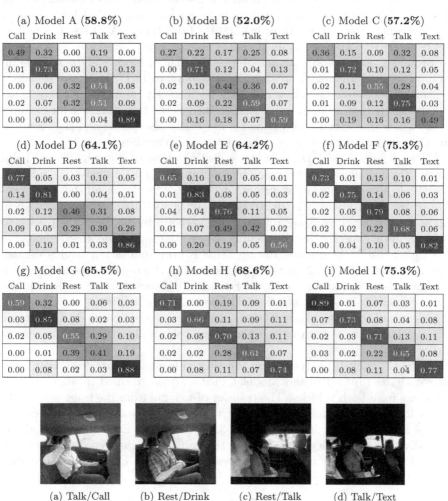

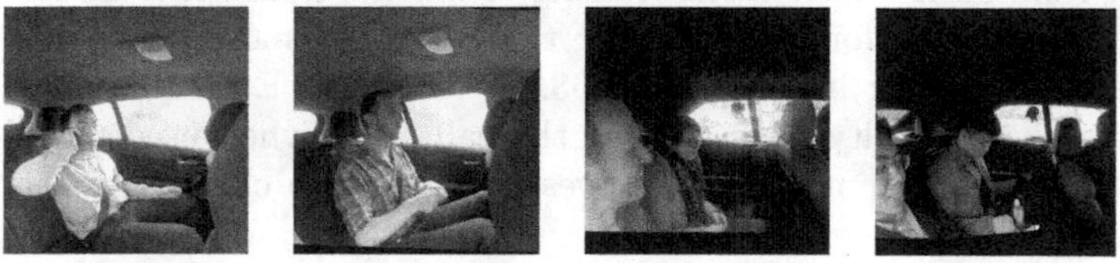

(a) Talk/Call (b) Rest/Drink (c) Rest/Talk (d) Talk/Text

Fig. 4. Misclassified images. Predicted label/true label.

all of the models. Figure 4c shows that the models find it difficult to distinguish between open and closed mouth states for certain individuals because there are many similarities between these two states, and given a small image it may not be possible to classify accurately.

The significance of using more data can be seen in the results for Model C; when the weights are used from Model A to help train Model C, there is an improvement of 5.2%, increasing the accuracy from 52.0% to 57.2%. The results for Model C in Table 1c, shows a large increase in accuracy for the talking class, up from 59% to 75%, albeit at the cost of more classes being misclassified talking - the texting class notably suffers the most from this. All other classes, except for the texting state, show a minor improvement from using Model A's weights.

Models D (Table 1d), E (Table 1e) and F (Table 1f) are A, B and C's respective counterparts and are trained using aligned data benefit from applying the proposed homography alignment process. These models show a substantial improvement, ranging from an increase of 5.3% to 18.1% in overall accuracy. The call state accuracy for these models is significantly improved. Although, the call labelled images are mistakenly misclassified as the talk state, see for example Fig. 4a where the presence of a phone is not easily discernible and the mouth is a much more prominent feature - the converse also applies.

Model F uses Model D's weights as initialisation of its weights and helps increase the overall average accuracy by a significant 11.1% from 64.2% in Model E, where it just uses ImageNet weights, to 75.3%. The classes all show improvement except for drinking. The drinking class images exhibit higher misclassification as the resting class. An example of this is shown in Fig. 4b as the passenger performs an unexpected bottle opening action, the model incorrectly classifies it as the closest looking state, resting. This is symptom of not having enough data, as noted in Sect. 3.2, the drinking state has the fewest images compared to other classes. In contrast to Model E, Model F shows a major increase in the text state, 56% to 82%, but this is still sometimes misclassified as the talk state. An example of this is shown in Fig. 4d, in some cases the images do not provide sufficient information, such as the phone not being fully visible in the camera field of view.

Model G (Table 1g) is trained on the aligned front seat data from both vehicles, Model H (Table 1h) and Model I (Table 1i) are trained in aligned, back seat views from both vehicles. The transfer of weights from the front seat model results in an increase for all states for the back seat model with the overall average accuracy rising by 6.7% from 68.6% to 75.3%. Even though Model G originally shows difficulty in discerning the call class, when used to help train the back seat model, it notably improves the call state classification accuracy from 71% to 89%.

4 Conclusions

In this paper, we propose a passenger state detection method that uses a convolutional neural network in combination with viewpoint warping using planar homography. This enables data which usually cannot be included at the training stage to be effectively used for re-training, and data re-purposed from driver monitoring to occupant state classification. The viewpoint normalisation and augmentation also allows the trained model to be re-trained with additional data to work between vehicle types. To evaluate the robustness of the proposed method, data was collected in two different vehicles at three different viewpoints, and used to demonstrate that viewpoint is a significant factor influencing accuracy. The results show that there is a benefit to using data from other vehicles and other viewpoints through transfer learning. Furthermore, we show that it is possible to usefully apply data from the driver monitoring to passenger state monitoring.

Being able to accurately classify passenger state, albeit to a limited number of distinct classes, opens up possibilities to build driving state monitoring systems for passenger-to-passenger and driver-to-passenger interactions. The current results are limited to well-lit vehicle cabins, and data was only captured with a relatively small number of people and two vehicle models. To assess the flexibility and robustness of the approach, future work will focus on further passenger states, data will be collected in a larger range of vehicle types, and under more demanding lighting environments, such as during night-time journeys.

Acknowledgement. This work was supported by Jaguar Land Rover and the UK-EPSRC grant EP/N012380/1 as part of the jointly funded Towards Autonomy: Smart and Connected Control (TASCC) Programme. We wish to thank all who volunteered to take part in the data collection.

References

1. Bay, H., Ess, A., Tuytelaars, T., Van Gool, L.: Speeded-up robust features (SURF). Comput. Vis. Image Underst. **110**(3), 346–359 (2008)
2. Bergasa, L.M., Nuevo, J., Sotelo, M.A., Barea, R., Lopez, M.E.: Real-time system for monitoring driver vigilance. IEEE Trans. ITS **7**(1), 63–77 (2006)
3. Chatfield, K., Simonyan, K., Vedaldi, A., Zisserman, A.: Return of the devil in the details: delving deep into convolutional nets. In: BMVC (2014)
4. Choi, I.H., Tran, T.B.H., Kim, Y.G.: Real-time categorization of driver's gaze zone and head pose using the convolutional neural network. In: Proceedings of HCI Korea, pp. 417–422. Hanbit Media, Inc. (2016)
5. Chuan, Z., Da Long, T., Feng, Z., Li, D.Z.: A planar homography estimation method for camera calibration. In: Proceedings of IEEE Computational Intelligence in Robotics and Automation, vol. 1, pp. 424–429. IEEE (2003)
6. Daza, I.G., Bergasa, L.M., Bronte, S., Yebes, J.J., Almazán, J., Arroyo, R.: Fusion of optimized indicators from Advanced Driver Assistance Systems (ADAS) for driver drowsiness detection. Sensors **14**(1), 1106–1131 (2014)
7. Gkioxari, G., Girshick, R., Malik, J.: Contextual action recognition with r* CNN. In: Proceedings of the IEEE ICCV, pp. 1080–1088 (2015)
8. Hartley, R., Zisserman, A.: Multiple View Geometry in Computer Vision. Cambridge University Press, Cambridge (2003)
9. Jaguar Land Rover: New ADAS technologies for 2017 Range Rover Sport (2016). http://media.landrover.com/node/10699. Accessed 18 June 2017
10. Ji, Q., Zhu, Z., Lan, P.: Real-time nonintrusive monitoring and prediction of driver fatigue. IEEE Trans. Vech. Tech. **53**(4), 1052–1068 (2004)
11. Ji, S., Xu, W., Yang, M., Yu, K.: 3D convolutional neural networks for human action recognition. IEEE Trans. PAMI **35**(1), 221–231 (2013)
12. Juan, L., Oubong, G.: SURF applied in panorama image stitching. In: Proceedings of Image Processing Theory Tools and Applications, pp. 495–499. IEEE (2010)
13. Krizhevsky, A., Sutskever, I., Hinton, G.E.: Imagenet classification with deep convolutional neural networks. In: NIPS, pp. 1097–1105 (2012)

14. LeCun, Y., Bottou, L., Bengio, Y., Haffner, P.: Gradient-based learning applied to document recognition. Proc. IEEE **86**(11), 2278–2324 (1998)
15. LeCun, Y.A., Bottou, L., Orr, G.B., Müller, K.-R.: Efficient backprop. In: Montavon, G., Orr, G.B., Müller, K.-R. (eds.) Neural Networks: Tricks of the Trade. LNCS, vol. 7700, pp. 9–48. Springer, Heidelberg (2012). doi:10.1007/978-3-642-35289-8_3
16. Mbouna, R.O., Kong, S.G., Chun, M.G.: Visual analysis of eye state and head pose for driver alertness monitoring. IEEE Trans. ITS **14**(3), 1462–1469 (2013)
17. McLaughlin, N., Del Rincon, J.M., Miller, P.: Data-augmentation for reducing dataset bias in person re-identification. In: AVSS, pp. 1–6. IEEE (2015)
18. Simonyan, K., Zisserman, A.: Two-stream convolutional networks for action recognition in videos. In: NIPS, pp. 568–576 (2014)
19. Simonyan, K., Zisserman, A.: Very deep convolutional networks for large-scale image recognition. arXiv preprint arXiv:1409.1556 (2014)
20. Vicente, F., Huang, Z., Xiong, X., De la Torre, F., Zhang, W., Levi, D.: Driver gaze tracking and eyes off the road detection system. IEEE Trans. ITS **16**(4), 2014–2027 (2015)
21. Yan, C., Coenen, F., Zhang, B.: Driving posture recognition by convolutional neural networks. IET Comput. Vis. **10**(2), 103–114 (2016)
22. Yan, C., Jiang, H., Zhang, B., Coenen, F.: Recognizing driver inattention by convolutional neural networks. In: CISP, pp. 680–685. IEEE (2015)
23. Yan, S., Teng, Y., Smith, J.S., Zhang, B.: Driver behavior recognition based on deep convolutional neural networks. In: ICNC-FSKD, pp. 636–641. IEEE (2016)
24. Zhang, Z.: A flexible new technique for camera calibration. IEEE Trans. PAMI **22**(11), 1330–1334 (2000)
25. Zhao, C., Zhang, B., He, J., Lian, J.: Recognition of driving postures by contourlet transform and random forests. IET Int. Trans. Syst. **6**(2), 161–168 (2012)

Demographic Classification Using Skin RGB Albedo Image Analysis

Wei Chen[1], Miguel Viana[1(✉)], Mohsen Ardabilian[1], and Abdel-Malek Zine[2]

[1] Laboratoire d'Informatique en Image et Systémes d'Information (LIRIS), École Centrale de Lyon, 36 Avenue Guy de Collongue, 69134 Écully Cedex, France
{wei.chen,miguel.viana,mohsen.ardabilian}@ec-lyon.fr
[2] Institut Camille Jordan, École Centrale de Lyon, 36 Avenue Guy de Collongue, 69134 Écully Cedex, France
abdel-malek.zine@ec-lyon.fr
https://liris.cnrs.fr, http://math.univ-lyon1.fr/

Abstract. Age, gender and skin type classification of demographics using common imaging techniques is costly and does not provide good performance. We propose an approach based on skin RGB albedo image analysis for demographic classification. The diffuse albedo uses inherent skin properties which prevail over illumination conditions variation despite being based on visual perception. The method was tested using skin samples from multiple facial regions to evaluate their performance for classification. Moreover, the application of a fusion algorithm using albedo data from each of the facial regions improved the overall performance resulting in rates above 90% accuracy in age, gender and skin type categories.

Keywords: RGB albedo · Image analysis · Skin reflectance · Gender recognition · Age recognition · Skin type recognition · Fusion · Machine learning

1 Introduction

Skin is one of the largest and most complex organs found in the human body. Its appearance is caused by several factors that can be clustered into two main ones: internal (endogenous) and external (exogenous). These factors determine the final skin structure and composition. The quantity of information provided by skin properties is enough to describe an individual and distinguish him from others. For example, it is possible to easily recognize the identities, sex and age range, lifestyle or health condition by observing people's faces. Human traits are perceived thanks to vision, which completely depends on the presence of light.

The electromagnetic radiation found between 350 nm and 750 nm is what we acknowledge as the visible light spectrum. Visual perception is based on reflected light radiation. Common representation techniques for imaging make use of color-spaces or colorimetry. These approaches portray some skin qualities like its tonality but are highly dependent on the illumination level used.

S. Battiato et al. (Eds.): ICIAP 2017, Part II, LNCS 10485, pp. 149–159, 2017.
https://doi.org/10.1007/978-3-319-68548-9_14

For example, in RGB color space the intensity for each color channel is defined as the integral of all energy sensed over the pixel.

State of the art techniques on subcutaneous prediction and evaluation endorse using reflectance based analysis. Several materials present unique traits in the spectral distribution of their reflectance. For skin, the analysis of its perception is known as optical biopsy and has applications in many areas of expertise such as demographics, medicine or cosmetics. Moreover, the acquisition of reflectance information is performed using non-invasive maneuvers. It is identified as the ratio between reflected and incident light energy thus implying that the information obtained is normalized.

1.1 Skin Anatomy

Skin complexity comes from its multiple layers, properties and particle presence. Although composition, size and structure is unique to each person some share range similarities [1]. The layers that are responsible for most of the appearance perceived from skin are those closest to the surface: epidermis and dermis. These two layers possess a combined thickness ranged between 0.5 mm and 4 mm with the dermis being approximately 24 times thicker. Epidermis is the outer-most layer which reveals the aspect and health condition of skin because it carries the main aesthetics of the body. Water presence in the epidermis is estimated around 10–20% and determines how bright or dry it is perceived. Epidermal thickness along with the presence of melanin cause the final aspect perceived by the skin through this layer.

The dermis layer is connected to the upper epidermal layer and consists of several cells, fiber and substances distributed into 2 sub-layers. These fibers belong to collagen and elastin structures. Up to 70% of dry skin weight comes from collagen distributed into thin and thick structures forming a complex rigid network that protect the blood vessels inside them. Hemoglobin is transported with the blood torrent and is the main responsible for blood vessel reddish tonality. Its concentration is closely related to blood circulation.

1.2 Skin Visual Perception

Because skin is a semi-translucent material, light trespasses its surface and interacts with the medium. Higher wavelengths reach deeper than the dermis layer. Pigments and structures are responsible for two different phenomena that define the reflectance spectra: absorption and diffraction [2]. Absorption corresponds to the loss of energy and can be mainly attributed to the presence of pigments and dried tissue in skin. On the other hand, diffraction relates to the trajectory deviation of incident light which causes that some wavelengths stay longer in the medium before exiting. This phenomenon is associated to the presence of collagen structures. The perceived skin aspect is mainly caused by the previous phenomena at the epidermis and dermis layers. Thinner skins present lighter tones while darker tonalities imply higher thickness and pigment presence. Due to the variations in skin anatomy among individuals we propose the usage of reflectance image analysis techniques for demographic classification.

1.3 Skin Related Demographics

As previously introduced, thickness and pigmentation are responsible for how skin is perceived. It is also one of the most demographic-related features as it varies from location to location, age range, ethnicity and gender. For example, female population usually present thinner skin than males (Poulsen et al. [15]) or younger people possess higher skin thickness than the elderly. In the case of an adult human it corresponds to almost 3.5 times thicker than infant skin tissue, but children do not exhibit similar skin thickness before reaching 5 years of age. It is known that epidermal thickness reaches it maximum at the age of 20 while for the dermis layer this event occurs at 30. Thicker layers mean a darker perceived skin tone.

Pigmentation presence is closely related to ethnic traits. However, the influence of age and solar radiation also intervene in the concentration of pigments in skin. Melanin is created as a response to the increase of subcutaneous metabolism. It is clear that sun exposure increases with age and thus the amount of skin damage and the growth in melanin presence. Males also present a higher level of melanin than women which is related to lighter skin tones across female population [8]. The elasticity of skin is related to the status of its collagen structures and as mentioned before, sun exposure and age wages on skin health. These are only some identifiable traits regarding demographics and the most defining properties of skin.

2 Related Work

Several studies indicate the presence of differences in pigmentation and skin structure between ethnicities, gender and age ranges [15]. To expand on these facts, we present the state of the art in reflectance image analysis to evaluate the feasibility of demographic classification using these techniques.

2.1 Skin Classification Based on Skin Aging and Light Exposure

Fitzpatrick's skin classification [7] is widely used in clinic and cosmetics as it establishes appearance as a response to possible effects of light exposure: sunburn and tanning capabilities of skin. Thus, it analyzes how different skin types react to light rather than associating ethnic traits to skin tonalities. Fitzpatrick identifies a total of 6 different skin types non directly related to color. More specifically, it considers the damage suffered on skin under UV exposure and melanin levels produced due to it. Fitzpatrick skin descriptions are provided in Table 1.

Under analysis of each Fitzpatrick's group, we identified a relation between types I to III and caucasian traits while IV and V corresponded to brownish tonalities. Finally, type VI identifies darker skin tones. However there is not a direct relation between the classification types and skin aging which can come from natural or photo-aging from UV exposure. Photo-aging causes dryness, hyper-pigmentation and yellowing of the skin.

Table 1. Fitzpatrick skin type classification

Type	Sunburn	Darkening	Unexposed color
I	Always	Never	White
II	Easily	Rarely, slight	White
III	Sometimes	Sometimes, moderate	White
IV	Hardly	Easily, moderate	White
V	Rarely	Easily, severe	Brown
VI	Never	Always, black	Black

2.2 Light-Skin Interaction Prediction: BSSRDF

To perform the association between light propagation optics and skin biological properties a light-skin interaction model is required. Most models are based on the Radiative Transfer Equation (RTE) theory by Chandrasekhar [3] and Ishimaru [11] which describes the light radiation variation as the emitted, reflected and transferred light at an arbitrary point in the medium. For skin, these are defined by the absorption and scattering phenomena.

The Bidirectional Subsurface Scattering Reflectance Distribution Function (BSSRDF) is one model to describe the interaction. It determines the outgoing location of a light beam that entered the medium, which can be different from the entry point. BSSRDF [5] takes into consideration the absorption and scattering events in propagation, and is defined by the ratio of outgoing to the incoming radiance:

$$S(x_i, \vec{\omega}_i, x_o, \vec{\omega}_o) = \frac{S(x_o, \vec{\omega}_o)}{S(x_i, \vec{\omega}_i)} \tag{1}$$

The BSSRDF describes the reflectance of the material as follows:

$$S(x_i, \vec{\omega}_i, x_o, \vec{\omega}_o) = \frac{1}{\pi} F_t(\eta, \vec{\omega}_i) R(||x_i - x_o||) F_t(\eta, \vec{\omega}_o) \tag{2}$$

where F_t is the Fresnel transmittance term and R is the reflectance of the material. On the other hand, the BRDF is a particularization of the BSSRDF where the light beam exits at the same point of entry. The BSSRDF can also describe the absorption and scattering phenomena in translucent materials.

2.3 State of the Art Skin Models

Several models can be applied to simulate the light-skin interaction that produces the reflectance signal depending on the accuracy and level of demand desired. Most models take into account the RTE to describe the different light interactions inside the medium.

The Monte Carlo model [16] simulates the physical phenomena in probabilistic and statistical manners and the propagation an interaction processes are described by the RTE. However the solution to the RTE is rather accurate. Accuracy increases when arbitrarily incrementing the number of process simulations. The result of the model provides statistical information regarding the absorption, reflection and transmission processes of the material. Figure 1 presents the Monte Carlo simulation steps to obtain these results.

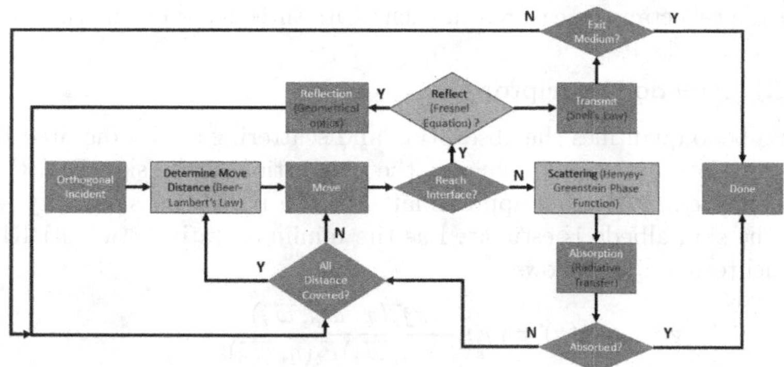

Fig. 1. Monte Carlo model for reconstructing light-skin interaction

On the other hand, the multipole model reduces the computation time required for the Monte Carlo model. Proposed by Jensen et al. [10] it simulates subsurface light transport in a high scattering medium like skin. The RTE is solved through approximating how light diffuses for each point in the medium. This presents a high difficulty level and can only be achieved by considering an isotropic medium. The last assumption is valid because skin is an anisotropic medium with strong scattering properties and thus light distribution becomes isotropic after multiple scattering events. If the medium were infinite, the solution to the equation would be the same as for Green's function. However this is not the case with skin and thus an expansion of the model is required to finally compute the reflectance in a multi-layered material.

Finally, the Kubelka-Munk model [12] presents a good computational speed and simplicity for skin analysis. It locally models light transport with a simple relation between optical medium properties and its reflectance. Given a medium slab with infinitely wide and length, absorption and scattering coefficients σ_a and σ_s respectively, the incident flux is modeled as 2 flux with opposite directions after passing a small distance. After traversing a dx thickness, each flux is partially absorbed and scattered. Downward flux I decreases due to absorption and scattering and increases because of the scattering of J as written in Eq. 3.

$$\begin{cases} dI = -(\sigma_a + \sigma_s)Idx + \sigma_s Jdx \\ -dJ = -(\sigma_a + \sigma_s)Jdx + \sigma_s Idx \end{cases} \tag{3}$$

3 Proposed Albedo Based Descriptor

Most demographic classification techniques are appearance based. Results present successful rates but values decrease for darker skin tonalities or bad illumination causing traits to be less perceivable. The usage of reflectance albedo provides color information of the skin while also overcoming the previous drawbacks. We present an RGB albedo descriptor for the depiction of skin properties based on the reflectance measured in each of its three color channels.

3.1 RGB Albedo Descriptor

The skin albedo quantifies the absorption and scattering ability dictated by the subcutaneous properties and demands the acquisition of the skin BSSRDF for that. Nevertheless, it is often approximated as the BDRF for simplicity and as a result, the skin albedo is estimated as the minimum ratio between BRDF f_r and Fresnel terms F_t as follows:

$$\alpha(x) = \min_{i} \frac{\pi f_r(x, \vec{\omega}_o, \vec{\omega}_i)}{F_t(\eta, \vec{\omega}_i) F_t(\eta, \vec{\omega}_o)} \tag{4}$$

where $\vec{\omega}_i$ and $\vec{\omega}_o$ are the incoming and outgoing light directions at surface point x, f_r is the surface BRDF, $F_t(\eta, \vec{\omega})$ is the Fresnel transmittance term with light direction $\vec{\omega}$, η is the relative refractive index between skin and air and is often assumed to be constant throughout all facial skin. The acquisition of skin BRDF can be performed using digital cameras, as proposed in the works of Weirich et al. [17]. From the skin diffuse albedo we extract our skin descriptor.

For each RGB channel, the albedo is divided into N equally distributed intervals between 0 and 0.4 because the skin RGB albedo generally stands in this range. After that, we compute the albedo distribution histogram as:

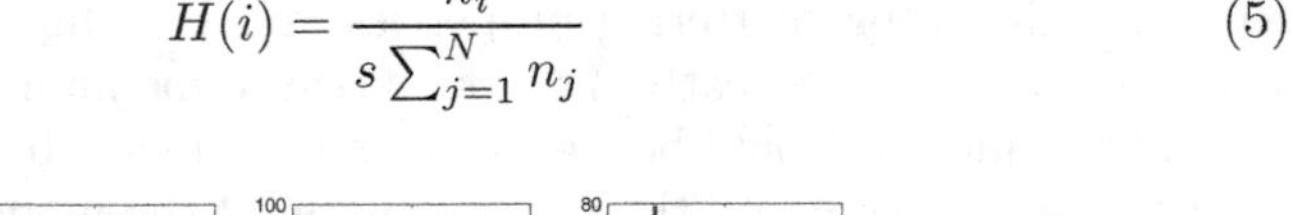

$$H(i) = \frac{n_i}{s \sum_{j=1}^{N} n_j} \tag{5}$$

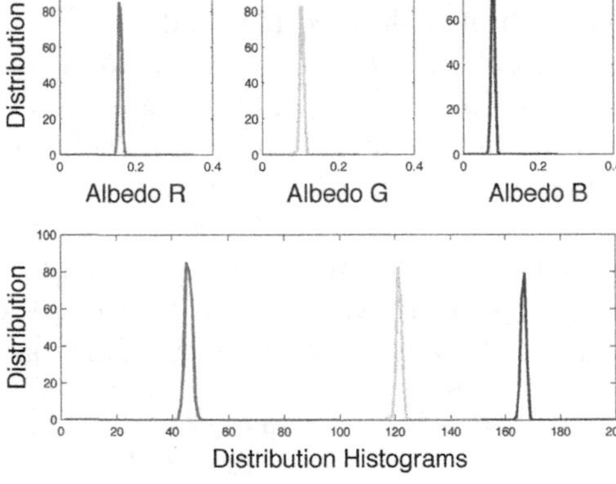

Fig. 2. R, G and B separate albedo histograms, and resulting concatenation (Color figure online)

where $H(i)$ is the value for the i-th bin, n_i is the quantity of albedo that corresponds in the $i-th$ interval, and s is the used step size between adjacent albedo intervals. The histogram satisfies the following properties:

$$\begin{cases} H(i) >= 0, i = 1, 2, \ldots, N \\ \sum_{i=0}^{N} H(i)s = 1 \end{cases} \tag{6}$$

Finally, by concatenating the R, G and B albedo histograms in order we obtain the proposed albedo-based descriptor as shown in Fig. 2.

4 Demographic Classification Results and Analysis

To evaluate the albedo-based descriptor performance, the MERL/ETH [18] database was used. It contains 156 subjects aged between 13 and 74 years with labeled gender. The diffuse albedo is obtained from 10 different facial regions as listed in Table 2. This section is divided into 2 parts. First, the data was clustered using k-means and compared with biological results. Then, we compare the performance for age, gender and Fitzpatrick skin type classification.

Table 2. Regions in MERL/ETH database

Region	Description	Region	Description
1	Philtrum	6	Nose
2	Forehead	7	Upper cheek
3	Eyebrows	8	Lips
4	Upper eyelid	9	Chin
5	Lower eyelid	10	Lower cheek

4.1 Gender, Skin Type and Age Clustering

The clustering process allows the establishment of relations between skin properties (age, gender and skin type) and RGB albedo information. We used the k-means algorithm for clustering subject albedo data. Initial cluster centers were randomly selected and the number of clusters was specified. Execution of the algorithm was performed enough times and final cluster center results were those that provided the lowest sum of point-to-center distances.

We obtained 3 different clusters for gender classification, as showed in Table 3. This table shows the value of the RGB albedo values related to each cluster center and the following conclusions were extracted. First, the proportion of women in cluster 1 is very low. Thus, women present higher reflectance meaning that their skin is thiner than men's. This is also described by the experiments in Giacomini et al. [6]. The small segment of women in cluster 1 present darker skin tones.

In regards to the skin type clustering, we defined 5 groups according to the work by Fitzpatrick. This is due to the combination of groups I and II

Table 3. Gender clustering results

Truth	Male	Female	R_m	G_m	B_m
Cluster 1	77	5	0.088	0.056	0.046
Cluster 2	39	17	0.120	0.066	0.056
Cluster 3	3	15	0.152	0.087	0.071

from Table 1 because subjects from I were scarce for clustering. Initially, lighter tonalities were associated to groups I to IV, brownish for V and dark for group VI. Groups I to III can manifest sunburns, while the IV to VI present difficulties to do that. The skin tanning level is perceivable in groups III to VI. After processing the algorithm, two clusters remain. The first one is composed by types I to III, while the second is conformed by types V and VI, which showed similar albedo properties. Group IV fell as an intermediate point in between, as displayed in Table 4. Results present high similarity to the work of Fitzpatrick on skin type categorization.

Table 4. Gender clustering results

Truth		I, II	III	IV	V	VI
Cluster 1		14	52	9	0	0
	Sub1	1	17	14	0	0
Cluster 2	Sub2	0	3	23	1	1
	Sub3	0	0	3	14	4

Finally, for age clustering four different groups were initially defined: under 25, 25 to 30, 30 to 40 and above 40. Over this initial assumption, two clusters appear for each of the previously defined ranges. Most concise data for the obtained clusters were related to masculine skin type III and IV respectively. Their data was analyzed to extract age effect. First, Table 5 shows the data related to type III skin male subjects, while Table 6 presents those for type IV.

Table 5. Age range clustering results for type III males

	0–25	25–30	30–40	40+	std_R	std_G	std_B
Cluster 1	12	13	6	1	9.54	11.61	11.52
Cluster 2	3	2	9	7	10.17	12.57	12.27

Table 6. Age range clustering results for type IV males

	0–25	25–30	30–40	40+	std_R	std_G	std_B
Cluster 1	4	6	6	2	7.88	10.49	10.62
Cluster 2	1	11	9	2	10.15	14.38	13.94

From this information we can denote the standard RGB reflectance (cluster centers) increase in variation with age which ultimately translates as that color homogeneity of classes decreases with age.

4.2 Gender, Skin Type and Age Classification

For the classification tasks, RBF-kernel Support Machine Vector (SVM) of age, gender and skin type was performed. Subjects from each class were randomly chosen for cross validation, training and testing also making up for over-sized or undersized classes. The cross validation consisted in a 10-fold approach partitioning data so that 7 sets were dedicated to training, 2 for testing and the one remaining was used for fusion training of the classifier. A total 10 rounds of this procedure were performed and the best result was kept. We analyzed data of the 6 most data-populated facial regions (1, 2, 6, 7, 8 and 9) and results obtained were as displayed in Table 7.

Table 7. Classification results using data from different regions

Region	Gender	Skin type	Age
1	86.875%	78%	69.06%
2	93.44%	81.25%	72.81%
6	85.94%	81.25%	68.44%
7	84.38%	66%	65.94%
8	70.94%	67.25%	50%
9	89.06%	68.75%	56.56%

Region 2 showed the better classification results while region 8 presented the worst performance rates overall under the proposed approach. Despite the improvable results in age, gender classification achieved good performance, specially for region 2. We used the class information provided by the SVM for late fusion of the data from the 5 best performing face regions. Each region is given a weight from the information provided by the SVM. Region fusion results are shown in Table 8.

Table 8. Classification results fusing data from different regions

Classification	Previous best	After
Gender	93.44%	96.88%
Skin type	81.25%	95%
Age	72.81%	90.63%

Fusion increases gender classification results by 3.44%. However, there is a significant increase in Fitzpatrick skin type (13.75%) and age range classification (17.82%), which resulted in overall categorization rates above 90% for all categories.

Finally, we analyzed the correlation between weights and average skin thickness, as described in Ha et al. [9]. For gender and age classifications, the weight increased along with skin thickness. This fact is supported by the works of Ortonne [13], where it is cited that sunlight exposure stimulates hyperpigmentation emergence. Males showed thicker skins which contain more melanin causing darker tonalities. However, Fitzpatrick skin type classification showed weight decrease when thickness increased. This was thought to be related to Fitzpatrick's association of higher pigmentation being responsible of low tanning capability of skin.

5 Conclusion

Typical imaging methods are highly dependent of illumination levels. Albedo based techniques provide a more stable and accurate solution to extract information. The proposed approach was based on the usage of skin reflectance RGB albedo for demographic classification under gender, age range and skin type. The proposed RGB albedo descriptor is produced concatenating the reflectance information measured over each one of the RGB channels. As a result, we obtain skin color normalized information not conditioned by light presence levels.

An initial clustering process was performed with albedo data to confirm skin properties hypothesis regarding thickness differences between gender, increase of pigmentation with age and for darker toned ethnicities. As for classification capabilities, different facial region albedo samples were used. Best results when individually analyzing the regional albedos provided close to 90% for gender, age range and Fitzpatrick skin type. In addition to these results, a fusion algorithm using each of the regional albedo was use to evaluate overall face performance for classification. Final results increased total classification capabilities of 96.88%, 95% and 90.63% for age, gender and skin type respectively. The previous numbers endorse our proposed approach for skin-based classification.

After evaluating the followed methodology and obtained results, we propose the inclusion of albedo information from additional color channels or spectral bands to increase performance in skin analysis and demographic classification.

References

1. Burns, T., et al.: Rook's textbook of dermatology (2004)
2. Bohren, C.F., Huffman, D.R.: Absorption and Scattering of Light by Small Particles. Wiley, Hoboken (2008)
3. Chandrasekhar, S.: Radiative Transfer. Courier Corporation, North Chelmsford (2013)
4. Donner, C., Jensen, H.W.: Light diffusion in multi-layered translucent materials. ACM Trans. Graph. (TOG) **24**(3), 1032–1039 (2005)

5. Donner, C., Jensen, H.W.: A spectral BSSRDF for shading human skin. Rend. Tech. **2006**, 409–418 (2006)
6. Giacomoni, P.U., Mammone, T., Teri, M.: Gender-linked differences in human skin. J. Dermatol. Sci. **55**(3), 144–149 (2009)
7. Fitzpatrick, T.B.: The validity and practicality of sunreactive skin types I through VI. Arch. Dermatol. **124**(6), 869–871 (1988)
8. Firooz, A., Sadr, B., Babakoohi, S., et al.: Variation of biophysical parameters of the skin with age, gender, and body region. Sci. World J. **2012**, Article ID 386936 (2012). doi:10.1100/2012/386936
9. Ha, R.Y., Nojima, K., Adams Jr., W.P., Brown, S.A.: Analysis of facial skin thickness: defining the relative thickness index. Plastic Reconstr. Surg. **115**(6), 1769–1773 (2005)
10. Jensen, H.W., Marschner, S.R., Levoy, M., Hanrahan, P.: A practical model for subsurface light transport. In: SIGGRAPH 2001: Proceedings of the 28th Annual Conference on Computer Graphics and Interactive Techniques, pp. 511–518. ACM, New York (2001). ACM Request Permissions
11. Ishimaru, A.: Wave Propagation and Scattering in Random Media, vol. 2. Academic Press, New York (1978)
12. Munk, F., Kubelka, P., Kubelka, P., et al.: Ein beitrag ztlr optik der farbanstriche. Z. Teeh. Physik. **12**, 501–593 (1931)
13. Ortonne, P.: Pigmentary changes of the ageing skin. Br. J. Dermatol. **122**, 21–28 (1990)
14. Malskies, C.R., Eibenberger, E., Angelopoulou, E.: The recognition of ethnic groups based on histological skin properties. In: Vision, Modeling, and Visualization, pp. 353–360 (2011)
15. Poulsen, T., Wulf, H.C., et al.: Epidermal thickness at different body sites: relationship to age, gender, pigmentation, blood content, skin type and smoking habits. Acta Derm. Venereol. **83**(6), 410–413 (2003)
16. Wang, L., Jacques, S.L.: Hybrid model of Monte Carlo simulation and diffusion theory for light reflectance by turbid media. JOSA A **10**(8), 1746–1752 (1993)
17. Weyrich, T., Matusik, W., Pfister, H., Ngan, A.: Measuring skin reflectance and subsurface scattering. Technical report TR-, Mitsubishi Electric Research Laboratories MERL (2005)
18. Weyrich, T., Matusik, W., Pfister, H., Bickel, B., Donner, C., Tu, C., McAndless, J., Lee, J., Ngan, A., Jensen, H.W., Gross, M.: Analysis of human faces using a measurement-based skin reflectance model. ACM Trans. Graph. (TOG) **25**, 1013–1024 (2006). ACM

Discriminative Dictionary Design for Action Classification in Still Images

Abhinaba Roy[1(✉)], Biplab Banerjee[2], and Vittorio Murino[1]

[1] Istituto Italiano di Technologia, Genova, Italy
{abhinaba.roy,vittorio.murino}@iit.it
[2] Indian Institute of Technology, Roorkee, India
bbanfcs@iitr.ac.in

Abstract. In this paper, we address the problem of action recognition from still images. Although used widespread, local features (SIFT, STIP) invariably engender two potential problems: the counts of such extracted features are not evenly distributed in different entities of a given category and many of such features are not paradigmatic of the visual concept the entities represent. In order to generate a discriminative dictionary taking the aforementioned issues into account, we propose a novel method for identifying robust and category specific local features which maximize the class separability to a possible extent. Specifically, we consider category independent region proposals to highlight local regions in still images. Further, the selection of potent local descriptors is cast as filtering based feature selection problem which ranks the local features per category based on a novel measure of distinctiveness. The underlying visual entities are subsequently represented based on the learned dictionary and this stage is followed by action classification using the random forest model. The framework is validated on the challenging Stanford-40 dataset and exhibits superior performances than the representative methods from the literature.

Keywords: Action recognition · Local features · Feature mining · Random forest

1 Introduction

Recognition of visual concepts is one of the most active research lines in computer vision. Other than the problem of generic object recognition from images [3], action recognition is another challenging research paradigm in this respect. With the growing amount of visual data available from various sources, intelligent analysis of human attributes and activities has gradually attracted the interest of the computer vision community. Action recognition based on local descriptors is another very popular paradigm which broadly depends upon three stages: (1) Extraction of local descriptors, (2) Codebook (dictionary) generation and feature encoding, and (3) Classification based on the encoded features. Although this framework has exhibited superlative performances for general

S. Battiato et al. (Eds.): ICIAP 2017, Part II, LNCS 10485, pp. 160–170, 2017.
https://doi.org/10.1007/978-3-319-68548-9_15

visual recognition tasks, the efficacy of such a model depends upon a number of factors, and effective codebook construction is undoubtedly the most noteworthy. The standard codebook construction process is based on vector quantizing local descriptors extracted from the available training data in which the cluster centroids define the codewords, the basic building blocks that are ultimately used to encode the underlying visual entities. Specifically, an entity is represented by a vector where the ith component can be either the number of local descriptors that fall in the ith cluster or a measure of proximity of local descriptors to the ith cluster centroid. Needless to mention, the quality of the extracted local descriptors affect representation power of the codewords which, in turn, has direct impact on the recognition performance. For instance, the descriptors extracted from background regions or the ones shared by many visual categories add little to the discriminative capability of the codebook in comparison to the ones specifically extracted from the objects of interest. However, it is impossible to ensure the selection of potentially useful local descriptors in advance since such feature extraction techniques are typically engineered and ad hoc. In other words, there are certain immediate advantages if the most discriminative local descriptors are used for the purpose of a cogent codebook construction, though the process is intrinsically complex in general. Selection of the discriminative local descriptors for effective codebook generation coping with action recognition from images is the very core topic of this paper. We propose a simple algorithm which gradually filters out unrepresentative descriptors before constructing a compact global codebook. The proposed method is generic in the sense that it can work with different types of local features irrespective of the underlying visual entities they refer to. Specifically, we represent each still image by a large pool of category independent region proposals [1]. Each region proposal is represented by convolutional neural network (CNN) features (4096 dimensions) obtained from a pre-trained network. We propose a sequential method for codebook construction which first clusters the local descriptors of each entity using the non-parametric mean-shift (MS) technique [7]. The cluster centroids thus obtained represent the reduced set of non-repetitive local features for the entity from now onwards. Another round of MS clustering on the new set of local descriptors calculated from all the entities of a given category is followed and the centroids thus obtained are employed to build a temporary codebook specific to each category. Further, we propose a ranking criteria to highlight potentially discriminative codewords from each category specific codebook and the global dictionary is built by accumulating these reduced set of codewords from all the categories.

We can summarize the main highlights of this paper as follows:

- The initial two level MS based clustering of the local descriptors on the entity and the category level largely reduces the effects of repetitive and uninteresting descriptors, yet selecting representative codewords from each locally dense region in the feature space. We further propose a novel measure to rank and select a subset of discriminative codewords per visual category under consideration The proposed ranking measure ensures that the selected set of

codewords are frequent in the entities of the same category while being sporadic in other visual categories.
- We evaluate the codebooks learned in this way for action categorization from still images. We observe that the learned codebooks, when used in conjunction with efficient feature encoding techniques, sharply outperform similar techniques from the literature.

The rest of the paper is organized as follows. We discuss a number of related works from the literature in Sect. 2. The proposed action recognition framework is described in Sect. 3. Experimental results are reported in Sect. 4, followed by concluding remarks and ideas of possible future endeavour.

2 Related Works

In this section, we primarily highlight two aspects of the proposed framework and discuss relevant techniques from the literature. First, action representation from images with a focus on local feature encoding based methods will be addressed and a discussion on the relevant codebook construction techniques will subsequently be followed.

2.1 Action Recognition from Still Images

Recognition of human actions and attributes [6] has been approached using traditional image classification methods [23]. In the standard dictionary learning based scenario, a typical framework extracts dense SIFT [16] from the training images and codebook is constructed by clustering the SIFT descriptors by k-means clustering. Further, efficient encoding techniques including bag of words (BoW), LLC, Fisher vector are used to represent the images before the classification stage is carried out in such a feature space [4].

Since the inherent idea of the BoW based frameworks is to learn recurring local patches, a different set of approaches directly models such object parts in images. Such techniques either initially define a template and try to fit it to object parts or iteratively learn distinctive parts for a given category.

Discriminative part based models (DPM) [8] are used extensively for this purpose and they served as the state of the art for a period. The hierarchical DPM model is used to parse human pose for action recognition in [22]. An efficient action and attributed representation based on sparse bases of local features is introduced in [24]. An expanded part model for human attribute and action recognition is proposed in [19]. The effects of empty cavity, ambiguity and pooling strategies are explored in order to design the optimal feature encoding for the purpose of human action recognition in still images in [25].

Very recently, the part learning paradigm has gained much attention because of its ability to represent mid-level visual features. Given a large pool of region proposals extracted from the images, such techniques iteratively learn part classifiers with high discriminative capabilities. Methods based on partness analysis [10], deterministic annealing for part learning [20] etc. are some of the representatives in this respect.

2.2 Efficient Codebook Construction for Classification

As already mentioned, the paper focuses on the identification of good local descriptors for building a better dictionary in order to enhance the action recognition performance. The dictionary learning strategies can be supervised or unsupervised in nature. A class of unsupervised dictionary learning strategies compute over-complete sparse bases considering the idea of alternate optimization. Such techniques iteratively update the dictionary components and sparse coefficients for the input samples using k-SVD and matching pursuit based methods. [21] proposes the LLC technique where a locality constraint is added to the loss function of sparse coding. [13] introduces an l_1-norm based sparse coding algorithm where feature-sign search is applied for encoding and Lagrange dual method for dictionary learning. Effective sampling strategies for the BoW model is the focus of discussion in [18] where several aspects including the codebook size, clustering techniques adopted etc. are exhaustively studied.

On the contrary, the supervised approaches include the class support in building the dictionary. Label consistent SVD [9], logistic regression based sparse coding [17] explicitly consider the class discrimination in designing the sparse bases for dictionary learning. Two different clustering based approaches for keypoints selection are introduced in [15] for the purpose of dictionary learning based generic scene recognition. Distance measures among the keypoints are modeled in an online fashion to filter out keypoints with low generalization capability.

We focus on the supervised dictionary learning paradigm and follow a sequential approach in building the dictionary. In contrast to similar methods from the literature, our framework is flexible, scalable and non-parametric. The MS based clustering technique is inherently capable of detecting all possible clusters in the feature space based on data density. It better captures the local aspects of the entities than the traditional k-means based vector quantization techniques which require a number of hyper-parameters to be optimized. In addition, our ranking measure can efficiently highlight codewords which are highly discriminative by exploring their frequency distributions in all the underlying categories. Further, the proposed ranking measure is generic in the sense that it is applicable to the broader domains of feature selection, part learning, ranking in a retrieval system to name a few.

3 Proposed Algorithm

We detail the proposed action recognition framework in this section. As already mentioned, the proposed framework consists of four major stages: (1) Extraction of local features (2) Discriminative dictionary construction (3) Feature encoding (4) Action classification.

For notational convenience, let us consider that $TR = \{X_i, Y_i\}_{i=1}^{N}$ constitutes N training examples belonging to L action categories where each X_i represents an image and Y_i is the corresponding class label. Entities in TR are represented by a set of local descriptors $F_i = \{F_i^1, F_i^2, \ldots, F_i^{\alpha_i}\}$ where $F_i^k \in \mathbb{R}^d$ and $d = 4096$ or $d = 162$, respectively, depending on whether the underlying X_i is an image. In addition, α_i represents the number of local descriptors extracted from X_i. Further, $\{C_1, C_2, \ldots, C_L\}$ represents the set of category specific codebooks learned by the proposed algorithm by exploiting the local features extracted from TR, whereas $C = [C_1 C_2 \ldots C_L]$ is the global codebook obtained by the concatenation of the local ones.

The framework is elaborated in the following sections.

3.1 Extraction of Local Features

We consider category independent region proposals to highlight local regions in still images.

Region proposal generation techniques highlight region segments in the image where the likelihood of the presence of an object part is high. This provides a structured way to identify interesting locations in the image and thus reduces the search space for efficient codeword generation. We specifically work with the objectness paradigm for region proposals generation from still images which is based on modeling several aspects regarding the characteristics of the objects in a Bayesian framework. Each region proposal is further represented by the CNN features. We prefer the ImageNet pre-trained VGG-F [5] model which has an architecture similar to AlexNet [11], and comprises of 5 convolutional layers and 3 fully-connected layers. The main difference of VGG-F and AlexNet is that VGG-F contains less number of convolutional layers and uses a stride of 4 pixels leading to better evaluation speed than the AlexNet architecture.

3.2 Discriminative Dictionary Learning

We first build category specific codebooks and then concatenate all the local codebooks to generate a global codebook.

Separate Dictionary Learning for Each Category. For a given $l \in \{1, 2, \ldots, L\}$, the dictionary learning process is summarized as follows:

1. For each training instance with the category label l, we first group the local descriptors using MS clustering and consider the cluster centroids as constituting the reduced set of local descriptors. MS is an iterative, non-parametric clustering method which does not require an estimation of the number of clusters as input. Instead, it relies on the kernel density estimate in the feature space to group samples which form dense clusters. Given $F_i = \{F_i^1, F_i^2, \ldots, F_i^{\alpha_i}\}$, the kernel density estimate at a point F_i^k is expressed as

$$f(F_i^k) = \frac{1}{\alpha_i h^d} \sum_{m=1}^{\alpha_i} K(\frac{F_i^k - F_i^m}{h}) \tag{1}$$

where K is a radially symmetric kernel function and h defines the width of the Parzen window to highlight the neighbourhood around F_i^k. A cluster is identified as the region where the data density is locally maximum. This

can alternatively be interpreted as the local regions where $\nabla f \approx 0$. ∇f can efficiently be calculated by iteratively shifting the centroids of the Parzen windows until the locally dense regions are reached [7].
Since all the descriptors in a dense region in the feature space highlight near similar local features, the mean-shift clustering is able to select one unique representative for all of them. Further, since mean-shift implicitly estimates the number of clusters present in the dataset, hence, the problem of over-merging is greatly reduced. On the other hand, spherical clustering techniques like k-means and fuzzy c-means create suboptimal codebooks as most of the cluster centroids fall near high density regions, thus under-representing equally discriminant low-to-medium density regions. MS resolves such problem by focusing on locally dense regions in the feature space. Let $\widehat{F}_i = \{\widehat{F}_i^1, \widehat{F}_i^2, \ldots, \widehat{F}_i^{\widehat{\alpha_i}}\}$ represents the new set of local descriptors for the ith training instance where each $\widehat{F}_i^k$ represents a cluster centroid.
2. Once $\widehat{F}_i$s are constructed for all the training instances with category label l, we vector quantize all such $\widehat{F}_i$s using MS clustering to build a temporary codebook $C_l = \{C_l^1, C_l^2, \ldots, C_l^{\beta_l}\}$ for the category with each C_l^k representing a codeword (cluster centroid). Similar to the previous stage, it is guaranteed that C_l is ensured to capture all the potential local features for the lth category.

$\{C_1, C_2, \ldots, C_L\}$ are constructed in the similar fashion for $l \in \{1, 2, \ldots, L\}$. It is to be noted that the labels of the codewords depend upon the action categories they refer to. Further, the sizes of the C_ls may differ from each other. The C_ls thus obtained are not optimal in the sense that they contain many codewords with low discriminative property. Such codewords need to be eliminated in order to build robust category specific codebooks. However, we need a measure to rank the descriptors based on their discriminative ability. In this respect, the following observations can be made:

- A potentially discriminative codeword is not frequent over many of the categories constituting the dataset.
- Most of its nearest neighbors in $\{C_1, C_2, \ldots, C_L\}$ share the same class label with the codeword under consideration.

We model the first observation in terms of the idea of conditional entropy whereas the second observation is replicated by the tf-idf score.

For a given codeword C_l^k, we find out the labels of its T nearest neighbours over the entire set of codewords in $\{C_1, C_2, \ldots, C_L\}$ and subsequently define the conditional entropy measure as:

$$H(Y|C_l^k) = -\sum_{l'=1}^{L} p(l'|C_l^k) \log_2 p(l'|C_l^k) \tag{2}$$

where $p(l'|C_l^k)$ represents the fraction of the retrieved codewords with label l'. For discriminative codewords, i.e. the ones which do not span many categories,

H is small whereas the value of H grows with the selection of codewords shared by many categories.

In addition to the H score, we also expect the nearest neighbours to be populated from the same category as of C_l^k. In order to impose this constraint, we define the tf-idf score for C_l^k as follows:

$$TI(C_l^k) = \frac{|C_{l'}^{k'}|C_{l'}^{k'} \in knn(C_l^k)\ AND\ l' = l|}{|C_{l'}^{k'}|C_{l'}^{k'} \in knn(C_l^k)|} \tag{3}$$

Both the measures are further combined in a convex fashion to define the ranking measure as follows:

$$Rank(C_l^k) = w_1 \frac{1}{H(Y|C_l^k)} + (1 - w_1)\ TI(C_l^k) \tag{4}$$

We repeat this stage for all the codewords in $\{C_1, C_2, \ldots, C_L\}$. As already mentioned, the $Rank(C_l^k)$ has high values for potentially discriminative and category specific codewords. We rank the codewords on the basis of the *Rank* scores and select top B codewords in a greedy fashion in order to define the final codebook $\widehat{C}_l$ for category l.

Global Dictionary Construction. The local codebooks obtained in the previous stage are concatenated in order to obtain a global codebook $\widehat{C} = [\widehat{C_1}\widehat{C_2}\ldots\widehat{C_L}]$ of size $L * B$.

3.3 Feature Encoding Using $\widehat{C}$

We represent each visual entity with respect to $\widehat{C}$ using efficient LLC based encoding technique due to its ability to deal with CNN features in case of action recognition in still images.

3.4 Classification

The final classification is performed using random forest ensemble classifier [2]. The decision tree learning algorithm used is information gain and bootstrap aggregation is employed to learn the ensemble model. Thus the forest reduces classifier variance without increasing bias. Random subspace splitting is used for each tree split and we consider $\sqrt{d}$ features for each split given d original feature dimensions. The generalization is performed by applying majority voting on the outcomes of the learned trees.

4 Experimental Details

4.1 Dataset

We consider the Stanford-40 [24] still image action recognition database to evaluate the effectiveness of the proposed framework.

Stanford-40 actions is a database of human actions with 40 diverse action types, e.g. brushing teeth, reading books, blowing bubbles, etc. The number of images per category ranges between 180 to 300 with a total of 9352 images. We use the suggested [24] train-test split with 100 images per category as training and remaining for testing.

4.2 Experimental Setup

The following experimental setup is considered in order to evaluate the performance of the proposed framework.

- MS clustering is used in conjunction with the Gaussian kernel. The adaptive bandwidth parameter (h) is fixed empirically as $\frac{D}{m}$ $(1 \leq m \leq 10)$, where D is the average pairwise distance of all the local descriptors extracted from all the visual entities of each category. The same setup is repeated for MS clustering in the entity and the category levels (Sect. 3.2).
- We extract 500 region proposals per image for the Stanford-40 dataset. Figure 1 depicts the extracted region proposals for a pair of images from the dataset. We further discard proposals which are largely overlapping to each other (overlap of $\geq$ 50%) in order to highlight potentially discriminative local patches in the images.
- The number of final distinctive codewords selected for each class based on the proposed ranking measure in Sect. 3.2 is set between $100 \leq B \leq 500$ and the best classification performances are reported. In LLC encoding, 100–200 nearest neighbors per local descriptor are considered to encode the images. We select the optimal hyper-parameters by cross-validation. Each image in

Fig. 1. Extraction region proposals from images of Stanford-40 using objectness

the Stanford-40 dataset is optimally represented by a sparse vector of length 8000×1 (100 neighbors in LLC).
- Each component tree in the random forest model is essentially a classification and regression tree (CART) [2]. We conduct experiments with random forest of different sizes (500–2000) and find that a random forest with 1000 CART trees exhibits superior performance.
- We compare the overall classification performance of the proposed technique with the representative techniques from the literature. All the experiments are repeated multiple times and the average performance measures are reported.

4.3 Performance Evaluation

Table 1 mentions the accuracy assessment of different techniques for the Stanford-40 dataset. The performances of the methods based on hand-crafted SIFT-like features are comparatively less (≈35.2%) [21] since the differences in human attributes for many of the action classes are subtle. Part learning based strategies obtain better recognition performance in this respect by explicitly modeling category specific parts. An overall classification accuracy of 40.7% is obtained with the generic expanded part models (EPM) of [19] which is further enhanced to 42.2% while the contextual information is incorporated in EPM. The best performance with shallow features obtained for this dataset is 45.7% by [24] which performs action recognition by combining bases of attributes, objects and poses. Further they derive their bases by using large amount of external information. The deep CNN models further enhance the state of the art performance in this respect thanks to the high level features they encode and the ImageNet pre-trained AlexNet reports a classification accuracy of 46% [11].

The proposed method further enhances the recognition performance to 49% while considering $B = 200$ codewords per visual category. We further observe a minor enhancement (≤0.75%) in the classification accuracy for $B \leq 275$ after which the performance degrades to some extent before being saturated to 47%. In this case, our method encapsulates the advantages of deep and shallow models effectively in a single framework. The CNN based region proposals are capable of encoding high level abstractions from the local regions. Since the images are captured in unconstrained environments, the backgrounds are uncorrelated in different images of a given category. The per category dictionary learning strategy reduces the effects of such background patches and the proposed ranking measure further boosts the proposals corresponding to the shared human attributes, human-objects interaction etc. for a given action category. In contrast to other techniques which are based on SVM classifier, our framework relies on the random forest model which does not explicitly require any cross-validation. Further, the ensemble nature of the classifier reduces the number of misclassified actions to some extent. We further observe that performance of the random forest model gradually improves with growing number of CART trees within the range 500–1000 and a random forest model with 1000 trees outputs the best performance.

Table 1. A summary of the performance of our classification framework for the Stanford-40 data in comparison to the literature

Method	Classification accuracy
ObjectBank [14]	32.5%
LLC with SIFT features [21]	35.2%
Spatial pyramid matching kernel [12]	34.9%
Expanded parts model [19]	40.7%
CNN AlexNet [11]	46%
Proposed framework (with a per class codebook with 200 codewords and LLC encoding)	49%

5 Conclusion

We introduce a novel supervised discriminative dictionary learning strategy for the purpose of action recognition from still images. We leverage the available training samples to optimally rank local features which are robust and discriminative. Further, we initially cluster the local features at the entity as well as category levels to eliminate the effects of features corresponding to non-recurrent or background locations. The proposed ranking paradigm holds wider applications in areas including feature selection, ranked set generation for retrieval etc. The effectiveness of the proposed dictionary learning is validated on challenging dataset (Stanford-40), on which, superior performance measures can be observed in comparison to popular techniques from the literature. We currently focus on the *learning to rank* paradigm for effective dictionary learning.

References

1. Alexe, B., Deselaers, T., Ferrari, V.: Measuring the objectness of image windows. IEEE Trans. Pattern Anal. Mach. Intell. **34**(11), 2189–2202 (2012)
2. Bishop, C.M.: Pattern recognition. Mach. Learn. **128** (2006)
3. Campbell, R.J., Flynn, P.J.: A survey of free-form object representation and recognition techniques. Comput. Vis. Image Underst. **81**(2), 166–210 (2001)
4. Chatfield, K., Lempitsky, V.S., Vedaldi, A., Zisserman, A.: The devil is in the details: an evaluation of recent feature encoding methods. In: BMVC, vol. 2, p. 8 (2011)
5. Chatfield, K., Simonyan, K., Vedaldi, A., Zisserman, A.: Return of the devil in the details: delving deep into convolutional nets. arXiv preprint arXiv:1405.3531 (2014)
6. Cheng, G., Wan, Y., Saudagar, A.N., Namuduri, K., Buckles, B.P.: Advances in human action recognition: a survey. arXiv preprint arXiv:1501.05964 (2015)
7. Comaniciu, D., Meer, P.: Mean shift: a robust approach toward feature space analysis. IEEE Trans. Pattern Anal. Mach. Intell. **24**(5), 603–619 (2002)
8. Felzenszwalb, P.F., Girshick, R.B., McAllester, D., Ramanan, D.: Object detection with discriminatively trained part-based models. IEEE Trans. Pattern Anal. Mach. Intell. **32**(9), 1627–1645 (2010)
9. Jiang, Z., Lin, Z., Davis, L.S.: Learning a discriminative dictionary for sparse coding via label consistent K-SVD. In: 2011 IEEE Conference on Computer Vision and Pattern Recognition (CVPR), pp. 1697–1704. IEEE (2011)

10. Juneja, M., Vedaldi, A., Jawahar, C., Zisserman, A.: Blocks that shout: distinctive parts for scene classification. In: Proceedings of the IEEE Conference on Computer Vision and Pattern Recognition, pp. 923–930 (2013)
11. Krizhevsky, A., Sutskever, I., Hinton, G.E.: Imagenet classification with deep convolutional neural networks. In: Advances in Neural Information Processing Systems, pp. 1097–1105 (2012)
12. Lazebnik, S., Schmid, C., Ponce, J.: Beyond bags of features: spatial pyramid matching for recognizing natural scene categories. In: 2006 IEEE Computer Society Conference on Computer Vision and Pattern Recognition (CVPR 2006), vol. 2, pp. 2169–2178. IEEE (2006)
13. Lee, H., Battle, A., Raina, R., Ng, A.Y.: Efficient sparse coding algorithms. In: Advances in Neural Information Processing Systems, pp. 801–808 (2006)
14. Li, L.J., Su, H., Fei-Fei, L., Xing, E.P.: Object bank: a high-level image representation for scene classification & semantic feature sparsification. In: Advances in Neural Information Processing Systems, pp. 1378–1386 (2010)
15. Lin, W.C., Tsai, C.F., Chen, Z.Y., Ke, S.W.: Keypoint selection for efficient bag-of-words feature generation and effective image classification. Inf. Sci. **329**, 33–51 (2016)
16. Lowe, D.G.: Distinctive image features from scale-invariant keypoints. Int. J. Comput. Vis. **60**(2), 91–110 (2004)
17. Mairal, J., Ponce, J., Sapiro, G., Zisserman, A., Bach, F.R.: Supervised dictionary learning. In: Advances in Neural Information Processing Systems, pp. 1033–1040 (2009)
18. Nowak, E., Jurie, F., Triggs, B.: Sampling strategies for bag-of-features image classification. In: Leonardis, A., Bischof, H., Pinz, A. (eds.) ECCV 2006. LNCS, vol. 3954, pp. 490–503. Springer, Heidelberg (2006). doi:10.1007/11744085_38
19. Sharma, G., Jurie, F., Schmid, C.: Expanded parts model for human attribute and action recognition in still images. In: Proceedings of the IEEE Conference on Computer Vision and Pattern Recognition, pp. 652–659 (2013)
20. Sicre, R., Jurie, F.: Discriminative part model for visual recognition. Comput. Vis. Image Underst. **141**, 28–37 (2015)
21. Wang, J., Yang, J., Yu, K., Lv, F., Huang, T., Gong, Y.: Locality-constrained linear coding for image classification. In: 2010 IEEE Conference on Computer Vision and Pattern Recognition (CVPR), pp. 3360–3367. IEEE (2010)
22. Wang, Y., Tran, D., Liao, Z., Forsyth, D.: Discriminative hierarchical part-based models for human parsing and action recognition. J. Mach. Learn. Res. **13**(Oct), 3075–3102 (2012)
23. Yang, W., Wang, Y., Mori, G.: Recognizing human actions from still images with latent poses. In: 2010 IEEE Conference on Computer Vision and Pattern Recognition (CVPR), pp. 2030–2037. IEEE (2010)
24. Yao, B., Jiang, X., Khosla, A., Lin, A.L., Guibas, L., Fei-Fei, L.: Human action recognition by learning bases of action attributes and parts. In: 2011 International Conference on Computer Vision, pp. 1331–1338. IEEE (2011)
25. Zhang, L., Li, C., Peng, P., Xiang, X., Song, J.: Towards optimal VLAD for human action recognition from still images. Image Vis. Comput. **55**, 53–63 (2016)

Enhanced Bags of Visual Words Representation Using Spatial Information

Lotfi Abdi[1,2(✉)], Rahma Kalboussi[1,2], and Aref Meddeb[1,2]

[1] National Engineering School of Sousse, University of Sousse, Sousse, Tunisia
lotfiabdi@hotmail.com, rahma.kalboussi@gmail.com,
Aref.Meddeb@infcom.rnu.tn

[2] Networked Objects Control and Communication Systems Laborator, University of Sousse, Sousse, Tunisia

Abstract. In order to achieve fast and robust Traffic Signs Recognition (TSR), we introduced a novel way to incorporate both distance and angle information in the BoVW representation. A novel approach for visual words construction was presented, which takes the spatial information of keypoints into account in order to enhance the quality of visual words generated from extracted keypoints. In this paper, we propose a new method for TSR system based on the Bags of Visual Words (BoVW) approach, using the distance and angle information in the BoVW representation. Second, we proposed a new computationally efficient method to model global spatial distribution of visual words by taking into consideration the spatial relationships of its visual words. Experimental results show that the suggested method could reach comparable performance of the state-of-the-art approaches with less computational complexity and shorter training time.

Keywords: Traffic sign recognition · Spatial information · Haar cascade · Bags of visual words · Driving safety · Intelligent transportation systems

1 Introduction

Extensive studies in human-machine interactivity are necessary to present the Traffic Sign Recognition (TSR) information in a careful way to inform the driver without causing distraction or confusion [3]. The visibility of traffic signs is crucial for the drivers' safety. In a real environment, it becomes a difficult task to recognize the traffic signs timely and accurately because the visibility of traffic signs may be decreased greatly by some unfavorable factors. For example, very serious accidents happen when drivers do not notice a stop sign. An automatic TSR system is helpful for assisting drivers and is essential for autonomous cars [4].

Road scenes are also generally very cluttered and contain many strong geometric shapes that could easily be misclassified as road signs. In conventional techniques, color, shape information, or geometric features of traffic signs are

S. Battiato et al. (Eds.): ICIAP 2017, Part II, LNCS 10485, pp. 171–179, 2017.
https://doi.org/10.1007/978-3-319-68548-9_16

utilized for performing detection and recognition with regard to the traffic signs in general. Road sign recognition systems usually have developed into two specific phases. First, in each frame, the detection stage identifies also the categories of signs based on shapes (circular, rectanglar, trianglar, etc.). The second task is to classify the detected signs and send the processing results (i.e., the types of signs and their locations) to the display and control units of an Advanced Drive Assistance System (ADAS) [1].

Fatigue, divergence of attention, and occlusion of signs due to road obstruction and natural scenes may lead to miss some important traffic signs, which can result in severe accidents. Guiding the driver attention to an imminent danger, somewhere around the car, is a potential application. After recognizing the traffic signs, a driver may be notified of the recognized traffic signs in a manner of audio or visual information. The recognition of road traffic signs correctly at the right time for that particular place is very important for any vehicle driver to ensure a safe journey for themselves and their passengers. The proposed system may make driving even more comfortable and may help the drivers receive important information regarding the signs, even before their eyes can actually see the sign, and in an easy and comprehensible way.

Several methods were recently proposed to incorporate spatial information to improve the BoVW model such as the spatial pyramid matching method [11], spatio-temporal interest point [7]. Other recent work have focused on the local stability of traffic sign regions [2]. Further, methods such a [8] proposed a novel classification techniques of TSR based on probabilistic latent semantic. The algorithm consists of two parts: (1) classify the shape of the traffic signs and (2) classify its actual class. In order to investigate the effect of coding methods and codebook size of local spatiotemporal features, [6] have proposed a coding method to alleviate the negative effect of quantization error by assigning the local spatiotemporal features to a few nearest visual words.

Correct and timely recognition of road traffic signs is crucial for any vehicle driver to ensure a safe journey for themselves and their passengers. Considering the processing time and classification accuracy as a whole, we have developed a novel technique to incorporate spatial information of visual words to improve accuracy while maintaining short retrieval time. In order to achieve fast and robust TSR, we introduced a novel way to incorporate both distance and angle information in the BoVW representation. A novel approach for visual words construction was presented, which takes the spatial information of keypoints into account in order to enhance the quality of visual words generated from extracted keypoints. This clearly demonstrated the complementarities of the additional relative spatial information provided by our approach to improve accuracy while maintaining short retrieval time.

In this paper, we propose a novel method for traffic sign recognition system based on the bag-of-visual-words approaches. We introduce a novel way to incorporate both distance and angle information in the BoVW representation. We proposed a new computationally efficient method to model the global spatial distribution of visual words by taking into consideration the spatial relationships of its visual words.

2 Traffic Signs Recognition

In both instances, the ability to recognize signs and their underlying information is highly desirable. This information can be used to warn the human driver of an oncoming change, or in more intelligent vehicle systems, to actually control the speed and/or steering of the vehicle. It is therefore necessary to classify the characteristics of the information given and find a way to represent the information according to these characteristics. To overcome this problem, we proposed a novel approch to integrate the spatial information to BoVW model.

2.1 Enhanced BoVW Using Spatial Information

This paper presents a new approach to integrate the spatial information to BoVW model, with explicit local and global structure models. The key idea is to consider the spatial distribution of visual words in an image. In [10], a pairwise spatial histogram is defined according to a discretization of the spatial neighborhood into several bins encoding the relative spatial position (distance and angle) of two visual words. Therefore, combining the frequency of occurrence and spatial information of visual words should be a promising direction for improving the image characterization.

To address this issue, we introduce a novel way to incorporate both distance and angle information in the BoVW representation. This method exploits spatial orientations and distances of all pairs of similar descriptors in the image. In the BoVWs model, a visual vocabulary $Voc = vi, i = \{1, \ldots, k\}$, then it is built by clustering these features into a certain number of K visual words. A given descriptor d_k is then mapped to a visual word V_i using euclidean distance Eq. 1 as follows:

$$v(d_k) = argminDist(v, d_k) \tag{1}$$

where $v \in Voc$, d_k is the kth descriptor in the ROI, $Dist(v, d_k)$ is the distance between the descriptor and the visual word based on the euclidean distance. For this reason, we consider the weighted sum of ROIs to implicity represent spacial information which is important for similarity measurement between images.

In the training stage, the SIFT features are extracted from all the training samples, using a dense grid. Since we are interested in the sign contents, only the descriptors that do not fall outside the sign contour are taken into account. Our system exploits the SIFT features, which have shown a high robustness to varied recording conditions. After the SIFT features are extracted for all the training samples, the number of feature points of each image is not entirely consistent, which will bring great difficulties to subsequent operations. Assignment of a visual feature to the vocabulary depends on the similarity metric. We propose a method that incorporates spatial information at feature level. This method exploits distances and spatial orientations of all pairs of similar descriptors in the image.

2.2 Similarity Measure

In order to improve similarity measurement between pairwise, we propose a simple and efficient method to infuse spatial information. We measure the spatial relationships between visual words using distance and orientation. This is done by creating an additional dictionary comprising of word pairs. These methods are inspired by [19], using a log-polar quantization of image spatial domain.

For each visual word, the average position and the standard deviation is computed based on all the occurrences of the visual word in the image. We consider the interaction between visual words by encoding their spatial distances, orientations and alignments. Figure 1 shows an example to better understand our approach. To encode spatial information, we use the distance (Fig. 1(a)) and orientation (Fig. 1(b)) information between pairs of patches in the image space.

More formally, we consider the set S_k of all the pairwise, where at least one patch in the pair belongs to the visual word w_k. A given pair $(P_i, P_j) \in S_k$ is characterized both by a pair of descriptors (d_i, d_j) and a pair of positions in the image space denoted (p_i, p_j) is illustrated in Fig. 1. Note that both d_i and p_i are vectors with $d_i \in R^D$ and $p_i \in R^2$.

After clustering the spatial information is implicitly included in the visual vocabulary. A pairwise spatial histogram (Fig. 1(d)) of similar patches is then defined considering a discretization of the image space into M bins denoted b_m,

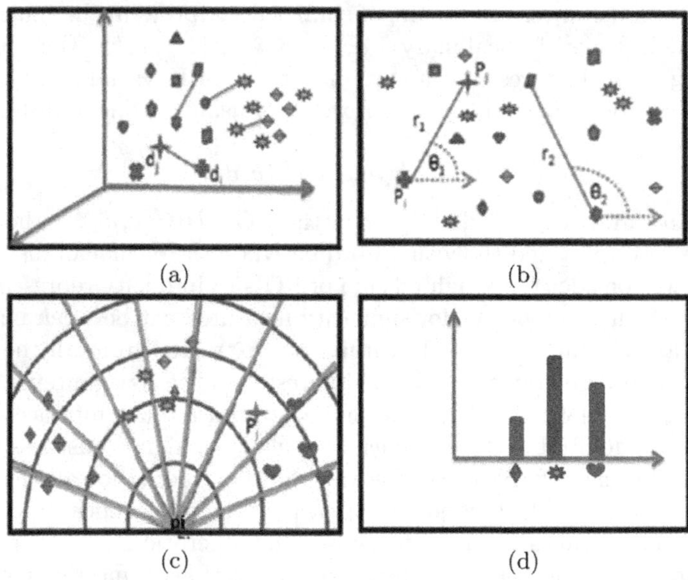

Fig. 1. Spatial histogram of similar pairwise using distance and orientation: (a) spatial distance of similar visual word, (b) spatial orientations of similar pairwise, (c) pairwise similarity distance-orientations information of similar patches, (d) pairwise spatial histograms

m = {1, ..., M} with an angle $\theta \in [0, \pi[$ split into M_θ angle bins and the radius $r \in [0, R]$ split into M_r radial bins so that $M = M_\theta.M_r$.

To classify a new feature, P_x, into the discovered classes, it is compared with the words included in the vocabulary using the distance described in Eq. 1. We assign the corresponding word i according to the nearest neighbor, but only if that distance is below a matching threshold, th_M.

$$W_i = argmin_{i \in [1,k]}(d(P_x, P_i)|jd(P_x,P_i) < th_M) \tag{2}$$

To represent a ROI in an image, we employ spatial pyramid to enhance voting and indexing criterion of the original inverted indexing technique. Based on the position of each visual words, this method exploits spatial orientations and distances of all pairs of similar descriptors in the image. It is relatively efficient during classification and can well present the spatial information contained in Spatial Pyramid features. In order to train our classifier, the framework of our proposed BoVW system is illustrated in Fig. 2.

For those purpose, a novel structural relationship between patches are defined for evaluating super-pixels' similarity. Particularly, simple spatial relations between visual words are considered the spatial locations of the words and the spatial relationship between the words were added to describe images in the BoW model. This histogram encodes spatial information (distance and orientation (Fig. 1) of pairwise similar patches, where at least one of the patches belongs to V_k. To have a global representation, we replace each bin of the BoVW frequency histogram with the spatial histogram associated to w_i. By this way, we keep the frequency information intact and add the spatial information. This modularity facilitates simple way to assemble the spatial histograms and to obtain the final representation.

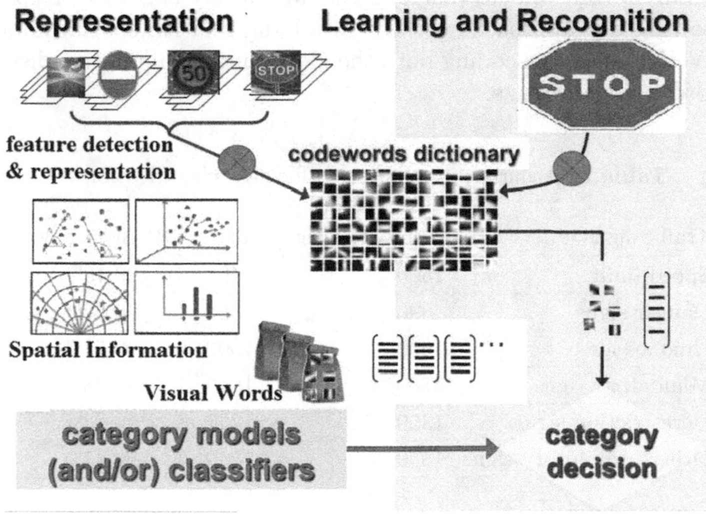

Fig. 2. Proposed approach BoVW for classification

3 Experiment Results

To evaluate the performance of the proposed algorithm, we implement the proposed AR-TSR method using the hardware environment of Core i7 640LM 2.13 GHz and the software environment of Windows 7, Visual Studio 2010 using OpenCV Library 2.4.8. In this paper, we focus on the detection of speed limit, unique signs and danger signs. We implement the suggested method in C++ and test the performance on the German Traffic Sign Detection Benchmark (GTSDB) dataset [15]. In the GTSRB dataset, there are 51,839 German traffic signs in 43 classes. These classes of traffic signs have been divided into six subsets speed limit sign subset, danger sign subset, mandatory sign subset, unique sign subset, derestriction sign subset and other prohibitory sign subset [17]. The solution of each image in the dataset is 1366×800. Here, we present quantitative a analysis of the system presented in this work.

3.1 Performance of the Proposed Method

The database used to train the classifiers has been designed using the ROIs obtained from the detection step and the model fitting methods presented in the previous sections. In order to evaluate the occlusion robustness of the suggested classification method, the content of the detected ROI is identified using the tree classifiers. This classifier is tested on static, low-resolution sign images. Some experiments have also been conducted to measure the performance, the GTSRB has been used, where Table 1 shows the classification rates of the linear SVM.

Furthermore, we evaluate the classification task on the detected signs returned by the previous detection module. As shown in Table 1, the overall classification accuracy is 99.31%. Note that only 3 (out of 1500) speed limit signs, while only 6 (out of 890) danger signs, are falsely classified. Experiments demonstrate that our approach succeeds in adding relative spatial information into the BoVW model by encoding both the global and local relative distribution of visual words over an image.

Table 1. Accuracy results for traffic sign classification

Traffic signs	Number of signs	TP	FP	Accuracy
Speed limit	1500	1497	3	99.13%
Danger signs	890	884	6	98.97%
Unique signs	1000	1000	0	99.51%
Mandatory signs	1700	1699	1	99.45%
Derestriction signs	1350	1347	3	99.32%
Other prohibitory signs	1850	1849	2	99.47%

3.2 Comparisons with Other State-of-the-Art Methods

In order to verify the discrimination performance and computation efficiency of the proposed feature for traffic sign detection, the experiments on the public available data set of traffic signs are implemented. Because the training and testing samples in the GTSRB and GTSDB dataset are split according to a fixed rule, an absolute performance comparison with other reported approaches is possible. We report these results in Table 2, where the results of the winning system from the IJCNN challenge and some reported results in the IJCNN 2011 are provided as references.

The performance results of the machine learning algorithms are all significantly different from each other. We have compared the suggested method with other state-of-the-art algorithms such as the committee of CNNs [5], Human Performance [16], Multi-Scale CNNs [14], Random Forests [8,18], wgy@HIT501 [9], VISICS [13], LITS1 [12,17], and Viola-Jones. The performance is analyzed in terms of detection and recognition accuracy.

According to the results for the GTSRB data set, shown in Table 2, this work achieves a 99.31% recognition accuracy, which is a comparable performance of 0.24% less then the work by [5], and a performance of 0.17% higher than the work by [16], and 0.69% higher than the work by [17], and 1.51% higher than the work by [14]. The accuracy of recognizing unique signs reach 99.31%, which is comparable with the best achieved one. The danger signs which have triangular shape have given the worst results compared with other traffic sign categories.

To prove the effectiveness of our proposed method, a comparison of its performance with the standard BoVW based on the GTSRB data set. The comparison results are reported in Table 3, where we examine the benefit of integration the spatial information to the standard BoVW.

As shown in Table 3, by combining spatial information with the traditional BoVW, our method outperforms the current state-of-the-art methods on the GTSRB databases. Comparisons with the standard BoVW model our method as a richer alternative for the usual BoW approach to build a visual vocabulary. Our method provides more representative semantic information of the traffic signs, including spatial information between visual words. We have shown that the spatial information between visual words is an important information

Table 2. Performance comparison with other traffic sign recognition methods.

Method	[5]	[16]	[14]	[18]	[8]	Our
Speed limit	99.47%	97.63%	98.61%	95.95%	98.82%	99.13%
Danger signs	99.07%	98.67%	98.03%	92.08%	96.85%	98.97%
Unique signs	99.22%	100.00%	98.63%	98.73%	100.00%	99.51%
Mandatory signs	99.89%	99.72%	97.18%	99.27%	96.86%	99.45%
Derestriction signs	99.72%	98.89%	94.44%	87.50%	97.93%	99.32%
Other prohibitory signs	99.93%	99.93%	99.87%	99.13%	98.27%	99.47%

Table 3. Comparison of our approach with the standard BoVW methods.

Method	Standard BoVW	Our
Speed limit	98.81%	99.13%
Danger signs	97.91%	98.97%
Unique signs	98.96%	99.51%
Mandatory signs	98.88%	99.45%
Derestriction signs	97.99%	99.32%
Other prohibitory signs	98.36%	99.47%

for category level recognition, the distribution of similar interest regions of the images is discriminative and can improve the performance of BoVW method significantly.

4 Conclusions

Driving is a complex, continuous, and multitask process that involves driver's cognition, perception, and motor movements. The way road traffic signs and vehicle information is displayed impacts strongly driver's attention with increased mental workload leading to safety concerns. We have developed a novel approach for visual words construction which takes the spatial information of keypoints into account in order to enhance the quality of visual words generated from extracted keypoints. In this paper, we have proposed a new computationally efficient method to model global spatial distribution of visual words and improved the standard BoVW representation. It clearly demonstrates the complementarity of the additional relative spatial information provided by our approach to improve accuracy while maintaining short retrieval time, and can obtain a better traffic signs classification accuracy than the methods based on the traditional BOVW model. Experimental results show that the suggested method could reach comparable performance of the state-of-the-art approaches with less computational complexity and shorter training time.

References

1. Abdi, L., Meddeb, A.: Deep learning traffic sign detection, recognition and augmentation. In: Proceedings of the Symposium on Applied Computing, pp. 131–136. ACM (2017)
2. Abdi, L., Meddeb, A.: In-vehicle augmented reality TSR to improve driving safety and enhance the driver's experience. Sig. Image Video Process. 1–8 (2017)
3. Abdi, L., Meddeb, A., Abdallah, F.B.: Augmented reality based traffic sign recognition for improved driving safety. In: Kassab, M., Berbineau, M., Vinel, A., Jonsson, M., Garcia, F., Soler, J. (eds.) Nets4Cars/Nets4Trains/Nets4Aircraft 2015. LNCS, vol. 9066, pp. 94–102. Springer, Cham (2015). doi:10.1007/978-3-319-17765-6_9

4. Azad, R., Azad, B., Kazerooni, I.T.: Optimized method for Iranian road signs detection and recognition system. arXiv preprint arXiv:1407.5324 (2014)
5. Cireşan, D., Meier, U., Masci, J., Schmidhuber, J.: Multi-column deep neural network for traffic sign classification. Neural Netw. **32**, 333–338 (2012)
6. Danelakis, A., Theoharis, T., Pratikakis, I.: A robust spatio-temporal scheme for dynamic 3D facial expression retrieval. Vis. Comput. **32**, 1–13 (2015)
7. Dawn, D.D., Shaikh, S.H.: A comprehensive survey of human action recognition with spatio-temporal interest point (STIP) detector. Vis. Comput. **32**, 1–18 (2015)
8. Haloi, M.: A novel pLSA based traffic signs classification system. arXiv preprint arXiv:1503.06643 (2015)
9. Houben, S., Stallkamp, J., Salmen, J., Schlipsing, M., Igel, C.: Detection of traffic signs in real-world images: the German traffic sign detection benchmark. In: The 2013 International Joint Conference on Neural Networks (IJCNN), pp. 1–8. IEEE (2013)
10. Khan, R., Barat, C., Muselet, D., Ducottet, C.: Spatial histograms of soft pairwise similar patches to improve the bag-of-visual-words model. Comput. Vis. Image Underst. **132**, 102–112 (2015)
11. Li, Y., Xu, J., Zhang, Y., Zhang, C., Yin, H., Lu, H.: Image classification using spatial difference descriptor under spatial pyramid matching framework. In: Tian, Q., Sebe, N., Qi, G.-J., Huet, B., Hong, R., Liu, X. (eds.) MMM 2016. LNCS, vol. 9516, pp. 527–539. Springer, Cham (2016). doi:10.1007/978-3-319-27671-7_44
12. Liang, M., Yuan, M., Hu, X., Li, J., Liu, H.: Traffic sign detection by ROI extraction and histogram features-based recognition. In: The 2013 International Joint Conference on Neural Networks (IJCNN), pp. 1–8. IEEE (2013)
13. Mathias, M., Timofte, R., Benenson, R., Van Gool, L.: Traffic sign recognition-how far are we from the solution? In: The 2013 International Joint Conference on Neural Networks (IJCNN), pp. 1–8. IEEE (2013)
14. Sermanet, P., LeCun, Y.: Traffic sign recognition with multi-scale convolutional networks. In: The 2011 International Joint Conference on Neural Networks (IJCNN), pp. 2809–2813. IEEE (2011)
15. Stallkamp, J., Schlipsing, M., Salmen, J., Igel, C.: The German traffic sign recognition benchmark: a multi-class classification competition. In: The 2011 International Joint Conference on Neural Networks (IJCNN), pp. 1453–1460. IEEE (2011)
16. Stallkamp, J., Schlipsing, M., Salmen, J., Igel, C.: Man vs. computer: benchmarking machine learning algorithms for traffic sign recognition. Neural Netw. **32**, 323–332 (2012)
17. Yin, S., Ouyang, P., Liu, L., Guo, Y., Wei, S.: Fast traffic sign recognition with a rotation invariant binary pattern based feature. Sensors **15**(1), 2161–2180 (2015)
18. Zaklouta, F., Stanciulescu, B., Hamdoun, O.: Traffic sign classification using K-d trees and random forests. In: The 2011 International Joint Conference on Neural Networks (IJCNN), pp. 2151–2155. IEEE (2011)
19. Zhang, E., Mayo, M.: Enhanced spatial pyramid matching using log-polar-based image subdivision and representation. In: 2010 International Conference on Digital Image Computing: Techniques and Applications (DICTA), pp. 208–213. IEEE (2010)

Exploiting Spatial Context in Nonlinear Mapping of Hyperspectral Image Data

Evgeny Myasnikov(✉)

Samara University, 34, Moskovskoye shosse, Samara 443086, Russia
mevg@geosamara.ru
http://www.ssau.ru

Abstract. Hyperspectral remote sensing image analysis is a challenging task due to the nature of such images. Therefore, dimensionality reduction techniques are often used as a step prior to image analysis. Although there are approaches, which exploit spatial information in image analysis, there is a lack of papers devoted to the problem of exploiting spatial information in dimensionality reduction methods. This paper is devoted to the problem of exploiting spatial context in nonlinear mapping method, which is one of the oldest and well-known dimensionality reduction techniques. To address this task, we use two possible approaches, based on window functions, and order statistics. We provide experimental results for several tasks of hyperspectral image analysis, namely classification, segmentation, and visualization. All the experiments were conducted using well-known hyperspectral images.

Keywords: Hyperspectral image · Spatial context · Nonlinear mapping · Segmentation · Classification · Clustering · Rand index

1 Introduction

A hyperspectral image is a three-dimensional array having two spatial dimensions, and one spectral dimension. Every pixel of a hyperspectral image is a vector containing hundreds of components corresponding to a wide range of wavelengths. Compared to grayscale and multispectral images, hyperspectral images offer new opportunities allowing to extract information about materials (components) located on images. Thanks to these unique properties, hyperspectral images are used in agriculture, medicine, chemistry and many other fields.

However, the high dimensionality of hyperspectral images often makes it impossible to directly apply traditional image analysis techniques to such images. For this reason, dimensionality reduction techniques are often used as a step prior to image analysis.

To reduce the dimensionality of hyperspectral images both linear and nonlinear dimensionality reduction techniques are used. Linear techniques including principal component analysis [1], independent component analysis [2], and projection pursuit are used more often. Nonlinear dimensionality reduction techniques (based on Isomap [3], locally linear embedding [4], laplacian eigenmaps [5],

S. Battiato et al. (Eds.): ICIAP 2017, Part II, LNCS 10485, pp. 180–190, 2017.
https://doi.org/10.1007/978-3-319-68548-9_17

nonlinear mapping [6]) are used less often, although it is known [3] that hyperspectral remote sensing images are subject to nonlinear effects due to nonlinear variations in reflectance, multipath light scattering, and the variable presence of water.

All considered dimensionality reduction techniques operate in a spectral space. But as it was outlined earlier, a hyperspectral image contains both spectral and spatial information. And the last type of information remains unused in traditional dimensionality reduction techniques. Although we can point out many papers that exploit spatial information in image analysis, and, particular, in hyperspectral image analysis (a recent review on this topic can be found in [8]), there is a lack of papers devoted to the problem of exploiting image spatial information in dimensionality reduction techniques. The purpose of this paper is to study the effectiveness of exploiting spatial information in the nonlinear mapping method, which is one of the oldest and well-known dimensionality reduction techniques.

The paper is organized as follows. In Sect. 2 we describe the dimensionality reduction approach used in this paper, and describe two schemes of exploiting spatial context. Section 3 contains the results of experiments for several tasks of image analysis, namely classification, segmentation (clustering), and visualization. The paper ends with conclusion.

2 Methods

2.1 Dimensionality Reduction

The nonlinear mapping method, that was adopted in this study as a dimensionality reduction technique, maps pixels of a hyperspectral image from a high dimensional hyperspectral space R^M into a low dimensional space R^L.

Let us denote the number of image pixels as N, the input pixel vectors in R^M as x_i, the output pixel vectors in R^L as y_i. Then the quality of a mapping can be written as a weighted sum of squared differences of distances between corresponding points in R^M and R^L:

$$\varepsilon = \mu \cdot \sum_{i,j=1,i<j}^{N} (\nu_{ij}(d(x_i, x_j) - d(y_i, y_j))^2). \tag{1}$$

Here $d()$ is a distance function, which is usually the Euclidean, μ and ν_{ij} are constants. In this study, we used $\mu = 1/\sum_{i<j} d^2(x_i, x_j), \nu_{i,j} = 1$. However, other values of constants can be used (see, for example, the Sammon's mapping [7]).

Usually, gradient descent based approaches are used to minimize the above error (1). But due to the high computational complexity of the basic gradient descent, we use the stochastic gradient descent approach [9]:

$$y_{ik}(t+1) = y_{ik}(t) + 2\alpha\mu \sum_{j=1}^{R} \nu_{i,r_j} \cdot \frac{d(x_i, x_{r_j}) - d(y_i, y_{r_j})}{d(y_i, y_{r_j})} \cdot (y_{ik}(t) - y_{r_j k}(t)). \tag{2}$$

In this equation r is the random vector of indices generated at each iteration t of the optimization process, r_j is the j-th element of this vector, R is the number of elements in r, α is the coefficient (step size) of the gradient descent.

On the whole, the nonlinear mapping method adopted in this paper consists in the initialization of output coordinates $y_i(0)$ following by the iterative optimization using (2) until coordinates $y_i(t)$ become stable. This simple algorithm allows us to find a suboptimal configuration of points y_i in the output space R^L. The computational complexity is $O(MLRN)$ per one iteration of the optimization algorithm that makes it possible to process hyperspectral images.

2.2 Exploiting Spatial Context in Dimensionality Reduction

The described above nonlinear mapping method (as many other dimensionality reduction techniques) do not provide a direct ability to account for a spatial context. In this paper, to overcome this limitation we embed spatial context information in dissimilarity measures. In particular, we follow the general idea of extending the feature space by the contextual information, extracted from spatial neighborhood of image pixels. The most obvious way is to extend the feature vector of a pixel by concatenating the values of neighbor pixels. The following equation shows such extension for a pixel with spatial coordinates (u, v) in the case of 4 pixel neighborhood (in contrast to the rest of the paper, here we use spatial coordinates of the pixel):

$$x^*_{u,v} = (x_{u,v}, x_{u-1,v}, x_{u,v+1}, x_{u+1,v}, x_{u,v-1}) . \quad (3)$$

Window Functions. Contrary to feature selection techniques, which allow one to automatically detect informative and noninformative components of extended feature vectors, for unsupervised dimensionality reduction techniques it is impossible to automatically adjust the impact of particular neighbors. To overcome this problem, we can use some weighting coefficients to control the impact of particular neighbors in a dissimilarity measure. For the Euclidean distance

$$d(x_i, x_j) = \sqrt{\sum_{m=1}^{M} (x_{im} - x_{jm})^2}, \quad (4)$$

the modification can be written in the form:

$$\rho(x^*_i, x^*_j) = \left(\sum_{k=1}^{K} w_k \sum_{m=1}^{M} (x^*_{i,kM+m} - x^*_{j,kM+m})^2 \right)^{1/2} . \quad (5)$$

Here K is the number of pixels in a spatial neighborhood, w_k are weighting coefficients.

Intuitively, we can suppose that greater values of coefficients should correspond to closer pixels, and smaller values should correspond to further ones. Also we can suppose that weighting coefficients don't depend on a direction.

In this case, such coefficients can be described by two-dimensional radial-symmetric window function. There were proposed a number of window functions that used widely in signal processing. Examples of such functions include Rectangle, Triangle, Gauss, Bartlet, Welch, Blackman, Hann, Nuttal window functions and others. In this paper, we use window functions based on the following equation (Fig. 1):

$$w(r) = \begin{cases} 1-(r/R)^p, & if(r \in [0;R])and(p \geq 0); \\ (1-r/R)^{-p}, & if(r \in [0;R])and(p < 0); \\ 0, & otherwise. \end{cases} \tag{6}$$

It should be noted that for particular values of p, this equation leads to some special windows: triangle window for $p = 1$, rectangle window for $p = \infty$.

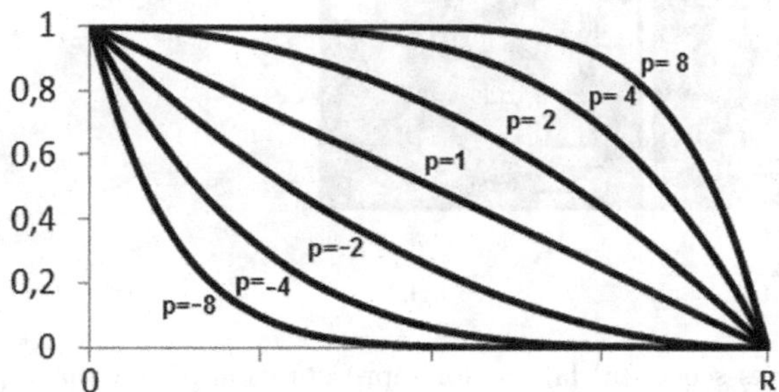

Fig. 1. Window functions.

Order Statistics. Another way of exploiting contextual information consists in using of the order statistics. Let us consider the sample $x_1, x_2, \dots x_K$. By sorting the sample and reindexing the values of the sample so that $x_{(1)} \leq x_{(2)} \leq \dots \leq x_{(K)}$, we obtain the set of the order statistics $x_{(i)}$. The first and the K-th order statistics are $x_{(1)} = min\{x_i\}$, $x_{(K)} = max\{x_i\}$ correspondingly.

To embed the contextual information, for each pixel x_i we consider a spatial neighborhood of radius R, and order pixels of the neighborhood by Euclidean distances to the pixel x_i in a spectral space. Then the feature space is extended by the first S order statistics (x_i is included in the neighborhood as $x_{(1)}$):

$$x_i^* = \left(x_{(1)}, x_{(2)}, \dots, x_{(S)}\right). \tag{7}$$

The modified dissimilarity measure takes the form:

$$\rho(x_i^*, x_j^*) = \left(\sum_{s=1}^{S} w_s \sum_{m=1}^{M} (x_{i,sM+m}^* - x_{j,sM+m}^*)^2\right)^{1/2}. \tag{8}$$

Here we use inverse weighting $w_s = 1/s$, and S is the number of the order statistics used as a spatial context.

3 Experimental Study

In our experiments, we used open and well-known hyperspectral remote sensing scenes [10]. Here we provide experimental results for Indian Pines scene (Fig. 2(a)), 145 × 145 pixels, 224 spectral bands acquired using AVIRIS sensor. Only 200 bands were selected by removing bands with the high level of noise and water absorption. This hyperspectral scene is provided with the groundtruth segmentation mask (Fig. 2(b)) that was used to evaluate the quality of classification and segmentation.

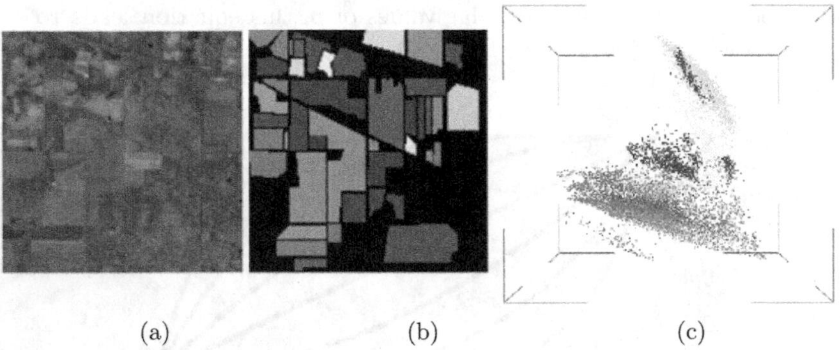

(a) (b) (c)

Fig. 2. Indian Pines scene: (a) false color representation of the image produced using nonlinear mapping; (b) ground truth image (classified pixels are shown with colors); (c) image pixels mapped into 3D space using nonlinear maping and spatial context (Color figure online).

3.1 Classification of Hyperspectral Image Data

In this section we present the evaluation of described approaches to exploiting spatial context in terms of a classification quality. We use two well-known classifiers for features obtained using the described nonlinear mapping algorithm, and exploiting spatial context. In the evaluation we use the following classifiers: k-nearest neighbors classifier (NN), and support vector machine (SVM). To perform experiments we divide the whole set of ground truth samples into a training subset, containing 60% of the sample, and a test subset, containing 40% of the sample.

Window Functions. Some results devoted to the study of window functions are shown in the Fig. 3. In all present cases the leftmost value of p corresponds to the window function

$$\delta(r) = \begin{cases} 1, if(r = 0), \\ 0, else, \end{cases} \tag{9}$$

which means that no spatial context is exploited in the nonlinear mapping. The rightmost value corresponds to the rectangular window function (*Rect*), which

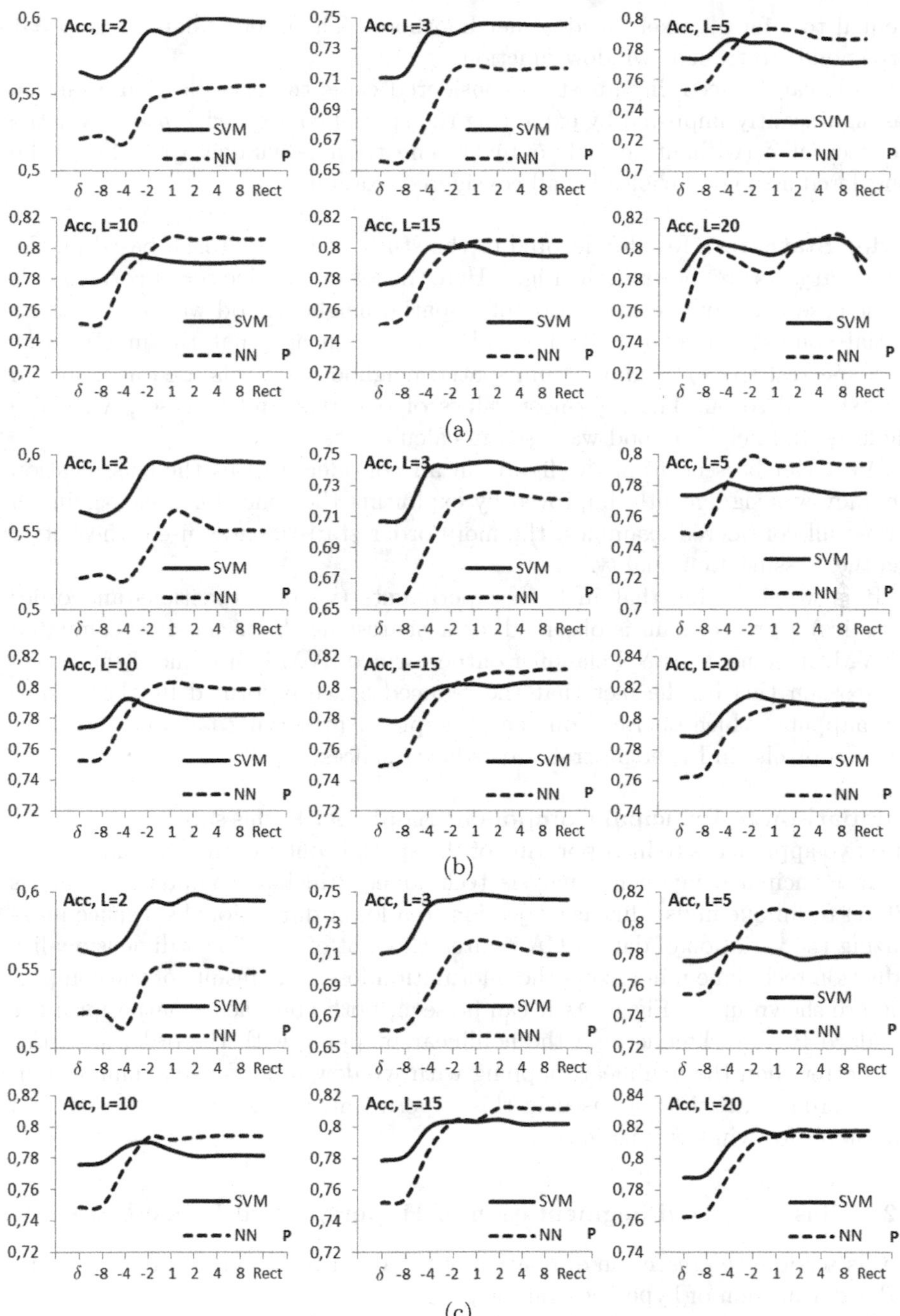

Fig. 3. The dependency of the classification accuracy (Acc) on the parameter p of the window function (6) for the specified dimensionality L of the reduced space R^L, and different neighborhood radius R: radius $R = 1$ (a), neighborhood radius $R = 2$ (b), neighborhood radius $R = 3$ (c). Parameter values δ and *Rect* correspond to special cases of window function: delta function (9) and rectangle window function.

is equal to 1 for all pixels inside a neighborhood. The intermediate value $p = 1$ corresponds to triangle window function.

As it can be seen, in almost all considered cases the classification accuracy was significantly improved by exploiting the spatial context with $p \geq 1$ over the accuracy at δ (without spatial context). The triangle window ($p = 1$) can be considered as a good choice for all considered examples.

Order Statistics. Results devoted to the study of the approach based on the order statistics are shown in the Fig. 4. Here the leftmost value $S = 1$ corresponds to the case when only central pixel of a spatial neighborhood was used, and no spatial context was exploited. The value $S = 2$ means that the first nearest (in a spectral space) neighbor from spatial neighborhood was used as a spatial context, and so on. The rightmost values of S correspond to cases, when the whole spatial neighborhood was used in calculations.

As it can be seen from the figure, in all considered cases the classification accuracy was significantly improved by exploiting the spacial context again. In almost all considered examples, the more order statistics was used, the better was the classification quality.

It is worth noting that in both experiments (window functions and order statistics) a greater gain is observed for a nearest neighbor classifier compared to SVM. In some cases NN classifier outperformed SVM in a reduced space. We can explain this by the fact that the reduced space is formed by the nonlinear mapping, which operates on the principle of preserving pairwise distances between pixels (in hyperspectral and reduced spaces).

Comparison to Principal Component Analysis. In this subsection we compare two approaches to incorporating of the spatial context, and the widely used linear Principal component analysis technique. Principal component analysis (PCA) technique finds a linear projection to a lower dimensional subspace maximizing the variation of data. PCA is often thought of as a linear dimensionality reduction technique minimizing the information loss. The results of the comparison are shown in the Fig. 5. As it can be seen, both approaches outperform the standard PCA technique, and the nonlinear mapping with the order statistics was better than the nonlinear mapping with window function (maximum number of order statistics was used in this experiment, and $P = 1$ was used as a parameter for window function).

3.2 Clustering and Segmentation of Hyperspectral Image Data

In this section we present an evaluation of the described approach for clustering and segmentation of hyperspectral image.

A segmentation method adopted in this paper is based on a clustering technique, and is quite straightforward. It consists of two steps. At first, the clustering of image pixels is performed in the reduced space. At this stage, a clustering algorithm partitions a set of image pixels into some number of subsets, according to pixels features. At second, an image markup procedure extracts connected

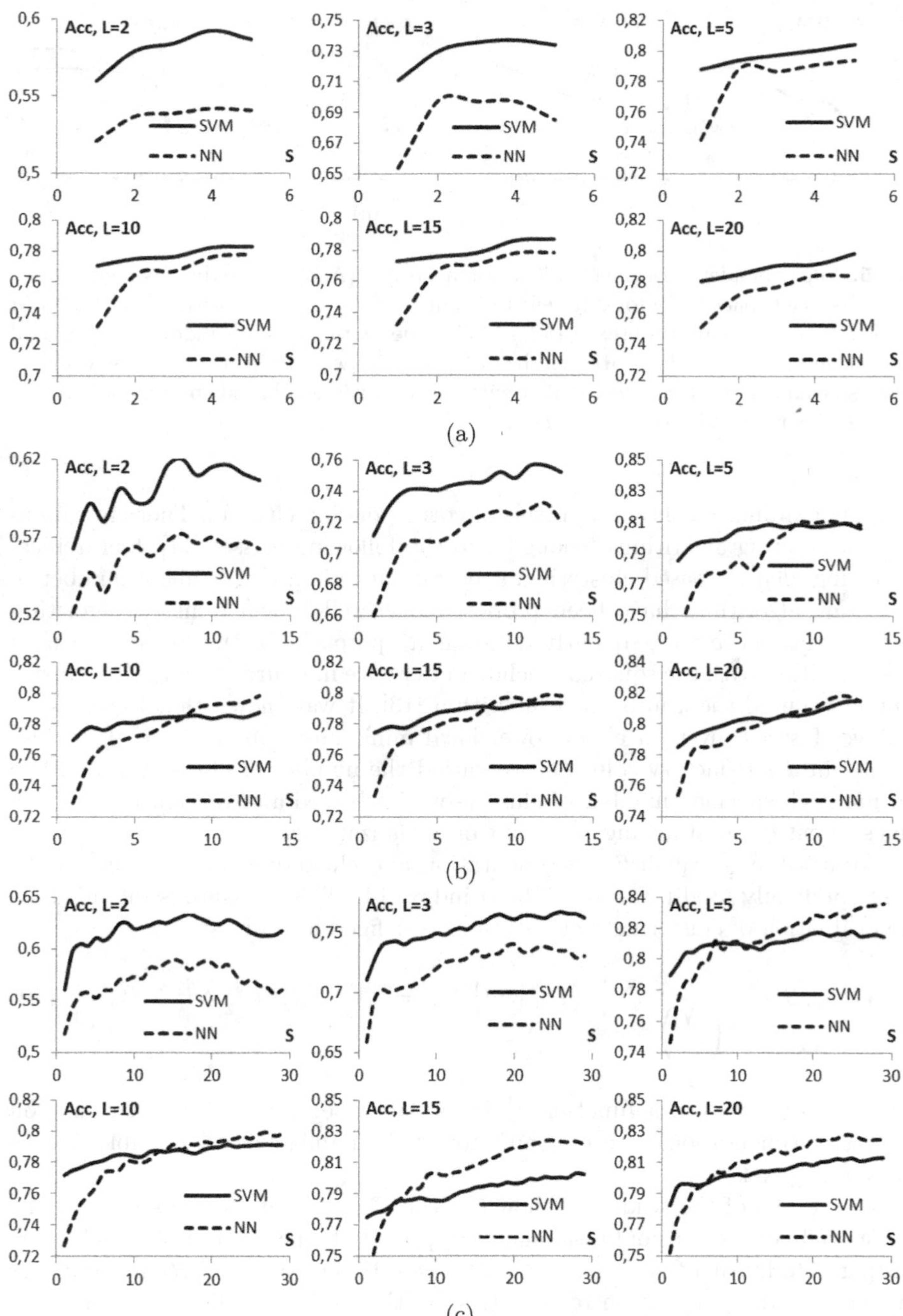

Fig. 4. The dependency of the classification accuracy (Acc) on the number S of the order statistics for the specified dimensionality L of the reduced space R^L, and different neighborhood radius R: radius $R = 1$ (a), neighborhood radius $R = 2$ (b), neighborhood radius $R = 3$ (c).

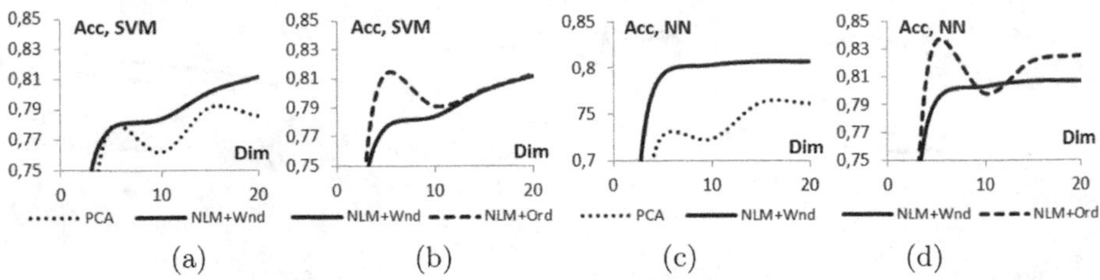

Fig. 5. The dependency of the classification accuracy (Acc) on the dimensionality Dim of the reduced space (R^L) for different techniques: SVM classifier with PCA and NLM + Window based spatial context (a), SVM classifier with NLM + window based spatial context and NLM + order statistics (b), NN classifier with PCA and NLM + Window based spatial context (c), NN classifier with NLM + window based spatial context and NLM + order statistics (d).

regions of an image containing pixels of corresponding clusters. There is a number of clustering algorithms belonging to the following classes [11]: hierarchical clustering, density-based clustering, spectral clustering, etc. While a number of clustering algorithms have been proposed, the well-known k-means algorithm [12] remains the most frequently mentioned approach. In this paper, we used this algorithm with the squared Euclidean distance measure. To initialize cluster centers we used the k-means++ algorithm [13]. It was shown that k-means++ achieves faster convergence to a lower local minimum than the base algorithm. To obtain a satisfactory solution, we varied the number of clusters from 10 to 50. For each specified number of clusters we initialized and ran clustering for 5 times to get the best arrangement out of initializations.

To measure the quality of segmentation and clustering, we used one of the most commonly used measures - Rand index [14]. This measure is intended for the evaluation of clustering methods, and is defined as follows:

$$RI(S_1, S_2) = \frac{1}{\binom{N}{2}} \sum_{i,j,i\neq j} \left(I(l_i^1 = l_j^1 \wedge l_i^2 = l_j^2) + I(l_i^1 \neq l_j^1 \wedge l_i^2 \neq l_j^2) \right) \quad (10)$$

Here $I()$ is the indicator function, l_i^k is the label (segment) of the i-th pixel on the k-th segmentation. The denominator is the number of all possible unique pairs of N pixels.

The results of the study are shown in the Fig. 6. As it can be seen, exploiting spatial context in nonlinear mapping provide higher values of Rand index compared to linear PCA technique in all cases. Higher values of RI mean better clustering and segmentation results. Besides that, using the first approach provided about 25% less number of segments on average (not shown on the figure), that means less oversegmentation.

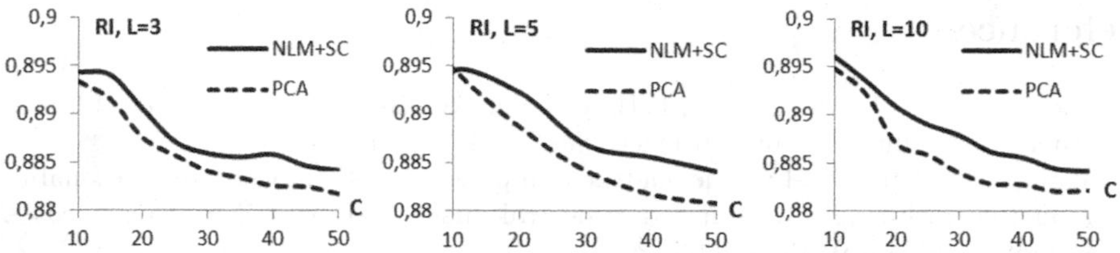

Fig. 6. The dependency of the Rand index (RI) on the number of clusters C for the specified dimensionality L of the reduced space (R^L), and dimensionality reduction techniques (nonlinear mapping with spatial context $NLM + SC$ and principal component analysis PCA).

3.3 Visualization of Hyperspectral Image Data

Described above approach to the dimensionality reduction can be used as a means of visual data analysis. By specifying the dimensionality $L = 2$ or $L = 3$ of the output space R^L, we can obtain 2D or 3D mappings of hyperspectral data. An example of 3D mapping of the test hyperspectral image is shown in the Fig. 2(c). We do not provide an evaluation of the quality of such mapping as this is out of the scope of the paper.

Besides that, dimensionality reduction techniques operating on the principle of preserving the pairwise distances between image pixels, can be applied to make hyperspectral image representation with special properties. An example of such representation is shown in the Fig. 2(a). Its distinguishing feature is that the distances between image pixels of the rendered image in color space approximate the distances between pixels in the source hyperspectral space. Similarly, the technique proposed in this paper can be used to produce false color representations of hyperspectral images using nonlinear mapping and spatial context.

4 Conclusion

In this work we proposed and evaluated the unsupervised nonlinear mapping technique, which exploit the spatial context of hyperspectral images. We evaluated two approaches to incorporating of the spatial context in the mapping technique. To evaluate the proposed technique, we considered three different tasks of hyperspectral image analysis, namely classification, segmentation, and visualization of hyperspectral images. Experimental study showed that proposed approaches can be successfully applied in hyperspectral image analysis.

Acknowledgements. The reported study was funded by RFBR according to the research project no. 16 – 29 – 09494 ofi_m, 16 – 37 – 00202–mol_a.

References

1. Richards, J.A., Jia, X., Ricken, D.E., Gessner, W.: Remote Sensing Digital Image Analysis: An Introduction. Springer, New York Inc. (1999)
2. Wang, J., Chang, C.-I.: Independent component analysis-based dimensionality reduction with applications in hyperspectral image analysis. IEEE Trans. Geosci. Remote Sens. **44**(6), 1586–1600 (2006)
3. Bachmann, C.M., Ainsworth, T.L., Fusina, R.A.: Improved manifold coordinate representations of large-scale hyperspectral scenes. IEEE Trans. Geosci. Remote Sens. **44**(10), 2786–2803 (2006)
4. Kim, D.H., Finkel, L.H.: Hyperspectral image processing using locally linear embedding. In: First International IEEE EMBS Conference on Neural Engineering, pp. 316–319 (2003)
5. Doster, T., Olson, C.C.: Building robust neighborhoods for manifold learning-based image classification and anomaly detection. In: Proceedings SPIE, vol. 9840, p. 984015 (2016)
6. Myasnikov, E.V.: Nonlinear mapping methods with adjustable computational complexity for hyperspectral image analysis. In: Proceedings of the SPIE, vol. 9875, p. 987508 (2015)
7. Sammon, J.W.: A nonlinear mapping for data structure analysis. IEEE Trans. Comput. **18**(5), 401–409 (1969)
8. Wang, L., Shi, C., Diao, C., Ji, W., Yin, D.: A survey of methods incorporating spatial information in image classification and spectral unmixing. Int. J. Remote Sens. **37**(16), 3870–3910 (2016)
9. Myasnikov, E.: Evaluation of stochastic gradient descent methods for nonlinear mapping of hyperspectral data. In: Campilho, A., Karray, F. (eds.) ICIAR 2016. LNCS, vol. 9730, pp. 276–283. Springer, Cham (2016). doi:10.1007/978-3-319-41501-7_31
10. Hyperspectral Remote Sensing Scenes. http://www.ehu.eus/ccwintco/index.php?title=Hyperspectral_Remote_Sensing_Scenes
11. Cariou, C., Chehdi, K.: Unsupervised nearest neighbors clustering with application to hyperspectral images. IEEE J. Select. Top. Sig. Process. **9**(6), 1105–1116 (2015)
12. Lloyd, S.P.: Least squares quantization in PCM. IEEE Trans. Inf. Theory **28**, 129–137 (1982)
13. David, A., Vassilvitskii, S.: K-means++: the advantages of careful seeding. In: Proceedings of the Eighteenth Annual ACM-SIAM Symposium on Discrete Algorithms, pp. 1027–1035 (2007)
14. Rand, W.M.: Objective criteria for the evaluation of clustering methods. J. Am. Stat. Assoc. **66**(336), 846–850 (1971)

Exploiting Visual Saliency Algorithms for Object-Based Attention: A New Color and Scale-Based Approach

Edoardo Ardizzone, Alessandro Bruno, and Francesco Gugliuzza(✉)

Dipartimento dell'Innovazione Industriale e Digitale (DIID), Università degli Studi di Palermo, Viale delle Scienze Ed. 6, 90128 Palermo, PA, Italy
{edoardo.ardizzone,alessandro.bruno15,francesco.gugliuzza}@unipa.it
http://www.diid.unipa.it/cvip

Abstract. Visual Saliency aims to detect the most important regions of an image from a perceptual point of view. More in detail, the goal of Visual Saliency is to build a Saliency Map revealing the salient subset of a given image by analyzing bottom-up and top-down factors of Visual Attention. In this paper we proposed a new method for Saliency detection based on colour and scale analysis, extending our previous work based on SIFT spatial density inspection. We conducted several experiments to study the relationships between saliency methods and the object attention processes and we collected experimental data by tracking the eye movements of thirty viewers in the first three seconds of observation of several images. More precisely, we used a dataset that consists of images with an object in the foreground on an homogeneous background. We are interested in studying the performance of our saliency method with respect to the real fixation maps collected during the experiments. We compared the performances of our method with several state of the art methods with very encouraging results.

Keywords: Visual saliency · Object-based attention · SIFT · Fixation maps · Dataset · Eye tracking

1 Introduction

The processing of visual information in human vision system begins in a thin layer made of neural tissue called retina. The architecture of our retina allows us to receive, every second, up to 10 billion bits information, while our cerebral cortex reaches about 10 billion neurons. Due to the lack of storage capability of our brain we cannot simultaneously perform complex analysis on all the input visual information [1]. One of the most important task of the human visual system is to detect the important visual subset, i.e. the salient subset. When a person performs any visual task (watching TV, driving a car) the eyes flick rapidly from place to place to inspect the visual scene. The movements of the

S. Battiato et al. (Eds.): ICIAP 2017, Part II, LNCS 10485, pp. 191–201, 2017.
https://doi.org/10.1007/978-3-319-68548-9_18

eyes, while observing a scene, are not random: each movement allows the central part of vision (fovea) to fall upon the region of interest of a picture (this is why vision is not uniform across our field of view and acuity decreases with eccentricity) [2].

There is an intimate connection between visual attention and eye movements; for this reason in the last decades, how, why and when we move our eyes is becoming a major topic in scientific research.

Two main factors drive visual attention: bottom-up factors and top-down factors. Bottom-up factors are derived solely from the visual scene [2]. Regions of interest attracting human attention are sufficiently discriminative with respect to surrounding features. This attentional mechanism is called exogenous [3]. Top-down attention, on the other hand, is driven by cognitive factors such as knowledge, expectations and current goals [4]. Other terms for top-down attention are endogenous [5], voluntary, or centrally cued attention.

The objective of Visual Saliency is to detect the most important regions of an image from a perceptual point of view, i.e. to imitate the behaviour of the human visual system. Visual Saliency studies are included in several research areas, such as Psychology, Neurobiology, Computer Science, Artificial Intelligence, Medicine; our work considers Visual Saliency from a Computer Science point of view. The objective is to build a saliency map revealing the salient subset in an image. It is usually a grayscale map and each pixel falls in the dynamic range [0, 255], where highest intensity values correspond to the most salient pixels of a picture.

In the same way as Visual Attention, Saliency approaches are based on Bottom-up factors and Top-down factors. More in detail, Visual Saliency methods can be grouped in three main approaches: Bottom-up, Top-down, Hybrid.

Bottom-up methods are stimulus-driven. These methods seek for the so-called "visual pop-out" saliency. Human attention, in these approaches, is considered as a cognitive process concentrating on the most unusual aspects of an environment while ignoring the common aspects. This consideration is implemented by several methods such as center-surround operation [6] and graph based activation maps [7].

Top-down methods for saliency detection are based on high level visual tasks such as Object Detection or Face Detection. In these methods the predefined task is given by the object class to be detected [8].

Hybrid methods are generally structured in two levels: a bottom-up layer gives rise to a noisy saliency map and a top-down layer filters out noisy regions in saliency maps created by the bottom-up layer.

The eye tracking technique has been typically adopted to examine human visual attention. The location of the eye fixation reflects attention, while the duration of the eye fixation reflects processing difficulty and amount of attention [9]. Specifically, fixation duration varies depending on types of information (e.g. texts or graphics) and types of tasks (e.g. reading or problem solving).

In our work we used an eye tracker to record the gaze path of thirty observers while viewing each image of a dataset [10]. Each image is show at full resolution for three seconds, separated by one second of viewing a gray screen (we adopted the same experimental approach of Torralba et al. [11]). The database [10] consists of several images with single objects in the foreground and homogeneous background color, but any dataset with a single main object (target) and a limited number of distractors in each image would have been appropriate as well. The viewers, while observing the images, sat at a distance of 70 cm from a 22 in. computer screen of 1920 × 1080 resolution.

We used the eye tracking data to create a ground truth made of fixation point maps showing where viewers look in the first three seconds of observation.

Our contributions in this work are three: a new available ground truth of fixation maps; a new saliency method, extending our previous study on visual saliency; a correlation study between saliency maps and the object attention process.

2 State of the Art

Models and approaches for visual saliency detection are inspired by human visual system mechanisms. As we reported in the first section of this paper, saliency methods can be divided in three main groups: Bottom-up, Top-down, Hybrid. In [6] the authors proposed a bottom-up approach based on multi-scale analysis of the image. In greater detail, multi-scale image features are used to create a topographical saliency map, then a dynamical neural network selects the attended locations with respect to the saliency values. The principle of center-surround difference is adopted in [12] for the parallel extraction of different feature maps. In [7] Harel et al. propose a saliency method (well known as GBVS) based on a biologically plausible graph-based model: the leading models of visual saliency may be organized into three stages: extraction, activation, normalization. Wang et al. [13] survey the corresponding literature on the low-level methods for visual saliency.

An effective method [14] for visual saliency detection based on multi-scale and multi-channel mean has been proposed by Sun et al. The image is decomposed and reconstructed by using wavelet transform and a bicubic interpolation algorithm is applied to narrow the filtered image in multi-scale. The saliency values are the distances between the narrowed images and the means of their channels. SIFT [15] keypoints density maps have been proposed in our previous works to extract saliency maps and texture scale [15–17].

In Top-down approaches [8,18], the visual attention process is considered task dependent, and the observer's expectations and wills analyzing the scene are the reason why a point is fixed rather than others. In [19] the authors perform saliency detection with a Top-down model that jointly learns a Conditional Random Field (CRF) and a visual dictionary.

Generally Hybrid systems for saliency use the combination of bottom-up and top-down stimuli. In many hybrid approaches [20,21], a Top-down layer is used to refine the noisy map extracted from the Bottom-up layer. For example the Top-down component in [20] is face detection. Chen et al. [21] used a combination of face and text detection and they found the optimal solutions through branch and bound technique. A well known state-of-the-art hybrid approach was proposed by Judd et al. [11] in addition to a database [22] of eye tracking data from 15 viewers. Low, middle and high-level features of this data have been used to train a model of saliency.

Yu et al. in [23] used a paradigm based on Gestalt grouping cues for object-based saliency detection.

In the last years several researchers focused their attention on deep learning approaches for saliency prediction because high-quality visual saliency model can be learned by using deep convolutional neural networks (CNNs). For instance, in [24] the authors introduced a neural network architecture, which has fully connected layers on top of a CNNs responsible for feature extraction at different scales.

The authors of [25] reported a comparative study that evaluates the performances of 13 state-of-the-art saliency models. A new metric is also proposed and compared with previous models. In [26] the authors give some formal definitions on three different type of approaches (Bottom-up, Top-down, Hybrid) and an overview on existing methods. Furthermore, the authors offer a description of publicly available datasets and the performance metrics used.

3 Proposed Methods

3.1 Eye Tracking Data Acquisition

We chose 5 different objects from the Object Pose Estimation Database (OPED) [10,27], and for each object we selected 19 views having a 130° fixed vertical angle and an horizontal angle ranging from 0° to 180° in 10° increments. The resulting 95 images have been thoroughly filtered to attenuate noise that could shift human attention and padded to fill the 22″ screen at 1920×1080 resolution. We showed the images to 24 users (males and females between 21 and 34 years old) placed at approximately 70 cm from the screen.

The acquisition procedure for each user was as follows: the Tobii EyeX [28] running at 60 Hz refresh rate was calibrated to the user, then each image was shown for 3 s while capturing all user's saccades and fixations, separated by 1 s of neutral gray screen, to keep the results consistent to those of previous works in literature [11] (Fig. 1).

Fixation Map Generation. Data acquired from the Tobii EyeX include three arrays of the same length: an array of positions looked at by the left eye, an array of positions looked at by the right eye, and an array of sampling times. The fixation points are calculated averaging data of both eyes and converting the result to screen coordinates, then full resolution maps are built by adding one to a pixel's value each time a user has looked at said pixel. The map are subsequently averaged using a Gaussian convolution kernel and normalized to [0, 1] (Fig. 2).

Fig. 1. The setup used for eye fixation data acquisition.

3.2 Proposed Saliency Map Generation Method

We aimed to improve the method we (Ardizzone et al.) developed in 2011 [15] by adding chroma information to the saliency map generation algorithm based on SIFT [29] Density Maps (SDMs). A SDM is built by counting the number of detected SIFT keypoints inside a sliding window of size k x k centered on each pixel of the image. To obtain a valid saliency map, the SDM is further processed by taking the absolute difference of each pixel with the most frequent value (mode) of the map, rescaling the values to [0, 1] and blurring the result with an average filter which has a window size that is half of that used to build the map (k).

Color-based saliency has been implemented in two ways by harnessing the power of HSV and CIE L*a*b* color spaces. We early found that the optimal SDM window size equation we used in [15]:

$$k = 2^{\lfloor \log_2 \left(\frac{\min(M,\ N)}{4} \right) \rfloor} \quad (1)$$

is unsuitable in object attention because it is calculated on entire image size, while the object only takes a small central portion of it, causing excessive loss of detail in the generated saliency maps. We overcame the problem by first taking the mean of the dimensions of the object bounding boxes in all images used during the data acquisition phase, then applying (1) to the calculated values.

HSV Color Space Saliency. In HSV an image is expressed using cylindrical coordinates, where hue is an angular dimension that goes from 0° to 360° and then back to 0°, while saturation and value are linear dimensions. 8-bit RGB images can be easily converted to HSV by projecting the RGB cube on a chromaticity plane in such a way that an hexagon is formed:

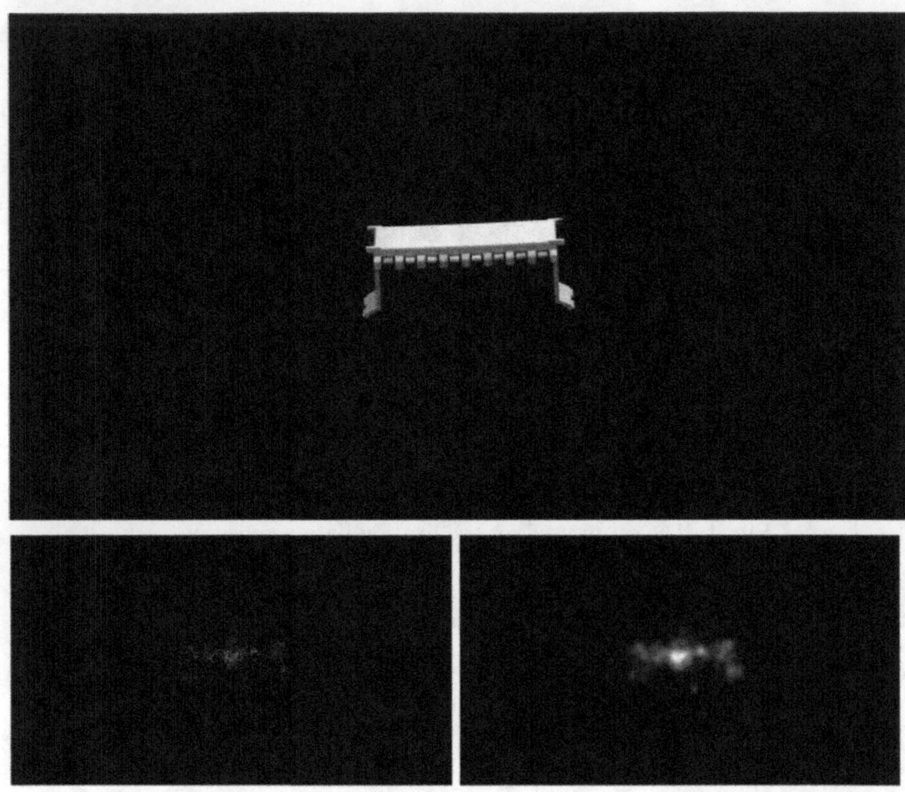

Fig. 2. An image from the OPED and its fixation map before and after Gaussian blurring.

$$
\begin{aligned}
C_{max} &= \max(R, G, B) \\
C_{min} &= \min(R, G, B) \\
\Delta &= C_{max} - C_{min} \\
H &= \begin{cases} 0 \text{ if } C_{max} = 0 \\ \left(60 \times \frac{G-B}{\Delta} + 360\right) \bmod 360 & \text{if } R = C_{max} \\ 60 \times \frac{B-R}{\Delta} + 120 & \text{if } G = C_{max} \\ 60 \times \frac{R-G}{\Delta} + 240 & \text{if } B = C_{max} \end{cases} \\
S &= \frac{\Delta}{C_{max}} \\
V &= \frac{C_{max}}{255}
\end{aligned}
\tag{2}
$$

We convert hue and saturation from polar coordinates to cartesian coordinates[1], in order to eliminate the discontinuity around zero in hue values:

$$\begin{aligned} X &= S \circ \cos(H) \\ Y &= S \circ \sin(H) \end{aligned} \tag{3}$$

then we rescale the X, Y, V channels to the [0, 1] range for convenience of processing and separately calculate statistically processed SDMs. The three maps are combined into the final saliency map (Fig. 3):

$$SM_{HSV} = \frac{1}{3}(SM_H + SM_S + SM_V). \tag{4}$$

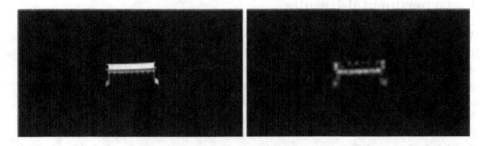

Fig. 3. An image from the OPED and its SIFT saliency map calculated in HSV space.

CIE L*a*b* Color Space Saliency. HSV space still shows some shortcomings, namely hue and saturation channels are dominated by noise when brightness is low; furthermore, it is not biologically inspired, and does not model the HVS color opponent process [30]. Therefore, we decided to implement SDM calculation also in the CIE L*a*b* space, which is perceptually uniform and designed with color opponency in mind [31].

The processing steps are essentially the same as the previous method: RGB → L*a*b* conversion (D65 illuminant used as reference), channel range rescaling, SDM calculation, statistical processing and fusion. Coordinate transformation has been omitted because this color space does not have mathematical discontinuities (Fig. 4).

[1] Not to be confused with CIE XYZ.

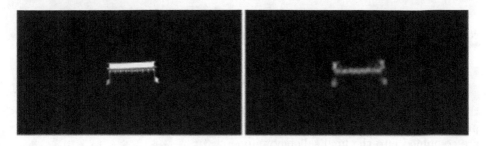

Fig. 4. An image from the OPED and its SIFT saliency map calculated in L*a*b* space.

4 Experimental Results

We generated saliency maps using various methods, as our legacy work [15], Itti-Koch-Niebur [6], GBVS [7], Judd [11], our two new color-based methods and a fixed centered Gaussian distribution as a baseline [11]. We ran tests on our 95 image dataset and its related fixation point and fixation map database, on an Intel Core i7-4770 computer with 4 cores (8 threads) and 16 GB of RAM. For the calculation of GBVS and Itti-Koch-Niebur saliency maps the GBVS Toolbox [32] has been used, as it includes an enhanced implementation of Itti's algorithm; Judd saliency maps were instead generated running Judd's code [22] with its original trained parameters. We binarized saliency and fixation maps at various percentiles [11,15] (between 0.95 and 0.5) and evaluated the performance of our method in terms of F-measure values:

$$P = \frac{n(M_D \cap M_R)}{n(M_D)}; R = \frac{n(M_D \cap M_R)}{n(M_R)}$$
$$F_1 = 2\frac{P \times R}{P + R} \tag{5}$$

where M_D is the binary version of the detected saliency map, while M_R is the binary version of the reference fixation map. We also calculated Normalized Scanpath Saliency (NSS) values, which is a well balanced, binarization-independent metric [33].

From Fig. 5, we note an opposite trend with respect to natural image saliency model performances reported in other works: in object attention, as saliency threshold reduces the F-measure tends to reduce as well, instead of increasing. The performances of both our models, instead, increase slightly with threshold until they reach a plateau at 90% saliency levels. Our CIE L*a*b*-based method always gets best results in both metrics, while the HSV-based method underperforms at high saliency levels with respect to GBVS and our previous work.

The execution time required for calculating a HSV or L*a*b* saliency map is about 12 s for a 1920 × 1080 image.

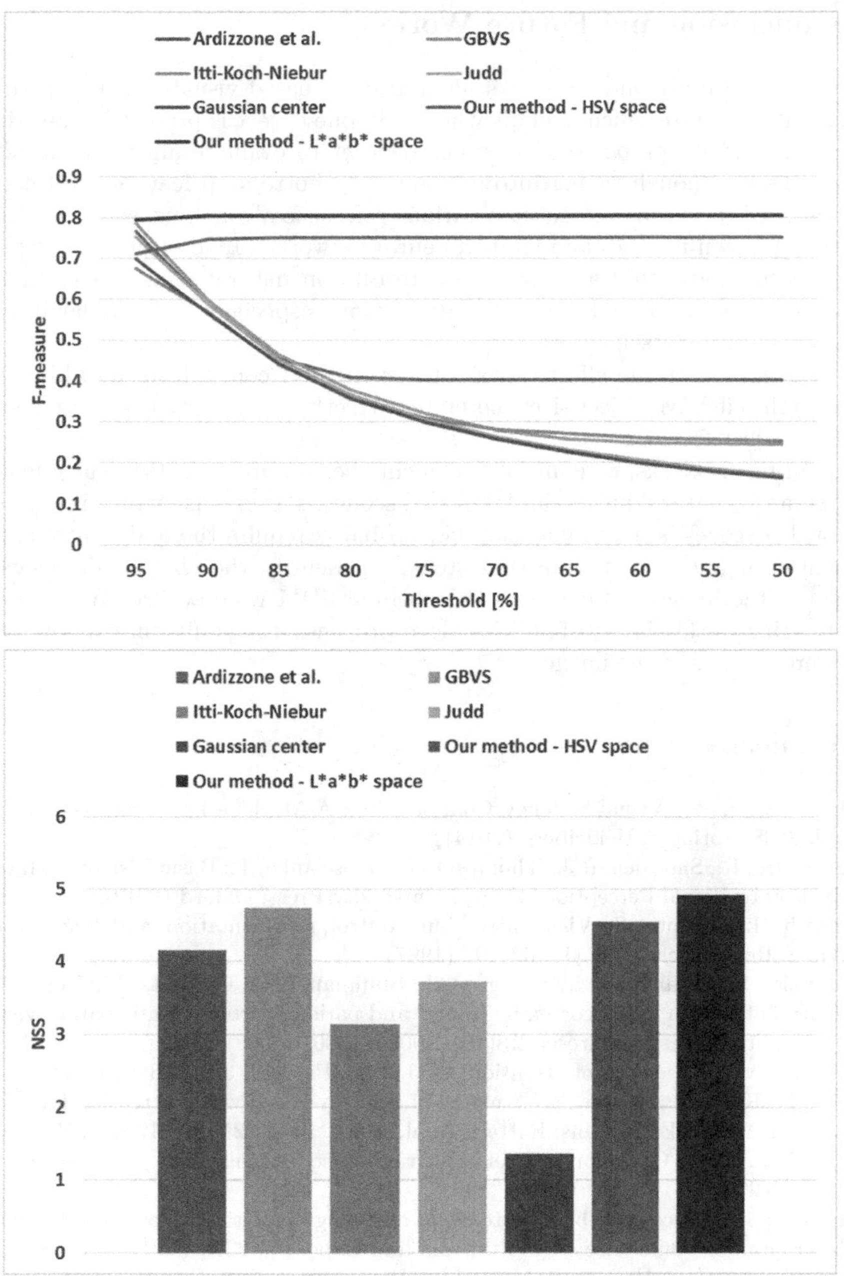

Fig. 5. Performance graphs of various saliency models in terms of F-measure vs. threshold (top) and NSS values (bottom).

5 Conclusion and Future Works

In this paper we presented a new scale and color-based visual saliency method to generate accurate saliency maps when only one object is present in the stimulus scene and we proposed a reference dataset to evaluate algorithm results. Our method, although taking into account only bottom-up features and being unsupervised, performs better than various reference algorithms, some of which exploit top-down features and trained neural networks (Judd et al.). We expect that this method would also give good results on natural images depicting a main object and a limited number of distractors, especially if dissimilar in size to the main object itself.

We believe that the effectiveness of our method comes from its ability to adapt to the effective object size, therefore correctly keeping track of the saliency of small object features.

In our future works, we plan to implement these improvements in our saliency algorithms for natural images and crowded scenes. The extension of this method to crowded scenes is not trivial and will probably require the addition of extra segmentation and object detection steps, to identify the size of all relevant objects in the image and compute the optimal SDM window size. We are also investigating the feasibility of multi-scale approaches using different window sizes on different parts of the image.

References

1. Li, J., Gao, W.: Visual Saliency Computation: A Machine Learning Perspective, vol. 8408. Springer, Heidelberg (2014)
2. Snowden, R., Snowden, R.J., Thompson, P., Troscianko, T.: Basic Vision: An Introduction to Visual Perception. Oxford University Press, Oxford (2012)
3. Egeth, H.E., Yantis, S.: Visual attention: control, representation, and time course. Annu. Rev. Psychol. **48**(1), 269–297 (1997)
4. Bressler, S.L., Tang, W., Sylvester, C.M., Shulman, G.L., Corbetta, M.: Top-down control of human visual cortex by frontal and parietal cortex in anticipatory visual spatial attention. J. Neurosci. **28**(40), 10056–10061 (2008)
5. Posner, M.I.: Orienting of attention. Q. J. Exp. Psychol. **32**(1), 3–25 (1980)
6. Itti, L., Koch, C., Niebur, E.: A model of saliency-based visual attention for rapid scene analysis. IEEE Trans. Pattern Anal. Mach. Intell. **20**(11), 1254–1259 (1998)
7. Harel, J., Koch, C., Perona, P., et al.: Graph-based visual saliency. In: NIPS, vol. 1, p. 5 (2006)
8. Luo, J.: Subject content-based intelligent cropping of digital photos. In: 2007 IEEE International Conference on Multimedia and Expo, pp. 2218–2221. IEEE (2007)
9. Tsai, M.J., Hou, H.T., Lai, M.L., Liu, W.Y., Yang, F.Y.: Visual attention for solving multiple-choice science problem: an eye-tracking analysis. Comput. Educ. **58**(1), 375–385 (2012)
10. http://www.cvl.isy.liu.se/research/objrec/posedb/
11. Judd, T., Ehinger, K., Durand, F., Torralba, A.: Learning to predict where humans look. In: 2009 IEEE 12th International Conference on Computer Vision, pp. 2106–2113. IEEE (2009)

12. Koch, C., Ullman, S.: Shifts in selective visual attention: towards the underlying neural circuitry. In: Vaina, L.M. (ed.) Matters of Intelligence. Synthese Library (Studies in Epistemology, Logic, Methodology, and Philosophy of Science), pp. 115–141. Springer, Dordrecht (1987). doi:10.1007/978-94-009-3833-5_5
13. Wang, L., Dong, S.L., Li, H.S., Zhu, X.B.: A brief survey of low-level saliency detection. In: 2016 International Conference on Information System and Artificial Intelligence (ISAI), pp. 590–593. IEEE (2016)
14. Sun, L., Tang, Y., Zhang, H.: Visual saliency detection based on multi-scale and multi-channel mean. Multimedia Tools Appl. **75**(1), 667–684 (2016)
15. Ardizzone, E., Bruno, A., Mazzola, G.: Visual saliency by keypoints distribution analysis. In: Maino, G., Foresti, G.L. (eds.) ICIAP 2011. LNCS, vol. 6978, pp. 691–699. Springer, Heidelberg (2011). doi:10.1007/978-3-642-24085-0_70
16. Ardizzone, E., Bruno, A., Mazzola, G.: Scale detection via keypoint density maps in regular or near-regular textures. Pattern Recogn. Lett. **34**(16), 2071–2078 (2013)
17. Ardizzone, E., Bruno, A., Mazzola, G.: Saliency based image cropping. In: Petrosino, A. (ed.) ICIAP 2013. LNCS, vol. 8156, pp. 773–782. Springer, Heidelberg (2013). doi:10.1007/978-3-642-41181-6_78
18. Sundstedt, V., Chalmers, A., Cater, K., Debattista, K.: Top-down visual attention for efficient rendering of task related scenes. In: VMV, pp. 209–216 (2004)
19. Yang, J., Yang, M.H.: Top-down visual saliency via joint CRF and dictionary learning. IEEE Trans. Pattern Anal. Mach. Intell. **39**(3), 576–588 (2017)
20. Tsotsos, J.K., Rothenstein, A.: Computational models of visual attention. Scholarpedia **6**(1), 6201 (2011)
21. Chen, L.Q., Xie, X., Fan, X., Ma, W.Y., Zhang, H.J., Zhou, H.Q.: A visual attention model for adapting images on small displays. Multimedia Syst. **9**(4), 353–364 (2003)
22. http://people.csail.mit.edu/tjudd/WherePeopleLook/index.html
23. Yu, J.G., Xia, G.S., Gao, C., Samal, A.: A computational model for object-based visual saliency: spreading attention along Gestalt cues. IEEE Trans. Multimedia **18**(2), 273–286 (2016)
24. Li, G., Yu, Y.: Visual saliency detection based on multiscale deep CNN features. IEEE Trans. Image Process. **25**(11), 5012–5024 (2016)
25. Toet, A.: Computational versus psychophysical bottom-up image saliency: a comparative evaluation study. IEEE Trans. Pattern Anal. Mach. Intell. **33**(11), 2131–2146 (2011)
26. Duncan, K., Sarkar, S.: Saliency in images and video: a brief survey. IET Comput. Vis. **6**(6), 514–523 (2012)
27. Viksten, F., Forssén, P.E., Johansson, B., Moe, A.: Comparison of local image descriptors for full 6 degree-of-freedom pose estimation. In: IEEE International Conference on Robotics and Automation, ICRA 2009, pp. 2779–2786. IEEE (2009)
28. Gibaldi, A., Vanegas, M., Bex, P.J., Maiello, G.: Evaluation of the Tobii EyeX Eye tracking controller and Matlab toolkit for research. Behav. Res. Methods **49**, 1–24 (2016)
29. Lowe, D.G.: Distinctive image features from scale-invariant keypoints. Int. J. Comput. Vis. **60**(2), 91–110 (2004)
30. Engel, S., Zhang, X., Wandell, B.: Colour tuning in human visual cortex measured with functional magnetic resonance imaging. Nature **388**(6637), 68–71 (1997)
31. Sharma, G.: Color fundamentals for digital imaging. In: Digital Color Imaging Handbook. CRC Press (2002)
32. http://www.vision.caltech.edu/~harel/share/gbvs.php
33. Bylinskii, Z., Judd, T., Oliva, A., Torralba, A., Durand, F.: What do different evaluation metrics tell us about saliency models? arXiv preprint arXiv:1604.03605 (2016)

Face Recognition with Single Training Sample per Subject

Taher Khadhraoui(✉) and Hamid Amiri

SITI Laboratory, National Engineering School of Tunis (ENIT),
University of Tunis El Manar, Tunis, Tunisia
khadhra.th@gmail.com, hamid.amiri@enit.rnu.tn

Abstract. This paper presents an approach called Patch uniform Local Binary Patterns (PuLBP) based Local Generic Representation (LGR) for face recognition. In fact, we insert a novel block that comports a uLBP in order to approximate both variation and reference subsets. Consequently, the focus will be on the difficult problem of a unique sample by person in a gallery set. More specifically, the major problem is if having solely one training person in every class is possible. The generation of virtual samples of every sample is one of the innovations of our technique. In a gallery set, each sample is used to generate the intra-personal variety of distinct individuals. We demonstrate the experimental results of our novel algorithm on many reference databases that include the FRGCv1, AR, the Georgia Tech (GT), the FEI and the Extended Cohn-Kanade.

1 Introduction

A multitude of new studies have focused on the way to ameliorate pattern recognition performance. Face recognition (FR) along with a Single Training Sample per Subject (STSS) is a significant difficulty because of the insufficient information got out of the training set in order to predict the likely facial variations within the query image. This case makes it impossible for many learning algorithms to solve this issue because those algorithms require training samples per subject in order to represent the query face image. One of the conventional learning techniques for FR is the LGR technique [26], that consists in producing an intra-class variation dictionary. In order to establish an effective and strong classification/recognition application, the amount of gallery cases per subject is one of the principal difficulties. Recognition with a unique gallery sample by individual/subject, needs data to predict the variety among samples of the subject, except if it is used along with a dictionary [11]. Moreover, in diverse applications, an amount of training samples by person may be available, covering a variety of expressions, diversity in illumination, or occlusion [21].

For many years, FR has been considered as a challenging task because of the variety of types of large face variations. For example, one may cite illuminations, expressions and disguises. Generally, the theory of sparse coding or representation has been employed for signal processing [9]. The Sparse-Representation-based Classification (SRC) theory of face recognition [25] is to stand for a query

S. Battiato et al. (Eds.): ICIAP 2017, Part II, LNCS 10485, pp. 202–212, 2017.
https://doi.org/10.1007/978-3-319-68548-9_19

face image just as a sparse linear compound of training samples from every class, from a dictionary, and hence representing it via just some non-zero coefficients existing within a sparse vector.

An SRC-algorithm-based face recognition through solving a l0-norm minimization problem was proposed by Fan et al. [5]. Indeed, the SRC algorithm non-zero coefficients should focus on the training cases having the same subject as a query sample. Besides, a new kernel sparse coding algorithm for efficient image classification and face recognition has been proposed by Geo et al. [7]. Wang et al. [20] presented a weighted SRC-algorithm-based locality that employed linearity and locality data. Khadhraoui et al. [10] put forward a multimodal biometric system based on the integration of 2D and 3D face modalities. Its main novelty was to apply a Relevance Vector Machine technique for score level fusion. An algorithm applying a sparse multiscale representation based on shearlets to extract the essential geometric content of facial features was proposed by Borgi et al. [2]. This article deals with the issue of available STSS. This subject is linked to a small sample size issue within object recognition. Even though the unique training sample has advantages, the creation of a face database is quick and simple, the storage conditions are modest.

Our algorithm is to identify the possible facial variations caused by a group of labeled faces (the variations group being considered as the label), that is able to classify the variety of any given query face. For this objective, a variant of uLBP called Patch uLBP (PuLBP), is employed as the face attributes. The very high effectiveness of PuLBP in facial recognition shows that it is able to efficiently generate discriminating facial geometric data. The extraction of nearby patches enables the construction of a local gallery dictionary, simultaneously a generic variation dictionary is constructed through the extraction of representative information out of external generic data in order to identify the likely possible facial variations in an automatic way. Every patch of the uLBP query sample is the only representation supplied at the corresponding location via the patch of uLBP gallery dictionary as well as the patch of uLBP variation dictionary. The half-quadratic optimization procedure is usually essential in order to find out a solution to the optimization problem. To conclude, the general residual representation of the uLBP query sample by every class is employed to execute the global classification.

This article is organized as follows: Sect. 2 describes the essential background of the Local Binary Pattern (LBP) versus the uniform LBP. A summary of the proposed approach is offered in Sect. 3. Section 4 presents the experimental results. The conclusion and future work are illustrated in Sect. 5.

2 Background of Local Binary Patterns

The initial Local Binary Patterns (LBP) operator was first offered by Ojala et al. [16] to characterize the image texture. The LBP is a new inexpensive image feature for texture classification/recognition and is successfully used with 2D face recognition [1]. In a formal way, The LBP result could be computed by setting a limit for the difference between the central pixel and its neighbors [1], and is expressed in decimals as:

$$LBP(x_c, y_c) = \sum_{k=1}^{n} s(i_n - i_c)2^n \tag{1}$$

where n runs over the 8 neighbors of the central pixel and i_c and i_n in are gray-level values of the central pixel and the neighbouring pixels respectively. The function $s(n)$ is defined as:

$$s(n) = \begin{cases} 1 \text{ if } n \geq 0, \\ 0 \text{ if } n < 0. \end{cases} \tag{2}$$

The elementary LBP has a 3×3 patch limitation is not big enough to capture the most important descriptors with the means of large scale structures. The LBP operator can be extended to a further extension named uniform [16]. The latter concept was presented in [16]. It was remarked that a number of binary codes came out as essential characteristics of the texture, and they represented for the big majority of possible models, superior to 90% [1]. These models were named "uniform models" since they had one common thing: At most the circular binary code had two transitions; either one-to-zero or zero-to-one.

To describe the uniformity of a neighborhood g_p concept, the uniformity measure $U(LBP_{(P,R)})$ is essential and is defined as:

$$U(LBP_{(P,R)}) = |s(g_{P-1} - g_c) - s(g_0 - g_c)| + \sum_{p=1}^{P-1} |s(g_p - g_c) - s(g_{P-1} - g_c)| \tag{3}$$

The LBP operator $LBP_{(P,R)}$ generates diverse 2^p output values, corresponding to binary 2^p codes [18]. It could be employed to represent spots, edges, fine lines, and corners. Ojala et al. called them uniform patterns [16], named $LBP^{u2}_{(P,R)}$. The uniformity measure $U(LBP_{(P,R)})$, as presented in Eq. (3), records the number of spatial transitions in the bit pattern as well as the uniform format, that includes at best two bit transitions, $U(LBP_{(P,R)}) \leq 2$. The operator uLBP $LBP^{u2}_{(P,R)}$ is determined as follows:

$$LBP^{u2}_{(P,R)}(x, y) \begin{cases} I(LBP_{(P,R)}(x, y)) \text{ if } U(LBP_{(P,R)}) \leq 2, \\ (P-1)P + 1 \text{ if } U(LBP_{(P,R)}) > 2. \end{cases} \tag{4}$$

where $u2$ expressed in Eq. (4), shows that the definition is linked to the uniform patterns with a value $U \leq 2$.

Using the uniform LBP code had two benefits. First is gaining memory and computation time. And second is that $LBP^{u2}_{(P,R)}$ can solely recognize significant local textures, as corners, spots, edges and fine line [18]. In fact, Ojala et al. [16] showed that the uLBP contained more than 90% of image information.

3 Overview of Proposed Approach

In order to succeed in determining the class label of the uz query face, an uLBP-LGR-based classification framework can be suggested. Figure 1 presents an overview of the suggested approach. This approach operates in two phases: offline and online.

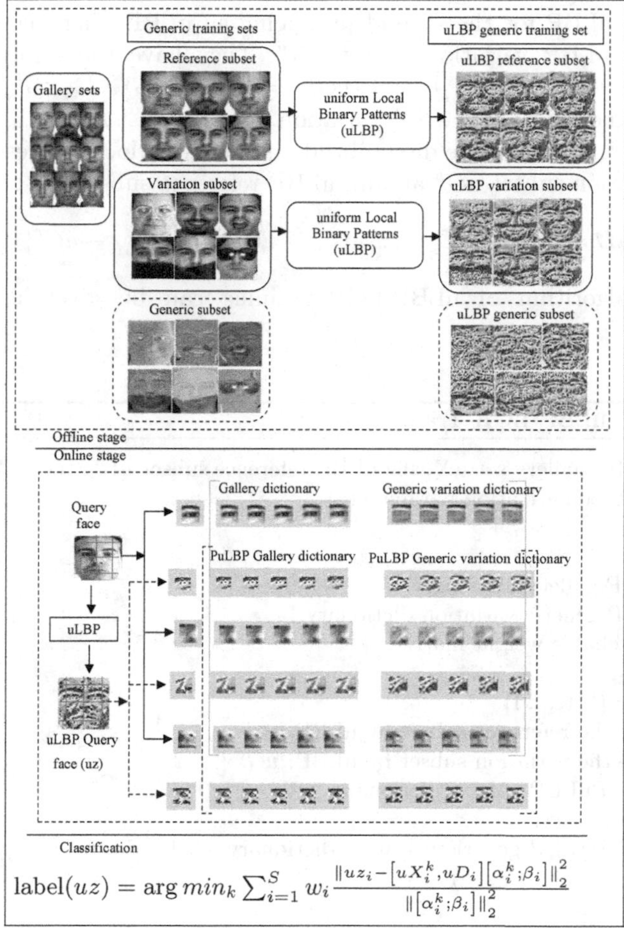

Fig. 1. Block diagram of proposed approach

To sum up, this can be depicted as a stage of feature extraction that would be based on uLBP and succeeded by an LGR based classification stage. The uLBP can efficiently capture the geometric elements in the facial image, to produce peculiar data of facial variation. This step is essential to ameliorate the recognition performance.

3.1 Offline Phase

In this phase, all the images from the generic training sets and gallery sets are composed as an STSS. The dataset gallery comprises the face images of the training. The generic training dataset is entirely divided into two sub categories or subsets that are: The reference subset G^r including neutral face images; and the variation subset G^v that includes variety in illumination, face expression and disguise (with sunglasses and scarf). Both sub categories are approximated by the uLBP for the sake of producing an uLBP reference subset uG^r as well as an uLBP variation subset uG^v. They have these forms: $uG^r = [uG_1^r, uG_2^r, \ldots, uG_m^r, \ldots, uG_M^r]$ and $uG^v = [uG_1^v, uG_2^v, \ldots, uG_m^v, \ldots, uG_M^v]$, where uG_m^v is the subset of the mth variation, $m = 1, 2, \ldots, M$. An uLBP generic subset is then fulfilled via the dissimilitude between the local representations of an uLBP variation subset uG^v and an uLBP reference subset uG^r.

$$uD = [uG_1^v - uG^r, \ldots, uG_m^v - uG^r, \ldots, uG_M^v - uG^r] \tag{5}$$

The full algorithm of uLBP-LGR-Training can be given as shown in Algorithm 1.

Algorithm 1. uLBP-LGR-Training

Input: The uLBP gallery set uX, the uLBP reference subset uG^r, the uLBP variation subset uG^v and regularization parameter λ.

Output:
uX: The PuLBP gallery dictionary;
uD: The PuLBP generic variation dictionary;
w: The most probable weight matrix.

1: Initialize $w = [1, 1, ..., 1]$
2: Approximate the reference subset by uLBP: uG^r
3: Approximate the variation subset by uLBP: uG^v
4: Calculate the PuLBP gallery dictionary: uX
$uX = [ux_1, ux_2, ..., ux_k]$
5: Calculate the PuLBP generic variation dictionary uD by Eq. (5)
6: Calculate $\{\alpha_i, \beta_i\}$ by Eq. (8)
7: While convergence
8: Update w by Eq. (15)
9: End

3.2 Online Phase

In this phase, the test face image is encoded or approximated by the uLBP. This test image is divided into several patches that are presented as $\{uz_1, uz_2, \ldots, uz_S\}$. Likewise, the partitioning of the gallery dictionary uX and the generic variation dictionary uD respectively gives, $\{uX_1, uX_2, \ldots, uX_S\}$ and

$\{uD_1, uD_2, \ldots, uD_S\}$. uX_i and uD_i are respectively the PuLBP gallery dictionary and the PuLBP generic variation dictionary linked to every local patch uz_i, $i = 1, 2, \ldots S$. On the basis of the two dictionaries uX_i and uD_i, every local patch uz_i is represented as follows:

$$uz_i = uX_i\alpha_i + uD_i\beta_i + e_i, i = 1, 2, \ldots S \tag{6}$$

where α_i and β_i are respectively vectors of the representation of uz_i over uX_i and uD_i, and where e_i stands for the residual representation. The patch of every dictionary (uX_i and uD_i) is the only representation assigned to the corresponding location of every uLBP query sample patch. Lastly, the overall uLBP test sample residual representation by each class is employed to accomplish the global classification. The identity of a test sample uz is calculated as follows:

$$\text{label}(uz) = \arg min_k \sum_{i=1}^{S} w_i \frac{\|uz_i - \left[uX_i^k, uD_i\right]\left[\alpha_i^k; \beta_i\right]\|_2^2}{\|\left[\alpha_i^k; \beta_i\right]\|_2^2} \tag{7}$$

$$\text{Subject to} \quad uz_i = uX_i\alpha_i + uD_i\beta_i + e_i, i = 1, 2, \ldots S$$

The full algorithm of uLBP-LGR-Classifying can be provided as presented in Algorithm 2.

Algorithm 2. uLBP-LGR-Classifying

Input:
uz: The uLBP query sample;
uX: The PuLBP gallery dictionary;
uD: The PuLBP generic variation dictionary;
w: The most probable weight matrix

Output: identity$\{uz\}$: Predicted class value

1: Partition uz, uX and uD into patches
2: Calculate $\{\widehat{\alpha_i}, \widehat{\beta_i}\}$ by Eq. (9)
3: Calculate the residual e_i by Lasso function [19]
4: Calculate uz_i by Eq.(6)
5: Output the class label of sample uz by Eq.(7)

3.3 Optimization and Classification

The half-quadratic optimization [15] has been widely used to explain the problem of minimization below as it is the efficient and unique solution to calculate $\{\alpha_i, \beta_i\}$.

$$\min_{\{\alpha_i, \beta_i\}} \sum_{i=1}^{S} l(\|e_i\|_2) + \lambda R(\alpha_i, \beta_i) \tag{8}$$

Clearly, to find significantly better solutions of the vectors α_i and β_i, proper regularization must be applied on α_i and β_i and the residual representation e_i makes possible the definition of the suitable loss function. The latter is defined on l_2-norm of e_i is represented by the $l(\|e_i\|_2)$ and some regularizers imposed on the representation coefficients are provided by $R(\alpha_i, \beta_i)$.

Now, the task is to calculate the regularize $R(\alpha_i, \beta_i)$ as well as the loss function $l(\|e_i\|_2)$. On the basis of the above analysis, one can simply get the below optimal regularization formulation:

$$\{\widehat{\alpha_i}, \widehat{\beta_i}\} = \min_{\{\alpha_i,\beta_i\}} \|uz_i - uX_i\alpha_i + uD_i\beta_i\|_2^2 + \lambda(\|\alpha_i\|_2^2 + \|\beta_i\|_2^2) \tag{9}$$

And then we compute the residual representation

$$e_i = \|uz_i - uX_i\widehat{\alpha_i} + uD_i\widehat{\beta_i}\|_2 \tag{10}$$

After computing the residual representation is in Eq. (10), a Correntropy Induced Metric (CIM) [12] is characterized as:

$$CIM(e_i) = (k_\sigma(0) - k_\sigma(e_i))^{1/2} \tag{11}$$

where $k_\sigma(x)$ is the kernel function.

The CIM is employed to measure the residual representation of every patch. Finally, the suggested uLBP-LGR model becomes:

$$\min_{\{\alpha_i,\beta_i\}} \sum_{i=1}^{S} (l - k_\sigma(\|e_i\|_2)) + \lambda(\|\alpha_i\|_2^2 + \|\beta_i\|_2^2) \tag{12}$$

$$\text{Subject to} \quad uz_i = uX_i\alpha_i + uD_i\beta_i + e_i, i = 1, 2, \ldots S$$

The augmented or increased minimization problem of Eq. (12) can be described as:

$$\min_{\{A,w\}} \sum_{i=1}^{S} (\frac{1}{2} w_i \|uz_i - uX_i\alpha_i - uD_i\beta_i\|_2^2 + \varphi(w_i)) + \lambda\|A\|_2^2 \tag{13}$$

where $A = [a_1, a_2, \ldots, a_i, \ldots, a_S]$ with $a_i = [\alpha_i; \beta_i]$, $w = [w_1, w_2, \ldots, w_S]$ and $\varphi(w_i)$ is the dual function. The above equation can be effectively optimized through the half-quadratic minimization [15], via the alternative update of A and w, when w is fixed, A shall be updated as:

$$\widehat{A} = \arg min_A \sum_{i=1}^{S} (w_i \|uz_i - uX_i\alpha_i - uD_i\beta_i\|_2^2) + \lambda\|A\|_F^2 \tag{14}$$

when A is fixed, the weights w can be solved by

$$\widehat{w_i} = \frac{1}{\sigma^2} exp(\frac{-\|uz_i - uX_i\alpha_i - uD_i\beta_i\|_2^2}{2\sigma^2}) \tag{15}$$

The weight w_i corresponds to the ith patch and is used for the control the portion of $\|e_i\|_2$ in the entire energy of Eq. (13).

3.4 Parameter Setting

In the experimental work, every image is resized and fixed as 80 × 80 pixels and the patch size is fixed as 20 × 20. It is employed for patch/block-based-methods comprising the SRC, the LGR and the suggested uLBP-LGR. Yet, the overlap between the neighboring patches is determined as 10 pixels which signifies that the test sample is divided into S = 25 patches. With the exception of the adjustment of the patch number and the size of plots, the existence of just two parameters set in the previously proposed uLBP-LGR is noted. The first one is the parameter of regularization that is fixed as $\lambda = 0.001$ and used in all experimentations. The second one is the parameter of scale employed in the kernel function $k_\sigma(x)$.

$$\sigma = \sqrt{\frac{1}{2S}\sum_{i=1}^{S}\|uz_i - uX_i\alpha_i + uD_i\beta_i\|_2^2} \tag{16}$$

The parameters of competing algorithms are tuned for improved results.

4 Experimental Results

This section, presents our approachs results as well as the evaluation of its performance on the standard face databases that present multiple variations (facial expressions, illumination and disguise). Our results demonstrate the amelioration of the uLBP-LGR over the LGR and other methods and offer a more comprehensive evaluation of the performance of the uLBP-LGR. Our approach is also compared to the state-of-the-art approaches, as the RRC [23], CRC [25], the Directed Acyclic Graph (SVM-DAG) [8], the RKR [24], the one against all (SVM-OAA), the nearest neighbor (NN), the BHDT [3] and the MetaFace [22].

4.1 The Experiments Comparison: FR on FRGCv1, OR, CK+, FEI and GT Databases

This experiment, sums up the results obtained via the application of our proposed approach on the benchmark face databases, including the FEI [6] (200 images), the FRGCv1 [17] (152 images), the Georgia Tech (GT) [4] (50 images), the ORL (40 images) face databases, the Extended Cohn-Kanade [13] (123 images) as well as the AR [14]. All the images are resized and fixed to 27 × 32. Table 1 displays the results obtained by applying our proposed approach on the five databases previously cited.

4.2 Running Time Comparison

All the experimentations were applied on MATLAB version 7.0.1 and the tests were performed on a PC with Intel(R) Core(TM) i3 Processor, clock frequency 2.2 GHz and 4Go RAM. In concrete applications, the training is usually an offline

Table 1. Recognition accuracy on the FRGCv1, ORL, CK+, FEI and GT databases

Method	FRGCv1	ORL	CK+	FEI	GT	AR
NN	-	69.94	-	-	-	48.10
SVM-OAA	59.21	87.50	98.37	96.00	28.00	88.00
SVM-DAG	60.53	87.50	98.37	96.00	28.00	82.00
BHDT	26.97	75.00	91.87	62.50	20.00	63.71
MetaFace	68.42	87.50	98.37	97.00	28.00	85.28
RKR	63.16	82.50	98.37	97.50	24.00	92.86
RRC	71.05	85.00	100	98.00	28.00	95.71
CRC	63.16	85.00	98.37	97.50	28.00	89.00
RSN	71.71	87.50	99.19	97.50	38.00	95.00
LGR	-	-	-	-	-	99.17
Our (uLBP-LGR)	81.00	-	99.66	98.74	30.33	99.58

Table 2. The average running time in seconds on FRGCv1, OR, CK+, FEI and GT databases

Method	FRGCv1	ORL	CK+	FEI	GT	AR
NN	-	0.7703	-	-	-	-
SVM-OAA	0.6415	0.0133	0.1146	0.1516	0.0212	-
SVM-DAG	0.0610	0.0113	0.0473	0.0786	0.0138	-
BHDT	0.0109	0.0019	0.0046	0.0057	0.0022	-
MetaFace	0.5042	0.6500	0.5238	1.0325	1.0684	-
RKR	0.0160	0.0160	1.2e−004	7.5e−005	0	-
RRC	0.0867	0.0102	0.1443	0.1758	0.1178	-
CRC	0.0027	7.7e−04	0.0017	0.0031	0.0012	-
RSN	0.0784	0.0094	0.1954	0.2341	0.1600	-
LGR	-	-	-	-	-	2.1e−04
Our (uLBP-LGR)	2.7e−04	-	2.2e−04	1.1e−05	8.3e−04	2.5e−04

stage and the recognition is usually an online one. Table 2 compares the average running time in seconds on FRGCv1, CK+, OR, FEI, GT and AR databases. Table 2 shows that uLBP-LGR is computationally less intense than the state-of-the-art approaches.

5 Conclusion

In this paper, we present a novel Patch uLBP-LGR approach for the difficulty task of face recognition with SSPP. The uLBP-LGR takes advantage of the benefits of patch uLBP and generic representation. A generic intra-class variation dictionary has been built on the basis of uLBP generic dataset. A CIM has been adopted to consider the highly non-gaussian distribution of the residual representation concerning various patches, as well as to make the measurement of each patch success. This makes possible a more robust evaluation for the benefit of the various patches in face recognition. The application of this algorithm on four reference face databases has proven the efficiency of the method based on LGR that does not cease to give a better recognition rate in comparison with the preceedingly cited methods of the state-of-the-art STSS.

The considerable numerical tests and the elaborated comparison with usual and state-of-the-art approaches have proved that the article's proposed uLBP-LGR approach is very competitive even in laborious classification tests. Besides, this method can be applied at a relatively smaller computational costs since it is based on the same effective structure that had been used in the LGR for the classification step.

References

1. Ahonen, T., Hadid, A., Pietikainen, M.: Face description with local binary patterns: application to face recognition. IEEE Trans. Pattern Anal. Mach. Intell. **28**(12), 2037–2041 (2006)
2. Borgi, M.A., Labate, D., El'Arbi, M., Amar, C.B.: Regularized shearlet network for face recognition using single sample per person. In: 2014 IEEE International Conference on Acoustics, Speech and Signal Processing (ICASSP), pp. 514–518. IEEE (2014)
3. Cevikalp, H.: New clustering algorithms for the support vector machine based hierarchical classification. Pattern Recogn. Lett. **31**(11), 1285–1291 (2010)
4. Chen, L., Man, H., Nefian, A.V.: Face recognition based on multi-class mapping of fisher scores. Pattern Recogn. **38**(6), 799–811 (2005)
5. Fan, Z., Ni, M., Zhu, Q., Sun, C., Kang, L.: L 0-norm sparse representation based on modified genetic algorithm for face recognition. J. Vis. Commun. Image Represent. **28**, 15–20 (2015)
6. de Oliveira Jr., L.L.: Captura e Alinhamento de Imagens: Um Banco de Faces Brasileiro. Relatório Final, Centro Universitário da FEI Projeto de Pesquisa (2005)
7. Gao, S., Tsang, I.W.-H., Chia, L.-T.: Kernel sparse representation for image classification and face recognition. In: Daniilidis, K., Maragos, P., Paragios, N. (eds.) ECCV 2010. LNCS, vol. 6314, pp. 1–14. Springer, Heidelberg (2010). doi:10.1007/978-3-642-15561-1_1
8. Hsu, C.-W., Lin, C.-J.: A comparison of methods for multiclass support vector machines. IEEE Trans. Neural Netw. **13**(2), 415–425 (2002)
9. Huang, K., Aviyente, S.: Sparse representation for signal classification. In: Advances in Neural Information Processing Systems, pp. 609–616 (2006)
10. Khadhraoui, T., Benzarti, F., Amiri, H.: Multimodal hybrid face recognition based on score level fusion using relevance vector machine. In: 2014 IEEE/ACIS 13th International Conference on Computer and Information Science (ICIS), pp. 211–215. IEEE (2014)
11. Khorsandi, R.S.: Sparse representation and dictionary learning for biometrics and object tracking (2015)

12. Liu, W., Pokharel, P.P., Príncipe, J.C.: Correntropy: properties and applications in non-gaussian signal processing. IEEE Trans. Sig. Process. **55**(11), 5286–5298 (2007)
13. Lucey, P., Cohn, J.F., Kanade, T., Saragih, J., Ambadar, Z., Matthews, I.: The extended cohn-kanade dataset (ck+): a complete dataset for action unit and emotion-specified expression. In: 2010 IEEE Computer Society Conference on Computer Vision and Pattern Recognition-Workshops, pp. 94–101. IEEE (2010)
14. Martinez, A.M.: The AR face database. CVC Technical report, 24 (1998)
15. Nikolova, M., Ng, M.K.: Analysis of half-quadratic minimization methods for signal and image recovery. SIAM J. Sci. Comput. **27**(3), 937–966 (2005)
16. Ojala, T., Pietikainen, M., Maenpaa, T.: Multiresolution gray-scale and rotation invariant texture classification with local binary patterns. IEEE Trans. Pattern Anal. Mach. Intell. **24**(7), 971–987 (2002)
17. Phillips, P.J., Flynn, P.J., Scruggs, T., Bowyer, K.W., Chang, J., Hoffman, K., Marques, J., Min, J., Worek, W.: Overview of the face recognition grand challenge. In: 2005 IEEE Computer Society Conference on Computer Vision and Pattern Recognition (CVPR 2005), vol. 1, pp. 947–954. IEEE (2005)
18. Pietikäinen, M., Hadid, A., Zhao, G., Ahonen, T.: Local binary patterns for still images. In: Pietikäinen, M., Hadid, A., Zhao, G., Ahonen, T. (eds.) Computer Vision Using Local Binary Patterns, vol. 40, pp. 13–47. Springer, London (2011). doi:10.1007/978-0-85729-748-8_2
19. Tibshirani, R.: Regression shrinkage and selection via the lasso: a retrospective. J. R. Stat. Soc.: Ser. B (Stat. Methodol.) **73**(3), 273–282 (2011)
20. Wang, C., Huang, K.: How to use bag-of-words model better for image classification. Image Vis. Comput. **38**, 65–74 (2015)
21. Yang, M., Van Gool, L., Zhang, L.: Sparse variation dictionary learning for face recognition with a single training sample per person. In: Proceedings of the IEEE International Conference on Computer Vision, pp. 689–696 (2013)
22. Yang, M., Zhang, L., Yang, J., Zhang, D.: Metaface learning for sparse representation based face recognition. In: 2010 IEEE International Conference on Image Processing, pp. 1601–1604. IEEE (2010)
23. Yang, M., Zhang, L., Yang, J., Zhang, D.: Regularized robust coding for face recognition. IEEE Trans. Image Process. **22**(5), 1753–1766 (2013)
24. Yang, M., Zhang, L., Zhang, D., Wang, S.: Relaxed collaborative representation for pattern classification. In: 2012 IEEE Conference on Computer Vision and Pattern Recognition (CVPR), pp. 2224–2231. IEEE (2012)
25. Zhang, L., Yang, M., Feng, X.: Sparse representation or collaborative representation: which helps face recognition? In: 2011 International Conference on Computer Vision, pp. 471–478. IEEE (2011)
26. Zhu, P., Yang, M., Zhang, L., Lee, I.-Y.: Local generic representation for face recognition with single sample per person. In: Cremers, D., Reid, I., Saito, H., Yang, M.-H. (eds.) ACCV 2014. LNCS, vol. 9005, pp. 34–50. Springer, Cham (2015). doi:10.1007/978-3-319-16811-1_3

Food Recognition Using Fusion of Classifiers Based on CNNs

Eduardo Aguilar(✉), Marc Bolaños, and Petia Radeva

Universitat de Barcelona & Computer Vision Center, Barcelona, Spain
{eduardo.aguilar,marc.bolanos,petia.ivanova}@ub.edu

Abstract. With the arrival of Convolutional Neural Networks, the complex problem of food recognition has experienced an important improvement recently. The best results have been obtained using methods based on very deep Convolutional Neural Networks, which show that the deeper the model, the better the classification accuracy is. However, very deep neural networks may suffer from the overfitting problem. In this paper, we propose a combination of multiple classifiers based on Convolutional models that complement each other and thus, achieve an improvement in performance. The evaluation of our approach is done on 2 public datasets: Food-101 as a dataset with a wide variety of fine-grained dishes, and Food-11 as a dataset of high-level food categories, where our approach outperforms the independent Convolutional Neural Networks models.

Keywords: Food recognition · Fusion classifiers · CNN

1 Introduction

In the field of computer vision, food recognition has caused a lot of interest for researchers considering its applicability in solutions that improve people's nutrition and hence, their lifestyle [1]. In relation to the healthy diet, traditional strategies for analyzing food consumption are based on self-reporting and manual quantification [2]. Hence, the information used to be inaccurate and incomplete [3]. Having an automatic monitoring system and being able to control the food consumption is of vital importance, especially for the treatment of individuals who have eating disorders, want to improve their diet or reduce their weight.

Food recognition is a key element within a food consumption monitoring system. Originally, it has been approached by using traditional approaches [4,5], which extracted ad-hoc image features by means of algorithms based mainly on color, texture and shape. More recently, other approaches focused on using Deep Learning techniques [5–8]. In these works, feature extraction algorithms are not hand-crafted and additionally, the models automatically learn the best way to discriminate the different classes to be classified. As for the results obtained, there is a great difference (more than 30%) between the best method based on hand-crafted features compared to newer methods based on Deep Learning, where the best results have been obtained with Convolutional Neural Networks (CNN) architectures that used inception modules [8] or residual networks [7].

S. Battiato et al. (Eds.): ICIAP 2017, Part II, LNCS 10485, pp. 213–224, 2017.
https://doi.org/10.1007/978-3-319-68548-9_20

Fig. 1. Example images of Food-101 dataset. Each image represents a dish class.

Food recognition can be considered as a special case of object recognition, being a very active topic in computer vision lately. The specific part is that dish classes have a much higher inter-class similarity and intra-class variation than usual Imagenet objects (cars, animals, rigid objects, etc.) (see Fig. 1). If we analyze the last accuracy increase in the ImageNet Large Scale Visual Recognition Challenge (ILSVRC) [9], it has been improved thanks to the depth increase of CNN models [10–13] and also to the fusion of CNNs models [11,13]. The main problem of CNNs is the need of large datasets to avoid overfitting the network as well as the need of high computational power for training them.

Considering the use of different classifiers, in general, trained on the same data, one can observe that patterns misclassified by the different models would not necessarily overlap [14]. This suggests that they could potentially offer complementary information that can be used to improve the final performance [14]. An option to combine the outputs of different classifiers was proposed in [15], where the authors used what they call a decision templates scheme instead of simple aggregation operators such as the product or average. As they showed, this scheme maintains a good performance using different training set sizes and is also less sensitive to particular datasets compared to the other schemes.

In this article, we integrate the fusion concept into the CNN framework, with the purpose of demonstrating that the combination of the classifiers' output, by using a decision template scheme, allows to improve the performance on the food recognition problem. Our contributions are the following: (1) we propose the first food recognition algorithm that fuses the output of different CNN models, (2) we show that our CNNs fusion approach has better performance compared to the use of CNN models separately, and (3) we demonstrate that our CNNs Fusion approach keeps a high performance independently of the target (dishes, family of dishes) and dataset validating it on 2 public datasets.

The organization of the article is as follows. In Sect. 2, we present the CNNs Fusion methodology. In Sect. 3, we present the datasets, the experimental setup and discuss the results. Finally, in Sect. 4, we describe the conclusions.

2 Methodology

In this section, we describe the CNN Fusion methodology (see Fig. 2), which is composed of two main steps: training K CNN models based on different architectures and fusing the CNN outputs using the decision templates scheme.

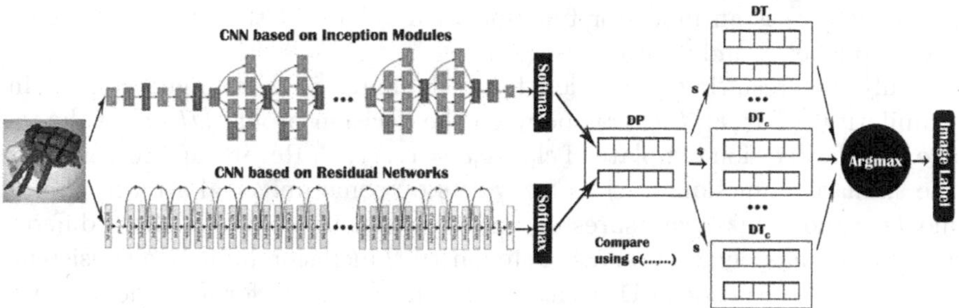

Fig. 2. General scheme of our CNNs fusion approach.

2.1 Training of CNN Models

The first step in our methodology involves separately training two CNN models. We chose two different kind of models winners of the ILSVRC in the object recognition task. Both models won or are based on the winner of the challenges made in 2014 and 2015 proposing novel architectures: the first based its design on "inception models" and the second on "residual networks". First, each model was pre-trained on the ILSVRC data. Later, all layers were fine-tuned by a certain number of epochs, selecting for each one the model that provides the best results in the validation set and that will be used in the fusion step.

2.2 Decision Templates for Classifiers Fusion

Once we trained the models on the food dataset, we combined the softmax classifier outputs of each model using the Decision Template (DT) scheme [15].

Let us annotate the output of the last layer of the k-th CNN model as $(\omega_{1,k}, \dots, \omega_{C,k})$, where $c = 1, \dots, C$ is the number of classes and $k = 1, \dots K$ is the index of the CNN model (in our case, K= 2). Usually, the softmax function is applied, to obtain the probability value of model k to classify image x to a class c: $p_{k,c}(x) = \frac{e^{\omega_{k,c}}}{\sum_{c=1}^{C} e^{\omega_{k,c}}}$. Let us consider the k-th decision vector D_k:

$$D_k(x) = [p_{k,1}(x), p_{k,2}(x), \dots, p_{k,C}(x)]$$

Definition [15]: A **Decision Profile,** DP for a given image x is defined as:

$$DP(x) = \begin{bmatrix} p_{1,1}(x) & p_{1,2}(x) & \dots & p_{1,C}(x) \\ & \dots & & \\ p_{K,1}(x) & p_{K,2}(x) & \dots & p_{K,C}(x) \end{bmatrix} \quad (1)$$

Definition [15]: Given N training images, a **Decision Template** is defined as a set of matrices $DT = (DT^1, \dots, DT^C)$, where the c-th element is obtained as the average of the decision profiles (1) on the training images of class c:

$$DT^c = \frac{\sum_{j=1}^{N} DP(x_j) \times Ind(x_j, c)}{\sum_{j=1}^{N} Ind(x_j, c)},$$

where $Ind(x_j, c)$ is an indicator function with value 1 if the training image x_j has a crisp label c, and 0, otherwise [16].

Finally, the resulting prediction for each image is determined considering the similarity $s(DP(x), DT^c(x))$ between the decision profile $DP(x)$ of the test image and the decision template of class c,$c = 1, \ldots, C$. Regarding the arguments of the similarity function $s(.,.)$ as fuzzy sets on some universal set with $K \times C$ elements, various fuzzy measures of similarity can be used. We chose different measures [15], namely 2 measures of similarity, 2 inclusion indices, a consistency measure and the Euclidean Distance. These measures are formally defined as:

$$S_1(DT^c, DP(x)) = \frac{\sum_{k=1}^{K}\sum_{i=1}^{C}\min(DT^c_{k,i}, DP_{k,i}(x))}{\sum_{k=1}^{K}\sum_{i=1}^{C}\max(DT^c_{k,i}, DP_{k,i}(x))},$$

$$S_2(DT^c, DP(x)) = 1 - \sup_u\{\left|DT^c_{k,i} - DP_{k,i}(x)\right| : c = 1, \ldots, C, k = 1, \ldots, K\},$$

$$I_1(DT^c, DP(x)) = \frac{\sum_{k=1}^{K}\sum_{i=1}^{C}\min(DT^c_{k,i}, DP_{k,i}(x))}{\sum_{k=1}^{K}\sum_{i=1}^{C}DT^c_{k,i}},$$

$$I_2(DT^c, DP(x)) = \inf_u\{\max(\overline{DT^c_{k,i}}, DP_{k,i}(x)) : c = 1, \ldots, C, k = 1, \ldots, K\},$$

$$C(DT^c, DP(x)) = \sup_u\{\min(DT^c_{k,i}, DP_{k,i}(x)) : c = 1, \ldots, C, k = 1, \ldots, K\},$$

$$N(DT^c, DP(x)) = 1 - \frac{\sum_{k=1}^{K}\sum_{i=1}^{C}(DT^c_{k,i} - DP_{k,i}(x))^2}{K \times C},$$

where $DT^c_{k,i}$ is the probability assigned to the class i by the classifier k in the DT^c, $\overline{DT^c_{k,i}}$ is the complement of $DT^c_{k,i}$ calculated as $1 - DT^c_{k,i}$, and $DP_{k,i}(x)$ is the probability assigned by the classifier k to the class i in the DP calculated for the image, x. The final label, L is obtained as the class that maximizes the similarity, s, the inclusion index, the consistency measure or the Euclidean distance between $DP(x)$ and DT^c: $L(x) = argmax_{c=1,\ldots,C}\{s(DT^c, DP(x))\}$.

3 Experiments

3.1 Datasets

The data used to evaluate our approach are two public datasets of very different images: Food-11 [17] and Food-101 [4], which are chosen in order to verify that the classifiers fusion provides good results regardless of the different properties of the target datasets, such as intra-class variability (the first one is composed of many dishes of the same general category, while the second one is composed of specific fine-grained dishes), inter-class similarity, number of images, number of classes, images acquisition condition, among others.

Food-11 is a dataset for food recognition [17], which contains 16,643 images grouped into 11 general categories of food: bread, dairy products, dessert, egg, fried food, meat, noodle/pasta, rice, seafood, soup and vegetable/fruit (see Fig. 3). The images were collected from existing food datasets (Food-101, UECFOOD100, UECFOOD256) and social networks (Flickr, Instagram). This dataset has an unbalanced number of images for each class with an average of 1,513 images per class and a standard deviation of 702. For our experiments, we used the same data split, images and proportions, provided by the authors [17]. These are divided as 60% for training, 20% for validation and 20% for test, that is 9,866, 3,430 and 3,347 images for each set, respectively.

Fig. 3. Images from the Food-11 dataset. Each image corresponds to a different class.

Food-101 is a standard to evaluate the performance of visual food recognition [4]. This dataset contains 101.000 real-world food images downloaded from foodspotting.com, which were taken under unconstrained conditions. The authors chose the top 101 most popular classes of food (see Fig. 1) and collected 1,000 images for each class: 75% for training and 25% for testing. With respect to the classes, these consist of very diverse and fine-grained dishes of various countries, but also with highly intra-class variation and inter-class similarity in most occasions. In our experiments, we used the same data splits provided by the authors. Unlike Food-11, and keeping the procedure followed by other authors [5,7,8], we validate and test our model on the same data split.

3.2 Experimental Setup

As usually, every CNN model was pre-trained on the ILSVRC dataset. Following, we adapted them by changing the output of the models to the number of classes for each target dataset and fine-tuned the models using the new images. For the training of the CNN models, we used the Deep Learning framework Keras[1]. The models chosen for Food-101 dataset due to their performance-efficiency ratio were InceptionV3 [18] and ResNet50 [13]. Both models were trained during 48 epochs with a batch size of 32, and a learning rate of 5×10^{-3} and 1×10^{-3}, respectively. In addition, we applied a decay of 0.1 during the training of InceptionV3 and of 0.8 for ResNet50 every 8 epochs. The parameters were chosen empirically by analyzing the training loss.

As to the Food-11 dataset, we kept the ResNet50 model, but changed InceptionV3 by GoogLeNet [12], since InceptionV3 did not generalize well over Food-11. We believe that the reason is the small number of images for each class not sufficient to avoid over-fitting; the model quickly obtained a good result on the training set, but a poor performance on the validation set. GoogLeNet and Resnet50 were trained during 32 epochs with a batch size of 32 and 16,

[1] www.keras.io.

respectively. The other parameters used for the ResNet50 were the same used for Food-101. In the case of GoogLeNet, we used a learning rate of 1×10^{-3} and applied a decay of 0.1 during every 8 epochs, that turned out empirically the optimal parameters for our problem.

3.3 Data Preprocessing and Metrics

The preprocessing made during the training, validation and testing phases was the following. During the training of our CNN models, we applied different preprocessing techniques on the images with the aim of increasing the samples and to prevent the over-fitting of the networks. First, we resized the images keeping the original aspect ratio as well as satisfying the following criteria: the smallest side of the resulting images should be greater than or equal to the input size of the model; and the biggest side should be less than or equal to the maximal size defined in each model to make random crops on the image. In the case of InceptionV3, we set to 320 pixels as maximal size, for GoogLeNet and ResNet50 the maximal size was defined as 256 pixels. After resizing the images, inspired by [8], we enhanced them by means of a series of random distortions such as: adjusting color balance, contrast, brightness and sharpness. Finally, we made random crops of the images, with a dimension of 299×299 for InceptionV3 and of 224×224 for the other models. Then, we applied random horizontal flips with a probability of 50%, and subtracted the average image value of the ImageNet dataset. During validation, we applied a similar preprocessing, with the difference that we made a center crop instead of random crops and that we did not apply random horizontal flips. During test, we followed the same procedure than in validation (1-Crop evaluation). Furthermore, we also evaluated the CNN using 10-Crops, which are: upper left, upper right, lower left, lower right and center crop, both in their original setup and also applying an horizontal flip [10]. As for 10-Crops evaluation, the classifier gets a tentative label for each crop, and then majority voting is used over all predictions. In the cases where two labels are predicted the same number of times, the final label is assigned comparing their highest average prediction probability.

We used four metrics to evaluate the performance of our approach, overall Accuracy (ACC), Precision (P), Recall (R), and F_1 score.

3.4 Experimental Results on Food-11

The results obtained during the experimentation on Food-11 dataset are shown in Table 1 giving the error rate (1 - accuracy) for the best CNN models, compared to the CNNs Fusion. We report the overall accuracy by processing the test data using two procedures: (1) a center crop (1-Crop), and (2) using 10 different crops of the image (10-Crops). The experimental results show an error rate of less than 10 % for all classifiers, achieving a slightly better performance when using 10-Crops. The best accuracy is achieved with our CNNs Fusion approach, which is about 0.75% better than the best result of the classifiers evaluated separately. On the other hand, the baseline classification on Food-11 was given by their authors, who obtained an overall accuracy of 83.5% using GoogLeNet models fine-tuned in the last six layers without any pre-processing and post-processing steps. Note that the best results obtained with our approach have been using the pointwise measures (S2, I2). The particularity of these measures is that they penalize big differences between corresponding values of DTs and DP being from the specific class to be assigned as the rest of the class values. From now on, in this section we only report the results based on the 10-Crops procedure.

Table 1. Overall test set error rate of Food-11 obtained for each model. The distance measure is shown between parenthesis in the CNNs Fusion models.

Authors	Model	1-Crop	10-Crops	N/A
[17]	GoogLeNet	-	-	16.5%
us	GoogLeNet	9.89%	9.29%	-
us	ResNet50	6.57%	6.39%	-
us	CNNs Fusion (S_1)	6.36%	5.86%	-
us	CNNs Fusion (S_2)	6.12%	**5.65%**	-
us	CNNs Fusion (I_1)	6.36%	5.89%	-
us	CNNs Fusion (I_2)	6.30%	**5.65%**	-
us	CNNs Fusion (C)	6.45%	6.07%	-
us	CNNs Fusion (N)	6.36%	5.92%	-

As shown in Table 2, the CNNs Fusion is able to properly classify not only the images that were correctly classified by both baselines, but in some occasions also when one or both fail. This suggests that in some cases both classifiers may be close to predicting the correct class and combining their outputs can make a better decision.

Table 2. Percentage of images well-classified and misclassified on Food-11 using our CNNs Fusion approach, distributed by the results obtained with GoogLeNet (CNN_1) and ResNet50 (CNN_2) models independently evaluated.

CNNs Fusion	CNNs evaluated independently			
	Both wrong	CNN_1 wrong	CNN_2 wrong	Both fine
Well-classified	3.08%	81.77%	54.76%	99.97%
Misclassified	96.92%	18.23%	45.24%	0.03%

Samples misclassified by our model are shown in Fig. 4, where most of them are produced by mixed items, high inter-class similarity and wrongly labeled images. We show the ground truth (top) and the predicted class (bottom) for each sample image.

Fig. 4. Misclassified Food-11 examples: predicted labels (on the top), and the groundtruth (on the bottom).

In Table 3, we show the precision, recall and F_1 score obtained for each class separately. By comparing the F_1 score, the best performance is achieved for the class Noodles_Pasta and the worst for Dairy products. Specifically, the class Noddles_Pasta only has one image misclassified, which furthermore is a hard sample, because it contains two classes together (see items mixed in Fig. 4). Considering the precision, the worst results are obtained for the class Bread, which is understandable considering that bread can sometimes be present in other classes (e.g. soup or egg). In the case of recall, the worst results are obtained for Dairy products, where an error greater than 8% is produced for misclassifying several images as class Dessert. The cause of this is mainly, because the class Dessert has a lot of items in their images that could also belong to the class Dairy products (e.g. frozen yogurt or ice cream) or that are visually similar.

Table 3. Some results obtained on the Food-11 using our CNNs Fusion approach.

Class	#Images	Precision	Recall	F1
Bread	368	**88.95%**	91.85%	90.37%
Dairy products	148	89.86%	**83.78%**	**86.71%**
Meat	432	94.12%	92.59%	93.35%
Noodles_Pasta	147	**100.00%**	**99.32%**	**99.66%**
Rice	96	94.95%	97.92%	96.41%
Vegetable_Fruit	231	98.22%	95.67%	96.93%

3.5 Experimental Results on Food-101

The overall accuracy on Food-101 dataset is shown in Table 4 for two classifiers based on CNN models, and also for our CNNs Fusion. The overall accuracy is obtained by means of the evaluation of the prediction using 1-Crop and 10-Crops. The experimental results show better performance (about 1% more) using 10-Crops instead of 1-Crop. From now on, in this section we only report the results based on the 10-Crops procedure. In the same way as observed in Food-11, the best accuracy obtained with our approach was by means of point-wise measures S2, I2, where the latter provides a slightly better performance. Again, the best accuracy is also achieved by the CNNs Fusion, which is about 1.5% higher than the best result of the classifiers evaluated separately. Note that the best performance on Food-101 (overall accuracy of 90.27%) was obtained using WISeR [7]. In addition, the authors show the performance by another deep learning-based approaches, in which three CNN models achieved over a 88% (InceptionV3, ResNet200 and WRN [19]). However, WISeR, WRN and ResNet200 models were not considered in our experiments since they need a multi-GPU server to replicate their results. In addition, those models have 2.5 times more parameters than the models chosen, which involve a high cost computational especially during the learning stage. Following the article steps, our best results replicating the methods were those using InceptionV3 and ResNet50 models used as a base to evaluate the performance of our CNNs Fusion approach.

Table 4. Overall test set accuracy of Food-101 obtained for each model.

Author	Model	1-Crop	10-Crops	N/A
[8]	InceptionV3	-	-	88.28%
[7]	ResNet200	-	88.38%	-
[7]	WRN	-	88.72%	-
[7]	WISeR	-	90.27%	-
us	ResNet50	82.31%	83.54%	-
us	InceptionV3	83.82%	84.98%	-
us	CNNs Fusion (S_1)	85.52%	86.51%	-
us	CNNs Fusion (S_2)	86.07%	86.70%	-
us	CNNs Fusion (I_1)	85.52%	86.51%	-
us	CNNs Fusion (I_2)	85.98%	**86.71%**	-
us	CNNs Fusion (C)	85.24%	86.09%	-
us	CNNs Fusion (N)	85.53%	86.50%	-

As shown in Table 5, in this dataset the CNNs Fusion is also able to properly classify not only the images that were correctly classified for both classifiers, but also when one or both fail. Therefore, we demonstrate that our proposed approach maintains its behavior independently of the target dataset.

Table 6 shows the top five worst and best classification results on Food-101 classes. We highlight the classes with the worst and best results. As for the worst class (Steak), the precision and recall achieved are 60.32% and 59.60%, respectively. Interestingly, about 26% error in the precision and 30% error in the recall is produced with only three classes: Filet mignon, Pork chop and Prime rib. As shown in Fig. 5, these are fine-grained classes with high inter-class similarities

Table 5. Percentage of images well-classified and misclassified on Food-101 using our CNNs Fusion approach, distributed by the results obtained with InceptionV3 (CNN_1) and ResNet50 (CNN_2) models independently evaluated.

CNNs Fusion	CNNs evaluated independently			
	Both wrong	CNN_1wrong	CNN_2wrong	Both fine
Well-classified	1.95%	73.07%	64.95%	99.97%
Misclassified	98.05%	26.93%	35.05%	0.03%

Table 6. Top 3 better and worst classification results on Food-101.

Class	Precision	Recall	F1
Spaghetti Bolognese	94.47%	95.60%	95.03%
Macarons	97.15%	95.60%	96.37%
Edamame	**99.60%**	**100.00%**	**99.80%**
Steak	**60.32%**	**59.60%**	**59.96%**
Pork Chop	75.71%	63.60%	69.13%
Foie Gras	72.96%	68.00%	70.39%

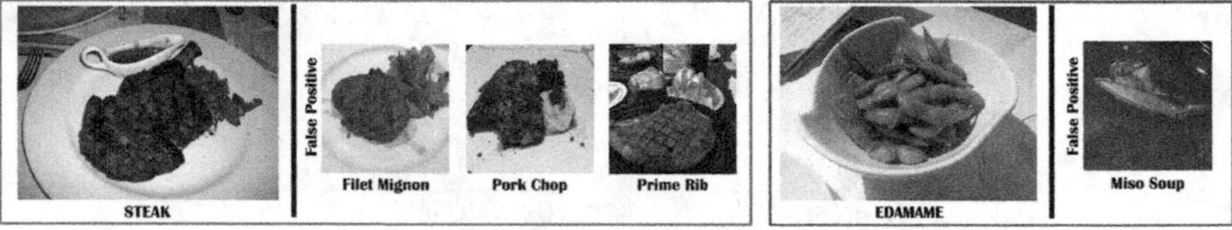

Fig. 5. Misclassified examples for the Food-101 classes that obtained the worst (steak) and best (edamame) classification results by F1 score (groundtruth label - bottom).

that imply high difficulty for the classifier, because it should identify small details that allow to determine the corresponding class of the images. On the other hand, the best class (Edamame) was classified achieving 99.60% of precision and 100% of recall. Unlike Steak, Edamame is a simple class to classify, because it has a low intra-class variation and low inter-class similarities. In other words, the images in this class have a similar visual appearance and they are quite different from the images of the other classes. Regarding the only one misclassified image, its visual appearance is close to the class Edamame as for the shape and color.

4 Conclusions

In this paper, we addressed the problem of food recognition and proposed a CNNs Fusion approach based on the concepts of decision templates and decision profiles and their similarity that improves the classification performance with respect to using CNN models separately. Evaluating different similarity measures, we show that the optimal one is based on the infinimum of the maximum between the complementary of the decision templates and the decision profile of the test images. On Food-11, our approach outperforms the baseline accuracy by more than 10% of accuracy. As for Food-101, we used two CNN architectures providing the best state of the art results where our CNNs Fusion strategy outperformed them again. As a future work, we plan to evaluate the performance of the CNN Fusion strategy as a function of the number of CNN models.

Acknowledgement. This work was partially funded by TIN2015-66951-C2, SGR 1219, CERCA, *ICREA Academia'2014*, CONICYT Becas Chile, FPU15/01347 and Grant 20141510 (Marató TV3). The funders had no role in the study design, data collection, analysis, and preparation of the manuscript. We acknowledge Nvidia Corporation for the donation of 3 Titan X GPUs.

References

1. Waxman, A., Norum, K.R.: WHO global strategy on diet, physical activity and health. Food Nutr. Bull. **25**, 292–302 (2004)
2. Shim, J.-S., Oh, K., Kim, H.C.: Dietary assessment methods in epidemiologic studies. Epidemiol. Health **36**, e2014009 (2014)
3. Rumpler, W.V., Kramer, M., Rhodes, D.G., Moshfegh, A.J., Paul, D.R.: Identifying sources of reporting error using measured food intake. Eur. J. Clin. Nutr. **62**, 544–552 (2008)
4. Bossard, L., Guillaumin, M., Van Gool, L.: Food-101-mining discriminative components with random forests. In: Fleet, D., Pajdla, T., Schiele, B., Tuytelaars, T. (eds.) ECCV 2014. LNCS, vol. 8694, pp. 446–461. Springer, Cham (2014). doi:10.1007/978-3-319-10599-4_29
5. Liu, C., Cao, Y., Luo, Y., Chen, G., Vokkarane, V., Ma, Y.: DeepFood: deep learning-based food image recognition for computer-aided dietary assessment. In: Chang, C.K., Chiari, L., Cao, Y., Jin, H., Mokhtari, M., Aloulou, H. (eds.) ICOST 2016. LNCS, vol. 9677, pp. 37–48. Springer, Cham (2016). doi:10.1007/978-3-319-39601-9_4
6. Yanai, K., Kawano, Y.: Food image recognition using deep convolutional network with pre-training and fine-tuning. In: ICMEW, pp. 1–6 (2015)
7. Martinel, N., Foresti, G.L., Micheloni, C.: Wide-slice residual networks for food recognition. arXiv Preprint (2016)
8. Hassannejad, H., Matrella, G., Ciampolini, P., De Munari, I., Mordonini, M., Cagnoni, S.: Food image recognition using very deep convolutional networks. In: Proceedings of the 2nd International Workshop on MADiMa, pp. 41–49 (2016)
9. Russakovsky, O., Deng, J., Su, H., Krause, J., Satheesh, S., Ma, S., Huang, Z., Karpathy, A., Khosla, A., Bernstein, M., Berg, A.C., Fei-Fei, L.: ImageNet large scale visual recognition challenge. Int. J. Comput. Vis. **115**, 211–252 (2015)
10. Krizhevsky, A., Sutskever, I., Hinton, G.E.: ImageNet classification with deep convolutional neural networks. In: Advances in Neural Information Processing Systems 25, pp. 1–9 (2012)
11. Zeiler, M.D., Fergus, R.: Visualizing and understanding convolutional networks. In: Fleet, D., Pajdla, T., Schiele, B., Tuytelaars, T. (eds.) ECCV 2014. LNCS, vol. 8689, pp. 818–833. Springer, Cham (2014). doi:10.1007/978-3-319-10590-1_53

12. Szegedy, C., Liu, W., Jia, Y., Sermanet, P., Reed, S., Anguelov, D., Erhan, D., Vanhoucke, V., Rabinovich, A.: Going deeper with convolutions. In: CVPR, pp. 1–9 (2015)
13. He, K., Zhang, X., Ren, S., Sun, J.: Deep residual learning for image recognition. In: CVPR, pp. 770–778 (2016)
14. Kittler, J., Hatef, M.: On combining classifiers. IEEE Trans. Pattern Anal. Mach. Intell. **20**, 226–239 (1998)
15. Kuncheva, L.I., Bezdek, J.C., Duin, R.P.W.: Decision templates for multiple classifier fusion: an experimental comparison. Pattern Recogn. **34**, 299–314 (2001)
16. Kuncheva, L.I., Kounchev, R.K., Zlatev, R.Z.: Aggregation of multiple classification decisions by fuzzy templates. In: EUFIT, pp. 1470–1474 (1995)
17. Singla, A., Yuan, L., Ebrahimi, T.: Food/non-food image classification and food categorization using pre-trained googlenet model. In: Proceedings of the 2nd International Workshop on MADiMa, pp. 3–11 (2016)
18. Szegedy, C., Vanhoucke, V., Ioffe, S., Shlens, J., Wojna, Z.: Rethinking the inception architecture for computer vision. In: CVPR, pp. 2818–2826 (2016)
19. Zagoruyko, S., Komodakis, N.: Wide residual networks. arXiv Preprint (2016)

MR Brain Tissue Segmentation Based on Clustering Techniques and Neural Network

Hayat Al-Dmour(✉) and Ahmed Al-Ani

Faculty of Engineering and Information Technology,
University of Technology Sydney, Ultimo, NSW 2007, Australia
{Hayat.Al-Dmour,Ahmed.Al-Ani}@uts.edu.au

Abstract. An effective fully-automatic brain tissue segmentation method based on the integration of clustering techniques and Neural Network (NN) is presented. The method aims to combine the strengths of a number of individual clustering techniques in order to achieve improved segmentation. The method starts by enhancing the image contrast through scaling of the pixel intensity. This is followed by pre-segmenting of the brain image through grouping neighboring pixels of similar intensities into objects. After that, the pixel intensity features are extracted for each group. Then, each object is partitioned using two (or more) different clustering techniques. The training labels are learned by feeding the extracted features and clustering methods' outputs into a NN model that is guided by the ground truth values (the target segmentation outputs). In the testing phase, the extracted features and clustering methods' outputs are fed to the trained NN to predict the class of each object. The efficiency of the proposed method is demonstrated on various MR brain images and compared with the base clustering techniques.

Keywords: Segmentation · K-means · Self-organized map · Neural network

1 Introduction

Amongst the important research studies for the past three decades is the segmentation of medical images including Magnetic Resonance (MR) brain images. Mainly, diagnosis, computer integrated surgery and treatment planning are the areas of application of this research domain [1]. The distinct advantage of the efficient and fully automated evaluation of MR brain images lies in the fact that it overcomes the human error during the investigation.

The MR brain image segmentation generally aims to divide the brain into three tissues, which are: white matter (WM), grey matter (GM) and cerebrospinal fluid (CSF) for functional visualization in the diagnosis of diseases like cancer and neurological disorders [2]. To analyze the original image easily, the segmentation process is applied to transform the original image representation into another meaningful format [3]. Fundamentally, boundaries between the brain tissues and assignment of a new label to every pixel in the image are

S. Battiato et al. (Eds.): ICIAP 2017, Part II, LNCS 10485, pp. 225–233, 2017.
https://doi.org/10.1007/978-3-319-68548-9_21

defined by image segmentation. The categorization and grouping of brain tissues are simplified by the assigned labels where a particular computed characteristic including texture, intensity, shape, or color represent pixels that contain the same label [3].

Development of an image segmentation, in terms of clustering has been the focus of many research studies in recent years, as it has been found that clustering can help in developing effective segmentation methods. The k-means, Fuzzy C-means (FCM), Gaussian Mixture Model (GMM), and Self-Organized Map (SOM) are some of the popular techniques of clustering. Each one of these methods has its own advantages and imitations [4]. For example, even though the k-means approach is faster and simple to use, it is highly influenced by the outlier points, and thus being sensitive to cluster initialization [5]. Alternatively, FCM relatively attains better outcome for overlapped classes and is less sensitive to initialization, but it has high computational cost [6]. Clearly, it proves to be challenging to have one clustering approach, which combines all the strength points of the existing approaches.

A completely-automatic segmentation approach for MR images is presented in this paper that integrates two or more clustering methods and a Neural Network model to retain the salient features of those clustering methods. In the beginning, a superpixel algorithm is applied to divide the image into objects. Then, two clustering methods are utilized to produce segmentation results. This is followed by training of a neural network model through combining the outcome of the clustering algorithm with the pixel intensity features.

The rest of this paper is organized as follows. The related work is described in Sect. 2. Section 3 presents details of the proposed method. Section 4 presents the experimental results, and the conclusion is given in Sect. 5.

2 Related Work

In recent years, many methods have been introduced in the area of medical image segmentation that include fully automatic methodologies for brain image segmentation. For instance, Kalaiselvi and Somasundaram [7] applied FCM for brain tissue segmentation, which is aimed to reduce the computational cost by using image histogram information to initialize the centroid values. The main drawback with that method is that it is iterative in nature, which makes the whole segmentation process computationally complex [8].

In [9], an automatic brain segmentation using K-means clustering was introduced to detect the shape and range of tumors. A median filter was applied to remove the artifacts and sharpen the edges of the image. Then, the k-means clustering method was utilized using random centroid values. To predict and calculate the tumor region in the image, a binary mask was applied to identify the class with high contrast values.

Ortiz et al. [10] proposed a fully unsupervised and automated method to segment MR brain images using Self-Organising Maps (SOMs) and Genetic

Algorithms (GAs). The method combined five steps: image acquisition and pre-processing, feature extraction, feature selection using genetic algorithm, voxel classification using SOM, and sharp map clustering.

To combine the advantage of two or more of the existing clustering methods, we develop an automatic segmentation technique that integrates the clustering methods using a neural network model. The next section presents our proposed method.

3 The Proposed Method

The proposed method is categorized into two main phases: training and testing. Below are the steps for each phase.

3.1 Training Phase

(1) *Pre-processing step*: This step mainly aims to improve the MR segmentation of the under-represented class, i.e., the cerebrospinal fluid (CSF) class. The first step comprises image scaling and resizing.

- Image scaling: The contrast of the image is improved by scaling the pixel intensity distribution to cover the range of 0 to 255. Equation 1 is applied to generate the new image.

$$\text{Scale pixel} = a + (\text{pixel} - c) * (\frac{b-a}{d-c}) \tag{1}$$

where a and b are the target minimum and maximum gray levels respectively, while c and d are the original gray levels.

- Image resizing: The dimension of the original and ground truth images are expanded from (256 × 256) to (512 × 512). This is made possible by duplicating each pixel into block of size (2 × 2). In contrast to WM and GM, CSF segmentation accuracy is poor in the majority of the existing algorithms. The CSF class comprises of a diminutive area with an average ratio of 2.10 in the brain image. Therefore, there is a possibility of integrating the CSF class with other classes while applying the superpixel algorithm.

(2) *Pre-segmentation*:

- Simple Linear Iterative Clustering (SLIC) Superpixels [11]: The superpixel algorithm is applied to over-partition the scaled image into objects. The first process involves manually defining the necessary number of superpixels (k). The initial superpixel cluster centers $C_j = [I_j, x_j, y_j]^T$ with $j = \{1, 2, \cdots, k\}$, where I_j is the pixel intensity and (x_j, y_j) is the coordinator of the center. The initial cluster centers are sampled on a regular grid spaced S pixels. In order to have approximately equal size of superpixel, the grid size is set to $\sqrt{\frac{N}{k}}$, where N is the total number of image pixels. Then, the centers are moved to the lowest gradient position in 3 × 3 neighborhood to prevent centering a superpixel on an edge or a noisy pixel. Grouping of the pixels is in terms of intensity similarity. After each iteration, the pixels that are assigned to the cluster are used for updating the cluster center values. This process is repeated until constancy is attained in superpixel centers.

- Remove the background: The over-partition image is divided into background and object regions based on Eq. 2.

$$g_{(x,y)} = \begin{cases} 1 \text{ if } f_{(x,y)} > Th \\ 0 \quad \text{otherwise} \end{cases} \tag{2}$$

where $g_{(x,y)}$ is the segmented binary image, $f_{(x,y)}$ is the original pixel value and Th is the threshold value, which is set to zero.

(3) *Feature Extraction*: For each object, the pixel intensity features are extracted to be input to two varied clustering techniques in a simultaneous manner. Thus, there is a difference in over-segmented objects' size. The five highest frequencies of pixel intensity are extracted for every object.

(4) *Clustering techniques*: The image is partitioned into groups through the number of base clustering techniques. The extracted features and number of classes are the inputs for each clustering method. It is important to note that there are three classes in the MR brain image excluding the background. k-means and Self-Organizing Map (SOM) are chosen to produce the segmented image. A brief description of each of these algorithms is given below.

 - k-means clustering: k-means is one of the most popular unsupervised clustering technique that separates the input data into groups based on their distance from each other. It is simple and fast to apply on images with large data points [12]. The k-means method is described in Algorithm 1.

$$J = \sum_{i=1}^{n} \sum_{j=1}^{k} \left\| x_i - c_j \right\|^2 \tag{3}$$

 where n is the number of data points $(x_1, x_2, \cdots, x_n)$, and k is the number of cluster centers.

$$c_j = \frac{1}{m_i} \sum_{x \in c_j} x \tag{4}$$

 c_j is the jth cluster center and m_i is the number of data point (x) in the jth cluster center.

 - Self-Organizing Map (SOM): SOM [13] is an unsupervised learning neural network. SOM has a feed-forward structure and does not require to specify the targeted output like other types of neural network, because it divides the data point into groups by learning from the data itself, instead of constructing a rule set. The map contains two layers: input and output (or competitive). In the input layer, each node corresponds to single input data, while the output layer is organized into a two dimensional grid of competitive neurons. Each input node is connected to the output node by adjustable weight vector and is updated in each iterative process.

Algorithm 1. k-means

Input: Data points (X) of size $n \times r$, and number of clusters (k).
Output: Cluster indices (C) of size $n \times 1$.
1 Randomly initialize cluster centers (C_j), where $C_j = \{c_1, c_2, \cdots, c_k\}$
2 **repeat**
  3 Calculate the distance between each data point of X and cluster center of C using Eq. 3
  4 Assign each data point(x_i) to cluster c_j which has the closest centroid
  5 Calculate the new cluster center using Eq. 4
until the cluster centers no longer change

The winner neuron, also know as Best Matching Unit (BMU), is determined by selecting the minimum Euclidean distance between the input data and weight vector at each iteration. SOM also utilizes neighborhood function. So, when the node wins a competition, the node neighbors are also updated. Let $X = \{x_1, x_2, \cdots, x_N\}$ be the input data of size $N \times 1$ and w_{ij} is the weight vector of the node x_i. The winner neuron c is computed by $c = \|x_i - w_{ij}\|$. The winner neuron and its neighbors are updated using Eq. 5.

$$w_{ij}(t+1) = w_{ij}(t) + H_{ci}\Big[x_i(t) - w_{ij}(t)\Big] \tag{5}$$

where w_{ij} and $w_{ij}(t+1)$ is the old and new adjusted weight for the node x_i respectively, t describes the iteration number of the training process, $x_i(t)$ is the input data at iteration t, and H_{ci} is the neighborhood function for the winner neuron c, which is calculated using Eq. 6.

$$H_{ci} = \alpha\left[\exp^{(-\frac{\|r_c - r_i\|^2}{2\sigma(t)^2})}\right] \tag{6}$$

where α is the learning rate, r_i and r_c are the positions of the node i, and the winner node c in the topological map (output space), respectively, $\|r_c - r_i\|$ is the distance between the i and winning neurons, σ is the search distance (number of neighborhood pixels).

(5) *Back Propagation Neural Network (BPNN)*: The simplest technique for organized training of multi-layered NN is BPNN. It adjusts the weight values by estimating the non-linear connection between the input and the output. The back propagation technique generalizes the Least Mean Square (LMS) algorithm by modifying the network weights for the purpose of minimizing the mean squared error between the network and targeted outputs. Two inputs are used to train the supervised network: the extracted features and outputs of the two clustering techniques. The BPNN is also provided with the targeted outputs (ground truth). We used the first five images of the dataset for training. Thus, a model for brain tissue segmentation is learned from the training dataset by combining the clustering techniques together.

3.2 Testing Phase

The method presents an effective combination of the two clustering methods and NN to achieve better segmentation, particularly for the CSF region. It aims to predict the final segmentation result by combining the two clustering method outputs. This combination is implemented using the trained NN model that receives two sets of inputs; i.e., extracted features and output of the two clustering techniques, and attempt to predict the class label of each superpixel.

The implementation of the testing process is explained in the following steps:

1. Pre-processing step: The test image is scaled to enhance the image contrast. Then, the image size is duplicated to help in classifying the CSF region.
2. Pre-segmentation step: The pre-processed image is divided into objects using the SLIC superpixel algorithm. Then, the background is excluded using a global threshold value.
3. Feature Extraction: For each object, we extract the five highest frequencies of pixel intensity.
4. Clustering Techniques: k-means and SOM are applied individually to divide the brain image into groups. The extracted features and number of classes are the inputs for each clustering algorithm.
5. Neural Network: The NN is used to predict the segmented image using the extracted features and outputs of the two clustering methods.

4 Experimental Results

The performance of the proposed method is evaluated using the Internet Brain Segmentation Repository (IBSR), which is made available by the Center for Morphometric Analysis, Massachusetts General Hospital (http://www.cma.mgh.harvard.edu/ibsr). The IBSR dataset contains a three dimensional T1-weighted MRI brain data set for 20 normal subjects and the corresponding manual segmentation, performed by a trained expert [14]. We utilize this manual segmentation of WM, GM, and CSF as the ground truth to evaluate our results.

We used the Jaccard Similarity (JS) metric to evaluate the spatial overlap between the ground truth and segmented images, which is computed as the ratio between the intersection and union of the segmented and ground truth images. The Jaccard Similarity (JS) is defined using Eq. 7.

$$JS = \frac{S_1 \cap S_2}{S_1 \cup S_2} \tag{7}$$

where S_1 is the ground truth image and S_2 is the segmented image.

Table 1 presents the results of Jaccard Similarity using k-means, SOM and our proposed method. It is observed that the proposed method achieved a higher degree of similarity for all three classes (WM, GM and CSF) compared to the other two clustering methods.

Table 1. Jaccard similarity values of k-means, SOM and our proposed method using slice number 20.

Class	k-means	SOM	Proposed
WM	0.712 ± 0.074	0.711 ± 0.075	$\mathbf{0.717 \pm 0.074}$
GM	0.632 ± 0.072	0.641 ± 0.097	$\mathbf{0.749 \pm 0.067}$
CSF	0.222 ± 0.148	0.240 ± 0.095	$\mathbf{0.553 \pm 0.207}$

We also used the Dice Similarity Coefficient (DSC), which is a statistical validation metric that was proposed to evaluate the accuracy of segmentation methods using Eq. 8.

$$DSC = 2 \times \frac{S_1 \bigcap S_2}{\mid S_1 \mid + \mid S_2 \mid} \tag{8}$$

Table 2. Dice similarity coefficient values of k-means, SOM and our proposed method using slice number 20.

Class	k-means	SOM	Proposed
WM	0.868 ± 0.054	0.867 ± 0.055	$\mathbf{0.870 \pm 0.054}$
GM	0.788 ± 0.073	0.792 ± 0.075	$\mathbf{0.863 \pm 0.046}$
CSF	0.469 ± 0.149	0.472 ± 0.153	$\mathbf{0.741 \pm 0.180}$

Table 2 shows the results of the Dice Similarity Coefficient obtained using the two based segmentation methods and our proposed method. The proposed method also achieved better segmentation than other methods using this measure.

The outcome of the segmentation method is also tested using Root Mean Square Error (RMSE). The RMSE is a statistical metric that finds the difference between the ground truth and segmented images. It is computed using Eq. 9.

$$RMSE = \sqrt{\frac{\sum_{i=1}^{W} \sum_{j=1}^{H} (S2 - S1)^2}{W \times H}} \tag{9}$$

where W and H are the width and height of the ground truth ($S1$) and segmented ($S2$) images.

Table 3 shows the experimental results of the proposed method compared to the k-means and SOM. It can be observed from Table 3 that the proposed method obtained better results compared to the other methods in terms of RMSE values.

Figures. 1(a) and (b) show the ground truth and segmented images respectively. The resulted image indicates a high degree of similarity with the manual segmented image.

Table 3. Root mean square error of k-means, SOM and our proposed method using slice number 20.

k-means	SOM	Proposed
0.217 ± 0.023	0.215 ± 0.025	$\mathbf{0.172 \pm 0.024}$

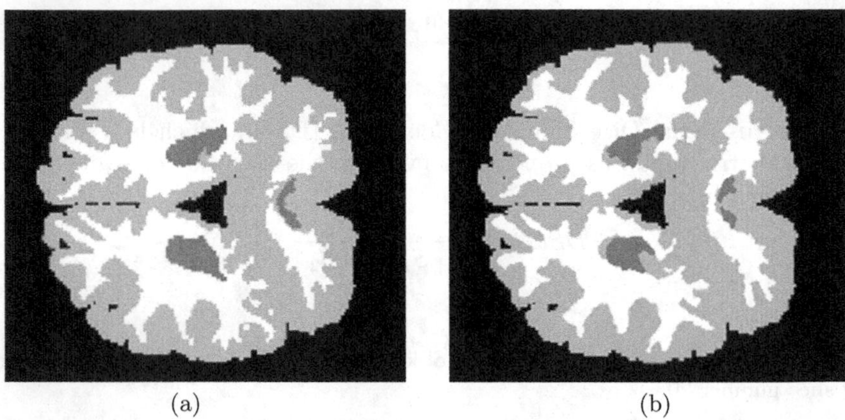

Fig. 1. Subject 202-3, slice 20: (a) ground truth and (b) proposed method.

5 Conclusions

This paper introduced a segmentation method for MR brain images based on clustering techniques and neural network. The method comprises two stages; training and testing. Both stages start with a pre-processing step to improve the image contrast, as the brain structure is not realized by unique intensities in MR images. Then, the brain image is partitioned into objects using the SLIC superpixels algorithm. The training stage involves the training of a NN model that is fed with features extracted from frequencies of pixel intensities and outputs of the two base clustering methods. Our method achieved better results compared to the based clustering according to the three measures of JS, DSC and MSE.

References

1. Bandyopadhyay, S.K.: A survey on brain image segmentation methods. J. Glob. Res. Comput. Sci. **2**, 4–7 (2011)
2. Attique, M., Gilanie, G., Mehmood, M.S., Naweed, M.S., Ikram, M., Kamran, J.A., Vitkin, A., et al.: Colorization and automated segmentation of human T2 MR brain images for characterization of soft tissues. PLoS ONE **7**, e33616 (2012)
3. Abdel-Maksoud, E., Elmogy, M., Al-Awadi, R.: Brain tumor segmentation based on a hybrid clustering technique. Egypt. Inform. J. **16**, 71–81 (2015)
4. Li, H., He, H., Wen, Y.: Dynamic particle swarm optimization and k-means clustering algorithm for image segmentation. Opt.-Int. J. Light Electron Opt. **126**, 4817–4822 (2015)

5. Celebi, M.E.: Improving the performance of k-means for color quantization. Image Vis. Comput. **29**, 260–271 (2011)
6. Cebeci, Z., Yildiz, F.: Comparison of k-means and fuzzy c-means algorithms on different cluster structures. Agrár./J. Agric. Inform. **6**, 13–23 (2015)
7. Kalaiselvi, T., Somasundaram, K.: Fuzzy c-means technique with histogram based centroid initialization for brain tissue segmentation in MRI of head scans. In: 2011 International Symposium on Humanities, Science & Engineering Research (SHUSER), pp. 149–154. IEEE (2011)
8. Gilanie, G., Attique, M., Naweed, S., Ahmed, E., Ikram, M., et al.: Object extraction from T2 weighted brain MR image using histogram based gradient calculation. Pattern Recogn. Lett. **34**, 1356–1363 (2013)
9. Dhanalakshmi, P., Kanimozhi, T.: Automatic segmentation of brain tumor using k-means clustering and its area calculation. Int. J. Adv. Electr. Electron. Eng. **2**, 130–134 (2013)
10. Ortiz, A., Gorriz, J., Ramirez, J., Salas-Gonzalez, D.: Improving MR brain image segmentation using self-organising maps and entropy-gradient clustering. Inf. Sci. **262**, 117–136 (2014)
11. Achanta, R., Shaji, A., Smith, K., Lucchi, A., Fua, P., Süsstrunk, S.: Slic superpixels compared to state-of-the-art superpixel methods. IEEE Trans. Pattern Anal. Mach. Intell. **34**, 2274–2282 (2012)
12. Xess, M., Agnes, S.A.: Survey on clustering based color image segmentation and novel approaches to FCM algorithm. Int. J. Res. Eng. Technol. **2013**, 346–349 (2013)
13. Kohonen, T.: Self-organizing maps. information sciences, 30th edn. Springer, Heidelberg (2001)
14. Worth, A.: The internet brain segmentation repository (IBSR) (2009)

Multi-branch CNN for Multi-scale Age Estimation

Marco Del Coco(✉), Pierluigi Carcagnì, Marco Leo, Paolo Spagnolo, Pier Luigi Mazzeo, and Cosimo Distante

National Research Council of Italy,Institute of Applied Science and Intelligent System,Ecotekne Campus, via Monteorni, 73100 Lecce, Italy
marco.delcoco@isasi.cnr.it

Abstract. Convolutional Neural Networks (CNNs) attracted growing interest in recent years thanks to their high generalization capabilities that are highly recommended especially for applications working *in the wild* context. However CNNs rely on a huge number of parameters that must be set during training sessions based on very large datasets in order to avoid over-fitting issues. As a consequence the lack in training data is one of the greatest limits for the applicability of deep networks. Another problem is represented by the fixed scale of the filter in the first convolutional layer that limits the analysis performed through the subsequent layers of the network.

This paper proposes a way to overcome these problems by the use of a multi-branch convolutional neural network with a reduced deep. In particular its effectiveness for age group classification has been proved demonstrating how, on the one hand, the reduced deep avoids the over-fitting issues, whereas, on the other hand, the multi-branch structure introduces a parallel multi-scale analysis capable to catch multiple size patterns.

Keywords: CNN · Deep learning · Age estimation

1 Introduction

Human beings are intrinsically used to estimate soft biometric of other people and, consequentially, they modulate their behavior to the age and gender of the ones involved in the interaction. It turns out that a more natural human-machine interaction could be achieved by introducing these biometric biases also in the human-machine interaction processes. This could improve acceptability and usability of assistive technologies [17] that are part of our everyday life [2], and in particular of social robots already largely used in clinical contexts [4]. In the light of the above the reliability of the soft biometric estimation algorithms becomes a key point in the human-machine scenario. Unfortunately this is still an open challenge since traditional approaches focused on the selection of specific biometric descriptors particularly suitable for a specific application

S. Battiato et al. (Eds.): ICIAP 2017, Part II, LNCS 10485, pp. 234–244, 2017.
https://doi.org/10.1007/978-3-319-68548-9_22

context but unsuitable for the use in unconstrained environments [3]. Where pose, illumination and occlusion changes have to be effectively addressed.

The recent explosion of the Convolutional Neural Network (CNN), jointly with the exponentially hardware advancements, has allowed to overcome the above limitations bringing to generalization capabilities. This is mainly due to the huge number of parameters that characterizes the CNN structure, as well as to the complete adaptability of the convolutional filtering side of the network that makes possible to discover a larger number of hidden structures. It is straightforward that the big amount of parameters requires a large amount of data for carrying out a learning phase free of over-fitting issues but, unfortunately, well annotated data are, nowadays, a limited resource due to the long and tedious phase for collecting and labeling them.

Another limit is that a CNN can adapt the inner parameters of each filter, but the filter size is chosen by the operator making the initial scale of analysis constrained to a fixed value.

A possible solution to avoid the overfitting issues, when available data are undersized, is to contain the network depth (i.e. the number of inner parameters), as suggested in [18]. This approach makes sense especially when the problem is quite simple (few output classes) with respect to classifications problems requiring many output classes (people or object recognition).

In this work the small sized network proposed in [18] is exploited for age group classification on the Adience face dataset [7]. In addition an innovative multi-branch structure (with different staring filter sizes) is proposed in order to achieve a better multi-scale analysis with respect to the one naturally accounted by classical CNN. In other words, the main contribution of this paper is that the proposed approach takes advantage of the parallel multi-scale analysis that is directly performed on the raw signal (see Fig. 1). In the proposed architecture, that of course has a larger number of parameters than the single branch one, overfitting issues are limited since the backpropagated errors are in parallel distribuited among the different branches.

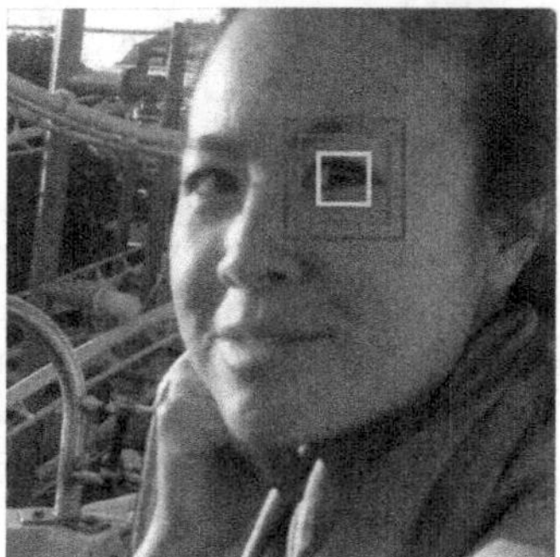

Fig. 1. Multiscale CNN processing: the multi-scale convolutional filters allow an analysis of different pattern sizes.

The rest of the paper is organized as follows: Sect. 2 aims to give a short overview of the leading approaches for age group classification presented in literature across last years; Sect. 3 presents the whole system giving a detailed description of the network structure; Sect. 4 deals with the presentation and discussion of experimental results; Sect. 5 is finally devoted to the conclusions and future works discussion.

2 Related Works

The problem of automatic age estimation from facial images attracted many researchers in the recent years. The solution has been treated in many ways, from the age group classification to the precise age estimation employing regression techniques. A detailed survey concerning both strategies is presented in [11].

Early researches focused the attention on the hand-crafted rule. First of all, facial features (i.e. nose mouth...) were detected in the face image and their size end mutual distances are computed; successively these measure were employed for the age group classification [15]. A similar approaches has been employed in order to model age progression in teenager people [19]. It is obvious that the narrow age range require a precise measurement of facial parts both in terms of size and distance, making the method unsuitable for *in the wild* application. Other approaches suffering the near frontal constraint are the ones based on the representation of age process as a subspace [9] or a manifold [10].

An alternative way is the use of local features for face aging representation. As an instance, Hidden-Markov-Model were used in [23] for representing face patch distributions. Alternatively, Gaussian Mixture Models (GMM) are employed in [22] for the same intent whereas [21] they were used again for representing the distribution of local facial measurements, but robust descriptors were used instead of pixel patches.

An improved versions of relevant component analysis [1] (exploited for distance learning) and locally preserving projections [12] (exploited for dimensionality reduction) are combined with Active Appearance Models (as image feature) in the work proposed in [5].

Another research line has been represented by the robust image descriptors. In [8], Gabor image descriptors were used jointly to a Fuzzy-LDA classifier which describe a face image as belonging to more than one age class.

Two of the main exploited approaches are Gabor filters and local binary patterns (LBP) that have been successfully employed in [6] along with a hierarchical age classifier. An interesting comparison of different descriptors, spacial reduction approaches and classification is shown in [3] where a detailed discussion of pose and scale influence is given.

The main limit of the listed methods is their inapplicability to constrained datasets, making their use in real scenarios, unpredictable.

The necessity of highly generalization capability led the research community toward a growing employment of CNN solutions, facilitated by the availability of high performance hardware (GPU equipped with thousand of cores and big amount of memory) and big dataset usually collected through the web.

One of the first solution has been represented by [16] where the LeNet-5 network has been successfully applied to the character recognition system. Another surprising application of CNN has been the classification task on *Imagenet* dataset, a dataset containing hundreds of categories in uncostrained environment [14].

Currently, one of the most interesting challenge is represented by face analysis on face datasets with significant pose variations, occlusions, and poor quality i.e. the Adience dataset. A preliminary solution has been presented in [7] by means of LBP descriptor variations [20] plus SVM. Successively, in [18], the authors suggest the use of a CNN made up by a reduced number of layers respect to the approach proposed in [14]; this is possible considering the limited number of classes in the age group problem (compared to the Image net one occuring in [14]) and allow to avoid the overfitting problem. Due to the notable results obtained this last work as been chosen as the main competitor for the proposed solution.

3 Multi-branch CNN

The most exploited CNN architecture is oriented to the modeling of *linear* and *deep* networks characterized by many parameters and great generalization capabilities. Unfortunately this kind of structure presents two main limits: the *overfitting problem* (occurring when the training data are not sufficient for the scope) and the *scale of analysis* that is constrained to the starting convolutional filter size and could be unsuitable to catch key patterns. This last aspect can be avoided by means of a brute force research of the best initial scale. Also in this case any assurance is given concerning the capacity of the network to capture all the important patterns across the image.

The proposed network is based on two main points:

- *A reduced network depth*: as suggested in [18] a reduction of the number of layers in the network helps to reduce the overfitting issue, especially in case such as gender or age group classification characterized by a small number of output classes.
- *A multi-branch architecture*: it is devoted to capture multiple scale appearance pattern keeping, at the same time, a sufficient robustness to the overfitting problem tanks to the independent back-propagation of the error.

The proposed network architecture, illustrated in Fig. 2, is made up by 2 main sides: the multi-branch one, aimed to decompose and filter the input image on multiple scales, and the single-branch one, aimed to collect the extracted features and to feed a series of full connected neural layers. RGB registered facial images of 256 $\times$ 256 pixels are centrally cropped to 227 $\times$ 227 pixels that are the inputs of the network. The i-th branch is referred as B(i), the j-th convolutional block (involving different layers) is referred as CL(j) and, finally, the k-th fully-connected layer is referred as FC (k).

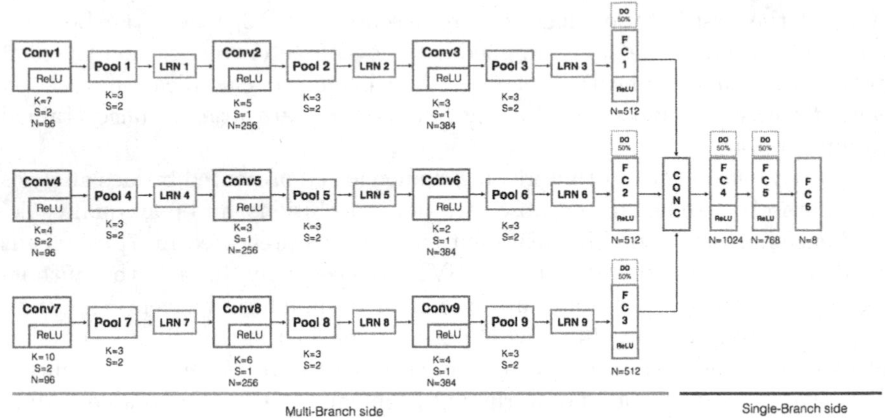

Fig. 2. Proposed network

B(1)-CL(1): it uses 96 filters of size 4 × 4 working directly on the input image. It is followed by a rectified linear operator (ReLU) and successively by a max pooling layer (aimed to take the maximal value of 3 × 3 regions with two-pixel strides) and a final local response normalization (LRN) layer.

B(1)-CL(2): this layer block processes the output of the previous one by means of 256 filter of size 96 × 3 × 3. As for the previous one, it is followed by a ReLU, a max pooling layer and a LRN layer working with the same parameters used before.

B(1)-CL (3): This last convolutional layer operates on the received input from the bottom one by applying a set of 384 filters of size 256 × 2 × 2 pixels followed by only a ReLU and a max pooling layer.

B(1)-FC (1): this first fully connected layer works on the received output of the last convolutional layer and contains 512 neurons. It is followed by a ReLU and a dropout layer.

B(2)-CL(4): it uses 96 filters of size 7 × 7 working directly on the input image. It is followed by a rectified linear operator (ReLU) and successively by a max pooling layer (aimed to take the maximal value of 3 × 3 regions with two-pixel strides) and a final local response normalization (LRN) layer.

B(2)-CL(5): this layer block processes the output of the previous one by means of 256 filter of size 96 × 5 × 5. As for the previous one, it is followed by a ReLU, a max pooling layer and a LRN layer working with the same parameters used before.

B(2)-CL (6): This last convolutional layer operates on the received input from the bottom one by applying a set of 384 filters of size 256 × 3 × 3 pixels followed by only a ReLU and a max pooling layer.

B(2)-FC (2): this first fully connected layer works on the received output of the last convolutional layer and contains 512 neurons. It is followed by a ReLU and a dropout layer.

B(3)-CL(7): it uses 96 filters of size 10 × 10 working directly on the input image. It is followed by a rectified linear operator (ReLU) and successively by a max pooling layer (aimed to take the maximal value of 3 × 3 regions with two-pixel strides) and a final local response normalization (LRN) layer.

B(3)-CL(8): this layer block processes the output of the previous one by means of 256 filter of size 96 × 6 × 6. As for the previous one, it is followed by a ReLU, a max pooling layer and a LRN layer working with the same parameters used before.

B(3)-CL (9): This last convolutional layer operates on the received input from the bottom one by applying a set of 384 filters of size 256 × 4 × 4 pixels followed by only a ReLU and a max pooling layer.

B(3)-FC (3): this first fully connected layer works on the received output of the last convolutional layer and contains 512 neurons. It is followed by a ReLU and a dropout layer.

CONCAT: a concatenation layer aimed to concatenate the output of the 3 fully-connected above mentioned.

FC (6): this fully connected layer, containing again 1024 neurons, receives the 1536-dimensional output of the previous fully connected. It is followed by a ReLU and a dropout layer.

FC (7): this fully connected layer, containing again 768 neurons, receives the 1024-dimensional output of the previous fully connected. It is followed by a ReLU and a dropout layer.

FC (8): this is the last step toward the mapping of the input image to the age categories (8 output layer).

As a final step, the output of the last fully connected layer is given as input to a soft max layer that produces an output in terms of probability for each class. The final prediction is the class corresponding to the maximum value of a soft-max layer.

4 Experimental Results

In order to evaluate the performance of the proposed architecture and its computational load, k-fold validation and tests on multiple hardware platforms have been carried out on the Adience dataset [7] (details in Table 1). The dataset consists of images automatically uploaded to Flickr from smart-phone devices.

Table 1. Adience face dataset: face distribution among different age classes.

Age group	0–2	4–6	8–13	15–20	25–32	38–43	48–53	60+	Total
(Label)	(0)	(1)	(2)	(3)	(4)	(5)	(6)	(7)	
Male	745	928	934	734	2308	1294	392	442	8192
Female	682	1234	1360	919	2589	1056	442	427	9411
Both	1427	2162	2294	1653	4897	2350	825	869	19487

The entire Adience collection includes roughly 26K images of 2,284 subjects. However it represents an highly challenging dataset (due to the presence of make-up, accessories and extremely rotated poses). In order to stress the performance of the proposed solution, the in-plane aligned version of the faces, originally used in [7], has been employed. In this way it is possible to highlight the generalization capabilities of the proposed solution, enhancing the advantage of a multiscale approach in unconstrained contexts. The training/testing procedure was based on a cross subject-exclusive validation that means that the same subject was not present, even with different images, in both the test and training sets used in a training/test round. More specifically the dataset under investigation was randomly split into k folds (the folding provided by [18]) with $k = 5$ has been employed. For each of the k validation steps, $k - 1$ folds were used for training, whereas the remaining fold was used for evaluating the estimation/validation capabilities.

The training was performed with a backforward optimization process aided by a stochastic gradient descend (SGD). More precisely a batch size of 50 elements and a momentum $\mu = 0.9$ were used.

Both training and testing procedures were performed using the Caffe open source framework [13] exploiting a Nvidia GTX 960 equipped with 1024 cores and 4 GB of video memory.

The outcomes were expressed in form of confusion matrix (computed as mean on 5 folds) both for the propose solution (Table 2) and the solution in [7] (Table 3). Moreover, the accuracies referred to the precise class and to the 1-class off were reported at the and of the tables and resumed in Fig. 3.

In order to be as fair as possible, the average accuracies have been computed as means of the confusion matrix diagonals, avoiding the unreliable results eventually due to the unbalancing of the classes.

Table 2. Accuracy performance on multi-branch network (proposed) approach: confusion matrix of *predicted* vs *real* classes; the last row reports the exact accuracy and the 1 class-off accuracy.

	0–2	4–6	8–13	15–20	25–32	38–43	48–53	60+
0–2	68.39%	28.32%	1.98%	0%	0.95%	0.08%	0.2%	0.08%
4–6	14.01%	64.4%	18.91%	0.32%	1.85%	0.32%	0.05%	0.14%
8–13	1.17%	24.48%	58.44%	4.6%	10.32%	0.75%	0.09%	0.14%
15–20	0.3%	5.79%	18.4%	15.74%	56.21%	3.26%	0.18%	0.12%
25–32	0.2%	1.49%	9.19%	8.26%	73.33%	7.03%	0.34%	0.14%
38–43	0.43%	2%	8.43%	4.26%	62.48%	18.3%	2.96%	1.13%
48–53	0.6%	1.81%	9.98%	1.57%	30.48%	40.24%	9.88%	5.54%
60+	0.23%	2.74%	7.77%	2.06%	13.83%	30.63%	16.46%	26.29%
Acc: 41.85%					Acc (1 out): 80.33%			

Table 3. Accuracy performance on single-branch network [18] approach: confusion matrix of *predicted* vs *real* classes; the last row reports the exact accuracy and the 1 class-off accuracy.

	0–2	4–6	8–13	15–20	25–32	38–43	48–53	60+
0–2	73.5%	23.01%	1.9%	0%	0.87%	0.44%	0.16%	0.12%
4–6	16.5%	61.49%	18.12%	0.55%	2.96%	0.18%	0%	0.18%
8–13	1.92%	21.95%	50.7%	7.46%	16.42%	1.31%	0.19%	0.05%
15–20	0.48%	5.91%	15.68%	15.98%	57.48%	3.74%	0.6%	0.12%
25–32	0.4%	1.66%	9.17%	7.76%	74.26%	5.84%	0.73%	0.18%
38–43	0.78%	2.65%	7.91%	2.7%	65.96%	14.91%	3.57%	1.52%
48–53	0.24%	2.53%	6.99%	3.37%	45.06%	28.43%	8.55%	4.82%
60+	0.8%	3.31%	6.63%	2.97%	23.2%	29.6%	12%	21.49%
Acc: 40.11%					Acc (1 out): 76.18%			

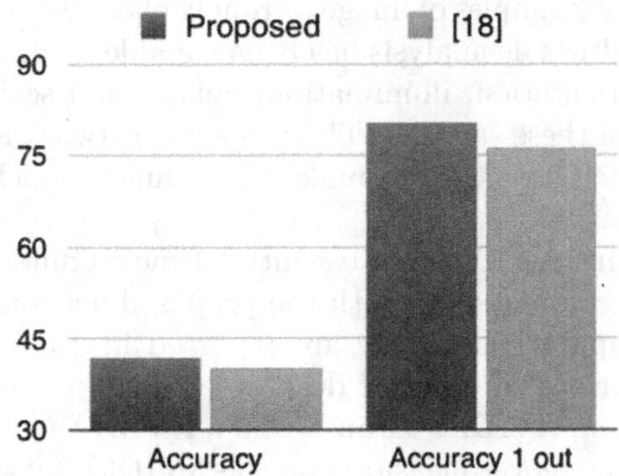

Fig. 3. Exact accuracy and the 1 class-off accuracy

A quick analysis of the results immediately highlights the superiority of the proposed approach respect to the single branch approach in [18]. Both the accuracies show that the use of multiple scales allows to improve the classification performance (+1.74% in exact class and +4.15% in 1 class-off class). Moreover, the greater increment on the 1 class-off accuracy proves as the proposed method allows a general polarization towards the right class.

Going deeper in the observation of confusion matrices it is possible to observe that, concerning the majority of the age classes, the proposed network guarantees an increment in accuracy. The most relevant improvement is obtained on the 60+ class (exact class = 4.8%, 1 class-off = 9.26) where the notable spread in age is well managed by the analysis on multiple scales. On the contrary, the 0–2 class is characterized by a narrow age gap and it is quite close to the successive one 4–6. This makes it suffering of the proposed approach that, catching more information, tends to spread the results leading to a sensible polarization toward

Table 4. Some example of sample wrongly classified by the single branch approach but correctly recognized by the proposed multi-branch structure. GT is the ground truth and the numbers are the labels referred in Table 1

GT	4	7	2	7
[18]	3	5	4	6
Proposed	4	7	2	7

the 4–6 class. Anyway, looking to the 1 class-off results for the first class, we can see as the proposed approach gets an accuracy of 96.71% against the 96.51%.

Table 4 shows some examples of images wrongly classified by the single branch approach that the multi-scale analysis has been capable to classify on the correct class. It is clear as orientation, illumination, quality and scale are some of the most evident issues on these samples. The proposed network is probably able to exploit its multi-branch structure to make the trained model more suitable to the aforementioned variations.

As an additional investigation, the execution time on different hardware configurations, has been computed for both the proposed network and the solution in [7]. Resulting computational times are reported in Table 5. More precisely the test have been performed on an i7 desktop pc equipped with 8 GB of RAM and, alternatively, on an NVIDIA Titan X and a NVIDIA GTX 970. The execution times have been moreover measured on the NVIDIA Jatson TX1 embedded system. It is clear as the proposed architecture introduces an increment of the computational time, that is, anyway, largely into the magnitude of the *real time processing*.

Table 5. Processing time: both the network was tested on 2 desktop configuration and (Nvidia Titan X and Nvidia GTX 970) and an embedded solution Nvidia Jatson TX1

	Execution time [ms]			Model size [MB]
	Jatson TX1	GTX 960	Titan X	
[18]	20.5	3.5	1.4	46
Proposed	70	12	5.3	243

5 Conclusions

Deep Learning, especially in computer vision issues, forcefully replaced classical approaches based on local feature extraction, allowing recognition to easily move toward unconstrained environments. Anyway the lack in data availability sets a serious limit to the development of very deep networks that are constrained to be mildly deep in order to avoid overfitting issues. In this work, the possibility to exploit a multi-branch structure capable to keep the overfitting under control, but meanwhile to take advantage of a multiple scale analysis, has been treated.

Results proved the validity of the suggested solution capable to outperform the leading state of the art solutions (at the best of our knowledge) increasing the average accuracy up to 4.15% for the 1-off class estimation.

References

1. Bar-Hillel, A., Hertz, T., Shental, N., Weinshall, D.: Learning distance functions using equivalence relations. In: ICML, vol. 3, pp. 11–18 (2003)
2. Carcagnì, P., Cazzato, D., Del Coco, M., Leo, M., Pioggia, G., Distante, C.: Real-time gender based behavior system for human-robot interaction. In: Beetz, M., Johnston, B., Williams, M.-A. (eds.) ICSR 2014. LNCS, vol. 8755, pp. 74–83. Springer, Cham (2014). doi:10.1007/978-3-319-11973-1_8
3. Carcagnì, P., Del Coco, M., Cazzato, D., Leo, M., Distante, C.: A study on different experimental configurations for age, race, and gender estimation problems. EURASIP J. Image Video Process. **2015**(1), 37 (2015)
4. Carcagnì, P., Cazzato, D., Del Coco, M., Distante, C., Leo, M.: Visual interaction including biometrics information for a socially assistive robotic platform. In: Second Workshop on Assistive Computer Vision and Robotics (ACVR) (2014)
5. Chao, W.L., Liu, J.Z., Ding, J.J.: Facial age estimation based on label-sensitive learning and age-oriented regression. Pattern Recogn. **46**(3), 628–641 (2013)
6. Choi, S.E., Lee, Y.J., Lee, S.J., Park, K.R., Kim, J.: Age estimation using a hierarchical classifier based on global and local facial features. Pattern Recogn. **44**(6), 1262–1281 (2011)
7. Eidinger, E., Enbar, R., Hassner, T.: Age and gender estimation of unfiltered faces. IEEE Trans. Inf. Forensics Secur. **9**(12), 2170–2179 (2014)
8. Gao, F., Ai, H.: Face age classification on consumer images with gabor feature and fuzzy LDA method. In: Tistarelli, M., Nixon, M.S. (eds.) ICB 2009. LNCS, vol. 5558, pp. 132–141. Springer, Heidelberg (2009). doi:10.1007/978-3-642-01793-3_14
9. Geng, X., Zhou, Z.H., Smith-Miles, K.: Automatic age estimation based on facial aging patterns. IEEE Trans. Pattern Anal. Mach. Intell. **29**(12), 2234–2240 (2007)
10. Guo, G., Fu, Y., Dyer, C.R., Huang, T.S.: Image-based human age estimation by manifold learning and locally adjusted robust regression. IEEE Trans. Image Process. **17**(7), 1178–1188 (2008)
11. Han, H., Otto, C., Jain, A.K.: Age estimation from face images: human vs. machine performance. In: 2013 International Conference on Biometrics (ICB), pp. 1–8. IEEE (2013)
12. He, X., Niyogi, P.: Locality preserving projections. In: NIPS, vol. 16 (2003)
13. Jia, Y., Shelhamer, E., Donahue, J., Karayev, S., Long, J., Girshick, R., Guadarrama, S., Darrell, T.: Caffe: convolutional architecture for fast feature embedding. arXiv preprint arXiv:1408.5093 (2014)
14. Krizhevsky, A., Sutskever, I., Hinton, G.E.: Imagenet classification with deep convolutional neural networks. In: Advances in Neural Information Processing Systems, pp. 1097–1105 (2012)

15. Kwon, Y.H., et al.: Age classification from facial images. In: 1994 IEEE Computer Vision and Pattern Recognition (CVPR), pp. 762–767 (1994)
16. LeCun, Y., Boser, B., Denker, J.S., Henderson, D., Howard, R.E., Hubbard, W., Jackel, L.D.: Backpropagation applied to handwritten zip code recognition. Neural Comput. **1**(4), 541–551 (1989)
17. Leo, M., Medioni, G., Trivedi, M., Kanade, T., Farinella, G.: Computer vision for assistive technologies. Comput. Vis. Image Underst. **154**, 1–15 (2017)
18. Levi, G., Hassner, T.: Age and gender classification using convolutional neural networks. In: IEEE Conference on Computer Vision and Pattern Recognition Workshops, pp. 34–42 (2015)
19. Ramanathan, N., Chellappa, R.: Modeling age progression in young faces. In: Computer Vision and Pattern Recognition (CVPR), vol. 1, pp. 387–394. IEEE (2006)
20. Wolf, L., Hassner, T., Taigman, Y.: Descriptor based methods in the wild. In: Workshop on Faces in 'Real-Life' Images: Detection, Alignment, and Recognition (2008)
21. Yan, S., Liu, M., Huang, T.S.: Extracting age information from local spatially flexible patches. In: IEEE International Conference Acoustics, Speech and Signal Processing (ICASSP), pp. 737–740 (2008)
22. Yan, S., Zhou, X., Liu, M., Hasegawa-Johnson, M., Huang, T.S.: Regression from patch-kernel. In: IEEE Computer Vision and Pattern Recognition (CVPR), pp. 1–8 (2008)
23. Zhuang, X., Zhou, X., Hasegawa-Johnson, M., Huang, T.: Face age estimation using patch-based hidden Markov model supervectors. In: 19th International Conference on Pattern Recognition (ICPR), pp. 1–4. IEEE (2008)

No-Reference Learning-Based and Human Visual-Based Image Quality Assessment Metric

Christophe Charrier[1(✉)], Abdelhakim Saadane[2], and Christine Fernandez-Maloigne[3]

[1] Normandie Univ, UNICAEN, ENSICAEN, CNRS, GREYC, 14000 Caen, France
christophe.charrier@unicaen.fr
[2] Université de Nantes, XLIM Lab, Nantes, France
[3] Université de Poitiers, XLIM Lab, Poitiers, France

Abstract. With the rapid growth of multimedia applications and technologies, objective image quality assessment (IQA) became a topic of fundamental interest. No-Reference (NR) IQA algorithms are more suitable to real-world applications where the original image is not available. In order to be more consistent with human perception, this paper proposes a new NR-IQA metric where the input image is firstly decomposed to several frequency sub-bands which mimic the human visual system (HVS). Then, the statistical features are extracted from these frequency bands and used to fit a multivariate Gaussian distribution (MVGD). Finally, the model obtained by training predicts the quality of the input image. Experimental results demonstrate the method effectiveness and show its robustness when tested by different databases. Moreover, the predicted quality is more consistent with human perception.

Keywords: Image quality assessment · No reference · Human visual system · Visual sub-bands · Multivariate gaussian distribution

1 Introduction

From the last two decades, a great number of researches have been conducted to design robust No-Reference Image Quality Assessment (NR-IQA) algorithms. As they don't use the reference image, these algorithms can be embedded in the development of new multimedia services. NR-IQA algorithms aim at predicting image quality from objective features extracted from distorted images. To reach this goal, these algorithms, either assume a priori knowledge of involved distortions, or look for a generic approach by directly assessing the image quality regardless the type of image distortion [1]. For specific approaches, metrics are mainly designed to quantify distortions induced by image encoders such as JPEG and JPEG2000. In studies [2,3], the block effect is estimated in the spatial domain while studies [4,5] quantify the same effect in the frequency domain. Assessment algorithms for blur effect are proposed in [6,7]. Distortions induced by JPEG2000 such as blur and ringing are considered in [8,9]. The blur effect

S. Battiato et al. (Eds.): ICIAP 2017, Part II, LNCS 10485, pp. 245–257, 2017.
https://doi.org/10.1007/978-3-319-68548-9_23

is characterized by an increase of the spread of edges while the ringing effect produces halos and/or rings near sharp object edges. As a result, the proposed metrics are generally based on the measurement of edges spreading. All of these metrics are interesting and some of them perform very well. However they remain limited by the distortions that they have to know.

Generic approaches aim to be universal and look to address all applications fields. Usually, the generic approaches follow two trends: Signal based approach that extracts and analyzes features in image signals and visually based approach that aims to mimic the human visual system (HVS) properties. Signal based approaches present a good trade-off between performance and complexity. Generally, these approaches require two steps. In the first one, relevant features are extracted while in the second, these features are pooled in order to produce the quality score of the image under test. The first step has been the subject of several investigations [10–12]. In contrast, the second step usually uses conventional combinations. To overcome this drawback, statistical modeling of natural images has been considered where the features extraction procedure is followed by a learning step. In the case of no-reference metrics which is the context of this paper, these learning methods can well map features and subjective assessments [13–17,23].

Since the result of evaluation ultimately depends on the final observer, the visually-based trend looks for designing metrics according to HVS behavior when assessing quality. Several metrics based on one or more HVS properties, are proposed in literature. The HVS sensitivity to image signals is used in [24,25]. The HVS sensitivity to spatial frequencies, luminance and/or structural information is pointed out in [26–28]. Nowadays, several HVS models are proposed in image quality assessment and their experimental results are very promising.

The goal of this paper is to combine the advantages of visually-based methods and the interesting results [29] of learning techniques in the context of NR-IQA metrics. More specifically, it uses the spatial frequency sensitivity of HVS to decompose the test image, extracts the statistical features of each visual sub-band and combines them using a multivariate Gaussian distribution, to assess the quality.

The rest of this paper is organized as follows: The proposed metric is presented in Sect. 2. The visual sub-band decomposition is described in Sect. 3. The experimental results are shown and discussed in Sect. 4.

2 The LEVIQI Index

2.1 General Framework

Figure 1 gives the overall framework of the proposed method.

The first block models the spatial frequency selectivity of the human visual system and performs a perceptual decomposition on both training and testing images. The steerable pyramid [30] is used to achieve the perceptual decomposition described in Sect. 2.2. The steerable pyramid is a multi-scale, multi-orientation ans self inverting image decomposition. With this decomposition, the image is divided into a set of sub-bands localized in scale and orientation. An example of the first level image decomposition where four oriented band-pass filters are used, is given Fig. 2.

The second block extracts the statistical features of each filtered visual channel. Instead of looking to define new features, this paper will exploit some features derived from well-known metrics.

The computed features and the DMOS (Difference of Mean Opinion Scores) values of training images are then used by the learning block to fit an IQA Multivariate Gaussian Distribution. Due to its simple parametric form, the MVGD is widely used for modeling vector-features signals and has been a sound choice in many image and video applications. Therefore, the resulting model, namely LEarning-based and Visual-based Image Quality Index (LEVIQI) is given by:

$$LEVIQI\,(x) = \frac{1}{(2\pi)^{k/2}\,|\Sigma|^{1/2}} \exp\left(-\frac{1}{2}\,(x-\beta)^T\,\Sigma^{-1}\,(x-\beta)\right) \tag{1}$$

where $x = (f_1, ..., f_k, DMOS)$ corresponds to the extracted features to which is added the DMOS of training distorted images. β and Σ denote the mean and covariance matrix of the MVGD model and are estimated using the maximum likelihood method. The features extracted from testing images with DMOS values lying between 0 and 100 with a step of 0.5, are fed into the learned LEVIQI to assess objective quality of image under test. The DMOS of the test image is the one that maximizes the distribution $p\,(x, \beta, \Sigma)$.

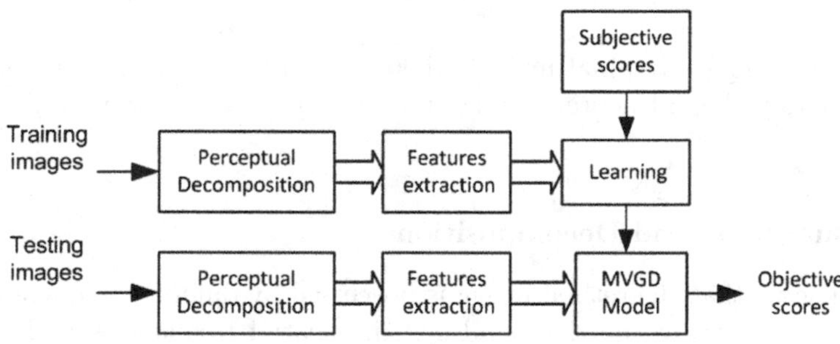

Fig. 1. Overall framework of the proposed method.

The probabilistic model is trained on the LIVE IQA database, for which, one has access to DMOS values. To ensure a robustness of results, multiple training sets were constructed. In each, the image database was subdivided into distinct training and test sets (completely content-separate). For each train set, 80% of the LIVE IQA Database content was chosen, inducing that the remaining 20% is considered for the test set. Specifically, each training set contained images derived from 23 original images, while each test set contained the images derived

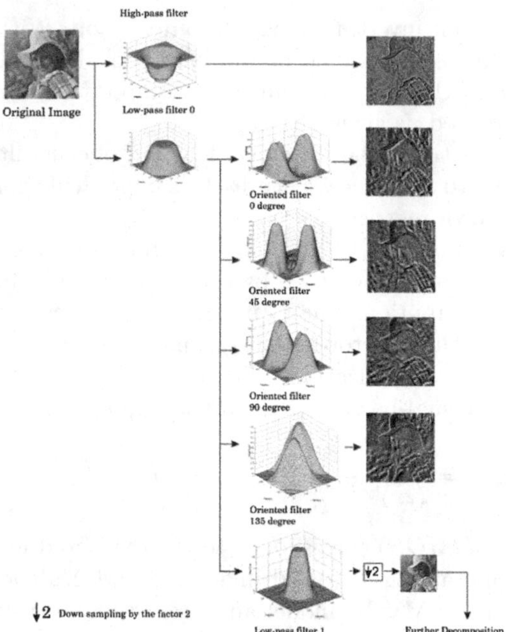

Fig. 2. Steerable pyramid-based image decomposition. Illustration of the first level image decomposition (radial selectivity) where four oriented band-pass filters are used (angular selectivity). The second level of decomposition is the result of the same process performed on low-pass filter 1 (low-pass filter 0 downsampled by the factor 2)

from the remaining 6 original images. 1000 randomly chosen training and test sets were obtained and the prediction of the quality scores was run over the 1000 iterations.

2.2 Visual Sub-band Decomposition

It is well known that the retinal image is processed by different frequency channels that are narrowly tuned around specific spatial frequencies and orientations. Numerous psychophysical experiments have been conducted to estimate the bandwidth of these channels. Quiet different values have been obtained by different types of experiments.

According to these experiments, most of the proposed decompositions suggest a spatial frequency bandwidth of approximately one octave and an orientation sensitivity that varies between 20° and 60° depending on the spatial frequency.

This study uses the decomposition of Fig. 3. Discussed in [31], this decomposition uses a radial frequency selectivity that is symmetric on a log-frequency axis with bandwidths nearly constant at one octave. It consists of one isotropic low-pass and three band-pass channels. The angular selectivity is constant and equal to 30°.

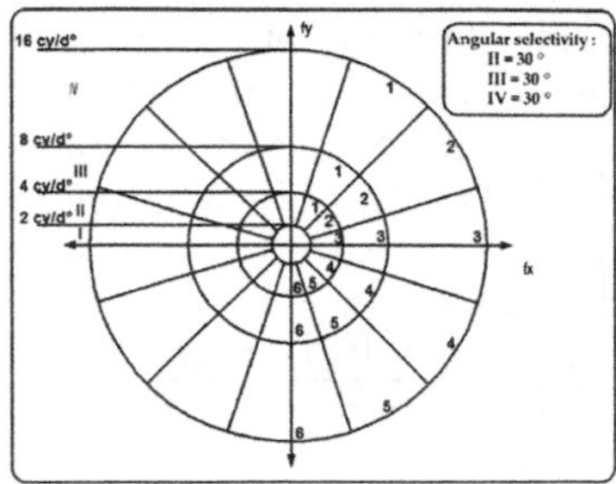

Fig. 3. Perceptual decomposition.

2.3 Selected Features

All features considered in this paper are extracted from nine commonly used learning-based NR-IQA metrics: (1) BRISQUE [18], (2) QAC [19], (3) BLIINDS-II [17], (4) NIQE [23], (5) DIIVINE [16], (6) BIQI [15], (7) IL-NIQE [20], (8) SSEQ [21] and (9) OG-IQA [22]. A reason of the choice of those trial algorithms is motivated by the fact that the code of all of them is publicly available.

Yet, since around 200 features are available from all the trial algorithms, only the most relevant are selected. Furthermore, some of them modelize similar visual characteristics, *e.g.*, luminance sensitivity, sub-band anisotropy, and so on. So, it is not necessary to select features which modelize similar characteristics.

All features are computed for all original images (and their associated degraded version) of the LIVE IQA database [33]. Then, the Spearman Rank-Order Correlation Coefficient (SROCC) between values of features and subjective DMOS is computed. Finally, only the 10 highest correlated attributes are considered to design LEVIQI under the constraint that the selected features do not modelize similar characteristics.

Table 1. Selected features to design LEVIQI

Feature from	Description	# features
BRISQUE	Shape parameter and the variance of GGD fit of the MSCN (mean subtracted contrast normalized) coefficients	2
BLIINDS2	Coefficient of frequency variation	1
BLIINDS2	Energy subband ratio measure	1
DIIVINE	Across scale and spatial correlations	2
SSEQ	Local spatial and spectral entropy features	2
IL-NIQE	Chromatic statistics	2

Table 1 presents the selected distinct features. Yet, these attributes have not been used as they have been defined in their associated NR-IQA schemes, but have been modified to adapt the visual sub-band decomposition.

GGD Fit Parameters: The first set of features, derived from BRISQUE, includes the shape parameter α and the variance σ^2 of generalized Gaussian distribution (GGD) fit of the MSCN (mean subtracted contrast normalized) coefficients for each sub-band. The MSCN coefficients refer to the transformed luminances $\hat{I}(i,j)$ given by:

$$\hat{I}(i,j) = \frac{I(i,j) - \mu(i,j)}{\sigma(i,j) + C}, \tag{2}$$

where (i,j) are spatial indices, C=1, and $\mu(i,j) = \sum_{k=-3}^{3} \sum_{l=-3}^{3} \omega_{k,l} I_{k,l}(i,j)$, and $\sigma(i,j) = \sqrt{\sum_{k=-3}^{3} \sum_{l=-3}^{3} \omega_{k,l} \left(I_{k,l}(i,j) - \mu(i,j)\right)^2}$, ω is a 2D circularly-symmetric Gaussian weighting function.

The shape α and the variance σ^2 parameter of the GGD are computed over all subbdands and then pooled by computing the lowest 10th percentile average of the local α scores and the local σ^2 scores across the sub-bands.

Coefficient of Frequency Variation: This feature, derived from BLIINDS2, is defined for each subband as:

$$\zeta = \frac{\sigma_{|X|}}{\mu_{|X|}} = \sqrt{\frac{\Gamma(1/\gamma)\,\Gamma(3/\gamma)}{\Gamma^2(2/\gamma)} - 1} \tag{3}$$

where $\sigma_{|X|}$ and $\mu_{|X|}$ are the standard deviation and mean of the frequency coefficient magnitudes $|X|$, respectively.

The feature ζ is computed for all sub-bands of the image and pooled by taking the highest 10th percentile and over all of the sub-band scores of the image.

Energy Sub-band Ratio Measure: This attribute, also derived from BLIINDS2, is used to capture local spectral signature. This feature is defined as:

$$R_{n,k} = \frac{\left| E_{n,k} - \frac{1}{n-1} \sum_{j<n} E_{j,k} \right|}{E_{n,k} + \frac{1}{n-1} \sum_{j<n} E_{j,k}} \tag{4}$$

where $E_{n,k} = \sigma^2_{n,k}$ is the average energy in the frequency sub-band n for a given radial band k, where $n \in \{1, 2, 3, \ldots, 6\}$ and $k \in \{1, 2, 3\}$. The mean of $(R_{n,k})_{n \in [2,\ldots,6]}$ is computed on the three radial bands k of the image and pooled by computing the highest 10th percentile average of the scores of the image.

Across Scale and Spatial Correlations: These two features are derived from features designed for DIIVINE. In order to capture the statistical dependencies between high-pass (HP) responses of natural images and their band-pass (BP) counterparts, a structural correlation is modeled as $\rho = (2\sigma_{xy} + C_2)/(\sigma_x^2 + \sigma_y^2 + C_2)$ where σ_{xy} is the cross-variance between the windowed regions from the BP and HP bands, and σ_x^2, σ_y^2 are their windowed variances respectively; C_2 is a stabilizing constant. The mean of the 18 correlation values (corresponding to the 18 sub-bands) is computed.

In order to capture spatial correlation statistics, the joint empirical distribution between coefficients at (i, j) and the set of spatial locations at a distance of τ is computed for each d_1^θ, $\theta \in \{0^o, 30^o, 60^o, 90^o, 120^o, 150^o\}$. The correlation between these two variables denoted X and Y is estimated by:

$$\rho(\tau) = \frac{E_{PXY(x,y)}\left[\left(X - E_{PX(x)}[X]\right)^T\left(Y - E_{PY(y)}[Y]\right)\right]}{\sigma_X \sigma_Y} \tag{5}$$

where $E_{PX(x)}[X]$ is the expectation of X with respect to the marginal distribution $p_X(x)$ (similarly for Y and (X,Y)). $\rho(\tau)$ is plotted for various distance across distortions and the obtained curve is fitted with a 3rd order polynomial. The coefficients of the polynomial and the error between the fit and the actual $\rho(\tau)$ form an 30 dimensional feature vector (5 values for each direction). The mean of $\rho(\tau)$ is computed over the three radial bands and pooled by computing the highest 10th percentile average of the scores.

Local Spatial and Spectral Entropy Features: The spatial entropy is computed on the obtained image after applying the first non-directional lower-pass filter applying.

Concerning the spectral entropy, a modified version of the attribute designed for SSEQ is defined. The variance across the six orientations per radial band $var(E[R_\theta])$ is computed. $(E[R_\theta])$ is the average of the Renyi entropy $R_\theta[n]$ per orientation for the three sub-bands (one per radial band) of orientation θ.

Finally, the mean of the three variance values per radial band is computed.

Chromatic Statistics: This attributed is derived from IL-NIQE and is computed on the color image before applying the DCP transform. From the RGB coordinates system, a logarithmic-scale opponent color space $\mathcal{RGB}$ is defined as $\mathcal{R} = \log R - \mu_R$, $\mathcal{G} = \log G - \mu_G$ and $\mathcal{B} = \log B - \mu_B$ where μ_R (resp. μ_G and μ_B) is the mean $\log R$ (resp. $\log G$ and $\log B$) over the image. Finally, the following linear color transform is applied on the $\mathcal{RGB}$ color space as $l_1 = (\mathcal{R}+\mathcal{G}+\mathcal{B})/\sqrt{3}$, $l_2 = (\mathcal{R}+\mathcal{G}-2\mathcal{B})/\sqrt{6}$, $l_3 = (\mathcal{R}-\mathcal{G})/\sqrt{2}$. The distributions of each channel l_1, l_2, l_3 conforms to a Gaussian probability law. In this paper, only the chromatic channels l_2 and l_3 are considered, since l_1 refers to a luminance channel. Finally, the two model parameters μ_C and σ_C^2 are estimated using a multivariate Gaussian model.

3 Performance Evaluation

3.1 Experimental Setup

Trial Databases: To provide comparison of NR-IQA algorithms, two publicly available databases are used: (1) TID2013 database [32] and (2) CSIQ database [34]. Since the LIVE database [33] has been used to train both the proposed metric and most of the trail NR-IQA schemes, it has not been used to evaluate performances of NR-IQA methods.

The TID2013 database contains images with multiple distortions. This database consists of 25 original images on which 24 different type of distortions have been applied using five degradation levels per distortion. A total of 3000 distorted images are generated. It is worth noting that seven new types of degradations have been introduced with respect to the 17 types of degradation existing in the previous version of the database, known as TID2008. The database contains 524340 subjective ratings from 971 different observers, and ratings are reported in the form of MOS.

The CSIQ Database consists of 30 original images, each is distorted using six different types of distortions at four to five different levels of distortion. A total of 866 distorted images have been generated. The database contains 5000 subjective ratings from 35 different observers, and ratings are reported in the form of DMOS.

Trial NR-IQA: To assess the performance of the proposed metric, nine commonly used NR-IQA schemes are used to compare LEVIQI with. The trial metrics are those mentioned in Sect. 2.3. Only NR-IQA schemes whose at least one feature has been adapted to design LEVIQI index, *i.e.*, BRISQUE, BLIINDS2, DIIVINE, SSEQ and ILNIQE are used to compare the performance of LEVIQI with.

Statistical Significance and Hypothesis Testing: Results obtained from the proposed metric are compared to results provided by all trial NR-IQA algorithms. To perform this evaluation, the Spearman Rank Order Correlation Coefficient (SROCC) is computed between the DMOS values and the predicted scores obtained from NR-IQA algorithms. In addition, to ascertain which differences between NR-IQA schemes performance are statistically significant, we applied an hypothesis test using the residuals between the DMOS values and the ratings provided by the IQA algorithms. This test is based on the t-test that determines whether two population means are equal or not. This test yields us to take a statistically-based conclusion of superiority (or not) of an NR-IQA algorithm.

3.2 Experimental Results

Table 2 gives SROCC mean values computed between predicted scores from NR-IQA schemes from which some extracted features are used to design LEVIQI and MOS values for the TID2013 Images database. When considering the whole database, LEVIQI overperforms BRIQUE, BLIINDS2, DIIVINE, SSEQ ad ILNIQE. In 67% of cases (16 subsets out of 24), scores predicted by LEVIQI allow to have a better correlation (higher SROCC value) with human judgments than that of any other tested NR-IQA. For the remaining cases (8 remaining subsets), the performance of LEVIQI is very close to the best values. If we consider multidistorted images associated to the seven last subset of Table 2, LEVIQI performs better than trial quality schemes except for 'Multiplicative Gaussian Noise' and 'Comfort Noise'. Yet, obtained SROCC values are highly competitive with the best quality index.

Table 2. SROCC values computed between predicted scores using NR-IQA schemes from which some attributes are extracted to design LEVIQI and MOS values for TID2013 Images database.

TID2013 subset	BRISQUE	BLIINDS2	DIIVINE	SSEQ	ILNIQE	LEVIQI
Additive Gaussian noise	0.852	0.722	0.855	0.807	0.876	**0.881**
Additive noise in color components	0.709	0.649	0.712	0.681	0.815	**0.817**
Spatially correlated noise	0.491	0.767	0.463	0.635	**0.923**	0.899
Masked noise	0.575	0.512	0.675	0.565	0.512	**0.779**
High frequency noise	0.753	0.824	0.878	0.860	0.868	**0.891**
Impulse noise	0.630	0.650	**0.806**	0.749	0.755	0.801
Quantization noise	0.798	0.781	0.165	0.468	**0.873**	0.860
Gaussian blur	0.813	0.855	0.834	**0.858**	0.814	0.851
Image denoising	0.586	0.711	0.723	0.783	0.750	**0.812**
JPEG compression	0.852	0.864	0.629	0.825	0.834	**0.867**
JPEG2000 compression	0.893	0.898	0.853	0.885	0.857	**0.901**
JPEG transmission errors	0.315	0.117	0.239	**0.354**	0.282	0.217
JPEG2000 transmission errors	0.360	0.620	0.060	0.561	0.524	**0.662**
Non eccentricity pattern noise	**0.145**	0.096	0.060	0.011	0.080	0.121
Local block-wise distortions	**0.224**	0.209	0.093	0.016	0.135	0.217
Mean shift	0.124	0.128	0.010	0.108	0.184	**0.211**
Contrast change	0.040	0.150	0.460	0.204	0.014	**0.521**
Change of color saturation	0.109	0.017	0.068	0.074	0.162	**0.471**
Multiplicative Gaussian noise	0.724	0.716	**0.787**	0.679	0.693	0.771
Comfort noise	0.008	0.017	0.116	0.033	**0.359**	0.315
Lossy compression of noisy images	0.685	0.719	0.633	0.610	0.828	**0.831**
Color quantization with dither	0.764	0.736	0.436	0.528	0.748	**0.892**
Chromatic aberrations	0.616	0.539	0.661	0.688	0.679	**0.699**
Sparse sampling and reconstruction	0.784	0.816	0.834	0.895	0.865	**0.899**
Cumulative subsets	0.367	0.393	0.355	0.332	0.494	**0.501**

Table 3. SROCC values computed between predicted scores from which some attributes are extracted to design LEVIQI and MOS values for CSIQ Images database.

CSIQ subset	BRISQUE	BLIINDS2	DIIVINE	SSEQ	ILNIQE	LEVIQI
JP2K	0.866	0.895	0.830	0.848	0.906	**0.911**
JPEG	0.903	0.901	0.799	0.865	0.899	**0.915**
Gaussian noise	0.252	0.379	0.176	**0.872**	0.850	0.862
Add. Gaussian pink noise	**0.925**	0.801	0.866	0.046	0.874	0.909
Gaussian blur	0.903	0.891	0.871	0.873	0.858	**0.931**
Global contrast decrement	0.029	0.012	0.396	0.200	0.501	**0.590**
Cumulative subsets	0.566	0.577	0.596	0.528	0.815	**0.821**

Similar results as those of Table 2 are given in Table 3 for CSIQ Images database. For this database also, The global performance (SROCC value of cumulative subsets) of LEVIQI is higher than that of trial metrics. More specifically,

LEVIQI performs better 4 times over 6. For the two remaining distortions 'Gaussian Noise' and 'Additive Gaussian Pink Noise', LEVIQI presents the second best performance.

In addition, Table 4 gives obtained results when a One-sided t-test is used to provide statistical significance of NR-IQA/LEVIQI quality scores on the 6 multidistortions subsets of TID2013 database (the 6 last subsets in Table 2). Each entry in this table is coded using six symbols. The position of each symbol corresponds to one subset (first position corresponds to 'Change of Color Saturation' subset, second position for 'Multiplicative Gaussian Noise' subset etc.). Each symbol gives the result of the hypothesis test on the subset. If the symbol equals '1', the NR-IQA on the row is statistically better than the NR-IQA on the column ('0' means worse, '-' is used when NR-IQAs are indistinguishable). Results confirm that, most of the time, LEVIQI is more consistent with human judgments than trial NR-IQA schemes from which some features have been extracted. As multidistortions subsets are not common to LIVE database, these results are more reliable.

Table 4. Statistical significance comparison of trial NR-IQA/LEVIQI quality scores on TID2013 database multidistortions subsets.

	BRISQUE	BLIINDS2	DIIVINE	SSEQ	ILNIQE
LEVIQI	1111111	1111111	10111-1	11111--	110-1--

In a similar way, Table 5 gives obtained results when a One-sided t-test is used to provide statistical significance of NR-IQA/LEVIQI quality scores on CSIQ database. One can notice that distortions present in CSIQ database are also present in LIVE database. For these learned distortions, LEVIQI exceeds in several cases all of the trial NR-IQA schemes.

Table 5. Statistical significance comparison of NR-IQA/LEVIQI quality scores on CSIQ database subsets.

	BRISQUE	BLIINDS2	DIIVINE	SSEQ	ILNIQE
LEVIQI	1-1-111	--11111	1111111	1101111	---111-

Finally, to compare the computational complexity of the proposed algorithm, we measured the average computation time required to assess an image of size 512×512, using a computer with Intel Core-I7 processor at 2.2 GHz. Table 6 reports the measurement results, which are rough estimates only, as no code optimization has been done on our Matlab implementations. LEVIQI is superior to DIIVINE and BLIINDS2 while inferior to BRISQUE and SSEQ and similar to ILNIQE.

Table 6. Comparison of computational time (in second/image)

NR-IQA	BRISQUE	BLIINDS2	DIIVINE	SSEQ	ILNIQE	LEVIQI
Time	1.33	87.13	27.33	2.69	11.85	12.38

4 Conclusion

In this paper, a machine learning-based and human vision-based quality index called LEVIQI has been proposed in the purpose of no reference quality evaluation. The model utilizes ten derivative relevant attributes from five well-known and highly competitive NR-IQA schemes. The selected attributes address human vision characteristics such as frequency sensitivity, chromatic sensitivity, anisotropy and contrast sensitivity. These attributes have been adapted to be computed on 18 sub-bands generated from three different radial bands and six different orientations, to simulate human perception sensitivity. Obtained results show that the performance of LEVIQI is competitive with top-performing NR-IQA schemes. The potential of this model to be used in real applications is the subject of investigations where real-time implementation of LEVIQI index is considered using efficient vectorization.

References

1. Zhang, S., Zhang, C., Zhang, T., Lei, Z.: Review on universal no-reference image quality assessment algorithm. Comput. Eng. Appl. **51**, 13–23 (2015)
2. Gao, W., Mermer, C., Kim, Y.: A de-blocking algorithm and a blockiness metric for highly compressed images. IEEE Trans. Circ. Syst. Video Technol. **12**(12), 1150–1159 (2002)
3. Pan, F., Lin, X., Rahardja, S., Lin, W., Ong, E., Yao, S., Lu, Z., Yang, X.: A locally adaptive algorithm for measuring blocking artifacts in images and videos. Sig. Process. Image Commun. **19**(6), 499–506 (2004)
4. Wang, Z., Bovik, A.C., Evan, B.: Blind measurement of blocking artifacts in images. In: International Conference on Image Processing, vol. 3, pp. 981–984. IEEE Press (2000)
5. Liu, S., Bovik, A.C.: Efficient DCT-domain blind measurement and reduction of blocking artifacts. IEEE Trans. Circ. Syst. Video Technol. **12**(12), 1139–1149 (2002)
6. Ferzli, R., Karam, L.J.: A no-reference objective image sharpness metric based on the notion of Just Noticeable Blur (JNB). IEEE Trans. Image Process. **18**, 717–728 (2009)
7. Sang, Q.B., Su, Y.Y., Li, C.F., Wu, X.J.: No-reference blur image quality assessment based on gradient similarity. J. Optoelectron. Laser **24**, 573–577 (2013)
8. Sazzad, Z.M.P., Kawayoke, Y., Horita, Y.: No reference image quality assessment for JPEG2000 based on spatial features. Signal Process. Image Commun. **23**, 257–268 (2008)
9. Barland, R., Saadane, A.: Reference free quality metric for JPEG-2000 compressed images. In: 8th International Symposium on Signal Processing and its Applications, pp. 351–354 (2005)

10. Wang, Z., Bovik, A.C., Sheikh, H.R., Simoncelli, E.P.: Image quality assessment: from error measurement to structural similarity. IEEE Trans. Image Process. **13**(1), 996–1006 (2004)
11. Shnayderman, A., Gusev, A., Eskicioglu, A.: An SVD-based grayscale image quality measure for local and global assessment. IEEE Trans. Image Process. **15**(2), 422–429 (2006)
12. Wee, C., Paramesran, R., Mukundan, R., Jiang, X.: Image quality assessment by discrete orthogonal moments. Pattern Recogn. **43**(12), 4055–4068 (2010)
13. Babu, R.V., Suresh, S., Perkis, A.: No-reference JPEG image quality assessment using GAP-RBF. Sig. Process. **87**(6), 1493–1503 (2007)
14. Gabarda, S., Cristobal, G.: Blind image quality assessment through anisotroy. JOSA **24**(12), B42–B51 (2007)
15. Moorthy, A.K., Bovik, A.C.: A two-step framework for constructing blind image quality indice. IEEE Sig. Process. Lett. **17**(5), 513–516 (2010)
16. Moorthy, A.K., Bovik, A.C.: Blind image quality assessment: from natural scene statistics to perceptual quality. IEEE Trans. Image Process. **20**(12), 3350–3364 (2011)
17. Saad, M., Bovik, A.C., Charrier, C.: Blind image quality assessment: a natural scene statistics approach in the DCT domain. IEEE Trans. Image Process. **21**(8), 3339–3352 (2012)
18. Mittal, A., Moorthy, A.K., Bovik, A.C.: No-reference image quality assessment in the spatial domain. IEEE Trans. Image Process. **21**(12), 46954708 (2012)
19. Xue, W., Zhang, L., Mou, X.: Learning without human scores for blind image quality assessment. In: IEEE Conference on Computer Vision and Pattern Recognition, pp. 995–1002 (2013)
20. Zhang, L., Zhang, L., Bovik, A.C.: A feature-enriched completely blind image quality evaluator. IEEE Trans. Image Process. **24**(8), 2579–2591 (2015)
21. Liu, L., Liu, B., Huang, H., Bovik, A.C.: No-reference image quality assessment based on spatial and spectral entropies. Sig. Proc. Image Commun. **29**, 856–863 (2014)
22. Liu, L., Hua, Y., Zhao, Q., Huang, H., Bovik, A.C.: Blind image quality assessment by relative gradient statistics and ada boosting neural network. Sig. Process.: Img. Commun. **40**, 1–15 (2016)
23. Mittal, A., Soundararajan, R., Bovik, A.C.: Making a "completely blind" image quality analyzer. IEEE Sig. Process. Lett. **20**(3), 209–212 (2013)
24. Chandler, D.M., Hemami, S.S.: VSNR: a wavelet-based visual signal-to-noise ratio for natural images. IEEE Trans. Image Process. **16**(9), 2284–2298 (2007)
25. Sheikh, H.R., Bovik, A.C.: Image information and visual quality. IEEE Trans. Image Process. **15**(2), 430–444 (2006)
26. Larson, E.C., Chandler, D.M.: Most apparent distortion: full-reference image quality assessment and the role of strategy. J. Electron. Imag. **19**(1), 011006-1–011006-21 (2010)
27. Pei, S.C., Chen, L.H.: Image quality assessment using human visual DOG model fused with random forest. IEEE Trans. Image Process. **24**(11), 3282–3292 (2015)
28. Le Callet, P., Saadane, A., Barba, D.: Frequency and spatial pooling of visual differences for still image quality assessment. Proc. SPIE **3959**, 595–603 (2000)
29. Gastaldo, P., Zunino, R., Redi, J.: Supporting visual quality assessment with machine learning. EURASIP J. Image Video Process. **2013**, 54 (2013)
30. Liu, Z., Tsukada, K., Hanasaki, K., Ho, Y.K., Dai, Y.P.: Image fusion by using steerable pyramid. Pattern Recogn. Lett. **22**, 929–939 (2001)

31. Daly, S.: A visual model of optimizing the design of image processing algorithm. In: Proceeding ICIP, Vol. II of III, pp. 16–20 (1994)
32. Ponomarenko, N., Jin, L., Ieremeiev, O., Lukin, V., Egiazarian, K., Astola, J., Vozel, B., Chehdi, K., Carli, M., Battisti, F., Jay Kuo, C.-C.: Image database TID2013: peculiarities, results and perspectives. Sig. Process. Image Commun. **30**, 57–77 (2015)
33. Laboratory for Image & Video Engineering, University of Texas (Austin), LIVE Image Quality Assessment Database (2002). http://live.ece.utexas.edu/research/Quality
34. Larson, E.C., Chandler, D.M.: Most apparent distortion: full-reference image quality assessment and the role of strategy. J. Electron. Imaging **19**(1), 1–21 (2010)

Performance Evaluation of Multiscale Covariance Descriptor in Underwater Object Detection

Farah Rekik(✉), Walid Ayedi, and Mohamed Jallouli

Computer and Embedded System Laboratory,
National Engineering School of Sfax, Sfax, Tunisia
farah.rekik@enis.tn, ayadiwalid@yahoo.fr,
mohjallouli@gmail.com

Abstract. Object detection is the fundamental process for the majority of the investigation projects in the submarine environment, and object detection is mainly based on image description done by the appropriate descriptor. In this paper we select and optimize parameters of multi-scale covariance descriptor for object detection in the submarine context. We adapt the descriptor parameters to be suitable to cope with the degradation of image quality in underwater environment, working on the homogeneity error tolerance and the precision degree of description. We justify the use of specific parameters values and well defined features. To perform our work we use support vector machine for data classification and Maris dataset as a benchmark.

Keywords: Multi-scale covariance descriptor · Object detection · Classification · Underwater pipe detection · Autonomous Underwater Vehicles

1 Introduction

Automatic detection and recognition of underwater object laying on the seafloor is an important project for international marine research. A great attention is paid to this area. The inspection tasks automation, recognition, detection or physicochemical parameters measurement, is strongly justified in this vast environment. Remotely operated vehicles were the first developed technique where the human is involved in the decision-making chain.

Problems to be solved then were the domain of seafloor technology and the transfer of energy and information via an umbilical link. Since the late 80 years, many research programs have emerged in the United States, Europe and Asia, to provide a solution based on autonomous vehicles. They are called Autonomous Underwater Vehicles (AUV).

These devices therefore have the ability to go in inaccessible areas and to do what other submarines could not do and go where they could not go.

Because of the inherent danger and time-sensitive nature of such missions, the next urgent priority is to embed intelligence in the AUV so that it can immediately react to the data it collects. By adapting its survey route in situ and efficiently allocating resources the AUV can collect the most informative data for the task at hand while simultaneously reducing costs [1].

S. Battiato et al. (Eds.): ICIAP 2017, Part II, LNCS 10485, pp. 258–266, 2017.
https://doi.org/10.1007/978-3-319-68548-9_24

To achieve this goal, two major obstacles must be overcome. First, an algorithm is needed that can perform the detection and recognition of underwater object in near real-time onboard an AUV with limited processing capabilities. Second, a plan is needed for specifying how the information gleaned from the detection results can be exploited to intelligently adapt the AUV route. To accomplish their mission, AUV used for this kind of project, are equipped by sonar system.

Compared to sonar, vision is not widely used in underwater research. This is due to degradation of image quality caused by absorption and scattering of light in water. But sonar suffers from several problems like cost resolution and complexity of use. Therefore there is a need for additional investigation to assess the actual potential of visual perception in underwater environments. Currently, the underwater video is increasingly used as a complementary sensor to the sonar especially for detection of objects or animals. However, the underwater images present some particular difficulties including natural and artificial illumination, color alteration and light attenuation. Therefore in the detection of underwater objects it is impossible to take only color as a detection criterion, but location, shape and color information must be combined.

Object detection is based on the extraction of discriminative features. This extraction is done by the meaning of a descriptor which describes the image through a characteristic vector using specific features which differ from one descriptor to another.

Our detection system is based on Multi-scale covariance descriptor (MSCOV) [2] and in this paper we will try to define the best parameters for better object detection.

The rest of the paper is organized as follows. Section 2 reviews an overview on AUV. In Sect. 3, we describe the method proposed for better underwater object detection. The experimental setup and experimental results are presented in Sect. 4. The paper concludes in Sect. 5.

2 Autonomous Underwater Vehicle

Autonomous Underwater Vehicles (AUV), also known as unmanned underwater vehicles, can be used to perform underwater survey missions such as detecting and mapping submerged wrecks, rocks, and obstructions that pose a hazard to navigation for commercial and recreational vessels. The AUV conducts its survey mission without operator intervention. When a mission is complete, the AUV will return to a pre-programmed location and the data collected can be downloaded and processed in the same way as data collected by shipboard systems.

Among the companies active in the field of underwater drones, some have AUVs equipped with sonar and video cameras. These drones are designed to detect and identify objects.

Object detection is a fundamental process for several submarine missions. Underwater surveillance and tracking require vision based control. Collision and obstacle avoidance are the basis of a safe UAV, so it's important to use a high-performance object detection and recognition system to ensure the safety of the submarine.

In this paper we are interested in the detection process and we will evaluate the descriptor parameter used for data description.

3 Image Descriptors

3.1 Global Approach

The structure of an object detection system is based on image description and data classification. This approach is described in Fig. 1.

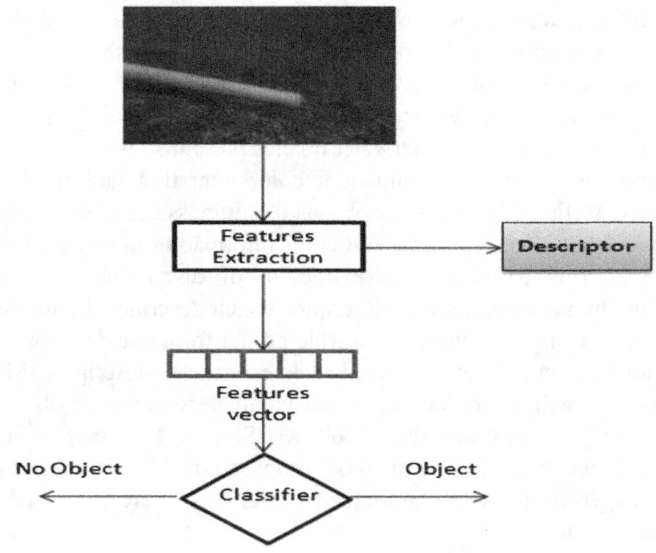

Fig. 1. Global approach for object detection

Object detection is based on the extraction of discriminative features. This extraction is done by the meaning of a descriptor which describes the image through a characteristic vector using specific features which differ from one descriptor to another. This vector is used to train the classifier.

3.2 Type of Descriptor

Global Descriptor

Global description consists on describing the whole image by their characteristics taken from each pixel. The color histogram is the best-known descriptor in this context. It represents the distribution of intensities or color components of the image. The most used global descriptors are statistics descriptors. They are determined following a frequency filtering, starting from the co-occurrence matrices, or from the first-order or high-order statistics.

Global descriptors are known for their speed and simplicity of implementation. The combination of several global characteristics can achieve good results. However, global description suffers from several problems. It implicitly assumes that the entire image is related to the object. Thus, any incoherent object would introduce noise affecting the description of the object. This limitation encourages the use of local descriptors, or even regions.

Local Descriptor
Local description is based on the identification of local points of interest with a vector of attributes, and on the use of local descriptors which characterize only a small part of the image. SIFT [3] is the most popular local descriptor.

The most interesting property of this descriptor is its robustness to the image transformation. The problem is that the objects are represented by a variable number of points of interest, whereas the classifiers require a vector with a fixed size as input. As for the descriptors by region, the feature vector is fixed, which is more suitable for classifiers.

Region Descriptor
This approach consists in decomposing the image into a set of fixed or variable size regions and then characterizing each of these regions. The decomposition is done in a predictable way in order to make the regions' characteristics homogeneous with each other. These descriptors have recently been successful in several applications. Covariance descriptor [4] is mainly used in human detection and re-identification. On the other hand, this descriptor has some limitations. Indeed, it implicitly assumes that the whole region is connected to the object to be modeled while the latter may have an incoherent shape.

MSCOV came to improve this descriptor by the adjustment of the trade-off between the local and the global description of the objects. This descriptor will be detailed in the next section.

3.3 Multiscale Covariance Descriptor

The descriptor adopted for this work is the multi-scale covariance descriptor. It is based on the quadtree structure which explains the multi-scale aspect. This structure is widely used for image representation in computer vision applications [5–7]. It is also used to store and index image characteristics and region of interest.

The quadtree represents a hierarchical structure constructed by recursive divisions of the image in four disjoint quadrants with the same size, according to homogeneity criterion, until a stop condition is reached. Figure 2 presents a quadtree applied to a frame of the dataset used in this work. Each image quadrant is represented by a quadtree node and the root node represents the whole image.

MSCOV descriptor characterizes a quadrant image through the characteristics stored in its associated node. Each node stores a features vector.

Feature vectors in each node are combined into a covariance matrix defined by (1).

$$C_r = \frac{1}{N_r - 1} \sum\nolimits_{c=1}^{N_r} (F_c - m)(F_c - m)^T \tag{1}$$

where Nr is the node number in the sub-tree of r, m the mean of the nodes features and Fc the feature vector of the node c descendant of r. This structure is nominated as

Fig. 2. Quadtree structure

"Image Quadtree Features" (IQF) [8]. Features are arranged in two groups. The structural characteristics which are related to image data location and the content characteristics that are derived from the color information as shown in Table 1.

Table 1. MSCOV features

	k	Features
Structure features	0	The x location of the corresponding image quadrant
	1	The y location of the corresponding image quadrant
	2	Node level
Content features	3	I grayscale intensity value (the luminance component)
	4	Cr color component value (the red chrominance component)
	5	Cb color component value (the blue chrominance component)
	6	Ix, the norm of the first order derivatives in x
	7	Iy, the norm of the first order derivatives in y
	8	Gradient, $o(x, y) = \arctan(\frac{I_y(x,y)}{I_x(x,y)})$
	9	Magnitude $mag(x, y) = \sqrt{I_x^2(x, y) + I_y^2(x, y)}$

The multi-scale covariance descriptor provides two main advantages. In fact, the decomposition into a quadtree makes it possible to capture the region of interest of the image (points of interest), and consequently it reduce the impact of noise and background information on the description of the object. Therefore the pre-treatment step is canceled. In addition, quadtree is used as a multi-level structure to extracts image features from different scales. It thus makes it possible to optimize the compromise between the local and the global description of the object.

MSCOV depends on two principal parameters: ε the homogeneity threshold, and $\propto$ the precision degree of description. The selection of ε depends on the background complexity. Indeed, background with low intensity variation can be easily discarded from image description using low ε values. In contrast, high intensity variation needs high ε values to be discarded. It depends on the nature of the image and the object to be described. Hence a tradeoff must be determined for better object description and therefore for better object detection. By varying $\propto$, the object is described from fine resolution level to coarse one. This parameter can considerably affect the object detection. The second factor that can also affect the detection rate is the choice of discriminative features for image description.

3.4 Support Vector Machine

Support vector machine (SVM) classifier is proposed by Vapnic [9]. It was very useful in pattern recognition. RBF is one of SVM kernel. We choose to adopt this kernel in our work with specific values of C and sigma parameter. In coming works we will compare it with other kernel in order to choose the best one with best parameters.

4 Experimental Result

This section is organized as follows. First we present the adopted evaluation metric. Then we describe the dataset used in the experimentation. Finally we present experiments that conduce to select and optimize parameters of MSCOV descriptor in the context of underwater object detection, and justify the choice of features for pipe detection.

4.1 Evaluation Metric

Precision, recall and F-measure are the appropriate metric to evaluate the detection accuracy. They have always been used for the evaluation of pattern detection algorithms. They are defined by (2) and (3).

$$Precision = \frac{TP}{TP + FP} \tag{2}$$

$$Recall = \frac{TP}{TP + FN} \tag{3}$$

where TP is the true positive, FN is the false negative and FP is the false positive. TP is the number of images real positive and predicted positive. FP is the number of image real positive and predicted negative. FN is the number of image real negative and predicted positive. F-Measure is also a measure of a test's accuracy. It is the harmonic-mean of precision and recall:

$$F_{Measure} = \frac{2.Recall.Precision}{Recall + Precision} \tag{4}$$

4.2 Maris Dataset

Maris dataset [9] is used to evaluate the performance of the proposed approach.

This dataset is acquired using a stereo vision system near Portofino (Italy). It provides images of cylindrical pipes with different color submerged at 10 m deep. The dataset include 9600 stereo images in Bayer encoded format with 1292 × 964 resolution, and it include positive (frames containing a pipe) and negative frame (frames presenting only the background) Fig. 3.

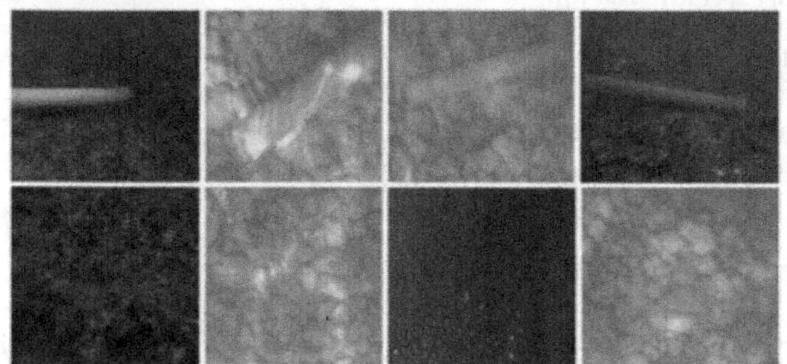

Fig. 3. Maris dataset simples (Color figure online)

From this dataset we extract arbitrary two sets as shown in Table 2.

Table 2. Maris dataset

	Data splits	Pipes/split	Non-pipe/split
Train set	2	300	300
Test set	2	300	300

4.3 Result

To select the best combination of $\propto$ and ε we compute F-measure using Train set and Test set described in Table 2 with different values of $\propto$ and ε.

From Fig. 4 we can conclude that detection performance is better when ε range from 10 to 15 which can explain the complexity of image background. According to the histogram the best result of F-Measure is 99.83%. It is reached when $\varepsilon = 15$ and $\propto = 2$.

To choose the best characteristic which walks with the submarine context and pipe detection, we evaluate the detection rate with different combination of content and structure features on Maris dataset. To reduce the number of combinations, it is possible to group the characteristics of the same context, as shown in Table 3.

The presented values in the following table are obtained after having made several tests by making random combinations of test and train subsets (pipe/non pipe) (Table 4).

Referring to the table we conclude that the best detection rate (99.91) is obtained when using F1, F2, F3, F4, F5 among all combination. It is the combination of all used structure features (location and level) and content features (shape and color).

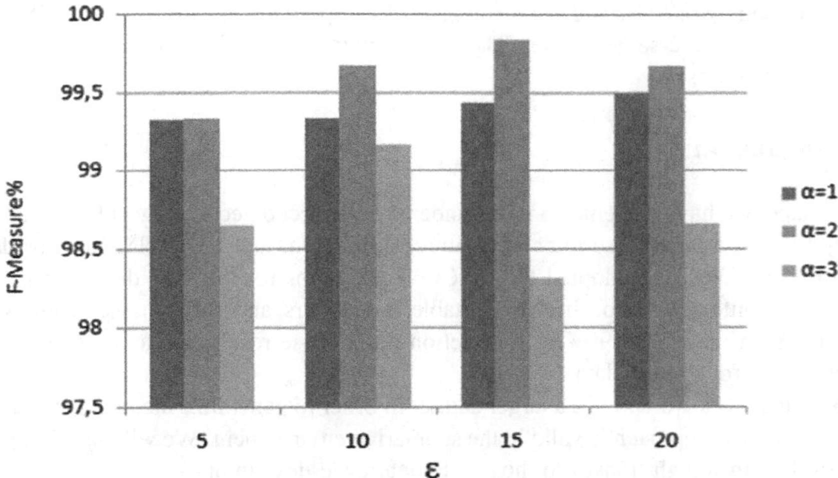

Fig. 4. F-measure for different combinations of ε and α

Table 3. New partition of MSCOV features

	F	k	Features
Structure feature	F1	0, 1, 2	Quadtree information's
Content features	F2	3, 4, 5	Luminance and chrominance components
	F3	6, 7	Norm of the first order derivatives in x and y
	F4	8	Gradient
	F5	9	Magnitude

Table 4. Detection rate and execution time for different features combination

Features	Detection rate %	Execution time ms
F1, F2	96.74	829630
F1, F3	92.58	782051
F1, F4	67.58	712023
F1, F5	66.33	645693
F2, F3	**99.08**	**309829**
F3, F4, F5	95.41	559557
F2, F3, F4, F5	98.16	463920
F3, F4, F5	93.41	429563
F1, F2, F3, F4, F5	**99.91**	**394016**

Although it has the best detection rate the execution time of F1 F2 F3 F4 F5 is not the optimal. The lowest execution time is obtained for F2 F3 combination with good detection rate (99.40). Therefore if we have a time constraint we can use this combination.

In previous work [10] we have compared this approach with PFC and MGS [11] algorithms using the same dataset. The experimental results show that it outperforms compared methods.

5 Conclusion

In this paper we have presented a novel underwater object detection algorithm based on multi-scale covariance descriptor for features extraction, and SVM classifier for data classification. We have adopted the MSCOV descriptor for the pipe detection in the submarine context by choosing the suitable parameters and the suitable features in order to reach more than 99% as a detection rate. Those results are token using train and test sets from Maris dataset.

In future work we will use a larger dataset in order to generalize the result and show that this adopted approach is valid in the submarine environment. We will also compare this result with hough transform, hog and covariance descriptors.

References

1. Williams, D.P.: On adaptive underwater object detection, April 2012
2. Ayedi, W., Snoussi, H., Smach, F., Abid, M.: The multi-scale covariance descriptor: performances analysis in human detection. In: 2012 IEEE Workshop on Biometric Measurements and Systems for Security and Medical Applications (BIOMS), pp. 1–5. IEEE, September 2012. Park, U., Jain, A.K., Kitahara, I., Kogure, K., Hagita, N.: Vise: visual search engine using multiple networked cameras. In: 18th International Conference on Pattern Recognition, vol. 3, pp. 1204–1207. IEEE (2006)
3. Tuzel, O., Porikli, F., Meer, P.: Region covariance: a fast descriptor for detection and classification. In: Leonardis, A., Bischof, H., Pinz, A. (eds.) ECCV 2006. LNCS, vol. 3952, pp. 589–600. Springer, Heidelberg (2006). doi:10.1007/11744047_45
4. Kim, D., Lee, D., Myung, H., Choi, H.T.: Object detection and tracking for autonomous underwater robots using weighted template matching. In: OCEANS, pp. 1–5 (2012)
5. Lin, S., Ozsu, M.T., Oria, V., Ng, R.: An extendible hash for multi-precision similarity querying of image databases. In: VLDB, pp. 221–230 (2001)
6. Malki, J., Boujemaa, N., Nastar, C., Winter, A.: Region queries without segmentation for image retrieval by content. In: Huijsmans, D.P., Smeulders, A.W.M. (eds.) VISUAL 1999. LNCS, vol. 1614, pp. 115–122. Springer, Heidelberg (1999). doi:10.1007/3-540-48762-X_15
7. Ayedi, W., Snoussi, H., Abid, M.: A fast multi-scale covariance descriptor for object re-identification. Pattern Recogn. Lett. **33**(14), 1902–1907 (2012)
8. Oleari, F., Kallasi, F., Rizzini, D.L., Aleotti, J., Caselli, S.: An underwater stereo vision system: from design to deployment and dataset acquisition. In: OCEANS 2015-Genova, pp. 1–6 (2015)
9. Sain, S.R.: The Nature of Statistical Learning Theory. Springer, New York (1996). doi:10. 1007/978-1-4757-2440-0
10. Rekik, F., Ayedi, W., Jallouli, M.: Evaluation of an object detection system in the submarine environment. In: WSCG 2017 (2017)
11. Kallasi, F., Rizzini, D.L., Oleari, F., Aleotti, J.: Computer vision in underwater environments: a multiscale graph segmentation approach. In: OCEANS 2015-Genova, pp. 1–6. IEEE, May 2015

Retinal Vessel Segmentation Through Denoising and Mathematical Morphology

Benedetta Savelli, Agnese Marchesi, Alessandro Bria(✉), Claudio Marrocco, Mario Molinara, and Francesco Tortorella

DIEI, University of Cassino and Southern Latium, Cassino, FR, Italy
{b.savelli,a.marchesi,a.bria,c.marrocco,m.molinara,tortorella}@unicas.it

Abstract. Automated retinal blood vessel segmentation plays an important role in the diagnosis and treatment of various cardiovascular and ophthalmologic diseases. In this paper, an unsupervised algorithm based on denoising and mathematical morphology is proposed to extract blood vessels from color fundus images. Specifically, our method consists of the following steps: (i) green channel extraction; (ii) non-local means denoising; (iii) vessel vasculature enhancement by means of a sum of black top-hat transforms; and (iv) image thresholding for the final segmentation. This method stands out for its simplicity, robustness to parameters change and low computational complexity. Experimental results on the publicly available database DRIVE show our method to be effective in segmenting blood vessels, achieving an accuracy comparable to that of unsupervised state-of-the-art methodologies.

Keywords: Color fundus images · Retinal vessel segmentation · Denoising · Mathematical morphology

1 Introduction

Retinal vessel structure is of great interest in ophthalmology since its characterization in terms of geometric features can provide indicative measures of the presence and severity of different cardiovascular and ophthalmologic diseases such as arteriosclerosis, diabetes, hypertension, and glaucoma [14]. The analysis of the change in morphological features of retinal blood vessels such as length, width, and branching angle, can help in the early detection of these diseases and in the identification of the optimal treatment in order to prevent visual loss [17]. Vessel segmentation is a crucial step for the assessment of these morphological properties. Because of the complexity of the vascular network, manual segmentation is time consuming and difficult: a feasible solution is the automation of the segmentation process. However, the automated segmentation of retinal blood vessels is a challenging task, mainly due to the width variability of vessels and to the low quality of retinal images that are, in general, noisy, unevenly illuminated, and of poor contrast [20]. In recent years, several methods have been proposed for blood vessel segmentation in color fundus images and they can be categorized as either supervised or unsupervised methods.

S. Battiato et al. (Eds.): ICIAP 2017, Part II, LNCS 10485, pp. 267–276, 2017.
https://doi.org/10.1007/978-3-319-68548-9_25

Supervised methods use hand-crafted or automatically extracted features to train a classifier in order to discriminate vessels and non-vessels pixels. A k-NN approach is used by Niemeijer et al. [23] and Staal et al. [27] to classify the feature vectors that are constructed by a multiscale Gaussian filter and by a ridge detector, respectively. Marín et al. [19] used gray-level and moment invariant-based features to build a 7-D feature vector and utilized a multilayer feed forward neural network for training and classification. Cheng et al. [7] used a random forest to fuse the information encoded in the hybrid context-aware features. Recent works use Deep Convolutional Neural Networks to automatically extract features from fundus images and segment the retinal vasculature [18,21,26].

Unsupervised segmentation methods rely on finding inherent properties of retinal blood vessels, which can be applied to distinguish vessel pixels in retinal images. According to the image processing methodologies, these methods can be classified into four sub-categories: matched filtering, vessel tracking, model-based, and mathematical morphology approaches.

In matched filtering techniques the retinal image is convolved with a 2-D kernel, and the matched filter response indicates the presence of the vessel. Matched filter detection was proposed by Chaudhuri et al. [6]. In this method, the authors used 12 rotated versions of a 2-D Gaussian shaped template for searching vessel segments along all possible directions. Azzopardi et al. [2] proposed a novel technique that uses a non-linear trainable filter called B-COSFIRE, designed to detect bar-shaped structures such as blood vessels. Vessels segmentation is achieved by adding up the responses of two B-COSFIRE filters, namely symmetric and asymmetric, followed by thresholding. Tracking methods, also mentioned as exploratory algorithms, start by locating the vessel points used for tracing the vasculature. This procedure can be done either manually or automatically by measuring some local image properties. Most of the methods reported in the literature use Gaussian functions to characterize vessel profiles [8,13]. This type of approaches can provide highly accurate vessel widths, but they are often unable to detect vessel segments that have no seed points. The model-based techniques apply explicit vessel models that can be regarded as vessel profile [16], parametric deformable [1], or geometric deformable models [28]. The vessel profile models approximate the vessel intensity profiles using mixture of Gaussians or the second-order derivative Gaussian, and other polynomial functions. The parametric and geometric deformable models are based on active contours models and on the theory of curve evolution geometric flows, respectively [11]. Mathematical morphology approaches exploit a priori information on the profile structure of vessels and involves the use of morphological operators. Zana and Klein [31] proposed a combination of morphological filters in order to enhance vessels and cross curvature evaluation to identify linearly coherent structures. Morphological processing for vessel segmentation was also reported in [22,29]. Fraz et al. [10], proposed a robust technique for the extraction of the retinal vessels map using a combination of centerlines detection and morphological operators. An hybrid technique was proposed by Yank et al. [30], where the blood vessels were enhanced and the background removed with a morphological top-hat operation, then fuzzy clustering was applied to extract the blood vessels.

In this paper we present an unsupervised method for retinal blood vessel segmentation in color fundus images based on denoising and mathematical morphology. The denoising step is required to prepare images for the further vessel enhancement phase. We use the non-local means denoising method that preserves image details and contrast properties and allows to obtain well-contrasted images with a smoothed background that are suited for the application of mathematical morphology. The use of morphological operators for vessel enhancement builds on the observation that a vessel can be defined as a dark pattern with Gaussian-shape cross-section profile, piece-wise connected, and locally linear [9]. Based on this assumption, in our method a sum of black top-hat transforms is used to highlight vessels with respect to these morphological features.

2 Dataset

The proposed method has been tested on images of the publicly available database DRIVE. It contains 40 color images of the retina, with 565×584 pixels and 8 bits per color channel, captured from a Canon CR5 non-mydriatic 3CCD camera at 45° field of view, and saved in JPEG-format. Besides the color images, the database includes masks with the delimitation of a field of view (FOV) of approximately 540 pixels in diameter for each image, and binary images with the results of manual vessel segmentation. The 40 images were divided into a training set and a test set by the authors of the database. The results of the manual segmentation are available for all the images of the two sets. We used the training images in the vessel enhancement step to verify the robustness of our approach to different parameters configurations, and the test images to evaluate the performance measures.

3 Method

The proposed approach consists of four phases. First, the green channel is extracted from RGB images, because it shows better vessels-to-background contrast [11,20]. Hence, from now on, we deal with grayscale images in which retinal vessels appear darker than the background. Then, the non-local means denoising algorithm is employed to remove the noise due to the image acquisition process [31]. Afterwards, a sum of black top-hat transforms is applied to emphasize the entire vasculature. A final thresholding is performed on the enhanced vessel map to obtain the desired segmentation. In the following section we describe in detail the last three steps of our approach. A diagram of the proposed approach is reported in Fig. 1.

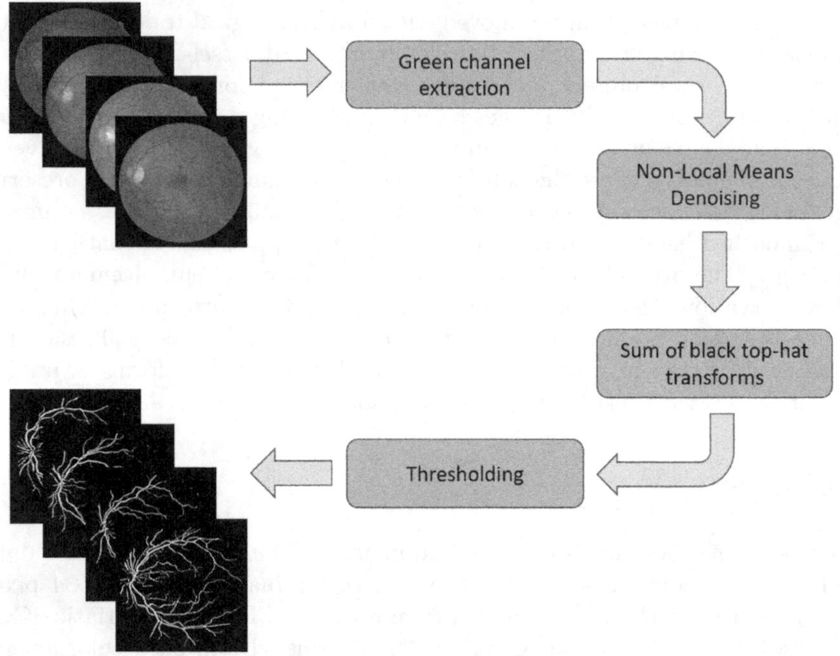

Fig. 1. Flow chart of the proposed approach.

3.1 Denoising: Non-local Means Algorithm

The success of an automatic retinal vessel segmentation algorithm is influenced by the presence of noise mainly caused by the digitization process [31]. Traditional denoising methods, like local smoothing and frequency domain filters, provide noise reduction and reconstruction of the main geometrical patterns of the image. These kind of filters are designed to remove only the high frequency noise of the image, but they often do not preserve fine details and structures that present the same frequency values as noise. To prevent this loss of information, in our work we apply the non-local means denoising. The main idea behind this denoising approach is that images contain a high level of redundancy, which means that it is possible to find similar intensity profiles either in adjacent or non-adjacent pixels. Thus, given a discrete noisy image f, for the pixel i, the denoised value $NL(f(i))$ is computed as a weighted average of all the pixels in the image:

$$NL(f(i)) = \sum_{j \in f} w(i,j) f(j) \tag{1}$$

where the family of weights $\{w(i,j)\}_j$ depends on the similarity between the pixels i and j, $0 \leq w(i,j) \leq 1$ and $\sum_j w(i,j) = 1$. Let N_i be a square neighborhood of size $n \times n$ centered on i, the restriction of f to a neighborhood N_i is defined by $f(N_i)$:

$$f(N_i) = \{f(j), j \in N_i\} \tag{2}$$

The measure of the similarity between the intensity gray level vectors $f(N_i)$ and $f(N_j)$ is obtained by means of a Gaussian weighted Euclidean distance $||f(N_i) - f(N_j)||^2_{2,\sigma}$, which is a L_2 norm convolved with a Gaussian kernel of standard deviation σ. The weights are defined according to the following equation:

$$w(i,j) = \frac{1}{Z(i)} e^{-\frac{||f(N_i)-f(N_j)||^2_{2,\sigma}}{h^2}} \tag{3}$$

where $Z(i)$ is a normalizing constant $Z(i) = \sum_j e^{-\frac{||f(N_i)-f(N_j)||^2_{2,\sigma}}{h^2}}$. This results in giving larger weights to those neighborhoods that are similar in terms of gray levels. The variable h in Eq. 3 is a parameter which controls the decay of the exponential function and consequently the decay of the weights. In order to reduce the computational cost, it shall be defined a $m \times m$ search window M on the image f, in which the similarity of the neighborhoods is evaluated. In this way, the denoised value $NL(f(i))$ is computed as a weighted average of all the pixels of the search window. In our application, we set all the parameters to the values suggested in the default implementation of the algorithm [5]. Therefore $h = 3$, $n = 3$ and $m = 21$.

3.2 Mathematical Morphology: Black Top-Hat

It can be assumed that retinal vessels are defined by a dark pattern with morphological properties like local linearity, connectivity, width, and a Gaussian-like cross-section profile [31]. These observations allow the use of mathematical morphology for retinal vessel enhancement. In particular, in this work, we apply a series of morphological black top-hat transforms. In general, let f be a grayscale image and b a structuring element (SE), the black top-hat transform of f by b at the ith pixel is defined as follows:

$$BTH(i) = f \bullet b - f \tag{4}$$

where $f \bullet b$ denotes the closing of f by b. We employ this operation because it enhances objects smaller than b and darker than their surroundings. Given the linearity property of vessels, we use a linear SE b_L, slightly longer than the width L of the largest vessels, to obtain the enhancement of all the vessels orthogonal to b_L. To target the orientation of a specific vessel, we use multiple $BTH(f, b_L^\alpha)$ for D different angles α equally spaced in the range $[0, \pi)$. The enhancement of the complete vessel tree is achieved with the sum of top-hats (STH)

$$STH(f, b_L) = \sum_\alpha (f \bullet b_L^\alpha - f) \tag{5}$$

In this case, the length of the SE b_L and the number of angles used are set respectively to $L = 16$ and $D = 12$. The choice of $L = 16$ is based on the visual inspection of the images by considering the width of primary large vessels, whereas D is chosen according to [6] where 12 different templates are applied to search for vessel segments along all possible directions. We also note that varying the parameter L in the range $[12, 20]$ and D in the range $[10, 16]$ did not affect the performance on the training images significantly, suggesting that the proposed method does not need a fine tuning of its parameters.

3.3 Image Thresholding

The enhanced image is then thresholded in order to obtain a binary image in which only vessel pixels are extracted, providing the final vessel segmentation. Given a threshold θ, all the pixels with an intensity gray level greater than θ are set to 255, whereas the others are set to 0. The value of θ is selected according to the desired accuracy level, i.e., the value that maximizes the segmentation accuracy on the training images of DRIVE dataset.

4 Experiments

To evaluate the performance of our algorithm we compared each resulting binary image from the test set with the corresponding ground truth, by computing the number of correctly classified pixels and misclassified ones. According to these values we calculated the classification accuracy as the number of correctly classified pixels with respect to the number of pixels in the image FOV. Furthermore, sensitivity and specificity were calculated. Experimental results in terms of accuracy (Acc), sensitivity (Se) and specificity (Sp) are reported in Table 1. We compared the effectiveness of the proposed method with other existing vessel segmentation methodologies, whose performances are also shown in Table 1.

Table 1. Sensitivity, specificity and accuracy of the compared methods obtained on the test images of DRIVE database.

	Method	Se	Sp	Acc
Unsupervised	**Proposed method**	**0.6806**	**0.9823**	**0.9440**
	Chaudhuri et al. [6]	0.3357	0.9794	0.8773
	Zana and Klein [31]	0.6971	0.9769	0.9377
	Mendonca and Campilho [22]	0.7344	0.9764	0.9463
	Lam et al. [16]	-	-	0.9472
	Fraz et al. [10]	0.7152	0.9768	0.9430
	Azzopardi et al. [2]	0.7655	0.9704	0.9442
Supervised	Niemeijer et al. [23]	0.7145	0.9801	0.9416
	Staal et al. [27]	0.7345	0.9773	0.9441
	Marín et al. [19]	0.7145	0.9801	0.9452
	Melinščak et al. [21]	0.7276	0.9785	0.9466
	Liskowski and Krawiec [18]	0.7763	0.9768	0.9495

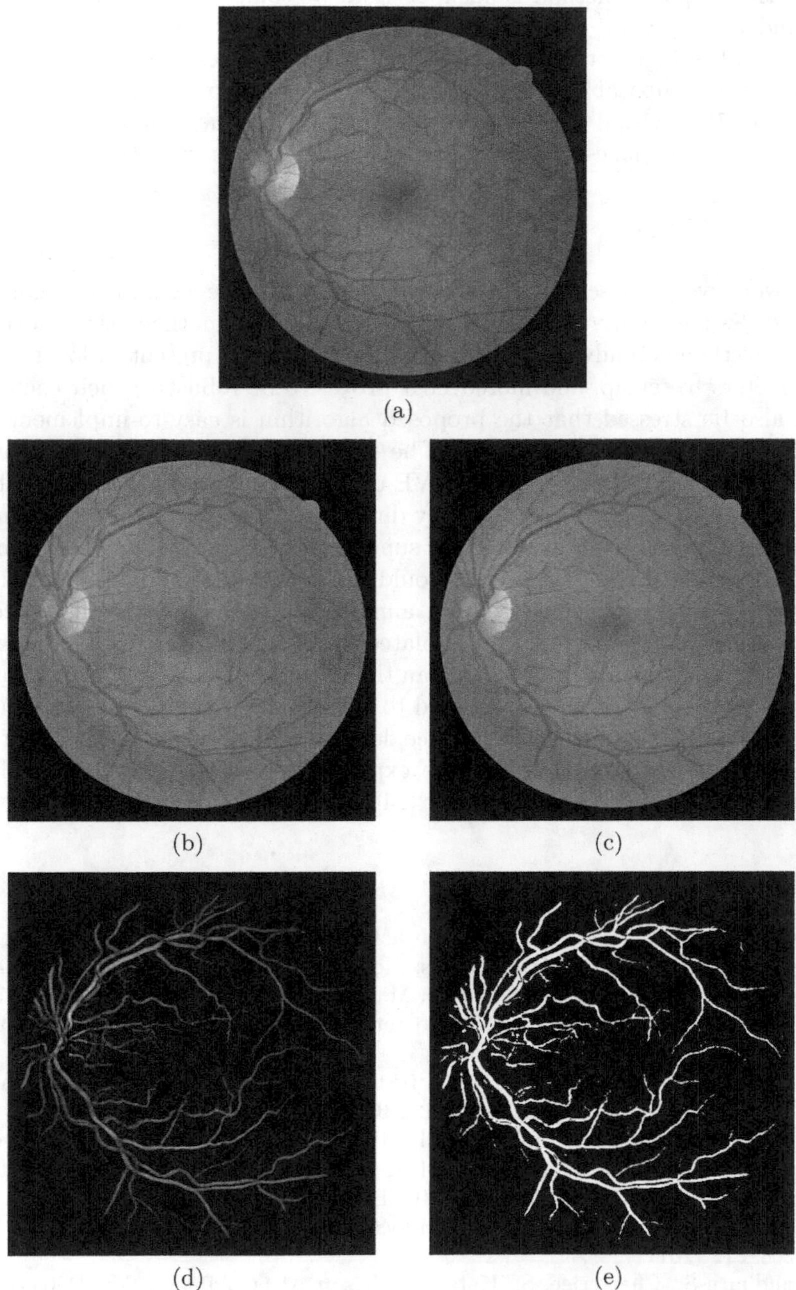

Fig. 2. Visualization of the output of each step of the proposed approach starting from (a) RGB retinal image, (b) green channel extraction, (c) non-local means denoising, (d) sum of black top-hat transforms, and (e) the segmentation obtained after thresholding.

Despite the proposed method is unsupervised, we compared it with both supervised and unsupervised techniques. Table 1 shows that the achieved accuracy $Acc = 94.40$ is higher or comparable than that of other unsupervised methods, and even comparable to that obtained by some reported supervised methods [19,23,27]. A visual step-by-step representation of the proposed framework applied to retinal images of DRIVE dataset is depicted in Fig. 2.

5 Conclusions

In this work we proposed an automated retinal vessel segmentation algorithm that provides for the use of a denoising technique and of mathematical morphology. One of the main advantage of this method is that it presents a low number of parameters to set up, and moreover it proves to be robust to their choice. It should also be stressed that the proposed algorithm is easy-to-implement and very time-efficient (<1 s per image). The performance was measured in terms of accuracy on the test set of the DRIVE dataset. The results shown in Table 1 demonstrate that we obtain an accuracy directly comparable with more complex methods for vessel segmentation, either supervised or unsupervised. These observations suggest that the performance could be improved in different ways: (i) by adding a post-processing method that aims at filling the gaps in the detected blood vessels and at removing small isolated areas misclassified as blood vessels; (ii) by constructing a feature vector from the unthresholded black top-hats as in [25] in order to apply methods designed to face the high imbalance between the two classes [3,4,12,15]. In addition, since denoising revealed to be crucial for our method, future directions also include experimenting other existing denoising techniques, such as the Bilateral Filter [24].

References

1. Al-Diri, B., Hunter, A., Steel, D.: An active contour model for segmenting and measuring retinal vessels. IEEE Trans. Med. Imaging **28**(9), 1488–1497 (2009)
2. Azzopardi, G., Strisciuglio, N., Vento, M., Petkov, N.: Trainable cosfire filters for vessel delineation with application to retinal images. Med. Image Anal. **19**(1), 46–57 (2015)
3. Bria, A., Marrocco, C., Molinara, M., Tortorella, F.: An effective learning strategy for cascaded object detection. Inf. Sci. **340**, 17–26 (2016)
4. Bria, A., Marrocco, C., Molinara, M., Tortorella, F.: A ranking-based cascade approach for unbalanced data. In: 2012 21st International Conference on Pattern Recognition (ICPR), pp. 3439–3442. IEEE (2012)
5. Buades, A., Coll, B., Morel, J.M.: Non-local means denoising. Image Process. Line **1**, 208–212 (2011)
6. Chaudhuri, S., Chatterjee, S., Katz, N., Nelson, M., Goldbaum, M.: Detection of blood vessels in retinal images using two-dimensional matched filters. IEEE Trans. Med. Imaging **8**(3), 263–269 (1989)
7. Cheng, E., Du, L., Wu, Y., Zhu, Y.J., Megalooikonomou, V., Ling, H.: Discriminative vessel segmentation in retinal images by fusing context-aware hybrid features. Mach. Vis. Appl. **25**(7), 1779–1792 (2014)

8. Chutatape, O., Zheng, L., Krishnan, S.M.: Retinal blood vessel detection and tracking by matched Gaussian and Kalman filters. In: Proceedings of the 20th Annual International Conference of the IEEE Engineering in Medicine and Biology Society, vol. 6, pp. 3144–3149. IEEE (1998)
9. Fang, B., Hsu, W., Lee, M.L.: Reconstruction of vascular structures in retinal images. In: 2003 International Conference on Image Processing, ICIP 2003, Proceedings, vol. 2, pp. II–157. IEEE (2003)
10. Fraz, M.M., Barman, S., Remagnino, P., Hoppe, A., Basit, A., Uyyanonvara, B., Rudnicka, A.R., Owen, C.G.: An approach to localize the retinal blood vessels using bit planes and centerline detection. Comput. Methods Programs Biomed. **108**(2), 600–616 (2012)
11. Fraz, M.M., Remagnino, P., Hoppe, A., Uyyanonvara, B., Rudnicka, A.R., Owen, C.G., Barman, S.A.: Blood vessel segmentation methodologies in retinal images-a survey. Comput. Methods Programs Biomed. **108**(1), 407–433 (2012)
12. Galar, M., Fernandez, A., Barrenechea, E., Bustince, H., Herrera, F.: A review on ensembles for the class imbalance problem: bagging-, boosting-, and hybrid-based approaches. IEEE Trans. Syst. Man Cybern. Part C (Appl. Rev.) **42**(4), 463–484 (2012)
13. Gao, X., Bharath, A., Stanton, A., Hughes, A., Chapman, N., Thom, S.: A method of vessel tracking for vessel diameter measurement on retinal images. In: 2001 International Conference on Image Processing, Proceedings, vol. 2, pp. 881–884. IEEE (2001)
14. Kanski, J.J., Bowling, B.: Clinical Ophthalmology: A Systematic Approach. Elsevier Health Sciences, Amsterdam (2011)
15. Lahiri, A., Roy, A.G., Sheet, D., Biswas, P.K.: Deep neural ensemble for retinal vessel segmentation in fundus images towards achieving label-free angiography. In: 2016 IEEE 38th Annual International Conference of the Engineering in Medicine and Biology Society (EMBC), pp. 1340–1343. IEEE (2016)
16. Lam, B.S., Gao, Y., Liew, A.W.C.: General retinal vessel segmentation using regularization-based multiconcavity modeling. IEEE Trans. Med. Imaging **29**(7), 1369–1381 (2010)
17. Leontidis, G., Al-Diri, B., Wigdahl, J., Hunter, A.: Evaluation of geometric features as biomarkers of diabetic retinopathy for characterizing the retinal vascular changes during the progression of diabetes. In: 2015 37th Annual International Conference of the IEEE Engineering in Medicine and Biology Society (EMBC), pp. 5255–5259. IEEE (2015)
18. Liskowski, P., Krawiec, K.: Segmenting retinal blood vessels with deep neural networks. IEEE Trans. Med. Imaging **35**(11), 2369–2380 (2016)
19. Marín, D., Aquino, A., Gegúndez-Arias, M.E., Bravo, J.M.: A new supervised method for blood vessel segmentation in retinal images by using gray-level and moment invariants-based features. IEEE Trans. Med. Imaging **30**(1), 146–158 (2011)
20. Mary, M.C.V.S., Rajsingh, E.B., Naik, G.R.: Retinal fundus image analysis for diagnosis of glaucoma: a comprehensive survey. IEEE Access **4**, 4327–4354 (2016)
21. Melinščak, M., Prentašić, P., Lončarić, S.: Retinal vessel segmentation using deep neural networks. In: VISAPP 2015, 10th International Conference on Computer Vision Theory and Applications (2015)
22. Mendonca, A.M., Campilho, A.: Segmentation of retinal blood vessels by combining the detection of centerlines and morphological reconstruction. IEEE Trans. Med. Imaging **25**(9), 1200–1213 (2006)

23. Niemeijer, M., Staal, J., van Ginneken, B., Loog, M., Abramoff, M.D.: Comparative study of retinal vessel segmentation methods on a new publicly available database. In: SPIE Medical Imaging, vol. 5370, pp. 648–656. SPIE (2004)
24. Paris, S., Kornprobst, P., Tumblin, J., Durand, F., et al.: Bilateral filtering: theory and applications. Found. Trends® Comput. Graph. Vis. **4**(1), 1–73 (2009)
25. Ricci, E., Perfetti, R.: Retinal blood vessel segmentation using line operators and support vector classification. IEEE Trans. Med. Imaging **26**(10), 1357–1365 (2007)
26. Savelli, B., Bria, A., Galdran, A., Marrocco, C., Molinara, M., Campilho, A., Tortorella, F.: Illumination correction by dehazing for retinal vessel segmentation. In: 2017 IEEE 30th International Symposium on Computer-Based Medical Systems (CBMS). IEEE (2017)
27. Staal, J., Abràmoff, M.D., Niemeijer, M., Viergever, M.A., Van Ginneken, B.: Ridge-based vessel segmentation in color images of the retina. IEEE Trans. Med. Imaging **23**(4), 501–509 (2004)
28. Sum, K., Cheung, P.Y.: Vessel extraction under non-uniform illumination: a level set approach. IEEE Trans. Biomed. Eng. **55**(1), 358–360 (2008)
29. Xiang, Y., Gao, X., Zou, B., Zhu, C., Qiu, C., Li, X.: Segmentation of retinal blood vessels based on divergence and bot-hat transform. In: 2014 International Conference on Progress in Informatics and Computing (PIC), pp. 316–320. IEEE (2014)
30. Yang, Y., Huang, S., Rao, N.: An automatic hybrid method for retinal blood vessel extraction. Int. J. Appl. Math. Comput. Sci. **18**(3), 399–407 (2008)
31. Zana, F., Klein, J.C.: Segmentation of vessel-like patterns using mathematical morphology and curvature evaluation. IEEE Trans. Image Process. **10**(7), 1010–1019 (2001)

Segmentation of Green Areas Using Bivariate Histograms Based in Hue-Saturation Type Color Spaces

Gilberto Alvarado-Robles[1], Ivan R. Terol-Villalobos[2], Marco A. Garduño-Ramon[1], and Luis A. Morales-Hernandez[1(✉)]

[1] Universidad Autónoma de Querétaro, 76806 San Juan Del Río, Querétaro, México
luis.morales@uaq.mx

[2] CIDETEQ, S.C., Parque Tecnológico Querétaro S/N, Sanfandila, 76700 Pedro Escobedo, Querétaro, México

Abstract. Image segmentation is the process where the pixels of the image with similar characteristics, like color, are clustered in homogeneous regions. However, this is a complex task, being the reason of the existence of many methods of color segmentation. Particularly, the study of vegetation mass in urban, residential and forestry zones is an interesting subject in image segmentation that has been increased in the last years. Therefore, it is necessary to focus on the research in image processing techniques to improve the results in this field. In this paper, an algorithm of vegetation detection is proposed. The use of a hue-based color space enables the detection of certain color tonality due to its low range of values that can be used to determine a range of similar colors (0° to 360°) with regard to the RGB color space (16,777,216 colors) that is a color space that is not uniform nor perceptually predictable. On the other hand, the use of chromatic histogram in combination with the morphological operators gives the advantage of detecting vegetation accurately despite its conditions and with the use of any camera to carry out this task.

Keywords: Color · Segmentation · HSI · Morphological operators · Watershed · Urban greening

1 Introduction

Image segmentation is the process in which the pixels in an image are clustered in homogeneous regions taking into account certain characteristics. In various areas of research, the segmentation of certain areas in an image has a great importance to study punctual zones on it, for this reason, various methods of segmentation have been proposed [17]. Some of these methods use mathematical morphology [22], and the applications of image segmentation range from thermography [21] to medicine [8], and recently green areas segmentation, defined as the zones covered by vegetation, has lately increased its interest as a research topic [5].

S. Battiato et al. (Eds.): ICIAP 2017, Part II, LNCS 10485, pp. 277–287, 2017.
https://doi.org/10.1007/978-3-319-68548-9_26

In image color segmentation numerous research works have been presented. For example, a color image segmentation using morphological clustering based on 2D-histograms was proposed by [11] having the disadvantage of fusing images that must be adjusted before this process. On the other hand, in [9] the authors have introduced a method that works with 2D-histogram multi-thresholding based on RGB color space (Red, Green, and Blue) and taking the three histograms RG, RB or BG in order to reduce the number of different values, and then fusing the resulting images. The main drawback of the method is the use of three histograms having many possible values for the fusing task. Another method uses orthogonal series to perform the image segmentation [24]; this is a fast algorithm but with the disadvantage of getting errors of segmentation in small zones. With regard to the vegetable zones detection, this subject has become a recurrent research topic. For instance, in [12], a work based on k-means clustering algorithm to detect individual trees in green areas, using the NDVI (Normalized Difference Vegetation Index), has been introduced. The disadvantage of this method is the use of a specific kind of camera. Another example is the classification of farmland images based on color features, that works with neural networks and different color spaces [16], this algorithm focuses only on green color zones; this algorithm focuses only on green color zones. Finally, the method used in [18] to detect green areas based on Hue threshold [4] has the disadvantage of being a semi-automatic method. As it was seen, color image segmentation and green areas detection have become a research topic of great interest. Although the methods aforementioned are functional, it is needed to develop an algorithm that works using any commercial camera, and regardless the difference in the green tone of vegetation, and with reliability.

In this paper, an automatic vegetable zone detection based on a color image segmentation method that uses bi-variable histograms of hue-based color spaces, [2] is proposed. The robustness of this method, in terms of variation of colors that vegetation can acquire during all seasons in the year, enables the detection of vegetable zones with accurate results.

2 Theoretical Framework

2.1 Color Spaces

A color space is a mathematical representation of a set of colors. The three most popular color models are RGB, YIQ (used in video systems), and CMYK (used in color printing). However, none of these color spaces is directly related to the intuitive notions of color; hue, saturation, and brightness. In fact, perceptual color spaces enable the simplification of programming, processing, and the end user manipulation. All aforementioned color spaces can be derived from the RGB information supplied by devices such as cameras [10].

The RGB color scheme encodes colors as combinations of the three primary colors: red, green, and blue. This scheme is widely used for transmission, representation, and storage of color images on both analogue devices such as television sets and digital devices. For this reason, many image processing and graphics programs use the RGB scheme as its internal representation for color images, and most language libraries use it as its standard image representation. RGB is an additive color system, which means that all colors are created by adding the primary colors. The RGB values are positive whose range is $[0, Cmax]$, where $Cmax = 255$ [10], it can be expressed with a cubic shape.

When humans see a color object, they tend to describe it by hue, saturation, and brightness. Hue is a color attribute that describes a pure color, whereas saturation gives a measure of the degree to which a pure color is vanished by white light. Hue is expressed as an angle around a hexagon that usually uses red as reference in 0. In Eq. 1 are shown the formulas to convert from RGB to HSV [20], where $max = max(R, G, B)$ and $min = min(R, G, B)$.

$$\begin{aligned} V &= max \\ S &= \frac{max - min}{max} \\ H &= \begin{cases} \frac{G-B}{max-min} & \text{if } R = max \\ \frac{B-R}{max-min} + 2 & \text{if } G = max \\ \frac{R-G}{max-min} + 4 & \text{if } B = max \end{cases} \end{aligned} \tag{1}$$

HSI color space decouples the intensity component of the color information (hue and saturation) in an image. As a result, the HSI model is an ideal tool for developing image processing algorithms based on color descriptions that are natural and intuitive to humans [13]. Equation 2 shows the formulas to convert from RGB to HSI.

$$\begin{aligned} I &= \left(\frac{1}{3}\right)(R+G+B) \\ S &= 1 - \left(\frac{1}{(R+G+B)}\right) min(R, G, B) \\ H &= \arccos\left[\frac{[\left(\frac{1}{2}\right)(R-G) - (R-B)]}{[(R-G)^2(R-B) + (G-B)]^{\frac{1}{2}}}\right] \end{aligned} \tag{2}$$

IHSL (Improved Hue, Saturation, and Luminance) is an improvement of HSL color space to overcome its limitations. There are two main advantages of the IHSL with respect to HSL. However, The most important problem are the instabilities that appear in saturation, not only in HSL color space but also in most of the other Hue-based color spaces. IHSL color space avoids these instabilities in saturation in high and low levels of luminance [1]. In Eq. 3 the formulas to convert from RGB to IHSL are illustrated as follows:

$$\begin{aligned} L &= 0.2126R + 0.7152G + 0.0722B \\ S &= max - min \\ H &= \begin{cases} \frac{G-B}{max-min} & \text{if } R = max \\ \frac{B-R}{max-min} + 2 & \text{if } G = max \\ \frac{R-G}{max-min} + 4 & \text{if } B = max \end{cases} \end{aligned} \tag{3}$$

2.2 Morphological Filter

A morphological filter ψ is an increasing and idempotent transformation in which the former characteristic implies that for two images f and g such that $f \leq g$, $\psi(f) \leq \psi(g)$, and the latter characteristic states $\psi|\psi(f)| = \psi(f)$ for any image f. Opening ($\gamma_{\lambda B}$) and closing ($\varphi_{\lambda B}$) are the basic morphological filters by a structural element λB, where B is the basic structuring element that contains the origin and λ is a homothetic parameter. The structural element can have different forms. Both of these filters are expressed in Eq. 4, where $\varepsilon_{\lambda B}$ and $\delta_{\lambda B}$ are the erosion and the dilation, respectively, defined as $\varepsilon_{\lambda B}(f)(x) = min\{f(y)\colon y\epsilon\lambda B\}$, and $\delta_{\lambda B}(f)(x) = max\{f(y)\colon y\epsilon\lambda B\}$, and max and min are the maximum and minimum value [22]:

$$\begin{aligned} \gamma_{\lambda B} &= \delta_{\lambda B}[\varepsilon_{\lambda B}(f)] \\ \varphi_{\lambda B} &= \varepsilon_{\lambda B}[\delta_{\lambda B}(f)] \end{aligned} \tag{4}$$

On the other hand, the opening ($\hat{\gamma}_{\lambda B}$) and closing ($\hat{\varphi}_{\lambda B}$) by reconstruction are developed using the geodesic dilation defined as $\delta_f^1(g) = f \wedge \delta_B(g)$ with $f \geq g$ and the geodesic erosion defined as $\varepsilon_f^1(g) = f \wedge \varepsilon_B(g)$ with $f \geq g$. These operators are applied until stability is reached in order to obtain the reconstruction (R) and the dual reconstruction (R^*) respectively [19]. Both operators are shown in Eq. 5.

$$\begin{aligned} \hat{\gamma}_{\lambda B} &= \lim_{x\to\infty} \delta_f^n[\varepsilon_{\lambda B}(f)] = R[f, \varepsilon_{\lambda B}(f)], \\ \hat{\varphi}_{\lambda B} &= \lim_{x\to\infty} \varepsilon_f^n[\delta_{\lambda B}(f)] = R^*[f, \varepsilon_{\lambda B}(f)] \end{aligned} \tag{5}$$

The alternating sequential filters are also increasing and idempotent transformations. They are formed by the composition of morphological openings and closings as shown in Eq. 6, with $\lambda_1 \leq \lambda_2 \leq ... \leq \lambda_n$ [19].

$$\begin{aligned} \psi_n(f) &= \varphi_{\lambda_n}\gamma_{\lambda_n}...\varphi_{\lambda_2}\gamma_{\lambda_2}\varphi_{\lambda_1}\gamma_{\lambda_1}(f) \\ \psi_n^*(f) &= \gamma_{\lambda_n}\varphi_{\lambda_n}...\gamma_{\lambda_2}\varphi_{\lambda_2}\gamma_{\lambda_1}\varphi_{\lambda_1}(f) \end{aligned} \tag{6}$$

2.3 Watershed with Dynamics

This operator consists in the interpretation of an image as a topographic surface, where the gray level at a point indicates the height at that point. A flooding is simulated starting with the minima of the image. When the water coming from the near minima are close to contact, a dam is formed. The whole dams represent the watershed [2]. In order to select the main maxima, we use a morphological tool known as dynamics or contrast extinction value introduced by Grimaud [7]. This measure of contrast maps each maximum with a value given by its contrast. The contrast of a maximum is the minimum descent necessary to move from the maximum to another higher maximum; the contrast of the highest maximum is defined as the difference between the maximum and minimum of the function.

3 Metodology

The proposed methodology of the paper is shown in Fig. 1.

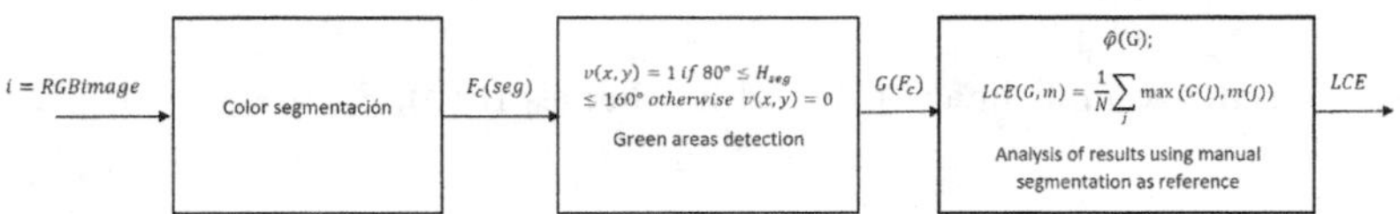

Fig. 1. Proposed methodology

Image i, shown in Fig. 3(a), is coded in RGB color space; therefore a color space conversion is realized by applying the Eqs. 1, 2 or 3. In this case, HSI color space is selected and each channel ($H_{g_{hsi}}$,$S_{g_{hsi}}$,$I_{g_{hsi}}$) is stored in individual images obtaining three grayscale images as shown in Fig. 3(b), (c) and (d). In Fig. 2 the proposed methodology for color segmentation is displayed.

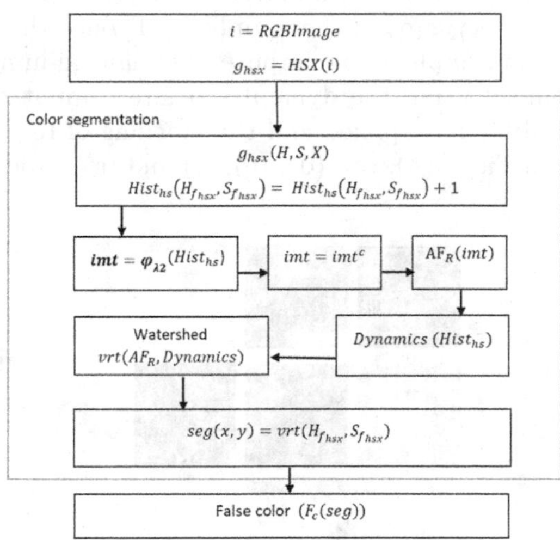

Fig. 2. Color segmentation methodology

Once the color space conversion of the picture is carried out, chromatic histogram is computed with coordinate placement of values using the matrices of H as horizontal axis and S as vertical axis $Hist_{hs}(H_{g_{hsi}}, S_{g_{hsi}}) = Hist_{hs}(H_{g_{hsi}}, S_{g_{hsi}}) + 1$, taking value of H as rows and S as columns.

Since the values in $Hist_{hs}$ could be higher than 255, that implies that it can not be represented in a RGB color image; therefore the Eq. 7 is computed by

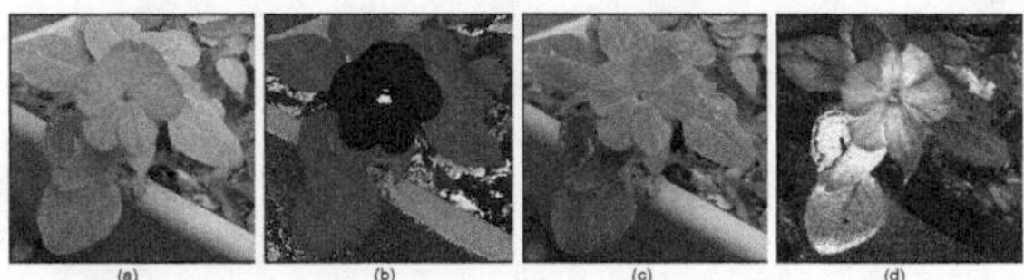

Fig. 3. Example image (a), and its chanels H (b), S (c) and I (d)

normalizing the range of values from 0 to 255 [2] in order to obtain the chromatic histogram shown in Fig. 4(a):

$$f_{hs}(x) = log \left[\frac{Hist_{hs}(x)}{max(Hist_{hs}(x))} \right] \quad (7)$$

The next step is to carry out the color segmentation $seg(i)$ using the computed histogram. Since the histogram contains a high amount of isolated spots, each spot representing a color in the image i, is processed with morphological operators. The following sequence of transformations is applied: (a) First a closing $\lambda = 2$ is applied, (b) the image is complemented, (c) an alternating filter by reconstruction $AF_R = \tilde{\varphi}_2\tilde{\gamma}_2\tilde{\varphi}_1\tilde{\gamma}_1$ is then applied, (d) once the histogram is filtered, the dynamics are applied to compute the main minima, (e) then, the watershed is determined using the dynamics before computed, (f) finally, an alternating sequential filter is applied and the labelling of regions is computed. Each step is shown in Figs. 4(b), (c), (d), (e), (f) and (g) respectively.

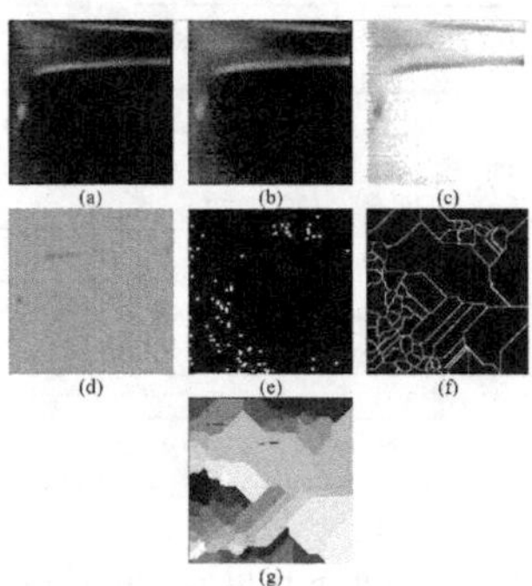

Fig. 4. Chromatic histogram (a), closing $\lambda = 2$ (b), complement (c), alternating sequential filter by reconstruction (d), dynamics of histogram (e), watershed (f) and slopes (g)

All preceding operators are applied in order to reduce the discontinuities of colors in the image. The labeling of regions in the watershed helps to obtain the color segmentation. This operation is defined by $seg(x, y) = vrt(H_{g_{hsv}}, S_{g_{hsv}})$. The computed image is shown in Fig. 5(a), then the false color image (F_c) is computed by taking at each value of gray as an average of colors in channels

RGB in image i (see Fig. 5(b)). Observe that the number of colors has been reduced in the segmented image $seg(i)$.

The detection of vegetation $G(i)$ is done by using the hue channel (H) that has a range from 0 to 360°. This is made by assigning values of 1 to values of H in a range of 80° to 160°. Using this criterion, colors out of the range are assigned as 0, thus a picture that keeps only green colors (in this case values in 1 are changed for pixels in image i). To reduce the noise a morphological reconstruction is applied R, then a closing by reconstruction $\hat{\varphi}_n$ is used to close small holes in green areas. The result is shown in Fig. 5(c).

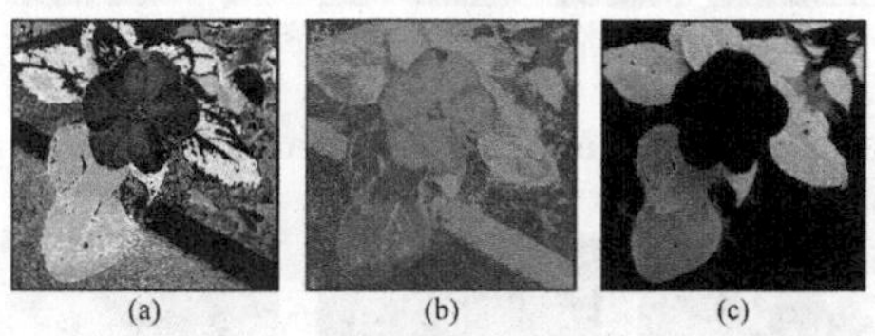

Fig. 5. Color segmentation (a), False color (b) and green areas detection (c) (Color figure online)

Finally, the automatic green areas detection is compared with manual segmentation $m(i)$ in order to determine the quality of vegetation detection [14], [15]. This is performed by computing the Local Consistency Error (LCE) [23] Eq. 8 where N is the number of pixels of the image i, $seg(i)$ is the automatic segmentation of i and $m(i)$ corresponds to manual segmentation.

$$LCE(seg, m) = \frac{1}{N} \sum max(seg(i), m(i)) \tag{8}$$

The value of LCE represents the coincidence between the automatic segmentation and the manual segmentation, and it ranges from 0 to 1 $[0, 1]$, where 0 indicates no similitude between $m(i)$ and $seg(i)$, and as LCE gets closer to 1, the similitude increases until the perfect match, then if LCE is close to 1, the segmentation is more accurate.

4 Results

In order to validate this methodology, an aerial picture is used. The image in Fig. 6(a) of dimensions 2880×1620 pixels is changed to HSI color space, as shown below. Figure 6(b) corresponds to channel H whereas Fig. 6(c) and (d) to channels S and I respectively.

After the color space conversion is done, the chromatic histogram is computed, obtaining the image shown in Fig. 7. Next, the morphological operators mentioned above are computed in order to get an image without discontinuities of colors. In Fig. 7, one observes the chromatic histogram (Fig. 7(a)), the dynamic of minima (Fig. 7(b)), and the labeled slopes (Fig. 7(c)).

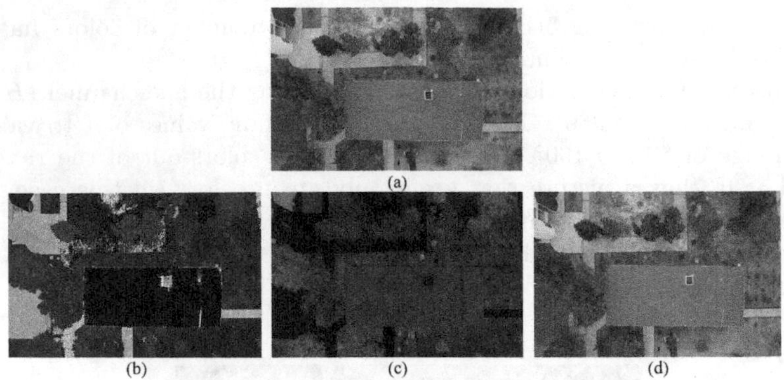

Fig. 6. Sample image (a) and HSI channels (b), (c) and (d)

Fig. 7. Chromatic histogram (a), dynamic minimum of histogram (b), labelled slopes (c)

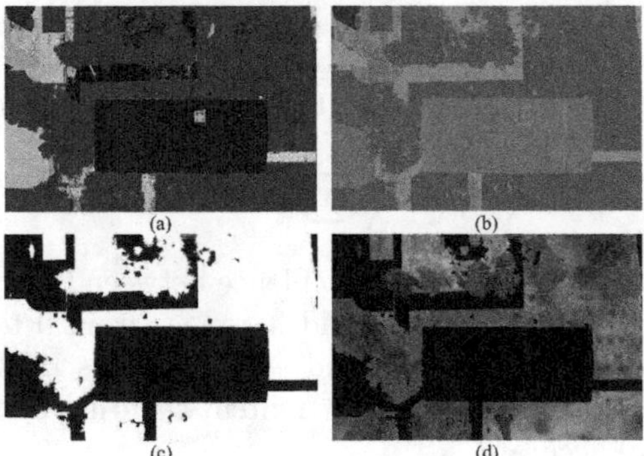

Fig. 8. Image Segmentation (a) and its false color (b), green areas detection binary (c), green areas with actual pixel (d) (Color figure online)

The segmentation is obtained by taking a value of the gray in $vrt(i)$ (Fig. 8(a)) to its corresponding coordinate in $seg(H_{g_{hsi}}, S_{g_{hsi}})$. The segmented image and its false color $seg(i)$ are shown in Fig. 8(a) and (b) respectively. It can be observed that this method groups green areas with a lower amount of green tones, enabling the easy detection of green areas. Once color segmentation and false color $seg(i)$ are computed, green areas detection is applied as shown in Fig. 8(c) and (d).

In Fig. 8 one observes that the algorithm enables the segmentation of different tones presented in the vegetation of the image as shown in Fig. 6(a). To analyze

Fig. 9. Manual Segmentation

Fig. 10. Sample aerial images (a) shadows in the green areas, (b) non-uniform in color tone vegetation and (c) normal conditions vegetation (Color figure online)

the results of this case of study the LCE is computed in order to measure quantitatively the detection of vegetation using manual segmentation as reference [3]. In Fig. 9 the manual segmentation $m(i)$ is shown. The computed value of LCE is 0.985, which means that green areas segmentation is accurate respect to manual segmentation, therefore, it is proved that this method is capable of segmenting green areas despite the differences of color in vegetation.

In order to validate this method, three more cases with particular characteristics were selected from the rest of the sample images; Fig. 10 illustrates the sample images that were selected from the database of the present work. Different Hue-based color spaces (HSV, HSI, and IHSL) were tested with the proposed segmentation method and compared with k-means clustering segmentation [6] that is used by [12] to green areas segmentation. Also the method of Hue threshold [4] used in green areas by [18] has been tested to compare the results. LCE values are computed in each case and The results obtained are shown in Table 1.

Table 1. Results of LCE in proposed method using different Hue-based color spaces, k-means clustering and Hue modification

Case	HSV	HSI	IHSL	K-means	H threshold
Image1	**0.975**	0.985	0.970	0.919	0.946
Image2	**0.978**	0.967	0.959	0.834	0.888
Image3	**0.987**	0.970	0.966	0.944	0.944
Image4	**0.983**	0.986	0.967	0.976	0.934
Average	**0.98**	0.977	0.965	0.918	0.930

From the results shown in Table 1, one observes that in general, the proposed method offers better results that k-means clustering and false color in green areas segmentation (the best results are in bold). In the case of HSI and HSV color spaces, both of them have an average LCE with a small difference between each other (0.03); therefore both are useful color spaces that can be selected to apply this algorithm.

5 Conclusion

In the present work, an algorithm for green areas segmentation that uses hue-based color spaces has been presented. It has been shown that the algorithm is able to segment vegetation in a large range of colors grouping them into a single region. This characteristic is important since vegetation color could vary depending of the season in the year or weather. During LCE analysis, it was demonstrated that the results of the algorithm are close to manual segmentation, and provides better results than k-means clustering and false color when green areas are considered. The advantage of the functionality of this system is that an additional sensor or image acquisition system is not required, which represents a lower cost. This is why it can be applied only with the use of a standard camera.

Acknowledgements. Alvarado-Robles expresses his gratitude to the 232143 PEI program, with fund number 1298, for the financing conferred to this work.

References

1. Angulo, J., Serra, J.: Color segmentation by ordered mergings. In: Proceedings of 2003 International Conference on Image Processing, ICIP 2003, vol. 2, pp. II–125. IEEE (2003)
2. Angulo, J., Serra, J.: Segmentación de imágenes en color utilizando histogramas bi-variables en espacios color polares luminancia/saturación/matiz. Computación y Sistemas **8**(4), 303–316 (2005)
3. Arbelaez, P., Maire, M., Fowlkes, C., Malik, J.: Contour detection and hierarchical image segmentation. IEEE Trans. Pattern Anal. Mach. Intell. **33**(5), 898–916 (2011)
4. Báez Rojas, J., Pérez, A.: Uso del sistema hsi para asignar falso color a objetos en imágenes digitales. Revista mexicana de física E **54**(2), 186–192 (2008)
5. Bendig, J., Yu, K., Aasen, H., Bolten, A., Bennertz, S., Broscheit, J., Gnyp, M.L., Bareth, G.: Combining UAV-based plant height from crop surface models, visible, and near infrared vegetation indices for biomass monitoring in barley. Int. J. Appl. Earth Obs. Geoinformation **39**, 79–87 (2015)
6. Chang, M.M., Sezan, M.I., Tekalp, A.M.: Adaptive Bayesian segmentation of color images. J. Electron. Imaging **3**(4), 404–414 (1994)
7. Grimaud, M.: New measure of contrast: the dynamics. In: San Diego 1992, pp. 292–305. International Society for Optics and Photonics (1992)
8. Ilunga-Mbuyamba, E., Avina-Cervantes, J.G., Garcia-Perez, A., de Jesus Romero-Troncoso, R., Aguirre-Ramos, H., Cruz-Aceves, I., Chalopin, C.: Localized active contour model with background intensity compensation applied on automatic MR brain tumor segmentation. Neurocomputing **220**, 84–97 (2017)

9. Kurugollu, F., Sankur, B., Harmanci, A.E.: Color image segmentation using histogram multithresholding and fusion. Image Vis. Comput. **19**(13), 915–928 (2001)
10. Les, T., Markiewicz, T., Jesiotr, M., Kozlowski, W.: Dots detection in HER2 FISH images based on alternative color spaces. Procedia Comput. Sci. **90**, 132–137 (2016)
11. Lezoray, O., Charrier, C.: Color image segmentation using morphological clustering and fusion with automatic scale selection. Pattern Recogn. Lett. **30**(4), 397–406 (2009)
12. Lin, Y., Jiang, M., Yao, Y., Zhang, L., Lin, J.: Use of UAV oblique imaging for the detection of individual trees in residential environments. Urban For. Urban Greening **14**(2), 404–412 (2015)
13. Luo, H., Lin, D., Yu, C., Chen, L.: Application of different HSI color models to detect fire-damaged mortar. Int. J. Transp. Sci. Technol. **2**(4), 303–316 (2013)
14. Martin, D.R., Fowlkes, C.C., Malik, J.: Learning to detect natural image boundaries using brightness and texture. In: Advances in Neural Information Processing Systems, pp. 1255–1262 (2002)
15. Martin, D.R., Fowlkes, C.C., Malik, J.: Learning to detect natural image boundaries using local brightness, color, and texture cues. IEEE Trans. Pattern Anal. Mach. Intell. **26**(5), 530–549 (2004)
16. Miao, R.H., Tang, J.L., Chen, X.Q.: Classification of farmland images based on color features. J. Vis. Commun. Image Representation **29**, 138–146 (2015)
17. Niu, S., Chen, Q., de Sisternes, L., Ji, Z., Zhou, Z., Rubin, D.L.: Robust noise region-based active contour model via local similarity factor for image segmentation. Pattern Recogn. **61**, 104–119 (2017)
18. Garduño Ramón, M.A., Sánchez-Gómez, J.I., Morales Hernández, L.A., Benítez-Rangel, J.P., Osornio-Ríos, R.A.: Methodology for automatic detection of trees and shrubs in aerial pictures from UAS. CONIIN 2014 (2014)
19. Serra, J., Vincent, L.: An overview of morphological filtering. Circuits Syst. Sig. Process. **11**(1), 47–108 (1992)
20. Shaik, K.B., Ganesan, P., Kalist, V., Sathish, B., Jenitha, J.M.M.: Comparative study of skin color detection and segmentation in HSV and YCBCR color space. Procedia Comput. Sci. **57**, 41–48 (2015)
21. Tan, J.H., Acharya, U.R.: Pseudocolours for thermography multi-segments colour scale. Infrared Phys. Technol. **72**, 140–147 (2015)
22. Terol-Villalobos, I.R., Mendiola-Santibáñez, J.D., Canchola-Magdaleno, S.L.: Image segmentation and filtering based on transformations with reconstruction criteria. J. Vis. Commun. Image Representation **17**(1), 107–130 (2006)
23. Unnikrishnan, R., Pantofaru, C., Hebert, M.: Toward objective evaluation of image segmentation algorithms. IEEE Trans. Pattern Anal. Mach. Intell. **29**(6), 929–944 (2007)
24. Zribi, M.: Unsupervised Bayesian image segmentation using orthogonal series. J. Vis. Commun. Image Representation **18**(6), 496–503 (2007)

Spatial Enhancement by Dehazing for Detection of Microcalcifications with Convolutional Nets

Alessandro Bria[1(✉)], Claudio Marrocco[1], Adrian Galdran[2], Aurélio Campilho[2,3], Agnese Marchesi[1], Jan-Jurre Mordang[4], Nico Karssemeijer[4], Mario Molinara[1], and Francesco Tortorella[1]

[1] DIEI, University of Cassino and Southern Latium, Cassino, FR, Italy
{a.bria,c.marrocco,a.marchesi,m.molinara,tortorella}@unicas.it

[2] INESC TEC, Institute for Systems and Computer Engineering, Technology and Science, Porto, Portugal
adrian.galdran@inesctec.pt

[3] Faculdade de Engenharia, Universidade do Porto, Porto, Portugal
campilho@fe.up.pt

[4] DIAG, Radboud University Nijmegen Medical Centre, Nijmegen, The Netherlands
{jan-jurre.mordang,nico.karssemeijer}@radboudumc.nl

Abstract. Microcalcifications are early indicators of breast cancer that appear on mammograms as small bright regions within the breast tissue. To assist screening radiologists in reading mammograms, supervised learning techniques have been found successful to detect microcalcifications automatically. Among them, Convolutional Neural Networks (CNNs) can automatically learn and extract low-level features that capture contrast and spatial information, and use these features to build robust classifiers. Therefore, spatial enhancement that enhances local contrast based on spatial context is expected to positively influence the learning task of the CNN and, as a result, its classification performance. In this work, we propose a novel spatial enhancement technique for microcalcifications based on the removal of haze, an apparently unrelated phenomenon that causes image degradation due to atmospheric absorption and scattering. We tested the influence of dehazing of digital mammograms on the microcalcification detection performance of two CNNs inspired by the popular AlexNet and VGGnet. Experiments were performed on 1,066 mammograms acquired with GE Senographe systems. Statistically significantly better microcalcification detection performance was obtained when dehazing was used as preprocessing. Results of dehazing were superior also to those obtained with Contrast Limited Adaptive Histogram Equalization (CLAHE).

Keywords: Spatial enhancement · Dehazing · Microcalcification detection · Convolutional neural networks · CAD

1 Introduction

Digital mammography is an effective and reliable method for early breast cancer detection, which is fundamental to increment the survival rate and improve

S. Battiato et al. (Eds.): ICIAP 2017, Part II, LNCS 10485, pp. 288–298, 2017.
https://doi.org/10.1007/978-3-319-68548-9_27

the life quality of the patients [1]. In the last few decades, Computer-Aided Detection (CADe) systems have been proposed to help radiologists in reading screening mammograms. Several studies have shown that CADe systems can improve the performance of individual radiologists [7] in detecting suspicious lesions in mammograms, such as microcalcifications (MCs) and masses. MCs are tiny deposits of calcium that appear on a mammogram as spots of size between 0.1 mm and 1 mm. They are of particular interest since they are usually associated with Ductal Carcinoma In Situ and invasive cancers [29]. Automatic MC detection is often based on supervised learning [4,5,14,21], where powerful binary classifiers are applied to determine whether a MC is present at a pixel location.

Among supervised techniques, Deep Learning approaches have recently acquired great popularity thanks to their outstanding performance in computer vision [17]. In particular, Convolutional Neural Networks (CNNs) have shown to be very effective for classification of image data and also received a large consensus in medical imaging problems [9,30], including MC detection [24,32]. A typical CNN architecture is a sequence of feed-forward layers where convolutional filters are interlaced with nonlinear activation functions and pooling. The convolutional layers determine a set of abstract features, whereas the last fully connected layers perform the classification. When training the CNNs, image preprocessing is a fundamental step. Among preprocessing techniques, those including contrast and spatial enhancement have shown to be particularly useful to improve the CNN performance. A preprocessing contrast-extracting layer was firstly used in [6], whereas a local contrast normalization layer was proposed in [12] with the aim of normalizing the responses across all features after each convolutional layer. A layer for brightness normalization was successively introduced in [16] and local plus global contrast normalization was used in [33] to normalize brightness and color variations of RGB images.

Preprocessing techniques are commonly applied in digital mammograms [19,20,22,23], and recently, the effect of contrast enhancement techniques on a CNN has been studied for medical imaging problems [18,25]. In this work, we propose a novel spatial enhancement method for MCs based on the removal of haze, an apparently unrelated phenomenon usually present in outdoor images that causes image degradation due to atmospheric absorption and scattering. Since CNNs automatically learn and extract low-level features that capture contrast and spatial information, spatial enhancement is expected to positively influence its classification performance. We show that applying an image dehazing approach on mammograms we enhance the contrast of MCs with respect to the surrounding tissue, thus obtaining statistically significantly better MC detection performance when dehazing is used as preprocessing for two different CNNs.

2 Dataset

For this study, we collected a database consisting of 1,066 mammograms acquired with GE Senographe systems (GE, Fairfield, Connecticut,

United States) in Radboud University Medical Center (Nijmegen, The Netherlands). All mammograms were acquired with standard clinical settings at a pixel resolution of 0.1 mm. A total of 7,579 individual MCs were annotated by an experienced reader who marked the center of each microcalcification based on the diagnostic reports. To feed the CNNs, we extracted a dataset of patches of size 12×12 pixels from the mammograms. The patches containing MCs (*positive* samples) were taken by centering the detector window at the groundtruth microcalcification centers, yielding the same number of samples as the individually labeled MCs. The background patches (*negative* samples) were randomly extracted from the remaining regions of the images, totalizing 27,017,503 samples.

3 Spatial Enhancement by Dehazing

3.1 Image Dehazing

The goal of image dehazing is to remove degradation in outdoor images caused by atmospheric absorption and scattering. This physical effect was modelled in [15] as being directly proportional to the distance of the object from the observer, according to the following light propagation law:

$$I(x) = t(x)R(x) + A(1 - t(x)), \tag{1}$$

where x represents a pixel location, $I(x)$ is the intensity captured, $R(x)$ is the radiance in a hypothetical haze-free scene, A is the predominant color of the atmosphere, and $t(x)$ is the transmission of light in the atmosphere. Following [31], the input image can be assumed to have intensities normalized in $[0, 1]$ and be white-balanced, so that the highest intensity in the image is white and A can be approximated by $A \approx (1, 1, 1)$. The haze degradation model simplifies in:

$$I(x) = t(x)R(x) + 1 - t(x), \tag{2}$$

from which, assuming that an estimate of $t(x)$ is available, we can factor the true radiance $R(x)$ by:

$$R(x) = 1 + \frac{I(x) - 1}{t(x)}. \tag{3}$$

Many methods have been proposed for an accurate and robust estimate of $t(x)$. Below we provide a detailed analysis of one of the most successful techniques, namely the Dark Channel Prior [10], that will reveal the direct link between haze removal and spatial enhancement of MCs.

3.2 Dark Channel Prior

The Dark Channel technique is probably the most popular method for image dehazing, partly due to its simplicity. It is based on the observation that most local patches in haze-free images contain some pixels with very low intensity in at least one color channel (the so-called Dark Channel Prior). Thus, given a pixel x and a local spatial neighborhood $\Omega(x)$ centered on it, the dark channel of the true radiance $R(x)$ contains mostly low values:

$$R^{\text{dark}}(x) = \min_{c\in\{R,G,B\}} \left(\min_{z\in\Omega(x)} R(z) \right) \to 0. \tag{4}$$

On the other hand, due to the additive degradation component in Eq. 1, the dark channel of the haze-degraded image $I(x)$ can be approximated by:

$$I^{\text{dark}}(x) = \min_{c\in\{R,G,B\}} \left(\min_{z\in\Omega(x)} I(z) \right) \approx A(1 - t(x)). \tag{5}$$

Using this prior in the simplified haze imaging model of Eq. 2, it is possible to directly estimate $t(x)$ as:

$$t(x) \approx 1 - \omega\ I_{\text{dark}}(x) \tag{6}$$

where $\omega \in (0, 1)$ is a parameter controlling the amount of contrast introduced in the final dehazed image. Due to the implicit local depth constancy made in Eq. 6, the estimated transmission map will usually suffer from a characteristic block artifact, that would lead to halos in the output image unless removed. This can be accomplished with different specialized refining filters, being the typical choice for this task the Guided Filter [11].

3.3 Spatial Enhancement of Microcalcifications

We applied the Dark Channel Prior on mammograms to selectively enhance the contrast of MCs with respect to the surrounding tissue. To show this, let us write $t(x)$ for a grayscale image:

$$t(x) = 1 - \omega \min_{z\in\Omega(x)} I(z) \tag{7}$$

which inserted into Eq. 3, and after simple algebraic manipulations, yields:

$$R(x) = 1 - \frac{1 - I(x)}{1 - \omega \left(\min_{z\in\Omega(x)} I(z) \right)} \tag{8}$$

The key factor in our case is the selection of a neighborhood $\Omega(x)$ slightly bigger than the MC size. In our case, since MCs have typical dimensions well below 1 mm and mammograms have a pixel resolution of 0.1 mm, we chose a squared neighborhood of size 11×11 pixels. This leads us to establish two key observations as explained in the following.

1. *The intensity of MCs is slightly reduced by dehazing.*
   If x belongs to a MC, then $\exists\epsilon \in \mathbb{R}^+, \epsilon \ll 1$ so that:

$$I(x) = 1 - \epsilon \tag{9}$$

since MCs have a high intensity in the image. Moreover, in the neighborhood $\Omega(x)$ there will be a background pixel that has the lowest intensity μ within $\Omega(x)$. Then, we can rewrite Eq. 8 as:

$$R(x) = 1 - \frac{\epsilon}{1 - \omega\mu} \tag{10}$$

Let $\Delta = I(x) - \mu$ be the difference between the intensity of the MC pixel under consideration and the lowest-intensity background pixel in the neighborhood $\Omega(x)$. Recalling that $0 < \omega < 1$, and after simple algebraic manipulations, we can bound $R(x)$ as:

$$I(x) > R(x) > \frac{\Delta}{\epsilon + \Delta} \tag{11}$$

Then, combining Eqs. 9 and 11 yields:

$$0 < I(x) - R(x) < 1 - \epsilon - \frac{\Delta}{\epsilon + \Delta} \tag{12}$$

which after simple algebraic manipulations rewrites as:

$$0 < I(x) - R(x) < \frac{\mu}{1 + \frac{\Delta}{\epsilon}} \tag{13}$$

Since $\mu, \epsilon, \Delta > 0$, this provides an upper bound to the difference in intensity between the MC pixels before and after dehazing. Specifically, since μ is the lowest-intensity background pixel in $\Omega(x)$, then $\Delta \gg \epsilon$ and the fraction in Eq. 13 yields a small value. In other words, independently from the choice of ω, the intensity of the MC pixels will only be slightly reduced by dehazing.

2. *The intensity of the background around MCs is greatly reduced by dehazing.* If x is a background pixel close to a MC so that part of the MC is within $\Omega(x)$, and $\Omega(x)$ is small, then we can approximate the lowest-intensity pixel in $\Omega(x)$ with $I(x)$:

$$\min_{z \in \Omega(x)} I(z) \approx I(x) \tag{14}$$

which combined with Eq. 8 yields:

$$R(x) \approx 1 - \frac{1 - I(x)}{1 - \omega I(x)} \tag{15}$$

This acts as a power-law gamma correction transform controlled by ω (see Fig. 1). Since $I(x)$ is supposed to have mid-low intensity, this transform will greatly darken $I(x)$. The closer ω to 1, the stronger the darkening of $I(x)$.

Following the above observations it is possible to conclude that the contrast between MCs and background tissue is enhanced by dehazing. This can be seen in Fig. 2 where we show a close-up of MCs before and after dehazing with $\omega = 0.9$ and $\Omega(x)$ of size 11×11 pixels. These parameters were fixed at the beginning of our experiments and were not varied afterwards.

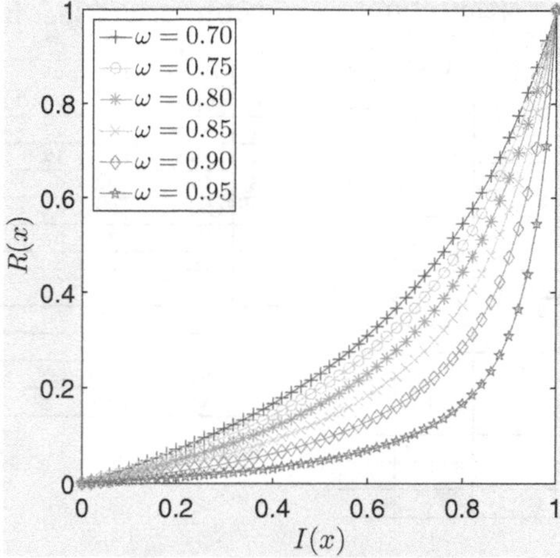

Fig. 1. Intensity transformations induced by dehazing on the background surrounding MCs for different values of ω. The closer ω to 1, the stronger the darkening of $I(x)$.

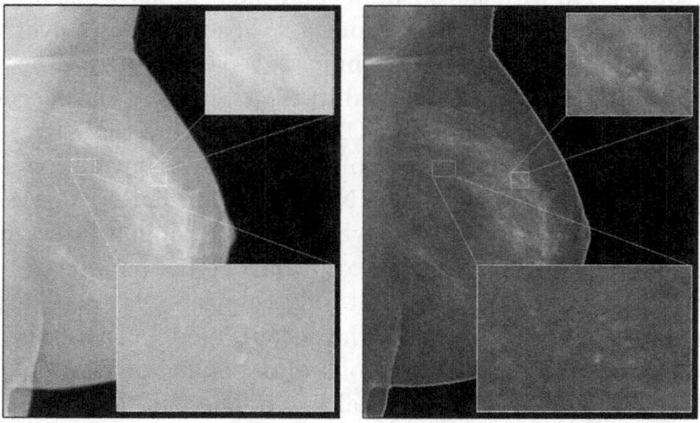

Fig. 2. A mammogram before (left) and after (right) dehazing. In the close-ups, two microcalcifications clusters are shown.

4 Convolutional Neural Networks

A CNN is an ensemble of neurons each featuring several weighted inputs and one output, performing convolution of inputs with weights and transforming the outcome according to a nonlinear activation function. Neurons are arranged in layers and usually share the same weights so as to produce a feature map and reduce the number of parameters. In a typical CNN architecture, convolutional layers are equipped with the Rectified Linear Units (ReLUs) and are intertwined with max-pooling layers. ReLUs apply a nonsaturating activation function $f(x) = \max(0, x)$ which allows the network to easily obtain sparse representations. Max-pooling layers aggregate the outputs of multiple neurons and

Table 1. AlexNet-based architecture

#	Type	Output	Kernel	Stride	Pad
0	Input	$1 \times 12 \times 12$			
1	Conv	$96 \times 12 \times 12$	3×3	1	1
2	LRN	$96 \times 12 \times 12$			
3	Maxpool	$96 \times 6 \times 6$	2×2	2	1
4	Conv	$256 \times 6 \times 6$	3×3	1	1
5	LRN	$256 \times 6 \times 6$			
6	Maxpool	$256 \times 3 \times 3$	2×2	2	1
7	Conv	$384 \times 3 \times 3$	3×3	1	1
8	Conv	$384 \times 3 \times 3$	3×3	1	1
9	Conv	$256 \times 5 \times 5$	3×3	1	1
10	Maxpool	$256 \times 3 \times 3$	3×3	1	1
11	FC	1,024	1×1		
12	Dropout	1,024			
13	FC	512	1×1		
14	Dropout	512			
15	FC	2	1×1		

Table 2. VGGnet-based architecture

#	Type	Output	Kernel	Stride	Pad
0	Input	$1 \times 12 \times 12$			
1	Conv	$32 \times 12 \times 12$	3×3	1	1
2	Conv	$32 \times 12 \times 12$	3×3	1	1
3	Maxpool	$32 \times 6 \times 6$	2×2	2	1
4	Conv	$32 \times 6 \times 6$	3×3	1	1
5	Conv	$32 \times 6 \times 6$	3×3	1	1
6	Maxpool	$32 \times 3 \times 3$	2×2	2	1
7	FC	256	1×1		
8	Dropout	256			
9	FC	256	1×1		
10	Dropout	256			
11	FC	2	1×1		

return the maximum, which results in less training time and lower network complexity. The final decision is made through one or more fully connected layers where each neuron is fed with the outputs of all the neurons of the previous layer. Dropout layers usually follow a fully connected layer to reduce overfitting. The term dropout indicates that, at each training stage, a fixed percentage of outputs coming from the previous layer is ignored in the training of the successive layer.

In this study, we implemented two CNNs inspired by the AlexNet [16] and the VGGnet [28]. The first model is composed by five convolutional and three fully connected layers. Local Response Normalization (LRN) layers follow the first and second convolutional layers, whereas max-pooling layers follow both LRN layers and the last convolutional layer. The ReLU nonlinearity is applied to the output of every convolutional and fully connected layer. The parameters of each layer are reported in Table 1. The second model consists of two stacks of two convolutional layers followed by one max-pooling layer. ReLU is used as activation function for each convolutional layer. The final layers are three fully connected layers. The parameters of each layer are shown in Table 2.

5 Experiments

We applied the two CNNs to the unprocessed mammograms and to the mammograms processed with dehazing and with CLAHE [26], which is a well-known method for spatial enhancement, also applied on mammograms [2]. The parameters of CLAHE were clip limit = 0.01 and block size = 8×8 pixels [34]. We used 2-fold cross validation to train and test the networks. In each cross validation step, the CNN was trained on the 50% of the samples and tested on the other 50%. Before training, positive and negative samples were balanced by means of data augmentation using flipping, rotation, and replication. Each network was trained to minimize the Softmax loss function by means of backpropagation and Mini-Batch Stochastic Gradient Descent, with mini-batches of 32 samples. Standardization was applied to the inputs by mean subtraction and normalization to unit variance. Weights of each learning layer were initialized using the algorithm of Glorot and Bengio [8]. The learning rate was set to the initial value of 10^{-3} and decreased during training by a factor of 10 every 6 epochs. Momentum and weight decay were set respectively to 0.9 and $5 \cdot 10^{-4}$. The dropout was performed with a probability of 0.5. For the LRN layers of the AlexNet we set the following parameters: $k = 1$, $n = 5$, $\alpha = 10^{-4}$, and $\beta = 0.75$. The learning was stopped after 30 epochs (1 epoch = 844, 297 iterations), i.e. when the loss function did not decrease significantly. We used the Caffe framework [13] for the implementation of both networks, and all the experiments were performed on a computer with 2 Intel Xeon e5-2609 processors, 256 GB of RAM and 2 GPU NVIDIA TitanX Pascal.

6 Results

The CNN-based microcalcification detectors without and with the two spatial enhancement methods have been evaluated in terms of Receiver Operating Characteristics (ROC) curve by plotting True Positive Rate (TPR) against False Positive Rate (FPR) for a series of thresholds on the CNN output associated to each sample. Furthermore, the mean sensitivity of the ROC curve in the specificity range on a logarithmic scale was calculated and compared. The mean sensitivity is defined as [24]:

$$\overline{S}(a,b) = \frac{1}{ln(b) - ln(a)} \int_a^b \frac{s(f)}{f} df \qquad (16)$$

where a and b are the lower and upper bound of the false positive fraction and were set, respectively, to 10^{-6} and 10^{-1} and $s(f)$ is the sensitivity at the false positive fraction f. Statistical comparisons were performed by means of bootstrapping [27] as in [3]. On the test set, average ROC curves were calculated over 1,000 bootstraps, and are reported in Fig. 3. Additionally, the mean sensitivity was calculated for each bootstrap and p-values were computed for

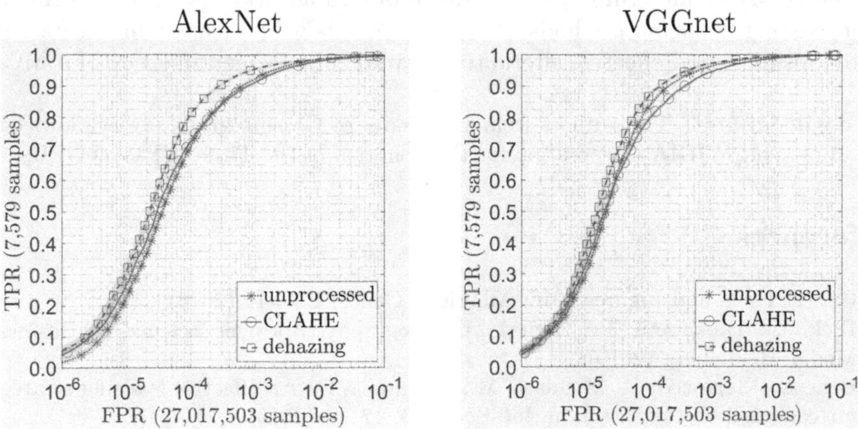

Fig. 3. ROC curves of the CNN detectors averaged from 1,000 bootstrap iterations.

Table 3. Comparative results of mean sensitivity $\overline{S}$ in the FPR range $[10^{-6}, 10^{-1}]$ for different methods (UN = unprocessed, DH = dehazing, CL = CLAHE).

CNN	$\overline{S}_{UN}$	$\overline{S}_{CL}$	$\overline{S}_{DH}$	$\overline{S}_{CL} - \overline{S}_{UN}$	$\overline{S}_{DH} - \overline{S}_{UN}$	$\overline{S}_{DH} - \overline{S}_{CL}$
AlexNet	68.47	69.64	73.29	**+1.17** ($p < 0.001$)	**+4.82** ($p < 0.001$)	**+3.65** ($p < 0.001$)
VGGnet	73.88	73.07	76.26	**−0.81** ($p < 0.001$)	**+2.38** ($p < 0.001$)	**+3.19** ($p < 0.001$)

testing significance. The statistical significance level was chosen as $\alpha = 0.05$ but, due to the number of comparisons $m = 3$, we applied the Bonferroni correction, so that performance differences were considered statistically significant if $p < 0.017$. Results are reported in Table 3. The mean sensitivities obtained on unprocessed images were 68.47 and 73.88 for the AlexNet- and the VGGnet-based CNNs, respectively. Results with dehazing were statistically significantly better than on unprocessed images (+4.82 with AlexNet and +2.38 with VGGnet) and also superior to those of CLAHE (+3.65 with AlexNet and +3.19 with VGGnet).

7 Conclusions

In the present study, we have established a novel connection between the problem of spatial enhancement of MCs in mammograms and the apparently unrelated problem of haze removal in outdoor images. We have shown that the performance of MC detection with CNNs can be greatly improved if mammograms are preprocessed with a dehazing technique to enhance local contrast of MCs. Indeed, it is known that the first layers of a CNN automatically learn and extract low-level features. We can suppose that improving the local contrast of MCs is beneficial for these layers that capture contrast and spatial information in the salient regions. Consequently, this may positively influence the learning task of the subsequent layers that are aimed at capturing more complex features.

Future works will be focused on analyzing the impact of image dehazing also on other CNN architectures and MC detectors. In addition, we will experiment other existing dehazing methods. If successful, this would lead to an entire new family of simple and effective alternative spatial enhancement methods for MCs.

Acknowledgment. The authors from University of Cassino gratefully acknowledge the support of NVIDIA Corporation for the donation of the Titan X Pascal GPUs.

References

1. Cancer Facts and Figures 2016. American Cancer Society (2016)
2. Bick, U., Diekmann, F.: Digital Mammography. Springer Science & Business Media, Heidelberg (2010)
3. Bria, A., Marrocco, C., Molinara, M., Tortorella, F.: An effective learning strategy for cascaded object detection. Inf. Sci. **340**, 17–26 (2016)

4. Bria, A., Marrocco, C., Karssemeijer, N., Molinara, M., Tortorella, F.: Deep cascade classifiers to detect clusters of microcalcifications. In: Tingberg, A., Lång, K., Timberg, P. (eds.) IWDM 2016. LNCS, vol. 9699, pp. 415–422. Springer, Cham (2016). doi:10.1007/978-3-319-41546-8_52
5. Bria, A., Marrocco, C., Molinara, M., Tortorella, F.: A ranking-based cascade approach for unbalanced data. In: 2012 21st International Conference on Pattern Recognition (ICPR), pp. 3439–3442. IEEE (2012)
6. Ciresan, D.C., Meier, U., Masci, J., Maria Gambardella, L., Schmidhuber, J.: Flexible, high performance convolutional neural networks for image classification. In: IJCAI Proceedings-International Joint Conference on Artificial Intelligence, Barcelona, Spain, vol. 22, p. 1237 (2011)
7. Eadie, L.H., Taylor, P., Gibson, A.P.: A systematic review of computer-assisted diagnosis in diagnostic cancer imaging. Eur. J. Radiol. **81**(1), e70–e76 (2012)
8. Glorot, X., Bengio, Y.: Understanding the difficulty of training deep feedforward neural networks. In: AISTATS, vol. 9, pp. 249–256 (2010)
9. Greenspan, H., van Ginneken, B., Summers, R.M.: Guest editorial deep learning in medical imaging: overview and future promise of an exciting new technique. IEEE Trans. Med. Imaging **35**(5), 1153–1159 (2016)
10. He, K., Sun, J., Tang, X.: Single image haze removal using dark channel prior. IEEE Trans. Pattern Anal. Mach. Intell. **33**(12), 2341–2353 (2011)
11. He, K., Sun, J., Tang, X.: Guided image filtering. IEEE Trans. Pattern Anal. Mach. Intell. **35**(6), 1397–1409 (2013)
12. Jarrett, K., Kavukcuoglu, K., LeCun, Y., et al.: What is the best multi-stage architecture for object recognition? In: 2009 IEEE 12th International Conference on Computer Vision, pp. 2146–2153. IEEE (2009)
13. Jia, Y., Shelhamer, E., Donahue, J., Karayev, S., Long, J., Girshick, R., Guadarrama, S., Darrell, T.: Caffe: convolutional architecture for fast feature embedding. arXiv preprint arXiv:1408.5093 (2014)
14. Jing, H., Yang, Y., Nishikawa, R.M.: Detection of clustered microcalcifications using spatial point process modeling. Phys. Med. Biol. **56**(1), 1–17 (2011)
15. Koschmieder, H.: Theorie der horizontalen Sichtweite: Kontrast und Sichtweite. Keim & Nemnich (1925)
16. Krizhevsky, A., Sutskever, I., Hinton, G.E.: Imagenet classification with deep convolutional neural networks. In: Advances in Neural Information Processing Systems, pp. 1097–1105 (2012)
17. LeCun, Y., Bengio, Y., Hinton, G.: Deep learning. Nature **521**(7553), 436–444 (2015)
18. Liskowski, P., Krawiec, K.: Segmenting retinal blood vessels with deep neural networks. IEEE Trans. Med. Imaging **35**(11), 2369–2380 (2016)
19. Maitra, I.K., Nag, S., Bandyopadhyay, S.K.: Technique for preprocessing of digital mammogram. Comput. Methods Programs Biomed. **107**(2), 175–188 (2012)
20. Marchesi, A., Bria, A., Marrocco, C., Molinara, M., Mordang, J.J., Tortorella, F., Karssemeijer, N.: The effect of mammogram preprocessing on microcalcification detection with convolutional neural networks. In: 2017 IEEE 30th International Symposium on Computer-Based Medical Systems (CBMS), pp. 118–121. IEEE (2017)
21. Marrocco, C., Molinara, M., Tortorella, F., Rinaldi, P., Bonomo, L., Ferrarotti, A., Aragno, C., lo Moriello, S.S.: Detection of cluster of microcalcifications based on watershed segmentation algorithm. In: 2012 25th International Symposium on Computer-Based Medical Systems (CBMS), pp. 1–5. IEEE (2012)

22. Mirzaalian, H., Ahmadzadeh, M.R., Sadri, S., Jafari, M.: Pre-processing algorithms on digital mammograms. In: MVA, pp. 118–121 (2007)
23. Molinara, M., Marrocco, C., Tortorella, F.: Automatic segmentation of the pectoral muscle in mediolateral oblique mammograms. In: 2013 IEEE 26th International Symposium on Computer-Based Medical Systems (CBMS), pp. 506–509. IEEE (2013)
24. Mordang, J.-J., Janssen, T., Bria, A., Kooi, T., Gubern-Mérida, A., Karssemeijer, N.: Automatic microcalcification detection in multi-vendor mammography using convolutional neural networks. In: Tingberg, A., Lång, K., Timberg, P. (eds.) IWDM 2016. LNCS, vol. 9699, pp. 35–42. Springer, Cham (2016). doi:10.1007/978-3-319-41546-8_5
25. Orlando, J.I., Prokofyeva, E., del Fresno, M., Blaschko, M.B.: Convolutional neural network transfer for automated glaucoma identification. In: 12th International Symposium on Medical Information Processing and Analysis, vol. 10160, p. 101600U (2017)
26. Pizer, S.M., Amburn, E.P., Austin, J.D., Cromartie, R., Geselowitz, A., Greer, T., ter Haar Romeny, B., Zimmerman, J.B., Zuiderveld, K.: Adaptive histogram equalization and its variations. Comput. Vis. Graph. Image Process. **39**(3), 355–368 (1987)
27. Samuelson, F.W., Petrick, N.: Comparing image detection algorithms using resampling. In: International Symposium on Biomedical Imaging, pp. 1312–1315 (2006)
28. Simonyan, K., Zisserman, A.: Very deep convolutional networks for large-scale image recognition. arXiv preprint arXiv:1409.1556 (2014)
29. Stomper, P.C., Geradts, J., Edge, S.B., Levine, E.G.: Mammographic predictors of the presence and size of invasive carcinomas associated with malignant microcalcification lesions without a mass. Am. J. Roentgenol. **181**(6), 1679–1684 (2003)
30. Tajbakhsh, N., Shin, J.Y., Gurudu, S.R., Hurst, R.T., Kendall, C.B., Gotway, M.B., Liang, J.: Convolutional neural networks for medical image analysis: full training or fine tuning? IEEE Trans. Med. Imaging **35**(5), 1299–1312 (2016)
31. Tarel, J.P., Hautiere, N.: Fast visibility restoration from a single color or gray level image. In: 2009 IEEE 12th International Conference on Computer Vision, pp. 2201–2208. IEEE (2009)
32. Wang, J., Yang, X., Cai, H., Tan, W., Jin, C., Li, L.: Discrimination of breast cancer with microcalcifications on mammography by deep learning. Sci. Rep. **6**, 27327 (2016)
33. Zeiler, M., Fergus, R.: Stochastic pooling for regularization of deep convolutional neural networks. In: Proceedings of the International Conference on Learning Representation (ICLR) (2013)
34. Zuiderveld, K.: Contrast limited adaptive histogram equalization. In: Graphics Gems IV, pp. 474–485. Academic Press Professional, Inc. (1994)

Towards Automatic Skin Tone Classification in Facial Images

Diana Borza[1(✉)], Sergiu Cosmin Nistor[2], and Adrian Sergiu Darabant[2]

[1] Computer Science Department, Technical University of Cluj Napoca, 28 Memorandumului Street, 400114 Cluj Napoca, Romania
Diana.borza@cs.utcluj.ro

[2] Computer Science Department, Babes Bolyai University, 58-60 Teodor Mihali, 400591 Cluj Napoca, Romania
sergiu.c.nistor@gmail.com, dadi@cs.ubbcluj.ro

Abstract. In this paper, we address the problem of skin tone classification in facial images, which has applications in various domains: visagisme, soft biometry and surveillance systems. We propose four skin tone classification algorithms and analyze their performance using different color spaces. The first two methods rely directly on pixel values, while the latter two divide the image into cells and classify the skin tone based on the color histograms of these cells. The proposed solutions were trained and evaluated on images from four publicly available databases and on images captured in our laboratory. The best accuracy (87.06%) is obtained using cell histograms of the Lab color space and support vector machine classifier.

Keywords: Skin tone · Color classification · Histograms · Support vector machines · Gaussian mixture models

1 Introduction

Skin color analysis is important to a variety of computer vision tasks, such as face detection and tracking, gesture analysis and human computer interaction. Although, automatic skin detection [1] is widely studied in the specialized literature, little work has been conducted on skin tone classification.

The notion of skin tone is very subjective, especially from a human interpretation point of view, as it doesn't have a clear and precise definition. Skin tone classification was pioneered by Felix von Luschan [2] in 1897 who defined a chromatic scale skin scale with 36 categories. The skin tone of a subject was determined by comparing its skin (from an area that was not exposed to sun) with 36 painted glass tiles. This method was initially used in anthropometric studies, but nowadays it is replaced with more accurate spectrophotometric methods [3]. In the field of dermatology, the Fitzpatrick scale [4] used six skin color classes (type I to type VI) to describe the sun-tanning behavior. Both of these taxonomies are subjective and often inconsistent: even trained practitioners give different results to the same skin tone.

S. Battiato et al. (Eds.): ICIAP 2017, Part II, LNCS 10485, pp. 299–309, 2017.
https://doi.org/10.1007/978-3-319-68548-9_28

The objective measurement of skin tone could bring benefits to a variety of real world applications: fashion, medicine, biometrics, surveillance systems, and the list can go on. Visagisme [5] is a relatively new concept in the field of fashion, eyewear, hairstyling, optic industry, etc., and its main purpose is to ensure the harmony between one's personality and appearance, with the aid of some tricks (color of the makeup, shape of the eyeglasses, hairstyle). The skin tone is an important factor in making these decisions. The skin color can also be used as a soft biometric trait. Although soft biometrics cannot be used to uniquely identify an individual, they are un-obtrusive, don't require any human cooperation and can be used to complement and increase the performance of traditional biometric systems. Finally, in the field of medicine, the skin tone can be used to quantify the UV radiation effects or skin lesions.

The rest of this work is structured as follows: in Sect. 2 we discuss the state of the art for skin color classification and in Sect. 3 we detail the proposed solution. The experimental results are discussed in Sect. 4. Section 5 presents the conclusions and directions for future work.

2 State of the Art

Color is a prominent and computation effective image representation feature, mainly due to its robustness towards geometrical transformations and partial occlusions. However, color classification is highly sensitive to illumination changes and to image capturing devices. In the case of skin color classification, the problem becomes more challenging as the skin tones are very similar to each other, and even trained human practitioners can label the same skin tone into different classes.

As the skin color is itself a subjective notion, not all the works classify the skin tones into the same number of classes. In [6] the skin tone is roughly differentiated into two classes: light and dark. The skin pixels are extracted from the face area by applying some threshold on the R, G, B channels. To classify the actual skin tone, several histogram distances between the test frame and two reference frames (one for light and one for black skin tone) are analyzed. This method achieves an accuracy of 87% on a subset of the Color FERET image database.

The task of skin detection has been extensively studied in the last decades. In [7], a detailed survey of skin detection methods, with emphasis on the color descriptor and skin modelling techniques is presented. Also, several skin color constancy and dynamic adaptation techniques that can improve the detection performance are discussed.

In [8] the skin color is classified into three classes: dark, brown and light, using 27 inference rules and fuzzy sets generated from the RGB values of each pixel. The method was trained and evaluated on images from the AR dataset and images from the Internet and it achieves a hit rate above 70%. The skin tone is classified into 16 color tones in [9, 10], but these methods imply that the user holds a color calibration pattern in each photo; this pattern is used for both color normalization and skin tone classification.

In [9] the skin pixels were selected from the face image by luminance filtering and a single Gaussian was used to model the distribution of each skin color sample. The classification was performed by choosing the label of the Gaussian with the smallest

Mahalanobis distance to the mean of the test subjects. In [10], the skin color classification was modeled so that it can handle more complex classification rules: a Gaussian Mixture Model (GMM) is fit to the face pixels' colors and the Kullback-Leibler (KL) distance is used to label the skin tone.

In this paper, we propose an automatic system that classifies the skin tone into three classes: dark, medium and light. The system is mainly intended for visagisme applications that detect the skin tone and suggests the make-up and the accessories that are most suited for the user. We argue that the 16 color tones model is unpractical and fails to bring useful information if not for the single reason that no human labeling will be good enough to accurately establish the ground truth. A model with six colors would probably be better as the six tones would closely match what we can distinguish visually amongst different regions and human races as predominant skin colors. However, some studies have shown [6] that natural, non-influenced classification of skin colors as performed by humans would contain only three classes: white/light, brown and black. Even with these three classes, the classification is subjective. Figure 1 shows some skin tone examples belonging to each skin tone class.

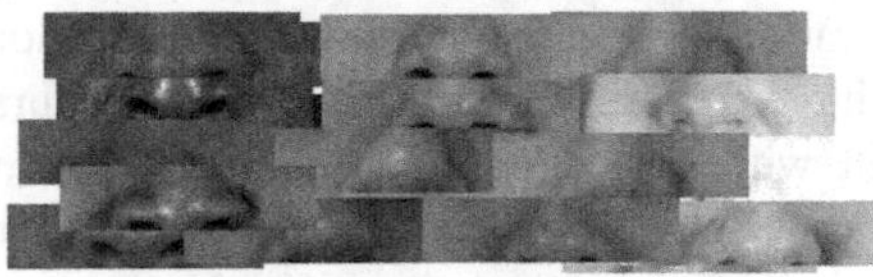

Fig. 1. Color samples belonging to each skin sample: starting from left or right: dark, medium and light skin tones

3 Proposed Solution

In this section, we describe in detail the proposed solutions for skin tone classification. There are three main steps that need to be addressed in the classification problem: determining the skin pixels from the face area, choosing the appropriate classification color descriptors and the design of the skin tone classifier. Based on the classification features, the proposed methods can be roughly classified as (1) pixel based (that use directly the pixel values for classification) and (2) patch based (that compute the color histograms of image regions and use these histograms to label the skin tone).

3.1 Region of Interest Selection

The first step that needs to be addressed is the selection of the skin pixels; the face contains several non-skin features (eyes, hair, eyeglasses) which could impact the classification result. We heuristically determined that the region below the eyes and right above the face center contains the fewest occlusions. Therefore, we first detect the face [11] in the input image and then we crop the face to a region of interest (ROI) defined by the rectangle $[0.2 \cdot w, 0.3 \cdot h, 0.6 \cdot w, 0.5 \cdot h]$, where w and h are the width and the height of the face region.

3.2 Color Descriptors

Color is a powerful image descriptor, and although it is invariant to geometrical transformations and partial occlusions, it has the main disadvantage of being highly dependent on the illumination conditions and the on the capturing devices. In [12] an extensive study of the invariance properties and the distinctiveness of color descriptors is presented. The invariance properties are derived from the diagonal model of illumination change, and several measures of invariance are defined: invariance to light intensity changes, light intensity shifts, light color changes and light color shifts.

The choice of the color space has a great impact on the color classification performance. Each color space describes and organizes the colors as tuples of numbers, such the colors are more easily distinguished and certain computations are more suitable. However, none of these color representations can be considered as a universal solution. In this paper, we have chosen to analyze three different color spaces: RGB, HSV and Lab. In the RBG color space, a color is expressed as the percent of blue, green and red components. The RGB histograms have no invariance properties. In the HSV color space the colors are encoded using a hexacone. The chromatic information is encoded by the Hue (H) and Saturation (S) components: hue represents the perceived dominant color and saturation is the relative purity of that color. The analysis shows that the certainty of the hue is inversely proportional to the saturation [12]. The Hue is scale and shift invariant with respect to light intensity. The value (V) component characterizes the brightness. Lab color space is a device independent color space and uses one channel for luminance and two chromaticity layers: a (for the green red axis) and b (for the blue yellow axis).

3.3 Pixel Based Classification

Distance to a Reference Frame

The straightforward approach to determine the skin tone is to compute, from the training data, a reference mean image for each skin tone class and label new examples based on the closest prototype. Figure 2 shows the mean images for each skin tone in the RGB color space.

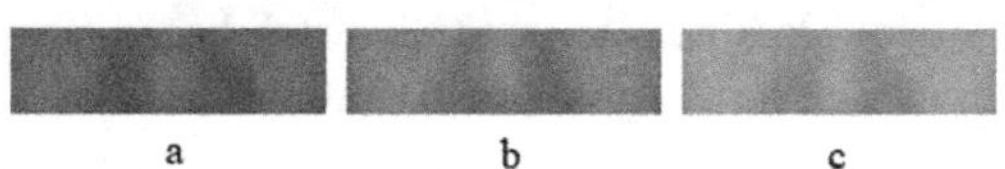

Fig. 2. Mean reference frames for each skin tone in the RGB color space: (a) dark, (b) medium, (c) light

All the images from the training data are scaled to a predefined size sz and the mean reference image is computed. In addition, the color components are normalized to the interval [0, 1] as in some color-spaces the channels have different magnitudes.

For the classification, the test image is preprocessed (scaled and normalized) and the Euclidian distance between the pixels from the test image and each reference frame is computed. The predicted skin label corresponds to the closest reference mean image:

$$pred = \underset{\substack{cls\in\{dark,\\ medium,\\ light\}}}{\arg\min} \sum_{i=0}^{rows-1} \sum_{j=0}^{cols-1} dist(I(i,j), \bar{I}_{cls}(i,j)),$$

where $I(i, j)$ is the pixel color from the test image at position (i, j) and $\bar{I}_{cls}(i, j)$ represents the color of the pixel (i, j) from the mean reference image of class *cls*. To compute the distance between two pixels we use the Euclidian distance.

$$d(p,q) = \sqrt{(p_0 - q_0)^2 + (p_1 - q_1)^2 + (p_2 - q_2)^2},$$

where p_i and q_i represent the ith color component of pixels p and q respectively.

Gaussian Mixture Models
Gaussian mixture models are a way of representing probabilistic density functions as a linear combination of several generative Gaussian models, as described in equation:

$$g(x \mid \mu, \Sigma) = \sum_{i=1}^{N} \omega_i \mathrm{N}(x \mid \mu_i, \Sigma s_i)$$

where x is a d-dimensional data vector, ω_i are the mixture weights and $\mathrm{N}(x \mid \mu_i, \Sigma_i)$ are the Gaussian densities. The GMM parameters are estimated from the training data using the Expectation-Maximization algorithm (EM).

Although GMM is an unsupervised learning algorithm, as we have the class labels for the training data, we initialize the parameters of the EM algorithm in a supervised manner, with the means of the classes from the training set.

The feature vector for the GMM classification is a 3-dimensional vector corresponding to the color components of a pixel. We compute the centroids for the pixels belonging to each skin tone class and the EM algorithm is initialized with these centroids. For the prediction step, each pixel from the selected region of interest is analyzed and the predicted skin tone with the highest frequency is the selected as final skin tone.

3.4 Patch Based Classification

For the patch based classification, the input ROI is split into several cells of size $(w \times w)$. We compute the color histogram of every cell and use it as a feature vector for the classification. We proposed two classification methods: the first method simply compares the histogram of each region with the corresponding histogram from the mean reference frame (Sect. 3.3). The second method uses a support vector machine classifier to determine the skin tone in the region based on the histograms.

Finally, the skin tone classification is computed by majority voting: for each cell position, the classifier gives a skin label and the final skin tone is selected as the class that occurs most often.

Figure 3 depicts the outline of the patch based classification methods.

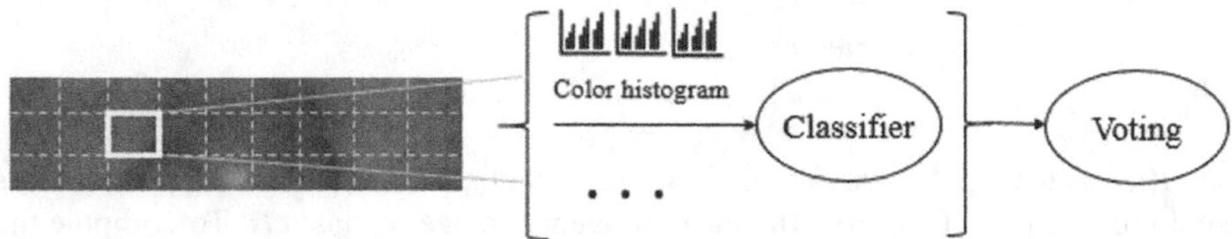

Fig. 3. Solution outline for patch based skin color classification

Histogram Differences

Determining the distance or similarity between two histograms is an important issue in the field of image processing and several measures for computing histogram distances have been proposed [13]. We analyze two types of differences: the correlation coefficient and the Bhattacharyya distance.

The correlation coefficient between two histograms is defined by as:

$$d(H_1, H_2) = \frac{\sum_I \left(H_1(I) - \overline{H_1}\right)\left(H_2(I) - \overline{H_2}\right)}{\sqrt{\sum_I \left(H_1(I) - \overline{H_1}\right)^2 \sum_I \left(H_2(I) - \overline{H_2}\right)^2}},$$

where H_1 and H_2 are the histograms under comparison.

The Bhattacharyya distance, takes values between 0 and 1, and it is a measure of the divergence between two histograms; it is formally defined as:

$$d(H_1, H_2) = \sqrt{1 - \frac{1}{\sqrt{\overline{H_1} \cdot \overline{H_2} \cdot N^2}} \sum_I \sqrt{H_1(I) H_2(I)}}$$

Support Vector Machine Classification

Support vector machine (SVM) classifiers are supervised learning algorithms that represent the training points in space, such that the examples from each class are separated by a margin as wide as possible. Test examples are transformed into the same space and assigned to the class based on which side of the separation margin they reside.

For the skin tone classification, the input vector for the SVM is the color histogram of a cell. A preprocessing step is applied: the input features are scaled such that each feature from the training set has zero mean and unit variance.

The histograms computed from each cell from the region of interest are fed to the SVM classifier and a skin tone is predicted for each cell. The final classification score is computed by majority voting on the color predictions of all the cells.

4 Experimental Results and Discussions

Data gathering is an important step of machine learning, as the training data determines what the classifier learns to recognize before being applied to unseen images. The training dataset was created by merging images from four publicly available databases [14–17]; the databases contain images captured in both controlled and natural environments. The 1999 Caltech Faces database [14] contains 450 frontal face images of 27 unique individuals captured in both indoor and outdoor scenarios. The Chicago Face Database [15] contains high-resolution, standardized images of 158 participants with ages between 18 and 40 years and extensive data about the subjects: race (Asia, Black, Hispanic/Latino), gender, facial attributes (feminine, attractive, baby-faced etc.). All the images were captured in controlled scenarios. The Minear-Park [16] face database contains frontal facial images from 575 individuals, with ages ranging from 18 to 93. Finally, the Brazilian Face Database [17] contains images from 100 male and 100 female individuals, with ages between 19 and 40 years. The original images from the database, and not the manually aligned subset is used. All the selected datasets contain frontal images, as the proposed solution is mainly envisioned for a visagisme application, where the user uploads a frontal image and various accessories and make-up suggestions are automatically provided based on its face appearance. There is no need for alignment of the images, as in the near-frontal case, the selected region of interest contains only skin pixels.

The training set was also enlarged by data augmentation in order to avoid overfitting. Each image from the dataset was subject to one of the following random transformations: contrast stretching, brightness enhancement and horizontal flips (to handle the cases when the light source is positioned sideway). We extracted subsets of each of the four databases such that the distribution of the skin tone classes is approximately equal. Each image was labeled by three independent labelers and the ground truth was established by merging their annotation results. Prior to the labeling process, we asked a visagist expert to point out several relevant images for each skin tone, and the labelers were instructed to classify the images based on their distance to the reference images. The final training dataset (after augmentation) consist of approximately 4000 images.

The classifier was evaluated on 200 images that were not used in the training process. Tables 1, 2 and 3 show the classification performance for each of the proposed methods. We evaluated the performances of the proposed solution using: RGB, HSV and Lab color spaces as color features.

Table 1. Pixel-based classification results

Colour space	Distance to reference frame			Gaussian mixture models		
	Precision	Recall	Accuracy	Precision	Recall	Accuracy
RGB	0.75	0.70	0.705	0.68	0.55	0.55
HSV	0.76	0.73	0.735	0.82	0.76	**0.76**
Lab	0.81	0.80	**0.80**	0.51	0.69	0.68

Table 2. Patch based classification results

Colour space	Hist. diff. correlation			Hist. diff. Bhattacharyya		
	Precision	Recall	Accuracy	Precision	Recall	Accuracy
RGB	0.62	0.60	0.605	0.75	0.69	0.69
HSV	0.55	0.45	0.55	0.66	0.68	0.68
Lab	0.73	0.72	**0.72**	0.74	0.69	**0.695**

Table 3. SVM based classification results

Color space	SVM based classification		
	Precision	Recall	Accuracy
RGB	0.88	0.84	0.8358
HSV	0.89	0.86	0.8557
Lab	0.89	0.87	**0.8706**

The best results for the pixel based comparison with the three mean reference images are obtained using the Lab color space (80% accuracy). The pixel based Gaussian Mixture Model classification achieves 76% accuracy for the HSV color space.

The classification based on the comparison of cell histograms with the corresponding histograms from the reference mean images for each class obtained the lowest classification accuracies. The best accuracy for this method (72%) is achieved using the Lab color space and the correlation metric between histograms.

Finally, the best results are achieved by the SVM classification (Table 3).

Table 4 shows the confusion matrix for the SVM based classification on the Lab color space. One can notice that the majority of "confusions" occurred between medium-light skin tones and dark-medium skin tones. This behavior is very similar to what we observed in the annotations of the ground truth by the three human labelers.

Table 4. Confusion matrix

Actual value	Predicted value			Accuracy
	Dark	Light	Medium	
Dark	51	1	11	80.95
Light	0	77	7	91.66
Medium	5	1	47	88.67

From the testing set, 25 samples were misclassified by the proposed solution and for 6 of these misclassified images there were also disagreements between the independent labelers (on 4 images the confusion dark-medium occurred and on 2 images the confusion medium-dark occurred). On these cases it was not possible to get an agreement between labelers on the real color class.

Some classification and miss-classification results are depicted in Fig. 4.

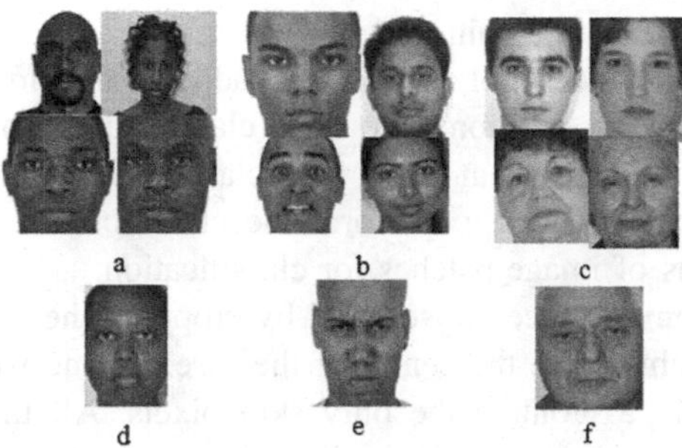

Fig. 4. Some classification results: (a) dark classified as dark, (b) medium classified as medium, (c) light classified as light, (d) dark classified as medium, (e) medium classified as dark, (f) light classified as medium

Next, we compare the proposed solution with other state of the art works that tackle the problem of skin tone classification. Table 5 compares the accuracy of the proposed method with the accuracy of other works from the specialized literature.

Table 5. Comparison with other state of the art works

Method	Skin tones	Accuracy
[6]	2	87%
[8]	3	70%
[9, 10]	16	–
Proposed solution	**3**	**87.06%**

As skin tone classification is a subjective task, not all the proposed methods classify the skin tones at the same granularity level. In [6] the skin tone is roughly distinguished into black and light and the obtained classification accuracy is 87%. In [8], the authors use fuzzy sets and 27 inference rules to classify the skin tone into the same skin tones as proposed in this paper. They obtain a hit rate above 70% for the best system setup. The test and training dataset of [8] consists of 200 images from the AR dataset [18] and images downloaded from the internet and the images and their annotations are not publicly available, therefore the proposed solution could not be tested on the same setup. Finally, in [9, 10] the skin tone is classified into 16 color classes and implies that the subject holds a color calibration target. This target arranges the primary and secondary colors and 16 patches representative of the range of human skin color into a known pattern, and it is used for both camera calibration and skin tone classification.

Our method does not impose any restrictions of the image capturing scenario and attains an accuracy rate of 87.06%.

5 Conclusions and Future Work

This paper addressed the issue of skin tone classification from facial images, a subject with applications in different domains, such as visagisme, soft biometry, surveillance systems etc. Our method does not require any additional color patterns or camera calibration and classifies the skin tone into three classes: dark, medium and light. We propose and analyze the performance of four classification methods: the first two method use the raw pixel values to perform the classification, while the latter two examine the histograms of image patches for classification.

The skin pixels from the face are selected by cropping the face image to the area below the eyes and right above the center of the face, a zone we heuristically determined it is more likely to contain the only skin pixels. All the proposed methods examine this ROI to classify the skin tone. The first method first computes the mean reference images for each skin tone; next the classification is simply performed based on the distance between the test image and each reference mean image. The second method uses Gaussian Mixture Models to determine skin tone. The last two methods split the face into cells and use histograms of these cells to perform the classification: the third method computes the distance between each cell from the test image with the corresponding cell from the mean reference image of each skin tone class. Finally, the last method uses a SVM classifier to classify the histogram of each region, and the skin tone is computed by majority voting. The best performance is achieved by the SVM classifier using the Lab color space.

The system was trained and tested on images from four publicly available face databases and on images captured in our laboratory. An accuracy of 87.06% and the extensive experiments we performed demonstrate the effectiveness of the proposed solution.

As a future work, we plan to use a more complex method to determine the facial skin pixels, and to combine the proposed method with the classification of other facial features, such as the eyes color and shape, the hair color etc.

References

1. Vezhnevets, V., Sazonov, V., Andreeva, A.: A survey on pixel-based skin color detection techniques. In: Proceedings Graphicon 2003, pp. 85–92 (2003)
2. von Luschan, B.: Beitrage zur Volkerkunde der deutschen Schutzgebiete. D. Reimer, Berlin (1897)
3. Thibodeau, E.A., D'Ambrosio, J.A.: Measurement of lip and skin pigmentation using reflectance spectrophotometry. Eur. J. Oral Sci. **105**(4), 373–375 (1997)
4. Fitzpatrick, T.B.: The validity and practicality of sun-reactive skin types I through VI. Arch. Dermatol. **124**(6), 869–871 (1988)
5. Juillard, C.: Brochure Méthode C. JUILLARD 2016. http://www.visagisme.net/Brochure-Methode-C-JUILLARD-2016.html. Accessed 13 April 2017
6. Jmal, M., Souidene, W., Attia, R., Youssef, A.: Classification of human skin color and its application to face recognition. In: The Sixth International Conferences on Advances in Multimedia, MMEDIA 2014 (2014)

7. Praveen, K., Makrogiannis, S., Bourbakis, N.: A survey of skin-color modeling and detection methods. Pattern Recogn. **40**(3), 1106–1122 (2007)
8. Boaventura, I.A.G., Volpe, V.M., da Silva, I.N., Gonzaga, A.: Fuzzy classification of human skin color in color images. In: IEEE International Conference on Systems, Man and Cybernetics, SMC 2006, vol. 6, pp. 5071– 5075 (2006)
9. Harville, M., Baker, H.H., Bhatti, N., Susstrunk, S.: Consistent image-based measurement and classification of skin color. In: Proceedings of the 2005 International Conference on Image Processing, ICIP 2005, Genoa, Italy, 11–14 September 2005, pp. 374–377 (2005)
10. Yoon, S., Harville, M., Baker, H., Bhatii, N.: Automatic skin pixel selection and skin color classification. In: 2006 IEEE International Conference on Image Processing, pp. 941–944. IEEE (2006)
11. Viola, P., Jones, M.: Rapid object detection using a boosted cascade of simple features. In: Computer Vision and Pattern Recognition (2001)
12. Van De Sande, K., Gevers, T., Snoek, C.: Evaluating color descriptors for object and scene recognition. IEEE Trans. Pattern Anal. Mach. Intell. **32**(9), 1582–1596 (2010)
13. Cha, S.H., Srihari, S.N.: On measuring the distance between histograms. Pattern Recogn. **35** (6), 1355–1370 (2002)
14. Caltech Faces. http://www.vision.caltech.edu. Image Datasets/faces, 1999
15. Ma, D.S., Correll, J., Wittenbrink, B.: The chicago face database: a free stimulus set of faces and norming data. Behav. Res. Methods **47**(4), 1122–1135 (2015)
16. Minear, M., Park, D.C.: A lifespan database of adult facial stimuli. Behav. Res. Methods Instrum. Comput. **36**(4), 630–633 (2004)
17. Thomaz, C.E., Giraldi, G.A.: A new ranking method for principal components analysis and its application to face image analysis. Image Vis. Comput. **28**(6), 902–913 (2010)
18. Martinez, A.M., Benavente, R.: The AR face database, CVC Technical Report #24 (1998)

Towards Detecting High-Uptake Lesions from Lung CT Scans Using Deep Learning

Krzysztof Pawełczyk[1,2], Michal Kawulok[1,2(✉)], Jakub Nalepa[1,2(✉)], Michael P. Hayball[1,3], Sarah J. McQuaid[4], Vineet Prakash[5], and Balaji Ganeshan[3,6]

[1] Future Processing, Gliwice, Poland
{krzysztof.pawelczyk,michal.kawulok,jakub.nalepa}@polsl.pl
[2] Silesian University of Technology, Gliwice, Poland
[3] Feedback PLC, London, UK
[4] Royal Surrey County Hospital, Guildford, UK
[5] Ashford and St. Peter's Hospitals, Ashford, UK
[6] University College London, London, UK

Abstract. Automatic detection of lung lesions from computed tomography (CT) and positron emission tomography (PET) is an important task in lung cancer diagnosis. While CT scans make it possible to retrieve structural information, PET images reveal the functional aspects of the tissue, hence combined PET/CT imagery allows for detecting metabolically active lesions. In this paper, we explore how to exploit deep convolutional neural networks to identify the active tumour tissue exclusively from CT scans, which, to the best of our knowledge, has not been attempted yet. Our experimental results are very encouraging and they clearly indicate the possibility of detecting lesions with high glucose uptake, which could increase the utility of CT in lung cancer diagnosis.

Keywords: PET/CT imaging · Lesion detection · Deep neural networks

1 Introduction

Cancer is one of the main causes of death worldwide, with 1.69 out of 8.8 million deaths caused by lung cancer in 2015. Therefore, improving lung cancer diagnostics from medical images is crucial, either as a part of the screening process or at a later stage to assess the effectiveness of treatment. An important, yet challenging, task is to differentiate between benign and malignant lesions. Some of the useful features concerned with the structure, shape and boundary smoothness can be observed relying on *structural imaging* (such as computed tomography—CT or magnetic resonance imaging—MRI). However, in many cases the physiological activity of the tissue must be captured to make this differentiation, which can be achieved with *functional imaging*, such as positron emission tomography (PET), functional MRI or dynamic contrast-enhanced imaging.

PET imaging allows for measuring the glucose uptake, which indicates the metabolism of the tissue to identify abnormally active lesions. In lung cancer

S. Battiato et al. (Eds.): ICIAP 2017, Part II, LNCS 10485, pp. 310–320, 2017.
https://doi.org/10.1007/978-3-319-68548-9_29

diagnostics, measuring the lesion's activity plays a key role in differentiating between benign and malignant tumours or nodules. This provides functional information, however as PET images are of poor spatial resolution and they do not reveal much of the anatomical details, they are usually complemented with the co-registered CT scans. This is necessary, as the high-uptake regions (*hot spots* in PET) include many false positives (FPs) which can be easily verified based on the anatomy and structure (e.g., the heart is usually a hot spot).

1.1 Contribution

Taking into account that different type of information is acquired with CT and PET imagery, these modalities are often fused to improve the diagnosis. In the research reported here, we explore the possibility of detecting the high-uptake lesions exclusively from CT scans. Identifying active tumour tissue from CT would be very useful as this is an important indicator of disease severity and response. To the best of our knowledge, there is no reported work on this problem and PET images are considered indispensable here.

Our contribution consists in proposing a deep convolutional neural network (CNN) to detect and segment the high-uptake lesions from CT scans. Deep neural networks (DNNs) [14] have been already successfully applied to solve a number of computer vision problems, including medical imaging challenges [15], and in many cases they reach beyond human performance. We validated the outcome obtained from CT scans using our CNN against the annotations performed by professional radiologists, asked to locate the high-uptake regions based on CT and PET data. Although the detection scores are below the scores obtained using the combined CT and PET modality, they are highly encouraging and they suggest that DNNs are capable of identifying high-uptake lesions from structural images. Finally, we compared the detection results of the CNN with those reported by our recent PET/CT lung lesion detection algorithm (LUNGCX) [17], which benefits from the information extracted from both modalities.

1.2 Paper Structure

This paper is structured as follows. Section 2 reviews the literature. The proposed CNN is described in Sect. 3 and the obtained experimental results are reported and discussed in Sect. 4. The paper is concluded in Sect. 5.

2 Related Literature

Detecting lung lesions (including nodules and tumours) is a deeply investigated problem of medical image analysis and there are numerous methods which operate from different image modalities. Here, we briefly outline the state of the art on detecting and segmenting lung lesions relying on CT and PET/CT images.

The two general tasks concerned with analysis of lung CT scans are: (i) detection of lung nodules aimed at early diagnosis of lung cancer and (ii) segmentation of the lesion region [11], helpful in differentiating between benign and malignant lesions, computer-aided surgery or planning radiation treatment. Clinical aspects of detecting lung nodules from CT scans are thoroughly discussed in a recently published survey [1]. Sensitivity of state-of-the-art methods vary significantly, spanning between 80% [22] and 98% [5] at several FPs per whole CT scan. This allows for improving the performance of inexperienced radiologists, and in some cases increases the detection sensitivity when used by experienced radiologists.

Most of the existing lung nodule detectors determine a set of candidates, i.e., the regions of dense tissue inside the lungs, which are further verified to filter out the FPs. In [5], the nodule candidates are extracted following a number of simple rules and preliminarily verified in each 2D slice. Subsequently, the candidates are combined in 3D to extract textural features, and classified using support vector machines (SVMs). A similar approach, employing shape descriptors, was proposed in [20]—all of the nodules whose size is at least 10 mm were reported to be correctly detected at 4 FPs per scan. Recently, a deep CNN with 5 convolutional and 3 max pooling layers was applied to detect lung nodules [6]. The reported sensitivity is 78.9% at 20 FPs per scan without using any FP reduction.

The differential diagnosis of malignant from benign lesions is difficult from CT scans, if a tumour is built of soft tissue without calcifications, as the metabolic information is not known to be manifested in CT scans. PET imaging allows for measuring the concentration of biologically active molecules (usually fluorodeoxyglucose, FDG) marked with positron-emitting isotopes. Hence, the high-uptake regions can be identified to indicate the tissue of high metabolism, which is a well-known marker in cancer diagnosis. The PET intensities are converted into standardised uptake values [3] and verified taking into account the anatomy, extracted from CT scans [7,17] or MRI [19]. The malignant lesions usually are seen as hot spots in PET due to increased metabolism of a tumour, but the hot spots do also appear in the healthy tissue (e.g., in the heart). In [7], the entire-body CT scan is divided into several sections using hidden Markov model, which subsequently makes it possible to classify each hot spot as normal or abnormal.

Not only are the CT scans used to verify the hot spots based on human body atlas, but they also allow for increasing the precision of delineating the lesions. The hot spots extracted from PET are treated as seeds for image segmentation performed with numerous techniques, including graph cuts [2], Markov random fields [8] or random walks [10]. The information extracted from PET may also be used during segmentation [25]. In [4], local maxima and saddle points are detected in a PET image to create a spatial-topological distance map, from which the tumours are segmented. In [23], the nodules are detected independently in both modalities—as hot spots in PET and using active contour filters in CT scans, and then a CNN with 3 convolutional layers is applied to extract the features of each candidate, which are classified with an SVM. Adding the CNN-based verification allowed for reducing the number of FPs from 72.8 to 4.9 per case, while the sensitivity dropped from 97.2% to 90.1%.

Deep CNNs have been successfully used for detecting and segmenting lung lesions both from CT and PET/CT modalities [6,23], but we have not encountered any reported attempts to bridge the gap between the results obtained from CT scans alone and from the combined PET/CT modalities. Since the deep networks allow for reaching beyond human performance in certain computer vision tasks, they may be helpful in detecting potentially high FDG uptake lesions from CT scans, improving the utility of CT in lung cancer diagnosis.

3 Detecting High-Uptake Lesions Using a CNN

Our algorithm for detecting active lesions is outlined in Fig. 1. At first, the 16-bit pixel values are normalised to the range of $\langle 0, 1 \rangle$ and the CT scan is split into 3D patches of size $5 \times 75 \times 75$ (2D patches of size 75×75 pixels are retrieved from 5 subsequent slices, as shown in Fig. 2). Such patches are used to train our deep CNN and afterwards each patch is classified by the trained network as *lesion* or *background*. For each patch, the CNN returns two responses (r_l and r_b) that express the similarity of the patch central pixel to the lesion and background classes, respectively. From these responses, the lesion similarity map is assembled to determine the lesion candidates, which are subject to two verification steps, based on (i) the maximal similarity within a blob and (ii) the blob's area.

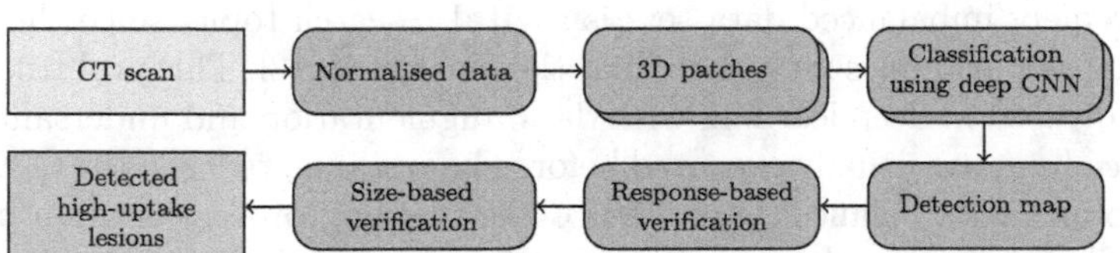

Fig. 1. Flowchart of the proposed DNN-based high-uptake lesions detection.

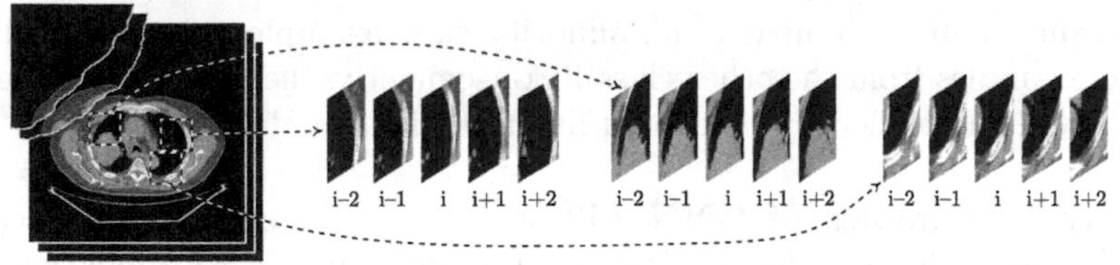

Fig. 2. Examples of 3D patches extracted for the *i*-th slice.

Network Architecture. The proposed network, whose architecture is presented in Fig. 3, is composed of two convolutional layers followed by three classical fully connected layers. According to [21], for small input images the pooling may be skipped to increase the performance, therefore there are no pooling layers in our CNN. Output of every hidden layer is adjusted by a rectified linear

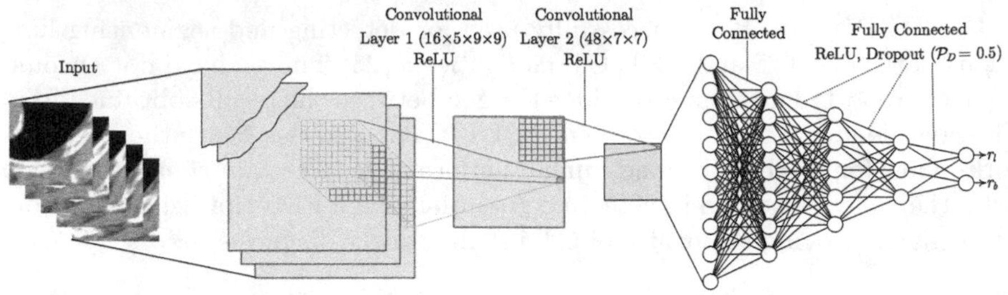

Fig. 3. Architecture of the proposed deep CNN.

unit (ReLU) [16] with an activation function $f(x) = \max(0, x)$. The main task of the first convolutional layer is to extract the low-level features from adjacent slices using a bank of 16 filters of the size $5 \times 9 \times 9$ pixels, applied with a stride of 2 pixels (the size of each patch is reduced from $5 \times 75 \times 75$ to 34×34). 3D patches are transformed into 2D ones of reduced dimensionality here. The second convolutional layer with the bank of 48 kernels of size 7×7 with a stride of 3 extracts higher-order features. The fully connected layers are a classical neural network with dropout (its probability is $\mathcal{P}_D$) which classifies the input vector.

Handling Extremely Imbalanced Data Sets. Training classification engines from extremely imbalanced data sets is a vital research topic, since the skewed distributions of examples may easily bias the classifiers [13]. This problem is commonly addressed in deep learning with data augmentation and undersampling—these procedures are usually executed before the training, to learn the CNN using a new, potentially "balanced" set. Hence the training process is often unaware of the underlying data characteristics and cannot adapt to retrieve a better-performing model. This was thoroughly discussed in a very recent work [24].

In our approach, we dynamically balance the data batches (each contains N samples) during the CNN training. After every $\mathcal{I}$ epochs of the CNN training, the classification accuracy η is quantified for each class separately (the validation set is balanced, and encompasses all minority-class examples, along with undersampled examples from the other class). Consequently, the lower η retrieved for a given class, the higher probability of including its samples in the batch.

Illustrative Examples of CNN Filters. Several examples of the learned filters (corresponding to the architecture presented in Fig. 3), along with the filtered images, are presented in Fig. 4 (the first layer responses are visualised before applying ReLU, while those for the second layer—after ReLU). For the first layer, each filter is composed of five 2D filters convolved with subsequent CT slices. It can be seen that these filters do not resemble the wavelets often reported in the literature, and they are rather "noisy"—possibly, some textural features are extracted in this way. Interestingly, the filters at the second layer are smoother and they are focused on extracting higher-order features, as expected. The extracted features in the presented example allow for correct segmentation

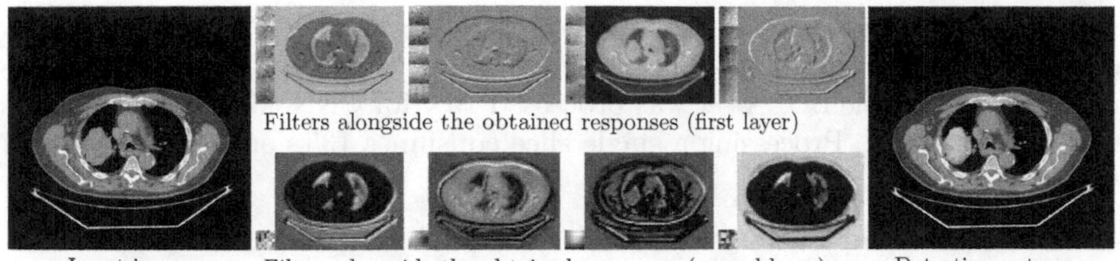

Fig. 4. Examples of filtered images obtained at the first and second convolutional layer. (Color figure online)

of an active lesion (yellow region in the detection outcome indicates true positives, while the blue—false negatives). We have observed that nearly half of the learned filters are uniformly gray when visualised, hence they average the signal—presence of such "dead" filters may mean that there are too many of them in the layer or that they should be of smaller dimensions [27]. Addressing that problem may improve our approach in the future.

Analysis of the Network Response. From the pixel-wise responses r_l and r_b, we compute the lesion similarity ($\mathcal{R} = r_l - r_b$) to create the lesion detection map. The pixels with non-negative similarities ($\mathcal{R} > 0$) are grouped spatially and each consistent region is considered a lesion candidate. These candidates are verified using two thresholds $\mathcal{T}_R$ and $\mathcal{T}_S$, imposed on the blob's maximum similarity ($\mathcal{R}_{max}$) and its area in pixels ($\mathcal{S}$), respectively. If $\mathcal{R}_{max} > \mathcal{T}_R$ and $\mathcal{S} > \mathcal{T}_S$, then the blob is labeled as the detected lesion.

4 Experimental Results

4.1 Experimental Setup

We validated our algorithm using two data sets, namely (i) our set with 90 CT scans of different patients (this includes the LUNGCX subset—44 scans used in [17]) and (ii) the LOLA set[1] with 55 CT scans without active lesions. For every study in our set (with active and non-active lesions), a single slice presenting the largest section of an active lesion was selected and manually segmented by an experienced radiologist based on both PET and CT (our algorithm *does not* exploit PET images). This set is extremely imbalanced—there are $7.3 \cdot 10^4$ pixels of active lesions, and $2.4 \cdot 10^7$ pixels of other tissues and background.

The algorithms were implemented in C++ with the Caffe framework [9] and validated on a computer with an Intel Xeon E5-2698 v3 processor (40M Cache, 2.30 GHz) with 128 GB RAM and NVIDIA Tesla K80 GPU 24 GB DDR5. The CNN internals were tuned to $\mathcal{P}_D = 0.5$, $N = 256$ and $\mathcal{I} = 500$, and we use the ADAM optimizer [12]. As there are no reported attempts to detect high-uptake

[1] LOLA set is available at https://lola11.grand-challenge.org.

lesions from CT, we compare our method with LUNGCX [17], exploiting the combined PET/CT modality. The reported results were obtained with 10-fold cross-validation. The LOLA set was processed 10× for CNN trained within each fold using our data. Processing a single slice consumes 135 s on average.

4.2 Analysis and Discussion

Quantitative Analysis. Table 1 presents the obtained detection scores. For our data set, we report precision and recall for the training and test sets (including the LUNGCX subset), averaged over 10 folds (entire slices are segmented). As the LOLA scans do not include active lesions, we only report the FP rate (i.e., the percentage of images with FP lesions). We show the scores obtained (i) without any verification ($\mathcal{R} > 0$), (ii) after response-based verification, (iii) with size-based verification, and (iv) after full verification. Clearly, the verification is critical, as it drastically decreases the FPs for LOLA (from 90.14% to 6.6%) and significantly improves the precision. The F-score differences are statistically important for all data sets at $p < .01$ (two-tailed Wilcoxon test). We also report the recall for four ranges of lesion size. Although the verification decreases the recall, mainly small lesions are affected—while this is obvious for size-based verification, the smaller active lesions also render lower $\mathcal{R}$, hence their vulnerability to the response-based cutoff.

In Figs. 5 and 6, we show the precision-recall curves obtained by varying the thresholds $\mathcal{T}_R$ and $\mathcal{T}_S$. The curves for the training set differ much across the folds, so increasing the amount of training data could be beneficial. For the test set, we apply either single or both thresholds (for each case only one threshold is being changed, while the other remains fixed), hence two curves for each case. The fixed threshold values were found independently for each fold, so as to obtain equal precision and recall for the training set, and these values were also applied to obtain the scores reported earlier in Table 1. While applying $\mathcal{T}_R$ after $\mathcal{T}_S$ improves the results (Fig. 6), the latter does not render any improvement, if

Table 1. Precision and recall obtained for our data set and FP rate for the LOLA set.

Verif	Precision	Recall	Recall for lesions of size (in mm)				W/out active FP rate (%)
			<15	(15, 25)	(25, 40)	>40	
Training set:							
None	.278 ± .102	1 ± .000	1 ± .000	1 ± .000	1 ± .000	1 ± .000	—
Resp.	.905 ± .052	.911 ± .050	.747 ± .115	.956 ± .053	.971 ± .060	1 ± .000	—
Size	.814 ± .060	.840 ± .074	.442 ± .242	.984 ± .034	1 ± .000	1 ± .000	—
Full	.925 ± .038	.808 ± .078	.379 ± .213	.951 ± .065	.971 ± .060	1 ± .000	—
Test set (full):							
None	.187 ± .072	.870 ± .064	.574 ± .259	.917 ± .171	1 ± .000	.981 ± .052	90.14 ± 8.14
Resp.	.741 ± .192	.610 ± .170	.269 ± .232	.758 ± .342	.622 ± .171	.867 ± .180	20.42 ± 9.68
Size	.589 ± .130	.680 ± .166	.315 ± .196	.758 ± .259	.811 ± .241	.944 ± .111	26.04 ± 9.11
Full	.789 ± .179	.570 ± .179	.176 ± .173	.708 ± .340	.622 ± .171	.852 ± .183	6.60 ± 3.52
Test set (LUNGCX subset):							
None	.261 ± .099	1 ± .000	—	1 ± .000	1 ± .000	1 ± .000	90.14 ± 8.14
Resp.	.852 ± .135	.677 ± .272	—	.700 ± .400	.593 ± .284	.852 ± .319	20.42 ± 9.68
Size	.779 ± .160	.872 ± .137	—	.700 ± .400	.824 ± .210	1 ± .000	26.04 ± 9.11
Full	.935 ± .134	.677 ± .272	—	.700 ± .400	.593 ± .284	.852 ± .319	6.60 ± 3.52

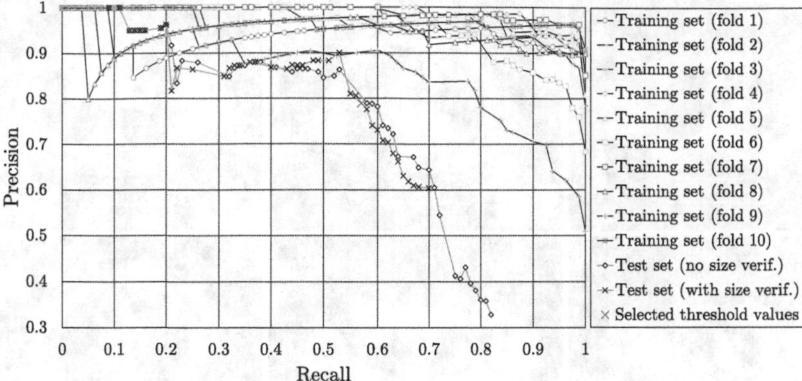

Fig. 5. Precision-recall curves obtained by varying the response threshold $\mathcal{T}_R$.

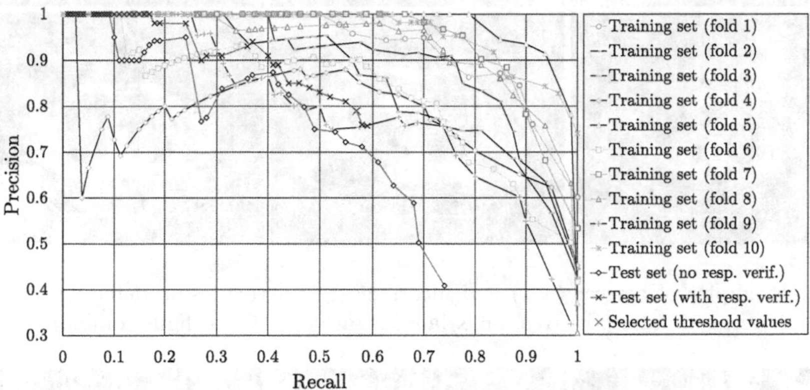

Fig. 6. Precision-recall curves obtained by varying the size threshold $\mathcal{T}_S$.

the former is applied (Fig. 5). Overall, we use $\mathcal{T}_R$ and $\mathcal{T}_S$, as we observed (from LOLA) that $\mathcal{T}_S$ is quite effective in reducing FPs for lungs without active lesions.

Qualitative Analysis. Figure 7 shows several examples of correct (Fig. 7a–e) and incorrect (Fig. 7f–i) detection. A very interesting example is presented in Fig. 7c—although the lesion is not very dense, our CNN identified it correctly and managed to differentiate it from another non-active lesion, which is present in the image. In Fig. 7f, the lesion was not detected and there has been one FP region found—naturally, we consider such cases as detection errors. Figure 7g and (h) show the outcome before and after the verification—several FPs were eliminated from (g), but the correctly detected lesion in (g) was also rejected.

Comparison with LUNGCX. We compared the proposed CNN classifier with our active lesion detection algorithm (LUNGCX) which operates on both PET and CT modalities [17]. In LUNGCX, the co-registered CT and PET series are identified at first, and the lungs (base and apex) are located from CT in the pixel-intensity histogram analysis. Then, for each lung-containing slice, we

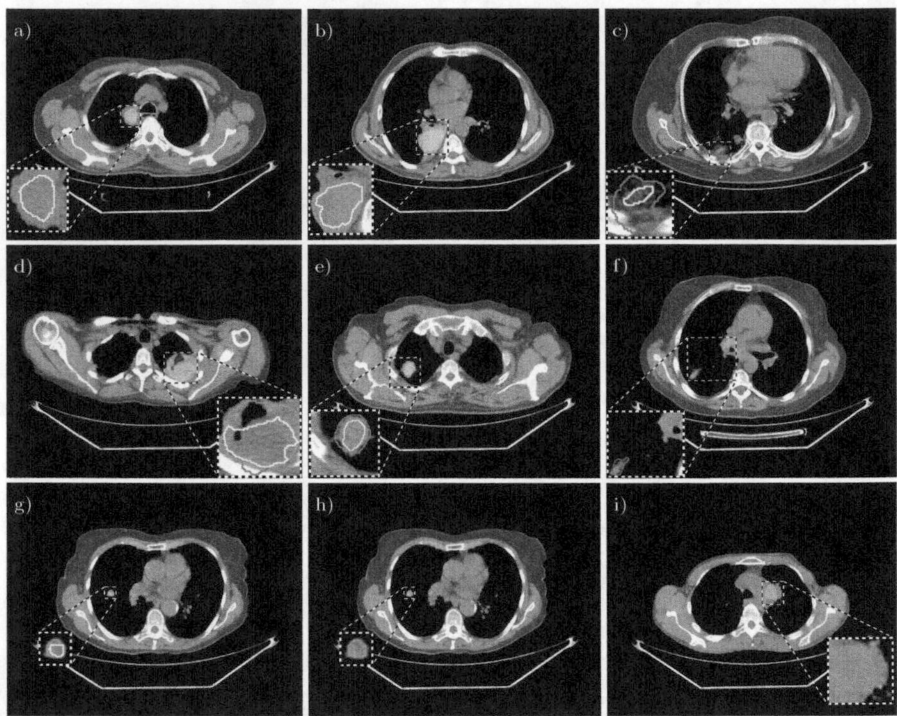

Fig. 7. Examples of correct (a–e) and incorrect (f–i) active lesion detection (yellow: true positives, red: false positives, blue: false negatives). (Color figure online)

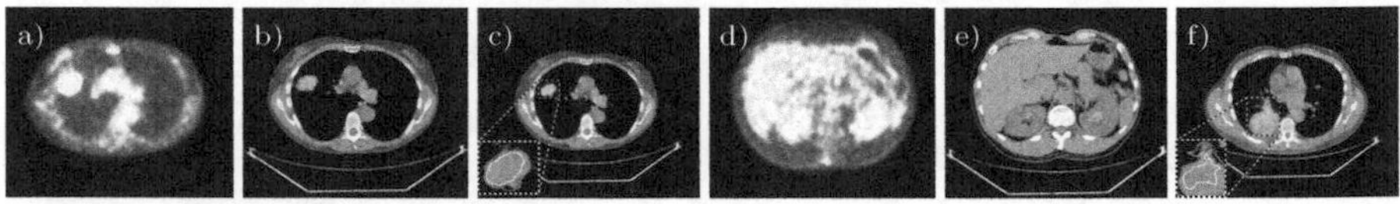

Fig. 8. Examples of active lesion detection from PET/CT with LUNGCX (a, b, d, e) and from CT using the proposed algorithm (c, f).

identify the lung tissue using thresholding, alongside its convex hull, as tumours may be associated with the lung wall or mediastinum. The active lesions (located only within the convex hulls of lungs) are extracted from the PET images [26].

Although the LUNGCX algorithm successfully identified active lesions in 41 (out of 44 patients in the LUNGCX subset) cases (93%), the example visualisations gathered in Fig. 8 show that the most avid PET regions are very patient-specific. In Fig. 8, we render the example of the problematic uptake in which the most active regions were found in the kidneys (Fig. 8d–e). Here, our CNN detected the active lesion correctly (Fig. 8f) and did not report any lesions for these high-uptake kidneys. The numerical results (Table 1) reveal that our CNN can produce results comparable with LUNGCX, and the analysis of the network response (see the Size variant) greatly improves the CNN recall measures.

5 Conclusions and Outlook

In this paper, we reported our attempt to employ deep learning for detecting high-uptake lesions from CT images, which is a challenging task, as it requires extraction of functional information from structural imaging. The obtained detection scores, though worse than those retrieved from the combined PET/CT [17], are encouraging and they indicate that our algorithm correctly differentiates between active and non-active lesions. It is clear from the experimental results that methods utilising deep CNNs can increase the CT diagnostic capacity.

Our ongoing research is aimed at improving the visualisation aspects to better understand which features are learned by the CNN and which image regions activate them, alongside comparing our method with other state-of-the-art techniques on larger data sets. Furthermore, we intend to focus on detecting small active lesions, being an important clinical goal, especially from low-dose CT. We are on the way to employ the existing methods for CT-based lesion detection (also exploiting the anatomical information) to pre-process the scans and narrow down the search in the pulmonary region. Also, we work on applying incrementally increased CNN architectures in our framework [18]. Overall, while the proposed method could be improved on many ways, it is an important step towards retrieving information on lesion activity from CT.

Acknowledgments. This work was supported by the National Centre for Research and Development under the grant: POIR.01.02.00-00-0030/15.

References

1. Al Mohammad, B., Brennan, P., Mello-Thoms, C.: A review of lung cancer screening and the role of computer-aided detection. Clin. Radiol. **72**, 433–442 (2017)
2. Ballangan, C., Wang, X., Fulham, M., Eberl, S., Feng, D.D.: Lung tumor segmentation in PET images using graph cuts. Comput. Methods Programs Biomed. **109**(3), 260–268 (2013)
3. Boellaard, R.: Mutatis mutandis: harmonize the standard!. J. Nucl. Med. **53**(1), 1–4 (2012)
4. Cui, H., Wang, X., Lin, W., Zhou, J., Eberl, S., Feng, D., Fulham, M.: Primary lung tumor segmentation from PET-CT volumes with spatial-topological constraint. Int. J. Comput. Assist. Radiol. Surg. **11**(1), 19–29 (2016)
5. Demir, Ö., Çamurcu, A.Y.: Computer-aided detection of lung nodules using outer surface features. Bio-Med. Mater. Eng. **26**(s1), S1213–S1222 (2015)
6. Golan, R., Jacob, C., Denzinger, J.: Lung nodule detection in ct images using deep convolutional neural networks. In: Proceedings of the IJCNN, pp. 243–250 (2016)
7. Guan, H., Kubota, T., Huang, X., Zhou, X.S., Turk, M.: Automatic hot spot detection and segmentation in whole body FDG-PET images. In: Proceedings of the IEEE ICIP, pp. 85–88 (2006)
8. Guo, Y., Feng, Y., Sun, J., Zhang, N., Lin, W., Sa, Y., Wang, P.: Automatic lung tumor segmentation on PET/CT images using fuzzy Markov random field model. Comput. Math. Methods Med. **2014** (2014)

9. Jia, Y., Shelhamer, E., Donahue, J., Karayev, S., Long, J., Girshick, R., Guadarrama, S., Darrell, T.: Caffe: convolutional architecture for fast feature embedding. arXiv preprint arXiv:1408.5093 (2014)
10. Ju, W., Xiang, D., Zhang, B., Wang, L., Kopriva, I., Chen, X.: Random walk and graph cut for co-segmentation of lung tumor on PET-CT images. IEEE Trans. Image Process. **24**(12), 5854–5867 (2015)
11. Keshani, M., Azimifar, Z., Tajeripour, F., Boostani, R.: Lung nodule segmentation and recognition using SVM classifier and active contour modeling: a complete intelligent system. Comput. Biol. Med. **43**(4), 287–300 (2013)
12. Kingma, D.P., Ba, J.: Adam: a method for stochastic optimization. CoRR abs/1412.6980 (2014)
13. Krawczyk, B.: Learning from imbalanced data: open challenges and future directions. Prog. Artif. Intell. **5**(4), 221–232 (2016)
14. LeCun, Y., Bengio, Y., Hinton, G.: Deep learning. Nature **521**(7553), 436–444 (2015)
15. Liskowski, P., Krawiec, K.: Segmenting retinal blood vessels with deep neural networks. IEEE Trans. Med. Imaging **35**(11), 2369–2380 (2016)
16. Nair, V., Hinton, G.E.: Rectified linear units improve restricted Boltzmann machines. In: Proceedings of the ICML, pp. 807–814 (2010)
17. Nalepa, J., Szymanek, J., McQuaid, S., Endozo, R., Prakash, V., Ganeshan, B., Menys, A., Hayball, M., et al.: PET/CT in lung cancer: an automated imaging tool for decision support. In: Proceedings of the RSNA, pp. 1–2 (2016)
18. Lorenzo, P.R., Nalepa, J., Kawulok, M., Ramos, L.S., Pastor, J.R.: Particle swarm optimization for hyper-parameter selection in deep neural networks. In: Proceedings of the GECCO, pp. 1–8. ACM, USA (2017)
19. Rundo, L., Stefano, A., Militello, C., et al.: A fully automatic approach for multimodal PET and MR image segmentation in gamma knife treatment planning. Comput. Methods Progr. Biomed. 77–96 (2017)
20. Setio, A.A., Jacobs, C., Gelderblom, J., Ginneken, B.: Automatic detection of large pulmonary solid nodules in thoracic CT. Med. Phys. **42**(10), 5642–5653 (2015)
21. Springenberg, J.T., Dosovitskiy, A., Brox, T., Riedmiller, M.: Striving for simplicity: the all convolutional net. ECCV 2014 arXiv preprint (2014)
22. Teramoto, A., Fujita, H.: Fast lung nodule detection in chest CT images using cylindrical nodule-enhancement filter. Int. J. Comput. Assist. Radiol. Surg. **8**(2), 193–205 (2013)
23. Teramoto, A., Fujita, H., Yamamuro, O., Tamaki, T.: Automated detection of pulmonary nodules in PET/CT images: ensemble false-positive reduction using a convolutional neural network technique. Med. Phys. **43**(6), 2821–2827 (2016)
24. Wang, S., Liu, W., Wu, J., Cao, L., Meng, Q., Kennedy, P.J.: Training deep neural networks on imbalanced data sets. In: Proceedings of the IJCNN, pp. 4368–4374 (2016)
25. Wang, X., Ballangan, C., Cui, H., Fulham, M., Eberl, S., Yin, Y., Feng, D.: Lung tumor delineation based on novel tumor-background likelihood models in PET-CT images. IEEE Trans. Nucl. Sci. **61**(1), 218–224 (2014)
26. Win, T., Miles, K.A., Janes, S.M., et al.: Tumor heterogeneity and permeability as measured on the CT component of PET/CT predict survival in patients with non–small cell lung cancer. Clin. Cancer Res. **19**(13), 3591–3599 (2013)
27. Zeiler, M.D., Fergus, R.: Visualizing and understanding convolutional networks. In: Fleet, D., Pajdla, T., Schiele, B., Tuytelaars, T. (eds.) ECCV 2014. LNCS, vol. 8689, pp. 818–833. Springer, Cham (2014). doi:10.1007/978-3-319-10590-1_53

Semi-automatic Training of a Vehicle Make and Model Recognition System

M.H. Zwemer[1,2(✉)], G.M.Y.E. Brouwers[2], R.G.J. Wijnhoven[2], and P.H.N. de With[1]

[1] Eindhoven University of Technology, Eindhoven, The Netherlands
m.zwemer@tue.nl
[2] ViNotion, Eindhoven, The Netherlands

Abstract. We propose a system for vehicle Make and Model Recognition (MMR) that automatically detects and classifies the make and model from a live camera mounted above the highway. Our system consists of a vehicle detection and MMR classification component. The vehicle detector is based on HOG features and can locate 98% of the vehicles with minimum false detections. We use a Convolutional Neural Network (CNN) for MMR classification on the vehicle locations. We propose a semi-automatic data-selection approach for the vehicle detector and the MMR classifier, by using an Automatic Number Plate Recognition engine for annotating new images, requiring minimal human annotation effort. In our results we show that our MMR classification has a top-1 accuracy of 98% for 500 vehicle models, where more than 500 training samples per model are desired to obtain accurate classification.

1 Introduction

There are thousands of surveillance cameras placed along highways which are mainly used for traffic management and law enforcement. Continuous manual inspection is not feasible, as it requires automatic visual interpretation. This enables detection and tracking of vehicles and classification into traffic classes. One specifically important concept is visual Make and Model Recognition (MMR). Make and model information of vehicles can be used to find stolen license plates when comparing the observed vehicle model with the model registered with the license plate. An additional application is to find specific vehicles after a crime when only a vehicle description is available (no license plate number). In such cases, make and model of the vehicle needs to be obtained visually.

Recognition of the vehicles in the above cases is now performed by an Automatic Number Plate Recognition (ANPR) system in a combination with a database lookup in the national vehicle registration database. Although this works for most cases, it is easy to circumvent this technique by altering the license plates. Moreover, it does not work for vehicles without a license plate, foreign vehicles or for motorcycles (when considering a frontal viewpoint).

We present an MMR system developed for the National Police, in which license plates are observed from a camera mounted in an overhead sign structure

S. Battiato et al. (Eds.): ICIAP 2017, Part II, LNCS 10485, pp. 321–332, 2017.
https://doi.org/10.1007/978-3-319-68548-9_30

on the highway where the focus is on a single lane (example video image in Fig. 3). The same camera is used to feed our recognition system. Due to bandwidth restrictions between the camera (online) and our training and testing facilities (offline), we have to optimize the gathering of training and testing samples. Therefore we propose a semi-automatic system to create a dataset.

The main contributions in this paper are the semi-automatic gathering of vehicle samples which are used for training our vehicle detector, an automatic procedure for acquiring make and model annotations for these samples and providing extensive insight in our MMR classification performance.

2 Related Work

Our vehicle recognition system consists of a detection and a classification stage, to localize and recognize vehicles in a full-frontal view. The first detection stage can be solved with different approaches. The full vehicle extent is detected using frame differencing by Ren and Lan [8] or background subtraction by Prokaj and Medioni [7]. Siddiqui *et al.* [12] and Petrović and Cootes [6] extend detections from a license-plate detector. Wijnhoven and de With [16] propose Histogram of Oriented Gradient (HOG) [2] to obtain contrast invariant detection. Recent work by Zhou *et al.* [18] reports on a Convolutional Neural Network (CNN) to obtain accurate vehicle detection. When the vehicle is detected, the vehicle region of the image is used as input for the classification task of MMR.

CNNs are state-of-the-art for image classification and originate by work from LeCun [5] and gained popularity by Krizhevsky [4] who used a CNN (AlexNet) to achieve top performance in the 1000-class ImageNet Challenge [9]. For MMR, Ren and Lan [8] propose a modified version of AlexNet to achieve 98.7% using 233 vehicle models in 42,624 images. Yang *et al.* [17] published a dataset which contains different car views, different internal and external parts, and 45,000 frontal images of 281 different models. They show that AlexNet [4] obtains comparable performance to the more recent Overfeat [10] and GoogLeNet [14] CNN models (98.0% vs. 98.3% and 98.4%, respectively). Siddiqui *et al.* [12] show that for small-scale classification problems, Bag of SURF features achieve an accuracy of 94.8% on the NTOU-MMR dataset[1] (containing 29 classes in 6,639 images).

Other work extends full-frontal recognition towards more unconstrained viewpoints. Sochor *et al.* [13] use a 3D box model to exploit viewpoint variation, Prokaj and Medioni [7] use structure from motion to align 3D vehicle models with images, and Dehghan *et al.* [3] achieve good recognition results but do not reveal details about their classification model.

In conclusion, detection methods involving background subtraction or frame differencing are sensitive to illumination changes and shadows. Therefore, we select Histogram of Oriented Gradients to obtain accurate detection. We have found that detection performance in this constrained viewpoint is sufficient whereas complex detection using CNNs [18] is considered too computationally

[1] NTOU MMR Dataset: http://mmplab.cs.ntou.edu.tw/mmplab/MMR/MMR.html.

expensive. We base our classification system on the AlexNet [4] classification model and focus on an extensive evaluation of the large-scale Make and Model Recognition problem. As shown by Yang *et al.* [17], AlexNet achieves state-of-the-art performance and the MMR problem does not benefit from more advanced CNN models such as GoogLeNet and Overfeat. Moreover, AlexNet is one of the fastest models at hand and suitable for a real-time implementation [1]. Our experiments are performed on our proprietary dataset, which contains 10x more images and car models than the public CompCar dataset [17]. We do not evaluate on the CompCar dataset because classification results are presented by Yang *et al.* [17] and we specifically focus on a large-scale evaluation.

3 System Description

The vehicle recognition system is shown in Fig. 1 and consists of two main components: detection and classification. The system is first trained offline and after training, it can fully automatically detect and recognize vehicles in a video stream from a camera mounted above the road. Both the detector and classification components are trained offline. To develop and train such a recognition system, it would be trivial to store long periods of raw video from the camera in the field and process this video data offline. However, this is not acceptable because typically only a low-bandwidth connection exists between the roadside setup and the backoffice. Therefore, the amount of video data transfer is rather limited. We obtain a low bandwidth when only transmitting a single image for each vehicle that passes the camera. To collect data for training the classification component, we use vehicle detection to select these images. However, we first need to train a vehicle detector. We start by downloading a limited amount of video (15 min) and manually annotating vehicles in these video frames. Using these images, we train our initial vehicle detector and then apply this detector to the roadside setup to collect images with vehicles and transmit these to our backoffice. We can now use these additional images to train an improved vehicle detector and to subsequently train our classification component. For both training purposes,

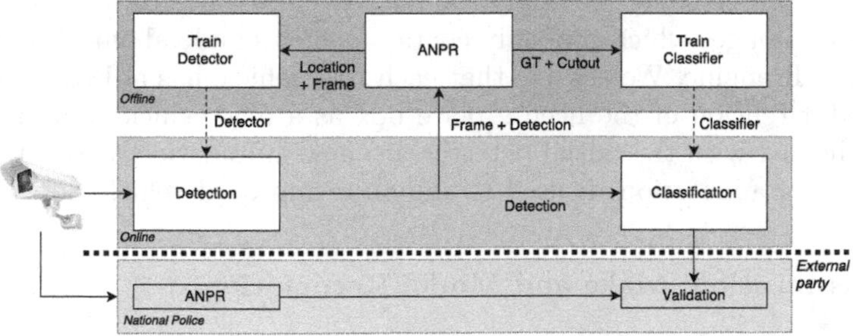

Fig. 1. System overview of the online make and model recognition and offline training of the detector and classifier. The blue bar is an additional validation stage carried out by an external party. (Color figure online)

we employ an Automatic Number Plate Recognition (ANPR)[2] engine. From the location of the number plate we will create additional vehicle annotations to improve the detector, while from the recognized license plate number we look up the vehicle make and model from a database. Next, we downscale each image to a lower resolution and only keep the make and model annotation and remove the license plate number to remove the identity of the vehicle. With this data, we train our vehicle recognition system, which has a privacy-friendly design because there is no identity information and license plates are not readable. Note that in the trained system, used in online operation, all images from the camera are directly down-sampled so that license plates are not readable, but the images are of sufficient resolution for classification. We will now discuss the detection and classification components in detail.

3.1 Detection: Vehicle Localization

Vehicle detection is performed by sliding a detection window over the image and classifying each window location into object/background. Our detector is trained on the grill of a vehicle covering the head lights and bumber, shown by the green rectangle in Fig. 3. Then, linear classification is realized, using HOG feature descriptions of the image as input. We compute HOG features of 12×5 cells of 4×4 pixels using 8 orientation bins ignoring the orientation sign, with L2 normalization of 1×1 blocks. For each cell, we add the gradient magnitude as an additional feature and train our linear classifier using Stochastic Gradient Descent [15]. The detection window is used on multiple, scaled versions of the input image and detections are merged by a mean-shift mode-finding merging algorithm. The detection process is performed every frame in the live video stream. Detections are tracked over time using Good Features to Track [11]. For each vehicle, the make and model classification is performed once when the vehicle is fully visible in the view.

Semi-automatic Training Data Collection for Detection
This approach is necessary because the detection performance of the initial detector is insufficient (missed cars and false detections). Because manual annotation of vehicles is cumbersome, we apply the initial vehicle detector at a low threshold to collect images which probably contain vehicles and validate these images using a ANPR engine. We assume that each real vehicle has a license plate and use a fixed extension of the license plate box as a new vehicle annotation. All images collected with the initial detector are now automatically annotated and the total set of annotations is used to train our improved vehicle detector.

3.2 Classification: Make and Model Recognition

Classification of make and model is performed once for each detected vehicle. The detection box is enlarged with a fixed factor to cover the grill, hood and

[2] ANPR Engine - CARMEN FreeFlow: http://www.arhungary.hu/.

windshield, shown as the blue rectangle in Fig. 3. This part of the image is scaled to a fixed low-resolution image of 256 × 256 pixels and used as the input of our MMR classifier combined with the make and model class label. We use the AlexNet classification model [4], which is a Convolutional Neural Network (CNN) consisting of 5 convolution layers and two fully-connected layers and a nonlinear operation between each layer. This large network is trained end-to-end by feeding our vehicle images and class labels and optimizing the network to predict the correct vehicle class for each image. Note that we predict the make and model combination, so that the number of classes equals the number of vehicle models. We use the AlexNet model pretrained on ImageNet and finetune it with our dataset. For each training image, multiple random subimages of 227 × 227 pixels are used. We train for 50,000 iterations using a batch size of 128. All other training parameters are equal to the original model [4].

Semi-automatic Make and Model Attribute Acquisition for Training Automatic finding of attributes is needed when classifying a large number of objects (order of 10^3–10^4), where sufficient samples are required for each class to distinguish intraclass variation from interclass variation. Moreover, not all vehicle models (classes) are equally popular, the distribution of models is extremely non-linear. To collect samples of rare vehicle models, it is required to on-the-fly annotate vehicles automatically. This is obtained when using our vehicle detector in the roadside setup. An ANPR engine processes every detection and the license-plate number is used to query a database with vehicle make and model information. Our setup is located in the Netherlands, enabling the use of the open-data interface of the Dutch Vehicle Authority (RDW), containing detailed information of all vehicles in the Netherlands[3]. This process allows for large-scale annotation of our dataset.

4 Evaluation of Proposed System

For evaluation, the vehicle detector is compared with the initial vehicle detector and furthermore, we provide insight in our make and model classification results.

4.1 Dataset

To train the initial vehicle detector, we have created a dataset by selecting frames which contain vehicles in 15 min of recorded video. This initial dataset contains 1,318 manually annotated vehicles. With our initial detector, we have collected images in the roadside setup over a period of four hours. The collected images are processed by the ANPR engine to remove false detections and correctly align vehicle boxes. In total, we collected 20,598 vehicle annotations. Half of this set is used to train the final detector, and half to evaluate the detector performance. The classification dataset was recorded during various weather conditions over a long interval of 34 days in which 670,706 images (100%) were collected. Examples

[3] RDW Open Data: https://opendata.rdw.nl/.

Fig. 2. Classification examples. Wind shield and license plates blurred for privacy.

of dark, strong shadows and rainy samples are shown in Fig. 2. All images are processed by the ANPR engine. In 649,955 of the images (97%), a license plate was found and the number could be extracted (other images contain too much noise for recognition). The make and model information was extracted from the database for 587,371 images (88%). Failure cases originate from non-Dutch license plates which are not registered in this database and license-plate numbers that are not read correctly (ANPR failure). In total we acquired 1,504 different vehicle models. The distribution of the number of samples per vehicle model is shown in Fig. 4, which follows a logarithmic behaviour. The top-500 models all have more than 30 samples. The last 700 models only have 1 or 2 samples and represent various high-end vehicles, old-timers and custom vehicles, such as modified recreational vehicles. The model which is seen most is the Volkswagen Golf, with a total of 20k samples (13% of the dataset).

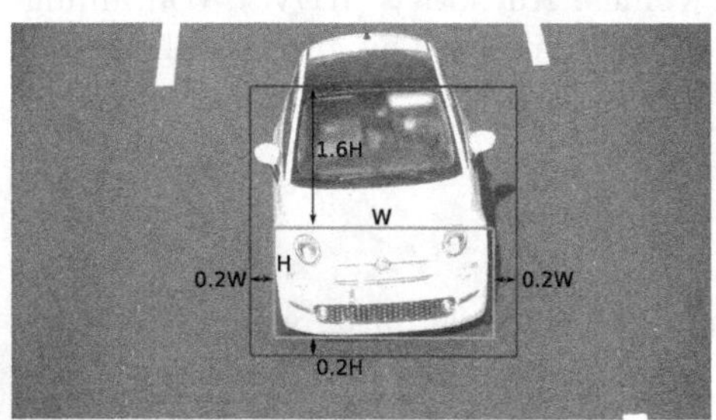

Fig. 3. Video frame, the detection box in green and classification ROI in blue. Windshield and license plate are blurred. (Color figure online)

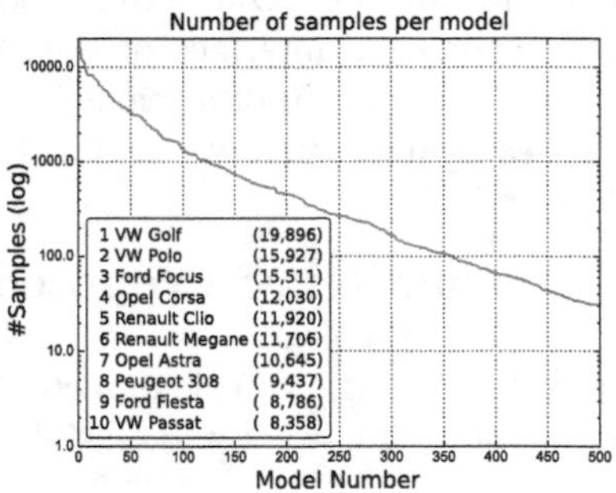

Fig. 4. Number of samples per model.

4.2 Evaluation Metrics

Detection performance is measured using recall and precision. A true positive TP rate is defined as a detection which has a minimum overlap (intersection over union) of 0.5 with the ground-truth box. Detections with lower overlap are false positives FP. Missed ground-truth samples are denoted as false negatives FN. The recall and precision are then computed by:

$$recall = TP/(TP + FN), \quad precision = TP/(TP + FP). \tag{1}$$

We summarize the recall-precision curve by a single value as the Area Under Curve (AUC), where perfect detection has a value of 100%.

The classification performance is measured by the top-1 accuracy, in which the number of correct classifications is divided by the total number of classifications performed. As a second metric, the performance per vehicle model is measured using recall and precision as in Eq. (1). A classification is a true positive TP if the classification label is equal to the ground-truth label, otherwise it is a false positive FP. A sample is a false negative FN for a ground-truth vehicle model if it is not correctly classified. Note that a false negative for one class results into a false positive for another class.

4.3 Vehicle Detection

This section evaluates our initial vehicle detector based on manual annotations and our final detector trained with the automatically collected vehicle annotations (Sect. 3.1). Figure 5 portrays the recall-precision curves for these detectors. The blue curve shows the performance of our initial detector and the red curve shows the results of the final detector. The initial detector already shows good performance, but regularly generates false detections. The final detector clearly outperforms the initial detector and is almost perfect with an AUC of 99%. The operation point has been empirically chosen to detect 98% of the vehicles, having negligible false detections, which is sufficient for the MMR application.

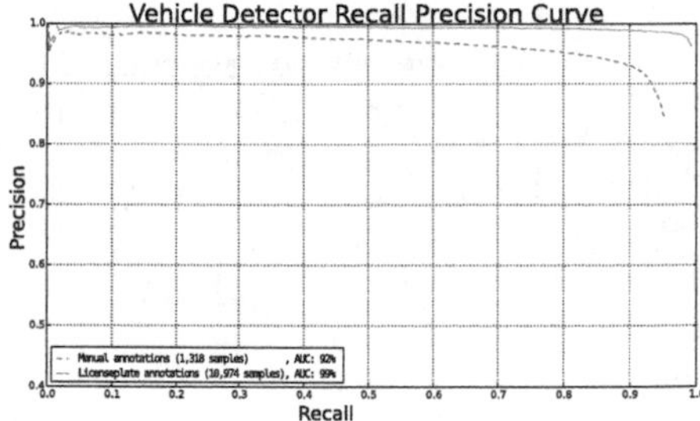

Fig. 5. Recall-precision curve of our initial and final vehicle detectors. (Color figure online)

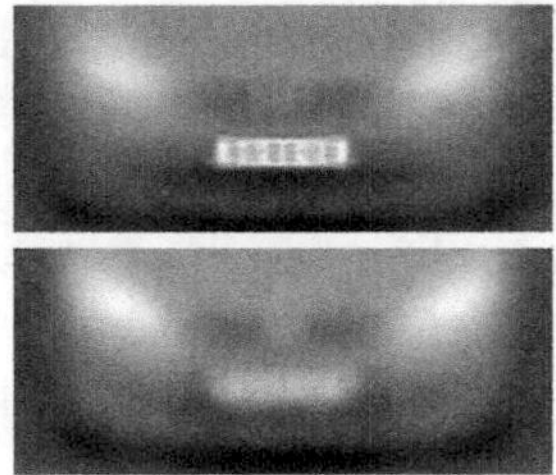

Fig. 6. Average automatically annotated detection box (top) and average detected result (bottom).

In Fig. 6 the average images of our training set and our detector output are shown. The top image shows the average image of the annotations that are used to train the detector (the output of the ANPR detector). It can be clearly seen that the image is aligned on the license plate. The bottom image shows the actual detections after training. Note that the detector does not focus only on the license plate but on the overall vehicle contour. This highlights that our process of automatic annotation is quite powerful and generalizes to the total vehicle characteristics.

4.4 Make and Model Classification

We investigate the classification performance in three main experiments. First, we investigate the relation between the number of classes and the classification performance. Second, we examine the effect of explicitly handling unknown classes. Third, we evaluate the per-class classification performance in relation to the amount of samples per class. Finally, miss-classifications are briefly discussed.

Training of our classification model is carried out by splitting up the vehicle classification dataset per day, to obtain a nice distribution of light and weather conditions. We randomly select 26 days (76%) for training and 6 days (18%) for testing and 2 days (6%) for monitoring the optimization process during training to avoid overfitting on the training set.

We investigate the classification performance when selecting an increasing number of classes in our model, incrementally adding the most frequent classes used. We distinguish the case where unconsidered classes are completely ignored during training ('no unknown' class) and where they are explicitly taken into account as an 'unknown' class.

As a first experiment, we investigate the 'no unknown' case. The classification performance for a low number of classes is constrained by the distribution of the data in the test set, e.g. for one model (VW Golf), the best possible accuracy is 13% because there are only so many samples in the dataset. The results are shown at the left in Fig. 7, where the classification performance (red, solid) approaches the theoretical boundary (green, dotted). Even for 500 classes, we

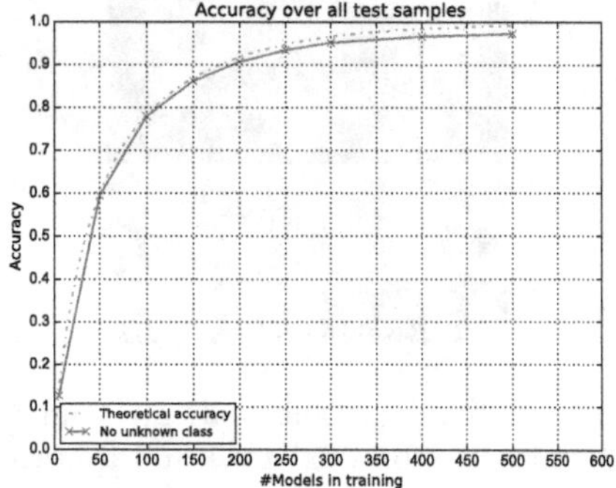

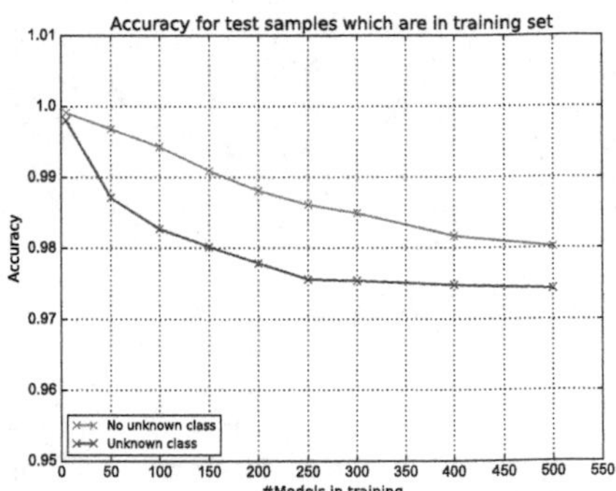

Fig. 7. Make and model classification accuracy. The left diagram evaluates over all test samples, the right diagram only over vehicle models that are incorporated in our classification model. The green dotted line denotes the theoretical accuracy (left), the red solid line is trained without 'unknown' label and the blue (bottom curve) with 'unknown' label. (Color figure online)

achieve an accuracy of 98%, showing that the classification model is able to handle this large-scale classification task. We now explicitly remove the bias from our results by normalizing to the theoretical accuracy, so that we only measure the performance of the classification model and completely ignore the statistics of the dataset. This is obtained by dividing the actual performance by the theoretical optimal performance. For example, for one model (VW Golf) the best possible accuracy is now 100%. The results are shown at the right in Fig. 7 by the red solid line. Although the accuracy over the complete range is high (>98%), it continuously decreases for a growing number of classes.

In a second experiment, we investigate the effect of explicitly taking the ignored classes into account. We expect a lower accuracy because the model has to deal with an extra class with a high amount of intraclass variation (contains all other vehicle classes). However, this case is very interesting because it learns the system to better classify the known models and also learns when a model is not recognizable. We compare this model to our 'no unknown' classification model without statistical bias in Fig. 7 by the blue bottom line. The model considering 'unknown' classes has a high accuracy, not much lower than the 'no unknown' model. More interestingly, the performance seems to saturate at 97.4%. However, with an increasing number of classes in our classification model, only a few samples are sorted into the additional (both for known and unknown) classes. Note that when training with all classes in the classification model, there is no 'unknown' class and both curves will have the same performance. In the third experiment, we provide more insight in the influence of the amount of training samples per model. We show the recall and precision per vehicle model versus the number of training samples available for that model in Fig. 8 for the top-500 vehicle models. Note that the plot is zoomed-in at sample sizes below 2,000. For the 66 models having more than 2,000 samples, both recall and precision approach unity (perfect classification). For classes with more than 500 samples, recall and precision are both exceed 95%. Using less than 200 model samples results in a performance drop. Outliers to this trend are annotated in the figure, which are further investigated and some examples are shown in Fig. 9. This figure shows an example TP classification and the highest FP classifications. We

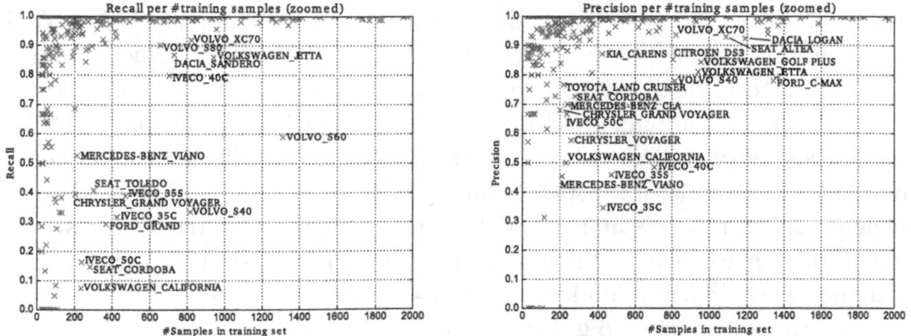

Fig. 8. Recall and precision for the amount of training samples available.

Fig. 9. TP classification (green) and the strongest FP classifications (red) for several models with low precision. The rate represents the distribution of the classifications. (Color figure online)

observe that for these cases, either the class labels are inconsistent (for example 'Citroen DS3' and 'DS 3'), or the classes are visually similar. For example, the Iveco model number relates to the wheel base and payload capacity which cannot be visually observed from the front of the vehicle and the more visually similar sedan models versus estate versions of a vehicle model. The proposed MMR system has been deployed as a live system in the Netherlands for the National Police. An evaluation of our system has been carried out by an external party (see blue region in Fig. 1). This independent evaluation uses our top-500 classification model and processed 4 different time periods with a total duration of 8 hours, under different weather and light conditions. A top-1 accuracy of 92.4% was measured. These results include very low-light conditions, where the vehicle is barely visible, which was not incorporated in our training process.

5 Conclusions

We have proposed a system for vehicle Make and Model Recognition (MMR) that automatically detects and classifies the make and model of each vehicle from a live camera mounted above the road. We have shown that with minimal manual annotation effort we can train an accurate vehicle detector (99% AUC), by using an Automatic Number Plate Recognition (ANPR) engine. Using this concept, we automatically detect vehicles and by extracting the license plate number we acquire the make and model information from a national database. We use a CNN for classification and experiment with the AlexNet model, leading to an MMR classifier with a top-1 accuracy of 98% for 500 vehicle models. The resulting classifier requires at least 500 training samples per model for accurate classification. An explicit unknown model class only leads to a small drop in performance (~0.5%), but makes the model aware of unrecognizable vehicles. This approach can be used to gather automatically more samples of rare vehicle models and new models.

Our choice of classifying the front of the vehicle has limitations. Differences between certain models are not always visible from the vehicle front. These models should be joined in a combined model description, or additional input data (e.g. a side view) is required to solve this classification task.

References

1. Canziani, A., Paszke, A., Culurciello, E.: An analysis of deep neural network models for practical applications. CoRR abs/1605.07678 (2016). http://arxiv.org/abs/1605.07678
2. Dalal, N., Triggs, B.: Histograms of oriented gradients for human detection. In: 2005 IEEE Computer Society Conference on Computer Vision and Pattern Recognition (CVPR 2005), vol. 1, pp. 886–893, June 2005
3. Dehghan, A., Masood, S.Z., Shu, G., Ortiz, E.G.: View independent vehicle make, model and color recognition using convolutional neural network. CoRR abs/1702.01721 (2017). http://arxiv.org/abs/1702.01721
4. Krizhevsky, A., Sutskever, I., Hinton, G.E.: Imagenet classification with deep convolutional neural networks. In: Advances in Neural Information Processing Systems (2012)
5. LeCun, Y., Boser, B., Denker, J.S., Henderson, D., Howard, R.E., Hubbard, W., Jackel, L.D.: Backpropagation applied to handwritten zip code recognition. Neural Comput. **1**(4), 541–551 (1989)
6. Petrovic, V.S., Cootes, T.F.: Analysis of features for rigid structure vehicle type recognition. In: BMVC, vol. 2, pp. 587–596 (2004)
7. Prokaj, J., Medioni, G.: 3-D model based vehicle recognition. In: Workshop on Applications of Computer Vision (WACV), pp. 1–7, December 2009
8. Ren, Y., Lan, S.: Vehicle make and model recognition based on convolutional neural networks. In: Proceedings of the IEEE ICSESS, pp. 692–695, August 2016
9. Russakovsky, O., Deng, J., Su, H., Krause, J., Satheesh, S., Ma, S., Huang, Z., Karpathy, A., Khosla, A., Bernstein, M., et al.: Imagenet large scale visual recognition challenge. Int. J. Comput. Vis. **115**(3), 211–252 (2015)
10. Sermanet, P., Eigen, D., Zhang, X., Mathieu, M., Fergus, R., LeCun, Y.: Overfeat: integrated recognition, localization and detection using convolutional networks. CoRR abs/1312.6229 (2013). http://arxiv.org/abs/1312.6229
11. Shi, J., Tomasi, C.: Good features to track. In: 1994 Proceedings of IEEE Conference on Computer Vision and Pattern Recognition, pp. 593–600, June 1994
12. Siddiqui, A.J., Mammeri, A., Boukerche, A.: Real-time vehicle make and model recognition based on a bag of surf features. IEEE Trans. Intell. Transp. Syst. **17**(11), 3205–3219 (2016)
13. Sochor, J., Herout, A., Havel, J.: BoxCars: 3D boxes as CNN input for improved fine-grained vehicle recognition. In: IEEE Conference on CVPR, June 2016
14. Szegedy, C., Liu, W., Jia, Y., Sermanet, P., Reed, S., Anguelov, D., Erhan, D., Vanhoucke, V., Rabinovich, A.: Going deeper with convolutions. In: The IEEE Conference on Computer Vision and Pattern Recognition (CVPR), June 2015
15. Wijnhoven, R.G.J., de With, P.H.N.: Fast training of object detection using stochastic gradient descent. In: Proceedings of the IEEE ICPR, pp. 424–427. IEEE Computer Society, Washington, D.C. (2010)

16. Wijnhoven, R.G.J., de With, P.H.N.: Unsupervised sub-categorization for object detection: finding cars from a driving vehicle. In: 2011 IEEE International Conference on Computer Vision Workshops (ICCV Workshops), pp. 2077–2083, November 2011
17. Yang, L., Luo, P., Change Loy, C., Tang, X.: A Large-Scale Car Dataset for Fine-Grained Categorization and Verification. ArXiv e-prints, June 2015
18. Zhou, Y., Liu, L., Shao, L., Mellor, M.: DAVE: a unified framework for fast vehicle detection and annotation. CoRR abs/1607.04564 (2016). http://arxiv.org/abs/1607.04564

A Computer Vision System for Monitoring Ice-Cream Freezers

Alessandro Torcinovich[1(✉)], Marco Fratton[2], Marcello Pelillo[1,3], Alberto Pravato[2], and Alessandro Roncato[1,2]

[1] DAIS, Ca' Foscari University, Via Torino 155, 30172 Venezia Mestre, Italy
840284@stud.unive.it
[2] PROSA S.r.l., via dell'Elettricità 3/d, 30175 Venezia-Marghera, Italy
[3] ECLT, Ca' Foscari University, S. Marco 2940, 30124 Venice, Italy

Abstract. In this paper, we describe a computer vision system aimed at monitoring the evolution of the content of a commercial ice-cream freezer. In particular, the system is able to detect the volume occupied by ice-creams in a basket and to track ice-cream sales. To this end, three modules have been developed performing the detection of the baskets and the products inside them, along with the tracking of the interactions with the freezer to take/drop products. The system comprises four cameras connected to an embedded mini-computer able to communicate with a telemetry system that sends information about the freezer context. Our proposed methods achieve promising results for the basket detection and the product tracking (accuracy around 70–80%) and good results in the volume estimation.

Keywords: Edge detection · Image segmentation · Optical flow estimation · Ice-cream freezers

1 Introduction

The need of a remote automated monitoring system for vending machines is mainly motivated by three facts. On a small scale, a more frequent analysis of vending machines content can help logistic activities plans, avoiding superfluous check-and-restock travels. On a larger scale, the collection of sales data coming from several vending machines allows more accurate sales strategies, regarding, for instance, what kind of products are sold in a particular geographical zone. Further, minor problems such as thief control and unwanted products detection, can be addressed as well.

To the best of our knowledge, there are not publicly documented projects with similar aims in the vending machines field, thus in this work we propose a novel system to address them. The effort has been focused on developing three modules devoted to perform two kinds of estimation, regarding: (a) the volume occupied by the ice-creams currently inside a basket (b) the amount of ice-creams sold in a particular time frame.

S. Battiato et al. (Eds.): ICIAP 2017, Part II, LNCS 10485, pp. 333–342, 2017.
https://doi.org/10.1007/978-3-319-68548-9_31

The former estimation has been addressed through the application of an interactive image segmentation approach, in particular a scribble-based one, to segment the *ice-cream zone*. To this end, Price et al.'s *geodesic graph cut* method [7] revealed to be the proper choice. However, since the system needs to work without the user aid, a further step was introduced to automatically draw the scribbles, based on the detection of the basket edges within which the ice-cream zone is.

As for the latter estimation, it has been developed a simple system able to detect the interaction between hands and ice-cream baskets, in particular the gestures related to a product being taken or being put into the basket. This has been accomplished through a motion tracker, based on a background subtractor and an optical flow estimator.

The data collection was performed by four fish-eye cameras, each placed above a basket. Fish-eye was required in order to capture as much visual information as possible.

The system have been tested with many configurations trying to emulate real use cases, and performed its required tasks with a good accuracy and efficiency.

2 Description of the Modules

In this section the algorithmic details of the developed modules are presented. At the beginning, the system checks for basket edges, then it performs periodical checks of the basket product level. At the same time, the freezer is constantly monitored to check for potential interactions and when they are found, they are asynchronously analyzed to detect gestures. The reader can refer to Fig. 1 for a sketch of the system pipeline outlining the interaction between the modules to perform the aforementioned tasks.

2.1 Basket Edges Detection (BED) Module

This module searches for peculiar parts of the baskets in the image, from now on referred as *basket edges*. In particular, the basket edges which we are interested in are the three visible *top borders* defining the rim of the basket, and the *bottom*. The basket is assumed to be in *overflowing state* if the top borders are not visible, and in *empty state* if the bottom is visible. It is important to note that the baskets are not fixed and the empty space between two of them can be up to 5 cm.

As for the top borders, our approach is inspired by the road-lines detection method explained in [1], since the two vertical top borders bear a lot of resemblance to road-lines.

First, the image is color-thresholded in order to filter out all non-white intensities. To this aim the image is converted to HSV color-space and then only the pixels with high V-values and low S-values are retained, more precisely all the pixels in the range $(\mu_l - \alpha_l\sigma_l, \mu_l + \beta_l\sigma_l)$, where μ_l and σ_l are the mean and standard error of the l-values in the image ($l \in \{S, V\}$), while $\alpha, \beta > 0$ are chosen

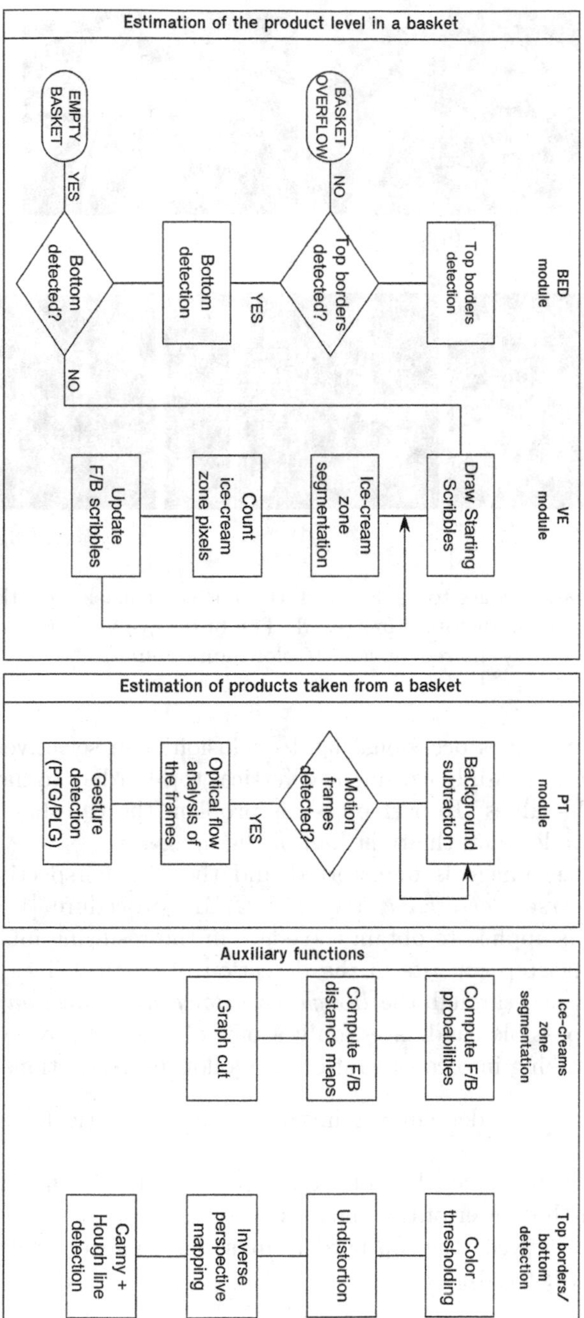

Fig. 1. A simplified sketch of the system pipeline

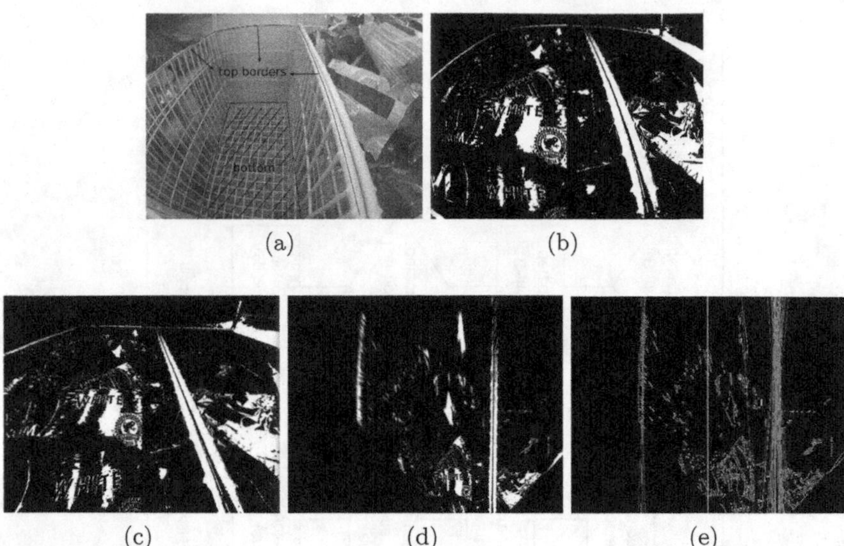

Fig. 2. (a) The basket edges to be detected. (b) The color mask after the color thresholding phase. (c) The undistorted image. (d) The bird's eye view. (e) Applying Hough transform on the Canny filtered version. (Color figure online)

heuristically. Sometimes occasional light occlusions can selectively darken one zone of the basket, possibly causing the vertical top borders to have very different associated V-values. To overcome this problem the thresholding is applied separately on the left and the right half of the image.

After this, the image is undistorted and then its perspective is canceled through an *Inverse Perspective Mapping (IPM)* procedure [1,3]. Further, a Canny filter [5] is applied to obtain the edges in the resulting image.

The following step consists in the detection of vertical lines in the image, that is accomplished through the *Hough Transform line detection* algorithm [9]. Among all the possible candidates, only a pair of lines has to be selected. The choice is made taking into consideration the following restrictions:

- The top borders under current investigation are vertical lines (constraint on θ).
- One of them is in the left half of the image, the other in the right half. They cannot be in the center of the image (constraint on ρ).
- The distance between the top borders must be consistent with the effective size of the basket (26 cm ca).

These constraints improve the search of a good pair of lines matching the top vertical borders.

After finding the two lines, their upper ends are joined to draw the third top border, then perspective and distortion are applied to. Figure 2 shows the main steps of the algorithm and the detected lines for a typical case.

The bottom is detected in a similar way, focusing only on the area bounded by the top borders, but horizontal lines—instead of vertical ones—are searched. In particular the algorithm gives separates responses for the upper and lower part of the bottom, depending on the number of detected horizontal lines in the two areas.

2.2 Volume Estimation (VE) Module

The VE module searches for the ice-cream zone within the basket edges found in the previous steps. Its implementation relies on the segmentation of the image through the approach described in [7], which basically proposes a revised version of the graph cut algorithm [4] with a cost function based on the geodesic segmentation [2]. The cooperation between the two algorithms results in a more accurate segmentation, aware of both the spatial and the colour intensity information. In summary, the so-called *Geodesic Graph Cut* works minimizing the following cost function:

$$E(\mathcal{L}) = \sum_{x_i \in P} R_{\mathcal{L}_i}(x_i) + \lambda \sum_{(x_i,x_j)\in N} B(x_i,x_j)|\mathcal{L}_i - \mathcal{L}_j| \tag{1}$$

where, $\mathcal{L} = (\mathcal{L}_i)$ is a binary vector of labels and $\mathcal{L}_i \in (\mathcal{F}, \mathcal{B})$ determines if a pixel is labeled as a foreground or background one. Here, $B(x_i, x_j)$ is a *boundary cost-term*, computed as:

$$B(x_i,x_j) = \frac{1}{1 + \|C(x_i) - C(x_j)\|^2} \tag{2}$$

$R_l(x_i)$ is instead a *region term*, computed as

$$R_l(x_i) = s_l(x_i) + M_l(x_i) + G_l(x_i) \tag{3}$$

where

$$s_l(x_i) = \begin{cases} \infty & \text{if } \mathcal{L}_i = l \\ 0 & \text{otherwise} \end{cases} \tag{4}$$

while $M_l(x_i) = P_{\bar{l}}(C(x_i))$ represents the probability that the color intensity associated to the pixel x_i belongs to the opposite label of l (if $l = \mathcal{F}$, then $\bar{l} = \mathcal{B}$).

The last term represents a normalized geodesic distance, in particular:

$$G_l(x_i) = \frac{D_l(x_i)}{D_{\mathcal{F}}(x_i) + D_{\mathcal{B}}(x_i)} \tag{5}$$

Figure 3 shows the pixel maps of the second and third terms, for both the foreground and background.

In order to speed up the computations the paper suggests to use approximate methods such as the Fast Gauss Transform [11] for the second term and the

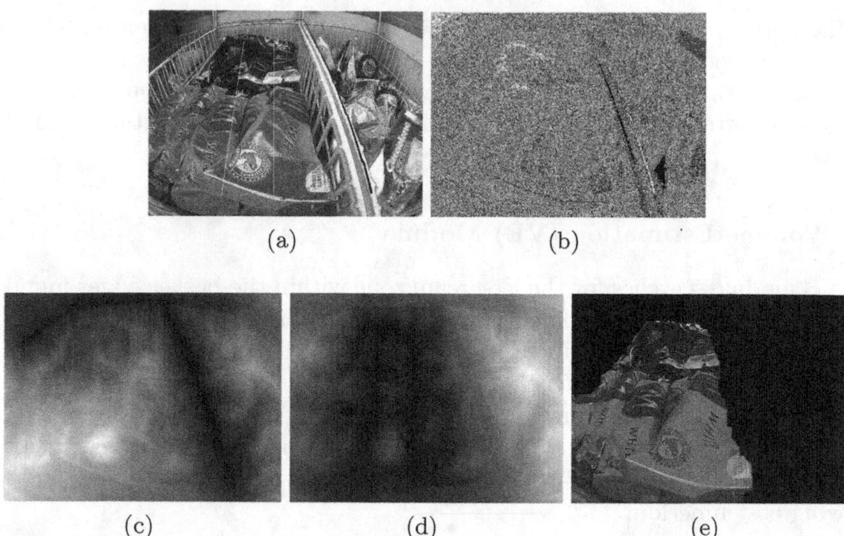

Fig. 3. (a–b) The probability pixel maps, foreground and background respectively, related to the second term in Eq. 3 (scribble colors are inverted to be more visible in the images). (c–d) The normalized distance maps related to the third term in the same equation. (e) An image segmented with the proposed method.

Fast Marching Method [2,8] for the third one. In addition, further accuracy improvements can be made (refer to [7] for more details).

The method requires the user to define an initial input representing a sample of background pixels and another one of foreground pixels. As previously noted, the output of the BED module is used to set the background strokes, while the foreground ones are drawn within the basket area. Once the first segmentation is performed, the ice-cream zone is dilated and the borders of the resulting area are taken as the new background strokes.

2.3 Product Tracking (PT) Module

The last module identifies significant interactions within the freezer and tries to guess what happened. The core of the module is based on two algorithms to detect movement. At first, a simple background subtractor [12] is run to capture movement. Whenever some activity is detected, the analysis is passed to optical flow detection algorithm until a period of overall inactivity is detected, where the background subtractor takes control again. The necessity to rely on two motion detection algorithms is due to the fact that the optical flow algorithm is resource-demanding, and cannot be run in real-time in the embedded system, however it is strictly needed because it is able to estimate the direction of the flow[1]. The

[1] Currently Farnebäck's method [6] is used, but in the future, to overcome this problem, it will be replaced with some faster technique like Tao's SimpleFlow [10].

frames between two inactivity periods are assumed to contain information about what kind of interaction has been performed within the freezer. Currently two gestures are detected, namely the *Product Taken* (*PTG*) and the *Product Left* (*PLG*) ones, denoting if an ice-cream has been taken from/put into the basket.

To detect a gesture, the pixels in a frame are first divided in four classes: (a–b) *up/down motion pixels* associated to motion vectors with a high norm and a direction pointing up/down, (c) *undefined motion pixels* associated to motion vectors with a low norm or a non vertical direction (or both), (d) *inactive pixels* where no motion is detected.

After classifying the pixels, the algorithm discriminates between *up/down motion frames*, simply checking what is the highest between up/down motion pixels. The next step consists in checking the highest amount of down pixels among all down frames against the respective amount among all up frames. If an ice-cream has been taken, it is expected that the max amount of up pixels, representing the hand and the ice-cream, is higher with respect to the max amount of down pixels representing the hand only, and the algorithm outputs a PTG response. The opposite can be stated for an ice-cream put into a basket.

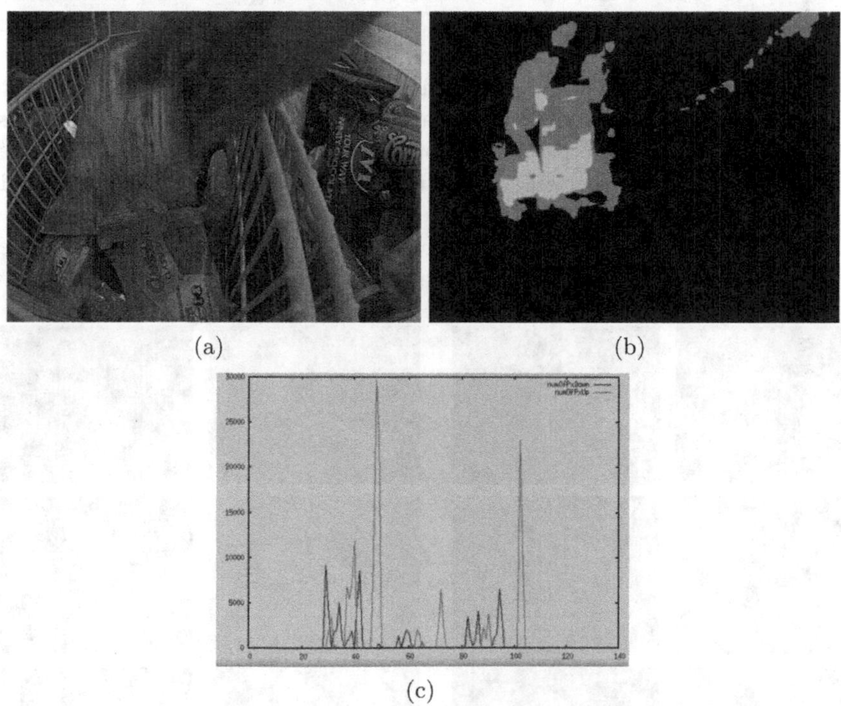

Fig. 4. (a) The optical flow of a video frame during an interaction with the freezer. (b) The corresponding pixel classification in up (green) and undefined (gray) motion pixels. Inactive pixels are colored in black, while down motion pixels are not present. (c) Plot representing the amount of down (red) and up (green) pixels in each frame of a video containing three PTGs. (Color figure online)

As an example, in Fig. 4(c) the amount of up/down pixels is shown for a sequence of three PTGs. Each gesture can be identified from the pattern of a down pixels peak, followed by an higher up pixels peak.

3 Experimental Results

A prototype of the ice-cream freezer was assembled, and we tested the three modules separately with multiple freezer contexts evolving over the time. All the tests were performed with several brands and different lighting conditions on a system with a dual core CPU with 3 GHz/core and a 2 GB RAM, however in the future further testing will be performed under more restrictive conditions. The running time of a volume analysis performed by the BED + VE modules stays under 12 s, while for the gesture detection, the time is equal to 130% of the video length containing the gesture.

The BED module testing was performed searching for the basket edges in 55 situations, comprising baskets with different product levels. A visual inspection of the results, revealed that the top borders detection was successful in $48/55 \approx 87.3\%$ of cases. In particular, sometimes the algorithm failed in the case of baskets with a top border next to the white inner wall of the freezer, because of the similarity in colour intensity. The bottom detection was performed only in case that the top borders have been found. Since the tests comprised an important percentage of overflow settings, the actual number of bottom detection tests decreased to 39 cases. The percentage of correct detections was $30/39 \approx 76\%$. Further testing revealed that also the bottom detection can suffer from similarity in colour intensity with the inner walls.

Some of the VE module tests are shown in Fig. 5. The initial scribbles were correctly drawn by the BED module and then the geodesic graph cut was

Fig. 5. Examples of VE analyses performed on an evolving context in the freezer.

iteratively applied. After tuning the number of dilations, the ice-cream zone was correctly detected with high precision without problems.

The PT module was tested analyzing 9 manually captured videos containing 66 PT/PL gestures. In the 73% of the cases the gesture was correctly detected and classified. The module was further tested with a setting in which a torch light was switched on and off periodically to simulate sudden lighting changes. The algorithm responded well, without erroneously detecting gestures.

4 Conclusions and Future Work

In this paper, we have described a suite of algorithms devoted to monitor the content of an ice-cream commercial freezer, to provide information about the content level and the products taken from the freezer, reaching good results for both the tasks. The project is still in a prototype stage and some of its aspects needs to be refined, in order to cover a wider majority of the real cases.

The BED module can be improved introducing the detection of basket dividers, which usually are freely movable and can be put to divide between multiple brands inside a basket. As for the PT module the algorithm as is, was tested using only one camera per basket, however being able to monitor the interactions in the fridge with two cameras could possibly increase the effectiveness of the algorithm and give us a more accurate position of the hand in the basket. This could be paired with a noise analyzer to detect the particular sound that a hand produces when scratching in search of ice-creams, providing us even more information about the interaction. The VE module, strongly depends on the drawn scribbles and assumes a progressive emptying of the basket. Indeed it can happen that the basket is subject to some "brutal" modification, like the re-fill or a total emptying, that can potentially invalidate the successive volume check. This can be dealt by pairing the module with the other two, especially with the product tracking module, making it capable to recognize these particular situations. The module currently work if the first check is performed with an adequately full basket. This aspect needs improvements in order that the first volume check can be run successfully with any possible basket status.

References

1. Aly, M.: Real time detection of lane markers in urban streets. In: 2008 IEEE Intelligent Vehicles Symposium, pp. 7–12. IEEE (2008)
2. Bai, X., Sapiro, G.: Geodesic matting: a framework for fast interactive image and video segmentation and matting (preprint). Technical report, DTIC Document (2008)
3. Bertozzi, M., Broggi, A.: Real-time lane and obstacle detection on the gold system. In: Proceedings of the 1996 IEEE Intelligent Vehicles Symposium, pp. 213–218. IEEE (1996)
4. Boykov, Y.Y., Jolly, M.P.: Interactive graph cuts for optimal boundary & region segmentation of objects in N-D images. In: Proceedings Eighth IEEE International Conference on Computer Vision, ICCV 2001, vol. 1, pp. 105–112. IEEE (2001)

5. Canny, J.: A computational approach to edge detection. IEEE Trans. Pattern Anal. Mach. Intell. **6**, 679–698 (1986)
6. Farnebäck, G.: Two-frame motion estimation based on polynomial expansion. In: Bigun, J., Gustavsson, T. (eds.) SCIA 2003. LNCS, vol. 2749, pp. 363–370. Springer, Heidelberg (2003). doi:10.1007/3-540-45103-X_50
7. Price, B.L., Morse, B., Cohen, S.: Geodesic graph cut for interactive image segmentation. In: 2010 IEEE Conference on Computer Vision and Pattern Recognition (CVPR), pp. 3161–3168. IEEE (2010)
8. Sethian, J.A.: A fast marching level set method for monotonically advancing fronts. Proc. Natl. Acad. Sci. **93**(4), 1591–1595 (1996)
9. Shapiro, L., Stockman, G.: Computer Vision. Prentice Hall, Upper Saddle River (2001)
10. Tao, M., Bai, J., Kohli, P., Paris, S.: Simpleflow: a non-iterative, sublinear optical flow algorithm. Comput. Graph. Forum **31**, 345–353 (2012). Wiley Online Library
11. Yang, C., Duraiswami, R., Gumerov, N.A., Davis, L.S., et al.: Improved fast gauss transform and efficient kernel density estimation. In: ICCV, vol. 1, p. 464 (2003)
12. Zivkovic, Z.: Improved adaptive Gaussian mixture model for background subtraction. In: Proceedings of the 17th International Conference on Pattern Recognition, ICPR 2004, vol. 2, pp. 28–31. IEEE (2004)

A Proposal of Objective Evaluation Measures Based on Eye-Contact and Face to Face Conversation for Videophone

Keiko Masuda, Ryuhei Hishiki, and Seiichiro Hangai(✉)

Department of Electrical Engineering, Faculty of Engineering,
Tokyo University of Science, 6-3-1 Niijuku, Katsushika, Tokyo 1258585, Japan
{masuda, hangai}@ee.kagu.tus.ac.jp
http://www.tus.ac.jp/en/fac/p/index.php?678

Abstract. In order to realize eye-contact and face to face communication, videophone or virtual conversational system which takes gaze line into consideration. Although many systems have been studied and reported, there is no objective measure for evaluating the quality of conversations. In this paper, we propose two objective measures such as the eye-contact conversation ratio (ECCR) and the face to face conversation ratio (FFCR) for evaluating the communication quality. By changing the position of camera from the above to the center of display, the ECCR increases from 24.3% to 25.3% in talking and decreases from 33.1% to 25.6% in listening. It is also found that the FFCR improved from 74.7% to 88.0% by centering a camera.

1 Introduction

In popular personal videophone system using PCs and Tablets, listening and talking with downcast eyes is inevitable. This is because a camera is installed above a display and there is no gaze line matching between two persons. In the teleconference system for multiple persons, half mirrors and cameras were used for realizing eye contact talks [1]. In another study, the picture plane was rotated for compensating the gaze direction, and the improvement of subjective perception was reported by the number of votes by 52 subjects [2]. Gaze correction method [3] and multi-viewpoint videos merging method [4] have been reported the improvement of eye-contact communication, too. However, there is no objective evaluation result in those studies.

Generally, in a natural conversation, eye-contact and face to face communication can be observed frequently, and those human behaviors should be taken into account by evaluating a system. In e-learning applications, eye mark recorder which recorded the fixation point movement on the view was applied to analyze the effectiveness of presentation methods [5].

In this paper, we define two objective measures, i.e., the ECCR and the FFCR, and show the experimental results using eye mark recorder [6] with changing the position of camera from the above to the center of display. We also discuss what makes a conversation natural using videophone and virtual conversational system.

S. Battiato et al. (Eds.): ICIAP 2017, Part II, LNCS 10485, pp. 343–350, 2017.
https://doi.org/10.1007/978-3-319-68548-9_32

2 Human Behaviors in Conversation

Mutual gaze during natural conversation is one of important interactions [7]. However, in personal videophone system, inconsistent gaze behavior, e.g., gaze at partner's clothes or out of display, is frequently observed.

Figure 1 shows the relationship between the gaze at partner's eye $G_{eye}(t)$, the gaze at partner's face $G_{face}(t)$, the Talk by subject $T_s(t)$, and the Talk by partner $T_p(t)$, and behavioral states. As each feature is represented by ON(1) or OFF(0), there are 16 behavioral states. In this figure, the duration of $G_{eye}(t) = 1$ and $G_{face}(t) = 1$, when $T_s(t) = 1$ or $T_p(t) = 1$, is the most important state. As shown in Fig. 1, we define Eye-Contact Conversation (ECC) state in which $G_{eye}(t) = 1$ and $T_s(t) = 1$ or $T_p(t) = 1$, and Face to Face Conversation (FFC) state in which $G_{face}(t) = 1$ and $T_s(t) = 1$ or $T_p(t) = 1$. The former is marked up by black bar and the latter is by gray bar in the bottom of figure.

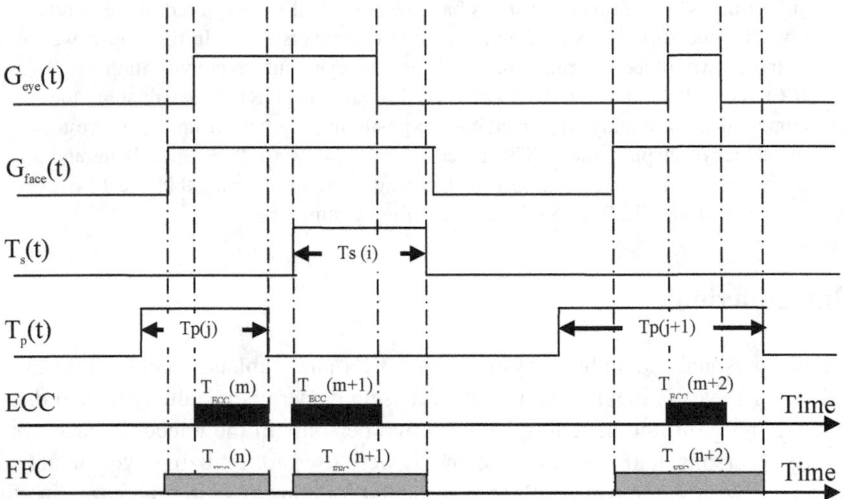

Fig. 1. Relationship between 4 features and 2 states

2.1 Eye-Contact Conversation Ratio (ECCR)

In order to estimate the eye-contact conversation objectively, we add up ECC durations and calculate ECC ratio by the following equation,

$$ECCR = \frac{\sum_{m=1}^{M} T_{ECC}(m)}{\sum_{i=1}^{I} T_s(i) + \sum_{j=1}^{J} T_p(j)} \times 100 \tag{1}$$

where, $T_{ECC}(m)$ is the m-th duration of ECC state, $T_s(i)$ is the i-th duration of subject's talk, and $T_p(j)$ is the j-th duration of partner's talk.

From the equation, ECCR presents eye-contact conversation ratio in both talking period and listening period.

2.2 Face to Face Conversation Ratio (FFCR)

As same as ECCR, we sum up FFC durations and calculate FFC ratio by the following equation,

$$FFCR = \frac{\sum_{n=1}^{N} T_{FFC}(n)}{\sum_{i=1}^{I} T_s(i) + \sum_{j=1}^{J} T_p(j)} \times 100 \tag{2}$$

where, $T_{FFC}(n)$ is the n-th duration of FFC state.

From the equation, FFCR presents face to face conversation ratio in both talking period and listening period. As shown in Fig. 1, FFC duration includes ECC duration, because eye is a part of face.

3 Experimental System

In order to evaluate the eye-contact conversation and the face to face conversation using videophone with different camera position, we have developed a videophone system, in which the camera position can be changed. Gaze point is recorded by the eye mark recorder which uses the infrared reflection of pupil/cornea, and the decision whether the gaze position of subject is at face or at eye or at others is made by analyzing the recorded images. Conversations are also recorded and separated into subject's talk and partner's talk after noise reduction.

In this section, the developed videophone system and the flow of signal processing are described.

3.1 Videophone System

The developed videophone system is shown in Fig. 2. A half mirror is located in front of a subject with 45° angle for realizing face to face conversation with the image of a partner. The flipped horizontal image is displayed on the monitor for avoiding left and right being reversed.

The height of the camera can be changed in any position. In this experiment, we use two positions such as center position and above position which simulates PC's camera. Two sets of systems are used in the experiment. Specification of each system is as follows,

Display size: 24.1 in. LCD
Videophone application: Skype
Left and Right Reverse: ManyCam[1]
Camera: 640pixels (H) by 480pixels (W), 24bit color, 30fps
Audio: fs = 44.1 kHz, 16bit

[1] https://manycam.com/.

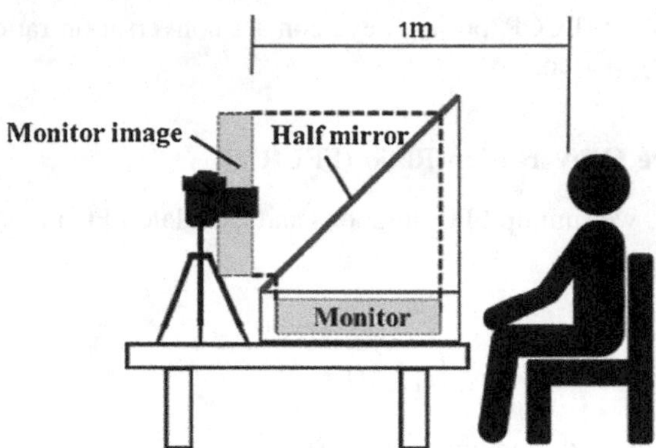

Fig. 2. Developed videophone system

A scene of the experiment using the developed videophone system is shown in Fig. 3.

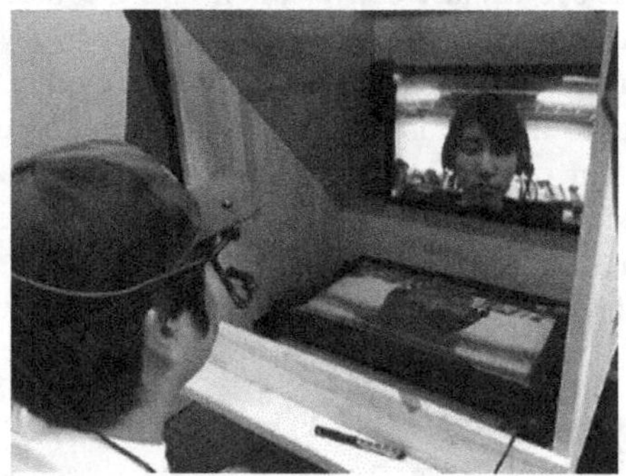

Fig. 3. Experimental setup

3.2 Gaze Point Estimation

The gaze points and audio information of the subject wearing the eye mark recorder (EMR-9) [6] are recorded.

The recorder measures the sight angle of the subject based on the infrared reflection image in the cornea and the pupil movement. The detection range is ±40° in horizontal and ±20° in vertical. The gaze points (Left: + , Right: □) and a parallax corrected gaze point: ○ are displayed on the image (640 × 480 pixels) taken by field of view cameras installed at the brim of a cap as shown in Fig. 4. The image was being recorded while conversation and analyzed together with recorded voice after the experiment to look into the location of the parallax gaze point.

Figure 4 shows examples of (a) "Eye-Contact Conversation in talking" scene, and (b) "Face to Face Conversation in listening" scene.

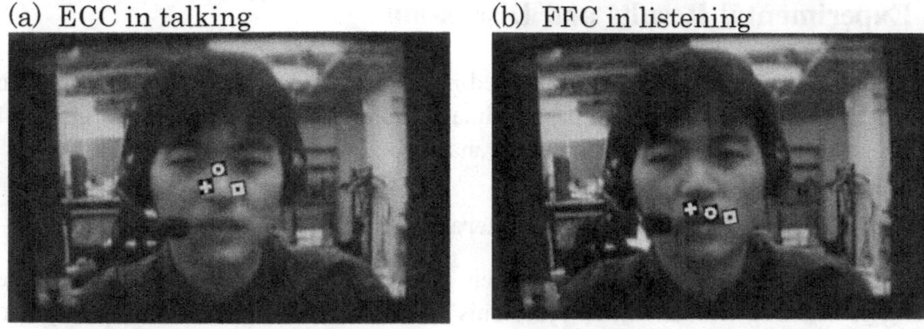

Fig. 4. Example of recorded image from EMR-9

Face area and eye area are manually determined by the face features such as skin color, eye-brow, eye, nose, mouth, and tin as shown in Fig. 5.

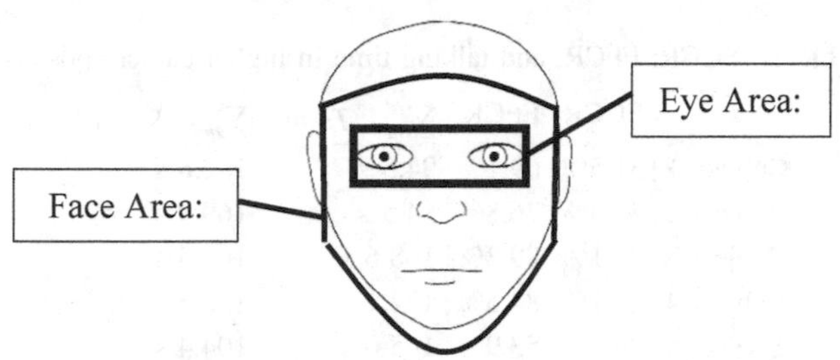

Fig. 5. Face area and eye area

3.3 Signal Processing Flow

Figure 6 shows the flow of signal processing to get 4 signals, i.e., $G_{eye}(t)$, $G_{face}(t)$, $T_s(t)$, and $T_p(t)$, and 4 states, i.e., T_{ECC} and T_{FFC} in talking/listening. In this study, $G_{eye}(t)$ and $G_{face}(t)$ are extracted manually, and the Talk by subject $T_s(t)$ and the Talk by partner $T_p(t)$ are extracted automatically based on the audio signal power.

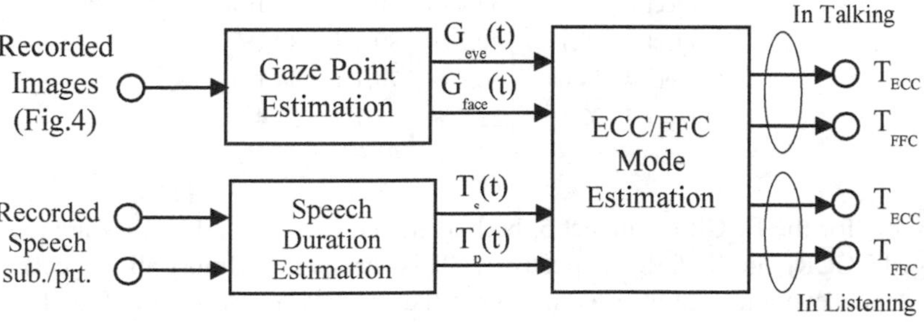

Fig. 6. Signal processing flow

4 Experimental Results and Discussion

10 male persons (Age: 22–24) were divided into 5 groups and made free conversation for 6 min. or more. After 1 min. passed, image including gaze point shown in Fig. 4 and speech were recorded for 5 min. and analyzed.

4.1 ECCR and FFCR in Higher Camera Position

It is expected that both a subject and their partner using PC based videophone are talking or listening with downcast eyes. This degrades the communication quality and leads to low ECCR and FFCR. Table 1 shows the ECCR and the FFCR in the conversation of 5 min. Total talking time of subject and Total talking time of partner are also indicated in seconds.

From Table 1, averaged ECCR of five subjects is 29.0% and the deviation is not large. On the other hand, averaged FFCR shows 74.7% even if a partner talks or listens with downcast eyes. Five FFCRs depend on subjects, and varies between 53.9% and 89.3%.

Table 1. ECCR, FFCR, and talking time in higher camera position

	ECCR	FFCR	$\sum_{m=1}^{M} Ts(m)$	$\sum_{n=1}^{N} Tp(n)$
Subject 1	31.5%	69.1%	94.5 s	192.8 s
Subject 2	30.3%	76.8%	84.5 s	169.9 s
Subject 3	28.1%	89.3%	168.6 s	103.3 s
Subject 4	26.1%	84.5%	124.1 s	135.5 s
Subject 5	29.0%	53.9%	165.0 s	104.4 s
Average	29.0%	74.7%	127.3 s	141.2 s

In order to inspect ECCR and FFCR in detail, we separate them into the ratios in talking and listening, and summarized as shown in Table 2. The suffix "T" and "L" shows "talking" and "listening" respectively.

Table 2. ECCR and FFCR in talking and listening in higher camera position

	$ECCR_T$	$ECCR_L$	$FFCR_T$	$FFCR_L$
Subject 1	27.9%	33.3%	61.2%	73.1%
Subject 2	26.1%	32.4%	70.0%	80.1%
Subject 3	20.7%	40.2%	89.4%	89.1%
Subject 4	14.6%	36.5%	74.1%	94.1%
Subject 5	32.1%	22.9%	51.7%	57.4%
Average	24.3%	33.1%	69.3%	78.8%

Except for the ECCR of subject 5, both averaged $ECCR_T$ and $FFCR_T$ are less than averaged $ECCR_L$ and $FFCR_L$, respectively. This means that almost all subjects watch the partner's eye and face in listening rather than in talking. Also, it is found that the subjects watch the partner's face rather than in the eye while talking.

4.2 ECCR and FFCR in Center Camera Position

Table 3 shows the ECCR, the FFCR, Total talking time of subject, and Total talking time of partner, and Table 4 shows the detail of ECCR and FFCR.

By comparing Table 3 with Table 1, the following are found,

(1) Averaged ECCR decreases by locating a camera to the center. This trend can be seen except for subject 1.
(2) Averaged FFCR increases by locating a camera to the center. This trend can be seen true for all subjects.

Table 3. ECCR, FFCR, and talking time in center camera position

	ECCR	FFCR	$\sum_{m=1}^{M} Ts(m)$	$\sum_{n=1}^{N} Tp(n)$
Subject 1	37.7%	83.9%	105.3 s	157.8 s
Subject 2	27.1%	93.2%	112.7 s	140.9 s
Subject 3	14.7%	97.2%	141.2 s	138 s
Subject 4	21.6%	93.8%	97.73 s	181.4 s
Subject 5	25.9%	71.9%	155.7 s	117.8 s
Average	25.4%	88.0%	122.5 s	147.2 s

Table 4. ECCR and FFCR in talking and listening in center camera position

	$ECCR_T$	$ECCR_L$	$FFCR_T$	$FFCR_L$
Subject 1	45.3%	32.6%	87.0%	81.9%
Subject 2	33.3%	22.1%	91.5%	94.7%
Subject 3	9.2%	20.3%	96.9%	97.5%
Subject 4	14.0%	25.7%	88.5%	96.7%
Subject 5	24.6%	27.5%	67.0%	78.5%
Average	25.3%	25.6%	86.2%	89.9%

By comparing Table 4 with Table 2, the following are found,

(3) Averaged $ECCR_T$ slightly increases by centering a camera. But, this is not a remarkable trend.
(4) Averaged $ECCR_L$ slightly decreases by centering a camera. This trend can be seen except for subject 5.
(5) Both $FFCR_T$ and $FFCR_L$ increase by centering a camera. This trend can be seen true for all subjects.

5 Conclusion

In order to improve the naturality of a conversation using videophone or virtual conversational system, we have proposed two objective measures, ECCR and FFCR, and developed a videophone system with half mirror. By changing the position of camera from above to center, FFCR increases from 74.7% to 88.0%. This means that the face to face conversation is affected by the gaze of the partner. Obviously, the conversation with mutual gaze increased by centering a camera, and the naturality of a conversation have been improved. However, because of wide eye area and no consideration of partner's gaze, ECCR in talking/listening does not change. For clarifying the true eye contact conversation ratio, the eye area and gaze of partner should be considered in future work. In addition, measures of affect or measure of emotion such as PANAS (Positive and Negative Affect Schedule) should be studied.

References

1. De Silva, L.C., et al.: A teleconferencing system capable of multiple person eye contact using half mirrors and cameras placed at common points of extended lines of gaze. IEEE Trans. Circ. Syst. Video Technol. **5**(4), 268–277 (1995)
2. Solina, F., Ravnik, R.: Fixing missing eye-contact in video conferencing system. In: Proceedings of ITI 2011, pp. 233–236 (2011)
3. Lu, J., Tao, X., Dong, L., Ge, N.: Chunk-wise face model based gaze correction in conversational videos with single camera. In: Proceedings of CITS 2016, pp. 1–5 (2016)
4. Ebara, Y., Nabuchi, T., Sakamoto, N., Koyamada, K.: Study on eye-to-eye contact by multi-viewpoint videos merging system for tele-immersive environment, In: Proceedings of AINA 2006, vol.2 (2006)
5. Ando, M., et al.: An analysis using eye-mark recorder of the effectiveness of presentation methods for e-learning. In: Proceedings of ICALT 2017, pp. 183–185 (2007)
6. http://www.eyemark.de/downloads/EMR9_Basic_Operations.pdf
7. Broz, F., et al.: Mutual gaze, personality, and familiarity: dual eye-tracking during conversation. In: Proceedings of IEEE RO-MAN 2012, pp. 858–864 (2012)

Multimedia

Wink Detection on the Eye Image as a Control Tool in Multimodal Interaction

Piotr Kowalczyk and Dariusz Sawicki(✉)

Warsaw University of Technology, Warsaw, Poland
p.kowalczyk@iem.pw.edu.pl,
Dariusz.Sawicki@ee.pw.edu.pl

Abstract. The selected problem of multimodal interaction for multimedia is discussed. The application for visual analysis of eye winking is presented. The proposed algorithm allows interpreting "eye gestures" as appropriate system events, which corresponds to executing system commands. The main aim of this paper is to develop replacing the usage of mouse keys by eye winking. Such a solution can work in multimodal interaction for many different applications. The work of surgeon in the operating room is a good example of situation, where busy hands do not allow controlling the computer system by traditional devices. Introduced interaction based on eye winking can help to communicate with system. The prototype has been built and set of experiments have been carried out in a group of 30 people. Tests have shown high acceptance of the proposed solution.

Keywords: Multimedia · Multimodal interaction · Eye winking recognition

1 Introduction

Multimedia covers a wide range of techniques from computer vision, machine learning, human-machine communications and artificial intelligence [1]. Computer vision and pattern recognition are treated as a tools which allow to develop devices providing multimedia entertainment. Today consumer electronics is commonly present in everyday life. However, multimedia is not just entertainment. Multimedia is also a multimodal interaction [2, 3], which is important in intelligent multimodal presentation as well as a working tool needed in the operating room. This last medical example is a specific sign of our time. On the one hand, it shows the need of using advanced graphical computer technology in a way that should be extremely simple – even intuitive. In this context effective control of a computer in the operating room [4] is one of the most difficult and also most interesting tasks of modern information technology. On the other hand this example, like others examples from multimodal interaction, breaks the tradition of combining functions in a pointing device. Since Ivan Sutherland and his sketchpad [5], the pointing device allows selecting the appropriate object in space and simultaneously generating the corresponding system event. In this way, the popular computer mouse, in addition to the ability to move the screen cursor, is equipped with keys (1 to 3). In multimodal interaction such functions can be controlled independently. This is important not only for the player, but also for the surgeon in the

S. Battiato et al. (Eds.): ICIAP 2017, Part II, LNCS 10485, pp. 353–362, 2017.
https://doi.org/10.1007/978-3-319-68548-9_33

operating room, who has a busy hands and looking at the computer screen would like to run a specific system event.

The main aim of this paper is to develop the simple method that allows generating system events using multimodal interaction. We focused on elements of body language which can be recognized using visual analysis.

2 The Task Analysis and the Main Assumptions of the Project

The typical (and known for the user) methods of generating system events is using the keys of mouse. If we assume that the hands are busy (e.g., in case of the surgeon), the control of the mouse keys can be replaced for example by movement of the head or facial expression.

The general survey of multimodal human computer interaction was presented by Jaimes and Sebe in [6], where many different methods of communications were analyzed. The image of face is used in many interactive solutions for multimedia. In the work of Mandal et al. [7] the video frames are analyzed. In the first step, the skin color is recognized, in the second the face/head pose and its movement is determined. Strumiłło and Pajor proposed a similar system based on face recognition [8]. The simple analysis of the closing and opening eyes is used in this solution. Surgeon's face analysis can be difficult due to the protective clothing required in the operating room. On the other hand, in such analysis there is a need to direct the face toward the camera or to provide a number of cameras so that in any head position one of the cameras can record the head/face properly The analysis of the closure of the eye seems to be a good option, provided, however, that it will be made from a close distance. Using eye tracking (oculography), we could isolate also eye blinks. Al-Rahayfeh, Faezipour [9] presented a survey of methods used in eye tracking and head position detections. A survey of eye tracking methods in different methods related to multimodal interaction was presented by Singh and Singh in [10]. Such a solution could be effective and elegant in considered task. However, the use of oculography in this task seems to be not only too complicated but also unnecessary.

In some multimedia applications infrared radiation is used. Kapoor and Picard [11] described a vision based system that detects head nods and head shakes using an IR camera equipped with IR LEDs. This idea allows analyzing the image independently of the room lighting, and more importantly, regardless of the head position.

After analyzing previous work on multimedia and multimodal interaction, we decided to use winking to generate events. The proposed solution should be able to correctly replace the keystrokes of a computer mouse by winks, on the following assumptions:

- Any position of the head of the user and direction of the eyesight are possible.
- Head and face of the user can be obscured (partially or almost completely).
- The solution should be simple, fast and effective. This applies particularly to the used algorithms for image processing.
- Analysis of the stage of eye should be performed in real time in order to measure eye closure time and distinguish natural blinking from intentional winking.

Simplifying the issue, we can use a camera to "observe" the face, extract the eye image from the face image, and after that identify the pupil and iris. Then, based on the corresponding image, we can recognize winking. In order to fulfill proposed requirements and to give the ability to work under any conditions and any head position, we decided to use an IR sensor and additionally we decided to mount sensors as close to the face/eyes as possible. This way the wearable technology [12] would be applied in our solution. Traditionally always accepted element (even in the operating room) are glasses. Therefore, wink detector can be "attached" to the glasses as a wearable sensor. It should ensure the correct operation in any eyesight direction.

3 Introduced Technical Solution

In the proposed solution, two modules that record reflection from the surface of the eye are used. Each module consists of micro camera recording infrared radiation and IR LED. The use of infrared radiation (instead of visible light) makes the analysis independent of external lighting conditions. In addition, it prevents the occurrence of glare, which could interfere with the user's work. Processing and analysis of images taken from a micro camera allows independently determining the state of the left and right eye. By the state of the eye to be understood one of two options – the eye is open or closed.

Cameras should be placed in such a way that they do not obscure the field of view and that they do record only image of eye – no other objects.

The quality of recorded images is a decisive parameter for the entire project, however small, low resolution cameras were also taken into consideration. We have experimentally determined that the micro camera should capture images with resolutions ranging from 4.8 thousand pixels (80×60) to 0.3 megapixels (640×480). Increasing the resolution improves the effectiveness recognition of the winking, while slightly increasing the computational cost. However, too large resolution does not improve effectiveness, only increases computational cost.

4 Influence of Infrared Radiation on the Eyeball

The influence of infrared radiation on the human body, especially on the eyes may be harmful. It is therefore necessary to select futures of IR LEDs that illuminate the eyes in such a way as to ensure work safety. It is a crucial condition for practical applications.

It is documented that the wavelength of the infrared radiation that is safe to the human eye should be greater than 1400 nm. Such waves do not penetrate the retina of the eye. It is further assumed that at the retinal level the emission should not exceed 100 W/m^2 [13]. We have used IR LEDs with a wavelength of 1550 nm and rated power that does not exceed the permissible level. In addition, experimentally, we have repeatedly reduced the power relative to the nominal value so that it is the smallest power at which the winking detection algorithm works correctly.

5 Algorithm of the Eye Winking Recognition

The proposed method of determining the state of the eye is based on the analysis of the degree of diffusion in the reflection of the radiation. This method allows performing accurate analysis while maintaining low computational cost. Usage of camera in this case is a typical application of camera as a sensor for registering and analyzing reflection [14].

The light that falls on any surface, including the surface of the body, is partially reflected [14, 15]. Depending on the type of surface, this reflection may be specular (directional) or diffuse (scattered). Infrared radiation emitted by the IR LED can be reflected from the eyeball when the eye is open or from the eyelid if the eye is closed. The eyeball is characterized by a smooth, glassy surface. From such a surface the light is reflected, primarily, directionally. On the images recorded by the micro cameras in this case, IR reflection emitted by the diode, is clearly visible. (Fig 1a). The reflection has small size but is very bright. Human skin, on the other hand, has very strong diffusing properties (Fig. 1b). The eyelid is illuminated, but there is no visible bright point of reflection.

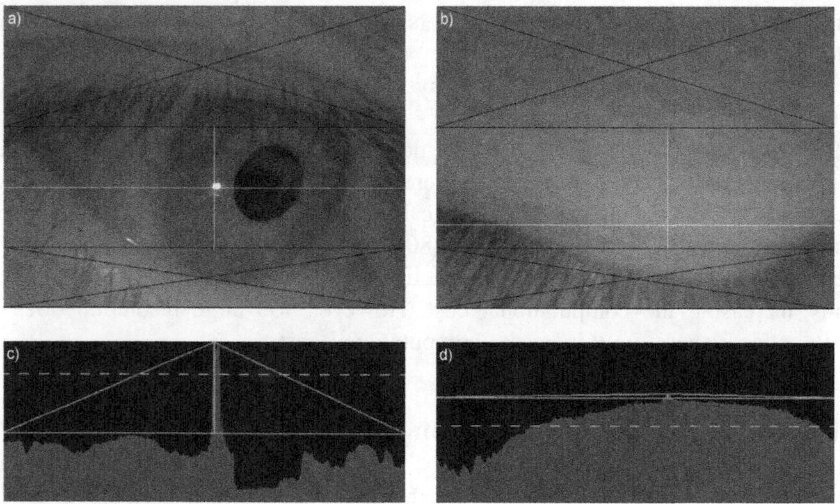

Fig. 1. The captured image of the right eye: (a) open (b) closed. The distribution of brightness in the horizontal line passing through the brightest pixel (c) open eye, (d) closed eye

It is therefore easy to determine the state of the eye based on the analysis of reflection. Strong diffusion means reflection from the eyelid: the eye is closed. The directional reflection from the eyeball means that the eye is open.

The algorithm for analysis of images recorded by micro cameras is implemented in three steps:

1. **Determination of the point of light reflection in the image.** This point is rarely one pixel in size – this is only possible with a very low resolution camera. Most often the point of reflection is represented as a group of bright, neighboring pixels. We analyze the difference in brightness between neighboring pixels and this way we build a group of pixels that creates a reflection area. Then, by specifying the center of gravity, we determine the exact point/pixel of the reflection. For several groups (what is possible while we notice diffuse reflection from wrinkled skin), we select the brightest group. This solution is accurate and, theoretically, allows eliminating other bright areas that could be mistakenly recognized as a reflection point. However, after conducting experiments on a set of about 5000 opened eye images and about 5000 closed eye images, we simplified the method. It turned out that simple finding the brightest pixel in the eye image, is effective enough and also much easier. The result of determining the brightest point of the image for the open eye is shown in Fig. 1c, and for the closed eye in Fig. 1d.
   The position of the reflection in the image is determined by the location of the IR LED and the camera. There is no need to look for reflection in the whole image. After experiments, in order to speeds up the analysis, we have limited the search area in the image. The excluded areas have been marked (crossed) in Figs. 1a and b.
2. **Determination of the brightness profile for the image.** Through the point selected in the first step, a horizontal line (cross-section) of one pixel width is carried out. For each point of this line the brightness of the image is determined. Then we generate a brightness graph on the section line. The pixel is brighter, the higher value it has and is represented in the brightness graph by a longer line. Because we analyze the image in one byte grayscale, so for each point of the section, a segment of up to 255 pixels high may appear. Examples of defined brightness profiles for the open and closed eye are illustrated respectively in Figs. 1c and d.
3. **Determination of the eye stage.** In this stage we analyze the brightness profile. The graph of specular reflection (Fig. 1c) is characterized by large differences in levels of luminance and rapid increase and decrease in luminance around the point of the highest level. In this case the eye is open. The graph for diffuse reflection (Fig. 1c) is characterized by a small differences in luminance levels and slow increase and decrease in luminance around the highest point. In this case the eye is close. The purpose of the analysis of brightness profile is only to isolate this two cases, rather than determining the reflection properties. Therefore, it is enough to test the differences in levels experimentally for many lighting environments and to specify luminance threshold. After that, using this threshold we look for local changing of luminance level around the point of maximum. Thus, the search starts from the maximum point (moving in the graph left or right) and finds the largest decrease in the neighborhood, taking into account local level differences. As the slope decreases, the process stops and the luminance level is determined for this point of the graph.

The proposed algorithm does not require complicated operations. After going through three stages, it is possible to unambiguously determine whether the user's eye is currently open or closed. The conducted experiments have shown practically 100% efficiency of determining the state of the eye in real time.

6 Analysis of Signals Generated from Eye Winks

In traditional control devices such as a mouse or keyboard, practically only single click (keyboard, mouse) or double click (mouse) is used. In both cases with very short time of individual click. Performing such activities using hands is not a problem, whereas analogous action with appropriate eye winking could be difficult for the user. The signals generated on the basis of winking, should be classified by the duration of the closure of the eye, and what is very important, the times of winking should be matched to the human capabilities. This will ensure the comfort of work. An additional problem is the filtering of natural eye blinks. To correctly identify winking, we have experimentally determined the longest time of natural blinking (t_0) and the shortest time of intentional winking (t_1).

We have separated two independent channels: for left and right eye. In each channel, the eye closure signal is recorded and its duration (t_x) is measured. We have proposed a simple algorithm for recognizing "eye gestures" and assigning the corresponding system event to them.

- **Simultaneous winking of both eyes – higher priority event.** Natural eye blinking is characterized by simultaneous occurrence of very short signal on both channels ($t_x < t_0$) – gesture B_0 in Fig. 2. This case is not further analyzed. When the user closes simultaneously his both eyes intentionally for a certain period of time ($t_x > t_0$), the signal B_1 is shown on both channels, as shown in Fig. 2. In this case, we have assumed that the B_1 gesture is used to enable/disable the detection system of winking.

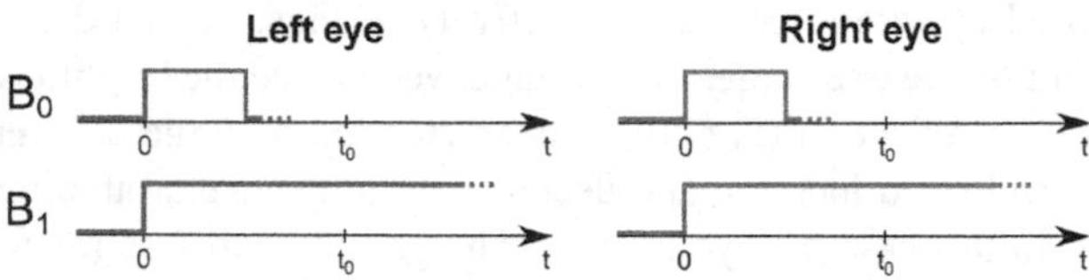

Fig. 2. Eye gestures: B_0 and B_1

- **Winking with one eye (the other is open) – event with lower priority.** Taking into account the time t_1, we have experimentally determined two additional times t_2 and t_3, such that $t_1 < t_2 < t_3$. If $t_x < t_1$ then the wink time is shorter than the shortest time of intentional wink and this case is not further analyzed. If $t_1 < t_x < t_2$, then there is a shorter intentional wink. If $t_2 < t_x < t_3$, then there is a longer intentional wink. If $t_3 < t_x$, then the eye is closed in intentional way (and it is not a wink!). The list of gestures for left and right eye is shown in Fig. 3.

The proposed method for time analyzing of the eye winking is the simplest possible, which gives the ability to separate up to 4 independent gestures for each eye.

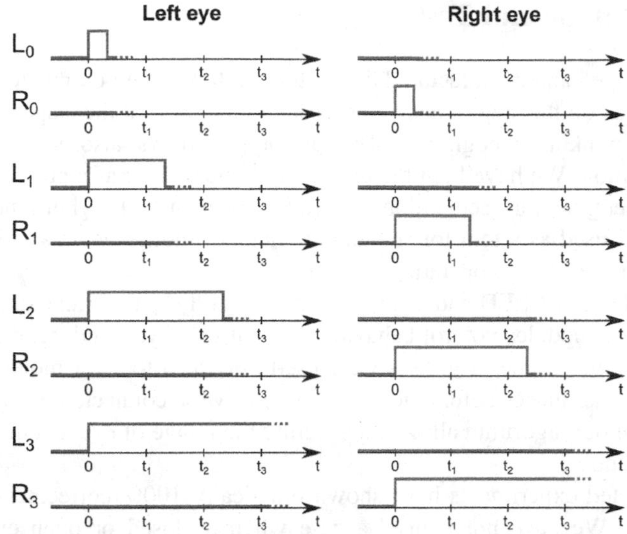

Fig. 3. Eye gestures: L_0 and R_0, L_1 and R_1, L_2 and R_2, L_3 and R_3

Experiments have shown that the proposed division into gestures is acceptable by the user and allows for quick and comfortable work. We have proposed how to assign system events to recognized gestures (Table 1.); based on known system commands generated using the mouse. Of course, the gestures described, generated on the basis of one or both eyes winking, can be arbitrarily translated into commands. This may be application-dependent and may be defined at the stage of software installation.

Table 1. Recognized eye gestures and corresponding events

Eye gesture	Closing time (t_x)		System event
	Left eye	Right eye	
B_0	$0 < t_x \leq t_0$	$0 < t_x \leq t_0$	None (this prevent blink recognition)
B_1	$t_0 < t_x$	$t_0 < t_x$	Switch on and off (of the interaction)
L_0	$0 < t \leq t_1$	$t_x = 0$	None
L_1	$t_1 < t_x \leq t_2$	$t_x = 0$	Single click of the left mouse button
L_2	$t_2 < t_x \leq t_3$	$t_x = 0$	Double click of the left mouse button
L_3	$t_3 < t_x$	$t_x = 0$	Hold down of the left mouse button
R_0	$t_x = 0$	$0 < t_x \leq t_1$	None
R_1	$t_x = 0$	$t_1 < t_x \leq t_2$	Single click of the right mouse button
R_2	$t_x = 0$	$t_2 < t_x \leq t_3$	Single click of the central mouse button or scroll (double click of the right mouse button is not popular)
R_3	$t_x = 0$	$t_3 < t_x$	Hold down of the right mouse button

7 The Prototype and Tests

We have developed and manufactured the prototype. It made it possible to carry out the necessary studies on the step of analyzing problem and also at the stage of developing algorithm for winking recognition. The prototype allows also testing the whole introduced solution. We have built the multimodal interaction: as a pointing device we have used an engine for recognition of head movements. The introduced winking recognition was used as a tool for simulating clicks of mouse buttons – this way user can create proper events in operating system.

Two modules of IR LED and camera are fixed rigidly to the frame of glasses worn on the head. The modules control behavior of the user's eyes (independently left and right eye). However frame of glasses were used only as a tool for mounting cameras and LEDs in proper places before the eyes. Cameras were connected by USB cables to computer where our algorithm allows interpreting the image of eye and turned winking into system events.

The conducted experiments have shown practically 100% correctness of the proposed solution. We have not noticed a case when a closed or open eye would be incorrectly identified. From technical point of view, the aim has been achieved. However, a much bigger problem is the acceptance of users – we deal with a completely new kind of interface. We have tested our eye control tool in a group of 30 participants. All participants used computer at work or home, but the proposed solution was new to all. Participants had to perform a set of simple tasks that are normally performed with the mouse: moving or selecting elements in the screen but using winking instead of mouse clicking. After the tests, participants were asked to evaluate the new solution. In the assessment we used methods consistent with the Standard ISO 9241-411:2012 [16]. The participants evaluated our solution using a 5-step subjective scale, in three independent subjects: Operation speed, Accuracy, General comfort (Fig. 4).

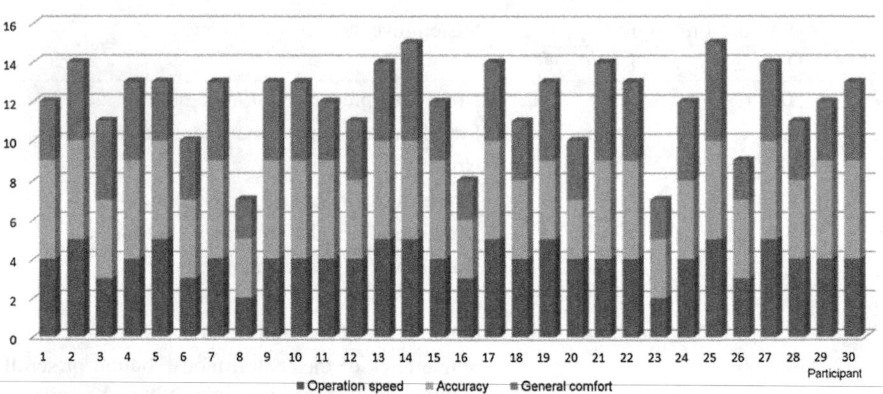

Fig. 4. The participants evaluation: Operation speed (from 1 = unacceptable to 5 = acceptable), the average result was 4.0 ($\delta = 0.83$); Accuracy (from 1 = very inaccurate to 5 = very accurate), the average result was 4.47 ($\delta = 0.73$); General comfort (from 1 = very uncomfortable to 5 = very comfortable), the average result was 3.5 ($\delta = 0.86$)

In the discussion, participants drew attention to the need of getting used to the new standard of communication. The high value of Operation speed (4.0) and Accuracy (4.7) shows the acceptance and correctness of our solution. Interesting, in this context, was lower level of General comfort (3.5), but the participants emphasized that this was the result of novelty of our solution and habits related to the standard mouse. One of the main problem for participants was to replace double clicking by long period of closed eye. The proposed algorithm allow for individual tuning the time conditions for each participant. The general opinion was that the solution seems to be slower, however the overall evaluation of all was positive. The participants pointed also out that there are people for whom winking at the right moment can be a very difficult task. This limitation seems to be the most serious problem of our solution.

On the other hand, the proposed method for identifying eye wink works properly in all cases. The proposed simple algorithm allowed interpreting the image of the eye and recognizing winking correctly. We have also tested the algorithm in different lighting environments. Additionally we have changed timing of events creation dependently of participant needs and we have noticed no problem in eye image interpretation. In all cases the replacement of mouse buttons by winking worked correctly.

8 Summary

The aim of the study was to present a new simple method for recognizing eye winking. We have used IR LEDs and cameras and built simple but effective algorithm for analysis of eye image. Initially, it was dedicated to control by head movement. However, as an independent module of interaction our solution can replace mouse buttons for creating proper system events in other situations. The introduced solution can be applicable in multimodal interaction for multimedia. As an additional control element for game players, as well as in professional systems for operating room or in control of the production line. In every situation, where operator hands are busy and cannot be used for keys click.

We have also built the prototype in which the proposed method was used. The solution was tested in different lighting environment. Simple algorithm of eye image interpretation allows selecting proper system events in any environment and with any head positions. The application shows the high usefulness of very simple method. The performed test on a group of participants showed correctness of proposed solution. The main advantage of the new solution was the ability to adjust the timing to the individual requirements of the users. For the users it is also important that the proposed solution will replace standard mouse clicks, and thus generate system events that are familiar to users.

In the future we plan, first of all, to change the connection to the computer – the wireless method will be used. In the future also, it is worth considering the possibility of automatically selecting the timing to the individual needs of users. This could be realized for example on the basis of a short test in which the user would be required to perform the winking in a predetermined manner in order to distinguish it from the standard blinking. More advanced methods (statistical model) will be also considered.

References

1. Shao, L., Shan, C., Luo, J., Etoh, M. (eds.) Multimedia Interaction and Intelligent User Interfaces: Principles, Methods and Applications (Advances in Computer Vision and Pattern Recognition). Springer (2010)
2. Turk, M.: A multimodal interaction: a review. Pattern Recogn. Lett. **36**, 189–195 (2014). doi:10.1016/j.patrec.2013.07.003
3. Muda, Z., Mohamed, R.E.K.: Adaptive user interface design in multimedia courseware. In: Proceeding of 2nd International Conference on Information and Communication Technologies. vol. 2, pp. 196–199 (2006) doi:10.1109/ICTTA.2006.1684369
4. O'Hara, K., et al.: Touchless interaction in surgery. Commun. ACM **57**, 70–77 (2014). doi:10.1145/2541883.2541899
5. Sutherland, I.E.: Sketchpad, a man-machine graphical communication system. Ph.D. thesis MIT, January 1963. Reprinted in Technical Report Number 574. University of Cambridge, Computer Laboratory. UCAM-CL-TR-574. September (2003)
6. Jaimes, A., Sebe, N.: Multimodal human computer interaction: a survey. Comput. Vis. Image Underst. **108**(1–2), 116–134 (2007). doi:10.1016/j.cviu.2006.10.019. (Special Issue on Vision for Human-Computer Interaction)
7. Mandal, B., Eng, H.-L., Lu, H., Chan, D.W.S., Ng, Y.-L.: Non-intrusive head movement analysis of videotaped seizures of epileptic origin. In: Proceeding of 34th Annual International Conference of the IEEE Engineering in Medicine and Biology Society, 28 August–1 September, pp. 6060–6063 (2012) doi:10.1109/EMBC.2012.6347376
8. Strumiłło, P., Pajor, T.: A vision-based head movement tracking system for human-computer interfacing. In: Proceeding of New Trends in Audio and Video/Signal Processing Algorithms, Architectures, Arrangements and Applications (NTAV/SPA), 28–29 September 2012, pp. 143–147 (2012)
9. Al-Rahayfeh, A., Faezipour, M.: Eye tracking and head movement detection: a state-of-art survey. IEEE J. Trans. Eng. Health Med. **1**, 2100212 (2013). doi:10.1109/JTEHM.2013.2289879
10. Singh, H., Singh, J.: Human eye tracking and related issues: a review. Int. J. Sci. Res. Publ. **2** (9), 1–9 (2012)
11. Kapoor, A., Picard, R.W.: A real-time head nod and shake detector. In: Proceeding of 2001 Workshop on Perceptive User Interfaces. November 2001, pp. 1–5 (2001) doi:10.1145/971478.971509
12. Calvo, A.A., Perugini, S.: Pointing devices for wearable computers. Adv. Hum Comput. interact. **2014**(527320), 10 (2014). doi:10.1155/2014/527320
13. Wolska, A.: Artificial Optical Radiation - Principles of Occupational Risk Assessment (in Polish). Central Institute for Labour Protection - National Research Institute, Warsaw, Poland (2013)
14. Dorsey, J., Rushmeier, H., Sillion, F.: Digital Modeling of Material Appearance. Morgan Kaufmann Publishers, Burlington (2007)
15. Baranoski, G.V.G., Krishnaswamy, A.: Light and Skin Interactions: Simulations for Computer Graphics Applications. Morgan Kaufmann, Burlington (2010)
16. Ergonomics of human-system interaction–Part 411: evaluation methods for the design of physical input devices. ISO Standard: ISO 9241–411:2012 (2012)

Adaptive Low Cost Algorithm for Video Stabilization

Giuseppe Spampinato(✉), Arcangelo Bruna, Filippo Naccari, and Valeria Tomaselli

Advanced System Technology, Imaging Group, ST Microelectronics, Stradale Primosole 50, 95100 Catania, Italy
{giuseppe.spampinato, arcangelo.bruna, filippo.naccari, valeria.tomaselli}@st.com

Abstract. Video stabilization is a technique used to compensate user hand shaking. It avoids grabbing the unintentional motion in a video sequence, which causes unpleasant effects for the final user. In this paper we present a very simple but effective low power consumption solution, suitable for cheap and small video cameras, which is robust to common difficult conditions, like noise perturbations, illumination changes, motion blurring and rolling shutter distortions.

Keywords: Video stabilization · Characteristics curves · IIR filters · FIR filters · Curves matching

1 Introduction

Different image stabilization techniques, both block-based and feature-based ones, have been proposed in literature. Most of these techniques are computationally expensive and thus not suitable for real time applications. We work under the assumption of limited (i.e. few pixels) misalignments between adjacent frames. Moreover, we take into account only vertical and horizontal shifts, because rotation is less disturbing for optical models with a wide field of view and the rotational compensation is expensive in terms of computations.

The proposed technique is primarily inspired by the use of characteristics curves and tends to overcome some limitations, often occurring in practical situations, especially in low-cost cameras. This is obtained by an accurate analysis of the characteristics curve by filtering out the deceptive information.

This paper is structured as follows: in Sect. 2 the prior art is shown; in Sect. 3 the proposed method is described. Then the experimental results are shown in Sect. 4, followed by some conclusions and future work in final Sect. 5.

2 Prior Art

Prior art digital video stabilization techniques can be grouped in two main categories: block based and feature based. The block based techniques split the image into blocks and then blocks of current frame are compared with blocks of previous frame to

S. Battiato et al. (Eds.): ICIAP 2017, Part II, LNCS 10485, pp. 363–372, 2017.
https://doi.org/10.1007/978-3-319-68548-9_34

calculate the related motion. The matching between blocks of the current and the previous frame is done on the basis of block matching criteria. Smaller the value of matching criterion better is the match between blocks. To retrieve the motion vector for each block, different matching metrics have been proposed, like the Partial Distortion Elimination (PDE) [1], Mean Absolute Difference (MAD) [2] or Universal Image Quality Index (UIQI) [3]. A compounding algorithm to detect the global motion vector is then applied. These techniques are usually robust, but really slow, because the whole image should be processed, in a block by block fashion.

The feature based techniques, on the contrary, allow retrieving directly the global motion vector through the analysis of a particular feature. Recent papers have mainly adopted Speeded Up Robust Features (SURF) [4], Scale Invariant Feature Transform (SIFT) [5] and Kanade-Lucas-Tomasi (KLT) techniques [6]. Even if there is no need to process the whole image, as the block based algorithms, the disadvantage of feature based methods is that they are strictly dependent on feature point extraction step [7], in terms of accuracy. Moreover, even if these kinds of algorithms are robust enough, they are too expensive to be used in low cost cameras. Their complexity is due, not only to the expensive feature calculation, but especially to the related feature matching.

Usually in feature based techniques, the optical flow analysis is executed [8–10]. The optical flow is a useful representation of the scene, consisting in the set of motion vectors, calculated between the previous and current frame. Main problem is to distinguish foreground motion vectors, caused by moving objects, which should be not considered for the video stabilization, from background motion vectors, which have to be used to determine the correct stabilization. To solve this problem usually the Random Sample Consensus (RANSAC) [11] is used. Since RANSAC is a not deterministic iterative method, the worst case could require too much iterations to converge, excessively slowing down the whole processing. In recent years some optimizations of RANSAC algorithm have been proposed [12], but not so relevant to drastically increase performances.

Looking at the complexity of the aforementioned methods, we have chosen to start from a simpler technique, based on motion estimation through integral projection curves [13, 14]. This technique works as follows: for simplicity, let us assume we have two gray-scale temporally adjacent frames, where M and N are the horizontal and vertical dimensions and p_{ij} is the pixel value in the position (i, j). The characteristics curves along the horizontal and vertical dimensions are respectively defined as:

$$C_h(j) = \frac{1}{N}\sum_{i=1}^{N} p_{ij} \tag{1}$$

$$C_v(i) = \frac{1}{M}\sum_{j=1}^{M} p_{ij} \tag{2}$$

The meaning of the curves can be easily understood by referring to the drawing of Fig. 1, where two schematic characteristics curves are reported. Figure 2 shows C_h curves for two successive frames F_1 and F_2. A shift of the curves along the x-axis represents an equivalent shift of the frames in the horizontal dimension (similar approach for y-axis and vertical shift). From Fig. 2 the horizontal shift is particularly evident. Hence, in order to properly evaluate the motion occurring between consecutive frames, the shift along the axes (off_h, off_v) of both C_h and C_v curves can be calculated as follows:

$$P_h(s) = 1\frac{1}{M-|s|}\sum_{j=\max(1,-s)}^{\min(M-s,M)}\left|C_h^{F_1}(j) - C_h^{F_2}(j+s)\right| \\ off_h = \{s' : P_h(s') = \min P_h(s)\} \tag{3}$$

$$P_v(s) = \frac{1}{N - |s|}\sum_{j=\max(1,-s)}^{\min(N-s,N)}\left|C_v^{F_1}(i) - C_v^{F_2}(i+s)\right| \\ off_v = \{s' : P_v(s') = \min P_v(s)\} \tag{4}$$

The term s is the search window size and represents the maximum retrievable displacement.

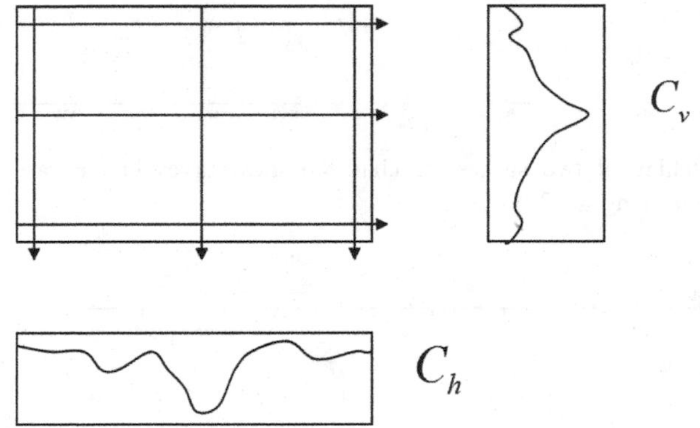

Fig. 1. A schematic representation of characteristics curves.

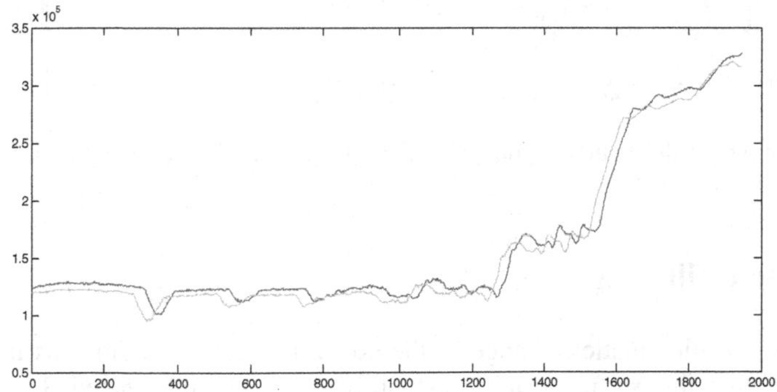

Fig. 2. Characteristics curves related to two adjacent frames.

The described method is low-cost and robust to noise, but it suffers in the cases of illumination changes and motion blur. The effect of illumination changes, in the calculated characteristics curves, is shown in the following example. Figure 3 represents the results of the ideal matching (−29), that is one curve should be shifted by 29 pixels to match with the other one. It is to note that the shape of the curves is similar, but the values are slightly different due to the scene illumination changes. Due to these differences, with the integral projection method, the matching is not correct, as indicated in Fig. 4. This effect is more evident along the edge (a strong edge is visible around the point 1600).

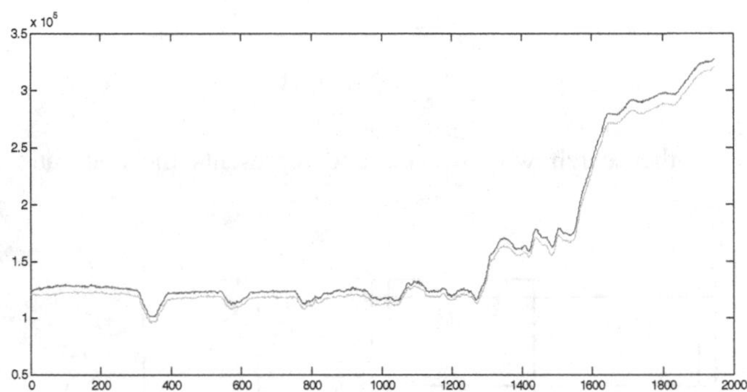

Fig. 3. Ideal matching of two successive characteristics curves in the case of illumination changes (perfect matching = −29).

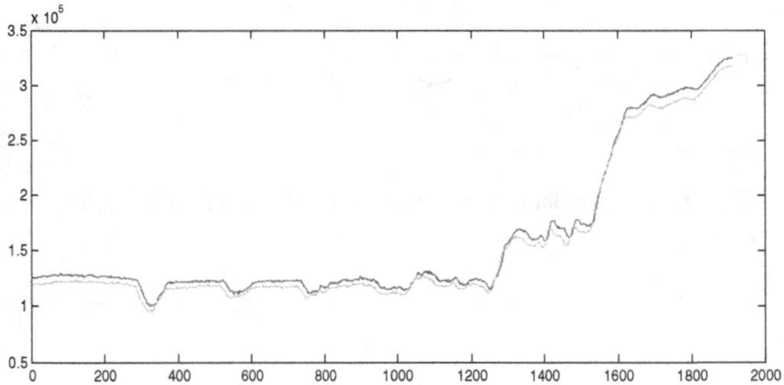

Fig. 4. Matching results with integral projection in the case of scene illumination changes (wrong matching = −25).

3 Proposed Filtering

The effect of the illumination change in the curve is basically a shift, as indicated in Fig. 3. Removing this shift, the matching problem can be eliminated. It can be performed deleting the DC component to the integral projection signal, by pre-filtering the characteristics curve with a High Pass Filter (HPF), thus enhancing curve peaks, obtained as follows:

$$C_x' = C_x - (C_x \otimes \ LPF) \tag{5}$$

$$C_y' = C_y - (C_y \otimes LPF) \tag{6}$$

A very simple LPF with 16 ones (i.e. [1111111111111111]/16) allows obtaining a simple HPF, with good results and low extra cost, subtracting the filtered data to the original one. The filter response is shown in Fig. 5. In the proposed example, with this simple filter, we obtain the perfect matching (−29), as shown in Fig. 6.

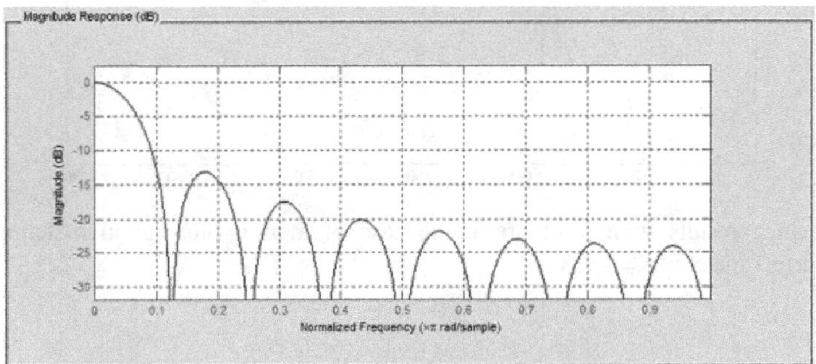

Fig. 5. The magnitude response of the HPF: 1 − [1111111111111111]/16.

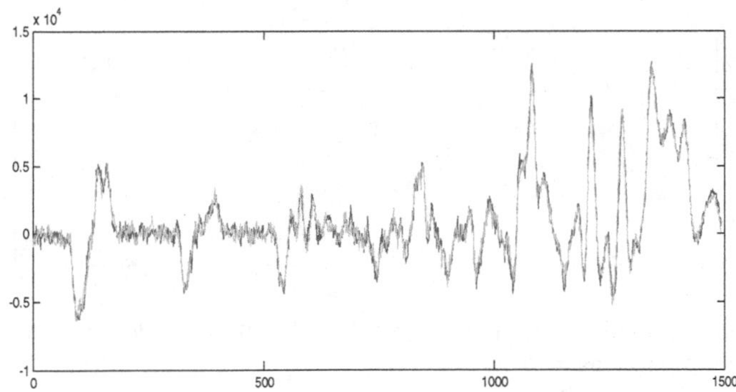

Fig. 6. Perfect matching results obtained applying a simple HPF to the characteristics curves in the case of scene illumination changes

Although this filter works better than prior art approaches in case of scene illumination changes, this is not the case when motion blur is present. To better understand this problem, we will show an example. In Fig. 7 the unfiltered characteristic curves are plotted. The matching is not perfect (+11 instead of +13 pixels). Using the HPF shown above, the matching is even worst (+7), as shown in Fig. 8. The mismatch is mainly due to the highest frequencies, considered as noise. By removing highest frequencies, hence applying a Low Pass Filter, the problem can be reduced.

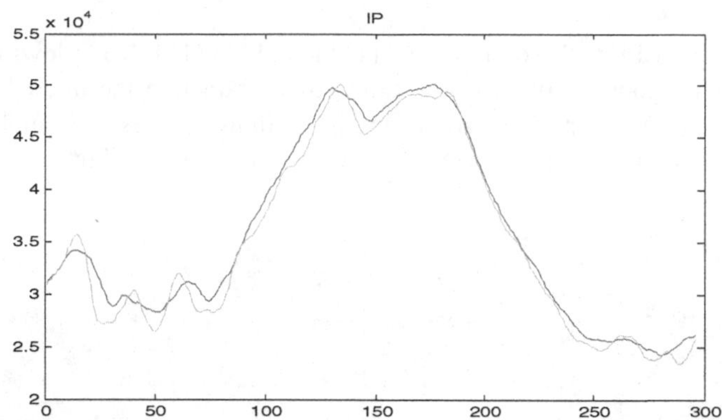

Fig. 7. Matching results with prior art in the case of motion blur (good matching = + 11, instead of perfect matching = + 13).

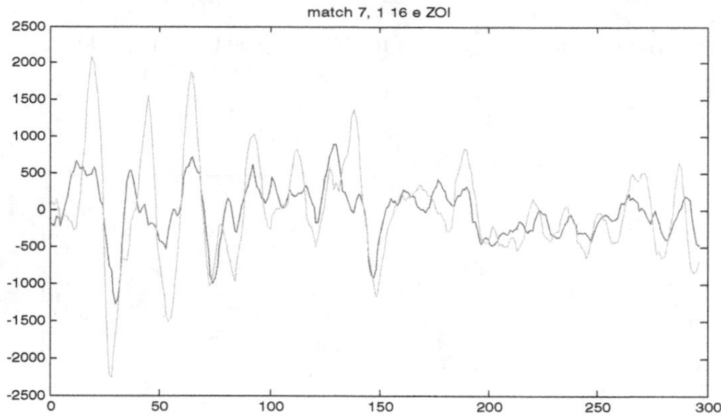

Fig. 8. Matching results obtained applying a simple HPF to the characteristics curves in the case of motion blur (bad matching = + 7).

Combining the two conditions, illumination changes and motion blur, a Band Pass Filter (BPF) seems to be the solution, so a Butterworth IIR BPF has been chosen. The general form of the IIR filter function is defined as:

$$H(z) = \frac{\sum_{l=0}^{L-1} b_l z^{-l}}{1 + \sum_{m=1}^{M} a_m z^{-m}} \tag{7}$$

The design problem consists in determining the coefficients b_l and a_m so that *H(z)* satisfies the given specifications. A second-order IIR filter has been chosen, to obtain a good tradeoff between implementation cost and results. The cutoff frequencies of the filter were fixed to $w1 = 0.01$ Hz and $w2 = 0.20$ Hz, chosen after having performed several simulations to obtain better results in the matching. The magnitude response of this filter is indicated in Fig. 9. With this filter, we obtain a good matching in both illumination changes and motion blur conditions, as indicated in Fig. 10 (related to the example in Fig. 4) and Fig. 11 (related to the example in Fig. 7).

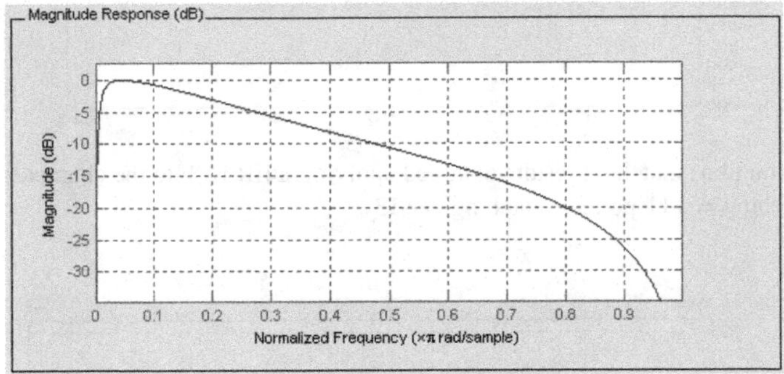

Fig. 9. The magnitude response of the filter 2th order BPF, Butterworth IIR, Direct form II, w1 = 0.01; w2 = 0.2.

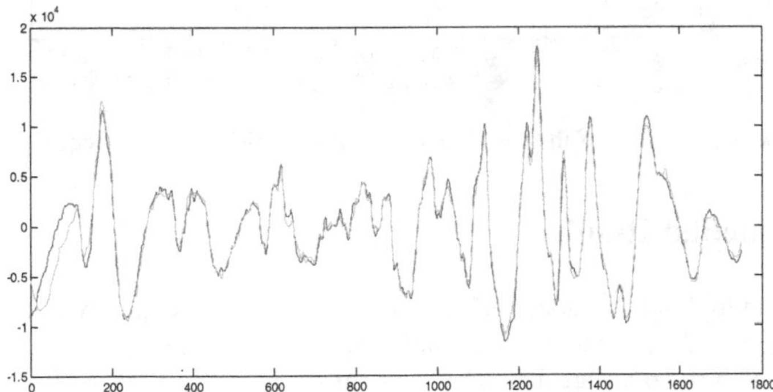

Fig. 10. Scene illumination changes: matching results obtained applying a IIR BPF to the characteristics curves (perfect matching = −29).

This IIR BPF is really a good solution, even if it requires several floating point multiplications. It should not be disregarded that IIR filters are often prone to coefficients quantization errors. The zeroes-poles graph of the IIR filter, represented in Fig. 12, shows that one pole is quite close to the unit circle, so this IIR filter is susceptible to finite precision effects, and hence optimization of this filter should be carefully chosen and tested.

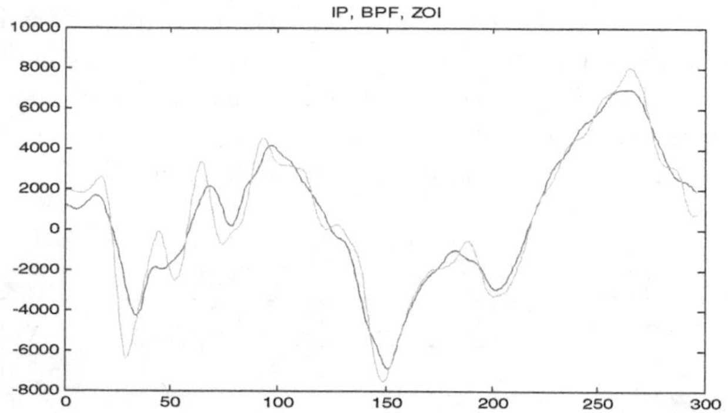

Fig. 11. Motion blur: matching results obtained applying a IIR BPF to the characteristics curves (obtained matching = +11 perfect matching = +13).

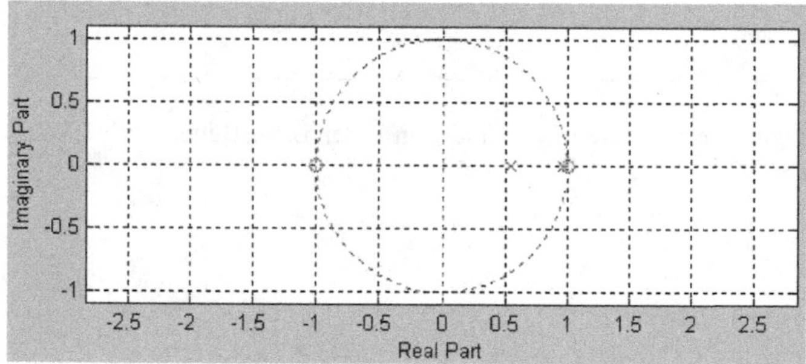

Fig. 12. Zeroes-poles graph of the IIR Butterworth BPF, order 2, cut-off frequencies 0.01–0.2.

4 Experimental Results

About one hundred video sequences have been used to test the quality of the results in both objective and subjective way. Tables 1, 2 and 3 show the results obtained respectively with the prior art method (Classical IP) and the proposed FIR (HPF) and IIR (BPF) methods. In these Tables, for each video sequence, the minimum (*MinX* and *MinY*), the maximum (*MaxX* and *MaxY*) and the standard deviation (*StdX* and *StdY*) were computed between the real motion vectors and the estimated ones (in pixel), for each axis (*X* and *Y*). We can note a reduction in the standard deviation error of about

11% for the FIR and about 16% for the IIR. In particular, the best improvement of IIR compared with the prior art method and FIR is obtained in the cases of weakness of these methods, represented by the presence of overall motion blur and artifacts due to the rolling shutter (respectively video 3 and 6).

Table 1. Previous art results.

Classical IP	MinX	MinY	MaxX	MaxY	StdX	StdY
1_intended_motion	−1.13	−1.47	1.16	1.18	0.47	0.52
2_no_intended_motion	−0.97	−1.08	1.04	1.13	0.33	0.38
3_motion_blur	−3.32	−2.91	4.15	2.98	0.64	0.73
4_no_intended_motion	−0.77	−0.70	0.84	0.65	0.32	0.31
5_intended_motion	−1.00	−1.35	1.39	1.37	0.44	0.52
6_rolling_shutter	−18.47	−1.86	3.74	16.72	1.49	1.29
7_no_intended_motion	−1.00	−1.28	1.07	1.22	0.40	0.48
8_intended_motion	−0.76	−0.49	0.77	0.54	0.31	0.23
Average	−3.43	−1.39	1.77	3.22	0.55	0.56

Table 2. Proposed FIR HPF results.

FIR	MinX	MinY	MaxX	MaxY	StdX	StdY
1_intended_motion	−1.13	−1.70	1.16	1.16	0.44	0.52
2_no_intended_motion	−1.11	−1.51	2.97	1.13	0.38	0.38
3_motion_blur	−4.07	−4.81	2.48	6.66	0.71	0.93
4_no_intended_motion	−1.19	−0.59	0.99	0.58	0.36	0.29
5_intended_motion	−0.99	−1.21	1.39	1.18	0.45	0.51
6_rolling_shutter	−2.85	−1.86	4.74	1.62	0.92	0.59
7_no_intended_motion	−1.03	−1.12	1.07	1.19	0.41	0.46
8_intended_motion	−0.53	−0.53	0.68	0.54	0.28	0.23
Average	−1.61	−1.67	1.93	1.76	0.49	0.49

Table 3. Proposed IIR BPF results.

IIR	MinX	MinY	MaxX	MaxY	StdX	StdY
1_intended_motion	−1.13	−1.47	1.16	1.13	0.44	0.48
2_no_intended_motion	−1.11	−1.08	1.04	1.13	0.34	0.38
3_motion_blur	−2.32	−2.50	1.18	5.23	0.58	0.72
4_no_intended_motion	−0.87	−0.65	0.99	0.58	0.34	0.30
5_intended_motion	−0.99	−1.35	1.39	1.18	0.43	0.49
6_rolling_shutter	−2.32	−1.66	4.54	1.50	0.85	0.54
7_no_intended_motion	−1.03	−1.11	1.07	1.22	0.40	0.46
8_intended_motion	−0.56	−0.54	0.64	0.55	0.28	0.23
Average	−1.29	−1.29	1.50	1.57	0.46	0.45

5 Conclusion and Future Work

A very low-cost algorithm for video stabilization has been developed, suitable for real-time processing. It achieves significant improvements in both subjective and objective manner compared with state of the art algorithms with similar complexity, reaching about 16% of improvement in standard deviation of the error. Moreover, apart to be robust to noise, like the prior art methods, it is also robust to illumination changes, motion blurring and rolling shutter distortions. Further investigation will involve a deeper research of better BPF to approximate the IIR chosen, to fully work in finite precision, and extension of the proposed technique to roto-traslation.

References

1. Jagtap, A.P., Baviskar, P.V.: Review of block based video stabilization. Int. Adv. Res. J. Sci. Eng. Technol. (2015)
2. Bhujbal, D., Pawar, B.V.: Review of video stabilization techniques using block based motion vectors. Int. Adv. Res. J. Sci. Eng. Technol., vol. 3, no. 3 (2016)
3. KovaazEvic, V., Pantic, Z., Beric, A., Jakovljevic, R.: Block-matching correlation motion estimation for frame-rate up-conversion. J. Sign. Process. Syst. **84**(2), 283–292 (2016)
4. Salunkhe, A., Jagtap, S.: Robust feature-based digital video stabilization. Int. J. Adv. Res. Electron. Commun. Eng. (IJARECE) (2015)
5. Patel, M., Parmar, N., Nilesh, M.: Comparative Analysis on Feature Descriptor Algorithms to Aid Video Stabilization and Smooth Boundary Reconstruction Using In-painting and Interpolation, Int. J. Comput. Appl., Vol. 140, No.4 (2016)
6. Spampinato, G., Bruna, A., Guarneri, I., Tomaselli, V.: Advanced feature based digital video stabilization. In: 6th International Conference on Consumer Electronics, ICCE Berlin (2016)
7. Rawat, P., Singhai, J.: Review of motion estimation and video stabilization techniques for hand held mobile video. Sign. Image Process. Int. J. (SIPIJ), vol. 2, no. 2 (2011)
8. Liu, S., Yuan, L., Tan, P., Sun, J.: Steadyflow: spatially smooth optical flow for video stabilization. In: IEEE Conference on Computer Vision and Pattern Recognition (CVPR) (2014)
9. Goldstein, A., Fattal, R.: Video stabilization using epipolar geometry. ACM Trans. Graph. **32**(5), 126 (2012)
10. Wang, Y.S., Liu, F., Hsu, P.S., Lee, T.Y.: Spatially and temporally optimized video stabilization. IEEE Tran. Vis. Comput. Graph. **19**(8), 1354–1361 (2013)
11. Fischler, M.A., Bolles, R.C.: Random sample consensus: a paradigm for model fitting with applications to image analysis and automated cartography. Commun. ACM **24**(6), 381–395 (1981)
12. Lebeda, K., Matas, J., Chum, O.: Fixing the locally optimized RANSAC. In: British Machine Vision Conference, Guildford, United Kingdom (2012)
13. Kim, M., Kim, E., Shim, D., Jang, S., Kim, G., Kim, W.: An efficient global motion characterization method for image processing applications. IEEE Trans. Consum. Electron. **43**(4), 1010–1018 (1997)
14. Koo, Y., Kim, W.: An image resolution enhancing technique using adaptive sub-pixel interpolation for digital still camera system. IEEE Trans. Consum. Electron. **45**(1), 118–123 (2005)

Remote Biometric Verification for eLearning Applications: Where We Are

Pietro S. Sanna(✉) and Gian Luca Marcialis

Department of Electrical and Electronic Engineering,
University of Cagliari, Cagliari, Italy
{pietro.sanna,marcialis}@diee.unica.it
http://pralab.diee.unica.it/en/Biometrics

Abstract. In recent years there has been a huge spread of online learning methods - eLearning - that have allowed learners of any type to enjoy well-structured lessons and articulated training course, with solid teaching methodologies. Many solutions and platforms have extremely useful and powerful features for the student; the innovative methodologies have brought unexpected results to all levels, multiple contexts, and areas of interest. In the e-learning there is an ongoing revolution that determines a strong impact on the training system on different fields: methodological, educational, social and technological. However, the identity of an individual - in eLearning - has always been the basis of some controversies. In this regard, a critical aspect concerns the User Identity verification and recognition. Currently there is an element of attention, the problem of examination is still to be done in the presence of the teacher in order to be permanently accredited. Presently are already used various user authentication and identification techniques, such as conventional User ID and password, challenging questions, and some biometric techniques. From this point of view, biometric technologies that have the feature of verifying their personal identity without being "stolen" or "lend" can provide a possible solution. Despite this fact, a few approaches have been proposed for eLearning applications. The purpose of this paper is to provide a summary of the state of the art of biometric methods applied to current eLearning solutions.

Keywords: eLearning · Biometrics · Pattern classification · Authentication · User identification · Learning Management System · MOOC · Gamification · Keystroke dynamics · Stylometry

1 Introduction

The development and dissemination of instruments and didactic methodologies in eLearning area has had a strong acceleration in recent years. Not only with regard to the technological aspects in the development of the systems and also in content management, but even more in the optimization of communication and online teaching models, in deepening cognitive engineering, in delivery typologies and in the defined standards, etc.

S. Battiato et al. (Eds.): ICIAP 2017, Part II, LNCS 10485, pp. 373–383, 2017.
https://doi.org/10.1007/978-3-319-68548-9_35

eLearning is in fact a process that combines content, technology and cognitive aspects, closely related to the teaching methodology.

The global industry of training and development is complex and changing, and the eLearning approach currently involves all educational scopes, from primary and secondary school to academic education - in turmoil on all eLearning issues - to long life learning aspects up to the professional and business upgrade.

Learners include broader age ranges, cultural and social. They are involved in the use of systems and eLearning contents increasingly rapid and far removed from the new emerging trend. Currently the interaction with the learner relies on the paradigm of speed and mobility, consequently, on the scarcity of the available time. In addition, digital technologies for teaching are undergoing profound changes, including social training, mobile learning micro-learning, the recent development of *MOOC* in academia and business, and more. Consider the very recent introduction of *gamification* techniques now spread, to new methods of Flipped Classroom or Digitized Classroom [1].

The lack of organization strategy in the definition of an online learning program, of expert human resources in monitoring the users participation and satisfaction, as well as the problems to deal with in the development and implementation of eLearning projects, among others, are still elements of discussion and resistance to change management.

In particular, the identity of an individual - in eLearning courses - has always been the basis of some controversies by those who are critical for distance education. A critical aspect on which we carried out an initial analysis concerns the User Identity Verification in eLearning, with particular reference to the arrangements for evaluating the learner on an online training course: using biometrics for identification and recognition.

The ability to authenticate and identify the learner throughout the whole phase of the examinations and forms of assessment in general is still matter of discussions. Moreover, all modes leading to invalidate the examination test, such as the use of unforeseen media (books, tablet, mobile phone, other web browser, etc.) or the interaction of people unrelated to the test, in the same environment or remotely, are among the basic issues when allowing a remote examination to the students.

Adopting biometric techniques to this aim, could be a possible solution beside conventional techniques such as *User ID* and *Password*. Some biometric traits have been proposed, such as voice recognition, video and synchronous facial monitoring, keystroke dynamics, and others.

In the following Sections, we synthesize the salient characteristics of eLearning environments, the related state of the art, and the proposed biometric traits for the user identification for the distance evaluation (test and examination) of students. We also take into account the limitations of a single trait, which leads to the use of multiple biometrics. The paper ends with some discussions about pros and cons of biometric authentication in eLearning applications, even by considering the current technological point of view.

2 eLearning Technologies

Observing the landscape of online education, in the broader sense (institutions, organizations, public and private market) it is possible to notice that there is large number of organizations that operate in this segment: large corporations offer online training services (e.g. *Coursera, Lynda, EDX, Codecademy, Litmos,* etc.) and institutions such as universities and research centers.

Technologies which allow training have gradually evolved, becoming increasingly complex. Mobile technologies, such as cloud and social tools have made possible new approaches to formative experiences.

Current technologies and instruments have to be "tuned" with learners' habits, minimizing the friction of the training process. Users know quite well popular software on the market, for which, for example, the use of an eLearning software should require any training. Moreover, users learn online on a daily basis (*e.g.*, how to prepare recipes, assembling furniture or learn languages), using standard and widely disseminated educational conventions. The role of gamification, augmented and virtual reality - that increases engagement - and eLearning motivation is unquestionably essential in this process [2].

An overly complex technology represents a serious obstacle for training and education; technology should be really exciting for users to learn better and faster. The history of distance learning starts in the early 1900s. One of the first forms of distance learning took place through correspondence courses with teaching materials, exercises, papers sent to students by postal services. Then, learners provided their feedback in the form of filled questionnaires and documents to the teachers for the examination.

With the era of technology and internet models, paradigms and contents, the distance learning approach changed significantly [3] *Computer-Based Training (CBT)* led to the evolution of learning management with audio/video modalities and then with CD-ROM technology. Afterwards, in 1990s, the *Web-Based Training (WBT)* had a wide diffusion with the Internet advent until years 2000, in which the organizations began to combine Instructor Lead Training (ILT) with WBT by defining the *Blended and Information Training.* Today, discussions about *Collaborative, Social and Talent-driven Learning* occur, as Innovative and new learning design to address knowledge gaps.

Clearly, Internet is a key technology element for optimal results, as well as the offline access. As already said, mobility allows a training process without solution of continuity, always and everywhere and in different conditions and with any device. Today's users live in symbiosis with their devices.

To date another key element is the micro-learning, with short and targeted contents, highly effective in blocking the learner's attention. This is a support and reinforcement of the traditional training.

In recent years, the number of eLearning systems strongly increased, with a number of new features and a variety of options to choose from. In particular, Learning Management System - LMS and Content management are evolving, by exploiting the standardization of contents (eg. SCORM standard and its evolutions), the support of multiple sources (eg. YouTube, TED, Slideshare, MOOCs, etc.) and the interaction capabilities of students.

On overall, the eLearning process combines content, technology and cognitive aspects, the latter closely related to teaching methodology.

As mentioned in the report "eLearning: La rivoluzione in corso e l'impatto sul sistema della formazione in Italia", Aspen, 2014, about technology, we can consider three main areas: (A) Learning Management System (LMS); (B) proprietary software; (C) applications for mobile systems ("apps") [1]. The area (B) relates to some great universities and research centres, which realize its own platform for teaching.

Generally they do not use web applications, but specific software shared among the different actors involved. The area (C) is the most recent and promising. The application market for mobile systems (tablets and smart phones) is expanding significantly, because apps are flexible tools, cheap, and do not require significant network performance. IT companies are pushing much the realization of mobile applications for education, such as educational games [1].

Currently, the area (A) is the most outstanding. Numerous commercial and open source solutions are available, each one having its strengths and weaknesses, methodologically different and technologically good. They exhibit very interesting features in content management, in specific functionalities, in teaching material management, exams, profiling of different user categories, and more.

All these functions are made possible through LMS that is mainly aimed at the management of learners and learning activities. A standard LMS is characterised by a minimal set of feature such as learner tools (communication, productivity, student involvement), support tools (administration, course delivery, curriculum design), and technical tools (hardware/software, licensing and pricing) [4].

Many evaluation tools have been developed to carry out the comparison of eLearning solutions and LMS, often dividing LMS in the groups of commercial and open Source, considering features as technical and support tools, adaptability, usability and international widespread, innovative functions and paradigms.

Furthermore, in the case of an LMS equipped with tools for content creation, as in the majority of cases, it is called *Learning Content Management System (LCMS)*, LMS belong to the broader category of Content Management System (CMS). According to the philosophy of education and learning Instructional Designers define, the product will assume the prerogative that will deviate slightly from the classic definition of LMS [7–11].

Some of the most popular LMS/LCMS platforms or with characteristics it is important to mention: *Moodle* considered being the best overall even if with some limitations. It is preferred in academia and PMI. *Claroline* available in over 30 languages worldwide, *BlackBoard*, which is a commercial VLE. And again, *Ilias, ATutor, Sakai*, with different adaptability features [4,5].

It may be noted that in 2012 there were more than 500 producers of eLearning Platforms (Bersin & Associates). The market is fragmented, in fact only five producers hold a market share of over 4% and *Moodle* hold a market share of 30% in the Public sector and Education.

Recently, the so-called *MOOCs solutions (Massive Open Online Courses)* increasingly affirmed worldwide as a major on-line training system [1,6]. In particular, MOOCs characterize a different paradigm:

- Short duration and pace of videos (multimedia teaching contents)
- Educational material organized in a flexible and dynamic by the teacher
- Delivery with intervention of experts in Virtual Classroom
- Frequent evaluation and self evaluation of learning
- Individual works (deliveries) evaluated by the teachers or discussed and evaluated between peers.

eLearning contents are usually designed and realized in order to be independents from the system used in delivery phase, and then interoperable, accessible and reusable. The content module is made up of Learning Objects - LO, short-term, rapid and reproducible at any time and with a definite teaching goal. To ensure such interoperability and traceability between platforms, they have been defined and perfected specific standards.

The most known and consolidate is the *Shareable Courseware Object Reference Model – SCORM* that is a set of technical standards developed for eLearning software products, it enables interoperability: the model determines how online Learning Content and LMS communicate each other.

A trend that may become a huge part of eLearning is known as *Tin Can* that was the Project Name; *Experience API* is the Name of the Standard. Tin Can was born with the aim of collecting information on learning of learners and it uses specific data structure called *LRS - Learning Record Store* - unique for each user. LRS allows to collect and transmit the set of events of every learner, and makes it possible to collect lot of data about the experiences of a user, online and offline in the background.

3 The Personal Verification Issue: The Biometric Perspective

As observed some years ago, eLearning courses represent a very challenging field for educators: students can operate in a highly independent manner that removes or significantly reduces the instructor control, by providing the opportunity to cheat [12,13]. It is also believed that in most online courses, students engage in a greater number of episodes of cheating than students who attend traditional courses [12,14,15].

We may note that from 2014, 5.5 million students took at least one online course. To ensure integrity in online education, the distance learning industry is working on strategies to verify the identity of distance learning students, with the aim to improve the certainty that students really did the assigned work to get his credits [16,18,19]. The rising status of online education requires new user authentication and identification parameters to preserve the credentials of institutional integrity and students code of conduct.

Determining the identity of students is a crucial issue which educators have to deal with on any eLearning module. This moves to the "front and center" of the classroom experience during testing and determining the originator of written exams. Moreover, determining user identity in the virtual classroom is also linked to user progress and student aid eligibility [12].

So, in recent years, various approaches and methodologies to User Authentication for online learning environment have been tested and some simple solutions are now integrated into a few online learning platforms. There is a strong interest in finding a reliable and cost-effective protocol to safely assess students engaging in learning [18,19].

Nowadays there is great variety of eLearning paradigms, and different assessment methodologies available, but one strategy will not fit all situations. Verifying user identity in every situation is not realistic, practical and cost effective, and again, the students satisfaction is a strong requirement whish shouldn't interfere with learning and a student's privacy [16].

With regard to this problem, biometric identification techniques are inherently designed to recognize particular characteristics of our bodies and our behavior, considering that body and behavioral dynamic changes although slight are inevitable. Consequently, biometric technologies require a reliable and robust operating environment [12].

Biometric traits fall into the *something you are* paradigm, in contrast to the widely used *something you know* (passwords and PINs) and *something you have* (smart cards, tokens) paradigms.

It is possible to consider two categories of biometric traits:

- Physiological characteristics, such as retina, iris or facial characteristics, fingerprints, hand or palm geometry.
- Behavioral characteristics such as signature, keystroke dynamics and gait, voice [22].

By considering the cost of online identification and verification, sophisticated systems and solutions can be integrated into the process [12,13]. A single biometric technology cannot meet all requirements; different biometrics have pros and cons, and each one has specific features and properties. Moreover, they exhibit a different degree of individuality [17].

In order to integrate a biometric trait for supporting the identity verification process of an online course: (i) the user verification identity must be performed continuously, during time (for example, during the examination); (ii) we can consider a closed environment, with a few of identities to be verified; (iii) users must be appropriately trained.

A first preliminary investigation was done in [20]. Facial recognition was proposed, because it allows a real-time verification of actual presence, low manufacturing costs and fair degree of reliability. A modular system implemented *detection and recognition operations*, able to verify the presence of learners beyond the screen and enable the authentication process.

At the same time, facial detection allowed to verify the contemporary presence of more individuals and unregistered people in the environment. Beside face, voice and keystroke dynamics were proposed. Voice Recognition was useful for detecting the presence of others who interact with the candidate, whilst the keystroke dynamics identified the candidate by typing frequency (continuous identity verification).

The combination of these three biometric traits significantly increases the percentage of recognition of candidates. Implementation costs are low and hardware equipment and sensors are usually already assembled on ICT devices [20].

Among the three cited traits, keystroke dynamics, that is, the way of typing, was particularly the focus of another work [13]. This investigation was carried out with the aim of developing a robust system to authenticate students taking online examinations. Keystroke dynamics focuses on the user's typing style by monitoring "the keyboard inputs thousands of times per second in an attempt to identify users based on habitual typing rhythm patterns" [13].

There is extensive evidence regarding the reliability of keystroke dynamics to accurately determine the user's identity. Keystroke dynamics is inexpensive compared to other biometric technologies: the capture device is the keyboard [13]. Two keystroke verification techniques were investigated:

- Static technique: it only works at specific time intervals.
- Continuous technique: it monitors the typing behavior throughout the interaction process; it is ideal to monitor environments where the fatigue or attention recognition, or continuous proof of the user must be performed.

Beside keystroke dynamics, the stylometry is a behavioural biometric trait which determines the authorship of manuscripts from the authors' linguistic styles. Typically statistical linguistic features are used at the word and syntax level. Keystroke and stylometry are appealing: not intrusive, inexpensive, continuing verification - dynamic verification.

Keystroke and stylometry were proposed in [21] for developing a robust system to authenticate students taking online examinations. This is also one the few works where experiments are carried out in order to quantify the effectiveness of the proposed approach.

Stylometry seems to be a useful addition to the process because the student could type the answers to the test while someone provides suggestions to the student that simply types the words of the coach without worrying about converting the linguistic style into his own [21].

The keystroke and the stylometry systems consisted of a data collector, a feature extractor, and a pattern classifier. Data were collected from 30 students of a spreadsheet modeling course (17 males and 13 females). A feature vector was extracted from keystroke and stylometry traits. The feature extractor parses each file creating both keystroke and stylometry feature vectors for later processing. Experiments were carried out by separating the two traits.

In the keystroke dynamics experiment, 239 features have been employed, with means and standard deviations of the timings of key press durations and transitions, and the percentage utilization of specific keys, as follows [21]:

– 78 duration features - 39 means and 39 standard deviations - individual letter and non-letter keys, and groups of letter and non-letter keys (see Fig. 1)
– 70 type - 1 transition features - 35 means and 35 standard deviations - transitions between any combination letters/non letters or groups of them (see Fig. 2)
– 70 type - 2 transition features - 35 means and 35 standard deviations - as type-1 transition features with different method of measurement (see Fig. 2)
– 19% features - non-letter keys and mouse clicks
– 2 keystroke input rates: total time to enter the text/total number of keystrokes and mouse events, total time to enter the text minus pauses greater than half second/total number of keystrokes and mouse events.

The following figures present the hierarchy trees for duration categories and transition categories [21]:

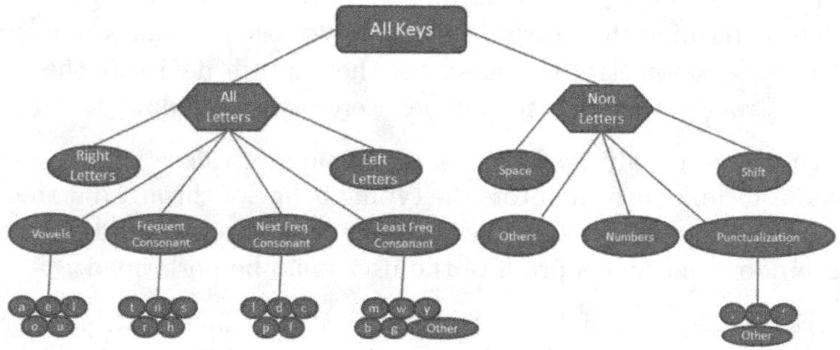

Fig. 1. Hierarchy tree - 39 duration categories - this figure quotes from [21]

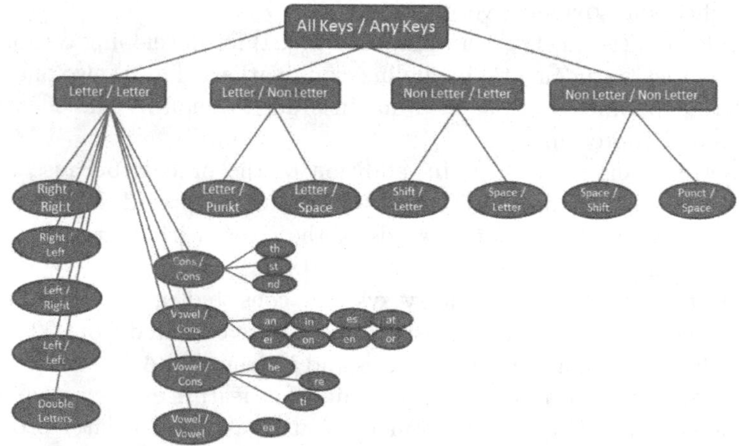

Fig. 2. Hierarchy tree - 35 transition categories (type 1 and type 2) - this figure quotes from [21]

In the stylometry experiment, a set of 228 linguistic features was used: 49 character-based, 13 word-based and 166 syntax-based features. "The features were normalized to be relatively independent of the text length – e.g. the number of different words (vocabulary)/total number of words was used rather than simply the number of different words" [21]. The choice of features shows typical differences in the student population, in fact it can be seen that some students have knowledge and use of extensive vocabulary and terminology, for others this is somewhat reduced.

For both keystroke dynamics and stylometry, the classification procedure is based on the *k-nearest-neighbor* algorithm with the Euclidean distance; it classified the unknown difference vectors, with a reference set composed of the differences between all combinations of the claimed user's enrolled vectors (within-person) and the differences between the claimed user and every other user (between-person) [21].

Two closed-system experiments were carried out on each of the keystroke and stylometry systems. It has operated on a relatively small database, data was collected by 30 students of the course. But with these requirements of the application that dimension may be acceptable: a small sets of subjects is the universe of interest, the remainder being outside of this universe. The following performance was achieved, given in terms of percentage of correctly verified users, on keystroke dynamics and stylometry (and combined) systems:

- **Keystroke System:**
  99.96% on the 3000 keystroke half test sample
  100% on the 6000 keystroke full test sample.

- **Stylometry System:**
  74% on test input of 500 words
  78% on test input od 1000 words.

In the two experiments the answers have been combined to obtain reasonably sized biometric samples; and each experiment has been based on four online short-answer tests of 10 questions each.

First experiment: each sample consisted of five test answers (half answers of each test), obtaining 8 overall samples per student, since each of the four tests contained ten questions for a total of 40 questions.

Second experiment: each sample consisted of ten answers (all the answers of a test) defining 4 samples per student.

Table 1 synthesizes design and results of both experiments. Observe that keystroke dynamics based system has achieved significantly superior performance over the stylometry based system. Keystroke dynamics and stylometry are both behavioral biometrics, but they work at different cognitive levels. The stylometry mainly involves aspects of syntax and semantics, and their is much more complex than that required by the keystroke dynamics [21].

Table 1. Design and results of experiments reported in [21]

Experiment	Data Samples	Keystr EER	Stylo EER
1	– 8 Samples/Student – 5 Answers Combined	0.04 %	~ 26 %
2	– 4 Samples/Student – 10 Answers Combined	0.00 %	~ 22 %

4 Discussions

Although biometrics have become very popular in the last years, some studies have proposed biometrics for eLearning applications, to achieve positive results in the identification of the user, in the continuous especially.

It is important to weigh up costs and ease of implementation, taking into account that now all the devices are equipped with a camera, microphone and keyboard, and then it seemed natural to choose biometric recognition techniques such as face, voice and keystroke dynamics.

In addition, biometrics can be used in eLearning to evaluate the degree of the candidate's attention. The different proposed combinations can provide excellent results in terms of efficiency and effectiveness.

Certainly eLearning technology and systems have achieved a high architectural and educational level. Current paradigms and technologies introduced new methodologies in collaborative learning, rapid and blended learning but still there is a gap in the management of continuous user identification process during the examination and assessment of learning.

This last aspect seems to be still a problem of not simple solution. Biometric techniques are sometimes invasive or costly or comfortably articulated during the implementation phase during an examination. There is also a psychological factor and the unevenness of the environments seems to make the process of continuous identification quite complex.

We believe that there isn't currently a criterion for completeness of biometric technologies, since we do not yet have truly specific mature technologies that are applicable to the eLearning world. It can be observed that discussed experiments, where done, were performed in a favorable, very specific context with verification in the continuous and during the examination phase by fully collaborative students. The results were certainly encouraging but significant efforts must still be done. The, positive, side effect was that the search for specific biometric verification methods to eLearning applications increased the interest on behavioral biometrics such as the keystroke dynamic and the stylometry. These traits, coupled with others, as the fingerprint and the face, are worthy to be considered as a good basis for developing specific approaches for personal verification to be performed continuously over time, in order to detect the presence and the identity of the student during the exam remotely.

References

1. Beltrametti, M., Lattanzi, R., Coppi, M., Genovese, P.V., Gilbert, P.: eLearning - la rivoluzione in corso e l'impatto sul sistema della formazione in Italia, Aspen (2014)
2. Leavoy, P., Biraghi, A.: The Importance of the Learner Experience in eLearning, Docebo (2017)
3. Campanella, S., Dimauro, G., Ferrante, A., Impedovo, D., Impedovo, S., Lucchese, M.G., Modugno, R., Pirlo, G., Sarcinella, L., Stasolla, E., Trullo, C.A.: eLearning platforms in the Italian Universities: the technological solutions at the University of Bari. WSEAS Trans. Adv. Eng. Educ. **5**(1) (2008)
4. Cavus, N., Zabadi, T.: A comparison of open source learning management systems. Soc. Behav. Sci. **143** (2014)
5. Ajlan, S.: A comparative study between eLearning features (2012)
6. CRUI: MOOCs MASSIVE OPEN ON-LINE COURSES - Prospettive e Opportunitá per l'Universitá italiana (2015)
7. Botturi, L., Canottoni, L., Lepori, B., Tardini, S.: Fast prototyping as communication catalyst for eLearning design (2008)
8. Dick, W., Carey, L., Carey, J.O.: The Systematic Design of Instruction, 8th edn. Pearson (2015)
9. Molenda, M.: In search of the elusive ADDIE model. Perform. Improv. **42**(5) (2003)
10. Morrison, G.R., Ross, S.M., Kemp, J.E.: Designing Effective Instruction, 4th edn. Wiley, New York (2004)
11. Abkulut, Y.: Implications of two well-known models for instructional designers in distance education: Dick-Carey versus Morrison-Ross-Kemp. Turk. Online J. Distance Educ.-TOJDE **8**(2) (2007)
12. Paullet, K., Douglas, D.M., Chawdhry, A.: Verifying user identities in distance learning courses: do we know who is sitting and submitting behind the screen? Issues Inform. Syst. **15**(1) (2014)
13. Monrose, F., Rubin, A.D.: Keystroke dynamics as a biometric for authentication. Future Gener. Comput. Syst. **16** (2000)
14. Frost, J., Hamlin, A., Barczyk, C.: A survey of professor reactions to scenarios of academic dishonesty in American universities. J. Bus. Inq.: Res. Educ. Appl. **6**(1), 11–18 (2007)
15. Grijalva, T.C., Kerkvliet, J., Nowell, C.: Academic honesty and online courses. Coll. Stud. J. **40**(1) (2006)
16. Jortberg, M.: Methods to verify the identity of Distant Learning Students. Axiom White Paper (2009)
17. Jain, A., Bolle, R., Pankanti, S.: Biometrics – Personal Identification in Network Society. Kluwer Academic Publishers, Dordrecht (1999)
18. Bailie, J.L., Jortberg, M.A.: Distance learning student authentication: verifying the identity of online students. ACXIOM - Third Annual Report on Identity in Distance Learning, White Paper (2008)
19. Bailie, J.L., Jortberg, M.A.: Online learner authentication: verifying the identity of online users. MERLOT J. Online Learn. Teach. **5**(2) (2009)
20. Mastronardi, G., Buonamassa, G., Patruno, T., Pierri, C.: Tecniche Biometriche Combinate nei Processi di eLearning e eVoting (2014)
21. Monaco, J.V., Stewart, J.C., Cha, S.-H., Tappert, C.C.: Behavioral biometric verification of student identity in online course assessment and authentication of authors in literary works (2013)
22. Gagbla, G.K.: Applying Keystroke Dynamics for Personal Authentication (2005). http://citeseerx.ist.psu.edu/viewdoc/download?doi=10.1.1.101.9517&rep=rep1&type=pdf

Towards Video Captioning with Naming: A Novel Dataset and a Multi-modal Approach

Stefano Pini, Marcella Cornia(✉), Lorenzo Baraldi, and Rita Cucchiara

Dipartimento di Ingegneria "Enzo Ferrari",
Università degli Studi di Modena e Reggio Emilia, Modena, Italy
stefanopi.93@gmail.com,
{marcella.cornia,lorenzo.baraldi,rita.cucchiara}@unimore.it

Abstract. Current approaches for movie description lack the ability to name characters with their proper names, and can only indicate people with a generic "someone" tag. In this paper we present two contributions towards the development of video description architectures with naming capabilities: firstly, we collect and release an extension of the popular Montreal Video Annotation Dataset in which the visual appearance of each character is linked both through time and to textual mentions in captions. We annotate, in a semi-automatic manner, a total of 53k face tracks and 29k textual mentions on 92 movies. Moreover, to underline and quantify the challenges of the task of generating captions with names, we present different multi-modal approaches to solve the problem on already generated captions.

Keywords: Video captioning · Naming · Datasets · Deep learning

1 Introduction

Video Captioning is a fundamental achievement towards machine intelligence, as it links together vision and language. The task has been gaining a lot of attention in the past few years, thanks to the spread of Deep Learning approaches [10,24] and large scale movie description datasets [15,21]. While current state-of-the-art approaches seem to have brought this task in a rather mature stage, and the computer vision community is focusing on novel techniques for including semantics [14] and enhance the quality of descriptions [18], one fundamental aspect is still missing: that of naming characters in captions with their proper names. Indeed, despite the availability of movie description datasets in which captions come with character names, it is usual practice to replace them with a "someone" tag, thus ignoring the naming task during the generation of the caption.

This choice has been well motivated by the need of training video captioning architectures which are not endowed with naming capabilities, in which the presence of proper names would have been more detrimental than useless. Each proper name, indeed, would have become a new entry in the dictionary, without being the network capable of identifying the relationship between the name and

S. Battiato et al. (Eds.): ICIAP 2017, Part II, LNCS 10485, pp. 384–395, 2017.
https://doi.org/10.1007/978-3-319-68548-9_36

Someone opens a door and glimpses someone and someone, who walk faster.
Lovejoy opens a door and glimpses Jack and Rose, who walk faster.

Fig. 1. Generating video captions with character names requires additional supervision besides the sole caption. To this end we augment the M-VAD dataset for movie description by annotating face tracks of movie characters and by associating them with their textual mentions.

the visual appearance of a person, thus making the presence of proper names a driver of noise rather than a useful information. Developing a video captioning network capable of generating captions with proper names in the correct place requires to address several tasks in a deep learning perspective, ranging from face detection and tracking, to face recognition with respect to a given set of names. Moreover, the generative recurrent architecture of such a model needs to be aware of the linguistic structure of the caption being generated, in order to identify the need of outputting a proper name, and then selecting it from a list of possible candidates by exploiting face identification features.

Training a video captioning architecture with naming capabilities by relying on the caption as the sole source of supervision would be particularly challenging, due to the lack of supervision between the textual and the visual domain. In each video, indeed, different faces can appear, only some of them mentioned in the ground-truth caption, and possibly some not even appearing in the cast of the movie, such in the case of background actors. A more grounded form of annotation is therefore needed, to link the visual appearance of a face to the list of characters playing the movie, and to link the mention of a character in a caption with the visual appearance of the corresponding face (Fig. 1).

In this paper, we present two contributions towards the development of video description architectures with naming capabilities: firstly, we augment one of the most popular movie description datasets, the Montreal Video Annotation dataset [21], with the annotations needed to train a neural architecture with naming capabilities. Each video of the dataset has been endowed with face tracks annotations, manually associated with characters of the movie and with textual mentions, thus closing the gap between the textual and the visual domain. Also, we release appropriate training, validation and test splits, which are carefully built by taking into account the specific challenges of the task. To our knowledge, this is the only publicly available dataset providing annotations for video captioning with naming. Secondly, we present a principled use case of the dataset by developing different multi-modal approaches which can solve the naming problem from already generated captions. Experimental results will enlighten and quantify many of the challenges associated with the task, and devise appropriate solutions.

2 Related Work

Our work is related to the generation of captions for video, and to the task of linking visual tracks to names.

The generation of natural language descriptions of visual content has received large interest since the emergence of recurrent networks, either for single images [7], user-generated videos [3] or movie clips [14,24]. First approaches described the input video through mean-pooled CNN features [25], or sequentially encoded by a recurrent layer [3,24]. Other works have then followed this kind of approach, either by incorporating attentive mechanisms [26] in the sentence decoder, by building a common visual-semantic emebedding [10], or by adding external knowledge with language models [23] or visual classifiers [14].

On a different note, the problem of linking people with their names has been tackled in different previous works [2,4,13,19,20], which rely on alignment of video to TV scripts. The goal is to track faces in the video and assign names to them. For example, [4,19] rely on the availability of aligned subtitles and script texts to associate TV or movie characters with their names. In [20] character appearances are modelled as a Markov Random Field, integrating face recognition, clothing appearance, speaker recognition and contextual constraints in a probabilistic manner. [2] propose a discriminative weakly supervised model jointly representing actions and actors in video. [13] tackle the problem of naming people in a video by including ambiguous mentions of people such as pronouns and nominals (i.e. co-reference resolution). [11] present a multiple instance learning based approach which focuses on recognizing background characters showing significant improvement over prior work.

Finally, in [16], authors address the problem of generating video descriptions with grounded and co-referenced people, by proposing a novel dataset and a deeply-learned model. This task significantly differs from the one tackled in this paper, as it aims at predicting the spatial location in which a given character appears, and at producing captions with proper names in the correct place.

3 The M-VAD Names Dataset

Providing the annotations needed to train novel movie description architectures capable of naming characters requires to annotate the visual appearance of each character of a movie, to track it through time, and to annotate the association between visual appearances, character names and textual mentions in the corresponding captions.

To this end, we collect new annotations for the Montreal Movie Description dataset [21]. The dataset consists of 84.6 h of video from 92 Hollywood movies, for a total of 46,523 video clips with corresponding captions, from which we retain only those that contain at least one character mention. For each movie, we provide manually annotated visual appearances of characters, their face and body tracks, and associations with textual mentions in the captions. The dataset, which we name *M-VAD Names*[1], is annotated in a semi-automatic way, by leveraging face recognition and clustering techniques, and then manually refined.

Face and Body Tracks Generation. To generate face and body tracks, we first detect faces on video frames by using the cascaded approach of [27], in which faces are detected by means of a multi-task Convolutional Network which jointly detects facial landmarks and predicts the face bounding box. After a face is detected, we track it in the following frames with the online tracker presented in [1], which learns an adaptive appearance model using Multiple Instance Learning. To account for tracking errors and camera shot changes, we also compute an appearance similarity measure between predictions in two consecutive frames, and we manually stop the tracking when the similarity is under a threshold. In practice, we found that a pixel-wise difference between consecutive detections is robust enough to discard the majority of errors.

For every face found and tracked for 16 consecutive frames, a new track is saved. This choice is also compatible with the temporal extent used in state-of-the-art action descriptors [22]. For every track, we store the face bounding box as well as the body bounding box, which can be used to compute action descriptors. This extended bounding box is generated with a linear transformation of the face bounding box, expanding it to obtain a square with a side of three times the height of the original bounding box with the face centered in the upper portion of the square. This approach is also similar to what has been done in [2].

Table 1. Overall statistics on the M-VAD Names dataset. Numbers of video clips in each split, and number of mentions, appearances and tracks.

	Number	Avg. per movie	Avg. per character
Train videos	17170	187	-
Test videos	2581	28	-
Validation videos	2708	29	-
Mentioned characters	1392	15	-
Annotated characters	908	10	-
Mentions	29862	325	22
Appearances	20631	224	24
Tracks	53665	583	62

Movie Character Annotations. Having detected face and body tracks, we associate them to characters from each movie with a coarse semi-automatic annotation strategy, which is then manually refined. Specifically, we employ a face

[1] The dataset is publicly available at: http://imagelab.ing.unimore.it/mvad-names.

recognition approach inspired to FaceNet [17] and trained on the MS-Celeb-1M dataset [6]. Each face track is projected into a 128-dimensional embedding space which is trained with the objective of giving very similar descriptors to similar faces (i.e. faces of the same character) and diverse descriptors to different faces (i.e. faces of different characters). At this point, an agglomerative clustering is used to cluster the face representations calculated in the previous phase for every movie. In this way, for every movie we obtain N clusters of similar faces associated to the relative video clips and tracks using an unsupervised method. Every cluster is then manually classified either as a character of the movie, as an unknown character (if it is not part of the cast, or it can not be not recognized by the human annotator), or as wrong (if the track does not contain a face due to errors in the face recognition and tracking phase). Every element of the cluster is manually checked and, if needed, moved to the appropriate cluster. At the end of this process, we obtain the annotation of the main characters of every movie and for every sequence of the dataset. This annotation is then used to associate tracks and appearances with textual mentions in the captions. To avoid human error, each annotation is checked by at least two different people.

Training, Validation and Test Splits. The authors of the M-VAD dataset provided official training/validation/test sets by splitting the original set of movies into three disjoint parts, so that captioning algorithms could be trained on a set of movies, and validated or tested on other movies, with only partially overlapped vocabularies. In the case of naming, it is instead important to have clips from the same movies in all splits, so that the visual appearance of a specific character can be learned and applied at testing time. For this reason, we propose and release an official split for the M-VAD Names dataset.

The split has been done randomly, but applying a series of soft constraints. We forced every movie to have 80% of the video clips into the training set, 10% into the validation set and 10% into the test set. Also, we applied the same splitting criterion to characters, to video clips with only one mention, and to video clips with two or more mentions. Ideally, these constraints ensure that the training set contains the 80% of the video clips of each movie, the same percentage of the characters, of clips with one mention, and of clips with two or more mentions. Finally, we gave priority to have at least a video clip for every character in every group set. We have considered the last constraint as the most important one to try to have ax example for every character in every set.

Statistics. Table 1 reports the overall statistics of the collected dataset. As it can be seen, the dataset contains a total of 22,459 video clips, which are divided into appropriate train, test and validation splits. The number of unique characters found in the screenplays is 1,392, while the overall number of mentions is 29,862. We found some errors in the captions of the M-VAD dataset for many films, so the correct values of unique characters and mentions would be lower. Finally, the unique annotated characters are 908. They appear a total of 20,631 times in the video clips and in 53,665 extracted tracks. Average per movie and

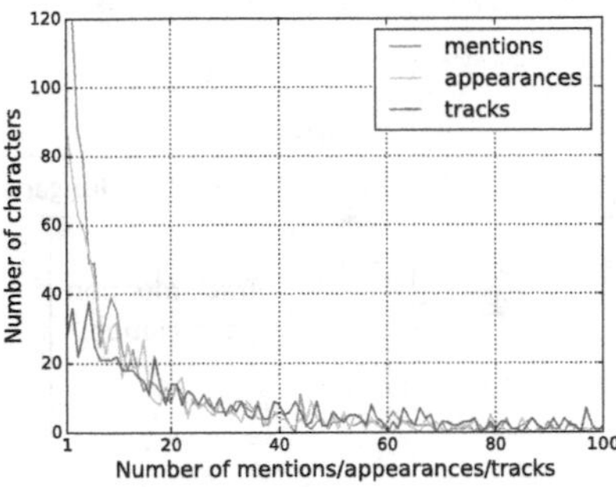

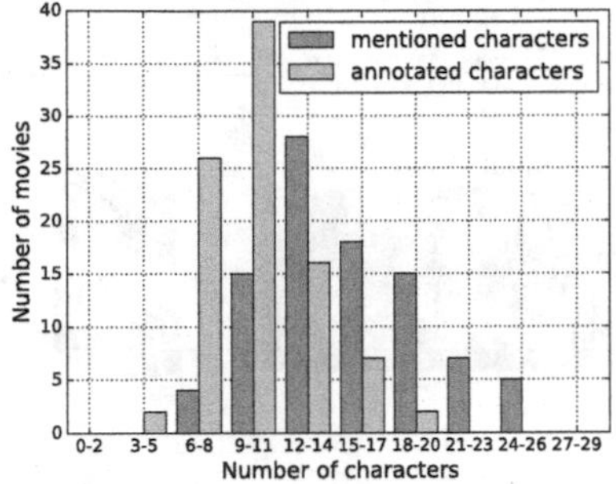

Fig. 2. Statistics on the dataset. Left: distribution of characters with respect to the number of mentions, appearances and tracks. Right: distribution of movies with respect to the number of characters mentioned in the captions and visually annotated.

per character values are reported, to give an estimate of how much data can be extracted from a single film or a single character. Figure 2 reports the distribution of characters with respect to the number of mentions, appearances and tracks on the left, and the distribution of movies with respect to the number of characters mentioned in the captions and visually annotated on the right. Due to errors in the M-VAD mentions and misses during the annotation pipeline, most of the films have less annotated characters than the mentioned ones, but the relative distribution is similar. For the same reasons, there is a peak in the graph on the left for the number of characters with one mention. As it can be observed, lots of characters have less than 20 mentions, annotation and tracks, and only few characters have more than 80.

4 Replacing the "Someone": An Approach to Video Description with Naming

The M-VAD Names dataset provides sufficient annotations to train generative architectures with naming capabilities, which for sure will need to solve the problem of associating textual mentions with character names at generation time. In this Section, we investigate a strictly related task which shares many of the challenges behind that of generating video description with proper names. Indeed, instead of proposing an architecture able to generate captions with names, we tackle the problem of replacing the "someone" tag in existing descriptions with the correct character name.

The task requires to identify for each video clip which character performs the action described in each textual mention in the relative caption. Therefore, it is not only necessary to identify characters present in the scene, but also to understand what each of them is doing. We start from ground-truth descriptions, where we replace each proper name with a "someone" tag, and extract the verb associated with each "someone" subject tag by using a NLP parser [8]. For each video clip, we also exploit face and body tracks from the M-VAD Names dataset,

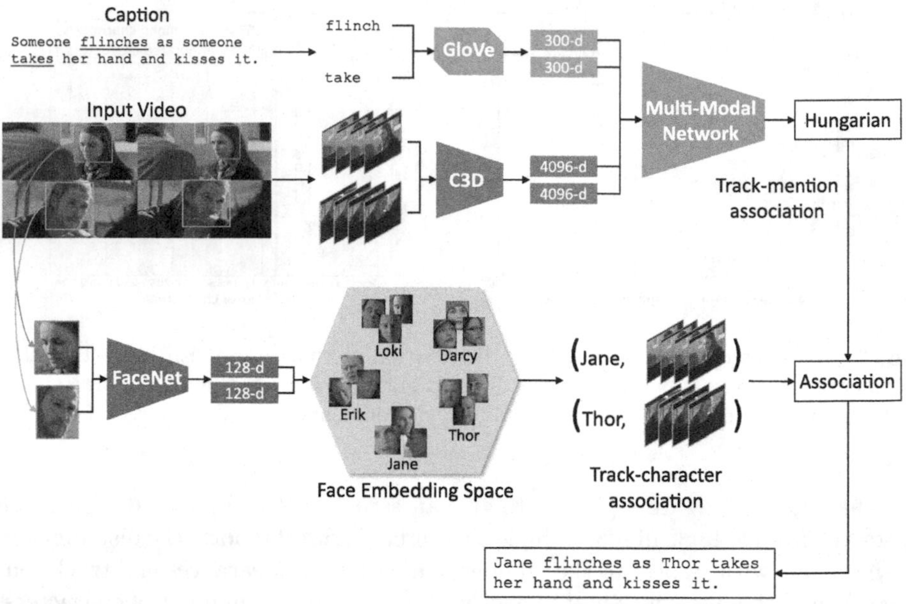

Fig. 3. Summary of our approach for replacing "someone" tags with the correct character names. A Multi-Modal network is trained to predict a matching between visual tracks and actions (expressed by the verbs found in the textual description), while a face recognition approach is used to complete the matching between the track and the character.

and the association with textual mentions. Starting from these information, we build a Multi-Modal network capable of associating a character track with the corresponding mention in the caption, by predicting the similarity between the performed action (i.e. the verb), and the visual track. A summary of the approach is shown in Fig. 3.

4.1 Multi-Modal Network

Given an input track and a verb, we preprocess them to extract appropriate feature vectors, and then feed them to a trained Multi-Modal network which can project visual and textual elements in a common multi-modal space. For tracks, which are 16 frames long, we resize them to 112×112 and extract a 4096-dimensional feature vector from the last but one fully convolutional layer of the C3D network [22], trained on the Sports-1M dataset [9], which encodes motion features computed around the middle frame of the input track. For each verb, instead, we extract a 300-dimensional feature vector by using the GloVe embedding [12] that provides a semantic vector representation for words.

The Multi-Modal network is composed of two different branches for each of the modalities. A visual branch $\phi_v(\cdot)$ takes the track descriptor as input and projects it into a multi-modal space, while a textual branch $\phi_t(\cdot)$ does the same for the textual counterpart. Training is performed by forcing the network to produce an embedding space such that a track and its corresponding verb have close projections, while a track and a verb which do not match should be far in the embedding space. We do this by introducing two distance terms, $p(\cdot)$ and $n(\cdot)$ to force projections of a verb and track to be close or far by at least a margin M:

$$p(\mathbf{t}, \mathbf{v}) = \|\phi_v(\mathbf{t}) - \phi_t(\mathbf{v})\|_2^2, \qquad n(\mathbf{t}, \mathbf{v}) = \max(M - \|\phi_v(\mathbf{t}) - \phi_t(\mathbf{v})\|_2^2, 0) \quad (1)$$

where $\mathbf{t}$ and $\mathbf{v}$ are respectively a visual track and a verb. In our case, we want each track to be classified as similar to its corresponding verb and dissimilar to a randomly chosen different verb. Additionally, we want to exploit the fact that the same verb could be associated to other tracks and to enforce tracks annotated as "wrong" (i.e. not containing a person) to be classified as dissimilar to that verb. Our final loss function is therefore:

$$L = \sum_{i=1}^{N} p(\mathbf{t}_i, \mathbf{v}_i) + n(\mathbf{t}_i, \mathbf{v}_i^-) + p(\mathbf{t}_i^+, \mathbf{v}_i) + n(\mathbf{t}_i^w, \mathbf{v}_i) \quad (2)$$

where, for each mention in the dataset, $\mathbf{t}_i$ represents the visual track associated with the mention; $\mathbf{v}_i$ the corresponding verb; $\mathbf{v}_i^-$ a randomly chosen verb from the dataset, different from $\mathbf{v}_i$; $\mathbf{t}_i^+$ a second visual track, associated with the same verb $\mathbf{t}_i$ (if present); $\mathbf{t}_i^w$ a "wrong" track.

At test time, we compute the distance between a track and a verb projected into the embedding space (i.e. $p(\mathbf{t}, \mathbf{v})$). Given a video clip, and all the distances between its verbs and its tracks, we compute the optimal assignment between tracks and verbs by using the Kuhn-Munkres algorithm.

4.2 Face Recognition

Once tracks have been assigned to textual mentions, tracks needs to be assigned to characters to complete the task of replacing the "someone". To this aim, similarly to what we did in Sect. 3, we use the 128-dimensional face representation provided FaceNet [17]. Then, we use these representations of the training faces of the characters as the training data of a K-Nearest Neighbours classifier with k equal to 5. In particular, an optimized version of it (kd-tree) is used. This classifier has been chosen to take advantage of the previously generated embeddings and to be sufficiently flexible in the case of characters whose aspect changes during the movie. Clustering classifiers and linear methods, in fact, are not always suitable to classify classes that can have multiple agglomerations in different areas of the space.

5 Experimental Evaluation

We evaluate the performance of the proposed pipeline with respect to different training strategies for the Multi-Modal network, and different classification

Table 2. Experimental results on the "replacing the someone" task, with different loss functions and face recognition approaches. Results are reported on the validation and test split of the M-VAD Names dataset, in terms of accuracy.

	Val. accuracy	Test accuracy
Random assignment	0.111	0.103
MMN Binary with two terms + Face AdaBoost	0.551	0.578
MMN Siamese + Face AdaBoost	0.569	0.593
MMN Triplet + Face AdaBoost	0.529	0.568
MMN Binary with two terms + Face SVM	0.667	0.670
MMN Siamese + Face SVM	0.683	0.685
MMN Triplet + Face SVM	0.630	0.650
MMN Binary with two terms + Face KNN	0.678	0.681
MMN Siamese + Face KNN	0.691	0.701
MMN Triplet + Face KNN	0.648	0.664
MMN Binary with four terms + Face KNN	0.710	0.691
Proposed MMN + Face KNN	**0.715**	**0.712**

approaches for the face recognition part. In particular, besides the proposed loss, we test a Siamese and a Triplet loss function, plus two Binary loss functions. The Siamese loss exploits only the first two terms of Eq. 2, while the Triplet loss is defined as follows:

$$L = \sum_{i=1}^{N} \max(\|\phi_v(\mathbf{t}_i) - \phi_t(\mathbf{v}_i)\|_2^2 - \|\phi_v(\mathbf{t}_i) - \phi_t(\mathbf{v}_i^-)\|_2^2 + M, 0). \qquad (3)$$

For the Binary loss functions, instead, we replace $p(\cdot)$ and $n(\cdot)$ with two binary cross-entropy terms, i.e. $p(\mathbf{t}, \mathbf{v}) = -\log d(\phi_v(\mathbf{t}), \phi_t(\mathbf{v}))$ and $n(\mathbf{t}, \mathbf{v}) = -\log(1 - d(\phi_v(\mathbf{t}), \phi_t(\mathbf{v})))$, where d is learnable classification function. In one version, we maintain all the four terms of Eq. 2, and in a second version, we keep only the first two terms, using the re-defined $p(\cdot)$ and $n(\cdot)$.

As mentioned, the architecture of our Multi-Modal network is composed by two different branches: the first one processes the track feature vectors coming from the C3D network, while the second processes the vectorized representations of verbs. In details, the track branch is composed by 4 fully connected layers with 1024, 256, 64 and 16 units respectively, while the verb branch is composed by 3 fully connected layers with 256, 64 and 16 units respectively, all with ReLU activations. Finally, in the case of the Binary loss functions, the outputs of these two branches are concatenated and fed into a fully connected layer with 1 unit and a sigmoid activation, which acts as the learnable classifier d. Weights of all layers are initialized according to [5].

During the training phase, we randomly sample a minibatch containing 32 training samples and we encourage the network to minimize one of the mentioned loss functions through the Stochastic Gradient Descent optimizer. SGD is applied with Nesterov momentum 0.9, weight decay 0.0005 and learning rate 10^{-2}. The margin M, after cross-validation, has been set to 1 for the Siamese loss, and to 0.2 for the Triplet loss.

5.1 Experimental Results

Experimental results are reported in Table 2. We show the results in term of the final association accuracy between textual mentions and character names, by comparing the proposed Multi-Modal network, described in Sect. 4.1, with the Siamese, Triplet and Binary alternatives. As it can be noticed, the proposed Multi-Modal network obtains the best results on both validation and test sets. Moreover, the performances obtained by the proposed loss and by the Binary loss with four terms confirm the advantage of exploiting wrong tracks and tracks associated with the same verb. For reference, we also show the result of a random replacement of all "someone" tags with a movie character randomly extracted from the character list of each movie.

Moreover, we report the accuracy results by using different face recognition approaches. In particular, we compare the K-Nearest Neighbours classifier (KNN) with the SVM and the Adaboost (with 30 Decision Trees) classifiers. As it can be observed, the KNN performs better than others on both validation and test sets.

6 Conclusion

In this paper we presented two contributions towards the development of movie description architectures capable of indicating characters with their proper names. First, we presented and released an extension of the M-VAD dataset, in which character faces and bodies are manually annotated and associated to textual mentions, thus closing the gap in supervision between the visual and the textual domain. Then, we proposed a multi-modal approach to tackle the task of naming characters on existing descriptions: our approach combines a Deep Network which jointly classifies visual tracks and textual actions, and state-of-the-art face recognition techniques. Experimental results enlighten and quantify the challenges associated with the task of developing novel video description architectures with naming capabilities.

References

1. Babenko, B., Yang, M.H., Belongie, S.: Visual tracking with online multiple instance learning. In: CVPR (2009)
2. Bojanowski, P., Bach, F., Laptev, I., Ponce, J., Schmid, C., Sivic, J.: Finding actors and actions in movies. In: ICCV (2013)
3. Donahue, J., Anne Hendricks, L., Guadarrama, S., Rohrbach, M., Venugopalan, S., Saenko, K., Darrell, T.: Long-term recurrent convolutional networks for visual recognition and description. In: CVPR (2015)

4. Everingham, M., Sivic, J., Zisserman, A.: "Hello! My name is.. Buffy"-Automatic Naming of Characters in TV Video. In: British Machine Vision Conference (2006)
5. Glorot, X., Bengio, Y.: Understanding the difficulty of training deep feedforward neural networks. In: International Conference on Artificial Intelligence and Statistics (2010)
6. Guo, Y., Zhang, L., Hu, Y., He, X., Gao, J.: MS-Celeb-1M: a dataset and benchmark for large-scale face recognition. In: Leibe, B., Matas, J., Sebe, N., Welling, M. (eds.) ECCV 2016. LNCS, vol. 9907, pp. 87–102. Springer, Cham (2016). doi:10.1007/978-3-319-46487-9_6
7. Hendricks, L.A., Venugopalan, S., Rohrbach, M., Mooney, R., Saenko, K., Darrell, T.: Deep compositional captioning: describing novel object categories without paired training data. In: CVPR (2016)
8. Honnibal, M., Johnson, M.: An improved non-monotonic transition system for dependency parsing. In: Conference on Empirical Methods in Natural Language Processing (2015)
9. Karpathy, A., Toderici, G., Shetty, S., Leung, T., Sukthankar, R., Fei-Fei, L.: Large-scale video classification with convolutional neural networks. In: CVPR (2014)
10. Pan, Y., Mei, T., Yao, T., Li, H., Rui, Y.: Jointly modeling embedding and translation to bridge video and language. In: CVPR (2016)
11. Parkhi, O.M., Rahtu, E., Zisserman, A.: Its in the bag: stronger supervision for automated face labelling. In: ICCV Workshops (2015)
12. Pennington, J., Socher, R., Manning, C.D.: GloVe: global vectors for word representation. In: Conference on Empirical Methods in Natural Language Processing (2014)
13. Ramanathan, V., Joulin, A., Liang, P., Fei-Fei, L.: Linking people in videos with "their" names using coreference resolution. In: Fleet, D., Pajdla, T., Schiele, B., Tuytelaars, T. (eds.) ECCV 2014. LNCS, vol. 8689, pp. 95–110. Springer, Cham (2014). doi:10.1007/978-3-319-10590-1_7
14. Rohrbach, A., Rohrbach, M., Schiele, B.: The long-short story of movie description. In: Gall, J., Gehler, P., Leibe, B. (eds.) GCPR 2015. LNCS, vol. 9358, pp. 209–221. Springer, Cham (2015). doi:10.1007/978-3-319-24947-6_17
15. Rohrbach, A., Rohrbach, M., Tandon, N., Schiele, B.: A dataset for movie description. In: CVPR (2015)
16. Rohrbach, A., Rohrbach, M., Tang, S., Oh, S.J., Schiele, B.: Generating descriptions with grounded and co-referenced people. In: CVPR (2017)
17. Schroff, F., Kalenichenko, D., Philbin, J.: Facenet: a unified embedding for face recognition and clustering. In: CVPR (2015)
18. Shetty, R., Rohrbach, M., Hendricks, L.A., Fritz, M., Schiele, B.: Speaking the same language: matching machine to human captions by adversarial training. arXiv preprint arXiv:1703.10476 (2017)
19. Sivic, J., Everingham, M., Zisserman, A.: Who are you?-learning person specific classifiers from video. In: CVPR (2009)
20. Tapaswi, M., Bäuml, M., Stiefelhagen, R.: Knock! Knock! Who is it? Probabilistic person identification in TV-series. In: CVPR (2012)
21. Torabi, A., Pal, C., Larochelle, H., Courville, A.: Using descriptive video services to create a large data source for video annotation research. arXiv preprint arXiv:1503.01070 (2015)
22. Tran, D., Bourdev, L., Fergus, R., Torresani, L., Paluri, M.: Learning spatiotemporal features with 3D convolutional networks. In: ICCV (2015)

23. Venugopalan, S., Hendricks, L.A., Mooney, R., Saenko, K.: Improving LSTM-based video description with linguistic knowledge mined from text. In: Conference on Empirical Methods in Natural Language Processing (2016)
24. Venugopalan, S., Rohrbach, M., Donahue, J., Mooney, R., Darrell, T., Saenko, K.: Sequence to sequence-video to text. In: ICCV (2015)
25. Venugopalan, S., Xu, H., Donahue, J., Rohrbach, M., Mooney, R., Saenko, K.: Translating videos to natural language using deep recurrent neural networks. In: North American Chapter of the Association for Computational Linguistics (2014)
26. Yao, L., Torabi, A., Cho, K., Ballas, N., Pal, C., Larochelle, H., Courville, A.: Describing videos by exploiting temporal structure. In: ICCV (2015)
27. Zhang, K., Zhang, Z., Li, Z., Qiao, Y.: Joint face detection and alignment using multi-task cascaded convolutional networks. arXiv preprint arXiv:1604.02878 (2016)

Biomedical and Assistive Technology

Bio-Inspired Feed-Forward System for Skin Lesion Analysis, Screening and Follow-Up

Francesco Rundo[1(✉)], Sabrina Conoci[1], Giuseppe L. Banna[2], Filippo Stanco[3], and Sebastiano Battiato[3(✉)]

[1] ADG Central R&D, STMicroelectronics, Catania, Italy
{francesco.rundo, sabrina.conoci}@st.com

[2] Medical Oncology Department, Cannizzaro Medical Hospital, Catania, Italy
gbanna@yahoo.com

[3] DMI IPLAB, University of Catania, Catania, Italy
{fstanco, battiato}@dmi.unict.it

Abstract. Traditional methods for early detection of melanoma rely upon a dermatologist who visually analyzes skin lesion using the so called ABCDE (*Asymmetry, Border irregularity, Color variegation, Diameter, Evolution*) criteria even though conclusive confirmation is obtained through biopsy performed by pathologist. The proposed method shows a bio-inspired feed-forward automatic pipeline based on morphological analysis and evaluation of skin lesion dermoscopy image. Preliminary segmentation and pre-processing of dermoscopy image by *SC-Cellular Neural Networks* is performed in order to get ad-hoc gray-level skin lesion image in which we compute analytic innovative hand-crafted image features for oncological risks assessment. At the end, pre-trained *Levenberg-Marquardt Neural Network* is used to perform ad-hoc clustering of such hand-crafted image features in order to get an efficient nevus discrimination (benign against melanoma) as well as a numerical array to be used for follow-up rate definition and assessment.

1 Introduction

The early detection of skin cancer, specifically, the so called *melanoma* is one of the main issues analyzed by medical oncologists and dermatologists as the probability of full oncological remission is strongly correlated to early detection. A robust and efficient approach for nevus discrimination is currently under investigation by physicians as well by bio-medical engineers with the target to develop a non-invasive "Point of Care (PoC)" for real-time skin cancer detection. Currently, both oncologists and dermatologists use heuristics approach to analyze skin lesion known with acronyms of "ABCDE" which means "Asymmetry, Border irregularity, Color variegation, Diameter, Evolution" applied to a classical dermoscopy image. Basically, the ABCDE criteria performs visual statistical assessment of the nevus based on physician experience and background. Clearly, this strategy suffers from clinician subjectivity as low sensibility and specificity; often it is required a nevus invasive biopsy to confirm diagnosis. In order to address that issue, the authors propose an automatic skin lesion pipeline based on analysis of dermoscopy images in order to discriminate benign lesions from

S. Battiato et al. (Eds.): ICIAP 2017, Part II, LNCS 10485, pp. 399–409, 2017.
https://doi.org/10.1007/978-3-319-68548-9_37

malignant ones trying to have a good trade-off between sensibility and specificity. The method has been successfully validated by using dermoscopy image dataset of PH^2 open database [3]. Several approaches have been proposed in the past for utomatic and robust detection of the skin cancer, including statistical hand-crafted features or soft computing approaches based on machine learning algorithms. In [9] the authors review several recent methods for melanoma detection basically based on usage of image features, neuro-fuzzy approaches, clustering methods (K-means, SVM, etc.) and some methods based on the study of melanotocytes distribution on the skin histopathologic images as well as based on probabilistic analysis (e.g. Bayesian Classifier). The overall analysis reports some very promising methods based on the usage of neural networks for classifying skin lesions or based on adaptive thresholding analysis. Among other methods based on classical pattern analysis approaches combined with classical statistical dermoscopy image features pointed-out acceptable results [9]. In [9, 10] the authors shows a method for classifying skin cancer based on usage of global and local features combined with different classification systems such as SVM, ANN, K-nearest, Naive-Bayes Algorithm showing drawbacks and advantages of the analyzed approaches. An interesting approach is proposed in [11] where some hand-crafted image features are combined together with a deep-learning algorithm able to extract other ad-hoc features. A final SVM engine is then used for performing finale classification by means of a *scores* based approach.

The results reported in [11] shows a limited sensibility/specificity for the analyzed image dataset. Other approaches [12, 13] makes use of "standard" image features coupled with a classification engine (i.e., SVM or K-nearest, etc.) with limited results.

The proposed method is based on combined approach of ad-hoc customized hand-crafted image features combined with a feed-forward neural network system which properly trained, is be able to perform a robust and efficient classification system.

2 Proposed Pipeline: System Description

In Fig. 1 the proposed system is synthetically detailed. Starting from preliminary dermoscopy acquisition made by dermatologist, till to the output of the algorithm consisting of numerical array containing representation of morphologic features applied to the pre-processed dermoscopic image as well as a classification of the skin lesion in terms of probability of oncological malignant progression.

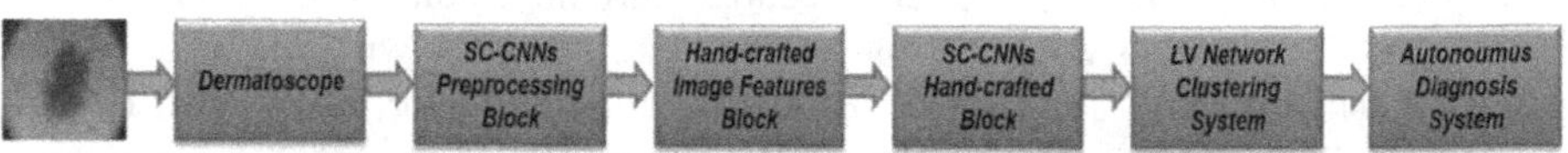

Fig. 1. The proposed feed-forward skin lesion analysis pipeline

2.1 Dermatoscope and SC-CNNs Pre-processing Block

The dermoscopy image is basically a RGB colour image acquired by classical medical optical dermatoscope. The medical dermatoscope provides also a zooming option for enlarging the image of the analyzed skin lesion. In Fig. 2 we show a classical dermoscopy image of a nevus:

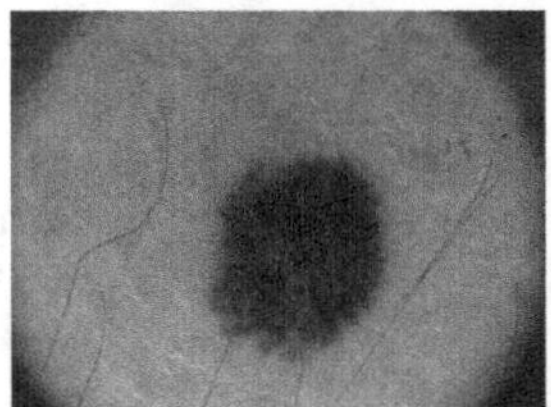

Fig. 2. A typical dermoscopy image of a skin nevus

In our pipeline, we avoid to use direct colour information; the method is applied only to luminance component of the source RGB dermoscopy image "*D(x, y)*" after classical conversion into YCbCr format [2]. The luminance component denoted as "$Y_{D(x,y)}$" is fed as "input" and "state" to a *State-Controlled Cellular Neural Networks (SC-CNNs)* having 2D matrix structure and size equal to source image $Y_{D(x,y)}$ of size $m \times n$.

The classical *Cellular Neural (or Nonlinear) Network (CNN)* was firstly introduced by Chua and Yang [14]. The CNN can be defined as high speed local interconnected computing array of analog processors [14]. The CNN processing is defined through the instructions provided by the so called *cloning templates* [14]. Each cell of the CNN array may be considered as dynamical system which is arranged into a topological structure, usually a 2D or 3D grid. The CNN cells interacts each other within its neighborhood defined by heuristically ad-hoc defined radius [14]. Each CNN cell has an *input*, a *state* and an *output* which is a functional mapping of the state (usually by means of *PWL* function). The CNN can be implemented with analog discrete components or VLSI technology so that it is able to perform high speed "near real-time" computations. Some stability results and consideration about the dynamics of the CNNs can be found in [14, 15]. An updated version of original Chua-Yang CNN model was introduced by Arena et al. in [15] and called "*State Controlled Cellular Neural Network (SC-CNN)*" as it explicates dependency of dynamic evolution of the cell to the "state" of the single cell directly. We refer to SC-CNNs in the following mathematical formulations. By assigning each normalized gray-level of the input source image ($Y_{D(x,y)}$ of size $m \times n$) to each cell of the SC-CNNs (input and state of each cell), several image processing tasks may be performed according to the defined cloning templates instructions [14]. The state equation of a ($m \times n$) SC-CNNs can be defined as follows:

$$C\frac{dx_{ij}(t)}{dt} = -x_{ij}(t) + \sum_{C(k,l)\in N_r(i,j)} A(i,j;k,l)y_{kl}(t) + \sum_{C(k,l)\in N_r(i,j)} B(i,j;k,l)u_{kl}(t) + \sum_{C(k,l)\in N_r(i,j)} C(i,j;k,l)x_{kl}(t) + I$$
$$N_r(i,j) = \{C(k,l), (\max(|k-i|, |l-j| < r\}$$
$$k \in [1,m], l \in [1,n]$$
$$y_{ij}(t) = 0.5\left(|x_{ij}(t)+1| - ||x_{ij}(t)-1||\right) \quad (1)$$

In (1) the $N_r(i, j)$ represents the neighborhood of each cell *C(i, j)* with radius *r*. The terms "x_{ij}", "y_{ij}", "u_{ij}", "*A(i, j; k, l)*", "*B(i, j; k, l)*", "*C(i, j; k, l)*", "*I*" are respectively: the state, the output and the input of the cell *C(i, j),* the cloning templates (*A, B, C*) and the bias (I) i.e. an heuristic ad-hoc defined constant.

The proposed SC-CNNs block can be used for several pre-processing operations such as lesion segmentation, edge detection, noise reductions, pixels averaging and so on [15]. In this paper we propose the usage of SC-CNNs for skin lesion pre-processing with novel ad-hoc cloning templates setup. Anyway, robust segmentation of the nevus is another challenge tasks because there are several issues related to noise removal (*hairs, angiomas, etc.*) without significant distortion of the original ROI of the lesion image. In this work we perform validation of our method by using dermoscopy images provided by PH^2 database in which segmented mask for each medical image is also provided [2] (Fig. 3).

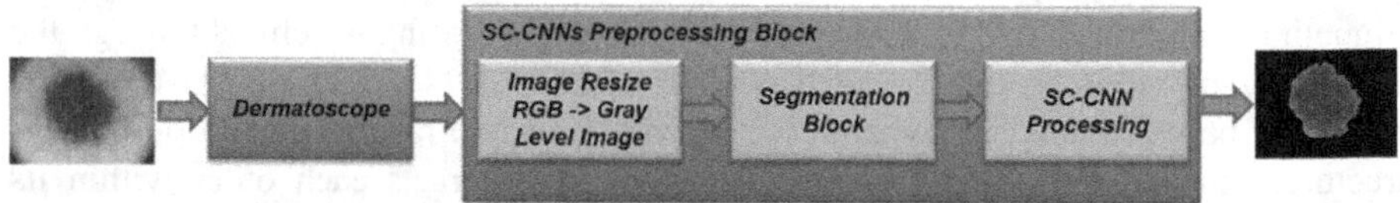

Fig. 3. Pre-processing block based on SC-CNNs

The SC-CNNs performs *transient* pre-processing of gray-level converting lesion image $Y_{D(x,y)}$ by using the following cloning templates [2, 8]:

$$A = \begin{bmatrix} 0 & 0 & 0 \\ 0 & 0 & 0 \\ 0 & 0 & 0 \end{bmatrix}; \ \mathrm{B} = \begin{bmatrix} 3 & 0.25 & 0.25 \\ 0.25 & 3 & 0.25 \\ 3 & 0.25 & 0.25 \end{bmatrix}; \ \mathrm{C} = \begin{bmatrix} 0 & 0 & 0 \\ 0 & 0 & 0 \\ 0 & 0 & 0 \end{bmatrix}; I = 0.7 \quad (2)$$

Above cloning templates is useful to configure SC-CNNs in order to perform ad-hoc adaptive time-transient increasing of gray-level $Y_{D(x,y)}$ image contrast reducing the image distortion produced by the typical *gel* used by dermatologist during medical dermoscopy. The dermoscopy gray level image processed by SC-CNNs will be denoted as $Y_{SC\text{-}CNN(x,y)}$. The following figure shows an example of dermoscopy image pre-processed by SC-CNN based block as described in Fig. 3:

The original dermosocpy image is also segmented by using same algorithm and image-mask used in PH^2 database as shown in Fig. 4.

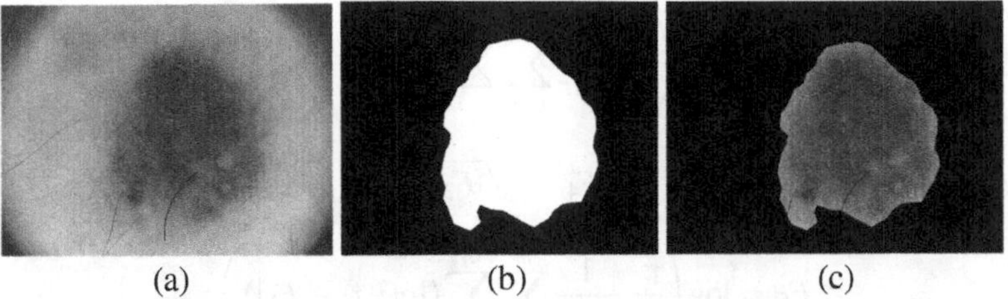

Fig. 4. (a) Original RGB dermoscopy image (b) image-mask provided by PH^2 database (c) SC-CNNs pre-processed segmented gray-level image.

2.2 The Hand-Crafted Image Features and SC-CNNs Hand-Crafted Blocks

At this level, the pre-processed gray-level dermoscopy image named $Y'_{SC-CNN(x,y)}$ is further processed by using ad-hoc morphological-heuristic set of hand-crafted features useful to reproduce the widely used *ABCDE* rule [1]. The image features are re-scaled via logarithmic function in order to compress the ranging of their values. Some of them are classical statistical indicators while other ones are heuristically defined encoding in some sense the *ABCDE* rule, with high precision and robustness. Moreover, some of the proposed hand-crafted image features tries to analyze the melanocytes distribution on the skin epidermis [1]. We denotes with *nr* and *mr* the dimension of the bounding box enclosing the signal (Fig. 5).

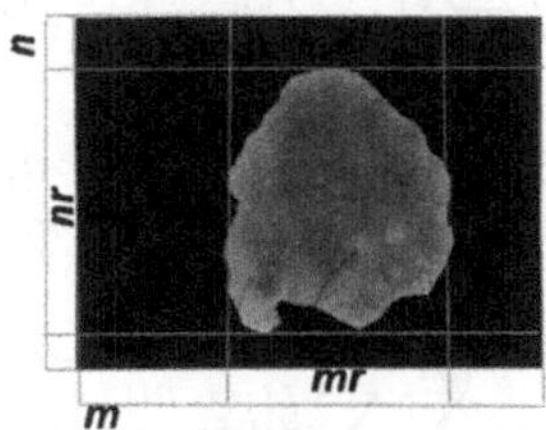

Fig. 5. Some geometric parameters for the segmented SC-CNNs gray-level image.

We indicate *p(i, j)* as gray-level intensity of the single image pixel while and $\Theta(p'(i,j))$ a measure of the frequency rate of the pixel $p'(i, j)$. In the following the overall list of employed feature F_i is reported

$$F_1 = \log\left(\frac{1}{m \cdot n}\sum_{i=1}^{mr}\sum_{j=1}^{nr} p(i,j)\right) \tag{3}$$

$$F_2 = \log\left(\frac{1}{mr \cdot nr}\sum_{i=1}^{m}\sum_{j=1}^{n}(|p(i,j) - F_1|)\right) \tag{4}$$

$$F_3 = \log\left(\frac{1}{mr \cdot nr}\sum_{i=1}^{m}\sum_{j=1}^{n}(p(i,j) - F_1)^2\right) \tag{5}$$

$$F_4 = \sqrt{|F_3|} \tag{6}$$

$$F_5 = \log\left(\frac{\pi}{2} \cdot \frac{1}{m \cdot n}\sum_{i=1}^{m}\sum_{j=1}^{n}(|p(i,j) - F_1|)\right) \tag{7}$$

$$F_6 = -\sum_{i=1}^{m}\sum_{j=1}^{n}\left(\Theta(p'(i,j)) \cdot log\left(\Theta\left(p'^{(i,j)}\right)\right)\right) \tag{8}$$

$$F_7 = \log\left(\sqrt{\left|F_3 - (F_6)^2\right|}\right) \tag{9}$$

$$F_8 = \log\left(\frac{1}{mr \cdot nr}\sum_{i=1}^{m}\sum_{j=1}^{n}\left(\frac{|p(i,j) - F_1|}{F_4}\right)^3\right) \tag{10}$$

$$F_9 = \log\left(\frac{1}{mr \cdot nr}\sum_{i=1}^{m}\sum_{j=1}^{n}\left(\left|i - \frac{m}{2}\right|\right) \cdot \left(\left|j - \frac{n}{2}\right|\right)\left(\frac{p(i,j) - F_1}{F_4}\right)^4\right) \tag{11}$$

$$F_{10} = \log\left(\frac{1}{mr \cdot nr}\sum_{i=1}^{m}\sum_{j=1}^{n}\left(\left|i - \frac{m}{2}\right|\right) \cdot \left(\left|j - \frac{n}{2}\right|\right)\left(\frac{|p(i,j) - F_1|}{F_4}\right)^5\right) \tag{12}$$

$$F_{11} = \log\left(\frac{1}{mr \cdot nr}\sum_{i=1}^{m}\sum_{j=1}^{n}\left(\left|i - \frac{m}{2}\right|\right) \cdot \left(\left|j - \frac{n}{2}\right|\right)\left(\frac{p(i,j) - F_1}{F_4}\right)^6\right) \tag{13}$$

$$F_{12} = \log\left(\frac{1}{mr \cdot nr}\sum_{i=1}^{m}\sum_{j=1}^{n}\left(\left|i - \frac{m}{2}\right| \cdot \left|j - \frac{n}{2}\right| \cdot p(i,j)\right)\right) \tag{14}$$

$$F_{13} = \log\left(\frac{1}{mr \cdot nr}\sum_{i=1}^{m}\sum_{j=1}^{n}\left(\left|i - \frac{m}{2}\right| \cdot \left|j - \frac{n}{2}\right| \cdot (p(i,j))^2\right)\right) \tag{15}$$

$$F_{14} = \log\left(\frac{1}{mr \cdot nr}\sum_{i=1}^{m}\sum_{j=1}^{n}\left(\left|i - \frac{m}{2}\right| \cdot \left|j - \frac{n}{2}\right| \cdot \left(p(i,j) \cdot (i-j)^2\right)\right)\right) \tag{16}$$

$$F_{15} = \log\left(\frac{1}{mr \cdot nr}\sum_{i=1}^{m-k}\sum_{j=1}^{n-k}\left(\left|i-\frac{m}{2}\right| \cdot \left|j-\frac{n}{2}\right| \cdot (|p(i,j) - p(i+k,j+k)|)\right)\right) \quad (17)$$

$$F_{16} = \log\left(\frac{1}{mr \cdot nr}\sum_{i=1}^{m-k}\sum_{j=1}^{n-k}\left(\left|i-\frac{m}{2}\right| \cdot \left|j-\frac{n}{2}\right|((|p(i,j) - F_1|) \cdot (|p(i+k,j+k) + - F_1|))\right)\right) \quad (18)$$

$$F_{17} = \log\left(m \cdot n \cdot \frac{1}{6} \cdot \frac{1}{mr \cdot nr} \cdot \left((F_8)^2 + \left(\frac{1}{4} \cdot (F_9 - 3)^2\right)\right)\right) \quad (19)$$

Note the feature F_{17} is a modified version of the "*Jarque-Bera index*" usually used in the field of financial markets able to point-out *kurtosis* and *skewness* of time-series. The authors adapted its meaning to their purpose related to 2D analysis of the *kurtosis* and *skewness* features for the analyzed nevus. In Eq. (17) we used k = 5 while in Eq. (18) we used k = 3.

$$F_{18} = \log\left(\begin{array}{l} \frac{1}{mr \cdot nr} \cdot \left(\sum_{i=1}^{round\left(\frac{m}{2}\right)}\sum_{j=1}^{n}\left(\left|i-\frac{m}{2}\right| \cdot \left|j-\frac{n}{2}\right|\right) \cdot \frac{|p(i,j) - p(i+1,j)|}{\sqrt{(p(i,j))^2 + (p(i+1,j))^2}}\right) \\ + \left(\sum_{i=1}^{m}\sum_{j=1}^{round\left(\frac{n}{2}\right)}\left(\left|i-\frac{m}{2}\right| \cdot \left|j-\frac{n}{2}\right|\right) \cdot \frac{|p(i,j) - p(i,j+1)|}{\sqrt{(p(i,j))^2 + (p(i,j+1))^2}}\right) \end{array}\right) \quad (20)$$

Above feature F_{18} is a modified version of the so called "*Cosine similarity*" indicator used to analyze the intrinsic similarity of the nevus. The features from F_{19} to F_{21} have been computed as results of further SC-CNNs post-processing performed in the "SC-CNNs Hand Crafted Block". The SC-CNNs programmed via "*edge detection*" cloning templates [14, 15] is used to detect edges of the gray-level segmented image $Y'_{SC-CNN(x,y)}$. Let denotes the so processed image as $Y^{pp}_{SC-CNN(x,y)}$. Considering same parameters as reported in Fig. 5 but related to image $Y^{pp}_{SC-CNN(x,y)}$, we compute the latest set of hand-crafted image features:

$$F_{19} = \log\left(\frac{1}{m \cdot n}\sum_{i=1}^{mr}\sum_{j=1}^{nr} p(i,j)\right) \quad (21)$$

$$F_{20} = \log\left(\pi \cdot \left(\frac{nr-1}{2}\right)^2\right) \quad (22)$$

$$F_{21} = \frac{\min(mr, nr)}{\max(mr, nr)} \quad (23)$$

2.3 The LV Network Clustering System

After we have collected the numerical values corresponding to morphological-heuristic e hand-crafted features, we are ready to perform a skin lesion classification needed for oncological risk assessment of the analyzed nevus i.e. we provide a classification of the analyzed skin lesion as "*benign nevus*" or "*melanoma*" (skin cancer). According to skin lesion membership of one of the above two classes, the algorithm provides follow-up rate according to ad-hoc heuristic risk assessment specifying a time-rate of periodic medical check-up of the analyzed lesion in order to monitoring time-evolution of the lesion by means of the numerical array of features $F_1 \ldots F_{21}$ as per above description.

The normalized set of proposed innovative image features $F_1 \ldots F_{21}$ are used as input of ad-hoc Feed-Forward *Levenberg-Marquardt* (LV) Optimized Neural Network previously pre-trained with carefully selected skin lesions training set containing lesions with different morphological features (color, border, geometry, irregularities, etc.) both benign and malignant(melanoma). We designed and trained that LV Feed-Forward Neural Network with one hidden layer: the neural network dimension is $21 \times 25 \times 2$. The output of LV Neural Network with value close to [1, 0] will be considered as benign nevus while [0, 1] will be considered as melanoma. The used neural network is trained with efficient optimized *Levenberg-Marquardt* back-propagation learning algorithm [15, 16] with stop criteria correlated to learning mean square error i.e. it Several works in literature have proposed the usage of *Artificial Neural Network or Convolutional Neural Networks* (CNNs) in order to address the target of early detection of skin cancer by analysis of dermoscopy images [5–7]. In our experiments in order to improve the generalization capability of the neural network as well as for avoiding the issue of local convergence of training algorithm, we have defined the training set carefully introducing selected dermoscopy images with different patterns (colors, borders, irregularities, benign lesions, melanoma, etc.). After that we have splitted training set in a part for training while remaining part has been used for testing/validation of the network. In our setup we have used MATLAB with Image and Neural Network toolboxes. The used dermoscopy images for training set have been taken partially from PH^2 database [10] and partially from other medical database provided us by oncologists. All the source images will be resized to 768×576 pixels by using same algorithm proposed in [8].

2.4 The Autonomous Diagnosis System

The numerical output of the LV Neural Network together with hand-crafted features is also used for determining the follow-up rate of the analyzed nevus, by using heuristic thresholds based rules as followed described. We define a set of thresholds for the proposed mainly discriminative hand-crafted image features i.e. the features F_{12}, F_{17}, F_{21}. We define those thresholds by means of averaging computation of the values heuristically computed during the training phase of the LV Neural Network according to oncologist advices:

$$Th_{F_{12}} = 12,99; Th_{F_{17}} = 104,00; Th_{F_{21}} = 6,49$$

The above thresholds are used to define ad-hoc follow-up rate according to nevus diagnosis performed by LV Neural Clustering System. Specifically, the Autonomous diagnosis system for "melanoma diagnosis" suggests "*Contact Physician as soon as possibile*" in case of both features F_{12}, F_{17}, F_{21} are greater than related thresholds while extend the follow-up to "1 month" otherwise (at least one of the above features is less of the corresponding thresholds) because even the diagnosis is "melanoma" with high probability, the exanimated nevus does not show high malignant medical indexes according to oncologists consultation. As per "melanoma diagnosis", in case of "benign nevus" diagnosis performed by LV Neural Clustering System, the Autonomous Diagnosis System suggests "*Follow-up rate* $\geq$ *1 year*" for both main features $(F_{12}$, F_{17}, $F_{21})$ within the corresponding thresholds while it reduces the time-range "*Follow-up rate* $\geq$ *6 months*" in case of at least one of the main features is greater than corresponding thresholds.

3 Experimental Results and Future Works

The proposed method has been validated by using the skin lesions database named "PH^2" kindly provided within the ADDI project [3, 10]. The PH^2 database consists of 176 dermoscopy images carefully classified in the related project webpage [3, 10]. The performance of the proposed pipeline has been compared with algorithms proposed in the ADDI project [3, 10] and validated with PH^2 database. In order to get easily the comparison between the proposed approach and other ones, we have computed same benchmark indicators i.e. *Sensibility (SE)* and *Specificity (SP)* and cost function "*C*" computed as reported in [3, 10]:

$$C = \frac{c_{10}(1 - SE) + c_{01}(1 - SP)}{c_{10} + c_{01}} \qquad c_{10} = 1.5c_{01};\ c_{01} = 1 \tag{24}$$

In (24) we defined c_{10} as the weight coefficient of incorrect classification of a melanoma as benign nevus (a false negative FN) while we indicating c_{01} as the weight coefficient for incorrect classification of benign nevus as melanoma (a false positive FP). From values reported in (24), it is clear we consider a missed classification of melanoma more dangerous with respect to a wrong classification of benign nevus as melanoma. Table 1 reports the performance benchmark comparison [3, 10].

The simple comparison between benchmarks reported in Table 1, shows the very promising performance of the proposed method with respect to other ones proposed in literature. An innovative combination between ad-hoc innovative hand-crafted image features with supervised neural networks properly trained has improved drastically the ability of the overall discrimination system in identifying the key nevus features for a robust classification of benign lesions against high suspected ones or melanoma cancer (see Table 2 for results). Often for improving the ability of their pipelines, some

Table 1. Benchmarks comparison for the proposed pipeline

Method	Performance indicators		
	Sensibility	Specificity	C
Proposed	**97%**	**95%**	**0.0682**
Global method (color)	90%	89%	0.1040
Global method (textures)	93%	78%	0.1300
Local method (color)	93%	84%	0.1060
Local method (textures)	88%	76%	0.1680
Global features C6 – kNN classifier	**100%**	71%	0.1160
Global features C1 – kNN classifier	88%	81%	0.1490
Global features C6 – SVM	84%	78%	0.183
Global features C6 – AdaBoost	96%	77%	0.117
Local features BoF	**98%**	79%	0.100

Table 2. Some instances of classified nevus

PH^2 image	***Classification by Proposed pipeline***	***Classification in PH^2 database***	***Follow-up rate***
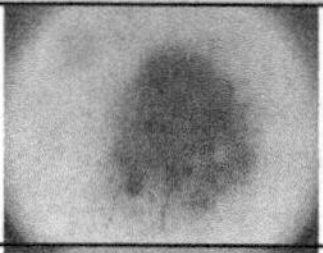	Non Melanoma (Benign Nevus)	Common Nevus (Benign Nevus)	>= 6 months
	Non Melanoma (Benign Nevus)	Common Nevus (Benign Nevus)	>= 1 year
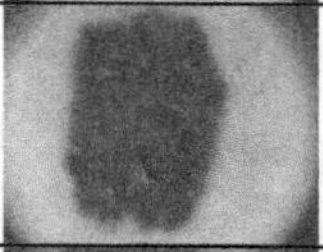	Non Melanoma (Benign Nevus)	Dysplastic Nevus (Benign Nevus)	>= 6 months
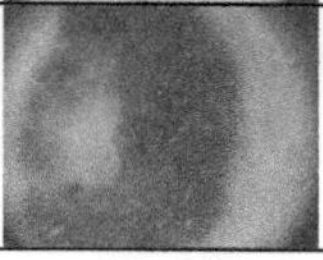	Melanoma (Cancer lesion)	Melanoma (Cancer lesion)	Contact physician as soon as possibile

authors increase sensibility reducing the overall checks of the skin lesion image statistics. The drawback of poor specificity of the pipeline means more nevus biopsy for the patients.

The time performance of the proposed method is acceptable as the proposed pipeline is able to analyze a single nevus in about 2.5 s (we are testing the pipeline in a PC Intel Core i5 3.10 GHz @ 64 bit with 4 Gbyte of RAM). The proposed pipeline can be ported easily from MATLAB environment to an embedded platform based on

STM32 [17]. We proposed a very efficient method for skin nevus analysis and oncological classification both for screening and follow-up of exanimated skin lesion. We are extending skin lesions image dataset in order to cover all the specific skin lesion features with aim to perform a more precise clustering including further class of "suspected" nevus often one of the main issue for dermatologists and oncologists.

References

1. Rundo, F., Banna, G.L.: A method of analyzing skin lesions, corresponding system, instrument and computer program product. EU Registered Patent App. N. 102016000121060, 29 November 2016
2. Gonzalez, R.C., Woods, R.E.: Digital Image Processing, 3rd edn. Prentice Hall, Upper Saddle River (2007)
3. Barata, C., Ruela, M., et al.: Two systems for the detection of melanomas in dermoscopy images using texture and color features. IEEE Syst. J. **99**, 1–15 (2013)
4. Celebi, M.E., Wen, Q., Hwang, S., Iyatomi, H., Schaefer, G.: Lesion border detection in dermoscopy images using ensembles of thresholding methods. Skin Res. Technol. **19**(1), e252–e258 (2013)
5. Fukushima, K.: Neocognitron: a self-organizing neural network model for a mechanism of pattern recognition unaffected by shift in position. Biol. Cybern. **36**(4), 93–202 (1980)
6. Fridan, U., et al.: Classification of skin lesions using ANN. Medical Technologies National Congress (TIPTEKNO) (2016)
7. Xie, F., Fan, F., et al.: Melanoma classification on dermoscopy images using a neural network ensemble model. IEEE Trans. Med. Imaging **36**(3), 849–858 (2016)
8. Battiato, S., Rundo, F., Stanco, F.: ALZ: adaptive learning for zooming digital image. In: IEEE Proceedings of International Conference on Consumer and Electronics (2007)
9. Binu Sathiya, S., Kumar, S.S., Prabin, A.: A survey on recent computer-aided diagnosis of Melanoma. In: 2014 International Conference on Control, Instrumentation, Communication and Computational Technologies (ICCICCT) (2014)
10. Mendonça, T., Ferreira, P.M., et al.: PH^2 - a dermoscopic image database for research and benchmarking. In: 35th International Conference of the IEEE Engineering in Medicine and Biology Society, 3–7 July 2013, Osaka, Japan (2013)
11. Majtner, T., Yildirim-Yayilgan, S., Hardeberg, J.Y.: Combining deep learning and hand-crafted features for skin lesion classification. In: 2016 Sixth International Conference on Image Processing Theory, Tools and Applications (IPTA) (2016)
12. Jamil, U., Khalid, S., Usman Akram, M.: Dermoscopic feature analysis for melanoma recognition and prevention. In: 2016 Sixth International Conference on Innovative Computing Technology (INTECH) (2016)
13. Rashad, M.W., Takruri, M.: Automatic non-invasive recognition of melanoma using Support Vector Machines. In: 2016 BioSMART Conference (2016)
14. Chua, L.O., Yang, L.: Cellular neural networks: theory. IEEE Trans. Circ. Syst. **35**(10), 1257–1272 (1988)
15. Arena, P., Baglio, S., Fortuna, L., Manganaro, G.: Dynamics of state controlled CNNs. In: IEEE Proceedings of International Symposium on Circuits and Systems, ISCAS 1996 (1996)
16. Hagan, M.T., Menhaj, M.: Training feed-forward networks with Marquardt algorithm. IEEE Trans. Neural Netw. **5**(6), 989–993 (1994)
17. STM32 32-bit ARM Cortex MCUs. http://www.st.com/en/microcontrollers/stm32-32-bit-arm-cortex-mcus.html?querycriteria=productId=SC1169

On the Estimation of Children's Poses

Giuseppa Sciortino[2(✉)], Giovanni Maria Farinella[1], Sebastiano Battiato[1], Marco Leo[2], and Cosimo Distante[2]

[1] IPLAB, Department of Mathematics and Computer Science, University of Catania, Catania, Italy
{gfarinella,battiato}@dmi.unict.it

[2] ISASI, Institute of Applied Sciences and Intelligent Systems, C.N.R National Research Council, Lecce, Italy
giuseppa.sciortino@isasi.cnr.it, {marco.leo,cosimo.distante}@cnr.it
http://iplab.dmi.unict.it/
http://www.isasi.cnr.it/

Abstract. Deep Learning architectures have obtained significant results for human pose estimation in the last years. Studies of the state of the art usually focus their attention on the estimation of the human pose of adults people depicted in images. The estimation of the pose of child (infants, toddlers, children) is sparsely studied despite it can be very useful in different application domains, such as Assistive Computer Vision (e.g. for early detection of autism spectrum disorder). The monitoring of the pose of a child over time could reveal important information especially during clinical trials. Human pose estimation methods have been benchmarked on a variety of challenging conditions, but studies to highlight performance specifically on children's poses are still missing. Infants, toddlers and children are not only smaller than adults, but also significantly different in anatomical proportions. Also, in assistive context, the unusual poses assumed by children can be very challenging to infer. The objective of the study in this paper is to compare different state of art approaches for human pose estimation on a benchmark dataset useful to understand their performances when subjects are children. Results reveal that accuracy of the state of art methods drop significantly, opening new challenges for the research community.

Keywords: Human pose estimation · Deep learning methods

1 Introduction and Motivations

Human Pose Estimation has obtained remarkable interest by the community in the last decades [1]. Thank to the advancements of deep learning and the availability larger labeled datasets, a boost on the pose estimation accuracy has been achieved. The main aim of human pose estimation is focused on finding joints (usually called keypoints) and parts (connections between joints) of people present in image to infer the body pose (Fig. 1a).

S. Battiato et al. (Eds.): ICIAP 2017, Part II, LNCS 10485, pp. 410–421, 2017.
https://doi.org/10.1007/978-3-319-68548-9_38

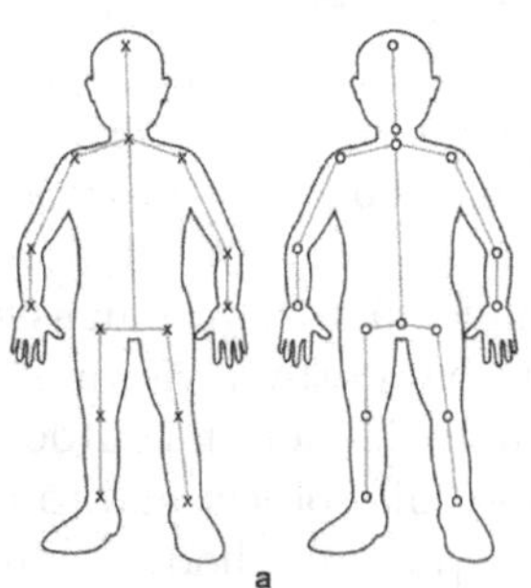

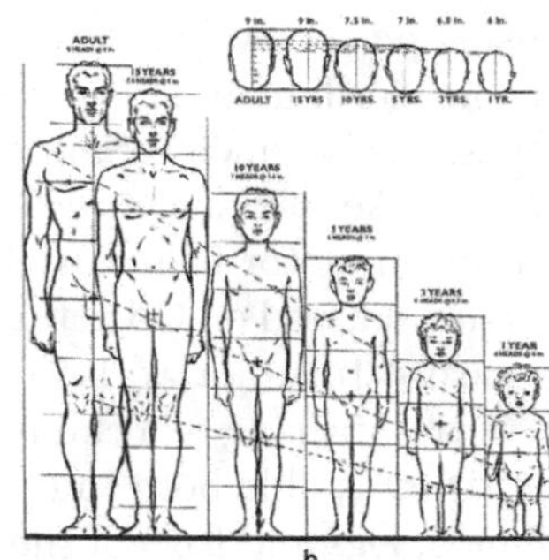

Fig. 1. a: Human skeleton, the human parts in green, the symbols "x" and "o" represent the joints, typically 14–16 joints are used to model the skeleton, 14 on the right and 16 on the left respectively. b: Age-related changes in human body proportions, image from book of Andrew Loomis "Figure Drawing For All It's Worth" (Color figure online)

The pose estimation is useful to analyze the human behaviors in many scenarios, such as in the context of Ambient Assisted Living (AAL) [2] for recognizing the user's activities of daily living, to the study and diagnosis of motor and psychological disorders, etc. The most of the available datasets used for human pose estimation [3–6] does not contain a sufficient number of images depicting children to evaluate the performances of the algorithms in this specific case. In the last years, MPII Human Pose Dataset [1] became a state of the art benchmark repository for the evaluation of human pose estimation algorithms. It contains 25K images related to 410 human activities performed in indoor and outdoor environments. Despite it is an excellent resource for human pose estimation algorithms, like other datasets does not contain enough images of children.

Human Pose and Body Proportions in Children. Although human pose estimation is a problem that has been studied for many years, most of the literature is focused on pose of adults, whereas it was sparsely studied in the case of children. It might seem that adults and children have the same body shape and therefore the same skeleton, but infants and children are not adults "miniatures", they are structurally different. The human body grows and develops continuously (not even uniformly) from birth through old age [7]. The proportion between body parts changes according to a predictable trend (Fig. 1b). For example, the length of head in adults is about one-seventh of the total body length, whereas in the infant is one-fourth. Proportionally, the trunk length is longer in children with respect to adults. The proportions of trunk and limbs change during the growth, and the lower limbs increase in length more rapidly than the upper limbs. The anatomy of the child's neck is one of the most particular aspects. The neck muscle strength increases with age, in children they are not generally properly developed and tend to appear as flattened due to the greater mass head perched on a thin neck. Indeed, in many images of children the neck joint is not visible. To the best of our knowledge there are no studies in the literature focusing on the evaluation of the state of art algorithms for pose estimation which consider human at early ages (children). In [8] is described a tools for the non-invasive assessment of autism spectrum disorder (ASD) considering four behavioral markers: visual tracking, disengagement of attention, sharing interest, and atypical motor behavior. The last marker is evaluated using a pose estimation algorithm based on Object Cloud Model [9] to detect an asymmetrical position of the arms. In [10] a work to simulate babysitter's vision is presented. The main aim is the tracking of a child-object in an indoor and outdoor environment. The algorithm is able both to track whole child-object and to track body parts (head, hands, legs and feet) of the child. In [11] a method to estimate body pose of infants in depth images by using random ferns is introduced. A pixelwise body part classifier is proposed and joints are located computing the center of mass of the points belonging to each body part. To the best of our knowledge no attempt has been made so far to establish a more representative dataset aiming to cover children's images for human pose estimation. A focused work to study child behaviors is reported in [12], where a study of typical autistic behaviors is performed. The annotations examine a set of representative attributes of the behavior (as stimming behavior category and intensity) but the ground truth of the joints is not available for this dataset. Some of these videos are used in the present work to build our experimental dataset.

In this paper we present a benchmark dataset in which the subjects are related to children and toddlers. On this dataset we compare different state of the art methods [13–16] to estimate the human pose and by performing an indept evaluation. The remainder of this paper is organized as follows. In Sect. 2, we briefly discuss some relevant related works for human pose estimation. In Sect. 3 we introduce the methods used for the comparative evaluation and present the benchmark dataset describing the collection and annotation processes. In Sect. 4, we detail the experiments and results. Section 5 concludes the paper.

2 Related Work

Human Pose Estimation. In this section we briefly review state of art methods for human pose estimation exploiting convolutional neural networks, by highlighting their main peculiarities. Nowadays, many limitation of classic approaches have been overcome through the widely use of convolutional neural networks. In the literature there are many models to solve articulated pose estimation exploiting deep learning architectures. To solve 2D human pose estimation from single image, some methods regress image to build heatmap or confidence maps [14,15,17–20], where heatmap represents the probability that the joints appear in a particular position of the image. Differently, in [21] Cartesian coordinates of human joints are directly predicted. Regressing heatmaps are preferred because the framework can be multimodal, indicating the existence of multiple joints. In [13] is presented a hybrid method that regresses on both heatmaps and cartesian coordinates.

Human Pose Estimation Multiperson. In the case of multi-person pose estimation, most of the approaches require a preliminary step where the person is detected [22–24]. This makes the results dependent on the correct detection of people within the images. In [25] the interdependence from people detection is considered, but the method requires additional initial assumptions. The authors of [16] propose a bottom-up method, which detects the joints as first step and then associates the different joints in parts to build the full skeleton. On the contrary in [26] a top-down strategy is presented. In the first stage the method predicts and scale the bounding box containing persons, then it regresses the locations of joints and finally performs the pose predictions.

3 Methods and Dataset

In this section we briefly describe the four methods we have considered to benchmark the problem of human pose estimation on images depicting children [13–16]. We also describe the dataset collected to perform the comparative evaluation of the different methods.

Recurrent Human Pose Estimation (RNN). In [14] a recurrent neural network model to estimate the human pose has been proposed. The approach is top-down based (i.e., joints are detected first and then pose is inferred without the need detecting the person as first step). The proposed architecture is composed by two main modules: Feed-Forward Module and Recurrent Module. The aim of the first module, is the detection of the "body joints" by regressing heatmaps (one for each joint) without knowing the body configuration or the association between couples of joints. The second module takes in input the heatmaps from the feed-forward module and it infers the contextual information. The first layer of feed-forward module use small filters, whereas larger filters are used in deep layers to learn the structures of the body. The whole network can be trained in an end-to-end fashion and outputs 16 joints (Fig. 1a). The network was trained and tested on MPII Human Pose [1] dataset and it has been tested on extended LSP [27] and on MPII [1] datasets obtaining good performances.

Human Pose Estimation with Iterative Error Feedback (IEF). Human pose Estimation with Iterative Error Feedback [13] introduces an iterative convolutional neural network to predict a body pose from 2D images. The authors present an iterative self-correcting model by feeding back-error prediction. Given a preliminary guess solution (i.e., Cartesian coordinates representations of joints positions) for each iteration the method applies a "bounded correction". In this way it predicts a direction in which to move a final solution. This correction is used to update the joints positions, and then the process is iterated. The IEF method takes in input the coordinates of any point that belongs to the torso as additional information. The architecture of the deep convolutional network consists of a pre-trained network (i.e., Imagenet [28]) where the first and the last layers were appropriately modified to adapt at the problem. The network is tested on two datasets, MPII [1] and LSP [27], obtaining good performances.

Convolutional Pose Machine (CPM). In [15] the authors proposed a method, Convolutional Pose Machine, to estimate human pose from 2D image for single-person. The method consists of a series of sequential convolutional networks. At every stage a network takes as input the belief map supplied as output from the previous network. The Convolutional Pose Machine is trained to learn image features and image-dependent spatial models. Each step corresponds to a sequential refinement. In the first stage a convolutional network is applied to obtain belief map from local evidence by using small receptive field. Successive stage use multiple layers to reach large receptive field in order to capture complex and long-range correlations between parts. CPM model takes in input additional information: a bounding box position containing the subject on which to estimate the pose. The model can be trained in end-to-end mode from scratch and outputs 14 joints. In evaluation section of the original paper, different training schemes are analyzed on LSP dataset. The performance are evaluate on three datasets (MPII [1], LSP [27] and FLIC [29]) obtaining good performances.

Realtime Multi-person 2D Pose Estimation Using Part Affinity Fields (MPP). In [16] a model to estimate 2D human pose for multi-person with bottom-up approach is presented. The method improves the performance with regard to computational time of the state of art. The proposed architecture consists of two-branch multi stage convolutional neural networks, and introduces the Part Affinity Field (PAF), namely a set of 2D vector fields related to body parts. The two branches work separately. The first predicts the confidence score maps for part detector and the second predicts PAFs for part association. Then their output are concatenated and the process is iterated. In the same manner, the model learns implicit spatial relationship between different people. The model is trained in end-to-end fashion and it outputs 14 joints, if all of them are detected. The performance are tested on the benchmark dataset MSCOCO [30] and on the MPII [1] outperforming the state-of-art. The authors provide an analysis of the computational time, the good performances allow have realtime method and to be applied for video sequence analysis.

3.1 Dataset

In this work we exploit a dataset of images of a particular age group: toddlers and children (approximately one-eleven years old). The images have been extracted from videos available on public domain websites and video portals. The dataset covers various activities typical of this age group as: 'learning to walk', 'sports' and 'play'. The collected images depict different variabilities, such as: indoor, outdoor, clothing type and include interaction with various objects and environments. The images are not splitted in activity categories since our purpose is not to recognize activities but to evaluate human pose estimation algorithms on child's images, to highlight and understand the differences in accuracy with regard to the state of art when used on adult's images. The available datasets in literature do not consider this sub-category and they contain very few images to carry out a specific evaluation. Our dataset covers a wide variety of poses. Children often assume poses more articulate than adults. Often due to privacy issues, the datasets regarding to child are not easily available on the web, and for this reason the images of our dataset are chosen from free videos available on youtube. Amateur and professional videos allowed to obtain a wide variety of subjects and varieties, due to the different acquisition sources.

Images Collection. As mentioned above, the dataset in [12] provides a list of 75 video URLs. These videos regarding to children behavioral disorders, are divided into three categories of stimming behaviors: arm flapping, head banging and spinning, that are typical behaviors observed in ASD. The videos in [12] were the starting point for our collection and research process. Some of them were discarded because they were not suitable for our purposes. For example some videos depict interaction between more persons, or the poses show strong truncations, same videos that show low quality. Other videos were searched and selected on youtube, using keywords like 'children', 'toddler', 'walking', 'learn to walk', 'children', 'video', 'talent show', 'gymnast', 'dancing', 'play', 'autism', and their possible combinations. In this way 150 videos were collected. The selected videos have been posted in youtube mainly by relatives or talent show. Afterwards, for each collected video all frames were extracted, without post-processing, preserving the original natural setting and a subset has been manually selected from several non-consecutive frames, trying to ensure different poses. We obtained a dataset with 1176 images related to 104 different subjects. This dataset is available for the research community upon request to the authors.

Images Annotation. The collected images were annotated by using a tool available on line [31]. The annotation process is designed by clicking control points in the image recording the positions, labels and visibility for each selected keypoints. The tool was slightly modified in order to mark up to 22 visible/occluded labeled keypoint locations: head (forehead, chin) ears, eyes, nose, mouth, neck, arms (shoulder, elbow, wrist), torso, legs (hip, knee, ankle). The annotations in our dataset are person-centric (i.e., right/left corresponding at right/left body parts of person) and are saved in an xml file for each image.

4 Comparative Evaluation

In this section we evaluate the performances of the methods described above on our dataset[1]. For the evaluation we considered the PCK measure [32], one of the most commonly used in the literature to measure the accuracy of detected joints. Generally, it is used a modified PCK measure, denoted PCKh, that considers a localized joint as correct if the distance between the predicted joint and correspondent ground-truth is less of 50% of head segment length. In this way, the PCKh measure is independent from the size of the bounding box considered by the measure PCK. The compared methods output different number of joints,

[1] The methods have been exploited considering pre-trained models without re-training or fine-tuning.

16 in [13,14] and 14 in [15,16] (see Fig. 1a). In our analysis, the joints related to eyes, nose, ears and mouth are not considered. In particular, for [15,16] shall not be considered chin and torso joints in evaluations because the joint relative to the chin and torso are not covered by their output (bin number 8 'Torso' and 10 'D Head' are missing in Figs. 3 and 4).

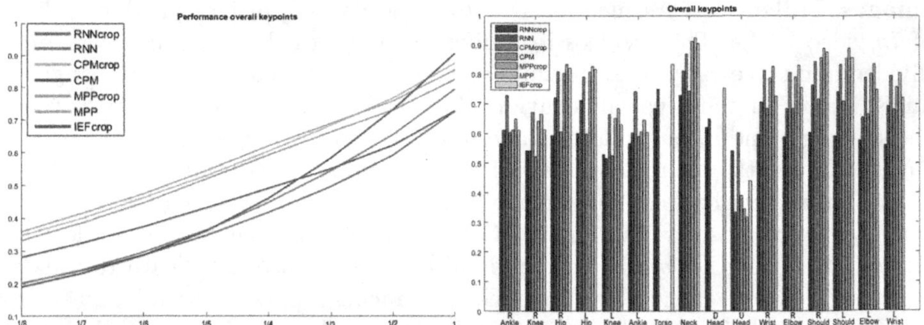

Fig. 2. Overall performance evaluation, for CPM and MPP methods considering 14 joint and 16 joints for IEF and RNN methods. With CPM crop, MPP crop and RNN crop we indicate the achieved performances on "cropped" images.

Fig. 3. Histogram performance evaluation per joint. In the x-axis the name of joints whereas in the y-axis the percentage of joints that are correctly detected by fixing the percentage to 50% of the head segment length for PCKh measure.

Performance Evaluation. Here we show the achieved performances for considered methods [13–16] on our dataset. Considering that CPM [15] and IEF [13] methods require additional input, we cropped every image on bounding box containing the person on which estimate the pose. The bounding box is roughly determined on ground truth annotation. Furthermore for IEF [13] method we considered that the center of the bounding box is the point belonging to torso. Figure 2 shows the performances of the considered methods depending on percentage of the head segment length taken into account in the evaluation measure. In the y-axis the percentage of joints that are correctly detected. For MPP, RNN and CPM methods are showed the performance achieved without input parameters. It is interesting to note that MPP method achieved the best performance when the full image is considered, without additional information about bounding box. In Fig. 3 are shown the performances for single joint. The performances are obtained by fixing the percentage to 50% of the head segment length for the evaluation measure PCKh.

In the analysis, the joints not present in the image (i.e., joints not present because the body part is not depicted in the image, namely truncations) are excluded from the evaluation for not penalizing the methods that output a fixed number of joints [13–15]. In Fig. 4 are presented the performances per single joints, where we split the joints in visible and occluded sets based on ground truth label. The histograms in Figs. 3 and 4 reveal that better performances for the joints belonging to the trunk (neck, shoulder, hip) than joints belonging to the limbs (elbow, wrist, upper head, knee, ankle), it is most evident in the case of occlusions (see Fig. 4 on the right). This could be due to convolutional layers that implicitly encode a configuration model. On the other hand, the deep networks are trained on dataset where the subjects are mainly adults. Therefore the deep networks may have learned the anatomical proportions of adults, in agreement with what has been described in the introduction section.

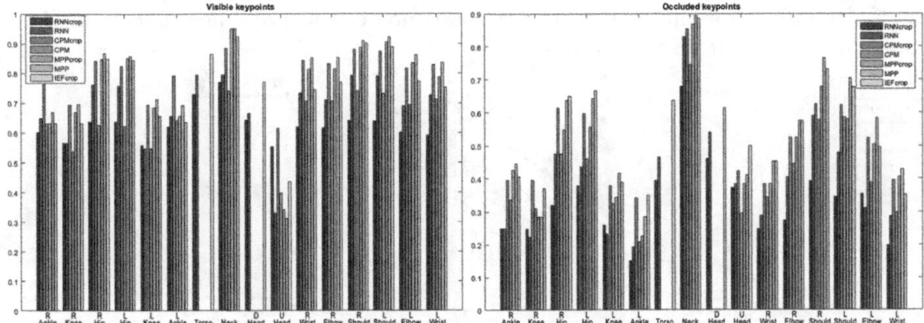

Fig. 4. Histogram performance evaluation on each joint 'visible' and 'occluded', on the right and on the left respectively. In the x-axis the name of joints, in the y-axis the percentage of joints that are correctly detected by fixing the percentage to 50% of the head segment length for 'PCKh' metric.

Occlusions and truncations can drastically affect performance. The best case is certainly represented by poses in which all the joints are visible and distinct. In this regard, we evaluate the performances of methods on subset of our datasets. We divide the dataset into four subsets:

Case 1: all the joint are present and visible in the image;
Case 2: all the joint are present but some are not visible;
Case 3: not all the joint are present in the image, those present are all visible;
Case 4: not all the joint are present in the image and some are not visible.

Based on these splitting we obtained four subsets. The first is about 20% of our dataset, the second 65%, the third 5%, and 10%. In Figs. 5 and 6 the performance are shown for each case depending on the head segment length taken into account.

To evaluate the accuracy of the entire pose for each image we calculated the average distance between each joint position and the respective ground truth joint position, if the average distance is less than 50% of head length segment the pose is considered as correct. In Table 1 the percentage of images that do not satisfy this relationship is showed. We observed that the highest performing

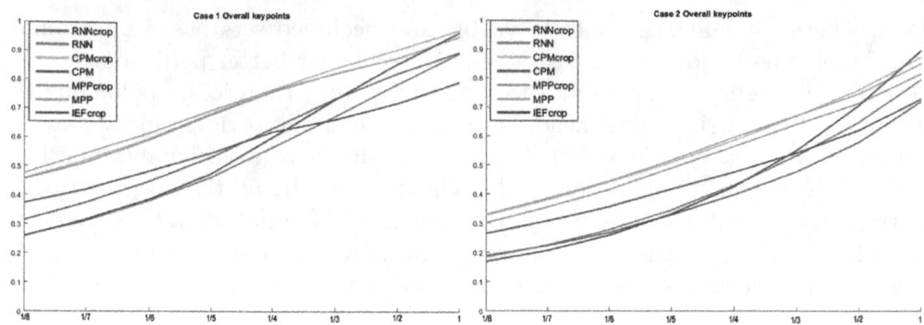

Fig. 5. Overall performance evaluation, Case 1: no occlusion and no truncation. Case 2: some occlusion but no truncations.

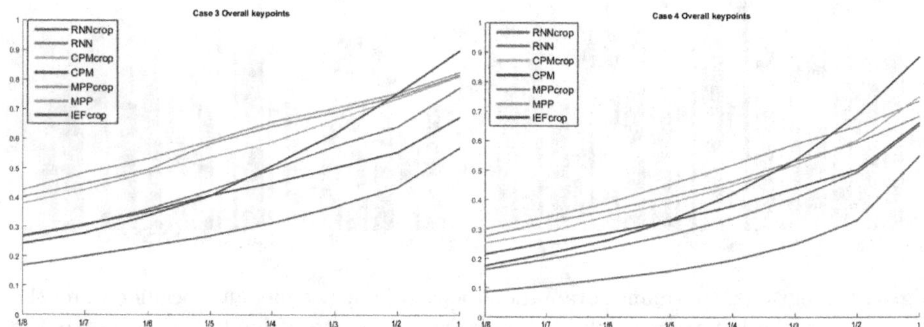

Fig. 6. Overall performance evaluation, Case 3: no occlusion but truncations are present, Case 4: occlusion and truncations are present.

Table 1. Percentage of worst pose based on average distance between all joints belonging to the pose and ground truth.

	RNNcrop	RNN	CPMcrop	CPM	MMPcrop	MMP	IEF
Case 1	23%	24%	4%	32%	26%	0%	7%
Case 2	58%	0.43%	27%	45%	49	10%	29%
Case 3	76%	39%	35%	47%	41%	0%	17%
Case 4	0.83%	0.60%	46%	56%	63	20%	25%

method is the one proposed in [16], although it is not the best in Case 1, it allows to have a good performances of the whole pose (despite occlusions and truncations), without requiring additional input parameters.

5 Conclusion

In this study we considered the problem of estimating pose of children from images. The aim is to verify whether the performance of state of the art methods on our dataset are comparable to those obtained on the benchmark datasets containing mainly images of adults. We have collected and annotated a dataset images containing children extracted from videos recorded in an uncontrolled environment and available on public domain websites and video portals. We have compared four well known methods on our dataset. Experiments point out that accuracy drops down for all methods in this application context. The results open new research challenges, especially for non-invasive assessment of behavioral or motor disorders of children for assistive technology. We expect that by retraining models the results can be improved. We plan to fine tune of the considered models and then extend our dataset in the next study.

References

1. Andriluka, M., Pishchulin, L., Gehler, P., Schiele, B.: 2D human pose estimation: new benchmark and state of the art analysis. In: IEEE Conference on Computer Vision and Pattern Recognition, pp. 3686–3693 (2014)
2. Leo, M., Medioni, G., Trivedi, M., Kanade, T., Farinella, G.M.: Computer vision for assistive technologies. Comput. Vis. Image Underst. **154**, 1–15 (2017)
3. Sigal, L., Balan, A.O., Black, M.J.: HumanEva: Synchronized video and motion capture dataset and baseline algorithm for evaluation of articulated human motion. Int. J. Comput. Vis. **87**(1–2), 4–27 (2010)
4. Ionescu, C., Papava, D., Olaru, V., Sminchisescu, C.: Human3.6M: large scale datasets and predictive methods for 3D human sensing in natural environments. IEEE Trans. Pattern Anal. Mach. Intell. **36**(7), 1325–1339 (2014)
5. Jhuang, H., Gall, J., Zuffi, S., Schmid, C., Black, M.J.: Towards understanding action recognition. In: IEEE International Conference on Computer Vision, pp. 3192–3199 (2013)
6. Everingham, M., Van Gool, L., Williams, C.K.I., Winn, J., Zisserman, A.: The PASCAL Visual Object Classes Challenge 2011 (VOC2011) Results. http://www.pascal-network.org/challenges/VOC/voc2011/workshop/index.html
7. Huelke, D.F.: An overview of anatomical considerations of infants and children in the adult world of automobile safety design. Annu. Proc./Assoc. Adv. Automot. Med. **42**, 93–113 (1998)
8. Hashemi, J., Spina, T.V., Tepper, M., Esler, A., Morellas, V., Papanikolopoulos, N., Sapiro, G.: Computer vision tools for the non-invasive assessment of autism-related behavioral markers. arXiv preprint arXiv:1210.7014 (2012)
9. Miranda, P., Falcão, A., Udupa, J.: Cloud models: their construction and employment in automatic MRI segmentation of the brain (2010)
10. Aljuaid, H., Mohamad, D.: Child video dataset tool to develop object tracking simulates babysitter vision robot. J. Comput. Sci. **10**(2), 296–304 (2014)
11. Hesse, N., Stachowiak, G., Breuer, T., Arens, M.: Estimating body pose of infants in depth images using random ferns. In: IEEE International Conference on Computer Vision Workshop, pp. 427–435 (2015)

12. Rajagopalan, S., Dhall, A., Goecke, R.: Self-stimulatory behaviours in the wild for autism diagnosis. In: IEEE International Conference on Computer Vision Workshops, pp. 755–761 (2013)
13. Carreira, J., Agrawal, P., Fragkiadaki, K., Malik, J.: Human pose estimation with iterative error feedback. In: IEEE Conference on Computer Vision and Pattern Recognition, pp. 4733–4742 (2016)
14. Belagiannis, V., Zisserman, A.: Recurrent human pose estimation. arXiv preprint arXiv:1605.02914 (2016)
15. Wei, S., Ramakrishna, V., Kanade, T., Sheikh, Y.: Convolutional pose machines. In: IEEE Conference on Computer Vision and Pattern Recognition, pp. 4724–4732 (2016)
16. Cao, Z., Simon, T., Wei, S., Sheikh, Y.: Realtime multi-person 2D pose estimation using part affinity fields. arXiv preprint arXiv:1611.08050 (2016)
17. Tompson, J.J., Goroshin, R., Jain, A., LeCun, Y., Bregler, C.: Efficient object localization using convolutional networks. In: IEEE Conference on Computer Vision and Pattern Recognition, pp. 648–656 (2015)
18. Tompson, J.J., Jain, A., LeCun, Y., Bregler, C.: Joint training of a convolutional network and a graphical model for human pose estimation. In: Advances in Neural Information Processing Systems, pp. 1799–1807 (2014)
19. Bulat, A., Tzimiropoulos, G.: Human pose estimation via convolutional part heatmap regression. In: Leibe, B., Matas, J., Sebe, N., Welling, M. (eds.) ECCV 2016. LNCS, vol. 9911, pp. 717–732. Springer, Cham (2016). doi:10.1007/978-3-319-46478-7_44
20. Jain, A., Tompson, J., Andriluka, M., Taylor, G.W., Bregler, C.: Learning human pose estimation features with convolutional networks. arXiv preprint arXiv:1312.7302 (2013)
21. Toshev, A., Szegedy, C.: DeepPose: human pose estimation via deep neural networks. In: IEEE Conference on Computer Vision and Pattern Recognition, pp. 1653–1660 (2014)
22. Pishchulin, L., Jain, A., Andriluka, M., Thormählen, T., Schiele, B.: Articulated people detection and pose estimation: reshaping the future. In: IEEE Conference on Computer Vision and Pattern Recognition, pp. 3178–3185 (2012)
23. Gkioxari, G., Hariharan, B., Girshick, R., Malik, J.: Using k-poselets for detecting people and localizing their keypoints. In: IEEE Conference on Computer Vision and Pattern Recognition, pp. 3582–3589 (2014)
24. Iqbal, U., Gall, J.: Multi-person pose estimation with local joint-to-person associations. In: Hua, G., Jégou, H. (eds.) ECCV 2016. LNCS, vol. 9914, pp. 627–642. Springer, Cham (2016). doi:10.1007/978-3-319-48881-3_44
25. Eichner, M., Ferrari, V.: We are family: joint pose estimation of multiple persons. In: Daniilidis, K., Maragos, P., Paragios, N. (eds.) ECCV 2010. LNCS, vol. 6311, pp. 228–242. Springer, Heidelberg (2010). doi:10.1007/978-3-642-15549-9_17
26. Papandreou, G., Zhu, T., Kanazawa, N., Toshev, A., Tompson, J., Bregler, C., Murphy, K.: Towards accurate multi-person pose estimation in the wild. arXiv preprint arXiv:1701.01779 (2017)
27. Johnson, S., Everingham, M.: Learning effective human pose estimation from inaccurate annotation. In: IEEE Conference on Computer Vision and Pattern Recognition, pp. 1465–1472 (2011)
28. Szegedy, C., Liu, W., Jia, Y., Sermanet, P., Reed, S., Anguelov, D., Erhan, D., Vanhoucke, V., Rabinovich, A.: Going deeper with convolutions. In: IEEE Conference on Computer Vision and Pattern Recognition, pp. 1–9 (2015)

29. Sapp, B., Taskar, B.: MODEC: multimodal decomposable models for human pose estimation. In: IEEE Conference on Computer Vision and Pattern Recognition, pp. 3674–3681 (2013)
30. Lin, T.-Y., Maire, M., Belongie, S., Hays, J., Perona, P., Ramanan, D., Dollár, P., Zitnick, C.L.: Microsoft COCO: common objects in context. In: Fleet, D., Pajdla, T., Schiele, B., Tuytelaars, T. (eds.) ECCV 2014. LNCS, vol. 8693, pp. 740–755. Springer, Cham (2014). doi:10.1007/978-3-319-10602-1_48
31. Bourdev, L., Malik, J.: The Human Annotation Tool. https://www2.eecs.berkeley.edu/Research/Projects/CS/vision/shape/hat/
32. Yang, Y., Ramanan, D.: Articulated human detection with flexible mixtures of parts. IEEE Trans. Pattern Anal. Mach. Intell. (PAMI) **35**(12), 2878–2890 (2013)

Optical Coherence Tomography Denoising by Means of a Fourier Butterworth Filter-Based Approach

Gabriela Samagaio, Joaquim de Moura(✉), Jorge Novo, and Marcos Ortega

Department of Computing, University of A Coruña, A Coruña, Spain
{gabriela.samagaio,joaquim.demoura,jnovo,mortega}@udc.es

Abstract. Optical Coherence Tomography (OCT) is affected by ubiquitous speckle noise that difficult the visualization and analysis of the retinal structures. Any denoising strategy should be able to remove efficiently the noise as well as preserves clinical information contained in the images. This information is crucial to analyses the retinal layer tissue that allows the posterior analysis and recognition of relevant diseases as macular edema or diabetic retinopathy.

To address this issue, a method based on the Fourier Butterworth filter combined with a contrast enhancement and a histogram regularization was developed in order to reduce the speckle noise in OCT retinal images. The proposed method was validated using 45 OCT retinal images organized into 3 groups of noise degree, comparing the results with the performance of representative methods of the state-of-the-art. The validation and comparison were made through three quantitative metrics: Signal-to-Noise Ratio (SNR), Contrast-to-Noise Ratio (CNR) and average Effective Number of Looks (ENL).

The experimental results showed that the proposed method offered satisfactory results, outperforming the results of the other methods by the achievement of a SNR of 7.04 dB and a CNR of 14.08 dB better than the second best filter, respectively, for the whole group of OCT retinal images.

Keywords: Optical Coherence Tomography · Retinal imaging · Speckle noise · Denoising · Fourier Butterworth filter

1 Introduction

Optical Coherence Tomography (OCT) is an imaging technique for the analysis of the eye fundus that provides a non-invasive and contactless cross-sectional retinal visualization with high resolution. It has been widely used in biomedical imaging, as in the extraction of medical information as spatial information but also in the measurement of biological tissues such as retinal layers [1–3] and other structures [4,5], in a micro scale way.

S. Battiato et al. (Eds.): ICIAP 2017, Part II, LNCS 10485, pp. 422–432, 2017.
https://doi.org/10.1007/978-3-319-68548-9_39

In OCT, speckle noise can be introduced in the capture process by many causes like the eye motion of the patient, multiple scattering or particular properties of each OCT system [6]. Therefore, the process of denoising of ophthalmic OCT images is essential to obtain better quality images that may lead to a better identification of the retinal structures as well as a better recognition and quantification of different retinal diseases, thus aiding the diagnosis made by the clinicians [7]. For speckle noise reduction, two major strategies are applied at different capture stages: (I) during image acquisition and (II) at a post-processing stage. The first strategy is usually performed by the device at the moment of the image capture, by the use of multiple uncorrelated recordings that are combined to produce the final image. The second stage requires a post-processing of the OCT scans based on computational algorithms [8]. A large variety of post-processing methods have been proposed for speckle noise reduction of OCT slices. In the literature, there are several OCT denoising techniques that are being adapted from other domains as active radar, synthetic aperture radar (SAR) or ultrasound, among others, since the granular noisy pattern is similar to the noise that appear in OCT scans [8]. Therefore, a substantial amount of signal processing research has been conducted to combat speckle noise and, in the last years, new versions of standard denoising filters [9] were adapted to the task of OCT image denoising [10,11]. Most of them were originally applied to other domains, such as the Lee filter [12], Kuan filter [13] and Frost filter [14], wavelet filter [15], general Bayesian filter [16] or anisotropic diffusion [17], which are the most widely discussed as spatial adaptive filters noise attenuation. Despite the efforts, it is being a challenge to develop an efficient speckle noise reduction algorithm that simultaneously preserves details such as texture and edge information, ensuring that the particular anatomy of those structures are preserved, beyond the image quality metrics, as signal-to-noise-ratio (SNR), contrast-to-noise-ratio (CNR) or average effective number of looks (ENL), that are typically used to measure any denoising strategy.

In this paper, we propose a new method for speckle noise reduction in OCT images based on the application of the Fourier Butterworth filter in combination with an image enhancement and histogram regularization. The method ensures, at the same time, image detail as well as preservation of particular morphological structures, allowing the possible recognition of pathologies in the internal retinal tissue. To validate the method, a comparative analysis with algorithms of the state-of-art was developed and implemented for an automatic calculation of quality images metrics, such as SNR, CNR and ENL and using three representative image sets of progressive noise complexity.

2 Methodology

To reduce the speckle noise in OCT retinal images the proposed method is based on the Fourier Butterworth filter. This low pass filter minimizes high frequency components which are essentially the aimed noise. Then, to enhance the image contrast, a post-processing stage based on histogram correction was applied to highlight the morphological structures as retinal layers and pathological structures.

Moreover, in order to develop a systematic comparison between the proposal with methods of the state-of-art, an automated iterative regions of interest (ROIs) extraction was implemented, ensuring an independent and objective process on the determination of quantitative metrics such as SNR, CNR and ENL.

2.1 Fourier Butterworth Image Filtering

Fourier Butterworth filter is a low-pass filter where images are transformed from the spatial to the frequency domain and vice-versa. It is characterized as a smooth filter where the high frequencies are cutoff, reducing the noise while preserving, simultaneously, the edges. The filter is governed by two main parameters: the critical frequency and the order filter [18]. The critical frequency determines the transition zone where amplitude of the frequency drops from 1 down to 0, while the order determines its steepness, such that a higher order produces a narrower transition zone [19].

Butterworth filter is applied in OCT images for speckle noise reduction following, as shown in the diagram of blocks in Fig. 1, the next steps: (1) fast Fourier transform (FFT), (2) Transfer Function $H(u, v)$, (3) inverse Fourier transform (IFFT). Based on spatial information of the input noisy image, a meshgrid array is created in order to process the information in frequency domain, which corresponds to the first process block of the diagram.

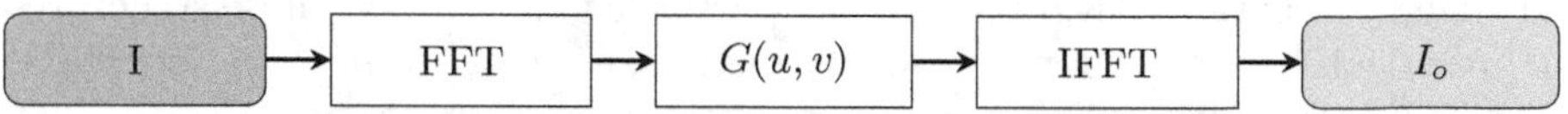

Fig. 1. Basic flow diagram of speckle noise reduction for OCT images based on Fourier Butterworth filter.

There, I is the input noise image and I_o is the output Fourier Butterworth filtered image. The second process block represents the transfer function $H(u, v)$ of the Butterworth low-pass filter of order n at a distance from the origin, defined by:

$$H(u, v) = \frac{1}{1 + (\frac{D(u,v)}{D_0})^2} \tag{1}$$

where, $D(u, v)$ is the distance from point (u, v) to the center of the filter and D_0 is the cutoff frequency. Having calculated $H(u, v)$ a multiply operation is applied with the Fourier transform of an original image, $F(u, v)$, as:

$$G(u, v) = H(u, v) \times F(u, v) \tag{2}$$

Finally, the resultant image, $G(u, v)$ is returned from the frequency to the spacial domain applying IFFT.

2.2 Post Processing

The Fourier Butterworth filter minimizes the presence of noise, Fig. 2(b). However, it is necessary to enhance the visualization of the morphological structures such as retinal layers that appear in the eye fundus in order to facilitate the posterior analysis. In an attempt to enhance these structures, two main steps were taken: a weighted multiplicative operation of the output Fourier Butterworth filter by the original image and a histogram correction.

Firstly, the resultant image of the Fourier Butterworth filter is multiplied by the original image with a weight $w = 1.2$, value empirically calculated based on extensive experimentation to yield the best results enhancing the boundaries of retinal layers, as well as preserving the details of the tissue of the morphological structures, as shown in Fig. 2(c).

Then, as a second step, a histogram correction was introduced to regularize the image contrast based on the histogram, as shown in Fig. 2(d), resulting in a more detailed and clear image.

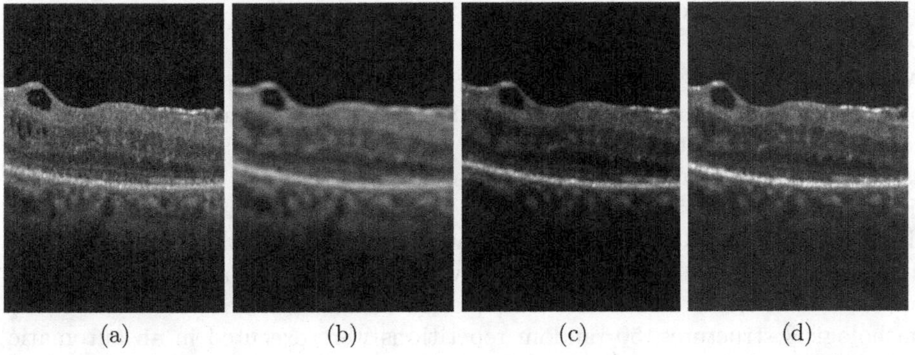

Fig. 2. Example of the post-processing stage of the OCT retinal images. (a) Original noisy image. (b) Output Fourier Butterworth filtered image. (c) Original image correction using the Fourier Butterworth output filter. (d) Contrast enhancement and histogram correction.

2.3 Automatic Validation Process

In order to validate and compare the proposed methodology with reference methods of the literature, quantitative image metrics were calculated, such as SRN, CNR and ENL, as some of the most used metrics in other denoising approaches as classification metrics for speckle noise reduction. However, the presence of pathological structures in the OCT images makes the scenario of image denoise quantification more complicated, since the pathological regions, black non-reflective spaces [20], can alter erroneously this metrics. In order to minimize this limitation, we implemented an automatic method for the extraction of a background and retinal layer ROIs that are needed for the calculation of the implemented metrics. Two main steps were implemented: an automatic segmentation of the retinal layers and the automatic ROIs extraction.

Automatic Segmentation of Retinal Layers: The calculation of the implemented metrics needs the selection of ROI windows in the background and inside the retinal layer tissue. To automatize this process, we need to identify the limits of these retinal layers. Based on the work of *Chiu et al.* [21], we implemented the identification of the boundaries of retinal tissues in OCT images. This automatic approach for segmenting retinal layers uses graph theory and dynamic programming to represent each OCT image as a graph of nodes, connecting optimum paths from both sides of the image. Firstly, the algorithm calculates dark-to-light gradient images, identifying adjacent layers and generating weights for the layer segmentations. The minimum weighted paths are found by the Dijkstra's algorithm [22] to progressively identify the main layers of the retina. This approach detects eight different layers. However, for this purpose, only the Inner Limiting Membrane (ILM) and the Retinal Pigment Epithelium (RPE) layers were used, as they delimit the retinal layer region.

Automatic ROIs Extraction: As second step, an iterative and random process of seed generation and ROI construction was implemented. The seed coordinates are needed to construct m^{th} ROI windows inside the retinal layers as well as 1 background ROI window. Based on these seeds, which are obtained from the original noisy image between the ILM and RPE retinal layers, it is possible to obtain the ROIs in exactly the same coordinates on the returned filtered images to measure the degree of improvement before and after the denoising process. In order to ensure that the quality metrics will not be influenced by pathological structures, 50 random repetitions were executed in an automatic way obtaining as a result the final quality metric means. Moreover, each set of random positions is used with all the methods, certifying that the approaches are all under the same conditions. The positions represent the upper left coordinates of mth ROIs, as shown in Fig. 3, and each ROI will be a window with a size of 10×20 pixels. The background ROI is obtained from the upper left part of the each OCT filtered image, having a total area equal to the sum of the mth ROIs areas, as:

$$A_b = \sum_{m=1}^{m} A_m \tag{3}$$

where, A_b is the background total area and A_m are the area of each mth ROIs.

The quality metrics, SNR, CNR and ENL, provide relevant information about the noise signal, contrast and homogeneity in images, respectively, and are defined as:

$$SNR_m = 20 \cdot log_{10}(\frac{\mu_m}{\sigma_b}) \tag{4}$$

$$CNR_m = \frac{\mu_m - \mu_b}{\sqrt{\sigma_m^2 + \sigma_b^2}} \tag{5}$$

$$ENL_m = \frac{\mu_m^2}{\sigma_m^2} \tag{6}$$

where, μ_b and σ_b are the mean and standard deviation of the background ROI, respectively, while μ_m and σ_m are the mean and standard deviation, of the mth ROIs, respectively. For this purpose, 8 ROIs were selected as m value, for the extraction of the retinal tissue regions, which is a representative and sufficient value to estimate accurately, the quality metrics, having finally:

$$SNR = \frac{1}{8}[\sum_{m=1}^{m=8} SNR_m] \quad CNR = \frac{1}{8}[\sum_{m=1}^{m=8} CNR_m] \quad ENL = \frac{1}{8}[\sum_{m=1}^{m=8} ENL_m] \tag{7}$$

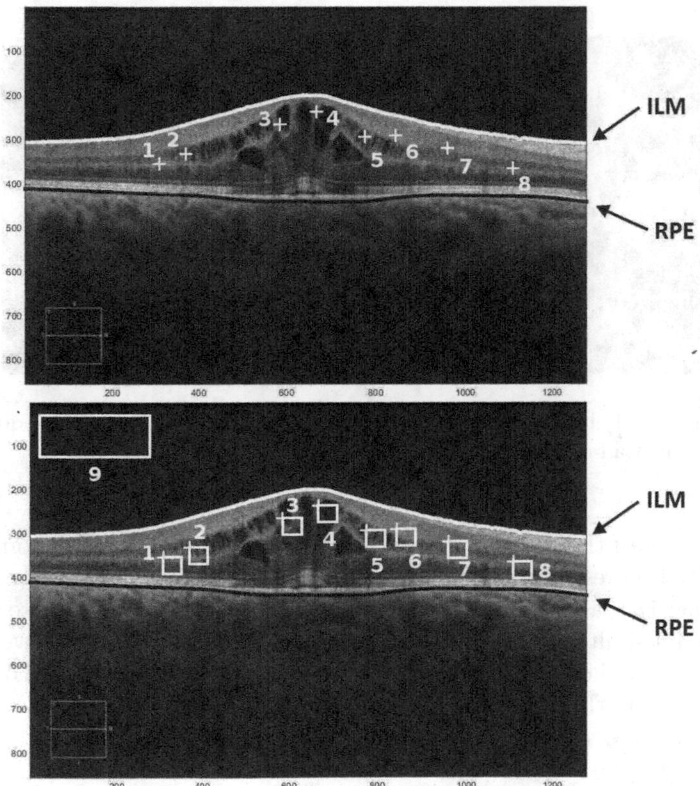

Fig. 3. Example of retinal layers delimitation and ROIs construction. 1st row, iterative random seeds (+) between ILM and RPE layers. 2nd row, ROIs construction where: 1–8, retinal layer ROIs; 9, the background ROI.

3 Results and Discussion

We used a dataset composed by 45 OCT retinal images acquired by two different OCT devices: CIRRUS^{TM} HD-OCT Zeiss OCT device and Spectralis® OCT confocal scanning laser ophtalmoscope from Heidelberg Engineering, two of the most representative and used of the market. We used these groups of images with different levels of noise and, therefore quality, each one of them with 15 images: (a) good quality images, (b) mid quality images and (c) bad quality images. Figure 4 shows representative examples of images included in each group. The vast majority, almost 90% of the images on the dataset belongs to unhealthy retinal patients, increasing the difficulty of the validation process due to the presence of dark regions, typically cystoid regions, that may influence in the metric results.

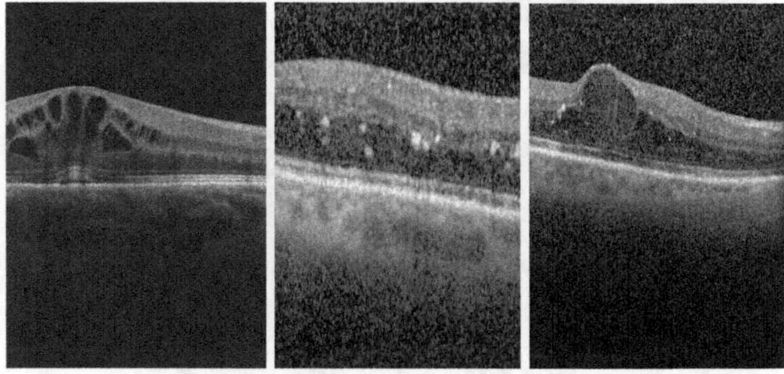

Fig. 4. Example of classification groups of OCT retinal images. (a) Good quality. (b) Mid quality. (c) Bad quality images.

In order to test the performance of the method and perform a fair comparison, we also tested representative denoising methods for OCT images of the state-of-the-art under the same conditions of the proposed methodology. These methods [10,11] are: Frost filters, Kuan filter, Kuwahara filter, Lee filter and wavelet filter. The proposed method was used with a cutoff frequency equal to 80 Hz and 5th order, parameters that were empirically established.

The results are synthesized in Fig. 5 which allows to do a quantitative comparison between the original and the filtered images of each quality group. The results show that the proposed method offers a better performance than the rest of the tested methods for all the groups of images in the dataset, having an increase of the SNR parameter from 22.9 dB to 35.6 dB in the group of good quality images and 9.3 dB better than the second best filter, the Frost filter. This demonstrates that the proposed filter suppresses noise in group (a) of images with good quality, but also, similarly positive results were obtained from (b) and (c) groups of images. Further, as expected in group (c) the proposed method have an improvement of 7.5 dB better than the next best filter, which is Kuan filter and 14.1 dB higher than the original images. Regarding the CNR metric, the proposed method also shows improvements of over 22.1 dB, 6.5 dB and 13.2 dB better than the second best method for (a) to (c), groups of images qualities, respectively. It is also possible to conclude that the image quality classification is coherent with the quantitative metrics, since the best group of original image quality has the highest values for SNR and CNR while the worth group of image quality has the lowest values.

Table 1 presents a general performance analysis and comparison using the entire dataset of 45 images. The three SNR, CNR and ENL were calculated for the original image, the state-of-the-art approaches and the proposed method. The SNR metric of the proposed method increases over 12.74 dB more when compared to the original noisy image and over 7.04 dB than the second best filter, the Kuan filter. Moreover, the proposed method also produced the best image contrast (CNR), near 17.81 dB better than the original image from 4.77 dB to 22.58 dB and 14.08 dB better than the Kuan filter (22.58 dB to 8.50 dB) also the second best method. This reinforces the effectiveness of the proposed method in noise removal conditions with the contrast enhancement of the images and the preservation of the retinal layer tissue details. Regarding the ENL metric, despite the proposal offers a satisfactory score, it is not the best of all of them. However, as this metric measures the homogeneity of the tissue, extremely high values are not desired as they may be derived due to more blurred output filtered images, losing, therefore, retinal tissue details (Fig. 6).

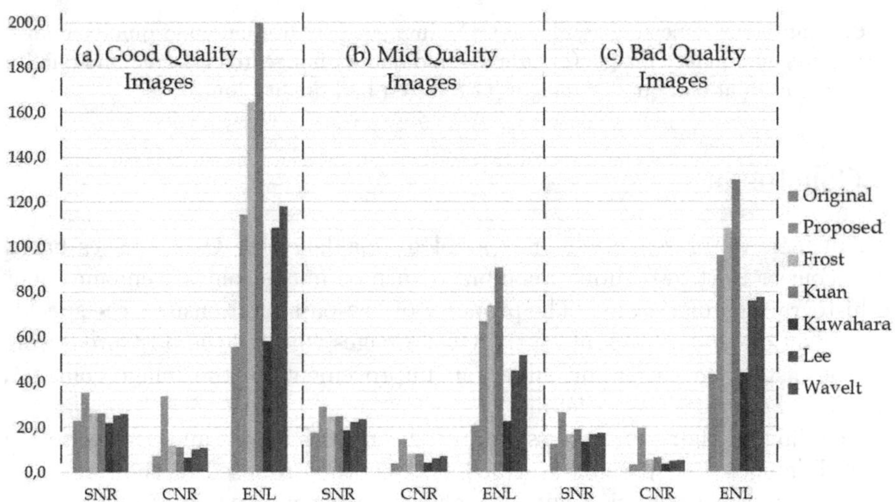

Fig. 5. Results of quality metrics for the tested denoising methods filters on the 3 groups of quality images: (a) Good quality, (b) Mid quality and (c) Bad quality images.

Table 1. Results of the quality metrics SNR, CNR and ENL using the entire dataset.

Method	All images		
	SNR (dB)	CNR (dB)	ENL
Original	*17.66*	*4.77*	*39.88*
Proposed	**30.41**	**22.58**	92.57
Frost	22.47	8.34	115.42
Kuan	23.36	8.50	**140.07**
Kuwahara	17.97	4.78	41.53
Lee	21.41	7.07	76.44
Wavelet	22.23	7.77	82.42

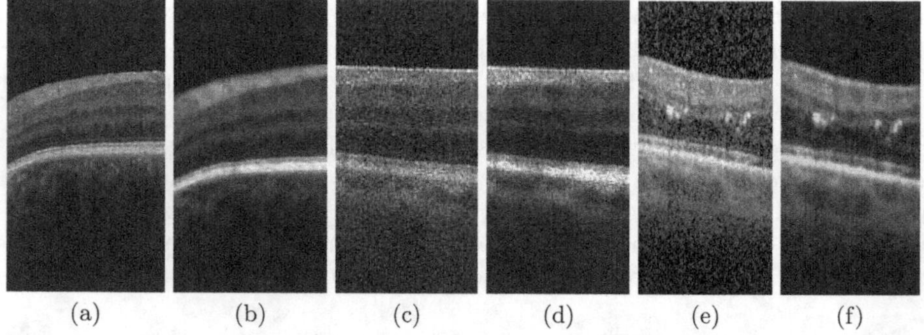

Fig. 6. Example of denoising results in OCT images. (a) Original good quality image. (b) Filtered good quality image. (c) Original mid quality image. (d) Filtered mid quality image. (e) Original bad quality image. (f) Filtered bad quality image.

4 Conclusions

In this paper, we present a new methodology for denoising OCT images based on the Fourier Butterworth filter combined with an image contrast enhancement and histogram regularization. The property of low-pass filter ensures the speckle noise reduction that is present on the OCT images whereas the post-processing with the histogram correction allows an improvement on the image contrast, enhancing the retinal tissue details.

According to the visual assessment of the results before and after filtering the OCT images, it is possible to conclude that the proposed method provides an efficient noise suppression with a higher contrast enhancing of morphological structures such as the boundaries of the different retinal layers, preserving clinical details as cysts without blurring excessively the output image in all the groups of original image qualities. Additionally, based on experimental quality metrics as SNR, CNR and ENL, we can offer an accurate method for noise suppression in comparison with other well-stated methods of the literature. The proposed method returns the best results in SNR, improving over 12.74 dB and

7.04 dB when compared with the original and the second best filtered images, respectively. Moreover, it has a great potential to improve the image contrast (CNR) with 17.81 dB better than original image and 14.08 dB better than the Kuan filter, the second best filter. In this way, the method is capable to improve the conditions of the OCT images, facilitating any posterior analysis of the retinal characteristics in the diagnosis of different retinal diseases.

Acknowledgments. This work is supported by the Instituto de Salud Carlos III, Government of Spain and FEDER funds of the European Union through the PI14/02161 and the DTS15/00153 research projects and by the Ministerio de Economía y Competitividad, Government of Spain through the DPI2015-69948-R research project.

References

1. Huang, D., Swanson, E., Lin, C., et al.: Optical coherence tomography. Science (New York, NY) **254**(5035), 1178 (1991)
2. Karri, S., Chakraborty, D., Chatterjee, J.: Transfer learning based classification of optical coherence tomography images with diabetic macular edema and dry age-related macular degeneration. Biom. Opt. Express **8**(2), 579–592 (2017)
3. González, A., Penedo, M.G., Vázquez, S.G., Novo, J., Charlón, P.: Cost function selection for a graph-based segmentation in OCT retinal images. In: Moreno-Díaz, R., Pichler, F., Quesada-Arencibia, A. (eds.) EUROCAST 2013. LNCS, vol. 8112, pp. 125–132. Springer, Heidelberg (2013). doi:10.1007/978-3-642-53862-9_17
4. de Moura, J., Novo, J., Ortega, M., et al.: Artery/vein classification of blood vessel tree in retinal imaging. In: Proceedings of the 12th International Joint Conference on Computer Vision, Imaging and Computer Graphics Theory and Applications, VISAPP, (VISIGRAPP 2017), vol. 4, pp. 371–377 (2017)
5. de Moura, J., Novo, J., Ortega, M., Charlón, P.: 3D retinal vessel tree segmentation and reconstruction with OCT images. In: Campilho, A., Karray, F. (eds.) ICIAR 2016. LNCS, vol. 9730, pp. 716–726. Springer, Cham (2016). doi:10.1007/978-3-319-41501-7_80
6. Hee, M., Izatt, J., Swanson, E., et al.: Optical coherence tomography of the human retina. Arch. Ophthalmol. **113**(3), 325–332 (1995)
7. Guedes, V., Schuman, J., Hertzmark, E., et al.: Optical coherence tomography measurement of macular and nerve fiber layer thickness in normal and glaucomatous human eyes. Ophthalmology **110**(1), 177–189 (2003)
8. Baghaie, A., D'Souza, R., Yu, Z.: Application of independent component analysis techniques in speckle noise reduction of retinal OCT images. Opt. Int. J. Light Electron Opt. **127**(15), 5783–5791 (2016)
9. Paes, S., Ryu, S., Na, J., et al.: Advantages of adaptive speckle filtering prior to application of iterative deconvolution methods for OCT imaging. Opt. Quantum Electron. **37**(13), 1225–1238 (2005)
10. Ozcan, A., Bilenca, A., Desjardins, A., et al.: Speckle reduction in OCT images using digital filtering. J. Opt. Soc. Am. A: **24**(7), 1901–1910 (2007)
11. Salinas, H., Fernández, D.: Comparison of PDE-based nonlinear diffusion approaches for image enhancement and denoising in OCT. IEEE Trans. Med. Imaging **26**(6), 761–771 (2007)
12. Lee, J.-S.: Speckle analysis and smoothing of synthetic aperture radar images. Comput. Graph. Image Process. **17**(1), 24–32 (1981)

13. Kuan, D., Sawchuk, A., Strand, T., et al.: Adaptive noise smoothing filter for images with signal-dependent noise. IEEE Trans. Pattern Anal. Mach. Intell. **2**, 165–177 (1985)
14. Frost, V., Stiles, J., Shanmugan, K., et al.: A model for radar images and its application to adaptive digital filtering of multiplicative noise. IEEE Trans. Pattern Anal. Mach. Intell. **2**, 157–166 (1982)
15. Mayer, M., Borsdorf, A., Wagner, M., et al.: Wavelet denoising of multiframe optical coherence tomography data. Biomed. Opt. Express **3**(3)
16. Wong, A., Mishra, A., Bizheva, K., et al.: General Bayesian estimation for speckle noise reduction in OCT retinal imagery. Opt. Express **18**(8), 8338–8352 (2010)
17. Puvanathasan, P., Bizheva, K.: Interval type-II fuzzy anisotropic diffusion algorithm for speckle noise reduction in OCT images. Opt. Express **17**(2), 733–746 (2009)
18. Lyra, M., Ploussi, A., Rouchota, M., et al.: Filters in 2D and 3D cardiac SPECT image processing. Cardiol. Res. Pract. Article ID 2014 (2014)
19. Prekeges, J., et al.: Nuclear Medicine Instrumentation. Jones & Bartlett Publishers, Burlington (2012)
20. Helmy, Y., Atta, H.: Optical coherence tomography classification of diabetic cystoid macular edema. Clin. Ophthalmol. **7**, 1731–1737 (2013)
21. Chiu, S., Li, X., Nicholas, P., et al.: Automatic segmentation of seven retinal layers in SDOCT images congruent with expert manual segmentation. Opt. Express **18**(18), 19413–19428 (2010)
22. Dijkstra, E.W.: A note on two problems in connexion with graphs. Numer. Math. **1**(1), 269–271 (1959)

Smartphone Based Pupillometry: An Empirical Evaluation of Accuracy and Safety

Davide Maria Calandra[1], Sergio Di Martino[2], Daniel Riccio[2(✉)], and Antonio Visconti[3]

[1] University of Naples "Federico II" - DM, Naples, Italy
davidemaria.calandra@unina.it
[2] University of Naples "Federico II" - DIETI, Naples, Italy
{sergio.dimartino,daniel.riccio}@unina.it
[3] Sober-Eye Inc., Menlo Park, CA, USA
av@sober-eye.com

Abstract. The Pupillary Light Reflex (PLR) is an involuntary reflex that changes the size of the pupil in response to varying light conditions. PLR analysis is widely employed in the evaluation of several neurological and ocular conditions and quantitative pupillometry requires the use of expensive ophthalmic instruments. In this paper, we describe an empirical evaluation we performed on the use of a commercially available smartphone (Apple iPhone 6s) to make quantitative measurements of PLR. Measurements were made with 30 healthy volunteers, equally distributed on three age ranges, including also different eye colors. Additionally, we also performed an assessment of the risks inducted by the use of the flash very close to the eye, showing that, if correctly used, it is by far below the constraints of international safety standard.

1 Introduction

It is well known that the pupil size adapts to changes in lighting conditions, as its radius constricts or dilates, according the intensity of light that falls on the retina. This phenomenon is called is Pupillary Light Reflex (PLR) and is an involuntary reflex. Interestingly, the amplitude and dynamic of PLR is not constant, but can be affected by many factors, such as brain injuries [3], assumptions of drugs and alcohol [2], cognitive activity [10], emotional reactions [7] and other neurological and ocular conditions [14]. Thus, in many scenarios, a measure of the PLR is a significant indicator of the state of the subject. For this reason, over the years, techniques to measure the PLR improved [18] both in precision and in non-invasivity, based on high quality, expensive cameras, characterized by high resolution and frame rate. However, to date, the measure of the PLR is confined to very specific contexts, like hospitals or specialized medical practices, able to afford the expenses for those systems.

On the other hand, the improvements in the quality, resolution and frame rate of mobile phones' cameras has raised the opportunity to measure the PLR

S. Battiato et al. (Eds.): ICIAP 2017, Part II, LNCS 10485, pp. 433–443, 2017.
https://doi.org/10.1007/978-3-319-68548-9_40

utilizing commercially available and inexpensive smartphone of the latest generation. This could give rise to a number of socially relevant use cases. Nevertheless, actually there is no sistematic work available in the literature assessing the potential accuracy of PLR measurements done using a smartphone.

To fill this gap, in this paper we report on the empirical evaluation we conducted on 30 healthy volunteers, in order to understand if recent smartphones can provide a reliable estimation of PLR, involving subjects of different ages and eye colors. To this aim, we reached out to the SOBER-EYE INC.[1] company, who developed a smartphone based pupilometer, allowing users to capture a video of their pupil to measure the PLR, using the flash as light stimulus. The main use case for this application is to self assess the level of impairment from drugs or alcohol. The company supported us in performing the empirical investigation, by providing us with access to their technology, the raw PLR videos and the reported data. A ground truth was obtained by manually labelling the recorded eyes videos, measuring the pupil size frame by frame.

By comparing the PLR data from Sober-Eye and the ground truth, we conducted three different studies. The first study was aimed to assess the accuracy in measuring the pupil radius: for each frame, the absolute difference between the pupil radius measured by the Sober-Eye technology and the corresponding one in the ground truth was calculated. The global accuracy was then obtained by averaging the error over all videos in the dataset. The goal of second and third experiments was to confirm, by means of the use of the smartphone, two specific claims reported in the literature [6], where it was reported that dark and light eyes have different PLRs, and that the amplitude of the PLR decreases with the age of the subjects.

In addition, we also investigated if the use of the phone camera flash as light source, in close proximity of the users eyes, presents risks in terms of light-induced retinal damage. Since different models of mobile phones have different types of flash and the implementation of smartphone PLR detection apps may also be different, a general blanket assessment of safety cannot be made. Therefore, the light hazard of all iPhone models compatible with the Sober-Eye set up (5s, 6, 6s, SE, and 7) was evaluated according to the requirements for optical radiation safety for ophthalmic instruments, specified by ISO15004-2.2 [9] and ANSI Z80.36 [1].

2 Related Work

Measuring pupil reaction to a light stimulus [12] is a critical part of the neurological examination process. The typical PLR to a single light pulse is shown in Fig. 1. Starting from an aperture A_{max}, after a few fractions of second the pupil starts to shrink, and after TA_{min} seconds it reaches the minimum aperture A_{min}, as discussed in [20].

Routine pupil examination with a conventional light source to estimate parameters like light reactivity, size and symmetry of the pupil is affected by many

[1] http://www.sober-eye.com/.

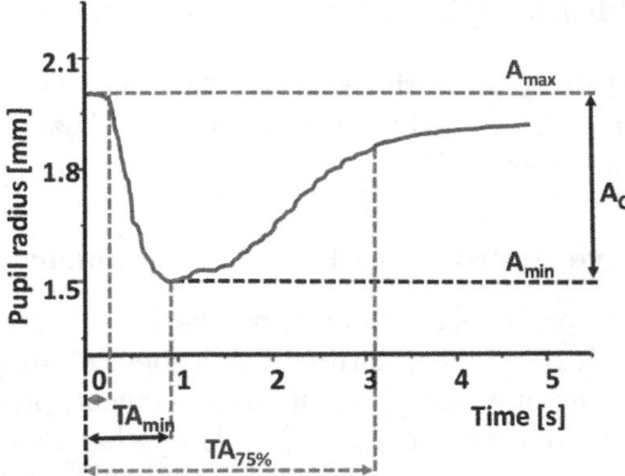

Fig. 1. Pupillary reflex to a single light pulse.

factors, such as ambient lighting, the examiner's experience and the intensity of the lighting stimulus. Thus, the use of imaging devices, namely pupillometers, for pupillary examination offers better measurement precision. One of the first study on automatic pupillometry based on infrared cameras was published in 1958 by Lowenstein and Loewenfeld [15], who argued their technique had high potential to be applied in pharmacological and physiological research. Thus, pupillometers significantly evolved from the first generation of devices that were time consuming and with very low precision, up to modern portable easy-to-use digital systems that enable less invasive examination with a more accessible and cost-effective method. Modern devices open to a wide range of applications like predicting neuroworsening [5], on-site evaluation of traumatic brain injury [16,22] or monitoring the effects of drugs [17]. In particular, as regarding the use of pupillometry for evaluating the effects of drugs, Martínez-Ricarte *et al.* [13] deeply discuss a wide casuistry including muscle relaxants, barbiturates, benzodiazepines and antidepressants.

Existing devices are mostly based on infrared cameras, so to be more robust to variations in ambient lighting. At the best of our knowledge, the mobile application for PLR proposed in [19] represents the only example of smartphone application, running on a Samsung Galaxy S4, for pupillometry working with a visible light camera. However, this smartphone application shows some important limitations. First of all, it requires a very invasive procedure to acquire a set of eye images, as the eyelid of the subject are hold opened by the examiner during the image acquisition process. Moreover, it just exploits five eye images to estimate the PLR, which can be too few to correctly capture the real dynamic of pupil parameters. Regarding the performance, even if almost comparable with those obtained with a PEN device, it was assessed on a limited sample, as the age of volunteers is mostly concentrated around 33 years with a very small variation (±5.3 years), while no variability was considered for the eye colour, being all subjects Koreans with dark eyes. At last, there was no investigation about the harmfulness of the Galaxy S4 flash for the human eye.

3 Assessing Mobile Phones for Pupillometry

In this section, we start by presenting the way the Sober-Eye app works. Then, we describe the procedure we used to measure the light hazard of the flash used in close proximity to the eye.

3.1 The Sober-Eye Approach to PLR Measurement

The Sober-Eye technology uses the smartphone flash as light source to produce a PLR response that is then video captured by the device camera and processed locally to extract the features useful for the measurement, processing the curve shown in Fig. 1. We used a version of the technology that also uses a cardboard enclosure (Fig. 2) that covers the users eyes to provide a near total darkness environment and to hold the phone at the appropriate distance (8 cm). Users hold the cardboard enclosure over their eyes for about 20 s before a PLR measurement is performed. During this time, the darkness of the enclosure stimulates a pupillary dilation (*mydriasys*). After the preset time, the camera flash turns on for about 4 s and the video stream is captured. The light impulse causes a pupil constriction (*myosis*). Then, the video is locally post-processed, in order to extract the features to measure the PLR. The app works by comparing a current measurement to a reference measurement (baseline), previously captured by the same user in sober conditions. Since the PLR dynamic is highly correlated with the reaction capabilities of a person, a significant deviation from the baseline is an indicator of possible impairment. The expected curve as response to a PLR is shown in Fig. 1.

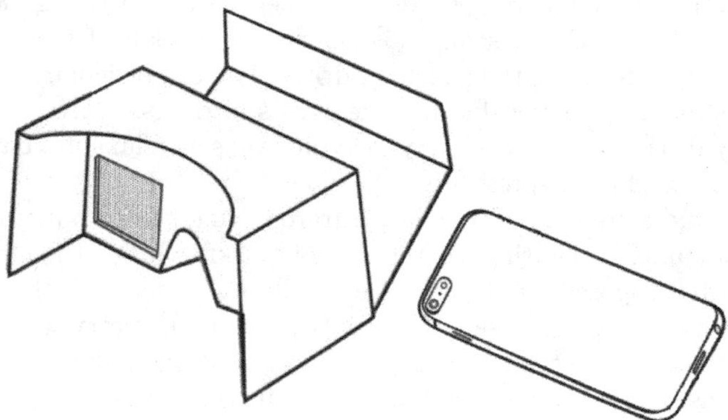

Fig. 2. Sober-Eye's cardboard enclosure.

3.2 Experimental Set-Up and Results

For the purposes of this investigation, we used Sober-Eye app in combination with the enclosure and an iPhone 6s. The app capture the video in HD (1920×1080) at

60 frames per second for about 4 s, producing 230 usable frames. As output values, the app returns the center coordinates for both iris and pupil and the related radius lengths.

3.3 The Dataset

Since no datasets of video sequences suitable to assess the use of smartphones as a pupillometer, we created our own collection of the videos. In particular, we created a dataset of 30 videos that were captured from 30 subjects by means of the Sober-Eye app. In order to consider also the variability of eye characteristics in the real population, we categorized the subjects by gender, age and eye color. We used 3 age ranges (18–30, 31–50, 51+) and two eye color categories (light/dark). Table 1 reports the distribution of subjects in each category. Unlike for age and gender, the distribution with respect to eye colors is less balanced. This is due to the geographical context where videos were gathered, since brown is the most frequent eye color in southern Italy.

Table 1. Number of persons in each group.

Total 30	Male			Female		
	18–30	31–50	51–	18–30	31–50	51–
Light	3	1	0	1	2	0
Dark	4	4	8	2	3	2

All videos have a spatial resolution of 1920 × 1080 pixels with 24 bits and were acquired at 60 fps by an iPhone 6s. Each video has been manually annotated by a human operator to build a ground truth to use to compare pupil radius measurements determined by Sober-Eye App. The annotation process was performed frame by frame and consisted in manually selecting three points on the pupil boundary with an angular distance of about 120 degrees each other. Circle fitting [21] was then applied to compute the pupil radius. In order to assess the correctness of the manual annotation process, we averaged the pupil and iris radius computed on the first frame over all videos, so obtaining a ratio of 0.5293. This result agrees with what reported by Ferrari et al. in [4], who claims that ratio between pupil and iris radius in healthy adults equals to 0.55 ± 0.056 in the frame preceding the light flash.

3.4 Testing the Accuracy of the Pupil Radius Measurement

In order to assess the performance of the Sober-Eye app, we performed three different studies. The first one aimed to assess the accuracy of the mobile app in automatically extracting and measuring the pupil radius. To this aim, for each frame of each video, we estimated the error by calculating the absolute difference in pixels between the pupil radius from the mobile app and the corresponding one in the ground truth. The global accuracy was then obtained by averaging the error over all videos in the dataset. The average error resulted to be very small, being 1.80 pixels with a standard deviation of 2.35. Given this very small error, we can affirm that a smartphone, equipped with the Sober-Eye technology and used in the cardboard enclosure, can be effectively used as pupillometer, as it is able to obtain the same performances of manual identification of the size of the pupil.

The goals of second and third study were to verify two specific claims reported by Spector in [6] by exploiting measurements performed by the Sober-Eye app. Indeed, in [6] the author stated that:

1. *a brown iris contracts less than a blue iris* and
2. *in old people and in patients with iris atrophy, the sphincter becomes rigid, hence the light reaction diminishes in extent.*

To verify the first claim, we partitioned the whole dataset in two groups of videos according to the eye color (light vs. dark) and we computed the average value of the A_{min} parameter (minimum radius of the pupil, see Fig. 1) for both groups. Interestingly, we found a difference in the two groups, with the average pupil radius for bright eyes of 14.39 pixels (std. dev. 1.95), and 13.58 pixels (std. dev. 1.23) for dark eyes. Thus, our experimental results conducted with the smartphone confirm that dark eyes contract less that light ones.

Regarding the second claim, we just considered the original partitioning of the dataset, according to age ranges. For this purpose, we considered the average value of the A_C parameter (difference between minimum and maximum size of the pupil, see Fig. 1) for the three groups separately, as this parameters relate to the amount of light reaction. Also in this case, results appear in line with the claim, as we obtained the following average value: A_C: $0.515, 0.503$ and 0.487 for young, adult and middle-aged groups, with the respective standard deviations of $0.040, 0.091$ and 0.078.

4 Light Hazard Evaluation

The use of the camera flash to stimulate a PLR response raise some concerns about the safety of using such light sources very close to the users' eyes. Indeed, the exposition of the pupil to a close light source is potentially harmful to the human eye. For this reason, it is important to verify the safety of the approach. Some studies already investigated the use of mobile phone flash in ophthalmic applications. For instance, in [11], the authors measured and calculated the light levels produced by the torch of an iPhone 4 in simulated conditions of indirect ophthalmoscopy and proved that retinal exposure from the smartphone was within the safety limits defined by the ISO standard ISO 15004-2.2 [9]. Hong et al. [8] evaluated the safety of iPhone 6 and 6 Plus in a similar application and concluded that the photobiological risk posed by iPhone 6 indirect ophthalmoscopy was at least 1 order of magnitude below the safety limits set by the ISO 15004-2.2. Both studies were related to the use of iPhones for fundus photography, where images of the retina are captured and analysed.

In order to evaluate the safety of the sober-eye approach, we conducted measurement in the specific sober-eye set-up with the flash at an 8 cm distance from the eye and on for 4 s. We followed the guidelines of the International Organization for Standardization (ISO15004-2.2), the standard sets the fundamental requirements and test methods to assess the light hazard of ophthalmic instruments [1,9].

The standard classifies ophthalmic instruments in the two following groups:

- *Group 1: ophthalmic instruments for which no potential light hazard exists.*
- *Group 2: ophthalmic instruments for which a potential light hazard exists.*

Since the iPhone is not considered a standard ophthalmic instrument, it must be classified as Group 2 and the light hazard must be evaluated with the exposure limits of Group 2 ophthalmic instruments.

The following iPhone models were tested: 5s, 6, 6s, 7 and 7 Plus. Apple's specifications indicate that iPhone 6, 6s, 6Plus, 6sPlus and SE have the same flash. Measurements confirm that the 6 and 6s have the same light spectra and radiant power. Likewise, the iPhone 7 and 7Plus flashes have similar light spectra and radiant power, while the iPhone 5s has a different flash. In conclusion, to cover all iPhone models from the 5s to the 7 Plus, it is sufficient to evaluate the safety of 3 types of flashes that we will refer in the following as *5s*, *6* and *7*.

4.1 Measurements

To assess the light hazard according to the ISO standard, it is necessary to measure the amount of radiant power per unit of area per wavelength reaching the retinal plane (retinal spectral irradiance E_λ), measured in $W/(cm^2\,nm)$. From this quantity, the ISO provides various formulas to calculate the hazard for different wavelength ranges (ultraviolet, visible and infrared).

Since all measured spectra (Fig. 4) showed no significant ultraviolet or infrared emission (below 400 and above 750 nm), as also verified in [11] and [8], to assess the hazard we need calculate only two values and compare them to the standard's limits.

1. *Weighted retinal visible and infrared radiation thermal irradiance* in the 380–1400 nm range (E_{VIR-R}) as defined by the standard, where E_λ is the measured retinal spectral irradiance and R_λ is a weighting function provided by the standard.
$$E_{VIR-R} = \sum_{380}^{1400} E_\lambda \times R(\lambda) \times \Delta\lambda$$
2. *Weighted retinal radiant exposure* in the 305–700 nm range (H_{A-R}), with $t = 4$ s and A_λ a weighting function provided by the standard.
$$H_{A-R} = \sum_{305}^{700} (E_\lambda \times t) \times A(\lambda) \times \Delta\lambda$$

The measurements have been carried out following the guidelines set by the standard. We used a Photo Research Inc. SpectraScan PR-670 spectroradiometer with a MS-75 lens and a Ocean Optics Reflectance standard WS-1. A screen covered with a light baffle and a 0.2 cm aperture was inserted between the iPhone and the Reflectance Standard C (4 cm from each) as shown in Fig. 3.

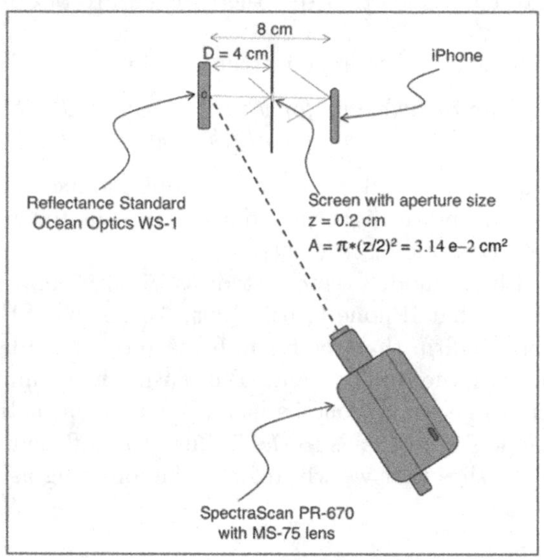

Fig. 3. Measurement setup.

In this set up, we directly measure the spectral radiance $L_{c\lambda}$ at the target C positioned at the corneal plane and then derive the retinal spectral irradiance E_{λ}:

$$L_{c\lambda} \qquad [W/(sr\ cm^2\ nm)]$$

since C is a reference Lambertian surface with reflectivity >99% in the spectrum range of interest (Ocean optics WS-1). The irradiance $E_{c\lambda}$ on C (corneal plane) is than calculated as follow:

$$E_{c\lambda} = \pi\ L_{c\lambda}\ W/(cm^2\ nm)$$

The iPhone spectral radiance L_{λ} is than calculated by multiplying $E_{c\lambda}$ by D^2/A, where A (3.14e−2 cm^2) is the area of the aperture and D (4.0 cm) is the distance from the aperture to the corneal plane.

$$L_{i\lambda} = D^2/A \quad E_{c\lambda} = 509\ E_{c\lambda}$$

Finally, from $L_{i\lambda}$, the retinal irradiance E_{λ} is calculated with the formula:

$$E_{\lambda} = A_p/f^2\ L_{i\lambda}$$

where $f = 1.7\,\text{cm}$ (focal length of human eye) $Ap = 3.85\text{e}{-1}\,\text{cm}^2$ (area of pupil)

$$E_\lambda = 0.13 L_{i\lambda}$$

4.2 Results

Figure 4 shows the spectral radiance of the tested iPhones. All tests were performed with the flashes at 100% intensity.

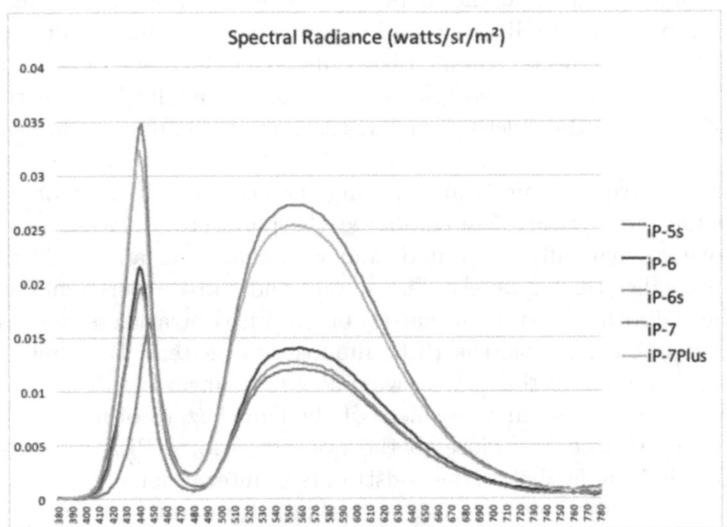

Fig. 4. Spectral radiance.

The limits set by the ISO stanstard [9] are 0.706 (W/cm^2) for E_{VIR-R} and 10 (J/cm^2) for H_{A-R}. In USA, the new standard ANSI Z80.36 [1] was adopted in 2016 and the limit for HA-R was reduced from 10 (J/cm^2) to 2.2 (J/cm^2). The measurements result and the limits are shown in Table 2 and all values for all the phones tested are significantly below both the ISO and ANSI limits.

In conclusion, the iPhone flashes 5s, 6 and 7 are safe to be used as light sources in the described set up per the ISO15004-2.2 and ANSI Z80.36 standards for ophthalmic instruments.

Table 2. Summary results.

	E_{VIR-r} (W/cm^2)	Limit (W/cm^2)	H_{A-R} $t = 4\,\text{sec}$ (J/cm^2)	ISO Limit (J/cm^2)	ANSI Limit (J/cm^2)
5S Type	0.040	0.706	0.033	10	2.2
6 Type	0.050	0.706	0.049	10	2.2
7 Type	0.096	0.706	0.084	10	2.2

5 Conclusions

Pupillary Light Reflex is a significant indicator of the state of a subject. To date, it is measured by means of very expensive ophthalmic instruments, in controlled environments. In this paper we described the results of the empirical evaluation we have conducted to assess if recent smartphones can be used to reliably measure the pupil size variation, when hit by the light, involving subjects of different ages and eye colors.

To this aim, we exploited the mobile app Sober Eye, which records a video stream and reports the pupil size variations over the time, running on an iPhone 6s. We involved 30 subjects, equally distributed over three age ranges, and with light and dark eyes. As for the ground truth, we manually labelled the pupil size for each frame of the videos. From this setting, we conducted three different experiments.

The first experiment aimed at assessing the accuracy of the mobile app in measuring the pupil radius. We compared the pupil size detected by the app with the one we manually annotated, and we found, over all the 30 videos, a mean error smaller than 2 pixels. The second and third experiments aimed at confirm the claims by Spector [6] that (I) brown iris contract less than blue iris, and (II) that in older people the PLR diminishes in extent. By using only the information extracted by the app we were able to confirm both the claims.

Finally, we performed an assessment of the light hazard of using the smartphone flash as light source, close to the eye, to perform PLR measurements, showing that it is by far below the constraints of international safety standard. As future research directions, we will be evaluating potential use of smartphone based pupilometry for screening or diagnostic purposes, by evaluating accuracy of measuring latency response, constriction speed and other relevant PLR features.

References

1. American national standard for ophthalmics - light hazard protection for ophthalmic instruments. Standard, American National Standards Institute (2016)
2. Brown, B., Adams, A.J., Haegerstrom-Portnoy, G., Jones, R.T., Flom, M.C.: Pupil size after use of marijuana and alcohol. Am. J. Ophthalmol. **83**(3), 350–354 (1977)
3. Capo-Aponte, J., Urosevich, T., Walsh, D., Temme, L., Tarbett, A.: Pupillary light reflex as an objective biomarker for early identification of blast-induced mTBI. J Spine **S4**, 2 (2013)
4. Ferrari, G.L., Marques, J.L., Gandhi, R.A., Heller, S.R., Schneider, F.K., Tesfaye, S., Gamba, H.R.: Using dynamic pupillometry as a simple screening tool to detect autonomic neuropathy in patients with diabetes: a pilot study. Biomed. Eng. Online **9**(1), 26 (2010)
5. Fountas, K.N., Kapsalaki, E.Z., Machinis, T.G., Boev, A.N., Robinson, J.S., Troup, E.C.: Clinical implications of quantitative infrared pupillometry in neurosurgical patients. Neurocrit. Care **5**(1), 55–60 (2006)
6. Spector, R.H.: The Pupils. In: Clinical Methods: The History, Physical, and Laboratory Examinations, 3 edn., pp. 300–304. Butterworths, Boston (1990)

7. Hess, E.H., Polt, J.M.: Pupil size as related to interest value of visual stimuli. Science **132**(3423), 349–350 (1960)
8. Hong, S.C., Wynn-Williams, G., Wilson, G.: Safety of iphone retinal photography. J. Med. Eng. Technol. **41**(3), 165–169 (2017)
9. Ophthalmic instruments - fundamental requirements and test methods - part 2: Light hazard protection. Standard, International Organization for Standardization (2016)
10. Kahneman, D., Beatty, J.: Pupil diameter and load on memory. Science **154**(3756), 1583–1585 (1966)
11. Kim, D.Y., Delori, F., Mukai, S.: Smartphone photography safety. Ophthalmology **119**(10), 2200–2201 (2012)
12. Larson, M.D., Behrends, M.: Portable infrared pupillometry: a review. Anesth. Analg. **120**(6), 1242–1253 (2015)
13. Martínez-Ricarte, F., Castro, A., Poca, M., Sahuquillo, J., Expósito, L., Arribas, M., Aparicio, J.: Infrared pupillometry basic principles and their application in the non-invasive monitoring of neurocritical patients. Neurología (Engl. Edn.) **28**(1), 41–51 (2013)
14. Meeker, M., Du, R., Bacchetti, P., Privitera, C.M., Larson, M.D., Holland, M.C., Manley, G.: Pupil examination: validity and clinical utility of an automated pupillometer (2005)
15. Lowenstein, O., Loewenfeld, I.E.: Electronic pupillography: a new instrument and some clinical applications. A.M.A. Arch. Ophthalmol. **59**(3), 352–363 (1958)
16. Pechman, D.M., Parikh, P., Vyas, K., Maranda, E.L., Habib, F.A.: Pupillometry in trauma: reducing variability associated with subjective assessment. J. Am. Coll. Surg. **217**(3), S52 (2013)
17. Pickworth, W.B., Murillo, R.: Pupillometry and eye tracking as predictive measures of drug abuse. In: Pharmacokinetics and Pharmacodynamics of Abused Drugs, 1 edn., pp. 127–142. CRC Press (2007)
18. Sari, J.N., Hanung, A.N., Lukito, E.N., Santosa, P.I., Ferdiana, R.: A study on algorithms of pupil diameter measurement. In: International Conference on Science and Technology-Computer (ICST), pp. 188–193. IEEE (2016)
19. Shin, Y., Bae, J., Kwon, E., Kim, H., Lee, T.S., Choi, Y.: Assessment of pupillary light reflex using a smartphone application. Exp. Ther. Med. **12**(2), 720–724 (2016)
20. Szczepanowska-Nowak, W., Hachol, A., Kasprzak, H.: System for measurement of the consensual pupil light reflex. Opt. Appl. **34**, 619–634 (2004)
21. Taubin, G.: Estimation of planar curves, surfaces, and nonplanar space curves defined by implicit equations with applications to edge and range image segmentation. IEEE Trans. Pattern Anal. Mach. Intell. **13**(11), 1115–1138 (1991)
22. Taylor, W.R., Chen, J.W., Hal, M., Gennarelli, T.A., Kelbch, C., Knowlton, S., Richardson, J., Lutch, M.J., Farin, A., Hults, K.N., Marshall, L.F.: Quantitative pupillometry, a new technology: normative data and preliminary observations in patients with acute head injury. J. Neurosurg. **98**(1), 205–213 (2003)

Pixel Classification Methods to Detect Skin Lesions on Dermoscopic Medical Images

Fabrizio Balducci(✉) and Costantino Grana

Dipartimento di Ingegneria "Enzo Ferrari",
Università degli Studi di Modena e Reggio Emilia,
Via Vivarelli 10, 41125 Modena, MO, Italy
{fabrizio.balducci,costantino.grana}@unimore.it

Abstract. In recent years the interest of biomedical and computer vision communities in acquisition and analysis of epidermal images increased because melanoma is one of the deadliest form of skin cancer and its early identification could save lives reducing unnecessary medical treatments. User-friendly automatic tools can be very useful for physicians and dermatologists in fact high-resolution images and their annotated data, combined with analysis pipelines and machine learning techniques, represent the base to develop intelligent and proactive diagnostic systems. In this work we present two skin lesion detection pipelines on dermoscopic medical images, by exploiting standard techniques combined with workarounds that improve results; moreover to highlight the performance we consider a set of metrics combined with pixel labeling and classification. A preliminary but functional evaluation phase has been conducted with a sub-set of hard-to-treat images, in order to check which proposed detection pipeline reaches the best results.

Keywords: Dermoscopic images · Image analysis · Annotation · Usability · Clustering

1 Introduction

The malignant melanoma is one of the most common and dangerous skin cancer, in fact 100,000 new cases with over 9,000 deaths are diagnosed every year by only considering the USA [7,8]; in this context automated system for fast and accurate melanoma detection are well accepted, also considering technical approaches like machine learning methods and the Convolutional Neural Networks.

The premise for this work was the development of an annotation tool for epidermal images: this has been the first step to create an heterogeneous data integrated system which architecture is depicted in Fig. 1.

This architecture presents the Digital Library with inner modules dedicated to image analysis and feature extraction and external tools to manage the *Annotation*, the *Information Visualization* and *Query* search capacities [3], for example by exploiting the DICOM (Digital Imaging and Communications in Medicine) metadata standard.

S. Battiato et al. (Eds.): ICIAP 2017, Part II, LNCS 10485, pp. 444–455, 2017.
https://doi.org/10.1007/978-3-319-68548-9_41

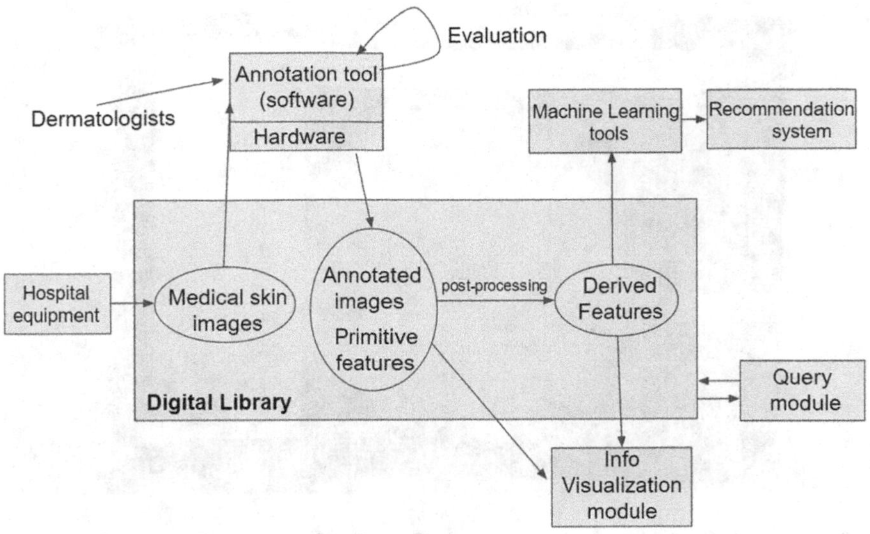

Fig. 1. The system architecture proposed for a medical digital library

The dataset coming from hospital equipment currently consists of 436 dermoscopic skin images in JPEG format with 4000×2664 or 4000×3000 resolution. In the view of the integrated system architecture, at this stage we focus on the exploit of state-of-art reliable methods for extracting the visual features we need, also introducing workarounds to improve results. The annotation software in Fig. 2 has been developed for domain experts like dermatologists which have peculiar working protocols and do not have much time to skill themselves on externals tools since usually overworked.

As hardware platform we chose the *Microsoft Surface Pro 3* which is a powerful non-invasive device that can be used in mobility in medical environments; it ensures that the recognized strokes will be only those that come from the specific *Surface Pen*, avoiding unwanted strokes coming from touch gestures or oversight movements.

This paper is organized as follows: Sect. 2 introduces some literature about epidermal and melanoma image analysis while Sect. 3 explains the skin lesion detection pipelines; Sect. 4 reports results obtained by the experimental sessions and finally, conclusions and future work are illustrated in Sect. 5.

2 Related Work

Nowadays standard video devices are commonly used for skin lesion inspection systems, in particular in the telemedicine field [20]; however, these solutions bring some issues, like poor camera resolution (melanoma or other skin details can be very small), variable illumination conditions and distortions caused by the camera lenses [14]. A complete and rich survey about skin lesion characterization is

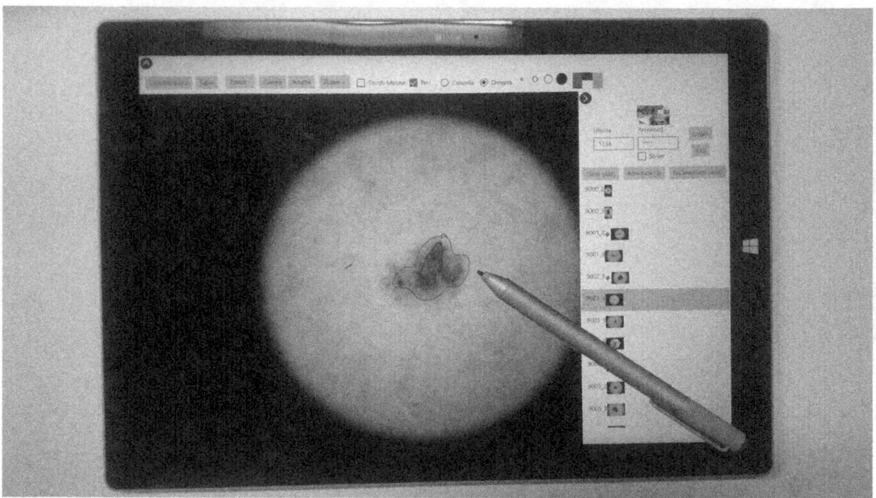

Fig. 2. Microsoft Surface is the hardware platform for the annotation tool (Color figure online)

in [5] with artifact removal techniques, evaluation metrics, lesion detection and preprocessing methods. The work of Seidenari *et al.* [19] provides an overview about the detection of melanoma by image analysis, Wighton *et al.* [21] presents a general model using supervised learning and MAP estimation that is capable of performing many common tasks in automated skin lesion diagnosis and Emre Celebi *et al.* [9] treats lesion border detection with thresholding methods as Fan *et al.* [10] that use saliency combined with Otsu thresholding. The work of Peruch *et al.* [17] faces lesion segmentation through Mimicking Expert Dermatologists' Segmentations (MEDS) and in Liu *et al.* [13] propose an unsupervised sub-segmentation technique for skin lesions.

In Codella *et al.* [6] manually pre-segmented images, already cropped around the region of interest, have been used in conjunction with hand-coded and unsupervised features to achieve state-of-the-art results in melanoma recognition task, with a dataset of 2,000 images. Learning approaches are exploited in Schaefer *et al.* [18] and deep learning techniques in Abbes and Sellami [2]; studies in [12,16] have been exploited in literature, while a combination of hand-coded features, sparse-coding methods, HOG and SVMs are used in Bakheet [4].

Finally, in 2016 a new challenge, called *Skin Lesion Analysis toward Melanoma Detection* [11], has been presented: the aim is to use one of the most complete dataset of melanoma images, collected by the *International Skin Imaging Collaboration* (ISIC) and obtained with the aggregation dataset of dermoscopic images from multiple institutions, to test and evaluate the automated techniques for the diagnosis of melanomas. The ISIC database has been also exploited in Yuan et al. [22] with a 19-layer deep convolutional neural network while classification and segmentation are achieved using deep learning approaches are also in Majtner et al. [15].

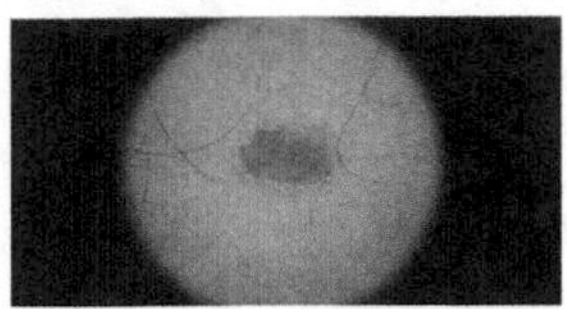

Fig. 3. A dermoscopic image annotated by a dermatologist (Color figure online)

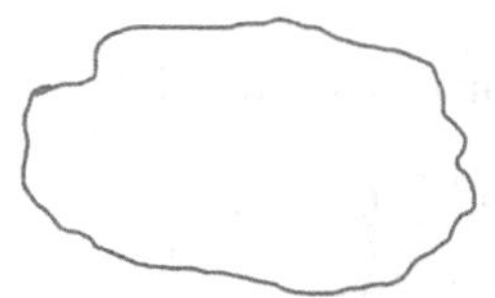

Fig. 4. The extracted annotation (primitive feature) (Color figure online)

Fig. 5. External black frame pixel mask (first version)

3 Visual Features and Skin Image Processing

In Fig. 3 is depicted a dermoscopic image where a dermatologist annotated the principal external contour of a skin lesion (red color): the primitive meta-data directly extracted by the annotation tool is in Fig. 4.

One of the targets of this system is to create and manage the image dataset for the automated analysis by combining low-level visual features representation, image processing techniques and machine learning algorithms: we want to obtain pipelines that detects skin lesions and automatically draws one or more contours by mimiking and increasing the dermatologist knowledge.

The drawn strokes introduce *primitive features* consisting of binary masks, coordinate points, color codes and pen sizes; the image processing functions extract *derived features* like contours, shapes, intersections, color features and numerical values.

Our method will exploit the manual annotations alongside image processing algorithms with the aim to evaluate how many pixels of the skin lesion could be automatically detected by the system; a pre-processing phase is necessary to remove thick skin hairs, because these artifacts influence shape and contours extraction [23].

3.1 Hair Removing

To accomplish the task we were inspired to the *DullRazor* [1] pipeline consisting of

1. *Detection* step that locates into an RGB image the slender and elongated structures that resemble the hairs on the skin by making an hair pixel mask
2. *Replacement* step that replaces each detected hair pixel with the interpolation of two lateral pixels chosen from a line segment built on a straight direction.

The *Detection* phase exploits the *generalized grayscale morphological closing* operation G_c: for each color channel c (Red-Green-Blue) the operator makes a set of morphologic closing by using different kernels with the aim to compare (by choosing the highest value c_p), for each pixel, which kernel better approximates a potential hair shape.

The value of G_c for each pixel p is calculated as:

$$\forall c \in r, g, b, \forall p, G_c = |b_c(p) - max(c_p)| \tag{1}$$

where $b_c(p)$ is the actual image pixel value and c_p measures how many pixels of the kernel are verified as hair structure for an image pixel p.

The kernel represents a sort of 'skeleton' that reconstructs pixel-to-pixel the hair shape (elongated, slight, weakly curvilinear) on the kernel-closed image and so the kernel structure is a very critical variable; we used four kernels for each of the possible directions (11 pixels horizontal at 0°/180°, 11 pixels vertical at 90°/270°, 9 pixels oblique at 45°/225° and 9 pixels oblique at 135°/315°) from which a potential hair could spread from a single pixel (located at the center of the kernel).

The final hair mask M is the union of the resulting pixel masks M_r, M_g, M_b, where each mask is obtained by a threshold on the generalized closing operator value previously calculated for each pixel.

Before the *Replacement* phase is performed, we must verify that each candidate hair pixel belongs to a valid thick and long structure so, for each direction previously described, a path is built having the pixel at the center until the non-hair regions are reached. The longest path is used to take the two interpolation pixels, selected on both the sides, perpendicular to the directions, at a fixed distance from the hair structure borders.

This pipeline is greatly affected by the variables used at each step, for example the kernel shape, the G_c threshold for the hair masks and the distance for the choose of the interpolation pixels.

3.2 Skin Lesion Detection

After the treatment that removes most of the hairs, it is possible to design the detection of the skin lesion area: we begin by using standard image processing techniques but with a fine-tuned *work pipelines* that will filter and refine results:

1. Otsu thresholding with pixel mask
2. Color clustering with pixel mask and cluster tolerance.

By considering the peculiar structure of a dermoscopic image, a binary mask M0 (Fig. 5) must be constructed to approximate the large black frame which surrounds the bright circular skin that contains the lesion. We execute a morphologic *erosion* on the original blurred image with a kernel size of 201 × 201: now the Otsu threshold (Fig. 8) is calculated considering the pixels not in M0 and, for future tests and comparisons, the resulting pixels will be in red color (in Fig. 6 the original image, in Fig. 7 with hairs removed).

For the color clustering technique we must develop an heuristic to differentiate if a cluster belongs to the bright skin or to the lesion skin by assign the label *skin* or *lesion* to each cluster; moreover it must be considered that a cluster can

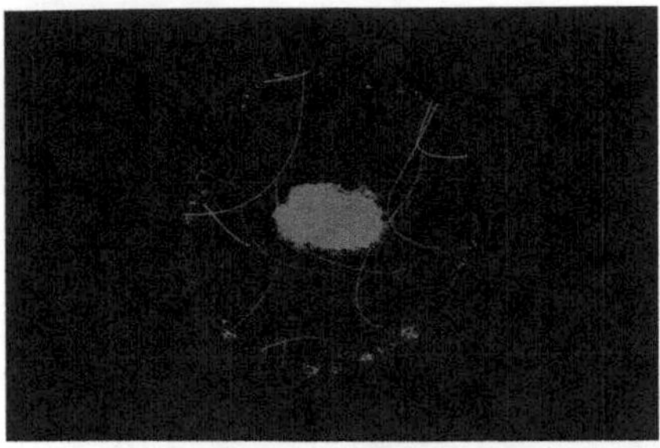

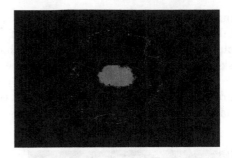

Fig. 6. Otsu with pixel mask, no hairs removed (Color figure online)

Fig. 7. Otsu with pixel mask, hairs removed (Color figure online)

easily intersect the two area types (some pixels on the lesion and others on the skin) especially on the borders.

To deal with this ambiguity we will develop a pixel *toleration area*: we execute a further morphologic *closing* and *erosion* on the otsu image by using two different kernels to obtain an *enlarged mask* (Fig. 9) that builds safety areas around the image borders (a side effect is that scattered hair pixels will be removed but groups may be enlarged).

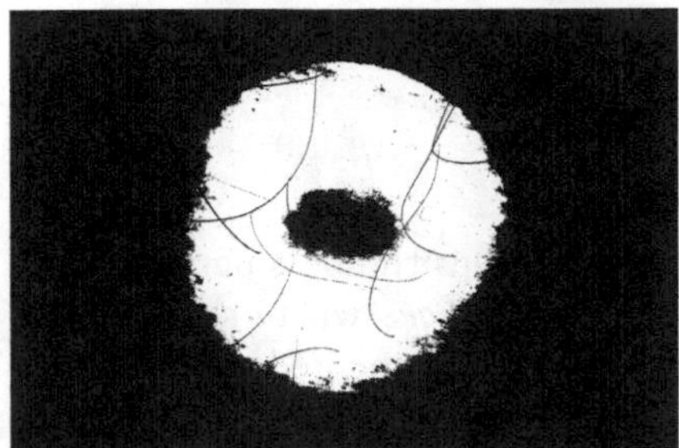

Fig. 8. Otsu from black binary mask, no hairs removed

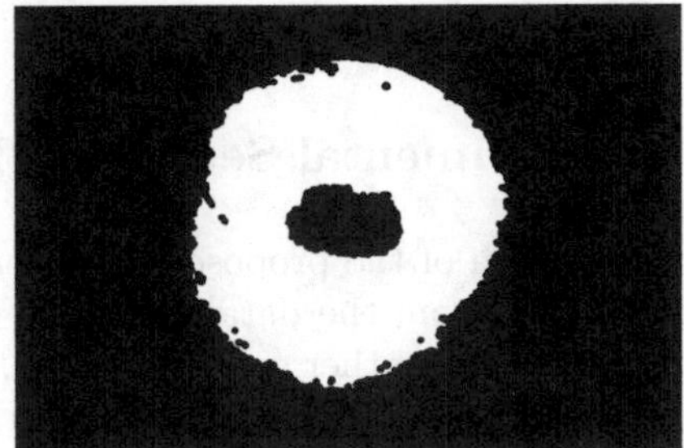

Fig. 9. Closed and eroded Otsu for the color cluster toleration area

The color cluster will be computed with OpenCV and, as for Otsu pipeline, by considering only the pixels not in the M0 mask; the number of clusters (K = 10) has been chosen empirically after a set of experimental sessions. Moreover the enlarged pixel mask in Fig. 9 has been divided into two sub-masks M1 and M2: the M1 mask represents the largest connected component that from now will replace M0, while M2 (the second largest connected component) will be considered as the tolerance area that approximates the central skin lesion pattern and that must contains the bigger and uniform clusters.

For each color cluster we count the number of pixels considered 'skin' or 'lesion' by using the M2 masks so if a cluster has more than 10% of its pixels out from this tolerance area it will be considered as simple skin (all of its pixels labeled 'skin'). It must be noticed that M1 must necessarily be used to exclude clusters that compose the concentric halo near the border between the bright skin and the black frame.

Examples of the global color cluster consisting by all the pixels labeled 'lesion' are in Figs. 10 and 11 respectively for the original image and for the one with hairs removed (for the future comparisons the pixels are in red).

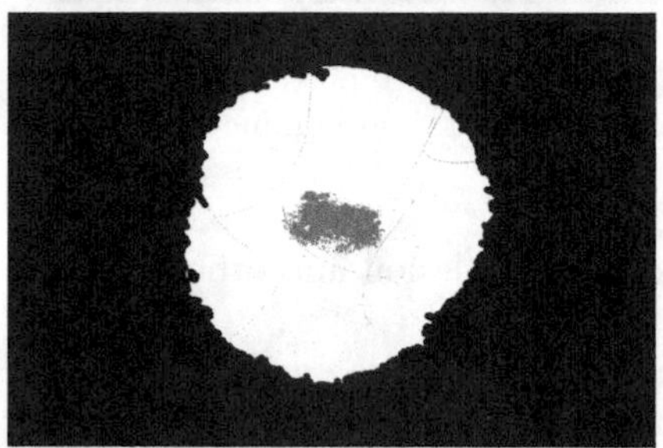

Fig. 10. Final global color cluster labeled 'lesion', no hairs removed (Color figure online)

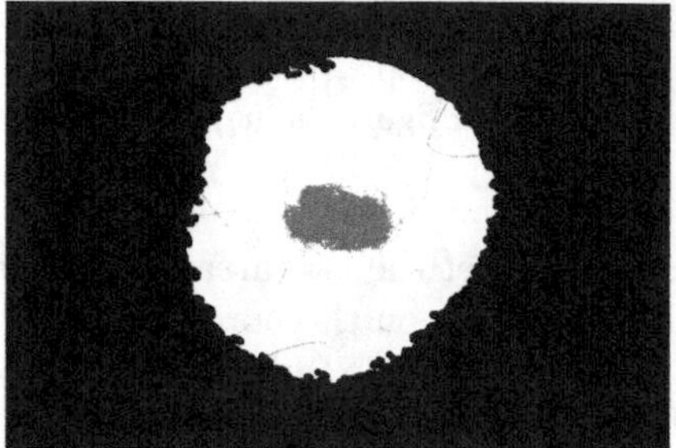

Fig. 11. Final global color cluster labeled 'lesion', hairs removed (Color figure online)

4 Experimental Setup and Results

To test which of the proposed detection pipelines reaches the best performance we will take from the dataset a small group of *hard images* where skin lesions differ from each other for size, colors, patterns, type (pigmented or not) and moreover for the presence of thick hairs at various sizes (Fig. 12).

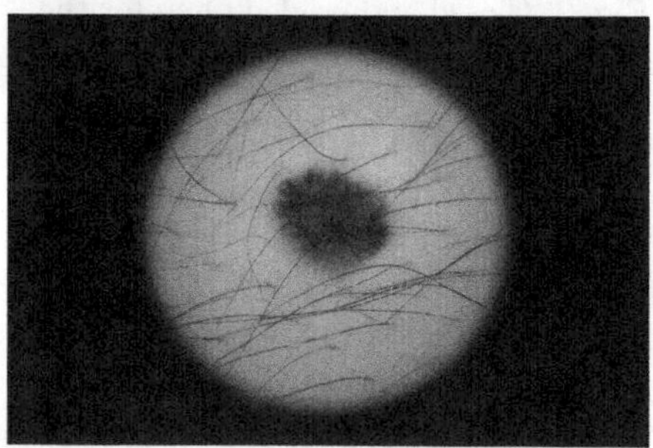

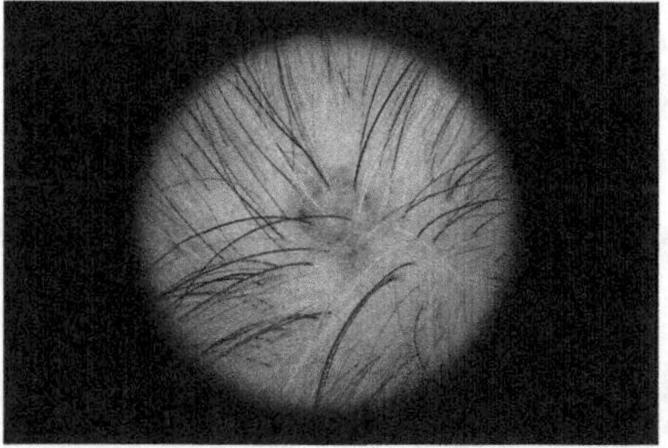

Fig. 12. Examples of hard dermoscopic images (Color figure online)

Each of the 17 chosen images are differentiated in 'original' version (Dataset 1) and 'hair removed' version (Dataset 2) for a total of 34 images. The experiment has a within-subjects design with two treatments and its structure is:

1. Dataset 1 - Otsu pipeline
2. Dataset 2 - Otsu pipeline
3. Dataset 1 - Color clustering pipeline
4. Dataset 2 - Color clustering pipeline

From a manual annotation we extract the green labeled area whose pixels must be considered as the *lesion ground truth* (Fig. 13). It is possible to consider, for each image in the datasets, the resulting labeled pixels (previously seen in red color) coming from the two detection pipelines and intersect them with the ground truth area to calculate measures like *Precision, Recall, f1-score* and *Accuracy*. To compute the four measures we consider a sort of *global goodness* for a detection pipeline, coming from the sum of the pixel of each classification, for each image of the dataset.

An example of the comparison between Figs. 13 and 6 for the Otsu pipeline is depicted in Fig. 14:

- true positives (TP): the pixel intersection (blue pixels)
- false positives (FP): red pixel that does not intersect the green ones
- false negatives (FN): green pixels that does not intersect the red ones
- true negatives (TN): pixels that are neither red nor green.

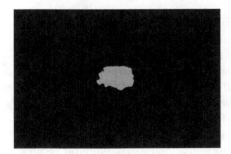

Fig. 13. Area derived from the manual annotation: ground truth (Color figure online)

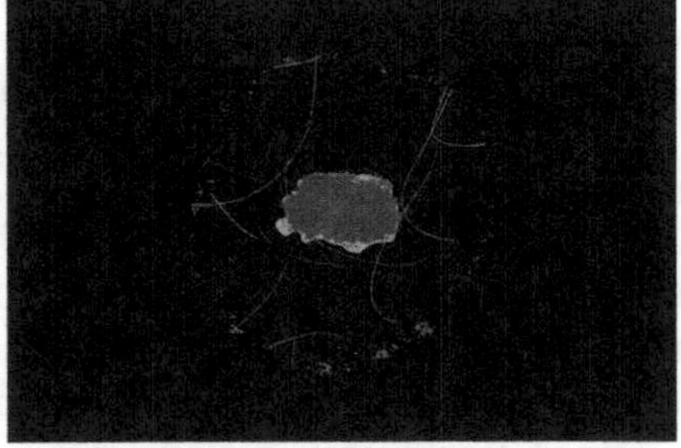

Fig. 14. Lesion detection comparison (blue pixels are the true positives) (Color figure online)

Experimental results are in Tables 1 and 2 for the Otsu pipeline while Tables 3 and 4 shows results for the Color clustering pipeline; it must be noticed that due to the peculiar image template and its size (pixel number) the *Accuracy* metric does not results significant, in fact the True Negatives (TN) represents most the large areas (as the black frame) that were considered and yet naturally excluded for pixel comparison during the detection pipelines.

Table 1. Results of Otsu pipeline on the original dataset

Otsu pipeline - Dataset 1				
Image	TP	FP	FN	TN
9001_2	217579	68714	33936	10335771
9012_2	70030	96724	2253	10486993
9013_3	364014	71399	14687	10205900
9017_2	87322	120510	5082	10443086
9030_2	296949	165099	3822	10190130
9051_2	7570	130537	219452	10298441
9067_2	106601	138285	10404	10400710
9084_2	52677	84071	3436	11859816
9108_2	214603	208246	1024	11576127
9110_2	176420	124390	12708	11686482
9112*a*_2	59915	47201	9617	11883267
9121_2	77964	41491	6187	11874358
9131_2	43994	109632	11788	11834586
9143_2	97002	113956	1150	11787892
9149_2	109552	61622	80655	11748171
9178*a*_2	645856	69300	71720	11213124
9190_2	818771	21707	82220	11077302
TOT.	3446819	1672884	570141	188902156
	Precision	*Recall*	*f1*	*Accuracy*
	0.67	0.86	0.75	0.99

Table 2. Results of Otsu pipeline on the hair removed dataset

Otsu pipeline - Dataset 2				
Image	TP	FP	FN	TN
9001_2	220376	43680	31139	10360805
9012_2	69964	80068	2319	10503649
9013_3	364010	66414	14691	10210885
9017_2	88283	106019	4121	10457577
9030_2	297509	73515	3262	10281714
9051_2	54816	129842	172206	10299136
9067_2	107926	116745	9079	10422250
9084_2	52832	57024	3281	11886863
9108_2	214525	158533	1102	11625840
9110_2	175242	92636	13886	11718236
9112*a*_2	59604	31391	9928	11899077
9121_2	77764	42813	6387	11873036
9131_2	44603	96025	11179	11848193
9143_2	96999	110616	1153	11791232
9149_2	114513	47094	75694	11762699
9178*a*_2	648944	36962	68632	11245462
9190_2	816557	8327	84434	11090682
TOT.	2476187	162813	1349042	190603958
	Precision	*Recall*	*f1*	*Accuracy*
	0.73	0.87	0.79	0.99

When comparing results between the two pipelines on each dataset we notice the detection improvement coming from the Color clustering especially in terms of the *Precision* metric, in fact it increases from 67% to 94% for the dataset 1 and from 73% to 94% for the second one. Also examining results between the two dataset for each pipeline we notice that all detection metrics always improve; only for the second dataset in Color Clustering pipeline, the *Precision* remains unchanged but *Recall* increases significantly): this demonstrates the need of hair removing treatment before each detection pipeline.

The *f1-score* metric provides the best interpretation of experimental results, in fact it increases in the Otsu pipeline (from 75% for dataset 1 to 79% for dataset 2) and in the Color clustering pipeline (from 79% for dataset 1 to 83% for dataset 2) and, as it is evident, it is always better for the second pipeline at the dataset equality. Finally, to explain these results it must be considered that, as previously explained, we considered images having specific characteristics and that, moreover, they represents only a part of the entire dermoscopic dataset.

Table 3. Results of Color clustering pipeline on the original dataset

Color clustering pipeline - Dataset 1				
Image	TP	FP	FN	TN
9001_2	126582	3625	111980	10413813
9012_2	38998	3192	25896	10587914
9013_3	299399	14928	64596	10277077
9017_2	0	0	84975	10571025
9030_2	116308	15577	170264	10353851
9051_2	0	0	215597	10440403
9067_2	88896	19563	19617	10527924
9084_2	0	0	50833	11949167
9108_2	190647	38220	10254	11760879
9110_2	133860	14066	44867	11807207
9112*a*_2	20617	1455	41831	11936097
9121_2	56131	739	20949	11922181
9131_2	0	0	49121	11950879
9143_2	86348	20269	3693	11889690
9149_2	82926	6518	97343	11813213
9178*a*_2	499044	17268	195287	11288401
9190_2	736431	7393	141939	11114237
TOT.	2476187	162813	1349042	190603958
	Precision	*Recall*	*f1-score*	*Accuracy*
	0.94	0.65	0.77	0.99

Table 4. Results of Color clustering pipeline on the hair removed dataset

Color clustering pipeline - Dataset 2				
Image	TP	FP	FN	TN
9001_2	183634	6196	54928	10411242
9012_2	57091	1600	7803	10589506
9013_3	299163	10266	64832	10281739
9017_2	59410	1423	25565	10569602
9030_2	278039	24196	8533	10345232
9051_2	0	0	215597	10440403
9067_2	82996	17204	25517	10530283
9084_2	30803	1022	20030	11948145
9108_2	198896	94849	2005	11704250
9110_2	156684	30981	22043	11790292
9112*a*_2	28484	120	33964	11937432
9121_2	56452	491	20628	11922429
9131_2	23825	3110	25296	11947769
9143_2	87097	21348	2944	11888611
9149_2	80332	268	99937	11819463
9178*a*_2	520811	3489	173520	11302180
9190_2	736402	1201	141968	11120429
TOT.	2880119	217764	945110	190549007
	Precision	*Recall*	*f1-score*	*Accuracy*
	0.93	0.75	0.83	0.99

5 Conclusions and Future Work

This work represents the second step (after the development and testing of the annotation tool) towards the development of a complex medical data management system used as an agent to support dermatologists in their decisions by exploiting all of the architecture modules while raw and structured data. The capacities must go from the images gathering and their annotation to the feature extraction and the analysis and visualization of their (meta)data, also exploiting Artificial Intelligence methods and CNN for advanced predictive performances.

Further tests and evaluations are needed, also considering the variety of lesion patterns and their stratification but the experiments presented shows encouraging results also with complex dermoscopic images, moreover the evaluation metrics proposed results adequate to verify and measure the best results: the color clustering technique featuring the pixel mask, the hair removing and the toleration areas reaches very positive performances.

References

1. DullRazor: a software approach to hair removal from images. Comput. Biolo. Med. **27**(6), 533–543 (1997)
2. Abbes, W., Sellami, D.: High-level features for automatic skin lesions neural network based classification. In: 2016 International Image Processing, Applications and Systems (IPAS), pp. 1–7, November 2016
3. Ahlberg, C., Williamson, C., Shneiderman, B.: Dynamic queries for information exploration: an implementation and evaluation. In: Proceedings of the SIGCHI Conference on Human Factors in Computing Systems, CHI 1992, pp. 619–626. ACM, New York, (1992). http://doi.acm.org/10.1145/142750.143054
4. Bakheet, S.: An SVM framework for malignant melanoma detection based on optimized HOG features. Computation **5**(1), 4 (2017)
5. Celebi, M.E., Wen, Q., Iyatomi, H., Shimizu, K., Zhou, H., Schaefer, G.: A state-of-the-art survey on lesion border detection in dermoscopy images. In: Dermoscopy Image Analysis, pp. 97–129. CRC Press (2015)
6. Codella, N., Cai, J., Abedini, M., Garnavi, R., Halpern, A., Smith, J.R.: Deep learning, sparse coding, and SVM for melanoma recognition in dermoscopy images. In: Zhou, L., Wang, L., Wang, Q., Shi, Y. (eds.) MLMI 2015. LNCS, vol. 9352, pp. 118–126. Springer, Cham (2015). doi:10.1007/978-3-319-24888-2_15
7. Codella, N.C.F., Nguyen, Q., Pankanti, S., Gutman, D., Helba, B., Halpern, A., Smith, J.R.: Deep learning ensembles for melanoma recognition in dermoscopy images. CoRR abs/1610.04662 (2016). http://arxiv.org/abs/1610.04662
8. Diepgen, T., Mahler, V.: The epidemiology of skin cancer. Br. J. Dermatol. **146**(s61), 1–6 (2002)
9. Emre Celebi, M., Wen, Q., Hwang, S., Iyatomi, H., Schaefer, G.: Lesion border detection in dermoscopy images using ensembles of thresholding methods. Skin Res. Technol. **19**(1), e252–e258 (2013)
10. Fan, H., Xie, F., Li, Y., Jiang, Z., Liu, J.: Automatic segmentation of dermoscopy images using saliency combined with Otsu threshold. Comput. Biol. Med. (2017). http://www.sciencedirect.com/science/article/pii/S001048251730080X
11. Gutman, D., Codella, N.C., Celebi, E., Helba, B., Marchetti, M., Mishra, N., Halpern, A.: Skin lesion analysis toward melanoma detection: a challenge at the international symposium on biomedical imaging (ISBI) 2016, hosted by the international skin imaging collaboration (ISIC). arXiv preprint arXiv:1605.01397 (2016)
12. Jaleel, J.A., Salim, S., Aswin, R.: Artificial neural network based detection of skin cancer. IJAREEIE **1**, 200–205 (2012)
13. Liu, Z., Sun, J., Smith, M., Smith, L., Warr, R.: Unsupervised sub-segmentation for pigmented skin lesions. Skin Res. Technol. **18**(1), 77–87 (2012)
14. Maglogiannis, I., Doukas, C.N.: Overview of advanced computer vision systems for skin lesions characterization. IEEE Trans. Inf Technol. Biomed. **13**(5), 721–733 (2009)
15. Majtner, T., Yildirim-Yayilgan, S., Hardeberg, J.Y.: Combining deep learning and hand-crafted features for skin lesion classification. In: 2016 Sixth International Conference on Image Processing Theory, Tools and Applications (IPTA), pp. 1–6, December 2016
16. Marín, C., Alférez, G.H., Córdova, J., González, V.: Detection of melanoma through image recognition and artificial neural networks. In: Jaffray, D.A. (ed.) World Congress on Medical Physics and Biomedical Engineering, Toronto, Canada, 7–12 June 2015. IP, vol. 51, pp. 832–835. Springer, Cham (2015). doi:10.1007/978-3-319-19387-8_204

17. Peruch, F., Bogo, F., Bonazza, M., Cappelleri, V.M., Peserico, E.: Simpler, faster, more accurate melanocytic lesion segmentation through MEDS. IEEE Trans. Biomed. Eng. **61**(2), 557–565 (2014)
18. Schaefer, G., Krawczyk, B., Celebi, M.E., Iyatomi, H.: An ensemble classification approach for melanoma diagnosis. Memet. Comput. **6**(4), 233–240 (2014)
19. Seidenari, S., Pellacani, G., Grana, C.: Early detection of melanoma by image analysis. In: Bioengineering of the Skin: Skin Imaging & Analysis, pp. 305–312. CRC Press (2006)
20. Singh, S., Stevenson, J., McGurty, D.: An evaluation of polaroid photographic imaging for cutaneous-lesion referrals to an outpatient clinic: a pilot study. Br. J. Plast. Surg. **54**(2), 140–143 (2001)
21. Wighton, P., Lee, T.K., Lui, H., McLean, D.I., Atkins, M.S.: Generalizing common tasks in automated skin lesion diagnosis. IEEE Trans. Inf. Technol. Biomed. **15**(4), 622–629 (2011)
22. Yuan, Y., Chao, M., Lo, Y.C.: Automatic skin lesion segmentation using deep fully convolutional networks with jaccard distance. IEEE Trans. Med. Imaging **PP**(99), 1 (2017)
23. Zagrouba, E., Barhoumi, W.: A prelimary approach for the automated recognition of malignant melanoma. Image Anal. Stereol. **23**(2), 121–135 (2011)

Feature Definition and Selection for Epiretinal Membrane Characterization in Optical Coherence Tomography Images

Sergio Baamonde(✉), Joaquim de Moura, Jorge Novo, José Rouco, and Marcos Ortega

VARPA Group, Departament of Computer Science,
University of A Coruña, A Coruña, Spain
{sergio.baamonde,joaquim.demoura,jnovo,jrouco,mortega}@udc.es

Abstract. Optical Coherence Tomography (OCT) is a common imaging technique for the detection and analysis of optical diseases, since it is a non invasive method that generates in vivo a cross-sectional visualization of the retinal tissues. These characteristics contributed to the use of OCT imaging in the analysis of pathologies as, for instance, vitreomacular traction, age-related macular degeneration or hypertension. Among its applications, OCT imaging can be used in the detection of any present epiretinal membrane section in the retina, a critical issue to prevent further complications caused by this pathology.

This work analyzed the main characteristics of the epiretinal membrane to define a complete and heterogeneous set of intensity and texture-based features. Those features were studied using representative selectors, as Correlation Feature Selection (CFS) and Relief-F, to identify the optimal subsets that offer the higher discriminative power. K-Nearest Neighbor (kNN), Naive Bayes and Random Forest were finally tested in a method for the automatic detection of the epiretinal membrane in OCT images. Previous works do not focus on automatic procedures and, on the contrary, depend on manual markers or supervised detections, while our method improves significantly this task by automating the search of the region of interest and the classification of the pixels belonging to that area.

The methodology was tested using a dataset of 129 OCT images. 120 samples were equally obtained from those scans, featuring both zones with and without epiretinal membrane. The best results were provided by the Random Forest classifier that, using a window size of 15 pixels, a quantity of 13 histogram bins and 28 features, achieved an accuracy of 93.89%.

This work is supported by the Instituto de Salud Carlos III, Government of Spain and FEDER funds of the European Union through the PI14/02161 and the DTS15/00153 research projects and by the Ministerio de Economía y Competitividad, Government of Spain through the DPI2015-69948-R research project.

S. Battiato et al. (Eds.): ICIAP 2017, Part II, LNCS 10485, pp. 456–466, 2017.
https://doi.org/10.1007/978-3-319-68548-9_42

Keywords: Computer-aided diagnosis · Retinal imaging · Optical Coherence Tomography · Epiretinal membrane · Feature selection · Classification

1 Introduction

Retinal image analysis is an important issue for the diagnosis of various optical diseases. To this end, it is necessary to identify precisely the pertinent structures of the eye fundus as, for instance, the optic disc [13] and the arterio-venular tree [12]. With this information, a characterization of cardiovascular complications [7] or pathologies such as diabetes [17] can be achieved.

Macular pucker, more commonly known as epiretinal membrane (ERM), is a fibrocellular tissue that can cause metamorphopsia, central vision decrease or blurred vision [1,10]. Moreover, epiretinal membranes are associated with different types of cysts (macular, paravascular, lamellar macular) [11], further contributing to the eyesight distortion or reduction.

Idiopathic ERMs are the most common, but retinal vascular diseases or changes in the vitreous humor [4] can induce a response from the immune system to protect the retina. This response causes, sometimes, that the retinal cells converge on the macular region, creating a transparent layer. This layer, that is scar tissue, causes tension on the retina by contraction, further increasing the chances of secondary ERMs to appear.

Optical Coherence Tomography (OCT) imaging [3] is frequently used to analyze the retinal morphology and detect the presence of ERM. ERM appears as a thin reflective layer on the retina [2], fact that can be used for its detection on OCT images. Irregularities on the retinal surface or retinal thickening can also indicate the presence of ERM on the patient.

The asymptomatic nature of this pathology makes necessary a reliable and accurate detection system. With an appropriate method, ERM can be early detected and treated before further complications appear. Those methods are usually based in the manual detection of the ERM by a specialist [14]. Similarly, the method of Wilkins *et al.* [16] uses real-time OCT images and, after an specialist establish manual markers on the image, ERM is detected by the use of information about the reflectivity and thickness of the retina on the selected points.

In this work, we aimed for the automation of the process by developing an algorithm that selects autonomously the region of interest (ROI) where the ERM can be present. We analyzed the main characteristics of the ERM and designed a complete and heterogeneous set of features that helped to characterize the regions where the ERM is present. Optimal subsets of those features were selected and used to train representative classifiers. We use those trained classifiers to identify automatically the points belonging to the region of interest and pinpoint the presence or absence of ERM in the selected area. This method aims to improve the general error tolerance of the process by avoiding the use of manual markers for ROI initialization and making them non-dependent of human interaction.

2 Methodology

The proposed system tries to identify automatically the Inner Limiting Membrane (ILM), which is the boundary between the retina and the vitreous body, area where the ERM can appear if the pathology is present on the patient. Then, using these identifications, we analyze all the points belonging to the ILM, generating a rectangular-shaped window for each point and calculating the relevant features of the constructed window. Finally, every feature vector is used to classify its associated point and obtain information about the presence of ERM in the ILM retinal layer.

2.1 Identification of the Region of Interest

In this work, we employed a new method based on the use of an active contour model (Snake) [8], which adjusts its shape to the ILM contour. A predefined number of points are initialized on the uppermost part of the OCT scan. These points adapt its contour to the shape of the ILM by using information about the intensity of this layer in contrast to the rest of the layers. We designed an adapted version of the Snake, since we restrict its movement to the vertical axis, allowing only downwards movement. All the points of the Snake are moved progressively, approaching to the ILM layer. Finally, if a point does not modify its energy after an iteration, that point is fixed and is not processed again. With this method, the Snake behaves like a cascade of points instead of a contracting closed shape.

The Snake finally reaches the ROI (defined by the ILM position) to identify the ERM presence. In order to obtain relevant information from the ROI, a large set of heterogeneous features is obtained from each point of the Snake. These features are measured in the surrounding area of each point of interest. This area is defined as a rectangular window where W_{size} is the width in pixels of the window and the height is $5 \times W_{\mathrm{size}}$ (Fig. 1), offering enough information of the layer tissue with respect to its surrounding area.

2.2 Feature Definition

Using the properties of the ILM with and without ERM presence, we selected a complete set of intensity and texture-based features of the windows obtained around the points of interest to be able to separate precisely the points with ERM from the normal ILM tissue.

The number of features varies between 223 and 263. This variability is caused by the use of the input parameter N_{bins} of the window features, depicted below. The used features can be classified in the following groups:

Window Features. Each window obtained from the Snake is divided in five different square-shaped windows. Then, we calculate the histogram for each sub-window. The number of bins was empirically selected, so the resulting number of features obtained oscillates between 35 and 85, depending on the

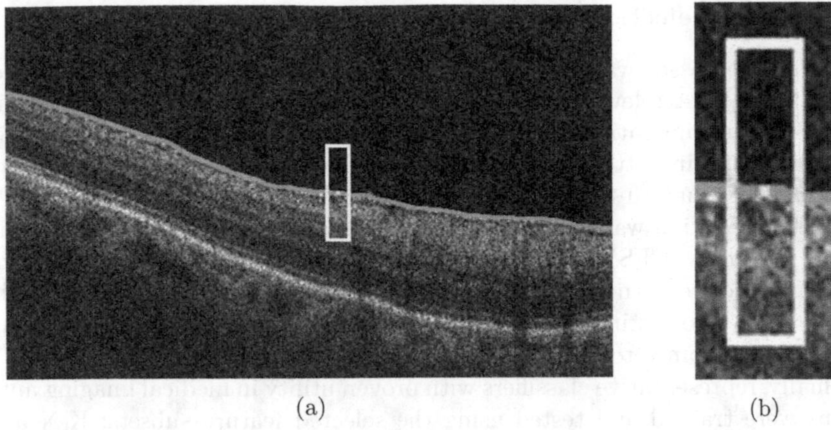

Fig. 1. Example of Region of Interest definition. (a): Snake situated on top of the ILM and window around a point of the ROI. (b): Zoom on the feature window

value of N_{bins}. This parameter is used to determine the number of bins for the histograms associated to each window. We tested configurations from 7 to 17 bins (in increments of 2). For each point we have 5 different sub-windows with an associated histogram, so the total number of features range from $5 \times 7 = 35$ to $5 \times 17 = 85$.

Intensity Features. 13 features are obtained with all the intensity information of the window: *maximum, minimum, mean, median, standard deviation, variance, first quartile, third quartile, skewness* and *maximum likelihood estimate* (for a normal distribution).

Gray-Level Intensity Histogram (GLIH). The histogram of the full window is calculated. From it, we obtain the following metrics: *obliquity, kurtosis, energy* and *entropy.*

Gray-Level Co-Ocurrence Matrix (GLCM). These features provide information about the spatial relationship of the pixels [15]. We use a distance of 2 pixels and 4 directions as proposed by Haralick *et al.* [6], for a total of 16 features.

Histogram of Oriented Gradients (HOG). Gradient orientation can be an useful feature, since it can contribute to the detection of the different patterns of gradients when the epiretinal membrane is found in contrast to its absence. Besides, HOG features are suitable to recognize gradient patterns in the ROI since they are invariant to scale, rotation or translation modifications. We used 9 HOG windows with 9 bins, obtaining a total of 81 features.

Local Binary Patterns (LBP). Local Binary Patterns also help to detect patterns of intensity changes in the selected window. Another advantage is their low sensitivity to intensity changes, since variation in illumination is usual in OCT images. We use a total of 64 features that will give a extended range of information.

2.3 Feature Selection and Classification

Once the feature set was specified, we proceeded with the analysis and selection of those most relevant features that contain meaningful data and provide the highest discriminative power. This way, we can optimize the classification process and obtain better results by avoiding the introduction of unneeded and redundant information to the classifiers.

Feature selection was performed using representative strategies. Correlation Feature Selection (CFS) algorithm [5], which works by selecting features correlated or predictive of the class that, otherwise, are irrelevant. Relief-F algorithm [9] is also used, consisting on the repeatedly sampling of a random instance and checking the distance to the nearest instances of the same or different class.

Finally, representative classifiers with proven utility in medical imaging applications were trained and tested using the selected feature subsets: K-Nearest Neighbor (kNN), Naive Bayes and Random Forest classifiers. For each classifier, we test three window sizes and six different number of bins for the window features mentioned beforehand.

3 Results and Discussion

The validation of the method was done by the use of a set of 129 OCT images. These scans were obtained with a tomograph CIRRUS^{TM} HD-OCT Zeiss, with Spectral Domain Technology. The resolution of the scans was 490×500 pixels without any preprocessing stage.

The scans were labeled by an expert clinician, identifying the areas where ERM is present and absent, respectively. With this groundtruth, we selected a set of 120 samples, divided in 60 samples with ERM presence and 60 with ERM absence. Furthermore, each group of samples can also be split in the following classes (Fig. 2):

1. *Membrane* class. Points with ERM presence on top of the ILM.
2. *Floating membrane* class. Points with ERM presence on the background.
3. *Non-membrane* class. Points situated on the first layer of the retina but without ERM presence.
4. *Background* class. Points not situated on the ILM layer, but on the background.

This way, we have labeled datasets for two and four class approximations. We performed experiments facing both approximations to test the capabilities of the designed method. To evaluate the results, we use the accuracy of the classifiers as our control metric. Table 1a and b present the results obtained using the 2-class approximation. Results are very similar across all configurations, only obtaining a slight improvement with $W_{size} = 17$ for the Random Forest classifier.

Table 2a and b show the results for the 4-class approximation. In this case, the improvement in the performance is more accentuated when using CFS algorithm, both with kNN and Random Forest classifiers for $W_{size} = 15$. Generally,

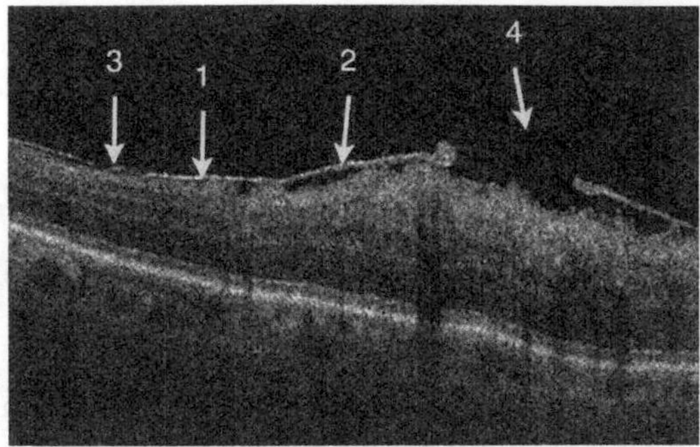

Fig. 2. Example of the different types of classes to be identified

Table 1. 2-class classification accuracy results

N_{bins}	kNN			Naive Bayes			Random Forest		
	W_{size}			W_{size}			W_{size}		
	13	15	17	13	15	17	13	15	17
(a) CFS algorithm									
7	81.80%	88.68%	87.44%	81.58%	89.71%	86.75%	83.79%	89.46%	90.45%
9	82.80%	87.46%	87.24%	80.85%	88.74%	87.26%	83.58%	89.22%	89.99%
11	84.22%	85.98%	88.95%	81.09%	88.73%	85.78%	83.81%	89.93%	89.73%
13	81.56%	86.19%	87.46%	80.61%	89.23%	85.29%	83.57%	88.72%	**90.48%**
15	81.29%	87.41%	85.48%	81.09%	89.70%	86.05%	82.10%	88.47%	90.23%
17	84.51%	88.19%	86.67%	80.61%	88.74%	85.54%	82.10%	90.21%	90.22%
(b) Relief algorithm									
7	81.14%	82.30%	85.73%	83.10%	82.34%	85.49%	81.88%	84.74%	86.00%
9	81.60%	82.04%	**89.22%**	84.30%	82.08%	86.29%	82.82%	82.54%	86.02%
11	81.65%	81.05%	87.48%	84.55%	82.08%	86.24%	82.64%	84.00%	86.53%
13	81.42%	85.76%	88.23%	83.84%	84.81%	87.01%	84.57%	85.51%	86.77%
15	83.32%	84.79%	84.29%	84.33%	82.85%	84.54%	84.80%	85.77%	83.83%
17	85.82%	85.78%	88.48%	85.53%	84.34%	87.04%	84.84%	85.54%	85.07%

the best accuracy is found for $W_{size} = 15$, while N_{bins} is a parameter with more discrepancy. Normally, extreme values on this parameter decrease the overall accuracy of the classifier. The best results are obtained for the 4-class approach, both with kNN and Random Forest classifiers, reaching an accuracy of 93.89% when using CFS algorithm for feature selection and Random Forest. Random Forest algorithm obtains a slightly higher accuracy than the kNN method. 28 features were selected for this configuration.

Regarding feature selection, we present the results using the 4-class approximation. The analysis and conclusions are analogous to the other 2-class approximation. About the selected features, Fig. 3 shows the ones provided by each selector in their best configuration, for a total of 28, grouped by type. The most

Table 2. 4-class classification accuracy results

N_{bins}	kNN			Naive Bayes			Random Forest		
	W_{size}			W_{size}			W_{size}		
	13	15	17	13	15	17	13	15	17
(a) CFS algorithm									
7	83.85%	89.77%	88.08%	86.60%	86.08%	88.77%	88.53%	92.47%	88.55%
9	85.34%	91.20%	88.32%	86.84%	89.23%	90.75%	85.86%	93.67%	87.57%
11	85.33%	92.17%	88.79%	88.11%	88.04%	90.23%	86.85%	91.96%	89.48%
13	85.37%	89.24%	88.05%	85.93%	88.05%	89.53%	88.09%	**93.89%**	89.25%
15	88.31%	91.45%	88.08%	85.64%	87.51%	88.02%	87.34%	91.71%	89.00%
17	86.14%	91.34%	88.28%	86.61%	87.55%	89.98%	86.63%	91.94%	89.24%
(b) Relief algorithm									
7	88.29%	86.54%	87.04%	86.58%	85.11%	83.88%	**92.19%**	90.43%	89.22%
9	87.55%	86.55%	86.09%	85.61%	83.38%	82.68%	90.24%	88.49%	88.74%
11	87.56%	87.81%	87.06%	87.04%	83.66%	83.17%	88.78%	87.29%	88.73%
13	90.98%	87.54%	88.32%	85.87%	84.65%	82.45%	89.97%	88.02%	89.48%
15	89.02%	87.03%	87.28%	87.80%	84.88%	83.16%	90.21%	88.97%	87.29%
17	88.79%	85.83%	85.86%	86.57%	84.15%	84.13%	89.48%	88.73%	89.97%

relevant features that were provided by both strategies belong to the window features. More precisely, information about the first bins on the third sub-window (this is, the central sub-window, which is the one where the ERM should be located) were included in the first positions. As we can observe, these features are selected because the core of the class differentiation is done by the use of information of the center of the window (luminosity values indicate presence or absence of ERM). In a lower degree, it is also relevant the information about the fourth sub-window (the second from the bottom) and the fifth sub-window (the bottom one). This is congruent with the theory that the information under the central sub-window contributes to the differentiation between *floating membrane* and *membrane* classes, the first having lower intensity values on those windows than the *membrane* class. The rest of the selected features are mostly HOG values and a few features from the LBP group with the CFS selection, as they provide higher information about intensity and patterns on the ROI and contribute to improve the discrimination between the ERM presence or absence. On the contrary, this information is better represented by mainly GLCM features using the Relief-F selected subset.

Figure 4 shows the accuracy of the classification stage with the most accurate configuration (Random Forest with $W_{\text{size}} = 15$ and $N_{\text{bins}} = 13$) when using progressive larger subsets of features with both selector strategies. In almost all situations, CFS selector performs better than the Relief-F selector.

Figure 5 presents a representative classification result of an OCT image using the most accurate configuration. As we can see, most of the points are classified correctly and the ERM presence is detected in almost all the ILM surface.

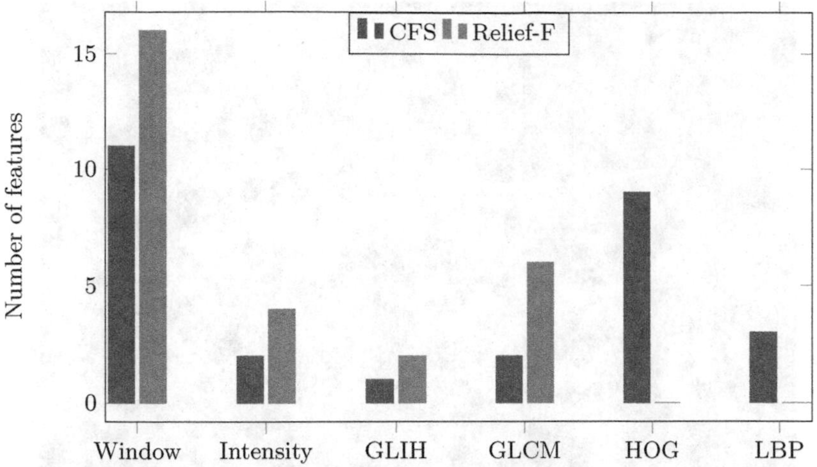

Fig. 3. Number of selected features, of a total of 28, for each feature selector method

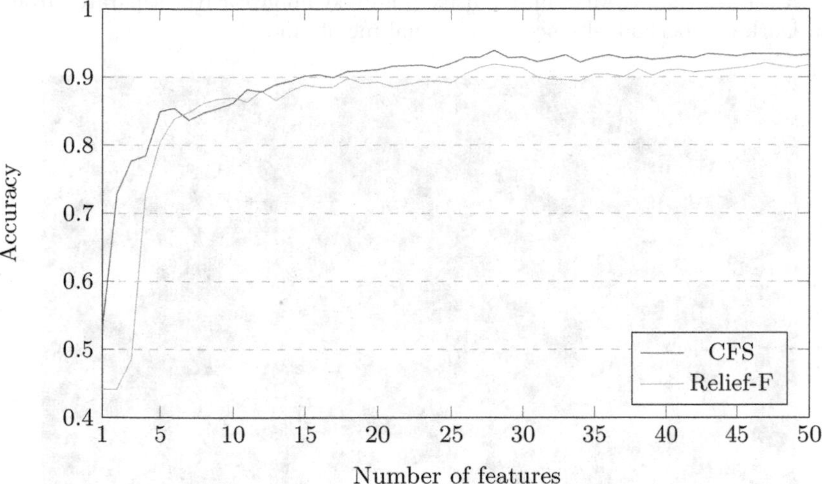

Fig. 4. Evolution of accuracy using Random Forest classifier and progressive larger feature subsets with the analyzed feature selectors

Differentiation between the ERM with or without separation from the ILM is also done correctly. In contrast, Fig. 6 show a common incorrect classification on the right points of the image. In this case, the Snake algorithm cannot be locally adjusted to the lower zone of the retina, being penalized the final adjustment and the points get detected at background positions.

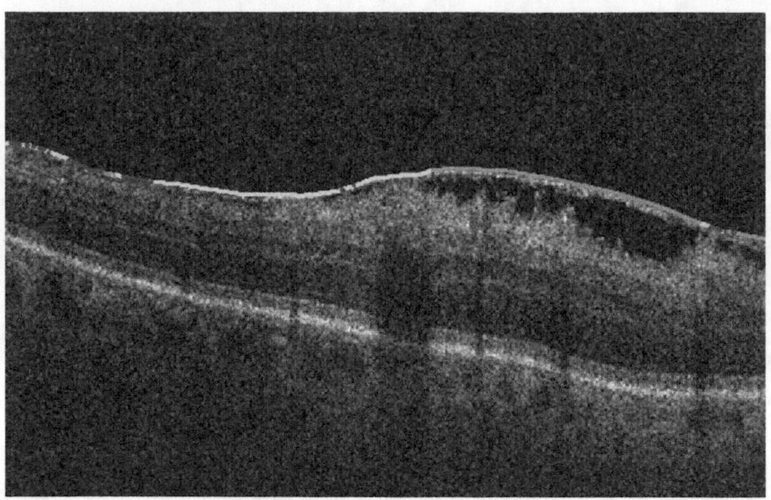

Fig. 5. Example of 4-class classification of an OCT image. Bright points represent the ERM on the ILM layer. Medium-intensity points symbolize ERM separated from the retina. Dark points show absence of epiretinal membrane

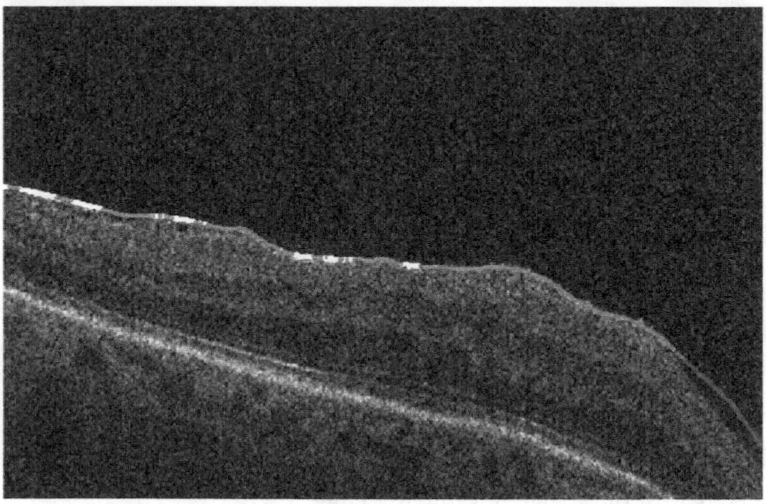

Fig. 6. Example of a region incorrectly identified and classified. Medium-intensity points on the right side symbolize background points, not contributing to the task at hand

4 Conclusions

The accurate identification of the presence of the ERM is an important issue in the retinal analysis as its early detection improves the chances of success of ERM removal surgery, avoiding the complications that its presence derive.

In this work, we proposed an automatic method to detect the ERM in OCT images. The method is fully automatic, instead of the few previous approaches that are based on manual detections by the specialist. Furthermore, we have achieved a higher level of tolerance to errors by using a deformable model to detect the ROI compared to the detection of this region based on manual markers.

The method firstly uses a deformable model (Snake) to initially identify the ILM retinal layer, ROI where the ERM is originated. Then, a complete and heterogeneous set of features were measured, based on the properties of the ERM. Representative feature selectors as CFS and Relief-F were used with the entire feature set to identify those that provide the highest discriminative power.

We defined a set of 223–263 features that were then filtered by a process of feature selection, obtaining 28 features with the method that provided the most accuracy afterwards (CFS). Different suitable classifiers were tested, for a total of 216 different configurations. For testing, we used a set of 120 samples, distributed equally between the different classes in both two and four class approximations.

The results were highly successful, obtaining an accuracy of 93.89% for the CFS algorithm, with a Random Forest classifier with W_{size} of 15 and a N_{bins} of 13.

For further works, an increase in the number of samples for training is planned in order to improve even further the accuracy of the classifiers. Furthermore, wrapper-based feature selection methods will be tested as well as a larger variability of classifiers.

References

1. Agarwal, A.: Gass' Atlas of Macular Diseases. Elsevier Health Sciences, Amsterdam (2011)
2. Brancato, R.: Optical coherence tomography (OCT) in macular edema. Doc. Ophthalmol. **97**, 337–339 (1999)
3. Do, D.V., Cho, M., Nguyen, Q.D., Shah, S.M., Handa, J.T., Campochiaro, P.A., et al.: Impact of optical coherence tomography on surgical decision making for epiretinal membranes and vitreomacular traction. Retina **27**, 552–556 (2007)
4. Foos, R.Y.: Vitreoretinal juncture; epiretinal membranes and vitreous. Invest. Ophthalmol. Vis. Sci. **16**, 416–422 (1977)
5. Hall, M.A.: Correlation-based feature selection for machine learning. The University of Waikato (1999)
6. Haralick, R.M., Shanmugam, K., et al.: Textural features for image classification. IEEE Trans. Syst. Man Cybern. **3**, 610–621 (1973)
7. Ikram, M.K., de Jong, F.J., Bos, M.J., Vingerling, J.R., Hofman, A., Koudstaal, P.J., et al.: Retinal vessel diameters and risk of stroke: the Rotterdam Study. Neurology **66**, 1339–1343 (2006)
8. Kass, M., Witkin, A., Terzopoulos, D.: Snakes: active contour models. Int. J. Comput. Vis. **1**, 321–331 (1988)
9. Kira, K., Rendell, L.A.: A practical approach to feature selection. In: Proceedings of the Ninth International Workshop on Machine Learning, pp. 249–256 (1992)

10. Medina, C.A., Townsend, J.H., Singh, A.D. (eds.): Manual of Retinal Diseases: A Guide to Diagnosis and Management. Springer, Cham (2016). doi:10.1007/978-3-319-20460-4
11. Meuer, S.M., Myers, C.E., Klein, B.E.K., Swift, M.K., Huang, Y., Gangaputra, S., et al.: The epidemiology of vitreoretinal interface abnormalities as detected by SD-OCT: the beaver dam eye study. Ophthalmology **122**, 787–795 (2015)
12. de Moura, J., Novo, J., Ortega, M., Charlón, P.: 3D retinal vessel tree segmentation and reconstruction with OCT images. In: Campilho, A., Karray, F. (eds.) ICIAR 2016. LNCS, vol. 9730, pp. 716–726. Springer, Cham (2016). doi:10.1007/978-3-319-41501-7_80
13. Novo, J., Penedo, M.G., Santos, J.: Optic disc segmentation by means of GA-optimized topological active nets. In: Campilho, A., Kamel, M. (eds.) ICIAR 2008. LNCS, vol. 5112, pp. 807–816. Springer, Heidelberg (2008). doi:10.1007/978-3-540-69812-8_80
14. Puliafito, C.A., Hee, M.R., Lin, C.P., Reichel, E., Schuman, J.S., Duker, J.S., et al.: Imaging of macular diseases with optical coherence tomography. Ophthalmology **102**, 217–229 (1995)
15. Ramamurthy, B., Chandran, K.R.: Content based medical image retrieval with texture content using gray level co-occurrence matrix and k-means clustering algorithms. J. Comput. Sci. **8**, 1070 (2012)
16. Wilkins, J.R., Puliafito, C.A., Hee, M.R., Duker, J.S., Reichel, E., Coker, J.G., et al.: Characterization of epiretinal membranes using optical coherence tomography. Ophthalmology **103**, 2142–2151 (1996)
17. Wong, T.Y., Klein, R., Sharrett, A.R., Schmidt, M.I., Pankow, J.S., et al.: Retinal arteriolar narrowing and risk of diabetes mellitus in middle-aged persons. JAMA **287**, 2528–2533 (2002)

Fully-Automated CNN-Based Computer Aided Celiac Disease Diagnosis

Michael Gadermayr[1](✉), Georg Wimmer[2], Andreas Uhl[2], Hubert Kogler[3], Andreas Vécsei[3], and Dorit Merhof[1]

[1] Aachen Center for Biomedical Image Analysis, Visualization and Exploration (ACTIVE), Institute of Imaging and Computer Vision, RWTH Aachen University, Aachen, Germany
Michael.Gadermayr@lfb.rwth-aachen.de

[2] Department of Computer Sciences, University of Salzburg, Salzburg, Austria

[3] Department of Pediatrics, St. Anna Children's Hospital, Medical University Vienna, Vienna, Austria

Abstract. While a significant amount of research has been on computer aided diagnosis of celiac disease, challenges remain especially due to difficult imaging conditions during endoscopy which frequently result in image degradations. To compensate for these degradations which often hide relevant disease markers, classification trials so far have been performed exclusively utilizing informative patches, which were manually selected by experienced physicians. In this work, we propose a novel fully-automated method to obtain decisions from computer aided diagnosis systems without any interaction, based on original endoscopic image data. For this purpose, we rely on a discriminative model based on convolutional neural networks trained with informative patch data. Additionally, we fit a probabilistic model utilizing original endoscopic image data to obtain realistic predictions for patches concerning their level of reliability. In our experiments, the state-of-the-art considering a classification on image as well as on patient level is outperformed.

1 Introduction

Celiac disease (CD) [1] is a common autoimmune disorder primarily affecting the small bowel. It is characterized by an inflammation affecting the mucosa of the duodenum and the pars descendens. During the course of the disease, the mucosa loses its absorptive villi and hyperplasia of the enteric crypts occurs, which leads to a diminished ability to absorb nutrients (Fig. 1). In the past, a significant amount of research has been performed in the field of computer aided CD diagnosis from endoscopic imaging data [2–6] with the target to provide a second opinion besides histological assessment of biopsies and/or to reduce the number of required biopsies [7]. In early literature, the focus in this field was mainly on handcrafted image representations including methods such as local binary patterns and mid-level representations such as Fisher vectors [3,7]. Recently, convolutional neural networks (CNNs) exhibited superior performance outperforming previous methods [4,6].

S. Battiato et al. (Eds.): ICIAP 2017, Part II, LNCS 10485, pp. 467–478, 2017.
https://doi.org/10.1007/978-3-319-68548-9_43

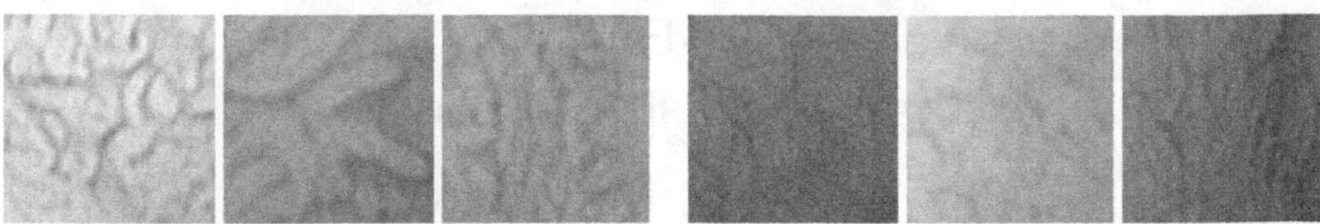

(a) A healthy mucosa is indicated by a pro-mounced villous structure as depicted.

(b) In case of a severe villous atrophy, the villous structure is completely missing.

Fig. 1. Informative patches clearly showing markers for distinguishing between healthy (a) and diseased subjects (b). In case of mild impairments, however, the discrimination is more difficult.

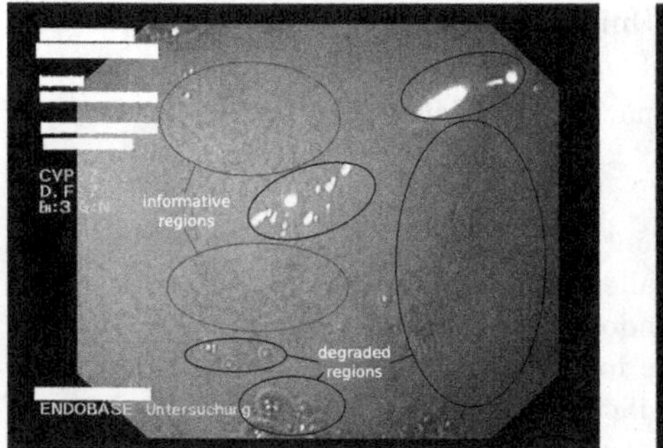

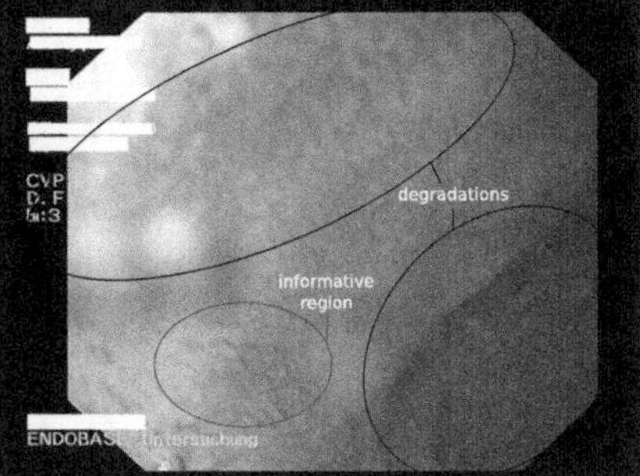

Fig. 2. Original endoscopic images showing several kinds of degradation (such as blur, low-contrast, bubbles and overexposure) partly or completely hiding the distinctive disease markers.

Most work in computer aided CD diagnosis [3,4,6,7] relies on 'informative' patches only, which were extracted by medical experts. These data is not only utilized for training the classification model, but also for evaluation. For a practical clinical system, however, this means that the physician needs to select a reliable patch before the automated classification approach can be applied. Thereby, manual effort is required and the obtained decision is no longer fully observer-independent.

A holistic (non-patch-based) approach cannot be applied effectively with limited training data, since disease markers are often visible locally only due to degradations (Fig. 2) and the patchy distribution of the disease markers [7].

A further issue is given by the fact that degraded images showing no sensible information are generally very similar (in feature-space) to patches exhibiting a diseased mucosa. The straight-forward application of patch-based approaches to randomly selected patches thereby leads to low as well as imbalanced outcomes considering sensitivity and specificity [8]. In previous work, it was suggested to select (and merge) discriminative patches from original images by computing a weighted linear combination of several basic quality measures to estimate a patch's quality [8]. These consist of metrics measuring (1) the average illumination, (2) the focus-level, (3) reflections, (4) noise and (5) contrast. Thereby, the classification rates were improved compared to a random patch selection. Nevertheless, a linear combination of five basic measures is unlikely to perfectly represent a patch's distinctiveness and is furthermore prone to changing imaging settings (e.g. if applying the narrow-band imaging technique [9]).

This work is partly motivated by an approach for histological cancer subtyping [10] which relies on a patch-wise classification followed by an aggregation of patch-wise decisions. That means, for each image, the final representation consists of a histogram collecting the classified subtype occurrences to determine the final image-wise subtype. This approach, however, demands more than two classes, as in case of dichotomization (considered in this work), the aggregation would be similar to majority voting, which is not effective for CD diagnosis [8]. The utilization of multiple instance learning [11] is inhibited by the rather small number of available patients for training.

In this work, the focus is first on learning a discriminative classification model based on CNNs utilizing informative (manually extracted) patches only. The aim of this stage is to obtain a model exhibiting a high discriminative power in case that distinctive information is available in the image. Then, based on automatically extracted image data, we estimate the probability of a correct patch-wise classification by means of a probabilistic model. Making use of 'real-world' training data partly containing non-discriminative information, the target of this stage is to obtain a reliable confidence estimation. Consequently, for an image to be evaluated, we obtain a decision as well as a confidence level for several patch-positions. In order to determine an image-wise decision, patch-based probabilities of each image are aggregated and a further classifier is trained to determine the final class of an original image. The training data utilized for fitting the final classifier can be increased without any manual effort when obtaining novel ground-truth labeled data during routine endoscopy (i.e. no further manual selection of informative patches is required in contrast to [8]).

2 Proposed Pipeline

The proposed pipeline (Fig. 3) consists of training a discriminative model (1) based on manually extracted informative patch data (Sect. 2.1), followed by fitting a probabilistic model (2) relying on real-world data (Sect. 2.2). The probabilistic patch-wise outputs are aggregated (3), a final classification model is trained (4) and evaluation is performed per image (5) and finally also a patient-wise classification (6) is proposed (Sects. 2.3–2.4).

The required training data consists of a data set containing ground-truth labeled informative patches (IP) which were extracted by medical experts and show distinct markers for diagnosis, as well as a data set containing ground-truth labeled original endoscopic images showing typical degradations leading to partly invisible disease markers. From this data set, for each image, a large number of not necessarily informative patches (AP) are automatically extracted as predefined position in an overlapping manner. The required processing steps are explained in detail in the following subsections:

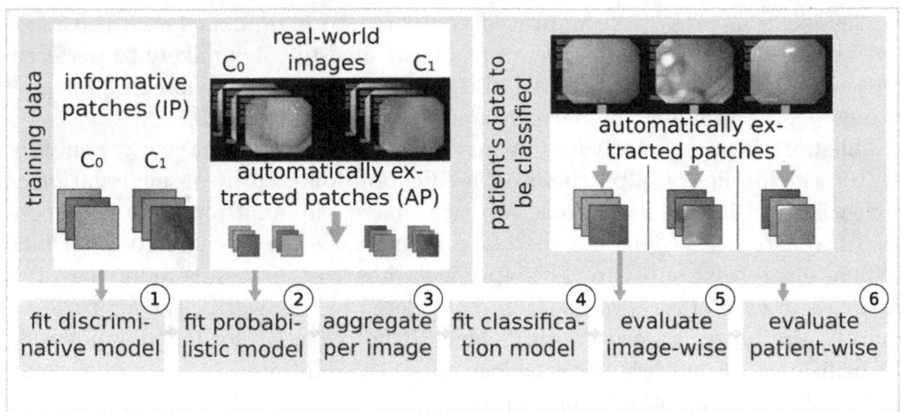

Fig. 3. Overview of the proposed classification pipeline.

2.1 Discriminative Patch-Based Model

First, the IP data set is utilized to train a discriminative classification model to distinguish between the two classes (CD, non-CD). To this end, we train a linear support vector machine (SVM) based on features extracted from CNNs, yielding exceptional performances in previous work [4]. The architectures as well as the training procedure of the utilized CNNs are described in detail in Sect. 2.5.

2.2 Probabilistic Patch-Based Model

The thereby obtained discriminative model is consequently applied to classify the AP data set. As the AP data set contains patches which distinctly differ from the IP data set (e.g. such as blur, low-contrast, bubbles and overexposure as shown in Fig. 2), the achieved patch-wise accuracies are expected to be significantly lower [8]. Additionally, an imbalance between specificity and sensitivity is expected due to higher similarities on average between unreliable data and class C_1 (CD) than between unreliable data and class C_0 (non-CD), [7].

In the next step, based on the AP data set, the probabilities of a correct classification are estimated. For this purpose, we reuse the discriminative model (CNN feature extraction followed by SVM classification) and apply it to patches of the AP data set resulting in classification outcomes (C_0 or C_1) for each patch as well as the distance to the decision boundary. In this space, providing the signed distances to the decision boundary, regression is performed by fitting a non-parametric model, specifically a Gaussian mixture model (GMM), to estimate the distribution of correctly and incorrectly classified samples for both classes. We estimate the distribution of correctly classified (d_c) and incorrectly classified (d_i) samples in order to determine the probability of a correct classification for a sample x by $p_c(x) = \frac{d_c(x)}{d_c(x)+d_i(x)}$. This is performed individually for samples classified as C_0 and C_1 obtaining $p_{c|C_0}(x)$ and $p_{c|C_1}(x)$.

The GMM is preferred to established methods such as Platt scaling and isotonic regression [12] due to the complete unawareness of the underlying distribution (p_c) which not necessarily shows a monotonic behavior. This is due to the difference in feature distribution between the IP and the AP data set and the similarity between samples of class C_1 and non-discriminative patches [8].

2.3 Image-Wise Classification

So far, we trained a discriminative as well as a probabilistic model based on informative (IP) and automatically extracted (AP) patch data. In order to obtain one final decision for an image to be evaluated, patches are automatically extracted. For each of the patches, the classification as well as the probabilistic outcome is estimated. Finally, all outputs for one image are merged by building a histogram based on the probabilistic output ($p_{c|C_0}$ and $p_{c|C_1}$, respectively) for both C_0 and C_1. These two histograms are concatenated and a further classification model is trained to distinguish between the classes on image-level. Training is again conducted based on the AP data set. As model, a k-nearest neighbor (kNN) classifier is utilized in combination with the histogram intersection as distance measure, which is common practice in histogram classification.

2.4 Patient-Wise Classification

The image-wise evaluation is performed by classifying a histogram obtained from patches of one endoscopic image. In order to obtain a patient-wise decision (i.e. a set of images are considered), the kNN output of all images of one patient is interpreted in a probabilistic way by considering the distribution of the nearest neighbors' class labels. Based on selecting the image with the highest confidence (i.e. the highest agreement of the nearest neighbors), one final class label is obtained. The advantages of this approach compared to simply collecting the outcomes of the patches of each image per patient into one histogram are given (a) by the higher number of training samples (the number of images is significantly larger than the number of patients) and (b) by the fact that images which could not be clearly assigned to a class (e.g. non-informative images) are not considered during evaluation.

2.5 Image Representations

For image representation, three CNN approaches are evaluated. Two of them exhibited excellent performances in previous work on the classification of manually selected patches [4] and the third one is a combination of the first two approaches:

Non-adapted CNN (NA-CNN) [13]: In the case of the first investigated representation, a convolutional neural network pretrained on the ImageNet challenge data[1], specifically the VGG-f network [13], is utilized. We chose this network

[1] http://www.image-net.org/challenges/LSVRC/.

because it provided the best outcomes for the classification of CD in [4]. The images are fed through the CNN and the activations of the last convolutional layer (4096 feature elements) are extracted as feature vectors.

Adapted CNN (A-CNN) [4]: The second representation (A-CNN) is based on the same network (VGG-f), however the (already pretrained) network is adapted to the classification of CD by training it on the IP data set. In previous work [4], this approach achieved excellent performances in classifying CD and outperformed the NA-CNN approach. Equally to the so called 'fully fine-tuned' VGG-f network in [4], the A-CNN network is trained for 5000 iterations utilizing stochastic gradient descent and the training images are randomly augmented (cropping, rotating and horizontally flipping). In fact, the only difference to the 'fully fine-tuned' VGG-f network in [4] is that the batches of images extracted for training consist of only 64 instead of 128 images due to the limited amount of available training samples. As for the NA-CNN approach, the activations of the last convolutional layer are extracted as feature vectors.

Combined CNN (C-CNN): The third descriptor C-CNN is a combination of NA-CNN and A-CNN. For this purpose, we concatenate the feature vectors of both approaches leading to an 8192 dimensional representation.

For further details as well as for evaluations of the utilized image representations, we refer to the original publication, where they were exhaustively assessed with respect the the classification of informative patches [4].

3 Experiments

3.1 Image Data Sets

The utilized material consists of images captured by physicians during routine gastrointestinal endoscopies at the St. Anna Children's hospital. Images where captured with Olympus endoscopes (GIF-Q165, GIF-N180) providing lossless-compressed (PNG) images with resolutions between 528 × 522 and 768 × 576 pixels.

For training and validating the proposed pipeline, an IP data set containing informative patches only (128 × 128 pixels) and an AP data set containing automatically extracted patches (72 patches (128 × 128) per image at fixed coordinates) is required. Due to the relatively small original endoscopic images, thereby an overlap between neighboring patches occurs. This, however, does not introduce any bias into the evaluation. Specifically, we extracted the patches in a fixed rectangular grid (nine coordinates in horizontal and eight in vertical direction) with an offset of 40 pixels. The grid is adjusted to capture real image information only and to exclude the outer parts providing meta information only (Fig. 2).

For the IP data set, an experienced physician manually selected between zero and four informative regions per endoscopic image exhibiting markers for an effective distinction between the two classes [7]. The IP data set is only utilized for training whereas the AP data set needs to be separated into a training and an evaluation set. To facilitate an unbiased evaluation, we take care that data of one patient (in the IP and the AP data set) is either used for training or for validation. We automatically generate four IP and corresponding AP data sets without overlaps in patient data (Table 1). To facilitate fusion on patient level (i.e. we obtain one decision per patient), the AP data sets contain four images for each patient. The discriminative model is trained on the IP data set after CNN feature extraction. For certainty estimation, the AP data set is randomly split into two equally sized, balanced sets (without overlaps of patient's data). One is utilized for certainty estimation and for providing the kNN's training samples and the other one for evaluation. This step is repeated for swapped training and evaluation sets. This policy is applied for all corresponding data sets (IP and AP) and the obtained mean classification rates (accuracy, sensitivity, specificity) as well as the standard deviations are finally reported.

We consider a binary classification between images showing healthy mucosa ('Marsh-0', in the following C_0) and CD ('Marsh-3', in the following C_1) [1,7]. The reason for working with this problem definition is given by the image data set available which contains rather few images for certain subclasses (e.g. 'Marsh-3C') when using the four-classes case ('Marsh-0' vs. 'Marsh-3A' vs. 'Marsh-3B' vs. 'Marsh-3C'). Furthermore, this two classes case is most relevant for clinical practice [7].

Table 1. Details on the data sets utilized for training and evaluation.

	IP_1	IP_2	IP_3	IP_4	AP_1	AP_2	AP_3	AP_4
# images class C_0	155	151	155	151	292×72	292×72	292×72	292×72
# images class C_1	157	149	154	152	92×72	92×72	92×72	92×72
# images overall	312	300	309	303	384×72	384×72	384×72	384×72
# patients class C_0	73	73	73	73	73	73	73	73
# patients class C_1	23	23	23	23	23	23	23	23
# patients overall	96	96	96	96	96	96	96	96

3.2 Evaluation and Implementation Details

As discriminative model to classify the extracted CNN features, a C-SVM (LIBLINEAR [14]) is applied exhibiting excellent performances [4]. The SVM cost factor (C) is evaluated utilizing inner cross validation (between 2^0 and 2^{10}) on the IP training data. The k value for kNN regression is also evaluated during inner cross-validation ($1 \leq k \leq 51$, step size 5). For the GMM, 64 components are specified to precisely estimate the distributions. Due to the low dimensionality, the large number of samples and the rather smooth distribution, this number is not critical and is thereby not optimized. The GMM was initialized randomly and optimized by means of expectation maximization. For the histograms building the final image representation, a linear binning with eight bins is utilized to obtain a trade-off between precision and robustness. For training the A-CNN, MatConvNet [15] is utilized.

3.3 Results

Figure 4 shows the classification results achieved with the proposed approach compared to the state-of-the-art quality-metric-based approach [8]. For image-wise classification, overall classification rates (accuracies) between 0.741 and 0.772 are obtained. C-CNN performs best on average, the proposed method exhibits a similar performance compared to the state-of-the-art. Regarding patient-based diagnosis, our proposed approach in combination with the C-CNN image representation exhibits the best accuracies (0.848).

Classifying on patch-level without any fusion (not shown in the table), we obtain accuracies between 0.60 and 0.65, which is clearly lower than the rates obtained for informative patches reported in [4] (accuracies above 0.9). Even with majority voting (similar to [10] in the two-classes case), accuracies are always below 0.7 for image-wise and patient-wise classification.

Figure 5 shows the rates obtained with selective classification, i.e. we perform image-wise classification with the restriction that only certain images are selected according to the confidence measure of the kNN classifier. This experiment is performed to facilitate a prediction if more (and more diverse) images would be available per patient. In these outcomes, a confidence > 0.7 indicates that at least 70% of the nearest neighbors must belong to one particular class so that a certain image is selected. By selecting images with a high confidence only, the accuracy increased up to 0.903 (NA-CNN) without obtaining a severe imbalance between sensitivity and specificity. Only the A-CNN feature suffers from a decreased sensitivity for high confidences and hence does not show comparable outcomes.

The precise figures as well as sensitivities and specificities of all experiments are provided in Table 2. For the experiment with selective classification, the fraction of selected samples (images) is provided in brackets.

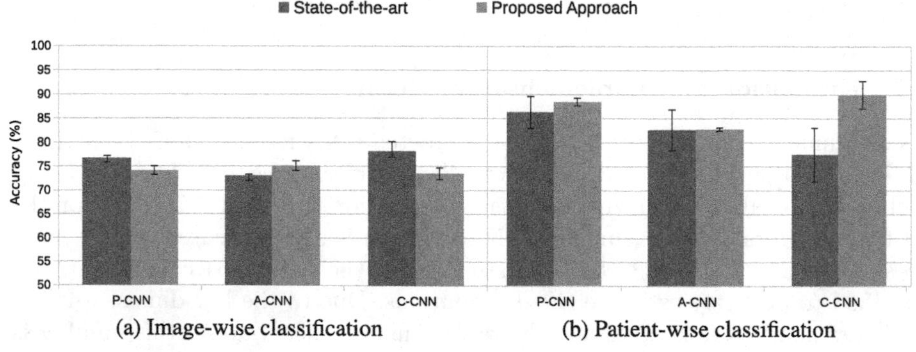

Fig. 4. Experimental results for image-wise (a) as well as patient-wise (b) classification.

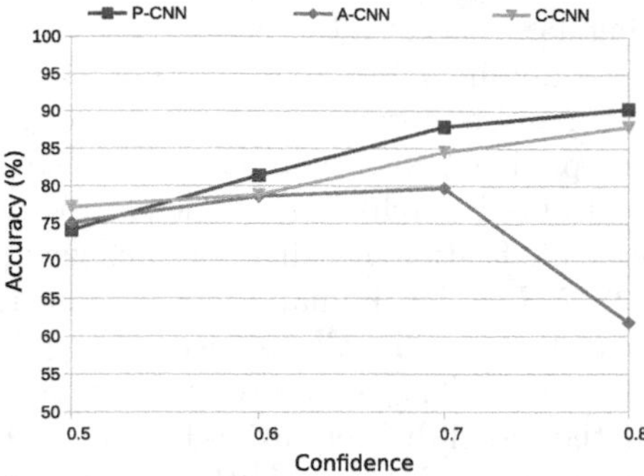

Fig. 5. Experimental results for image-wise classification with varying confidence levels. The confidence level here corresponds to the distribution of class labels of the kNN's selected neighbors. The left column (confidence 0.5) corresponds to the typical setting (Fig. 2 (a)), where each image is classified.

Table 2. Classification results achieved with the proposed and the metric-based reference approach [8] for image- and patient-wise classification. Additionally to the sensitivity (sens), the specificity (spec) and the accuracy, we provide the accuracies' standard deviations for classification without a minimum confidence (upper part). In case of introduced confidences (image-wise classification), we provide the fraction of the selected images in brackets.

Approach	NA-CNN			A-CNN			C-CNN		
	Sens	Spec	Accuracy	Sens	Spec	Accuracy	Sens	Spec	Accuracy
Image-wise [8]	.770	.764	.767 ± .005	.754	.707	.730 ± .003	.836	.726	**.781** ± .020
Patient-wise [8]	.831	.801	**.816** ± .029	.765	.803	.784 ± .037	.627	.851	.739 ± .049
Proposed image-wise	.788	.695	.741 ± .009	.714	.787	.751 ± .010	.809	.735	**.772** ± .012
Proposed patient-wise	.858	.811	.835 ± .007	.728	.842	.785 ± .003	.837	.859	**.848** ± .025
Selective classification	NA-CNN			A-CNN			C-CNN		
	Sens	Spec	Accuracy	Sens	Spec	Accuracy	Sens	Spec	Accuracy
Proposed (conf. > 0.6)	.855	.774	**.814** (75%)	.753	.819	.786 (83%)	.818	.759	.788 (90%)
Proposed (conf. > 0.7)	.905	.854	**.879** (55%)	.736	.857	.797 (62%)	.866	.823	.845 (72%)
Proposed (conf. > 0.8)	.876	.929	**.903** (34%)	.295	.941	.618 (19%)	.882	.876	.879 (50%)

3.4 Discussion and Conclusion

We proposed a method to obtain decisions from computer aided diagnosis systems without any interaction based on original endoscopic image data. A previous method [8], representing the state-of-the-art relying on a linear combination of basic quality measures, was outperformed for patient-based classification for each image representation. Furthermore, the variability in performance of our proposed approach is lower. Another advantage of the proposed method compared to the quality-metric-based approach is that no handcrafted metrics are required for assessing the quality. Consequently, no conceptual changes need to be performed in the case of changing the imaging conditions. Furthermore, an augmentation of training data, specifically the AP data set, can be performed by adding endoscopic images obtained during endoscopies (i.e. no manual selection of informative patches is required for enlarging the training data set as in [8]).

Interestingly, the adapted A-CNN, exhibiting the best accuracies in patch-wise classification for idealized patch data [4], leads to inferior outcomes for image- and patient-wise classification utilizing non-idealized patch data. We suppose that by training a CNN to discriminate between informative patches with clearly visible mucosal structures, the CNN-features lose distinctive power for rating non-informative patches. The structure of the mucosal villi is the most important criterion for the visual differentiation between a healthy mucosa and CD. The mucosal inflammation in CD causes either a mild villi atrophy, a marked villi atrophy or an entirely absence of villous structure, depending on the severity of the disease. Also, in non-informative patches, villous structures are often not visible due to image degradations (e.g. out of focus, low contrast, etc.) and so the CNN probably misinterprets those non-informative patches without visible villi to be affected by CD. As a consequence, the network output cannot be utilized effectively to estimate probabilities for correct classifications. Therefore, training CNNs utilizing informative CD image data actually turned out to be a disadvantage for the classification of non-informative images. However, this disadvantage can be compensated by combining adapted CNNs with non-adapted CNNs (C-CNN), which provided the best accuracies for image- and patient-wise diagnosis. In case of a selective classification, we notice that NA-CNN outperforms C-CNN which is probably caused by the increasing imbalance between sensitivity and specificity for the A-CNN at higher rates of confidence.

In this study, we put emphasis on including highly realistic endoscopic images and not only rather idealized data into the image data sets for evaluation. For this reason, the proposed approach could be implemented into a clinical system without any significant changes. Considering the obtained classification accuracies, a point of criticism could be that the obtained accuracies are still far away from 100%. However, we identified several scenarios which can lead to distinctly higher accuracies in the clinical routine: Novel endoscopic devices (exhibiting higher resolutions and new modalities) could potentially be applied in order to improve the classification performance even further. Furthermore, the data for evaluation exhibits a high degree of correlation between a patient's images. We had to include patient's image data although the images show similar regions of the mucosa in order to keep the amount of data relatively high. Based on the outcomes with increased confidence levels (Table 2), it can be assumed that by utilizing uncorrelated image material the rates could be improved significantly. Finally, even video material (endoscopic video frames) could be utilized additionally to the images taken by the physicians to increase data diversity.

To conclude, we proposed an effective approach for fully-automated CNN-based classification of endoscopic images for CD diagnosis. Notably, the best performances are not obtained with the adapted CNN, but with a network trained for the image-net challenge as well as with combinations of the two evaluated networks. Fusing data on patient-level, an accuracy of 0.85 can be obtained and the state-of-the-art is thereby clearly outperformed. Experiments provide evidence that the rates could be increased further if more data per patient would be available, creating incentive for incorporating video material in future work.

Acknowledgments. This work was supported by the German Research Foundation (DFG) under grant no. ME3737/3-1.

References

1. Oberhuber, G., Granditsch, G., Vogelsang, H.: The histopathology of coeliac disease: time for a standardized report scheme for pathologists. Eur. J. Gastroenterol. Hepatol. **11**, 1185–1194 (1999)
2. Quantitative image analysis of celiac disease. World J. Gastroenterol.**21**(9) (2015)
3. Kwitt, R., Hegenbart, S., Rasiwasia, N., Vécsei, A., Uhl, A.: Do we need annotation experts? A case study in celiac disease classification. In: Golland, P., Hata, N., Barillot, C., Hornegger, J., Howe, R. (eds.) MICCAI 2014. LNCS, vol. 8674, pp. 454–461. Springer, Cham (2014). doi:10.1007/978-3-319-10470-6_57
4. Wimmer, G., Vécsei, A., Uhl, A.: CNN transfer learning for the automated diagnosis of celiac disease. In: Proceedings of the International Conference on Image Processing Theory, Tools and Applications (IPTA 2016) (2016)
5. Gadermayr, M., Liedlgruber, M., Uhl, A., Vécsei, A.: Problems in distortion corrected texture classification and the impact of scale and interpolation. In: Petrosino, A. (ed.) ICIAP 2013. LNCS, vol. 8156, pp. 513–522. Springer, Heidelberg (2013). doi:10.1007/978-3-642-41181-6_52
6. Wimmer, G., Gadermayr, M., Kwitt, R., Häfner, M., Merhof, D., Uhl, A.: Evaluation of i-scan virtual chromoendoscopy and traditional chromoendoscopy for the automated diagnosis of colonic polyps. In: Peters, T., Yang, G.-Z., Navab, N., Mori, K., Luo, X., Reichl, T., McLeod, J. (eds.) CARE 2016. LNCS, vol. 10170, pp. 59–71. Springer, Cham (2017). doi:10.1007/978-3-319-54057-3_6
7. Hegenbart, S., Uhl, A., Vécsei, A., Wimmer, G.: Scale invariant texture descriptors for classifying celiac disease. Med. Image Anal. (MIA) **17**(4), 458–474 (2013)
8. Gadermayr, M., Uhl, A., Vécsei, A.: Quality based information fusion in fully automatized celiac disease diagnosis. In: Jiang, X., Hornegger, J., Koch, R. (eds.) GCPR 2014. LNCS, vol. 8753, pp. 666–677. Springer, Cham (2014). doi:10.1007/978-3-319-11752-2_55
9. Gadermayr, M., Hegenbart, S., Kwitt, R., Uhl, A., Vécsei, A.: Narrow band imaging versus white-light: what is best for computer-assisted diagnosis of celiac disease? In: Proceedings of the IEEE International Symposium on Biomedical Imaging (ISBI 2015) (2016)
10. Hou, L., Samaras, D., Kurc, T.M., Gao, Y., Davis, J.E., Saltz, J.H.: Patch-based convolutional neural network for whole slide tissue image classification. In: Proceedings of the International Conference of Computer Vision (CVPR 2016) (2016)
11. Xu, Y., Zhu, J.Y., Chang, E.I.C., Lai, M., Tu, Z.: Weakly supervised histopathology cancer image segmentation and classification. Med. Image Anal. **18**(3), 591–604 (2014)
12. Niculescu-Mizil, A., Caruana, R.: Predicting good probabilities with supervised learning. In: Proceedings of the International Conference on Machine learning (ICML 2015)

13. Chatfield, K., Simonyan, K., Vedaldi, A., Zisserman, A.: Return of the devil in the details: delving deep into convolutional nets. In: Proceedings of the British Machine Vision Conference (BMVC 2014) (2014)
14. Fan, R.E., Chang, K.W., Hsieh, C.J., Wang, X.R., Lin, C.J.: LIBLINEAR: a library for large linear classification. J. Mach. Learn. Res. **9**, 1871–1874 (2008)
15. Vedaldi, A., Lenc, K.: MatConvNet - convolutional neural networks for matlab. In: Proceeding of the ACM International Conference on Multimedia (MM 2015) (2015)

An Investigation of Deep Learning for Lesions Malignancy Classification in Breast DCE-MRI

Stefano Marrone[1], Gabriele Piantadosi[1], Roberta Fusco[2], Antonella Petrillo[2], Mario Sansone[1], and Carlo Sansone[1(✉)]

[1] DIETI - University of Naples Federico II, Naples, Italy
{stefano.marrone,gabriele.piantadosi,mario.sansone, carlo.sansone}@unina.it

[2] Department of Diagnostic Imaging, National Cancer Institute of Naples 'Pascale Foundation', Naples, Italy
roberta.fusco@unina.it, antonellapetrillo2@gmail.com

Abstract. Dynamic Contrast Enhanced-Magnetic Resonance Imaging (DCE-MRI) is gaining popularity as a complementary diagnostic method for early detection and diagnosis of breast cancer. However, due to the large amount of data, DCE-MRI can hardly be inspected without the use of a Computer Aided Diagnosis (CAD) system. Among the major issues in developing CAD for breast DCE-MRI there is the classification of regions of interest according to their aggressiveness. For this task newer hand-crafted features are continuously proposed by domain experts. On the other hand, deep learning approaches have gained popularity in many pattern recognition tasks, being able to outperform classical machine learning techniques in different fields, by learning compact hierarchical representations of an image which well fit the specific task to solve. The aim of this work is to explore the applicability of Convolutional Neural Networks (CNN) in automatic lesion malignancy assessment for breast DCE-MRI data. Our findings show that while promising results in treating DCE-MRI can be obtained by using transfer learning, CNNs have to be carefully designed and tuned in order to outperform approaches specifically designed to exploit all the available data information.

Keywords: Deep convolutional neural network · DCE-MRI · Breast · Cancer

1 Introduction

Breast cancer represents about 12% of all tumour new cases and about the 25% of all cancers in women. With these numbers, it is the most common women tumour worldwide and the second overall, with almost 2 million new diagnosed cases/year [5]. The key factor to improve the breast neoplasm prognosis is early detection, especially for cancer (malignant tumours). The world health organisation indicates the x-ray mammography as the standard diagnostic tool [26],

S. Battiato et al. (Eds.): ICIAP 2017, Part II, LNCS 10485, pp. 479–489, 2017.
https://doi.org/10.1007/978-3-319-68548-9_44

for its high resolution and low operational costs. However, among its main disadvantages, there is the use of ionising radiations (x-rays) and its low specificity, especially for radiographically dense breast tissue (as in under-forty women) or when the patients have scars or breast implants.

In recent years, Dynamic Contrast Enhanced-Magnetic Resonance Imaging (DCE-MRI) has demonstrated a great potential in screening different tumours tissues, gaining an increasing popularity as an important complementary diagnostic methodology for early detection of breast cancer [12]. DCE-MRI advantages include its ability to acquire 3D high resolution dynamic (functional) information, not available with conventional RX imaging [23] and its limited invasiveness, since it does not make use of any ionising radiations or radiocontrast. It has been successfully used for under-forty women and for high-risk patients [4], both for assessing therapy effects and for staging newly diagnosed breast cancer [17].

DCE-MRI consists of 4-dimensional data, obtained by combining different 3D volumes acquired before (pre) and after (post) the intravenous injection of a paramagnetic contrast agent (usually Gadolinium-based), as depicted in Fig. 1a. Each voxel is associated with a Time Intensity Curve (TIC) representative of the temporal dynamics of the acquired signal (see Fig. 1b) that reflects the absorption and the release of the contrast agent, following the vascularisation characteristics of the tissue under analysis [22].

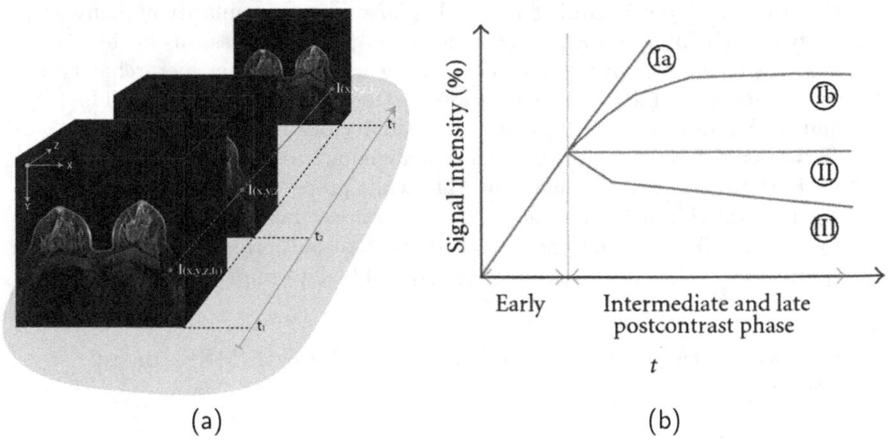

Fig. 1. DCE-MRI and time intensity curves. (a) A representation of the four dimensions (3 spatial + 1 temporal) of a typical breast DCE-MRI scan; (b) some examples of time intensity curves.

While the use of DCE-MRI has proved to improve breast cancer diagnosis [14], it is a very time-consuming and error-prone task that involves analysis of a huge amount of data [16]. A visual assessment of the lesion malignity could be performed referring to the Fig. 1b. *Type I* corresponds to a straight (Ia) or curved

(Ib) line where the contrast absorption continues over the entire dynamic study (typical of healthy tissues or benign neoplasms); *Type II* represents a plateau curve with a sharp bend after the initial upstroke (typical of probably malignant lesions); finally, *Type III* shows a washout time course (typical of malignant lesions). It follows that radiologists can hardly inspect DCE-MRI data without the use of a Computer Aided Detection/Diagnosis (CAD) system designed to reduce such amount of data, allowing them to focus attention only on regions of interest.

In particular, in this paper we focus on a CAD system for the automatic lesion diagnosis in order to better assist the physician decision by reducing the inter/intra-operator variability.

When lesion diagnosis is performed by means of classifier systems [8], many features have been proposed so far, roughly grouped in Clinical [9] (as age, parental and history), Dynamic [7,9] (directly extracted from the TIC), Textural [6] (as variance, kurtosis and skewness), Pharmacokinetic [15] (extracted by means of mathematical models of contrast agent and tissue), Spatio-temporal [25] (as DFT coefficient map, margin and radial gradients) and Morphological [7] (as lesion eccentricity, compactness and perimeter).

While newer hand-crafted features are continuously proposed by domain experts, in the last years deep learning approaches have gained popularity in many pattern recognition tasks, being able to outperform classical machine learning techniques in different fields [10]. Among these approaches, we can cite Convolutional Neural Networks (CNN) that are composed of different convolutional layers stacked in a deep architecture (as in Fig. 2) meant to automatically learn the best data representation as composed by simpler concepts. They usually perform better than classifiers trained on hand-crafted features because are able to learn a compact hierarchical representation of an image which well fits the specific task to solve.

Deep approaches have been used in brain DCE-MRI, both for lesion and anatomical segmentation [18,24], in prostate tissues analysis, by using deep auto-encoders for tumours grading and diagnosis [13,20], while breast DCE-MRI cancer lesion detection was never faced.

In a recent publication, Antropova et al. [2] first proposed to apply Deep Learning for the breast cancer lesion diagnosis task, by feeding a pre-trained CNN with the ROI extracted from DCE-MRI slices containing a lesion. They adopted the AlexNet architecture pre-trained on the ImageNet dataset [11] as feature extractor, by using a Support Vector Machine (SVM) for the malignant/benignant classification task. To the best of our knowledge, it is the first time that a Deep Learning approach was used in breast DCE-MRI analysis. This notwithstanding, the results reported in [2] refer to a single training modality (the CNN used as feature extractor); moreover, they are given by using cross-validation (CV) for performance evaluation, which can provide in this case an overestimation of the actual performance. Finally, no comparisons with other approaches on the same data are reported.

The aim of this work is then to assess CNN capability for automatic lesion classification in breast DCE-MRI, in order to have broader results on its applicability and effectiveness for a very specific task such as the considered one. In addition to the training modality proposed in [2], we also explored the fine-tuning of a pre-trained AlexNet and the complete training from scratch of the same net. All the results have been also compared with those achieved by other proposals, by using a Leave-One-Patient-Out CV evaluation in order to ensure fair and more reliable findings.

This rest of the paper is organized as follows: Sect. 2 gives some information on the proposed methodology and on the literature proposals used to validate the approach considered in this study and presents the dataset used for comparing the results. In Sect. 3, we present the results obtained by the different CNN-based approaches, together with those achieved by other methods under comparison. Finally, in Sect. 4, we discuss these results, by providing some conclusions.

2 Materials and Methods

2.1 Dataset

Patients. The dataset is constituted of 42 women breast DCE-MRI 4D data, (average age 40 years, in range 16–69) with benign or malignant lesions histopathologically proven: 42 regions of interest (ROIs) were malignant and 25 were benign for a total of 67 ROIs.

Data Acquisition. All patients underwent imaging with a 1.5 T scanner (Magnetom Symphony, Siemens Medical System, Erlangen, Germany) equipped with breast coil. DCE T1-weighted FLASH 3D coronal images were acquired (TR/TE: 9.8/4.76 ms; flip angle: 25°; field of view 370 × 185 mm × mm; matrix: 256 × 128; thickness: 2 mm; gap: 0; acquisition time: 56 s; 80 slices spanning entire breast volume). One series (t_0) was acquired before and 9 series (t_1-t_9) after intravenous injection of 0.1 mmol/kg of a positive paramagnetic contrast agent (gadolinium-diethylene-triamine penta-acetic acid, Gd-DOTA, Dotarem, Guerbet, Roissy CdG Cedex, France). An automatic injection system was used (Spectris Solaris EP MR, MEDRAD, Inc., Indianola, PA) and injection flow rate was set to 2 ml/s followed by a flush of 10 ml saline solution at the same rate.

Gold Standard. An experienced radiologist (A.P.) delineated suspect ROIs using T1-weighted and *subtractive* image series. Starting from DCE-MRI acquired data, the subtractive image series is defined by subtracting t_0 series from t_4 series. In subtractive images, any tissue that does not absorb contrast agent is suppressed. Manual segmentation stage was performed in OsiriX [21], that allows the user to define ROIs at a sub-pixel level. Per each ROI the lesion was histopathologically proven. The evidence of malignity was used as Gold Standard (GS) for the ROI Classification problem.

2.2 Lesion Diagnosis in DCE-MRI

The breast lesion diagnosis issue has been mainly designed in the pattern recognition framework. The most recent literature proposals mostly differ in the feature vector used to describe the classification subject. In the following we will briefly review those we considered for comparison; then, in the next sub-section the CNN-based approach is described. The choice of the comparing approaches is motivated by the attempt of covering the most part of the feature taxonomy presented in the introduction.

In [9] Glaßer et al. proposed to use a Decision Tree trained on Clinical and Dynamic features in order to consider both the patient high-level information and contrast agent perfusion parameters derived from the signal temporal dynamic.

Fusco et al. [7] suggested to use both Dynamic and Morphological features, combining them by using a Multiple Classifier System, in order to take into account the contrast agent concentration and the lesion shape.

In [19], trying to exploit both spatial and spatio-temporal information, a Random Forest classifier (made up of 10 Random Trees each one using a random subset of features with no limitation on its maximum depth) was trained on the spatio-temporal version of Local Binary Pattern in Three Orthogonal Planes (LBP-TOP [1]).

2.3 CNN for Lesion Diagnosis

As stated in the introduction, Antropova et al. [2] investigated deep learning for the lesion diagnosis task using DCE-MRI data. They proposed to apply transfer learning from a pre-trained CNN (see Fig. 2).

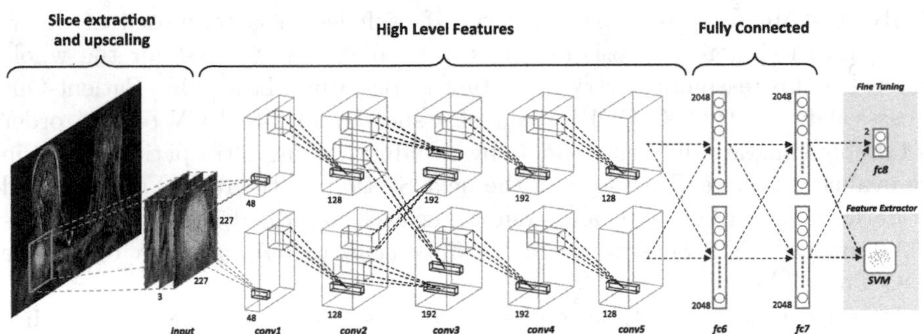

Fig. 2. The considered convolutional neural network.

Their approach can be summarised as follows:

1. Take only the second post-contrast series from the 4D DCE-MRI data.
2. Depending on the size of the tumour, a tile around the lesion is extracted from each lesion slice. The tile size varies between 1 and 1.5 times the maximum diameter of the observed lesion.

3. Images are up-sampled to yield 256×256 pixel ROIs.
4. AlexNet [11] pre-trained on the ImageNet [3] database is used to extract a feature vector from the last internal convolutional layer (*fc7* in Fig. 2). 4096 features per each slice are extracted.
5. Performance are evaluated using a 10-fold cross-validation, by considering an SVM as a classifier (no information about kernel or others hyper-parameters were provided).

To better investigate the capability of the Antropova et al. [2] proposal, we also explored two other training modalities, i.e., fine-tuning and the training from scratch of AlexNet.

Fine-tuning of a pre-trained CNN consists in replacing the last trained fully connected layers with the same untrained ones. Retraining the so modified network will softly adapt the weights of the pre-trained layers to the new task and will strongly train the new fully connected layer according to the new findings. Fine-tuning allows us to use a reduced number of training images and to achieve a stable result in fewer epochs. We chose the best number of epochs needed to avoid overfitting by considering the loss function values during the training phase.

In order to perform training from scratch, we deployed exactly the same network architecture proposed by AlexNet in [11], so performing a totally new supervised training. This approach is, usually, higher demanding and needs a greater amount of images to achieve a valuable solution.

2.4 Performance Evaluation

To obtain a fair performance evaluation, k-fold cross-validation (CV) is commonly used. In our case, however, even if each lesion is composed of different slices, the lesion diagnosis task has to predict a single class for the whole lesion. For this reason, it is very important to perform a Leave-One-Patient-Out Cross-Validation (LOPO-CV) instead of a slice-based k-fold CV one, in order to reliably compare different models by avoiding mixing intra-patient slices in the evaluation phase. Therefore, in the next Section all the results are reported by performing a LOPO-CV and comparing each described approach in terms of Accuracy (ACC), Sensitivity (SEN), Specificity (SPE) and Area Under the ROC Curve (AUC).

It is worth noting that, since classification is always performed at the slice level, a combining strategy has to be applied in order to provide a unique class label for each lesion. Among the possible combining strategies of the results provided by the classifier at slice level, we chose to investigate the following ones:

Majority Voting: The class of the lesion corresponds to the most voted class over all the slices.

Weighted Majority Voting: As for the majority voting, but each slice contribution is weighted by its class probability provided by the classifier.

Weighted Majority Voting by Slice Area: As in the previous case, but each slice contribution is proportionally weighted by using its area.

Naïve Bayes: Predicted classes of the slices are combined according to the Naïve Bayes approach.

Max Prob. Value per Slice: The class of the lesion corresponds to the class of the slice with the highest class probability values as provided by the classifier.

Biggest Slice: The class of the lesion is the same of the biggest slice.

3 Experimental Results

This Section shows the results obtained by applying deep approaches for lesion diagnosis. All the results are here presented without any remark. A discussion is reported in the next Section.

Table 1 compares all the CNN-based approaches so far presented, by performing the training to the best of our capability in order to achieve a fair comparison. Antropova et al. [2], in fact, propose to use AlexNet as a feature extractor, but do not provide enough information about the SVM hyper-parameters settings. So, we performed an optimization of the classification stage: The best results were obtained by using a SVM with a polynomial kernel of degree equal to 3 and C=1.

The fine-tuning of AlexNet has been performed as discussed in Sect. 2; in this case, the best results were achieved with 75 epochs and a learning rate of 10^{-5}.

Table 1. Comparing different AlexNet training modalities, by varying the slice combining strategy. Average values obtained in Leave-One-Patient-Out CV over 42 patients are reported.

Training modality	Combining strategy	ACC [%]	SEN [%]	SPE [%]	AUC [%]
Feature Extraction	Majority voting	76.19	78.26	73.68	76.43
	Weighted majority voting	73.81	78.26	68.42	75.97
	WMV by slice area	76.19	78.26	73.68	76.20
	Naïve Bayes	76.19	82.61	68.42	72.77
	Max prob. value per slice	71.43	78.26	63.16	74.83
	Biggest slice	76.19	73.91	78.95	76.43
Fine tuning	Majority voting	71.43	82.61	57.89	72.65
	Weighted majority voting	69.05	82.61	52.63	73.23
	WMV by slice area	69.05	82.61	52.63	71.40
	Naive Bayes	66.67	82.61	47.37	71.17
	Max prob. value per slice	69.05	82.61	52.63	73.00
	Biggest slice	73.81	86.96	57.89	72.43

Finally, we also investigate whether training by scratch of AlexNet could improve the lesion diagnosis with respect to the previously described approaches. In this case, we found that the learning rate strongly influences the learning

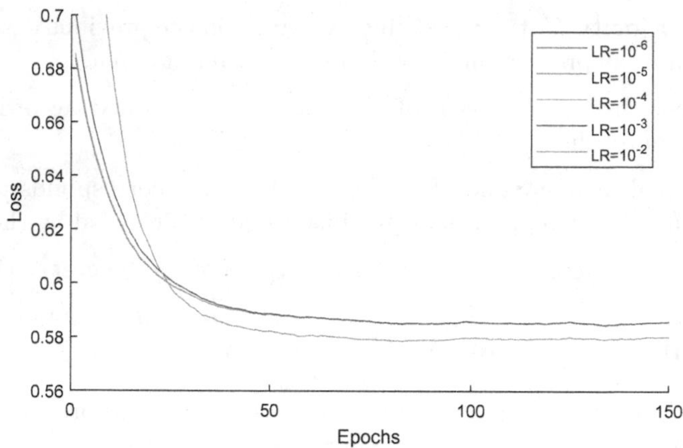

Fig. 3. Training from scratch loss function values for different learning rates (LR), by varying the number of epochs.

curve. Looking at the Loss function evaluation per each epoch (Fig. 3), we can see that the network suffers from under-fitting. The loss function settles to a certain value and, even if further training steps are performed, the net does not learn more.

For the training from scratch approach in the considered case, a learning rate of 10^{-2} and 100 epochs are enough to reach the best working point of the net. Unfortunately, very poor results can be achieved, with an accuracy equal to 54.76% and an AUC value of 68.55%.

Table 2 compares the best results obtained by a deep approach with those obtained by applying the methods proposed in [7,9,19].

Table 2. Comparison of the best results obtained by a CNN-based lesion diagnosis approach with those achieved by other state-of-the-art approaches. Average values obtained in Leave-One-Patient-Out CV over 42 patients are reported.

Methodology	ACC [%]	SEN [%]	SPE [%]	AUC [%]
LBP-TOP (Piantadosi et al. [19])	83.33	95.14	66.67	88.41
Best CNN	76.19	73.91	78.95	76.43
Dyn. & Morph. + MCS (Fusco et al. [7])	69.05	78.26	57.89	68.08
Decision trees (Glaßer et al. [9])	64.29	95.65	26.32	60.98

Finally, in order to closely replicate the results reported in [2], we also performed a slice based 10-fold CV of the SVM (using the same parameters that gave us the results reported in the previous tables) fed with the 4096 features extracted by using AlexNet, obtaining 91.75% and 96.90% in terms of Accuracy and AUC, respectively.

All the results have been obtained by using the MatLab Neural Network Toolbox from Mathworks, over our University's computing infrastructure (SCoPE - www.scope.unina.it) where three DELL R720 are equipped with two Intel(R) Xeon(R) CPU E5-2680 v2 @ 2.80GHz, 128GB RAM and a cluster of five NVIDIA Tesla K20m GPU is available.

4 Discussion and Conclusions

The aim of this paper was to investigate automatic lesion malignancy classification in breast DCE-MRI by means of Convolutional Neural Networks (CNN), analyzing a literature proposal and comparing it with different training modality and combining strategies. The evaluations were performed on ROIs manually segmented by an experienced radiologist (A.P.) for all the patients in our database. Obtained results were compared with previous findings in the literature, by using a Leave-One-Patient-Out cross-validation (LOPO-CV) in order to ensure a fair comparison and more reliable findings.

Antropova et al. [2] presented the first use of deep learning for the lesion classification task in DCE-MRI data. They propose to apply transfer learning from a pre-trained CNN. Results presented in table 1 show that a CNN pre-trained on natural images (ImageNet dataset) used as a feature extractor performs better than considering a fine-tuning modality, in terms of Accuracy, Sensitivity, Specificity and AUC. The same table also shows that the most effective slice combining technique is to consider as lesion class the one predicted by the slice containing the biggest ROI. This is reasonable, since the biggest ROI in a lesion is likely to bring the majority of the lesion malignancy information. Reported results confirm that the training from scratch approach is not feasible with the reduced number of biomedical images usually available.

Table 2 compares the best CNN-based approach with other approaches presented so far in the literature, showing that, even if deep learning can outperform two of them [7,9], it cannot overcome the method based on the LBP-TOP descriptor [19]. This result seems to suggest that while CNNs show promising results in treating biomedical images, they have to be carefully designed and tuned in order to outperform approaches specifically designed to suitably exploit data information for the specific task. It is worth recalling, in fact, that the presented CNN, differently from the LBP-TOP descriptor, do not use neither dynamic nor spatio-temporal information. Moreover, training times of our best CNN-based solution are about two orders of magnitude graeater with respect to those needed by the other approaches.

Our results also confirm that it is important to compare all the approaches on a patient base, in order to obtain a more reliable outcome. The results obtained by using a 10-fold CV are in fact significantly higher when compared with those obtained by using a LOPO-CV. The former are however intrinsically unfair because of exploiting inter-patient knowledge, so biasing the cross-validation results.

As a final remark, we would like to highlight that a problem concerning the use of a CNN is that there is no clear physiological interpretation of the classifier operational mode, thus it can be very difficult to motivate results from the physician point of view. Another limit of this study is the population size; our findings should be then confirmed on a larger dataset. Future works will focus on these aspects and on the study of net design choices that can be able to suitably exploit dynamic or spatio-temporal information coming from DCE-MRI data.

Acknowledgments. The authors gratefully acknowledge the availability of the Calculation Centre SCoPE of the University of Naples Federico II and thank the SCoPE academic staff for the given support.

References

1. Ahonen, T., Hadid, A., Pietikäinen, M.: Face description with local binary patterns: application to face recognition. IEEE Trans. Pattern Anal. Mach. Intell. **28**, 2037–2041 (2006)
2. Antropova, N., Huynh, B., Giger, M.: SU-D-207B-06: predicting breast cancer malignancy on DCE-MRI data using pre-trained convolutional neural networks. Med. Phys. **43**(6), 3349–3350 (2016)
3. Deng, J., Dong, W., Socher, R., Li, L.J., Li, K., Fei-Fei, L.: ImageNet: a large-scale hierarchical image database. In: IEEE Conference on Computer Vision and Pattern Recognition, CVPR 2009, pp. 248–255. IEEE (2009)
4. El-Kwae, E.A., Fishman, J.E., Bianchi, M.J., Pattany, P.M., Kabuka, M.R.: Detection of suspected malignant patterns in three-dimensional magnetic resonance breast images. J. Digit. Imaging: Official J. Soc. Comput. Appl. Radiol. **11**, 83–93 (1998)
5. Ferlay, J., Héry, C., Autier, P., Sankaranarayanan, R.: Global burden of breast cancer. In: Li, C. (ed.) Breast Cancer Epidemiology, pp. 1–19. Springer, Heidelberg (2010)
6. Fusco, R., Sansone, M.: Segmentation and classification of breast lesions using dynamic and textural features in dynamic contrast enhanced-magnetic resonance imaging. In: 25th International Symposium on Computer-Based Medical Systems (CBMS), pp. 1–4. IEEE (2012)
7. Fusco, R., Sansone, M., Petrillo, A., Sansone, C.: A multiple classifier system for classification of breast lesions using dynamic and morphological features in DCE-MRI. In: Gimel'farb, G., Hancock, E., Imiya, A., Kuijper, A., Kudo, M., Omachi, S., Windeatt, T., Yamada, K. (eds.) SSPR /SPR 2012. LNCS, vol. 7626, pp. 684–692. Springer, Heidelberg (2012). doi:10.1007/978-3-642-34166-3_75
8. Fusco, R., Sansone, M., Filice, S., Carone, G., Amato, D.M., Sansone, C., Petrillo, A.: Pattern recognition approaches for breast cancer DCE-MRI classification: a systematic review. J. of Med. Biol. Eng. **36**, 449–459 (2016)
9. Glaßer, S., Niemann, U., Preim, B., Spiliopoulou, M.: Can we distinguish between benign and malignant breast tumors in DCE-MRI by studying a tumor's most suspect region only? In: Proceedings of CBMS 2013–26th IEEE International Symposium on Computer-Based Medical Systems, pp. 77–82. IEEE (2013)
10. He, K., Zhang, X., Ren, S., Sun, J.: Deep residual learning for image recognition. In: Proceedings of the IEEE Conference on Computer Vision and Pattern Recognition, pp. 770–778 (2016)

11. Krizhevsky, A., Sutskever, I., Hinton, G.E.: ImageNet classification with deep convolutional neural networks. In: Advances in Neural Information Processing Systems, pp. 1097–1105 (2012)
12. Lehman, C.D., Gatsonis, C., Kuhl, C.K., Hendrick, R.E., Pisano, E.D., Hanna, L., Peacock, S., Smazal, S.F., Maki, D.D., Julian, T.B., DePeri, E.R., Bluemke, D.A., Schnall, M.D.: MRI evaluation of the contralateral breast in women with recently diagnosed breast cancer. Technical report (2007)
13. Liao, S., Gao, Y., Oto, A., Shen, D.: Representation learning: a unified deep learning framework for automatic prostate MR segmentation. In: International Conference on Medical Image Computing and Computer-Assisted Intervention, vol. 16(Pt 2), pp. 254–261 (2013)
14. Marrone, S., Piantadosi, G., Fusco, R., Petrillo, A., Sansone, M., Sansone, C.: Automatic lesion detection in breast DCE-MRI. In: Petrosino, A. (ed.) ICIAP 2013. LNCS, vol. 8157, pp. 359–368. Springer, Heidelberg (2013). doi:10.1007/978-3-642-41184-7_37
15. Marrone, S., Piantadosi, G., Fusco, R., Petrillo, A., Sansone, M., Sansone, C.: A novel model-based measure for quality evaluation of image registration techniques in DCE-MRI. In: CBMS 2014, pp. 209–214. IEEE (2014)
16. Nodine, C.F., Kundel, H.L., et al.: Using eye movements to study visual search and to improve tumor detection. Radiographics **7**(6), 1241–1250 (1987)
17. Olsen, O., Gøtzsche, P.C.: Cochrane review on screening for breast cancer with mammography. Lancet **358**(9290), 1340–1342 (2001)
18. Pereira, S., Pinto, A., Alves, V., Silva, C.A.: Brain tumor segmentation using convolutional neural networks in MRI Images. IEEE Trans. Med. Imaging **35**(5), 1240–1251 (2016)
19. Piantadosi, G., Fusco, R., Petrillo, A., Sansone, M., Sansone, C.: LBP-TOP for volume lesion classification in breast DCE-MRI. In: Murino, V., Puppo, E. (eds.) ICIAP 2015. LNCS, vol. 9279, pp. 647–657. Springer, Cham (2015). doi:10.1007/978-3-319-23231-7_58
20. Reda, I., Shalaby, A., El-Ghar, M.A., Khalifa, F., Elmogy, M., Aboulfotouh, A., Hosseini-Asl, E., El-Baz, A., Keynton, R.: A new NMF-autoencoder based CAD system for early diagnosis of prostate cancer. In: Proceedings - International Symposium on Biomedical Imaging 2016-June, pp. 1237–1240 (2016)
21. Rosset, A., Spadola, L., Ratib, O.: OsiriX: an open-source software for navigating in multidimensional DICOM images. J. Digit. Imaging **17**, 205–216 (2004)
22. Tofts, P.S.: T1-weighted DCE imaging concepts: modelling, acquisition and analysis. Magneton Flash Siemens **3**, 30–39 (2010)
23. Twellmann, T., Saalbach, A., Müller, C., Nattkemper, T.W., Wismüller, A.: Detection of suspicious lesions in dynamic contrast enhanced MRI data. In: Annual International Conference of the IEEE Engineering in Medicine and Biology Society, vol. 1, pp. 454–457 (2004)
24. Wang, Y., Sun, Z., Liu, C., Peng, W., Zhang, J.: MRI image segmentation by fully convolutional networks. In: 2016 IEEE International Conference on Mechatronics and Automation, pp. 1697–1702. IEEE, August 2016
25. Zheng, Y., Englander, S., Baloch, S., Zacharaki, E.I., Fan, Y., Schnall, M.D., Shen, D.: STEP: spatiotemporal enhancement pattern for MR-based breast tumor diagnosis. Med. Phys. **36**, 3192–3204 (2009)
26. Zoorob, R., Anderson, R., Cefalu, C., Sidani, M.: Cancer screening guidelines. Am. Fam. Phys. **63**(6), 1101–1112 (2001)

A Smartphone-Based System for Detecting Falls Using Anomaly Detection

Vincenzo Carletti(✉), Antonio Greco, Alessia Saggese, and Mario Vento

Department of Information Engineering,
Electrical Engineering and Applied Mathematics, University of Salerno,
Fisciano, Italy
{vcarletti,agreco,asaggese,mvento}@unisa.it
http://mivia.unisa.it

Abstract. As reported by the World Health Organization, falls are a severe medical and financial issue; they represent the second leading cause of unintentional injury death, after road traffic injuries. Therefore, in recent years, the interest in realizing fall detection systems is considerably increased. Although the overall architecture of such systems in terms of its basic components is consolidated, the definition of an effective method to detect falls is a challenging problem due to several difficulties arising when the system has to work in the real environment. A very recent research trend is focused on the realization of fall detection systems running directly on a smartphone, so as to avoid the inconvenience of buying and carrying additional devices. In this paper we propose a novel smartphone-based fall detection system that considers falls as anomalies with respect to a model of normal activities. Our method is compared with other very recent approaches in the state of the art and it is proved to be suitable to work on a smartphone placed in the trousers pocket. This result is confirmed both from the achieved accuracy and the required hardware resources.

Keywords: Smartphone-base falls detection · Embedded pattern recognition · Anomaly detection · One-class classification

1 Introduction

The World Health Organization (WHO) defines a fall as an *event which results in a person coming to rest inadvertently on the ground or floor or other low level.* Globally, falls are a major public health problem [1–3]. An estimated 424.000 fatal falls occur each year, making them the second leading cause of unintentional injury death, after road traffic injuries. Though not fatal, every year approximately 37.3 million falls are severe enough to require medical attention. Such falls are responsible for over 17 million DALYs lost (the Disability-Adjusted Life Year is a measure of overall disease burden, expressed as the number of years lost due to ill-health). The largest morbidity occurs in people aged 65 years or older, young adults aged 15–29 years and children aged 15 years or younger. But, more

S. Battiato et al. (Eds.): ICIAP 2017, Part II, LNCS 10485, pp. 490–499, 2017.
https://doi.org/10.1007/978-3-319-68548-9_45

generally, elderly, children, disabled individuals, workers, athletes and patients with visual, balance, gait, orthopaedic, neurological and psychological disorders are all at risk from falls. Furthermore, fall-related injuries have relevant financial costs and as highlighted by a recent analysis conducted in Canada, the implementation of effective prevention strategies could create a net savings of over USD 120 million each year. For this reason, there is a growing interest of the scientific community in realizing effective fall detection systems [4–6].

A fall detection system can be defined as a device whose main objective is to alert when a fall event has occurred. Its main purpose is to distinguish whether a given event is a fall or a normal Activity of Daily Living (ADL). The general structure of a fall detection system is shown in Fig. 1; it can be considered a typical pattern recognition system, composed of a batch of sensors, a feature extraction stage and a detection (or classification) stage. An additional component is generally used to notify the detection of fall to other people that will promptly act to help the fallen one.

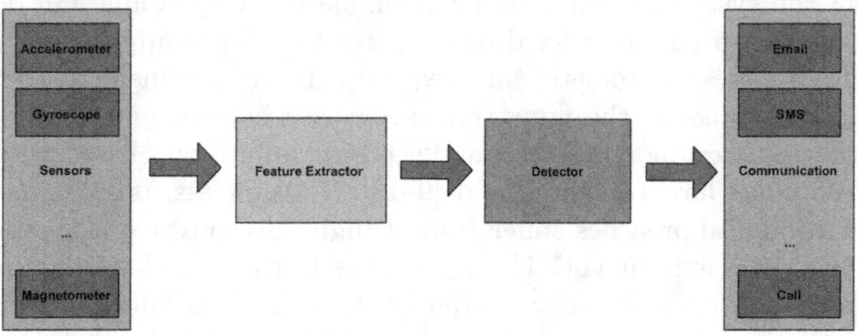

Fig. 1. Common structure of a fall detection system.

In a real-life scenario, a fall detection system potentially reduces some of the adverse consequences of a fall and, more specifically, it can have a direct impact on the rapid provision of assistance after a fall. But, to be effective, this system has to provide the best trade-off between false negatives and false positives. Indeed, if the system does not detect a fall, the safety of the monitored person could be jeopardized. On the other hand, if the system reports an excessive number of false positives, the users could perceive it as ineffective and useless.

Reviewing the current scientific literature [4–8] it is possible to identify two classes of fall detection systems: *context-aware systems* and *wearable systems*. In Context-aware systems the sensors are deployed in the environment to detect falls, such as cameras, acoustic sensors, pressure sensors, infrared sensors, lasers and Radio Frequency Identification (RFID) tags. The advantage is that the targeted person does not need to wear or carry any special device. But they are suitable for those situations where the environment to monitor is well defined and circumscribed such as hospitals, nursing house and other indoor environment. When the activities performed by the user take place both indoors and outdoors and involve going from one place to another, the approach of this type of systems becomes unsuitable due to the restrictions imposed on the mobility of the person. Wearable fall detectors have been developed in response to the problems affecting context-aware systems. They are specialized sensor-based devices that detect falls by analyzing users motion. Most of these devices are composed by a tri-axis accelerometer, and optionally additional sensors (such as gyroscopes and magnetometers), a dedicated elaboration units and wireless connectivity. The availability on recent smartphones of all the required elements makes them a suitable solution to realize a self-sufficient fall detection system [5]. Despite that, the use of a smartphone have some challenging issues that can affect the accuracy and the precision of the system. For instance, the *quality of the sensors* vary among different models of the same manufacturer and can not be sufficient to detect falls in some real situations; the *placement of the smartphone* (waist, pocket, hand, arm and so on) influences the measures collected from the sensors (indeed, moving the smartphone from the shirt pocket to the arm the accelerations can be very different); the *power consumption* sets some limits to the number of sensors, the data sampling rate and the complexity of the algorithm used to perform the task. As discussed in [5], to deal with these challenges, the majority of the systems running on smartphone use very simple fall detection algorithms that are based on fixed or adaptive thresholds applied on the data collected from the accelerometer and require a fixed placement of the phone. Only recently, thanks to the improved hardware resources and battery life of the new smartphones, more complex and accurate approaches based on machine learning techniques have been proposed [9–13]. Nevertheless, as discussed in [14], all the traditional approaches suffer from a high false positive rate when they are used in a real environment. The reason lies in the fact that these methods suppose to be able to collect enough data concerning real falls. This is a weak assumption because it is not easy to collect real falls, especially when the system has to monitor complex activities such as running, biking or skating. An alternative, proposed in [14,15], is to consider fall detection as an anomaly detection problem, where the detector is trained only on the ADLs and considers all the anomalies as falls. In this paper we propose a smartphone-based fall detector using Self-Organizing Maps (SOMs) to deal with the constraints depending on the limited hardware resources of a smartphone. In order to proof the effectiveness and the efficiency of the proposed method, we have compared SOMs with other well-known classifiers at the state of the art: k-Nearest Neighbor and Support Vector Machine. In particular, the performance comparison considered both the accuracy and the resource requirements of each method and has been conducted directly on a smartphone.

2 The Proposed Method

Anomaly detection refers to the problem of finding patterns in data that do not conform to the expected behaviour [16]. The aim of the anomaly detection is to define a methodology through which establish a region representing the normal behaviour and classify any observation in the data which does not belong to this normal region as an anomaly. In the case of fall detection, the system has to consider as normal the data representing the ADLs for a given activity and as anomalies (or falls) any unknown situation. Defining such a region that includes every possible ADL is not a trivial problem, indeed, the boundary between normal and abnormal instances is often not precise. In particular, abnormal observations very close to the region boundary can be considered as normal, and vice-versa. This problem is further enhanced by the noise: if the data representing ADLs are affected by noise they could be very similar to abnormal data and so not easy to distinguish. Furthermore, in many domains the ADLs can evolve and a current notion of normal instances might not be sufficiently representative in the future. For instance, in the case of sport activities, the experience of a user can increase, so he would be able to perform more difficult actions that where not considered in the model of the ADLs initially provided to the system.

It is important to point out that the system has to run in real-time using very limited hardware resources and battery in order to allow the normal operation of the smartphone. Therefore, on the one hand, the system has to be robust with respect to the noise and the small variations that can affect the normal instances. But, on the other hand, the model of ADLs used by the detector has to be as much compact as possible in order to require a low computational cost and to provide a decision with a low amount of memory. In addition, the hardware constraints could not allow to update the model directly on the smartphone, and, if the classifier requires an expensive training procedure, it could not be performed on the smartphone.

The previous considerations have a high impact on the features and the classifier that we can use to realize the system. As for the features, they have to be easy to compute in real-time and provide a compact representation. Similarly to Medrano [14] and Micucci [15], we used only the accelerations obtained from the three-axis accelerometer of the smartphone. Even though it could not be the best choice, we have decided to avoid the use of other sensors available on the smartphone, such as the gyroscope and the magnetometer, so as to reduce at minimum the energy required by the system. Indeed, the more sensors we use, the higher the battery consumption is. Moreover, differently from the other sensors, all the smartphones are equipped with good quality accelerometers because they are very cheap.

In more details, the three-axes accelerometer provides three raw measures (x, y and z axes) for each sample, as shown in Fig. 2. According to most of the current scientific literature, we have used a sampling rate of *50 Hz* that is adequate to our purpose and compatible with the highest sampling rate allowed by the majority of the smartphone (85 Hz). So that, the signal produced by the accelerometer is arranged in non-overlapped time windows of 50 samples. Each window is characterized by a feature vector of 150 features (50 features for each axis) representing the raw measures collected from the sensor. As discussed in [15] raw features can provide a good accuracy without requiring further elaborations to extract the feature vector; thus, we reduce the time and the battery required by the feature extraction stage.

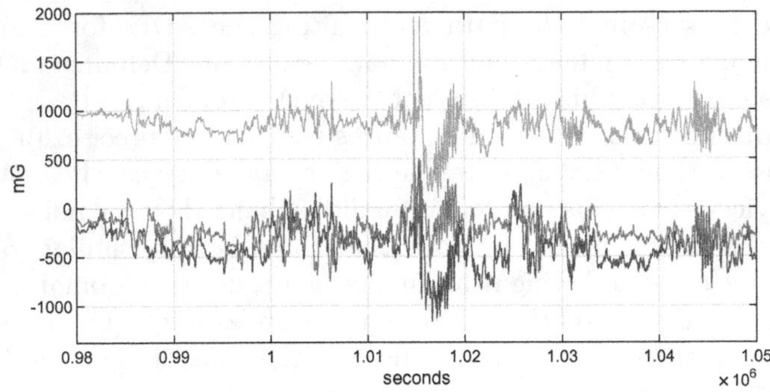

Fig. 2. Common fall pattern measured by the accelerometer of a smartphone. The accelerations on the three axes x, y and z are shown in red, blue and yellow, respectively. (Color figure online)

Concerning the selection of a suitable classifier, the anomaly detection is nicely addressed using *One-Class Classifiers* (OCCs) [16–19]. Indeed, the problem is to determine, in the data space, the boundary to distinguish between normal and abnormal patterns, i.e. ADLs and falls. In Micucci et al. [15], the authors compared the performance of a Support Vector Machine (SVM) and a k-NN, used both as one-class and binary classifiers. Differently from them, we propose to use Self-Organizing Maps since they are able to generate a very compact model and the decision stage has a very low computational cost.

SOMs are feed-forward neural networks composed of only two levels of neurons. Their main purpose is to produce a discrete representation of the input space, where the instances lies. This is performed by adopting the *competitive learning*, an unsupervised method that makes the network able to divide the input space in different partitions, each of them represented by one ore more prototypes (i.e. output neurons). In a multi-class problem each partition is composed of instances belonging to the same class, but a class can be represented by one or more partitions. A new instance is classified on the basis of the class the closest prototype belongs to. Alternatively, the class of the new instance can be decided similarly to the k-NN, by considering the class of the k nearest prototypes. In the case of anomaly detection, all the partitions shaped by the SOM are related to the normal class; thus, when a new instance comes, the classifier checks whether it belongs or not to the normal region by considering the average distance between the new instance and the k nearest prototypes. If such a distance is higher than a given *rejection threshold*, the instance is considered abnormal (i.e. a fall).

Therefore, recalling the architecture shown in Fig. 1, we propose a fall detection system where the features are obtained by collecting raw data from the accelerometer, the detection is performed by a SOM that is trained out of the smartphone using only ADLs and the communication is carried out by using the basic function of a phone, such as calls and messages.

3 Experiments

As previously discussed, the limited hardware resources of a smartphone (or other wearable systems) introduces additional constraints; thus, the most accurate method could not be the most suitable to run in real-time on a smartphone. In our experiments, we have firstly compared the accuracy of SOM, k-NN and ν-SVM on a test environment in Matlab by considering two performance measures: the *sensitivity* (SE) (a.k.a the recall) which represents the true positive rate, i.e. the number of falls that are correctly recognized, and the *specificity* (SP) that is the true negative rate or, in our case, the number of properly classified ADLs. It is important to note that if we have a high sensitivity and a low specificity the system will generate many false alarms. Differently, if the specificity is high and the sensitivity is low the system will have many miss alarms. So that, it is clear that the aim is to tune the classifier in order to have the best trade-off between the two indices.

Successively, once the best parameters for each classifier have been assessed, we have verified the hardware and battery requirements on a Huawei P8 lite equipped with Android 6.

Table 1. Area under the curve (AUC) of k-NN considering different values of k.

	$k = 1$	$k = 3$	$k = 5$	$k = 7$
Medrano	**0.976**	0.976	0.975	0.974
Sisfall	**0.706**	0.705	0.704	0.705

Table 2. Area under the curve (AUC), sensitivity (SE) and specificity (SP) of ν-SVM considering different values of ν.

		$\nu = 0.1$	$\nu = 0.2$	$\nu = 0.3$	$\nu = 0.4$	$\nu = 0.5$	$\nu = 0.6$	$\nu = 0.7$	$\nu = 0.8$	$\nu = 0.9$	$\nu = 1.0$
Medrano	*AUC*	0.978	0.998	0.978	0.977	0.977	0.975	0.975	0.974	0.974	**0.974**
	SE	0.132	0.198	0.398	0.589	0.641	0.812	0.839	0.901	0.924	**0.948**
	SP	1.000	1.000	1.000	1.000	0.992	0.969	0.961	0.939	0.973	**0.922**
Sisfall	*AUC*	0.975	0.984	0.985	0.985	0.985	0.985	0.984	0.984	0.981	**0.956**
	SE	0.421	0.511	0.549	0.601	0.619	0.580	0.581	0.563	0.638	**0.827**
	SP	1.000	1.000	1.000	1.000	1.000	1.000	0.998	0.997	0.995	**0.919**

3.1 Datasets

The experiments have been conducted on two very recent publicly available datasets, Medrano et al. [14] and Sucerquia et al. [20],

The former has been specifically realized to investigate the use of anomaly detection techniques for fall detection in smartphone-based solution. It is composed of ADLs and simulated falls recorded with the built-in three-axes accelerometer of a smartphone (a Samsung Galaxy Mini running Android). and considers eight different typologies of falls: forward falls, backward falls, left and right-lateral falls, syncope, falls sitting on empty chair, falls using compensation strategies to prevent the impact, and falls with contact to an obstacle before hitting the ground. Each fall has been repeated three times (for a total of 24 fall simulations per subject) and has been completed on a soft mattress in a laboratory environment. During the falls, participants wore a smartphone in both their two pockets. On the other hand, the ADLs have not been collected in a laboratory environment, but under real-life conditions. In particular, participants carried a smartphone in their pocket for at least one week to record everyday behaviour. On average, about 800 ADLs records have been collected from each subject. The dataset is divided in entries, each of them representing a time window of 6 seconds around a peak.

The SisFall dataset is composed of simulated activities of daily living and falls. This dataset has been generated in a laboratory environment with the collaboration of 38 volunteers divided in two groups: elderly people and young adults. Both the groups contains men and women. Differently from Medrano, SisFall has not been used by Micucci et al. [15] and is composed also of more interesting situations such as running and jogging people. The duration of each entry varies from 12 seconds, for simple ADLs, to 100 seconds, for more complex activities.

Both the datasets have been divided in ADLs and simulated falls because only the former have been used to train the systems, while the latter have been used to validate and test the classifiers. In more details, the *training set* consists of the 40% of the ADLs while the *validation set* is composed of the 30% of the ADLs and the 50% of the falls. The remaining part has been used as *test set*.

3.2 Results

The accuracy and the computational requirements of each classifier can vary on the base of how we set its parameters. We have assessed the best setup of each classifier by analyzing the *Area Under the Curve* (AUC) for different combination of their parameters. Additionally, only for the ν-SVM, we have considered the combined evolution of the specificity and sensitivity curves so as to avoid the overfitting.

Before discussing the validation of the ν-SVM, it is important to recall the meaning of the parameter ν in order to properly understand the results in Table 2. According to the technical report of Schölkopf [21], ν allows to control the model complexity; it defines how much the process of definition of the separating hyperplane will lie upon the training instances. In particular, low values of ν state that the hyperplane will be completely defined on the training set, while high values of ν assert that the hyperplane will take the training data just as an indication. Then in the former case, we risk to overfit the training set, while in the latter we risk to have an excessive generalization with a more complex model. In our results (see Table 2) it is possible to understand that even if the AUC is high for most of the values of ν, the best value for ν on both the datasets is 1.0, because it provides the best trade-off between the miss and false alarms. Moreover, the high values of specificity and sensitivity guarantee that the system is not too specialized on the training set.

Then, Table 1 shows the validation results of k-NN. The accuracy of the k-NN is not improved when k grows; we have selected the lowest value of k so has the best trade-off between the accuracy and the computational cost of the prediction.

As for the SOMs, the aim is to have as less neurons as possible so as to reduce the complexity of the model used on the smartphone. Analysing the validation results in Table 3, it is possible to note that increasing the number of neurons or the value of k the accuracy of the classifier is not substantially affected. Thus, we have selected 9 neurons and $k = 3$.

Table 3. Area under the curve (AUC) for different values of the number of prototype vectors and the parameter k on both the datasets. The best values for each combinations of the parameters are highlighted in bold.

	k	*Neurons*							
		9	16	25	36	49	64	81	100
Medrano	$k = 1$	0.947	0.960	0.964	0.969	0.972	**0.976**	**0.976**	0.974
	$k=3$	0.961	0.971	0.971	0.970	0.972	0.974	0.973	**0.974**
	$k=5$	0.936	0.962	0.972	0.970	0.973	**0.974**	0.973	0.973
	$k=7$	0.879	0.951	0.973	0.971	0.973	**0.975**	0.974	0.973
SisFall	$k = 1$	0.943	0.931	0.944	**0.950**	0.943	0.941	0.949	0.949
	$k=3$	**0.964**	0.959	0.959	0.958	0.960	0.957	0.964	0.955
	$k=5$	**0.978**	0.971	0.968	0.966	0.960	0.960	0.966	0.960
	$k=7$	**0.987**	0.977	0.971	0.969	0.959	0.962	0.966	0.962

The results achieved by using the best setup for each dataset are shown in Table 4. As regards the Medrano Dataset, even if the ν-SVM has achieved the best accuracy, there is a small difference among the three classifiers. The reason lies in the fact that the ADLs composing the dataset are very simple so the falls are easy to distinguish. The results are completely different when more complex ADLs are considered, as in the case of the SisFall dataset. Indeed, in activities like jogging and running the accelerations measured by the system can be very similar to those representing falls. In these situations, that are very frequent in a real environment, the k-NN achieved a very low specificity and an acceptable sensitivity. This is a symptom of the fact the classifier often decides for a fall and confuses it with ADLs. The ν-SVM, instead, is more precise, but the recall is still low with a high value of false negatives. So it is clear that the best accuracy is achieved by the SOMs. Furthermore, if we join the results in Table 4 with those in Table 3, it is evident that the best accuracy is obtained with only 9 prototypes;

Table 4. Comparison of the performance on Medrano Dataset and SisFall.

	Medrano			SisFall		
	k-NN	SVM	SOM	k-NN	SVM	SOM
Sensitivity	0.952	**0.952**	0.944	0.842	0.824	**0.912**
Specificity	0.944	**0.954**	0.918	0.410	0.929	**0.981**

so we can suppose that SOMs will have very low resources requirement during the detection stage.

The comparison performed on the smartphone confirms the feasibility of the proposed method. In fact, the results in Table 5 show the efficiency of SOMs with respect to the other two classifiers. K-NN and ν-SVM need complex models to achieve a good accuracy, so they require also more time and battery to detect falls. Differently from them, SOMs are suitable to our purpose, since they are faster and cheaper due to a simpler model. They require just 0.04 seconds to classify 1 second of measures with a good battery consumption.

Table 5. Performance comparison in terms of hardware resources on a Huawei P8 lite.

	k-NN	SVM	SOM
Time	9 s	9 s	0.04 s
Memory	40 Mb	40 Mb	28 Mb
Battery (10 min)	$\geq$2%	$\geq$2%	1%

4 Conclusion

In this paper we have proposed and realized a smartphone-based fall detection system that can be comfortably placed in the trousers pocket. Due to the constraints imposed by the limited hardware resources and the battery of a smartphone we have adopted Self-Organizing Maps to distinguish normal activities and falls. They have the advantage to produce a very compact model and, consequently, to limit the computational cost required during the detection. The suitability of SOMs have been confirmed by the experimental evaluation, which demonstrated the effectiveness of the proposed method both in terms of accuracy and hardware requirements. To the best of our knowledge, the relevance of these results is enhanced by the fact that this is the first performance comparison of fall detection machine learning approaches performed on a smartphone.

References

1. Sadigh, S., Reimers, A., Andersson, R., Laflamme, L.: Falls and fall-related injuries among the elderly: a survey of residential-care facilities in a swedish municipality. J. Commun. Health **29**(2), 129–140 (2004)

2. Kalace, A., Fu, D., Yoshida, S.: Who Global Report on Falls Prevention in Older Age. World Health Organization Press, Ginevra (2008)
3. Igual, R., Medrano, C., Plaza, I.: Challenges, issues and trends in fall detection systems. BioMed. Eng. Online **12**(1), 1–24 (2013)
4. Delahoz, Y.S., Labrador, M.A.: Survey on fall detection and fall prevention using wearable and external sensors. Sensors **14**(10), 19806–19842 (2014)
5. Habib, M.A., Mohktar, M.S., Kamaruzzaman, S.B., Lim, K.S., Pin, T.M., Ibrahim, F.: Smartphone-based solutions for fall detection and prevention: challenges and open issues. Sensors **14**(4), 7181–7208 (2014)
6. Koshmak, G., Loutfi, A., Linden, M.: Challenges and issues in multisensor fusion approach for fall detection: review paper. J. Sens. (2016)
7. Brun, L., Saggese, A., Vento, M.: Dynamic scene understanding for behavior analysis based on string kernels. IEEE Trans. Circ. Syst. Video Technol. **24**(10), 1669–1681 (2014)
8. Leo, M., Medioni, G., Trivedi, M., Kanade, T., Farinella, G.M.: Computer vision for assistive technologies. Comput. Vis. Image Underst. **154**, 1–15 (2017)
9. Choi, Y., Ralhan, A.S., Ko, S.: A study on machine learning algorithms for fall detection and movement classification. In: 2011 International Conference on Information Science and Applications, pp. 1–8, April 2011
10. Abbate, S., Avvenuti, M., Bonatesta, F., Cola, G., Corsini, P., Vecchio, A.: A smartphone-based fall detection system. Pervasive Mob. Comput. **8**(6), 883–899 (2012). Special Issue on Pervasive Healthcare
11. Albert, M.V., Kording, K., Herrmann, M., Jayaraman, A.: Fall classification by machine learning using mobile phones. PLoS ONE **7**(5), 1–6 (2012)
12. Shi, Y., Shi, Y., Wang, X.: Fall detection on mobile phones using features from a five-phase model. In: 2012 9th International Conference on Ubiquitous Intelligence and Computing and 9th International Conference on Autonomic and Trusted Computing, pp. 951–956, September 2012
13. Ali Fahmi, P.N., Viet, V., Deok-Jai, C.: Semi-supervised fall detection algorithm using fall indicators in smartphone. In: Proceedings of the 6th International Conference on Ubiquitous Information Management and Communication, pp. 122:1–122:9 (2012)
14. Medrano, C., Igual, R., Plaza, I., Castro, M.: Detecting falls as novelties in acceleration patterns acquired with smartphones. PLoS ONE **9**(4), 1–9 (2014)
15. Micucci, D., Mobilio, M., Napoletano, P., Tisato, F.: Falls as anomalies? An experimental evaluation using smartphone accelerometer data. J. Ambient Intell. Humanized Comput. **8**(1), 87–99 (2017)
16. Chandola, V., Banerjee, A., Kumar, V.: Anomaly detection: a survey. ACM Comput. Surv. **41**(3), 15:1–15:58 (2009)
17. Bartkowiak, A.M.: Anomaly, novelty, one-class classification: a comprehensive introduction. Int. J. Comput. Inf. Syst. Ind. Manag. Appl. **3**(1), 61–71 (2011)
18. Pimentel, M.A.F., Clifton, D.A., Clifton, L., Tarassenko, L.: Review: a review of novelty detection. Sig. Process. **99**, 215–249 (2014)
19. Khan, S.S., Madden, M.G.: One-class classification: taxonomy of study and review of techniques. Knowl. Eng. Rev. **29**(3), 345–374 (2014)
20. Sucerquia, A., Lpez, J.D., Vargas-Bonilla, J.F.: SisFall: a fall and movement dataset. Sensors **17**(1), 198 (2017)
21. Platt, J., Schlkopf, B., Shawe-Taylor, J., Smola, A.J., Williamson, R.C.: Estimating the support of a high-dimensional distribution. Technical report, November 1999

CNN-Based Identification of Hyperspectral Bacterial Signatures for Digital Microbiology

Giovanni Turra[1,2], Simone Arrigoni[2], and Alberto Signoroni[1(✉)]

[1] Information Engineering Department, University of Brescia, Brescia, Italy
alberto.signoroni@unibs.it

[2] Futura Science Park, Copan Italia S.p.A., Brescia, Italy

Abstract. The rapidly increasing diffusion of Full Microbiology Laboratory Automation plants is reshaping the way microbiologists perform diagnostic tasks. A huge stream of digital visual data is expected to be produced daily in the coming years in the emerging field of Digital Microbiology Imaging. In this context, we want to assess the suitability and effectiveness of a Deep Learning approach to solve the diagnostically relevant but visually challenging task of directly identifying pathogens on bacterial growing plates. In particular, starting from hyperspectral acquisitions in the VNIR range and spatial-spectral processing of cultured plates, we approach the identification problem as the classification of computed spectral signatures of the bacterial colonies. In a highly relevant clinical context (urinary tract infections) and on a database of HSI images, we designed and trained a Convolutional Neural Network for pathogen identification, assessing its performance and comparing it against conventional classification solutions.

Keywords: Hyperspectral imaging · Spatial-spectral processing · Convolutional neural networks · Pattern recognition · Clinical microbiology

1 Introduction

The present work is situated at the intersection of three significant innovation trends and looks to exploit the different opportunities they offer and propose a solution for direct pathogen identification on bacteria culturing plates. At first, the concrete possibility of (deeply) learning salient visual features determined, in recent years, the success of Deep Learning (DL) architectures. For visual recognition tasks, DL models are normally implemented with Convolutional Neural Network (CNN) [1] for low-to-high level visual feature learning. This is currently influencing, if not significantly impacting, several application domains. In the biomedical field a transition from handcrafted to learned feature-based approaches can bring significant benefits, especially when high data throughput and visual content variability are involved [2]. However, data dimensionality (biomedical data are often 3D or higher dimensional) introduces further challenges for DL solutions only partially addressed so far.

S. Battiato et al. (Eds.): ICIAP 2017, Part II, LNCS 10485, pp. 500–510, 2017.
https://doi.org/10.1007/978-3-319-68548-9_46

The second trend we consider, is the increasing attention on small-scale applications of hyperspectral imaging (HSI) in several domains, such as industrial quality controls (especially food, pharma and chemical [3,4]), cultural heritage preservation [5] and a number of biomedical applications [6]. What currently contributes to the proliferation and diversification of small-scale HSI applications, in addition to the classical Remote Sensing (RS) ones, is the increasing technological variety and ever lower cost of acquisition equipment [7,8] and, as for DL, the continued increase in computational power and storage/transmission capabilities of computing hardware and networks. In many situations, where visual analysis is limited by spatial-spectral resolution trade-offs, the alternative or concurrent use of HSI acquisition systems can play a determinant role for improved data interpretation. However, the restricted number of non-RS available datasets still hinder the popularity of HSI data acquisition and analysis research for non-RS applications.

The third evolution we consider defines the application context of our work. This is related to a recent digitization trend significantly impacting the field of Clinical Microbiology (CM), the escalating diffusion of Full Laboratory Automation (FLA). An FLA system is capable of handling all phases of bacterial colony culturing, from the processing of various human collected specimens through seeding and streaking on culturing plates (Petri dishes), to automatic incubation and further processing for subsequent analysis [9,10]. All relevant phases of bacteria colony growing can be captured by digital cameras, visualized on diagnostic workstations, stored/communicated and processed. This determined the advent of Digital Microbiology [11,12] and a fundamentally new way of work for microbiologists.

In Digital Microbiology Imaging (DMI), image-based decision making can be automated for certain tasks or support the work of the microbiologist for others. One of the most impacting capabilities (not yet provided by commercial products) would be reliable and fast identification of bacterial species by direct image analysis and machine learning solutions. Early identification of bacteria species is needed to determine the correct therapy for the patient with potentially significant impact on life expectation. In addition, early identification is one of the most powerful ways to contrast the worldwide threat related to antibiotic resistance. This is especially true if one considers very general and massive diagnostic investigation procedures such as screening for urinary tract infection (UTI) pathogen identification [13]. UTI are widespread and serious health problems that interest many millions of people every year around the world, accounting for a significant part of CM labs' workload [14]. Unfortunately, *presumptive identification* by visual inspection of UTI pathogens on the most diffused culturing media (e.g. blood agar) can be a very complex and ambiguous task, even for highly skilled microbiologists (examples of different pathogens exhibiting high visual similarity are showcased in Fig. 1). This is the reason why, despite their higher cost, *chromogenic* media [15] have gained widespread market diffusion, thanks to their ability to mark different colonies with different colors (through the use of pathogen-specific enzyme substrates). However, these

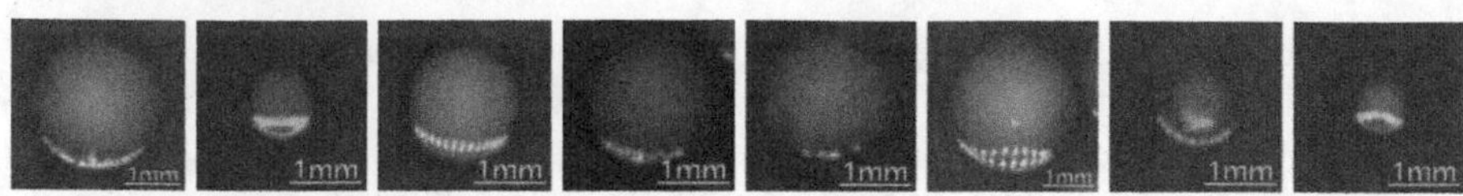

Fig. 1. Examples of different UTI bacteria colonies grown on blood agar media.

media have several limitations in terms of the number of pathogens that can be differentiated [16]. HSI technology could provide support where three-chromatic imaging does not give enough spectral information for reliable discrimination. Therefore UTI identification is a good case study for HSI-based bacteria identification because UTI represents a diagnostic context involving, for a single laboratory, hundreds of analyses per day, so a technology investment can be rapidly amortized.

There are still very few examples of DL-based approaches for RS applications [17–20] and, to our knowledge, still none for other fields, including biomedicine. Moreover, though both conventional machine learning [21–23] and DL solutions [24] have already been implemented for DMI analysis tasks, and hyperspectral classification has already been explored in CM [25–29], the present work is the first attempt to combine HSI, CM-FLA and CNN for direct bacterial identification purposes. In this work, we want to exploit the enhanced spectral information coming from HSI acquisitions to prove the feasibility of reliable bacteria species discrimination based on a DL approach. We raise the complexity of the problem compared to our preliminary study [27] by increasing the cardinality of pathogens, building a larger HSI UTI dataset (made available online) and by exploiting an improved acquisition setup. Unlike typical Remote Sensing techniques, that seek to increase the spatial consistency of the spectral classification at a pixel level [30], our pathogen recognition takes place on each single bacterial colony growing on the agar substrate. To this end, we propose a new (with respect to [27]) spatial-spectral distance measure to extract Colony Spectral Signatures (CSS). We designed and trained a 1D-CNN acting on CSS for pathogen identification and compared it against other conventional machine learning approaches, well selected and designed for the same purpose [29]. In particular, classification accuracy, computational efficiency and scalability comparisons are proposed along with examples and further considerations.

2 Proposed Method

A general scheme of the proposed HSI processing and classification workflow for rapid UTI bacteria discrimination is given in Fig. 2. In describing the various stages of our system, we give more emphasis on the novel CNN-based solution for CSS discrimination. Details about other parts, HSI database and conventional handcrafted feature-based classification solutions can be found in [29].

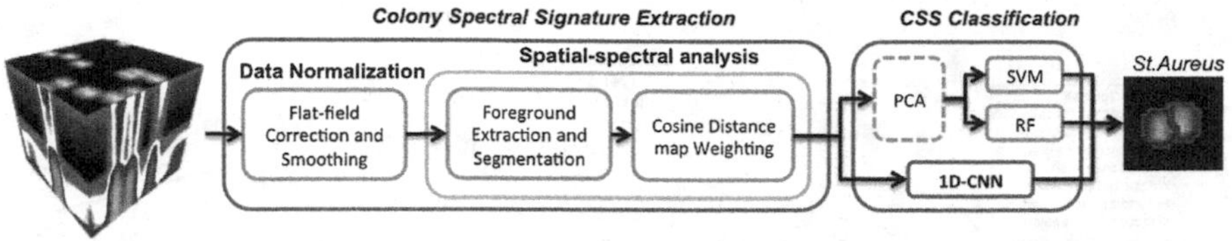

Fig. 2. Processing and classification pipeline.

Hyperspectral Acquisition System. The HSI target is a 90 mm diameter Petri plate. The main parts of the acquisition system are: (1) *HSI camera* - a linear VNIR camera (Specim Spectral Camera V10E) with spectral range between 400 and 1000 nm, tele-centric fore lenses (Specim OLE23, focal length 23 mm); spatial resolution has been doubled with respect to [27] (640 × 600 pixels) maintaining scanning time under 15 seconds (compatible with FLA needs). (2) *Illumination system* - the light of two 150W halogen lamps is conducted by two 13 mm-diameter optical fibers, spread by cylindrical lenses and finally reflected to the inner side of a semi-cylindrical dome. This configuration avoids total reflection effects on translucent colonies. (3) *Conveyor system* - a conveyor sliding system, mounting a shuttle which accommodates both the plate and a calibration bar (coated with $BaSO_4$ optopolymer), allows push-broom plate acquisition and a per-sample radiometric calibration.

Colony Spectral Signature (CSS) Extraction. Flat-field calibration was applied to the hypercube to derive a normalized (with respect to a white calibration bar) relative reflectance measure $R_{i,\lambda}$

$$R_{i,\lambda} = \frac{S_{i,\lambda} - D_{i,\lambda}}{W_{i,\lambda} - D_{i,\lambda}} \tag{1}$$

where $S_{i,\lambda}$ is the acquired reflectance, $W_{i,\lambda}$ and $D_{i,\lambda}$ are the white calibration and the dark current spatial(i)-spectral(λ) profiles. A signal preserving Savitzky-Golay [31] denoising (window size of 7) is then applied. Since illumination power from the halogen sources decreases at the spectrum extrema, corresponding bands were cut off, preserving the ones with highest SNR in the range from 430 nm to 780 nm (for a total of 125 spectral bands). Then, a threshold-based foreground extraction is performed on the spectral band at wavelength 520 nm. This produces a reliable isolation of the grown colonies because, at this specific wavelength, the contrast between relative reflectances of pathogens and blood agar is greater than in other bands. At this point, spatial distance transform is calculated for each colony using a spectral *cosine distance map*, computed as:

$$1 - \frac{\mathbf{u} \cdot \mathbf{v}}{||\mathbf{u}||_2 ||\mathbf{v}||_2}, \tag{2}$$

between each pixel signature $\mathbf{v}$ and the agar footprint $\mathbf{u}$, obtained by averaging the spectral signatures of background pixels. We use the resultant map as an

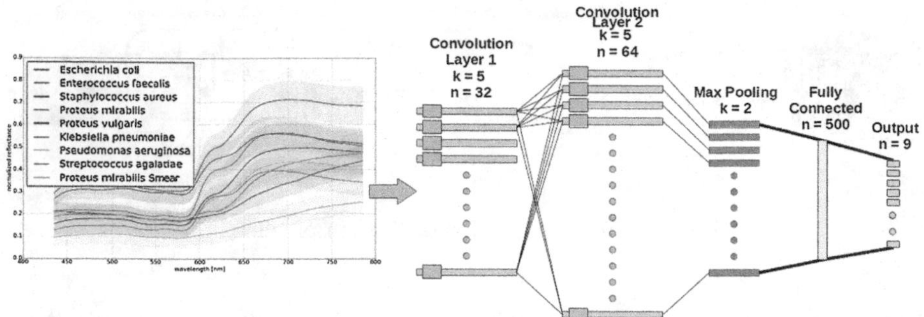

Fig. 3. Average spectral signatures of UTI bacteria, their standard deviations and CNN structure selected.

elevation map for a reliable watershed segmentation of bacterial colonies. For each detected colony we then extract a representative spectral signature (a 125-dimension vector) where we set colony pixel weighting factors proportional to the previously computed cosine distance map:

$$\mathbf{CSS_{colony}} = \sum_{p \in P} w_p \cdot \mathbf{R}_p \qquad \in \quad \mathbb{R}^{125} \tag{3}$$

with P the set of colony pixels, w_p the weighting factor for the pixel p and $\mathbf{R}_p$ the relative reflectance spectrum of the pixel p. Representative CSSs for each pathogen (the list is given in Sect. 3) are shown in Fig. 3 (left), along with their standard deviation (shadowed).

Classification Methods. CNN architectures [1] have been related to models of the visual cortex [32] and are characterized by locally overlapped connections (receptive fields) and shared weights implemented within a stacked hierarchy (from low to high level visual tasks) of convolutional feature extraction layers alternated with pooling layers (usually exploiting a max pooling rule). This is followed by one or more fully connected classification layers. Non-linear activation function layers are typically employed following convolutional ones, and the whole network produces a differentiable score function allowing the network parameters to be learned (weights and biases of the convolutional and fully connected layers). Unlike spatial-spectral 3D-CNN configurations [33], which are more susceptible to overfitting and therefore needing dedicated regularization strategies, we exploit a 1D-CNN configuration similar to that considered in [33,34] for RS hypercubes. However, instead of considering single pixel spectra, we take advantage of the proposed spatial-spectral processing so that the CNN sees the extracted CSS as inputs while producing the colony-based class scores as output. Our network topology, see Fig. 3 (right), contains 2 convolutional layers, 1 pooling layer, 1 fully-connected layer and a final probability-based (softmax) classifier layer, for a total of 1,905,496 network parameters to learn. The first convolutional layer evaluates 32 feature maps from the 125-dimensional CSS input, for each map a 5-tap filter is trained to produce same size output. The structure of the second convolution layer is similar and is composed of 64 feature maps (again 5-tap filters). Parametric Rectified Linear Units (PReLU) were used as activation functions [35]. After the two convolutional layers, a max pooling layer halves the size of feature maps given as input of a fully connected layer composed of 500 units, eventually followed by 9 output units. The selection of the above CNN structure was based on the evaluation of many possibilities by changing the number of convolutional and fully connected layers, learning rate and learning decay value (see Sect. 3). We implement the whole structure in Python 2.7 and TensorFlow 1.0 [36].

For comparison purposes we selected two conventional classification approaches among those which have been shown to be effective in handling reflectance spectral data in this and other HSI analysis contexts: SVM and Random Forests. *Support Vector Machine* (SVM) [37] is a popular non-parametric technique for binary classification. It is a suitable tool in cases of data not regularly distributed or data with an unknown distribution. We implement SVM according to multi-class one-against-all structure, with Radial Basis Function kernels configured through iterated model selection for each pathogen binary classifier.

Random Forests (RF) [38] is an ensemble learning method that operates by constructing a multitude of decision trees. They predict (through a bagging approach) deep insights into the structure of data. Each tree is built on different samples with randomness in the growing phase to ensure dissimilarity. Class with most votes (among all the trees in the forest) determines the prediction. The use of randomness and averaging improves the predictive accuracy and contrasts overfitting. We also tested both SVM and RF combined with information preserving dimensionality reduction obtained by Principal Component Analysis (PCA) [39], used to reduce spectral redundancy with 99.9% of retained variance.

3 Results and Discussion

We built and analyzed a database of 16642 colonies streaked and grown on Petri dishes (5% sheep blood agar plates, BBL, BD Diagnostics, Sparks, MD) from 106 HSI volumes acquired after 18 hours of incubation in O_2. Target pathogens in our analysis, all belonging to the American Type Culture Collection (ATCC), and covering over 85% of UTI species of interest, are: *E.coli* (5539 colonies), *E.faecalis* (1958), *S.aureus* (2355), *P.mirabilis* (2315), *P.vulgaris* (654), *K.pneumoniae* (542), *Ps.aeruginosa* (1529) and *Str.agalactiae* (1750). Representative colony examples (from RGB images) and corresponding average spectral signatures are shown in Figs. 1 and 3 respectively. The whole dataset has been licensed for research use and can be accessed on http://www.microbia.org.

Classification Accuracy. Bacterial species classifiers based on CNN, as well as SVM and RF (with or w/o PCA) have been implemented and compared on the experimental dataset. In Table 1, classification performance in terms of average accuracy are reported.

Table 1. Classification accuracy (avg and std). With asterisk configurations considered in Fig. 4.

	Accuracy
CNN*	**0.997** ± 0.001
SVM*	0.995 ± 0.001
RF	0.938 ± 0.002
PCA+SVM	0.984 ± 0.002
PCA+RF*	0.971 ± 0.002

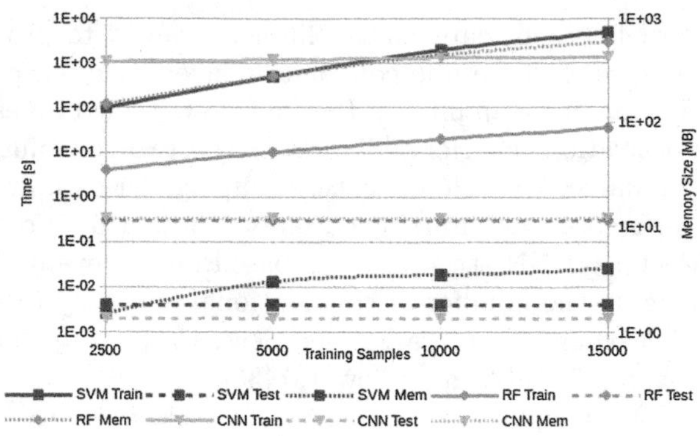

Fig. 4. Computational and memory footprint performance: training times–*solid line*, testing times–*dashed line*, and memory footprint–*dotted line* of the classifiers, versus the number of training samples

The selected CNN model, after 50,000 training iterations, reached an accuracy of 99.7% becoming our best option. A learning rate of 0.01 and learning decay of 0.005 were selected after many different tests, resulting in the following observations: (a) by growing the number of convolutional and/or FC layers we obtained minor improvements with more than double the training/testing time; (b) comparable classification results can be obtained with learning rates between 0.005 and 0.01, while using 0.05 leads to a lack of convergence for all tested configurations except a suboptimal one with single Conv and FC layers; (c) with dropout of 0.75 and momentum of 0.9 (commonly adopted values) we preserve both the network structure and highest accuracy levels with training and test timings fully acceptable to guarantee FLA compatible near real-time classification (see below).

Accuracy assessments are based on a 70/30 random split in training and validation sets for each class in the database and repeated five times following a Shuffle & Split cross-validation approach. For the CNN solution it is particularly significant to assess how the method behaves as the dimension of the training set decreases. We therefore considered different percentages of the training set and, in Fig. 5(a) we track accuracy performance as a function of the number of learning iterations. Several curves are used to show increased accuracy when increasing the training set dimension (we are able to reach accuracy already greater than 99% by using only 15% of the training set).

Though slightly inferior with respect to CNN, SVM also reached comparable performance, showing an accuracy peak of 99.5% without PCA, while we obtained >1% accuracy drop by adopting dimensionality reduction. It is therefore possible to create hyper-surfaces (thanks to RBF kernels) to accurately separate the analyzed classes.

RF was used differently to normal. When making predictions on the test dataset, we tried to exploit every tree in the forest in order to leverage the

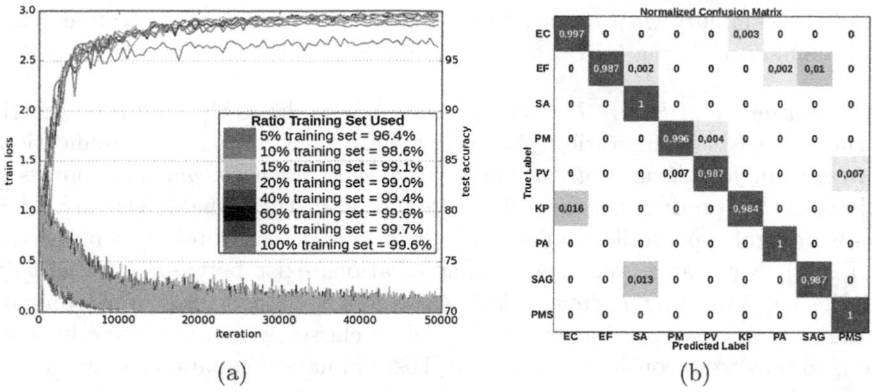

Fig. 5. (a) CNN dataset evaluation and (b) Confusion matrix with 99.7% accuracy.

full forest and benefit from averaging the prediction. A decision is taken only if 70% of the forest agrees. This may decrease the overall accuracy but it increases classification precision, and reduces wrong predictions for test samples that bring new factors that were not in the training set (as not yet considered species or other undesirable alterations). Using this configuration, RF obtains its own best performance using PCA (97.1%, i.e. +3.3*pp* with respect to the baseline).

Computational and Scalability Assessment. Fast CSS classification is needed especially in the context of FLA. Classifiers present strong discrepancies in terms of computational efficiency and scalability features according to the dataset dimension. This section analyses training time (Fig. 4 solid line), testing time (Fig. 4 dashed line) and classifier memory footprint (Fig. 4 dotted lined) versus the number of training samples. We used a standard PC for all classifiers (Intel Core i5-3470 CPU 4×3.2GHz, 16 GB RAM) except CNN (Intel Core i7-5930K CPU 12×3.5 GHz, 32 GB RAM, GeForce GTX TITAN X). SVM and CNN are the slowest proposed solutions by almost two order of magnitude as long as RF on training times. SVM has a bigger slope compared to CNN that maintains similar values throughout the dataset size (meaning a better scalability). RF generates many decision trees (1000 in configurations applied to our dataset) and in order to reduce the training time it is possible to prune some branches (or to limit the branching depth) in a possible trade-off with the accuracy. The cardinality of the dataset has little influence on the classification (testing) time. RF is the slowest (while CNN is the fastest) because any sample must flow along every decision tree. Classifier memory footprint rises for SVM and RF and it remains constant for CNN. In absolute, RF requires much more space than CNN and SVM. Though not the best in terms of accuracy, RF demonstrates a high level of precision (low false positives) and facilitates extrapolation of additional information. On the other hand, SVM shows great accuracy. CNN proved to be the best solution in terms of accuracy, memory footprint and testing time, as well as scalability with respect to the dataset dimension, while training time can be limited by using GPU and specific hardware.

Considerations and Future Directions. Figure. 5(b) shows the confusion matrix for the best CNN configuration. We observe only a few mutual misclassifications between *Ent. faecalis* and *Str. agalact.*, *Esch. coli* and *Kleb. pneum.*, couples of pathogens that produce colonies which are hardly distinguishable visually. They are also roughly spectrally similar (though in average separated by a bias term, see Fig. 5). Noticeably, very few misclassifications exist between *Proteus vulg.* and the not swarming *Proteus mirab.* (also almost impossible to discriminate visually) while, in a previous work [27], these classes were not distinctly separated, so they were considered as joined. Discrimination capability between two species of the same bacterial genus, as for *Proteus*, is of high application value and this is evidence of the improved CSS extraction introduced in this work.

According to accuracy of classification, complexity of the structure, memory footprint, training and testing times, the CNN-based method is seen as the best analyzed bacterial identification pipeline. However, near perfect species differentiation reveal the need and opportunities to further increase the number of considered pathogens as well as the size and variability of the dataset (e.g. including plates coming from clinical specimens). Even if, on same experimental setting, conventional classification methods reached high classification performance as well, we can expect that DL-based approaches will be more appropriate in presence of scalability needs and variability factors that will be considered in order to bring this HSI technology closer to clinical application.

4 Conclusion

We verified the possibility of applying a deep learning approach to UTI bacteria identification by using HSI technology operating in the VNIR spectrum. Our CNN-based solution obtained highest classification accuracies on a large laboratory dataset, notwithstanding the significant number of analyzed pathogens and the fact that pathogen spectral signature differentiation is challenging and made even harder by spectral mixing with the growing media. There are also notable differences in term of scalability (both training, testing and memory used) driving our CNN implementation selection above alternate methods. Improvements over previous works have also been obtained thanks to a better data acquisition setup and a more reliable CSS assessment. This study suggests that further investigations are desirable by making our deep learning pipeline functional in a real clinical lab environment. Future activities should take into account an even higher number of UTI-relevant pathogens and clinical laboratory validations.

Acknowledgments. This work was partially supported by Italian Ministry of University and Research: Smart Factory Cluster, Adaptive Manufacturing, CTN01 00163 216730. The authors would to express their sincere thanks to the scientific and technical staff of Copan Group S.p.A. (Brescia, Italy) for their essential support in the creation of the hyperspectral image database.

References

1. LeCun, Y., Bengio, L., Haffner, P.: Gradient-based learning applied to document recognition. Proc. IEEE **86**(11), 2278–2324 (1998)
2. Litjens, G., Kooi, T., Bejnordi, B.E., Setio, A.A.A., Ciompi, F., Ghafoorian, M., van der Laak, J.A., van Ginneken, B., Sánchez, C.I.: A survey on deep learning in medical image analysis. arXiv preprint (2017). arXiv:1702.05747
3. Serranti, S., Cesare, D., Marini, F., Bonifazi, G.: Classification of oat and groat kernels using nir hyperspectral imaging. Talanta **103**, 276–284 (2013)
4. Gowen, A., Feng, Y., Gaston, E., Valdramidis, V.: Recent applications of hyperspectral imaging in microbiology. Talanta **137**, 43–54 (2015)
5. Liang, H.: Advances in multispectral and hyperspectral imaging for archaeology and art conservation. Appl. Phys. A **106**(2), 309–323 (2012)
6. Lu, G., Fei, B.: Medical hyperspectral imaging: a review. J. Biomed. Optics **19**(1), 010901 (2014)
7. Wug Oh, S., Brown, M.S., Pollefeys, M., Joo Kim, S.: Do it yourself hyperspectral imaging with everyday digital cameras. In: IEEE CVPR (2016)
8. Lambrechts, A., Gonzalez, P., Geelen, B., Soussan, P., Tack, K., Jayapala, M.: A CMOS-compatible, integrated approach to hyper-and multispectral imaging. In: IEEE International Electron Devices Meeting, pp. 10.5.1–10.5.4 (2014)
9. Bourbeau, P., Ledeboer, N.: Automation in clinical microbiology. J. Clin. Microbiol. **51**(6), 1658–1665 (2013)
10. Novak, S., Marlowe, E.: Automation in the clinical microbiology laboratory. Clin. Lab. Med. **33**(3), 567–588 (2013)
11. Rhoads, D., Sintchenko, V., Rauch, C., Pantanowitz, L.: Clinical microbiology informatics. Clin. Microbiol. Rev. **27**(4), 1025–1047 (2014)
12. Doern, C.D., Holfelder, M.: Automation and design of the clinical microbiology laboratory. In: Manual of Clinical Microbiology, Eleventh Edition, pp. 44–53. American Society of Microbiology (2015)
13. Flores-Mireles, A.L., Walker, J.N., Caparon, M., Hultgren, S.J.: Urinary tract infections: epidemiology, mechanisms of infection and treatment options. Nat. rev. Microbiol. **13**(5), 269–284 (2015)
14. Wilson, M.L., Gaido, L.: Laboratory diagnosis of urinary tract infections in adult patients. Clin. Infect. Dis. **38**(8), 1150–1158 (2004)
15. D'Souza, H., Campbell, M., Baron, E.: Practical bench comparison of BBL chromagar orientation and standard two-plate media for urine cultures. J. Clin. Microbiol. **42**(1), 60–64 (2004)
16. Perry, J., Freydire, A.: The application of chromogenic media in clinical microbiology. J. Appl. Microbiol. **103**(6), 2046–2055 (2007)
17. Chen, Y., Zhao, X., Jia, X.: Spectral-spatial classification of hyperspectral data based on deep belief network. IEEE J. Sel. Top. Appl. Earth Obs. in Remote Sens. **8**(6), 2381–2392 (2015)
18. Makantasis, K., Karantzalos, K., Doulamis, A., Doulamis, N.: Deep supervised learning for hyperspectral data classification through convolutional neural networks. In: 2015 IEEE IGARSS, pp. 4959–4962 (2015)
19. Zhang, L., Zhang, L., Du, B.: Deep learning for remote sensing data: a technical tutorial on the state of the art. IEEE Geosci. Remote Sens. Mag. **4**(2), 22–40 (2016)
20. Ma, X., Wang, H., Geng, J., Wang, J.: Hyperspectral image classification with small training set by deep network and relative distance prior. In: 2016 IEEE IGARSS, pp. 3282–3285 (2016)

21. Ferrari, A., Signoroni, A.: Multistage classification for bacterial colonies recognition on solid agar images. In: 2014 IEEE International Conference on Imaging Systems and Techniques (IST), pp. 101–106. IEEE (2014)
22. Ferrari, A., Lombardi, S., Signoroni, A.: Bacterial colony counting by convolutional neural networks. In: 2015 37th Annual International Conference of the IEEE Engineering in Medicine and Biology Society (EMBC), pp. 7458–7461 (2015)
23. Andreini, P., Bonechi, S., Bianchini, M., Mecocci, A., Di Massa, V.: Automatic image analysis and classification for urinary bacteria infection screening. In: Murino, V., Puppo, E. (eds.) ICIAP 2015. LNCS, vol. 9279, pp. 635–646. Springer, Cham (2015). doi:10.1007/978-3-319-23231-7_57
24. Ferrari, A., Lombardi, S., Signoroni, A.: Bacterial colony counting with convolutional neural networks in digital microbiology imaging. Pattern Recogn. **61**, 629–640 (2017)
25. Yoon, S., Lawrence, K., Park, B.: Automatic counting and classification of bacterial colonies using hyperspectral imaging. Food Bio. Tech. **8**(10), 2047–2065 (2015)
26. Leroux, D., Midahuen, R., Perrin, G., Pescatore, J., Imbaud, P.: Hyperspectral imaging applied to microbial categorization in an automated microbiology workflow. In: Clinical and Biomedical Spectroscopy and Imaging IV, vol. 9537, pp. 1–9. SPIE (2015)
27. Turra, G., Conti, N., Signoroni, A.: Hyperspectral image acquisition and analysis of cultured bacteria for the discrimination of urinary tract infections. In: Engineering in Medicine and Biology Society, EMBC, pp. 759–762. IEEE (2015)
28. Kammies, T., et al.: Differentiation of foodborne bacteria using nir hyperspectral imaging and multivariate data analysis. Appl. Microbiol. Biotech. **100**(21), 9305–9320 (2016)
29. Arrigoni, S., Turra, G., Signoroni, A.: Hyperspectral image analysis for rapid and accurate discrimination of bacterial infections: a benchmark study. Comput. Biol. Med. **88**, 60–71 (2017)
30. Bernard, K., Tarabalka, Y., Angulo, J., Chanussot, J., Benediktsson, J.A.: Spectral-spatial classification of hyperspectral data based on a stochastic minimum spanning forest approach. IEEE Trans. Image Process. **21**(4), 2008–2021 (2012)
31. Savitzky, A., Golay, M.: Smoothing and differentiation of data by simplified least squares procedures. Anal. Chem. **36**(8), 1627–1639 (1964)
32. LeCun, Y., Bengio, Y., Hinton, G.: Deep learning. Nature **521**(7553), 436–444 (2015)
33. Chen, Y., Jiang, H., Li, C., Jia, X., Ghamisi, P.: Deep feature extraction and classification of hyperspectral images based on convolutional neural networks. IEEE Trans. Geosci. Remote Sens. **54**(10), 6232–6251 (2016)
34. Li, J., Bruzzone, L., Liu, S.: Deep feature representation for hyperspectral image classification. In: 2015 IEEE International Geoscience and Remote Sensing Symposium (IGARSS), pp. 4951–4954 (2015)
35. He, K., Zhang, X., Ren, S., Sun, J.: Delving deep into rectifiers: surpassing human-level performance on imagenet classification. ArXiv e-prints (2015)
36. Abadi, M., et al.: Tensorflow: a system for large-scale machine learning. In: Proceedings of the 12th USENIX Conference on OSDI 2016, pp. 265–283 (2016). http://dl.acm.org/citation.cfm?id=3026877.3026899
37. Cortes, C., Vapnik, V.: Support-vector networks. Mach. Learn. **20**(3), 273–297 (1995)
38. Breiman, L.: Random forests. Mach. Learn. **45**(1), 5–32 (2001)
39. Shaw, G., Manolakis, D.: Signal processing for hyperspectral image exploitation. IEEE Sig. Process. Mag. **19**(1), 12–16 (2002)

Description of Breast Morphology Through Bag of Normals Representation

Dario Allegra[1(✉)], Filippo L.M. Milotta[1], Diego Sinitò[1], Filippo Stanco[1], Giovanni Gallo[1], Wafa Taher[2], and Giuseppe Catanuto[3]

[1] Department of Mathematics and Computer Science, University of Catania, Catania, Italy
{allegra,milotta,dsinito,fstanco,gallo}@dmi.unict.it
[2] International Fellowship Querci Della Rovere, London, UK
misswtaher@yahoo.com
[3] Multidisciplinary Breast Unit, Azienda Ospedaliera Cannizzaro, Catania, Italy
giuseppecatanuto@gmail.com

Abstract. In this work we focus on digital shape analysis of breast models to assist breast surgeon for medical and surgical purposes. A clinical procedure for female breast digital scan is proposed. After a manual ROI definition through cropping, the meshes are automatically processed. The breasts are represented exploiting "bag of normals" representation, resulting in a 64-d descriptor. PCA is computed and the obtained first 2 principal components are used to plot the breasts shape into a 2D space. We show how the breasts subject to a surgery change their representation in this space and provide a cue about the error in this estimation. We believe that the proposed procedure represents a valid solution to evaluate the results of surgeries, since one of the most important goal of the specialists is to symmetrically reconstruct breasts and an objective tool to measure the result is currently missing.

Keywords: 3D scanning · Breast surgery · Histogram of normals · Principal Component Analysis

1 Introduction and Motivation

In the last decade 3D scanners have been employed in architecture, engineering, biology, cultural heritage as well as diagnostic medicine and reconstruction surgery [1–9]. These devices allow doctors to get a detailed virtual model of a human body. The opportunity to acquire body parts shape, including soft tissues like the female human breast, has motivated our conjunct study with the medical specialists in breast reconstruction.

Our main aim is to find a discriminative parametrization of female breast shape i.e., a small set of parameters to meaningfully describe it. This kind of mathematical representation gives the possibility to easily define accurate metric for breast difference evaluation. This result is very attractive for breast surgeon, since it can be used to develop new tools to assess the symmetry after a breast

S. Battiato et al. (Eds.): ICIAP 2017, Part II, LNCS 10485, pp. 511–521, 2017.
https://doi.org/10.1007/978-3-319-68548-9_47

reconstruction. It could also be an effective strategy to create clear and well-defined breast shape categories.

Currently, the surgeons are routinely used to acquire pictures of the patients, or rather a 2D projection of them. The only way to evaluate the surgery is still based on a photographic comparison using pictures taken before and after the surgery. Nevertheless, 3D scanners capture and store more information, like volume estimation, curvature and so on. The use of these data would enable the specialists to plan and asses the surgery in a more accurate way.

The 3D scanner acquisition of human body parts requires a certain time and skills. Long scanning time, tends to increase the patient stress as well as the amount of noise due to the breath and involuntary micro-movements. Modern hand-held scanners, reduces these problems by allowing low acquisition time. Furthermore it guarantees a sufficiently high quality of the data. Actually, extremely high resolution and accuracy are pointless to capture general shape. Moreover, dense points clouds would affect the processing time. For this reasons we propose to perform dataset acquisition with a fast and low-cost hand-held 3D scanner: Structure Sensor [10]. High portability of hand-held scanners simplifies the operator job, that can easily turn around the patient.

The 3D data have to be processed and simplified to capture just the information that surgeons need for their analysis. In the proposed approach we consider normals orientation to build a compact representation of breast model. To further simplify processed 3D data, Principal Component Analysis (PCA) [11] has been employed. PCA is a popular and valuable approach to reduce the high dimensionality of the datasets and capture just the most significant features. Feature reductions through PCA has already been used in the parametrization process of the human body parts [12,13]. Concerning the breast shapes, other authors proposed to analyse them either using linear measurements, stationary laser scanner, MRI, X-rays or thermoplastic moulding [14–19]. Compared to our previous work [9], in this paper we do not employ the planar projections. In [9] the 3D meshes are projected in 2D space and then Thin-Plate Splines [20] is used to estimate the non-linear transformation which change each breast projection in the average one. Our contribution in the field can be summarized in the following points:

- The acquisition of 3D breast models to build a proper dataset and perform significant experiments. At the best of our knowledge there are not available dataset like this.
- The idea to exploit 3D normals to create a compact representation of 3D breast models.
- Time and cost optimization by employing a hand-held 3D scanner.

The remainder of this paper is structured as follows: employed devices and proposed method are described in Sect. 2. Details on the dataset are provided in Subsect. 2.1, while the proposed parametrization method is detailed in Subsect. 2.2. Experimental results are given in Sect. 3. A final discussion, with some consideration for future works, ends the paper.

2 Materials and Methods

The study we conducted is mainly focused on digital shape analysis of breast models to assist breast surgeons for medical and surgical purposes. Our idea is based on three key points: minimally invasive for the patient, use of low cost devices, easy data visualization-&-understanding for people with a medical background.

We employed a 3D scanner with structured infrared light technology that allows us to acquire the information about depth of thousands of points at the same time. The Structure Sensor (Fig. 1) is a hand-held scanner proved to be empirically able to acquire up to 12 m, although it is recommended a distance in the range 0.4 and 3.5 m. Its maximum accuracy is 0.5 mm, but worsens when the volume of the area scanned is large. Since the scanner uses infrared rays, it is recommended for indoor usage only. The device is calibrated, that means each 3D model will show its real size. The sensor itself is not able to acquire RGB colour mode information, however it is possible to plug into an iPad and uses the tablet camera to this purpose.

To acquire a breast model, we propose a clinical procedure in which the female patients hold the hands behind and above the head. In this way the operator can move around the breast with the Structure Sensor (which is clipped onto the iPad). Although texture information have been acquired, this has not been used for the present investigation. An example of the model acquired with Structure Sensor is shown in Fig. 2.

Once the model is acquired, it is automatically pre-processed through a 3D processing software (Meshlab [21]), in order to remove noise, isolated vertices and faces. Mesh editing is followed by a manual definition, through cropping, of the Region of Interest (ROI). ROI extraction is a critical part of the proposed procedure. We adopt a simple approach that has been proved to be replicable and reasonable precise. We manually selected the ROI exploiting four anatomical reperees suggested by the breast surgeons (Figs. 2 and 3). In our acquisitions, we scanned both left and right breasts but all of them have been, when needed, vertically mirrored in order to make the dataset right-left side invariant, as shown in Fig. 3.

Each model is saved with the standard OBJ format, which describes the information on vertices, faces and face normals. The average number of vertices

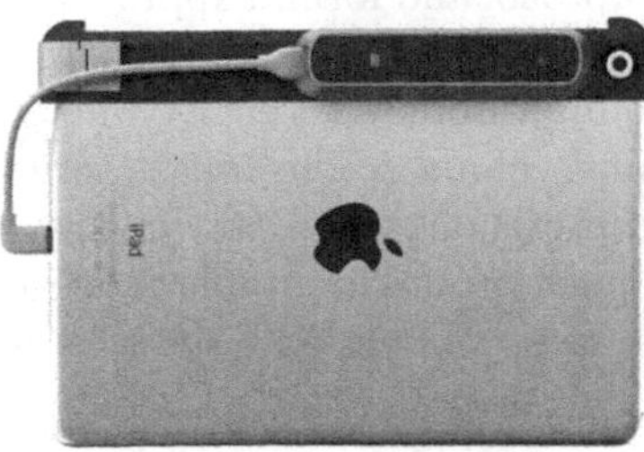

Fig. 1. A Structure Sensor clipped onto an iPad. We used the same setting in our acquisitions.

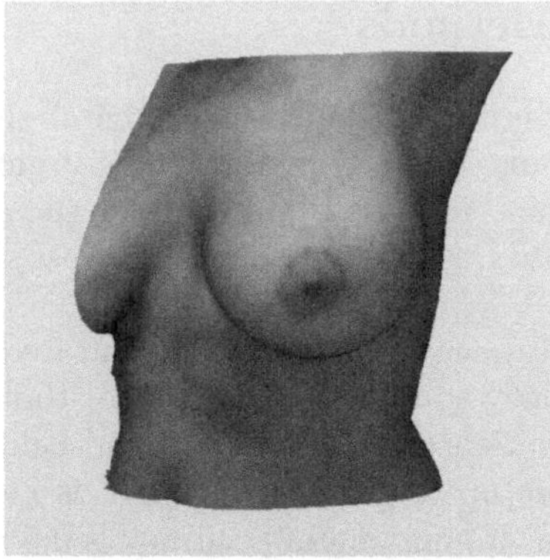

Fig. 2. Example of a textured mesh as it is acquired by the Structure Sensor.

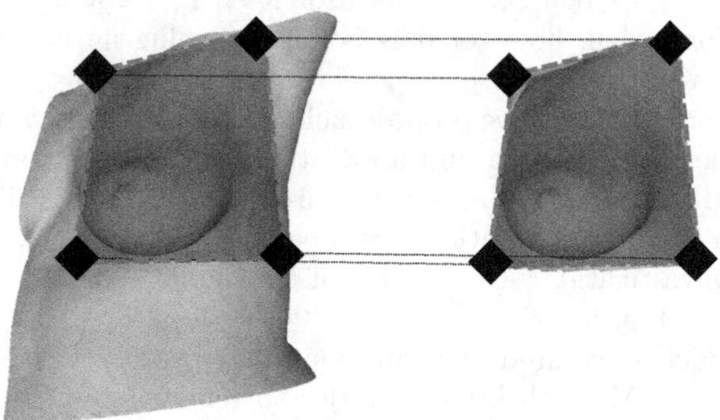

Fig. 3. Definition of ROI through 4 anatomical reperees suggested by the breast surgeons.

is ~1,500, while the average number of faces is ~4,000. These models resolution is not extremely high but it is enough to capture information about breast shape, which is the point of this work.

2.1 3D Breast Dataset

After review of the study protocol and formal approval by the internal ethic committee of ASLT (Associazione Santantonese per la Lotta ai Tumori) we gathered a dataset with breasts acquired from different volunteers, aged between 25 and 65, with different shapes and volumes. The breast surgeons put a label on each model, describing size and ptosis of the breast. The severity of ptosis is characterized by evaluating the position of the nipple relative to the infra-mammary fold. Supervised by the doctors, we created a dataset in the following way:

- Main Dataset: is made up of 31 breasts, 17 left and 14 right. To guarantee a proper dataset variability, we have included breasts of different size and ptosis.

Then, in order to test the strength of the proposed methodology, we selected a patient and acquired her breast several times in pre-operation and post-operation conditions. Hence, two more groups of meshes is distinguishable:

- Pre-operation - Group 1: is made by 52 meshes, 26 left and 26 right. Notice that this set of meshes has been acquired by two operators, namely a junior and an expert one, so it can be used to investigate how the proficiency of the operator may change the parametrization
- Post-operation - Group 2: is made by 16 breasts, 8 left and 8 right. These models come from the same patient of Group 1 after a surgery on both breasts.

2.2 Shape Parametrization

In this subsection we present the method employed to process the 3D models in order to parametrize the breast shape with a minimum number of parameters. Our idea is to describe each 3D model as histograms of normals. Since normal vectors define the orientation of each model vertex/face, the proposed algorithm starts with a registration step. Actually, although the acquisition device is calibrated, it doesn't have a system to get the correct orientation into the real space (e.g., gravity sensors). Hence, the meshes have initially to be oriented along the same direction and its centroid moved on the origin of a 3D Cartesian coordinate system. As second step, the normal space is clustered and the occurrences for each cluster counted. This descriptor is finally reduced by Principal Component Analysis. The summarized pipeline of proposed method is shown in Fig. 4.

Breast Registration. As mentioned before, the acquired data have to be roto-translated since the built descriptor depends on normals orientation. This process is automatically performed as described in [9]. First of all the mesh centroid is moved into the origin of a cartesian coordinate system. Subsequently, the average normal is computed to find the rotation matrix, in order to align it along the Z axis. This means, we use the unit vector $(0, 0, 1)$ as reference. Finally, to get the matrix, a closed form named Rodrigues's rotation formula [22] is employed. Specifically, given two vectors u_1 and u_2, formula computes the rotation to align u_1 to u_2. In our case $u_1 = averageNormal$ and $u_2 = (0, 0, 1)$.

Bag of Normals. After each mesh has been correctly oriented in our coordinate system, we can proceed to obtain a representation of the normals distribution over a suitably quantized grid. Firstly, all the normals are normalized. We divided each normal u for $||u||$, in order to get a unit vector. By performing this process, the three components of normal vector (u_x, u_y, u_z) fall in the range $[-1, 1]$. We linearly quantize the space of each component into 4 levels, in order to obtain $4 \times 4 \times 4 = 64$ different cluster. Finally, each mesh is represented by counting the occurrences in each cluster. This histogram with 64 bins is then in turn normalized to get the final bag of normals descriptor.

Principal Component Analysis (PCA). PCA is a popular statistical method that is commonly used for finding patterns in data of high dimension or reducing such dimensionality. This reduction is more interesting when one wants to extract

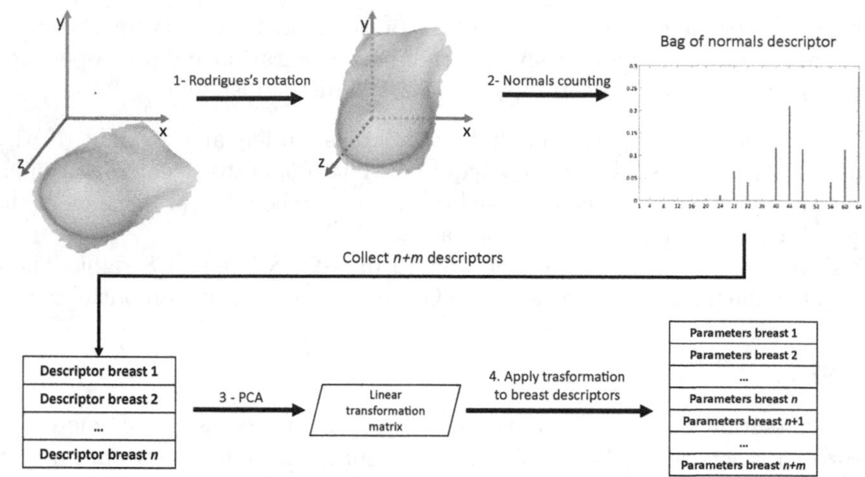

Fig. 4. Pipeline of the proposed method. Note that PCA is applied on n breast descriptors. Then, the "learnt" transformation matrix is used as model to extract parameters of all the 3D meshes. Additional details are reported in Sect. 3.

the main characteristics of complex data. PCA is applied on datasets which are described by several attributes. It is able to find a linear transformation which move the data into another space where the transformed attributes are uncorrelated. The aim is to identify the "Principal Components", or rather a reduced set of attributes which represent the original data [11].

We applied PCA on the 64-d descriptors obtained at the previous step in order to describe each 3D breast with a very small set of parameters, namely 2. This procedure allows us to represent each 3D model as a point in 2D coordinate system where axes are the first two Principal Components. This kind of representation, allows to visually asses the results of a surgery intervention by observing the change of position of a breast in the 2D space. It the next section we report the results that show that 2 components are enough to represent breast shape.

3 Results

We computed PCA on the 31 models in the main dataset. Exploiting only the first 2 principal components we obtained a variance retain of $48.04 + 29.35 = 77.39\%$ (Fig. 5(a)) and the models can be represented in a chart, as shown in Fig. 5(b). The breast surgeons confirmed us the evidence of Fig. 5(b): the first 2 principal components seem enough to distinguish characteristic traits of the labelled models, since models are clearly separated in the obtained result. However, since there are not official metrics to describe breast shape, we currently cannot associate each component to a specific geometrical property.

In order to further assess the soundness of the proposed method we plotted the models of Group 1, exploiting the PCA computed only on the main

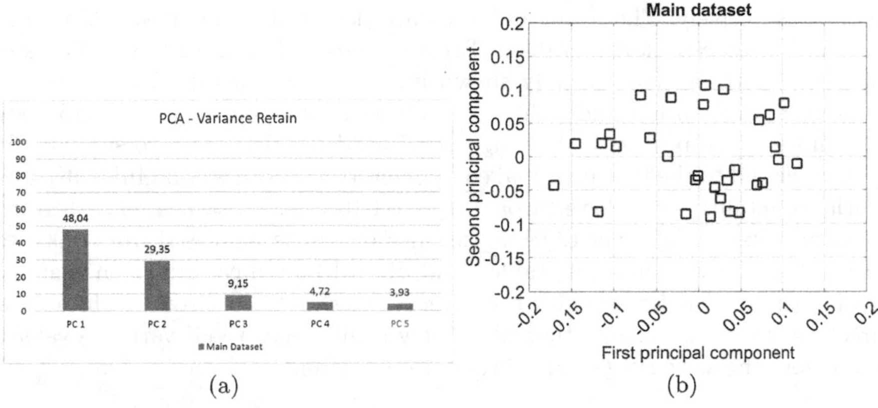

Fig. 5. PCA computed on the Main Dataset. (a) Variance Retain of the first 5 principal components. The sum of the first 2 principal components is 77.39%. (b) Plot of the 31 models in the Main Dataset using the first 2 principal components.

dataset (Fig. 6(a)). The left breast is clearly distinguishable from the right one, as expected. Once more, using the same principal components, we plotted also the models from Group 2 (Fig. 6(b)). We remark that 3D models in Groups 1 and 2 includes the right and left breast of the same patient, before and after a

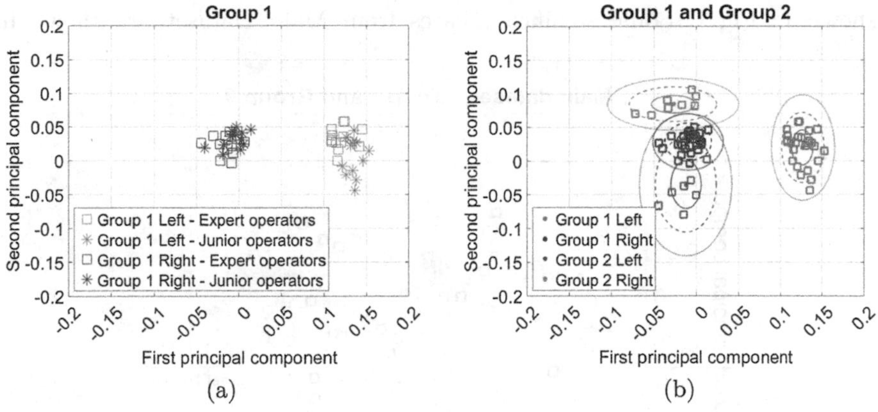

Fig. 6. Plots of the models in Group 1 and Group 2 using the first 2 principal components of PCA computed on the Main Dataset. (a) Visual comparison of the principal components of Group 1 between models acquired by the two groups of operators, properly juniors and experts. (b) Comparison of the principal components between Group 1 (pre-surgery) and 2 (post-surgery). Error in the parametrization has been highlighted through error ellipses added on each set of models. Starting from the ellipsis centroid (the mean value of the set), each concentric error ellipsis contains the 68% (σ), the 95% (2σ) and the 99% (3σ) of the elements, respectively.

surgery, respectively. The mean and standard deviation of models in Groups 1 and 2 have been reported in Table 1. Error ellipses including the 68%, 95% and 99% of the data are contextually shown in Fig. 6(b). The Euclidean distances between centroids of left and right breast clusters for Group 1 and Group 2 are 0.137 and 0.12, respectively. Although the Euclidean distances are similar, the distance related to the first principal component (the most meaningful, with 48% of variance retain) is way lower: from 0.136 to 0.014. The distance related to the second principal component (29% of variance retain) is increased from 0.008 to 0.119. So, the results shown in Table 1 and Fig. 6(b) are a confirmation that the right breast and left breast after the surgery (meshes from Group 2) have now a first principal components that has pretty similar mean and variance values, while before the surgery (Group 1) they were different.

Table 1. Mean and Standard Deviation of models in Group 1 and 2. L stands for Left, R for Right. Each entry is a pair in which the values are related to the first and second principal component, respectively.

	Group 1		Group 2	
	L	R	L	R
Mean	(0.1269, 0.0198)	(−0.0098, 0.0281)	(−0.0124, −0.0352)	(−0.0273, 0.0843)
Stand. Dev.	(0.0127, 0.0262)	(0.0138, 0.0139)	(0.0181, 0.0353)	(0.0258, 0.0129)

The comparative chart with the components of all the digitized breasts is shown in Fig. 7. Some significant cases from Main Dataset are shown in

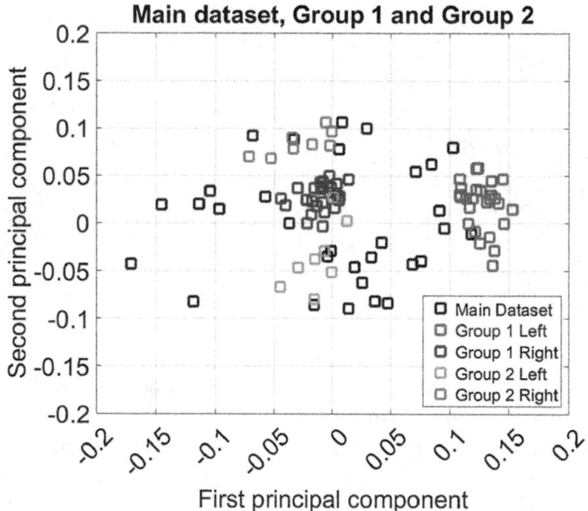

Fig. 7. Comparison of the first 2 principal components (X and Y axis, respectively) between different datasets. PCA computed on the main dataset, comparison between the main dataset, Group 1 and Group 2.

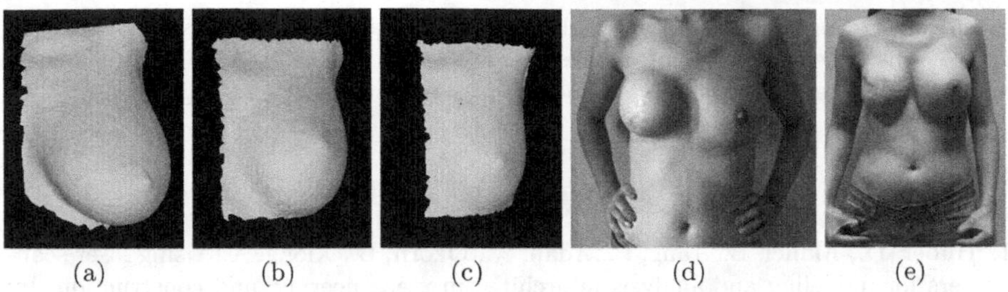

Fig. 8. Significant acquired models. (a–c) Models from Main Dataset with principal components (−0.17; −0.04), (0; −0.02) and (0.11; −0.01), respectively. They are in the most left, central e right position of the plot of Fig. 5(b). A clear difference about the shape of the breasts can be noticed. (d–e) Patient of Group 1 (pre-surgery) and Group 2 (post-surgery), respectively; note that we considered right breast the one corresponding to the right arm of the patient.

Figs. 8(a)–(c), while the patient scanned in Groups 1 and 2 is shown in Figs. 8(d) and (e). The breast surgeons confirmed us that the positions of models from these latter sets are coherent with respect to the one of the models from Main Dataset. These results show that the first principal component is strong enough to characterize the shape of a breast, and through the standard deviation computations on Group 1 and 2 we can also give a cue about the error in this estimation.

4 Conclusions

In this work we have focused on digital shape analysis of breast models to assist breast specialists for medical and surgical purposes. We fixed three key points for our proposed solution: minimally invasive for the patient approach, use of low cost devices, easy data visualization-&-understanding for people with a medical background. We proposed a clinical procedure in which the female patients hold the hands behind and above the head, while an operator can digitize her breast with a 3D scanner. After a manual ROI definition through cropping, the meshes are automatically processed. The breasts are represented exploiting bag of normals representation, resulting in a 64-d descriptor. A reference dataset has been used to compute PCA on a set of discriminative and different breasts, and the obtained first 2 principal components have been used to plot the breasts into a 2D space. We empirically proved that breasts subjected to a surgery change their representation in this space, and through the variance computations on Group 1 and 2 we also gave a cue about the error in this estimation. We believe that the proposed procedure, assessed by the surgeon, represents a valid solution to evaluate the results of surgeries, since one of the most important goal of the specialists is to symmetrically reconstruct breasts, but an objective tool to measure the result is currently missing. As future works, we planned to augment the ROI extraction phase, which is a critical part of the proposed procedure and requires professionals with a proper know-how of 3D object editing.

Acknowledgment. The authors would like to thank the "Azienda Ospedaliera Cannizzaro", the "Associazione Santantonese per la lotta ai tumori (ASLT)" and the female volunteers for their contribution as models.

References

1. Huber, D., Akinci, B., Tang, P., Adan, A., Okorn, B., Xiong, X.: Using laser scanners for modeling and analysis in architecture, engineering, and construction. In: Conference on Information Sciences and Systems (CISS), pp. 1–6, March 2010
2. Stoll, J., Novotny, P., Howe, R., Dupont, P.: Real-time 3D ultrasound-based servoing of a surgical instrument. In: International Conference on Robotics and Automation (ICRA), pp. 613–618, May 2006
3. Bottino, A., De Simone, M., Laurentini, A., Sforza, C.: A new 3-D tool for planning plastic surgery. IEEE Trans. Biomed. Eng. **59**(12), 3439–3449 (2012)
4. Treleaven, P., Wells, J.: 3D body scanning and healthcare applications. Computer **40**(7), 28–34 (2007)
5. Dai, Y., Tian, J., Dong, D., Yan, G., Zheng, H.: Real-time visualized freehand 3D ultrasound reconstruction based on GPU. IEEE Trans. Inf. Technol. Biomed. **14**(6), 1338–1345 (2010)
6. Stanco, F., Tanasi, D., Allegra, D., Milotta, F.L.M., Lamagna, G., Monterosso, G.: Virtual anastylosis of Greek sculpture as museum policy for public outreach and cognitive accessibility. J. Electron. Imaging **26**(1), 011025 (2017)
7. Laing, R., Leon, M., Isaacs, J.: Monuments visualization: from 3D scanned data to a holistic approach, an application to the city of Aberdeen. In: International Conference on Information Visualisation, pp. 512–517, July 2015
8. Nguyen, C.V., Fripp, J., Lovell, D.R., Furbank, R., Kuffner, P., Daily, H., Sirault, X.: 3D scanning system for automatic high-resolution plant phenotyping. In: International Conference on Digital Image Computing: Techniques and Applications (DICTA), pp. 1–8, November 2016
9. Gallo, G., Allegra, D., Atani, Y.G., Milotta, F.L.M., Stanco, F., Catanuto, G.: Breast shape parametrization through planar projections. In: Blanc-Talon, J., Distante, C., Philips, W., Popescu, D., Scheunders, P. (eds.) ACIVS 2016. LNCS, vol. 10016, pp. 135–146. Springer, Cham (2016). doi:10.1007/978-3-319-48680-2_13
10. Structure Sensor Website. http://structure.io/. Accessed Apr 2017
11. Pearson, K.: On lines and planes of closest fit to systems of points in space. Phil. Mag. **2**(6), 559–572 (1901)
12. Allen, B., Curless, B., Popovi, Z.: The space of human body shapes: reconstruction and parameterization from range scans. In: International Conference on Computer Graphics and Interactive Techniques, pp. 587–594 (2003)
13. Gallo, G., Guarnera, G.C., Catanuto, G.: Human breast shape analysis using PCA. In: Proceedings of the Third International Conference on Bio-inspired Systems and Signal Processing (BIOSIGNALS) (2010)
14. Smith Jr., D.J., Palin Jr., W.E., Katch, V.L., Bennett, J.E.: Breast volume and anthropomorphic measurements: normal values. Plast. Reconstr. Surg. **78**(3), 331–335 (1986)
15. Farinella, G.M., Impoco, G., Gallo, G., Spoto, S., Catanuto, G.: Unambiguous analysis of woman breast shape for plastic surgery outcome evaluation. In: 4th Conference Eurographics Italian Chapter (2006)

16. Catanuto, G., Gallo, G., Farinella, G.M., Impoco, G., Nava, M.B., Pennati, A., Spano, A.: Breast shape analysis on three-dimensional models. In: Third European Conference on Plastic and Reconstructive Surgery of the Breast (2005)
17. Galdino, G.M., Nahabedian, M., Chiaramonte, M., Geng, J.Z., Klatsky, S., Manson, P.: Clinical applications of three-dimensional photography in breast surgery. Plast. Reconstr. Surg. **110**(1), 58–70 (2002)
18. Nahabedian, M.Y., Galdino, G.: Symmetrical breast reconstruction: is there a role for three-dimensional digital photography? Plast. Reconstr. Surg. **112**(6), 1582–1590 (2003)
19. Lee, H.Y., Hong, K., Kim, E.A.: Measurement protocol of womens nude breasts using a 3D scanning technique. Appl. Ergon. **35**, 353–360 (2004)
20. Bookstein, F.L.: Principal warps: thin-plate splines and the decomposition of deformations. Trans. Pattern Anal. Mach. Intell. **11**(6), 567–585 (1989)
21. Cignoni, P., Callieri, M., Corsini, M., Dellepiane, M., Ganovelli, F., Ranzuglia, G.: Meshlab: an open-source mesh processing tool. In: Eurographics Italian Chapter Conference, vol. 2008, pp. 129–136 (2008)
22. Weisstein, E.: Rodrigues' Rotation Formula. http://mathworld.wolfram.com/rodriguesrotationformula.html. Accessed Apr 2017

Measuring Refractive Properties of Human Vision by Showing 4D Light Fields

Megumi Hori[1], Fumihiko Sakaue[1], Jun Sato[1(✉)], and Roberto Cipolla[2]

[1] Department of Computer Science, Nagoya Institute of Technology, Nagoya 466-8555, Japan
hori@cv.nitech.ac.jp, {sakaue,junsato}@nitech.ac.jp

[2] Department of Engineering, University of Cambridge, Cambridge CB2 1PZ, UK
cipolla@eng.cam.ac.uk

http://www.cv.nitech.ac.jp, http://www.eng.cam.ac.uk

Abstract. In this paper, we propose a novel method for measuring refractive properties of human vision. Our method generates a special 4D light field and present it to human observers, so that the observers will see different images according to their dioptric properties, i.e. the amount of near sightedness and far sightedness. Thus, we can measure the visual properties of observers just by showing the light field images. The proposed method does not require expensive measurement systems nor well trained medical doctors unlike the existing eyesight measurement methods. The real image experiments and the synthetic image experiments show the efficiency of the proposed method.

1 Introduction

In recent years, many people are suffering from eyesight problems. Thus, it is important to measure the refractive properties of human vision easily and accurately. Traditionally, the refractive properties of human vision have been measured by using autorefractors [3,12]. Although they can measure precise dioptric properties of human vision, they are very expensive and must be operated by well trained medical doctors. Thus, we in this paper propose a novel method for measuring dioptric properties of human vision easily and quickly.

Our estimation method is very different from the standard computer vision techniques. In computer vision, we usually obtain camera images of objects and estimate their properties, such as shape, reflectance and refractivity, from images. On the contrary, our method measures the refractive properties of eyesight by showing images to observers. In this method we generate a special 4D light field and present it to human observers. Then, the human observers will see specific images according to the refractive properties of their eyesight. As a result, we can measure the refractive properties of their eyesight from the observed images. For example, a person who has nearsightedness with the dioptric power of $-0.50D$ will observe a flower, while a person whose dioptric power is $+0.10D$ will observe a dog.

S. Battiato et al. (Eds.): ICIAP 2017, Part II, LNCS 10485, pp. 522–533, 2017.
https://doi.org/10.1007/978-3-319-68548-9_48

Since the proposed method does not need to capture images nor compute them, it does not require computation time and is very efficient.

We first present related work in Sect. 2, and show the advance of the proposed method. We next explain the refractive properties of human vision in Sect. 3. Then, we propose a novel method for measuring the refractive properties of human vision by presenting 4D light field in Sect. 4. The experimental results in Sect. 5 show the efficiency of the proposed method.

2 Related Work

Since the pioneering work of a light field display by Levoy and Hanrahan [11], the light field displays and their applications have been studied extensively [6–10,18,19]. Recently, Huang et al. [7] proposed a novel displaying method, which enables weak sighted people to see visual information correctly without wearing eyeglasses. Their method generates 4D light fields, so that the distorted visual system of a weak sighted person observes in-focus images. The visual acuity can also be improved by displaying images convolved with the inverse PSF of human vision [1,2]. These vision correction methods require the measurement of the refractive properties of human vision before using them. However, the precise measurement of the refractive properties of human vision requires very expensive measurement systems, such as autorefractors [3,12], and only well trained medical doctors can operate them. Thus, we need more simple and low cost methods for measuring the precise refractive properties of human vision.

For measuring the refractive properties of transparent objects, the ray tracing has often been used in the existing methods [5,14–17]. These methods use the relationship between the input light rays and the output light rays of transparent objects for analyzing their refractive properties. However, it is very difficult to obtain light rays which passed the human visual systems, and thus we cannot use these existing methods for measuring the refractive properties of human vision. For extracting the dioptric properties of human vision, Pamplona et al. [13] used a light field display. They showed that it is possible to measure refractive properties of human vision by controlling light rays of the light field display so that two light rays meet at the retina of observers. Although their method is efficient and can measure the refractive properties interactively, it requires the observers to control the light rays iteratively, so that their spots on the retina coincide. Thus, people who do not have skills to control the light field display, such as infants, cannot use their method.

Therefore, we in this paper propose a novel method for measuring the refractive properties of human vision just by showing a 4D light field to human observers. Our method generate a special 4D light field, so that the human observers see different images according to the refractive properties of their vision. Thus, we can measure their refractive properties just by checking what are observed by them. Hence, our method does not need any skills to control the light field display.

3 Refractive Properties of Human Vision

If we have a normal vision, the light rays which come into the eye intersect on the retina as shown in Fig. 1(a), and sharp images are obtained on the retina. If the eye lens is not controlled properly and is thicker than usual, the input light rays intersect in front of the retina and are spread on the retina as shown in Fig. 1(b). As a result, the observed images are blurred. This is called nearsightedness. Similarly, if the eye lens is thinner than usual, the input light rays also spread on the retina, and observed images are blurred as shown in Fig. 1(c). This is called farsightedness. In general, the refractive properties of human vision are described by using a dioptric power D, which is the reciprocal of the maximum or minimum focal length f controllable by the eye lens of the human observer as follows:

$$D = \frac{1}{f} \tag{1}$$

For example, the dioptric power of a near sight observer with the maximum visible distance of 2 m is $\frac{1}{-2} = -0.5D$, and that of a far sight observer with the minimum visible distance of 0.5 m is $\frac{1}{0.5} = +2.0D$.

The visual acuity can be measured by several methods. The most popular method is the Landolt ring method [4], in which a ring with a gap is shown to the human observer to check the visibility of the gap. The Landolt ring method is very simple and easy to check the grade of eyesight. However, since it is too simple, we cannot obtain the detail characteristics of the eyesight. For example, we cannot distinguish nearsightedness and farsightedness just from the Landolt ring method. This is because the Landolt ring method only measures the amount of blur on the human retina.

For measuring the refractive properties of human vision more precisely, autorefractors [3,12] are often used by medical doctors and eyeglass designers. Although the autorefractor can measure precise refractive properties of eyesight, it is very expensive and must be operated by well trained medical doctors or eyeglass designers, and cannot be used by ordinary people. Thus, for using flexible eyesight correction systems such as [7], we need more simple and low cost methods for measuring precise refractive properties of human vision. In the following sections, we show that by controlling the 4D light field, we can measure the detail characteristics of human eyesight very easily.

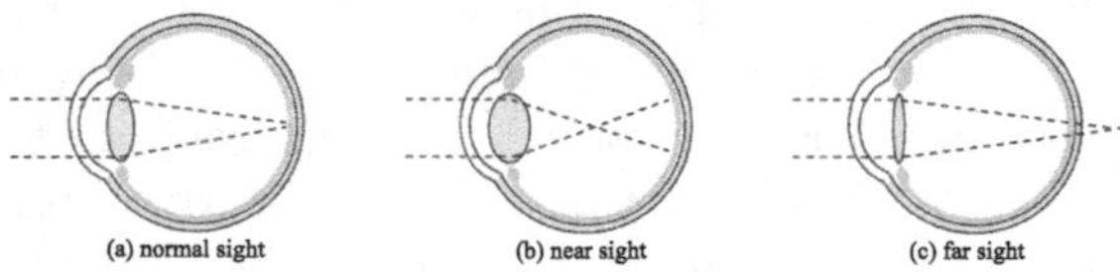

Fig. 1. Eyesight

4 Estimating Refractive Properties from 4D Light Field

4.1 Observation of Light Field in Human Vision

For measuring the refractive properties of human vision by showing images, we use a light field display. The light field display consists of a 2D display and a micro lens array, and can generate arbitrary intensity of light toward arbitrary orientations as shown in Fig. 2.

Suppose a farsighted person observes a standard 2D display which is close to the person. Then, because of the farsightedness, light rays emitted from different points on the display converge to an identical point on the retina as shown in Fig. 3(a). Since the different points on the display emit different colors in general, the point on the retina observes the sum of these colors as shown in Fig. 3(a). As a result, the farsighted person observes blurred images.

On the contrary, if we use a 4D light field display, we can emit different light in each orientation at each position on the display. Thus, we can emit light fields, so that the light rays, which converge to an identical point on the retina, have the same color as shown in Fig. 3(b). As a result, the farsighted person can observe clear sharp images, even if the display is closer than the minimal local length of the person.

We next consider the observation model of light emitted from a 4D light field display. Let us consider a light $L(s,t,u,v)$ emitted from a point (s,t) on the display into the orientation (u,v) by using the light field display. Then, since the

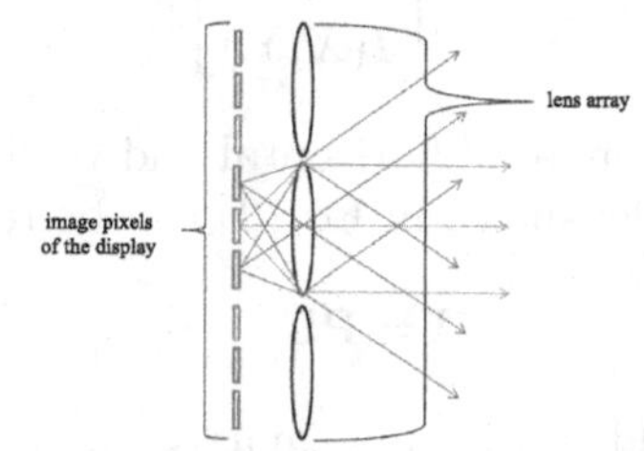

Fig. 2. Light field display

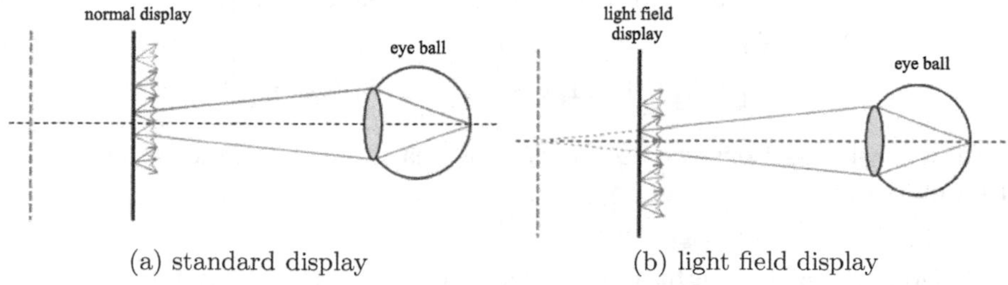

(a) standard display (b) light field display

Fig. 3. Observation of light field (Color figure online)

observed intensity $I(x, y)$ is the sum of light which comes into a point (x, y) on the retina, it can be described as follows:

$$I(x, y) = \sum_{s=1}^{S} \sum_{t=1}^{T} \sum_{u=1}^{U} \sum_{v=1}^{V} \delta(x, y, s, t, u, v) L(s, t, u, v) \tag{2}$$

where, S and T are the number of horizontal and vertical points on the light field display, and U and V are the number of horizontal and vertical orientations of the light field display. δ denotes a 6th order tensor, whose component $\delta(x, y, s, t, u, v)$ is 1 if a light ray $L(s, t, u, v)$ passes through a point (x, y) on the retina, and is 0 if it does not pass through the point (x, y) as follows:

$$\delta(x, y, s, t, u, v) = \begin{cases} 1 \; if \; L(s, t, u, v) \rightarrow I(x, y) \\ 0 \qquad otherwise. \end{cases} \tag{3}$$

Let us consider a vector $\mathbf{L}$ which consists of light rays $L(s, t, u, v)$, and a vector $\mathbf{I}$ which consists of the observed intensities $I(x, y)$ as follows:

$$\mathbf{L} = \begin{bmatrix} L(1, 1, 1, 1) \\ \vdots \\ L(S, T, U, V) \end{bmatrix} \tag{4}$$

$$\mathbf{I} = \begin{bmatrix} I(1, 1) \\ \vdots \\ I(X, Y) \end{bmatrix} \tag{5}$$

where, X and Y are the number of horizontal and vertical points on the retina. Then, the observation model shown in Eq. (2) can be rewritten as follows:

$$\mathbf{I} = \mathbf{PL} \tag{6}$$

where, $\mathbf{P}$ is a light field projection matrix, and is described by using $\delta(x, y, s, t, u, v)$ as follows:

$$\mathbf{P} = \begin{bmatrix} \delta(1, 1, 1, 1, 1, 1) & \cdots & \delta(1, 1, S, T, U, V) \\ \vdots & \ddots & \vdots \\ \delta(X, Y, 1, 1, 1, 1) & \cdots & \delta(X, Y, S, T, U, V) \end{bmatrix} \tag{7}$$

In the following sections, we use Eq. (7) for measuring the refractive properties of human vision.

4.2 Measuring Eyesight Characteristics from 4D Light Field

We next describe a method for measuring the refractive properties of human vision by using the light field display. In our method, we present a specially designed light field from a light field display, so that human observers see different images according to their refractive properties.

Suppose we have N different eyesight characteristics. Then, these N eyesight characteristics are represented by N different light field projection matrices $\mathbf{P}_i$ $(i = 1, \cdots, N)$. Now, let us consider N observers O_i $(i = 1, \cdots, N)$, whose eyesight characteristics are described by $\mathbf{P}_i$ $(i = 1, \cdots, N)$. We show N different images $\mathbf{I}_i$ $(i = 1, \cdots, N)$ to these N observers O_i $(i = 1, \cdots, N)$. Then, the observation of these N observers can be described as follows:

$$\begin{bmatrix} \mathbf{I}_1 \\ \vdots \\ \mathbf{I}_N \end{bmatrix} = \begin{bmatrix} \mathbf{P}_1 \\ \vdots \\ \mathbf{P}_N \end{bmatrix} \mathbf{L} \tag{8}$$

Thus, the light field $\mathbf{L}$, which provides N different images $\mathbf{I}_i$ $(i = 1 \cdots, N)$ toward N observers O_i $(i = 1 \cdots, N)$, can be derived as follows:

$$\mathbf{L} = \begin{bmatrix} \mathbf{P}_1 \\ \vdots \\ \mathbf{P}_N \end{bmatrix}^{-} \begin{bmatrix} \mathbf{I}_1 \\ \vdots \\ \mathbf{I}_N \end{bmatrix} \tag{9}$$

where, $(^-)$ denotes the pseudo inverse of a matrix. This means N different visual acuities of human observer can be identified just by showing the light filed $\mathbf{L}$ to the observer and checking the image observed by the observer.

However, unfortunately the light field $\mathbf{L}$ derived from Eq. (9) cannot be displayed by the real light field display, since the light field $\mathbf{L}$ derived from Eq. (9) includes negative intensity values in general. To cope with this problem, we derive a light field $\mathbf{L}$ by solving the following conditional minimization problem:

$$\min_{\mathbf{L}} \sum_{i=1}^{N} ||\mathbf{I}_i - \mathbf{P}_i \mathbf{L}||^2 \quad \text{subject to} \quad 0 \leq L(s,t,u,v) \leq I_{max} \tag{10}$$

where, I_{max} denotes a maximum intensity value of the light field display. Then, the refractive properties of a human observer can be measured by showing the light filed $\mathbf{L}$ derived from Eq. (10) to the observer.

5 Experiments

5.1 Real Image Experiments

We next show the results from real image experiments. In our experiments, we put a micro lens array on the display of a Nexus 7 tablet, and used them as a light field display. The resolution of the display of the tablet was 1920×1200 pixels, but we used only 80×80 pixels for a light field display. The resolution of position of the light field display was 20×20, and the resolution of orientation of the light field display was 4×4. We used a CCD camera with the resolution

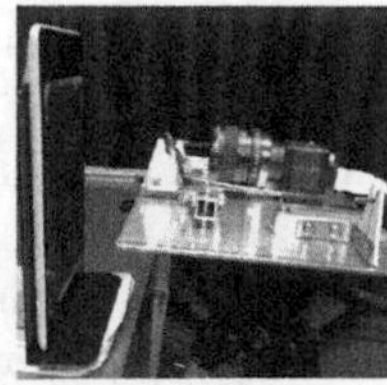

Fig. 4. Experimental setup.

(a) normal sight observer (b) near sight observer

Fig. 5. Target images for a normal sight observer and a near sight observer.

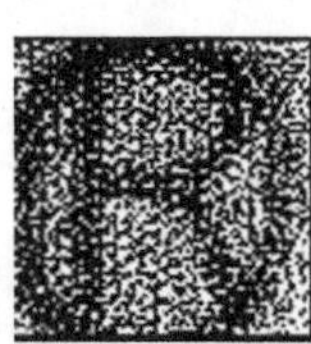

Fig. 6. 4D light field derived from Fig. 5.

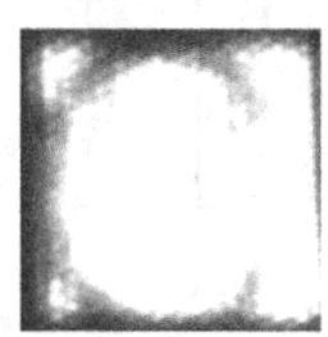

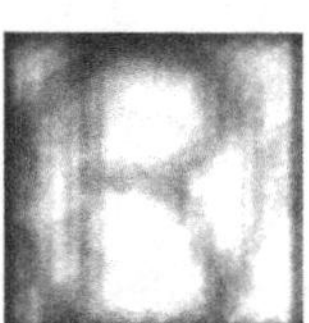

(a) normal sight observer (b) near sight observer

Fig. 7. Images observed by a normal sight observer and a near sight observer.

of 1292×964 as an observer, and it observed the light field display as shown in Fig. 4. The distance from the display to the camera was 10 cm.

For simulating two different eyesight characteristics, we put eyeglasses of nearsightedness in front of the camera, and took images by the camera with and without eyeglasses. The focus of the cameras was set up, so that it focuses on the display when we put eyeglasses in front of the camera. Thus, the camera with eyeglasses is a normal sight observer, and the camera without eyeglasses is a near sight observer.

We next derived a 4D light field from Eq. (10), so that the normal sight observer observes the image shown in Fig. 5(a) and the near sight observer observes the image shown in Fig. 5(b). The derived 4D light field is shown in Fig. 6, in which the 4D light field is represented as a 2D image.

The light field shown in Fig. 6 was presented by using the light field display, and was observed by the normal sight observer and the near sight observer. The images observed by these two observers are shown in Fig. 7(a) and (b) respectively. As shown in Fig. 7(a) and (b), the images observed by the normal sight observer and the near sight observer are different from each other, and are close to the target images shown in Fig. 5(a) and (b). These results show that observers with different eyesight characteristics see different images, and their refractive properties can be measured just from observed images by using the proposed method.

5.2 Accuracy Evaluation

We next evaluated the accuracy of identification of refractive properties in the proposed method by using synthetic light fields. In this experiment, the distance from the lens center to the retina was 24 mm, and the distance from the display to the lens center was 250 mm. Thus, the focal length of the normal sight observer was $f = 21.8978$ mm in this case. Then, the focal length of the near sight observer is $f - \alpha$, and the focal length of the far sight observer is $f + \alpha$ respectively, where α is the magnitude of nearsightedness or farsightedness. The light field display had the resolution of 30×30 in position and 5×5 in orientation, and thus its total resolution was $30 \times 30 \times 5 \times 5$. The resolution of the image on the retina was 30×30 pixels.

The accuracy of identification of refractive properties was evaluated by using the similarity between a target image I and an observed image $\hat{I}$. The similarity was measured by the zero-mean normalized cross correlation (ZNCC) as follows:

$$ZNCC = \frac{\sum_{x=1}^{X}\sum_{y=1}^{Y}(I(x,y) - \bar{I}(x,y))(\hat{I}(x,y) - \bar{\hat{I}}(x,y))}{\sqrt{\sum_{x=1}^{X}\sum_{y=1}^{Y}(I(x,y) - \bar{I}(x,y))^2}\sqrt{\sum_{x=1}^{X}\sum_{y=1}^{Y}(\hat{I}(x,y) - \bar{\hat{I}}(x,y))^2}}$$

where, $\bar{I}$ and $\bar{\hat{I}}$ are typical values of I and $\hat{I}$ respectively. By using ZNCC, we evaluated the accuracy of identification in the case of three eyesight characteristics and five eyesight characteristics respectively.

We first evaluated the accuracy of identification in the case of three eyesight characteristics. The focal length of these three eyesight characteristics was $f - 0.10$, f and $f + 0.10$, which correspond to near sight, normal sight and far sight observes respectively. The target images for these three eyesight characteristics are shown in Fig. 8(a)–(c). The 4D light field for identifying these three eyesight characteristics was derived from the proposed method and presented to the observer. The focal length of the observer was changed from $f - 0.15$ to $f + 0.15$, and the light field was observed at each focal length. Figure 9 shows observed images when the focal length of the observer was $f - 0.10$, $f - 0.05$,

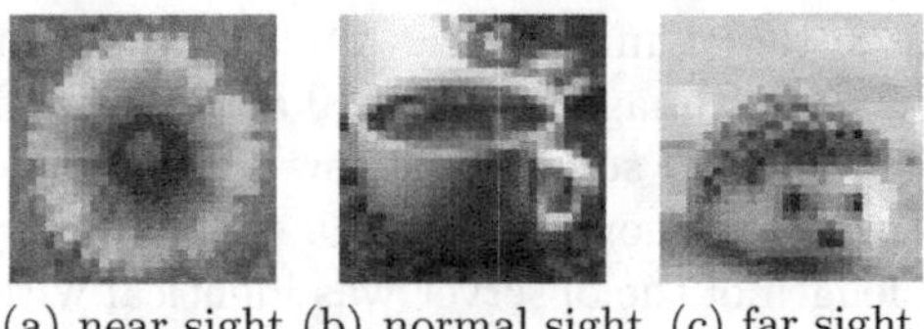

(a) near sight (b) normal sight (c) far sight

Fig. 8. The target images for the near sight, normal sight and the far sight observers.

(a) near sight (b) normal sight (c) far sight

Fig. 9. The observed images of the near sight, normal sight and the far sight observers.

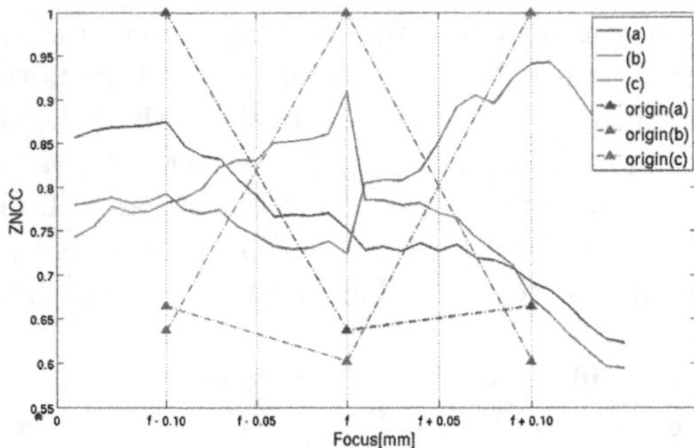

Fig. 10. The similarity between observed images and target images. (Color figure online)

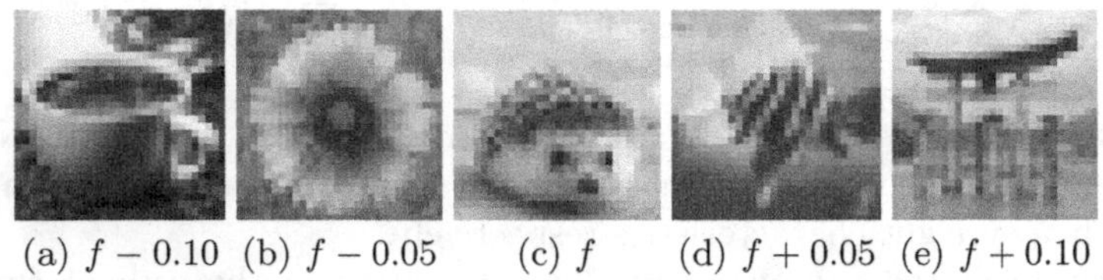

(a) $f - 0.10$ (b) $f - 0.05$ (c) f (d) $f + 0.05$ (e) $f + 0.10$

Fig. 11. The target images of $f - 0.10$, $f - 0.05$, f, $f + 0.05$ and $f + 0.10$.

f, $f + 0.05$ and $f + 0.10$. As shown in these images, the observed images at $f - 0.10$, f and $f + 0.10$ are almost identical with target images at $f - 0.10$, f and $f + 0.10$ respectively. The similarity ZNCC between observed images and target images was computed by changing the focal length of the observer from $f - 0.15$ to $f + 0.15$. The result is shown in Fig. 10. The horizontal axis is the focal length of the observer, and the vertical axis is the similarity between the observed image and the target image. The blue solid line in Fig. 10 shows the similarity between the target image at $f - 0.10$ and observed images at various focal length. The green and red solid lines show those of the target image at f and $f + 0.10$ respectively. As shown in Fig. 10, the similarity takes the largest value when the focal length of the observer was identical with that of the target image. For the reference of readers, we also show the ground truth of the similarity by the dash-dot lines in Fig. 10. From these results, we find that we can identify the refractive properties of observers properly by using the proposed method.

We next evaluated the accuracy of identification in the case of five eyesight characteristics. The focal length of these five eyesight characteristics was $f - 0.10$, $f - 0.05$, f, $f + 0.05$ and $f + 0.10$. The target images for these five eyesight characteristics are shown in Fig. 11. The 4D light field for identifying these five eyesight characteristics was computed from the proposed method and presented to the observer. Figure 12 shows observed images when the focal length of the observer was $f - 0.10$, $f - 0.05$, f, $f + 0.05$ and $f + 0.10$. Again, the observed

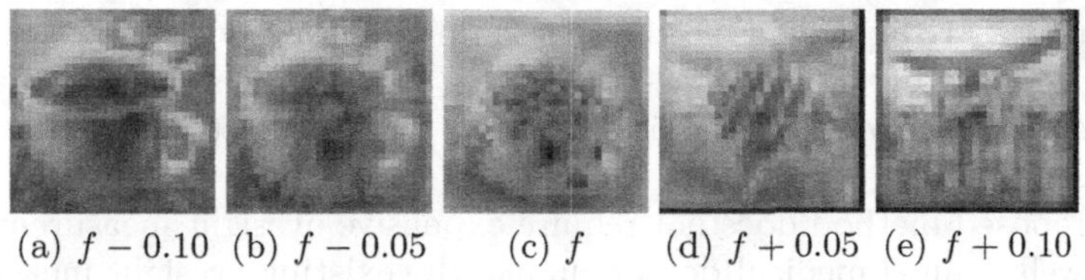

(a) $f - 0.10$ (b) $f - 0.05$ (c) f (d) $f + 0.05$ (e) $f + 0.10$

Fig. 12. The images observed by an observer with the focal length of $f - 0.10$, $f - 0.05$, f, $f + 0.05$ and $f + 0.10$.

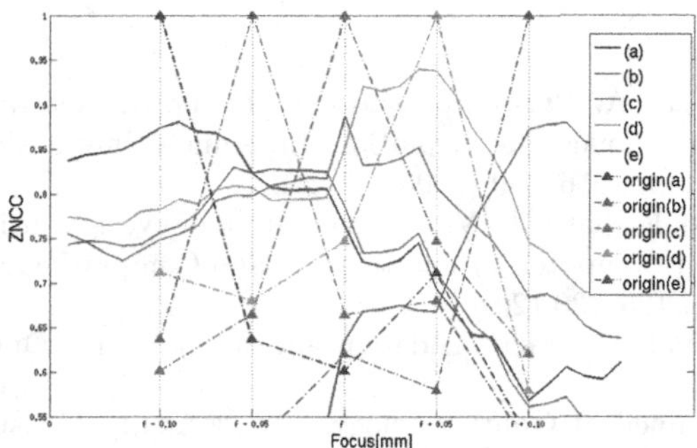

Fig. 13. The similarity between observed images and target images. (Color figure online)

images are almost identical with target images, when the focal length of the observer is identical with that of target images. The similarity ZNCC between observed images and target images was computed by changing the focal length of the observer from $f - 0.15$ to $f + 0.15$. The result is shown in Fig. 13. The horizontal axis is the focal length of the observer, and the vertical axis is the similarity between the observed image and the target image. The blue solid line in Fig. 13 shows the similarity between the target image at $f - 0.10$ and observed images at various focal length. The green, red, light blue and purple solid lines show those of the target image at $f - 0.05$, f, $f + 0.05$ and $f + 0.10$ respectively. As shown in Fig. 13, the similarity takes the largest value when the focal length of the observer was identical with that of the target image. From these results, we find that the refractive properties of observers can be identified properly by using the proposed method. The proposed method requires just a tablet or smart phone and a lens array sheet, and thus it is quite simple and efficient. The result of the measurement can be used for controlling the display, so that near sight and far sight observers can see clear images on the display.

6 Conclusion

In this paper, we proposed a novel method for measuring the refractive properties of human vision just by showing 4D light fields to observers. For this objective, we generated a 4D light field, so that the human observers see different images according to their eyesight characteristics. By using our method, we can identify the refractive properties of observers just by showing the light field to the observers.

The proposed method does not require expensive eyesight measurement systems nor well trained medical doctors unlike the existing eyesight measurement methods. The efficiency of the proposed method was shown by the real image experiments as well as the synthetic image experiments.

References

1. Alonso, M., Barreto, A.: Pre-compensation for high-order aberrations of the human eye using on-screen image deconvolution. In: Annual International Conference of the IEEE EMBS, pp. 556–559 (2003)
2. Aoki, M., Sakaue, F., Sato, J.: Virtual correction of eyesight using visual illusions. In: Conference on Computer Vision, Imaging and Computer Graphics Theory and Applications, pp. 125–130 (2016)
3. Dave, T.: Automated refraction: design and applications. Optom. Today 28–32 (2004)
4. International Council of Ophthalmology. Visual Acuity Measurement Standard. Ital. J. Ophthalmol. (1988)
5. Ding, Y., Li, F., Ji, Y., Yu, J.: Dynamic fluid surface acquisition using a camera array. In: Proceedings of ICCV (2011)
6. Hirsch, M., Wetzstein, G., Rasker, R.: A compressive light field projection system. In: ACM Transactions on Graphics (SIGGRAPH 2014) (2014)
7. Huang, F.C., Wetzstein, G., Barsky, B., Raskar, R.: Eyeglasses-free display: towards correcting visual aberrations with computational light field displays. In: ACM Transactions on Graphics (SIGGRAPH 2014), vol. 33, no. 4 (2014)
8. Huang, F., Chen, K., Wetzstein, G.: The light field stereoscope immersive computer graphics via factored near-eye light field displays with focus cues. In: ACM SIGGRAPH 2015 (2015)
9. Isaksen, A., MacMillan, L., Gortler, S.J.: Dynamically reparameterized light fields. In: Proceedings of the 27th Annual Conference on Computer Graphics and Interactive Techniques (SIGGRAPH), pp. 297–306 (2000)
10. Lanman, D., Luebke, D.: Near-eye light field displays. In: ACM SIGGRAPH Asia (2013)
11. Levoy, M., Hanrahan, P.: Light field rendering. In: ACM SGGRAPH 1996 (1996)
12. Medina, A.: Electronic refractor. Spanish Patent (1979)
13. Pamplona, V.F., Mohan, A., Oliveira, M.M., Raskar, R.: NETRA: interactive display for estimating refractive errors and focal range. ACM Trans. Graph. **29**(4), 77 (2010)
14. Qian, Y., Gong, G., Yang, Y.: 3D reconstruction of transparent objects with position-normal consistency. In: Proceedings of Conference on Computer Vision and Pattern Recognition (2016)
15. Shan, Q., Agarwal, S., Curless, B.: Refractive height fields from single and multiple images. In: Proceedings of Conference on Computer Vision and Pattern Recognition (2012)
16. Tsai, C., Veeraraghavan, A., Sankaranarayanan, A.: What does a light ray reveal about a transparent object? In: Proceedings of ICIP (2015)

17. Wetzstein, G., Roodnick, D., Heidrich, W., Raskar, R.: Refractive shape from light field distortion. In: Proceedings of ICCV (2011)
18. Wetzstein, G., Lanman, D., Heidrich, W., Raskar, R.: Layered 3D: tomographic image synthesis for attenuation-based light field and high dynamic range displays. ACM Trans. Graph. **30**(4), Article No. 95 (2011)
19. Wetzstein, G., Lanman, D., Hirsch, M., Rasker, R.: Tensor displays: compressive light field synthesis using multilayer displays with directional backlighting. In: ACM Transactions on Graphics (SIGGRAPH 2012) (2012)

Crossing the Road Without Traffic Lights: An Android-Based Safety Device

Adi Perry, Dor Verbin, and Nahum Kiryati(✉)

School of Electrical Engineering, Tel Aviv University, 69978 Tel Aviv, Israel
nk@eng.tau.ac.il

Abstract. In the absence of pedestrian crossing lights, finding a safe moment to cross the road is often hazardous and challenging, especially for people with visual impairments. We present a reliable low-cost solution, an Android device attached to a traffic sign or lighting pole near the crossing, indicating whether it is safe to cross the road. The indication can be by sound, display, vibration, and various communication modalities provided by the Android device. The integral system camera is aimed at approaching traffic. Optical flow is computed from the incoming video stream, and projected onto an influx map, automatically acquired during a brief training period. The crossing safety is determined based on a 1-dimensional temporal signal derived from the projection. We implemented the complete system on a Samsung Galaxy K-zoom Android smartphone, and obtained real-time operation. Promising experimental results provide pedestrians with sufficiently early warning of approaching vehicles. The system can serve as a stand-alone safety device, that can be installed where pedestrian crossing lights are ruled out. Requiring no dedicated infrastructure, it can be powered by a solar panel and remotely maintained via the cellular network.

Keywords: Pedestrian crossing · Traffic analysis · Optical flow · Blind and visually impaired · Resource-limited computer vision · Android device

1 Introduction

Safe road crossing requires conflict-free sharing of road surface. Pedestrian crossing lights stop traffic to ensure secure crossing. Other right of way regulations, such as zebra-crossings, are less reliable since drivers do not consistently yield to pedestrians [19]. Thus, for safe crossing in the absence of crossing lights, pedestrians need to identify occasional traffic gaps [7].

The duration of a traffic gap allowing safe crossing is determined by the crossing time. US data [1] indicates a representative crossing speed of about 1.2 m/s and urban road widths of 5.5 m (one-way single lane) or 8.5 m (two-way or two-lane one-way) with shoulders. The corresponding crossing times are 4.5 s and 7 s respectively. The necessary traffic gap, equivalent to the needed advance warning of an approaching vehicle, must be at least that long.

S. Battiato et al. (Eds.): ICIAP 2017, Part II, LNCS 10485, pp. 534–544, 2017.
https://doi.org/10.1007/978-3-319-68548-9_49

Recognizing an approaching vehicle 7 s ahead of contact time is possible for people with normal or corrected vision and solid cognitive function. For blind and visually impaired people, as well as for elderly persons, children [15] and others, sufficiently early recognition of an approaching vehicle can be implausible. Blind people are trained to rely on hearing and follow the "cross when quiet" strategy [18]. Auditory-based performance is less than ideal [4,9] and further difficulties arise where background urban noise masks the low noise emitted by a modern car [6], or when the pedestrian has less than perfect hearing.

Image and video processing methods to assist visually handicapped people were reviewed in [13]. Many techniques convert visual data to auditory or haptic signals. These include systems to locate street crossings [11,16]. Converting traffic light status to an auditory or other indication was considered in [3]. Sensor-rich "smart canes" have been developed [5] for close-range obstacle detection and path planning. However, the problem of road crossing at unsignalized crosswalks has remained open. A noteworthy exception is the work of [2], aimed at providing a robotic system with street-crossing capability.

We recently presented [17] a system positioned near a road crossing, alerting pedestrians of approaching traffic and detecting traffic-gaps suitable for safe crossing. The experimental results in [17] were obtained using a powerful laptop computer connected to a good-quality external USB camera with an added narrow field of view lens. The feasibility of real-time implementation on a low-cost and power-frugal platform remained an open question. In this paper we meet this challenge using an off-the-shelf Android device, establishing the potential for widespread deployment and stand-alone solar-powered operation. To complete the feasibility proof, we demonstrate successful autonomous operation at actual road crossings.[1]

2 Stationary Location-Specific Solution

Consider a vehicle approaching a crosswalk at 50 km/h. Detecting this vehicle 7 s in advance, implies detection at about 100 m from the crosswalk. At that distance, the frontal viewing angle of a typical car is about 1° horizontally and slightly less vertically. In many suburban and country roads, speeds up to 100 km/h are not unusual, corresponding to roughly 0.5° at the required detection distance.

Assuming a straight road towards the crosswalk, the tiny-looking approaching vehicle appears near the vanishing point in the picture. Reliable detection requires predetermination of the point of first appearance, and a narrow field of view (FOV) around that point. With some exaggeration, this can be illustrated as watching the scene through a straw. One might wish for a smartphone application, warning the holder of approaching vehicles. However, typical mobile phones have wide FOV lenses. Worse, a hand-held device would require careful pointing towards the point of first appearance, a challenge for anyone, and a formidable undertaking for people with visual, cognitive or motoric impairments.

[1] Video: www.eng.tau.ac.il/~nk/crosswalk-safety.html.

Fig. 1. The system, implemented on a Samsung Galaxy K-zoom smartphone, attached to a lighting pole near a road crossing. The camera faces the approaching traffic. Note the extended position of the zoom lens.

We suggest a stationary and location-specific solution, where the system is attached, for example, to a traffic sign or a lighting pole, facing the approaching traffic. The absence of substantial ego-motion improves robustness, facilitates training and adaptation, and reduces the computational requirements, thus minimizing cost and power consumption. Beyond its technical advantages, the location-specific approach does not require a personal device, and is thus accessible to all members of society.

The essential system elements are a video camera with a narrow field of view lens, a processing unit, and a user interface. Optional elements include a compact solar panel and a rechargeable battery, enabling off-grid operation, and a communication unit (cellular, WiFi, etc.), allowing remote maintenance as well as cooperation with nearby instances of the system, e.g. in two-way multi-lane road crossings. As discussed in the sequel, we successfully implemented the system on the Samsung Galaxy K-Zoom smartphone, see Fig. 1.

Indication of incoming traffic, or lack thereof, can be delivered to the user as an audio, tactile or visual signal, or via a local networking interface (such as Bluetooth or WLAN) to a smartphone or a dedicated receiver.

3 Principles of Operation

Vehicles approaching a crosswalk emerge as tiny blobs at the point of first appearance. A key design choice has been to provide the pedestrian with the earliest possible warning of an approaching vehicle, rather than the latest warning that would allow crossing. This implies that the task at hand is early detection, rather than precise estimation of the time to contact (TTC) [10,12].

We transform the incoming video to a scalar temporal signal, referred to as *Activity*, quantifying the overall risk-inducing motion in the scene. Vehicles approaching the camera lead to pulses in the Activity signal, so that substantial activity ascent reflects an approaching vehicle. Timely detection of incoming traffic, with sparse false alarms, calls for reliable separation between meaningful Activity ascent and random-like fluctuations.

Fig. 2. Influx map, computed by averaging the optical flow field over a training period. Left: arrow-style vector display. Right: color coded vector representation, corresponding to the color-key shown top-right. The motion direction in each pixel is represented by hue, its norm by saturation. (Color figure online)

3.1 Activity Estimation

Motion in the scene leads to optical flow. Parts of the optical flow field may correspond to vehicles approaching the camera. Other optical flow elements are due to harmless phenomena, such as traffic in other directions, crossing pedestrians, animals on the road, moving leaves and slight camera vibrations. The suggested activity signal A_n, where n is the temporal index, is computed as the projection of the optical flow field $\boldsymbol{u}_n(x, y)$ onto an *influx map*. The influx map $\boldsymbol{m}(x, y)$ is a vector field, representing the location-dependent typical course of traffic approaching the camera. The effective support of the influx map largely corresponds to incoming traffic-lanes. Mathematically,

$$A_n = \sum_{x,y} \boldsymbol{m}(x, y) \cdot \boldsymbol{u}_n(x, y). \tag{1}$$

The scalar activity signal quantifies the overall hazardous traffic in the scene.

Assuming a one-way street, the influx map can be obtained by temporal averaging the optical flow field over a training period, as illustrated in Fig. 2. Temporal averaging cancels randomly oriented motion elements due to vibrations, wind and similar effects, while maintaining the salient hazardous traffic motion patterns in the road area.

The influx-map vectors can be intensified to emphasize motion near the point of first appearance, thus increasing the advance warning time. Note that the point of first appearance can be identified during training, e.g. by back-tracking the influx-map vectors to the source. Minor adjustment of the influx map computation procedure is necessary for handing two-way roads, where consistent outbound traffic is also expected.

We have tacitly presented optical flow and influx map computation in *dense* form, meaning that their spatial resolution is uniform and commensurate with the resolution of the video frame. However, *sparse* computation of the optical flow and influx map, at spatially variant density, is beneficial with regard to both performance and computational cost. Foveated computation of Eq. 1, dense near the point of first appearance and sparse elsewhere, has the desirable effect of emphasizing motion in that critical region.

3.2 Detecting Approaching Vehicles

Based on the raw Activity signal, we wish to generate a binary indication, suggesting whether risk-inducing traffic is approaching ('TRAFFIC' state), or a sufficient traffic gap allows safe crossing ('GAP' state).

As a vehicle approaches the camera, the peak of the corresponding Activity pulse is usually evident. However, *early* warning requires detection at an early stage of ascent, where the signal to noise ratio is poor.

Reliable, timely detection cannot be achieved by direct thresholding of the Activity signal. The signal is therefore evaluated within a sliding temporal window of N samples. This improves the effective signal to noise ratio, consequently increasing the detection probability and reducing the false alarm rate. However, sliding window computation inherently introduces detection latency. The extent N of the sliding window must therefore be limited, to maintain the advance warning period required for safe crossing.

Suppose that the sliding window falls either on a traffic gap ('GAP' hypothesis θ_0), modelling the Activity signal as $A_n = \nu_n$, where ν_n is zero-mean additive white Gaussian noise, or at the precise moment a vehicle emerges ('TRAFFIC' hypothesis θ_1), such that $A_n = s_n + \nu_n$, where s_n is the clean Activity pulse shape for the specific vehicle and viewing conditions.

We discriminate between the two hypotheses using the Likelihood Ratio Test

$$\mathrm{LRT} = \frac{\frac{1}{(2\pi)^{N/2}\sigma^N} \prod_{n=0}^{N-1} \exp[-\frac{(A_n - s_n)^2}{2\sigma^2}]}{\frac{1}{(2\pi)^{N/2}\sigma^N} \prod_{n=0}^{N-1} \exp[-\frac{A_n^2}{2\sigma^2}]} \gtrless \lambda, \tag{2}$$

where λ is a discriminative threshold. This leads to the detection rule:

$$\tilde{y} \equiv \sum_{n=0}^{N-1} A_n s_n \gtrless \frac{1}{2}\left(2\sigma^2 \ln \lambda + \sum_{n=0}^{N-1} s_n^2\right) \equiv \gamma, \tag{3}$$

where $\tilde{y}$ denotes the correlation of the Activity signal with the pulse template. The pulse shape depends on the particular installation, and can be learned during training. For a given installation, specific vehicle characteristics lead to some variability. The right hand side is an application-dependent threshold denoted γ. We determine the threshold using the Neyman-Pearson criterion, tuned to obtain a specified false-alarm probability.

Given either hypothesis, the random variable $\tilde{y}$ is Gaussian with variance $\sigma^2 \sum_{n=0}^{N-1} s_n^2$. The probability of false alarm is determined by the tail distribution of the 'GAP' hypothesis θ_0:

$$P_{FA} = Pr\left(\tilde{y} > \gamma | \theta_0\right) = 1 - \phi\left(\frac{\gamma}{\sigma\sqrt{\sum s_n^2}}\right), \tag{4}$$

where $\phi(x)$ is the cumulative distribution of the standard normal distribution. Let $Q(x)$ denote its tail probability $Q(x) = 1 - Q(-x) = 1 - \phi(x)$. The threshold γ can thus be expressed in terms of the admissible false-alarm rate, the noise and the Activity pulse template. The decision rule can be reformulated as

$$\frac{\sum_{n=0}^{N-1} A_n s_n}{\sqrt{\sum_{n=0}^{N-1} s_n^2}} \gtrless Q^{-1}\left(P_{FA}\right)\sigma, \tag{5}$$

which is the familiar matched filter result, correlating the input sequence with the Activity pulse template we wish to detect.

So far, we have discussed the most demanding detection condition, the earliest detection of an incoming vehicle. The Activity pulse increases as the vehicle approaches; thus, once the correlator output $\tilde{y}$ crosses the threshold, it generally exceeds the threshold as long as the vehicle has not reached the crossing. When a vehicle appears late (e.g. leaving a parking space near the crossing), the Activity pulse rapidly ascends, and the correlator promptly crosses the threshold.

3.3 Setting Detector Parameters

The detector is tuned according to the noise variance and the Activity pulse template. During training, after the influx map had been computed, the system produces Activity measurements. These measurements are used to automatically estimate the Activity pulse template and the noise variance.

The system locates the salient local maxima of the Activity signal during training, and sets a temporal window containing each maximum, where most of the window precedes the maximum. We set the Activity pulse template as the median of the Activity signal windows, see Fig. 3.

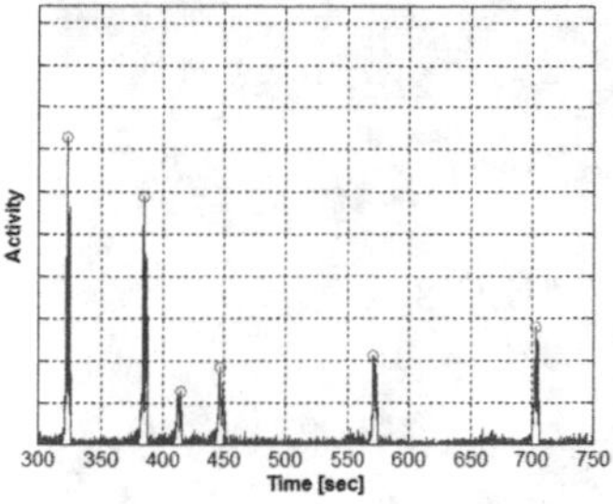

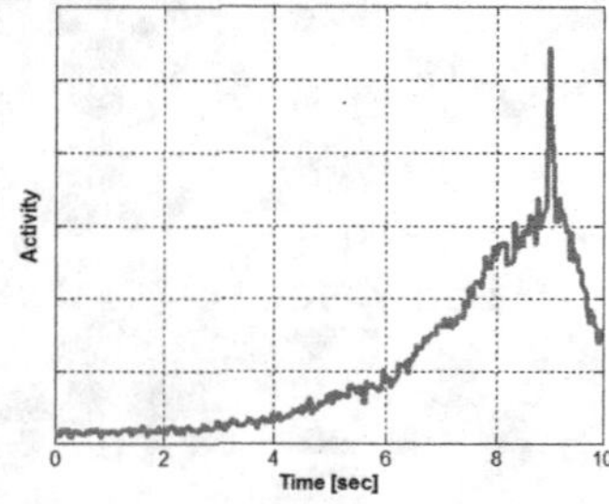

Fig. 3. Generating the activity pulse template. Left: The Activity signal over a small part of the training period, with the salient local maxima detected. Right: The Activity pulse template, generated as the median of the Activity signal windows

4 Examples and Evaluation

Figure 4 shows the activity signal computed during six minutes at a test site. The color represents the prediction issued to the pedestrian, red signifying TRAFFIC and blue meaning GAP. TRAFFIC is typically indicated about 8–10 s before the Activity signal peak, i.e. before the car reaches the camera, allowing secure crossing with sufficient margin. Multimodal activity pulses, e.g. at $t \approx 160$ s and at $t \approx 250$ s are each due to several cars arriving in sequence. Corresponding snapshots are shown in Fig. 5. The false alarms at $t \approx 80$ s and $t \approx 205$ s are not due to random noise: the first was caused by a person on the road, the second

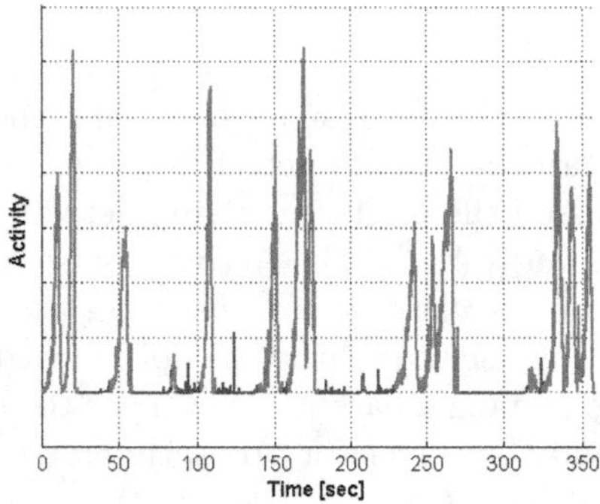

Fig. 4. The activity signal computed over six minutes, including the events shown in Fig. 5. TRAFFIC and GAP indications are represented by red and blue respectively. (Color figure online)

Fig. 5. Video frames corresponding to two multiple car events.

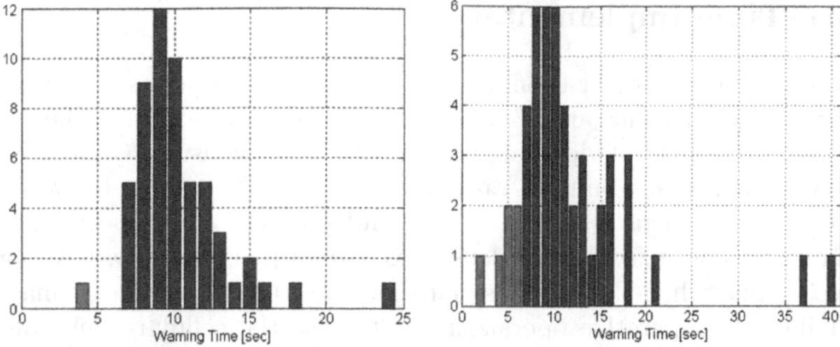

Fig. 6. Left: histogram of advance warning times provided by the real-time implementation during two hours at the test site shown in Fig. 5. Right: histogram of advance warning times obtained during one hour at the test site shown in Fig. 7.

Fig. 7. Irregular events on the road. Left: a slow, distant bicyclist. Right: a person suddenly appearing on the road.

by a cat. The random noise level is rather low, as seen circa $t = 300s$ in Fig. 4. Figure 6(left) is the histogram of advance warning times provided by the system during two hours of testing.

In the archetypal case, an approaching vehicle emerges close to the point of first appearance, thus maximizing the achievable warning time. However, in various urban or suburban settings, a vehicle can exit a parking place close to the camera, or turn into the monitored road from an adjacent driveway. Thus, while the typical visibility period of an incoming vehicle is often greater than 10 s, in unusual cases it can be as short as 1 s, providing neither the system nor an alert human with sufficient warning. Figure 6(right) is the histogram of the advance warning times obtained during one hour at the challenging installation shown in Fig. 7, characterized by substantial occlusion near the point of first appearance and a residential afternoon activity pattern, including parking, biking and strolling. Short warning times usually correspond to bikes and very slow vehicles, that impose minimal hazard. Irregular events acquired by the system, leading to unusually short or long advance warnings, are exemplified in Fig. 7. Additional results, including ROC curves, can be found in [17].

5 Real-Time Implementation

To show that low-cost realization and wide deployment are possible, we implemented the system using an off-the-shelf Android device. Since the solution requires a narrow field-of-view lens, we selected the Samsung Galaxy K-zoom smartphone, that has a built-in variable zoom lens. Note that the variability of the optical zoom lens is not necessary, and the system can be readily built and operated using a fixed-zoom lens. Fixed-zoom lenses are widely available as low-cost smartphone accessories. Furthermore, unless remote management and distributed cooperative operation are sought, the cellular communication parts of the smartphone are redundant. In addition, although the K-zoom has a GPU capable of reducing computation time and increasing the frame rate, our implementation only utilized the smartphone's CPU. Thus, the system can be implemented on extremely low-cost Android tablet-like devices. In many applications, the screen of the Android device is not in use, and can be shut down or removed altogether to further reduce the power consumption and cost.

Other than its variable zoom lens, the Samsung Galaxy K-zoom, introduced in 2014, is similar to smartphones of its generation, with Android v4.4.2 (KitKat) as its native operating system. Figure 1 shows an installation of the system at a test site, attached to a lighting pole, with its zoom lens extended.

The target system software was developed using OpenCV4Android SDK version 2.4.10, FFmpeg and JavaCPP. The most demanding computational task is optical flow estimation. During training, dense optical flow estimation is required for generating the influx map. For that purpose, we used the OpenCV4Android implementation of Farnebäck's algorithm [8], skipping frames to improve the estimation of the tiny optical flow vectors near the point of first appearance. Following the identification of that point, a subset of 2000 image points was sampled using a Gaussian density function centered at the point of first appearance. In the online phase, sparse optical flow is evaluated only at those points, using the OpenCV4Android implementation of the Lucas-Kanade [14] algorithm. The Activity signal is computed by projecting the sparse optical flow field onto the corresponding sparse subset of the influx map. A rate of 8 frames per second is easily obtained on the K-zoom smartphone. To reduce power consumption, the rate can be decreased to two frames per second without significant performance degradation. The template matching process (3) is carried out at 30 frames per second using temporal interpolation. The results shown in Figs. 4 and 6(left) were obtained using the real-time smartphone-based implementation.

Two way streets, as encountered for example in the test site shown in Fig. 1, raise two issues. First, warning about traffic approaching the crossing from the opposite direction can be obtained by installing an additional instance of the system with its camera pointing in that direction. Cellular, WiFi or Bluetooth communication between the two systems allows issuing a mutually agreed indication to the pedestrian. Second, in a two way road, as the camera monitors incoming traffic, distancing vehicles may be seen in the field of view. To distinguish incoming from outgoing traffic based on optical flow, one might compute vector field measures such as divergence. In practice, assuming that the camera is installed higher than typical vehicles, the distinction can be based on the sign of the vertical component of the optical flow vector: incoming vehicles are associated with descending optical flow vectors. In the Android implementation, upwards-pointing vectors in the influx map are simply nullified. Indeed, the system works well at the test site shown in Fig. 1 using that method.

6 Discussion

The global number of visually impaired people worldwide has been estimated to be 285 millions [20], and many more have age-related disabilities limiting their road-crossing skills. Reliable mobile personal assistive devices for road crossing without traffic lights are yet to be invented, but in any case are likely to be beyond the means of needy people worldwide. The goal of this research has been the development of an effective, easy to use, low cost solution. It is facilitated by the stationary, location-specific approach, avoiding system ego-motion altogether. System installation is simple, as the location-dependent influx map and the Activity pulse template are automatically obtained by training.

Compared to our previous work [17], the breakthrough is real-time stand-alone implementation using an off-the-shelf Android device. The retail price of a suitable Android tablet, including all necessary components, is less than US$40 at the time of writing, and the marginal cost of a fixed narrow field lens is negligible. The mass-production manufacturing bill of materials, especially for a screen-less version of the system, is therefore likely to be widely affordable.

Interestingly, in their future work section, [2] imagined a stationary solution for the unsignalized road crossing problem. Our approach is substantially more robust, computationally lighter and better exploits stationarity when compared to the tracking-based method applied by [2] in their robotic system.

We evaluated the system at several test locations, obtaining promising results. Additional work is needed to ensure reliable operation at night or in the presence of rain or snow.

References

1. American Association of State Highway and Transportation Officials. A Policy on Geometric Design of Highways and Streets. Washington, DC (2004)
2. Baker, M., Yanco, H.: Automated street crossing for assistive robots. In: Proceedings of the 9th International Conference on Rehabilitation Robotics. IEEE (2005)
3. Barlow, J., Bentzen, B., Tabor, L., Signals, A.P.: Synthesis and Guide to Best Practice. National Cooperative Highway Research Program (2003)
4. Barton, B., Ulrich, A., Lew, R.: Auditory detection and localization of approaching vehicles. Accid. Anal. Prev. **49**, 347–353 (2012)
5. Buchs, G., Maidenbaum, S., Amedi, A.: Obstacle identification and avoidance using the 'EyeCane': a tactile sensory substitution device for blind individuals. In: Auvray, M., Duriez, C. (eds.) EUROHAPTICS 2014. LNCS, vol. 8619, pp. 96–103. Springer, Heidelberg (2014). doi:10.1007/978-3-662-44196-1_13

6. Emerson, R., Naghshineh, K., Hapeman, J., Wiener, W.: A pilot study of pedestrians with visual impairments detecting traffic gaps and surges containing hybrid vehicles. Transp. Res. Part F Traffic Psychol. Behav. **14**(2), 117–127 (2012)
7. Emerson, R., Sauerburger, D.: Detecting approaching vehicles at streets with no traffic control. J. Vis. Impair. Blindness **102**(12), 747–760 (2008)
8. Farnebäck, G.: Two-frame motion estimation based on polynomial expansion. In: Bigun, J., Gustavsson, T. (eds.) SCIA 2003. LNCS, vol. 2749, pp. 363–370. Springer, Heidelberg (2003). doi:10.1007/3-540-45103-X_50
9. Hassan, S.: Are normally sighted, visually impaired, and blind pedestrians accurate and reliable at making street crossing decisions? Invest. Ophthalmol. Vis. Sci. **53**(6), 2593–2600 (2012)
10. Horn, B., Fang, Y., Masaki, I.: Time to contact relative to a planar surface. In: Proceedings of the IEEE Intelligent Vehicles Symposium, pp. 68–74 (2007)
11. Ivanchenko, V., Coughlan, J., Shen, H.: Crosswatch: a camera phone system for orienting visually impaired pedestrians at traffic intersections. In: Miesenberger, K., Klaus, J., Zagler, W., Karshmer, A. (eds.) ICCHP 2008. LNCS, vol. 5105, pp. 1122–1128. Springer, Heidelberg (2008). doi:10.1007/978-3-540-70540-6_168
12. Lee, D.: A theory of visual control of braking based on information about time-to-collision. Perception **5**, 437–459 (1976)
13. Leo, M., Medion, G., Trivedi, M., Kanade, T., Farinella, G.: Computer vision for assistive technologies. In: Computer Vision and Image Understanding (2016, in press)
14. Lucas, B., Kanade, T.: An iterative image registration technique with an application to stereo vision. In: Proceedings of the 7th IJCAI, vol. 2, pp. 674–679 (1981)
15. Morrongiello, B., Corbett, M., Malinovic, M., Beer, J.: Using a virtual environment to examine how children cross streets: advancing our understanding of how injury risk arises. J. Pediatr. Psychol. **41**(2), 265–275 (2016)
16. Murali, V., Coughlan, J.: Smartphone-based crosswalk detection and localization for visually impaired pedestrians. In: Proceedings of the International Conference Multimedia Expo Workshops, pp. 1–7. IEEE (2013)
17. Perry, A., Kiryati, N.: Road-crossing assistance by traffic flow analysis. In: Agapito, L., Bronstein, M.M., Rother, C. (eds.) ECCV 2014. LNCS, vol. 8927, pp. 361–374. Springer, Cham (2015). doi:10.1007/978-3-319-16199-0_26
18. Sauerburger, D.: Developing criteria and judgment of safety for crossing streets with gaps in traffic. J. Vis. Impair. Blindness **93**(7), 447–450 (1999)
19. Schroeder, B., Rouphail, N.: Event-based modeling of driver yielding behavior at unsignalized crosswalks. J. Transp. Eng. **137**(7), 455–465 (2011)
20. World Health Organization. Global data on visual impairments (2010). http://www.who.int/blindness/GLOBALDATAFINALforweb.pdf

Information Forensics and Security

A Novel Statistical Detector for Contourlet Domain Image Watermarking Using 2D-GARCH Model

Maryam Amirmazlaghani(✉)

Computer Engineering and IT Department, Amirkabir University of Technology, Tehran, Iran
mazlaghani@aut.ac.ir

Abstract. In this paper, we propose a novel watermark detector in contourlet domain using likelihood ratio test (LRT). Since the accuracy of an LRT based watermark detector is dependent on the efficiency of the applied statistical model, first, we study the statistical properties of the contourlet coefficients. Using different tests, we demonstrate that the marginal distribution of contourlet coefficients is heavy-tailed and heteroscedasticity exists in these coefficients. All of the previously proposed models for contourlet coefficients assume that these coefficients are identically distributed, so they can not capture the characteristics of the contourlet coefficients. To overcome this problem, we propose using two dimensional generalized autoregressive conditional heteroscedasticity (2D-GARCH) model for contourlet coefficients that provides an efficient structure for the dependencies of these coefficients. Based on using 2D-GARCH model, a novel LRT based heteroscedastic watermark detector is designed in contourlet domain. Experimental results confirm the efficiency of the proposed watermark detector under different types of attacks and its outperformance compared with alternative watermarking methods.

Keywords: Image watermarking · Contourlet transform · Likelihood ratio test · 2D-GARCH model

1 Introduction

Digital watermarking which embeds information into an original media by modifying its content has been used in many applications such as copyright protection, broadcast monitoring, content authentication, and covert communication [1]. Distribution of digital data on the internet increases the importance of the intellectual property rights, and copyright protection. In this paper, we focus on image watermarking for copyright protection. Imperceptibility and robustness are two major requirements for copyright protection watermarking approaches. In copyright protection, the main goal is to distinguish whether a specific watermark is present in the received media or not. So, to check the validity of the

S. Battiato et al. (Eds.): ICIAP 2017, Part II, LNCS 10485, pp. 547–557, 2017.
https://doi.org/10.1007/978-3-319-68548-9_50

watermark for copyright protection, the detection of the watermark is sufficient [2–4]. In some other applications, the watermark decoding may be required, i.e., the hidden message (watermark) should be extracted and decoded [5,6].

A watermarking scheme consists of two main steps: watermark embedding and watermark detection. According to the embedding method, watermarking techniques can be classified into two categories: spread spectrum based [2–4,7] and quantization based techniques [8]. Spread spectrum watermarking provides high level of security and robustness because of spreading the watermark throughout the spectrum of the image. So, it is more popular. In spread spectrum schemes, usually the watermark is embedded in a transformed domain such as the discrete cosine transform (DCT) [9], the discrete wavelet transform (DWT) [3,7] and the contourlet transform [2–4].

Contourlet transform which is an efficient extension of the wavelet transform can capture the smooth contours of images using a double filter bank. Contourlet transform consists of two main steps: Applying Laplacian pyramid to capture point discontinuities and using directional filter bank to obtain directional information [10]. This transform provides some good properties such as multiresolution, nearly critical sampling, and directionality [10]. Moreover, contourlet transform has spreading property that is important in watermarking. It means that inserting the watermark into a special subband causes in spreading out it in all subbands at the step of reconstracting the watermarked image [4].

In the recent years, many contourlet domain image watermarking methods have been proposed [2,4,6,11,12]. Using correlation detector is a common way to detect transform domain watermarks [11,12]. But, this detector is optimal only when the distribution of data is Gaussian. But, the non-Gaussian nature of contourlet coefficients has been studied in many works such as [2,4,13]. So, in this situation, watermark detection can be formulated as a binary hypothesis test and optimal watermark detector can be achieved using Bayesian log-likelihood ratio test (LRT). To design a contourlet domain LRT based watermark detector, the contourlet coefficients should be modeled statistically. More accurate statistical model for contourlet coefficients results in more efficient watermark detector. In the literature, to derive contourlet domain watermark detectors, different statistical models have been proposed such as generalized Gaussian (GG) [14], Bessel K form [2], alpha-stable [4], and Normal Inverse Gaussian (NIG) [6].

In this paper, we propose a novel contourlet domain watermark detector. We embed the watermark in contourlet domain using an additive spread spectrum approach. Then, to design an efficient LRT based watermark detector, first, we study the statistical properties of the contourlet coefficients. All of the previously proposed models for contourlet coefficients suppose that these coefficients are identically distributed. In this work, we demonstrate that heteroscedasticity exists in contourlet coefficients and they are not identically distributed. In this way, we propose to apply a heteroscedastic model called two dimensional generalized autoregressive conditional heteroscedasticity (2D-GARCH) model for the contourlet coefficients. 2D-GARCH model is the two dimensional extension of GARCH model [15] and GARCH model has been introduced for financial time series by Bollersrev in [16]. Moreover, we demonstrate the efficiency of 2D-GARCH model in capturing the heteroscedasicity of contourlet coefficients and the dependencies among them. Then, we design an LRT based contourlet domain watermark detector using 2D-GARCH model. Experimental results confirm the high efficiency of the proposed detector and its robustness under different kinds of attacks. Also, we compare our proposed watermark detector with some alternative detectors and show the outperformance of our method.

This paper is organized as follows. In Sect. 2, we review the contourlet transform, 2D-GARCH model and discuss the compatibility of contourlet coefficients and 2D-GARCH model. Watermark embedding process and 2D-GARCH based watermark detector are explained in Sect. 3. Section 4 reports the simulation results. Finally, Sect. 5 concludes the paper.

2 Statistical Modeling of Contourlet Coefficients

In this section, first, we review the contourlet transform and the 2D-GARCH model. Then, we verify the efficiency of 2D-GARCH to model the contourlet coefficients.

2.1 Contourlet Transform

Wavelet transform can not recognize the smoothness of the images contours. To overcome this problem, many directional image representations have been proposed recently. Contourlet is one of them proposed in [10] that provides an efficient representation for typical images with smooth contours. Contourlet transform of an image constructed in two steps. First, Laplacian pyramid filter is applied that results in subband decomposition. Then, the directional filter bank is used. Combining these two steps makes Pyramidal Directional Filter Bank (PDFB). Contourlet transform is computationally efficient because of using iterated filter banks. It can provide a desired number of direction in each subband and also it is close to critically sampled [10]. So, we use contourlet transform.

2.2 2D-GARCH Model

Suppose $x(i,j)$ represents a two-dimensional zero mean stochastic process. $x(i,j)$ follows a 2D-GARCH (p_1, p_2, q_1, q_2) model, where (p_1, p_2, q_1, q_2) denotes the order of the model, if

$$x(i,j) = \sqrt{h(i,j)}\varepsilon(i,j) \tag{1}$$

$$h(i,j) = \alpha_0 + \sum_{k\ell \in \Lambda_1} \alpha_{k\ell} x(i-k, j-\ell)^2 + \sum_{k\ell \in \Lambda_2} \beta_{k\ell} h(i-k, j-\ell), \tag{2}$$

where

$$\Lambda_1 = \{k\ell \,|0 \leqslant k \leqslant q_1, 0 \leqslant \ell \leqslant q_2 \,, (k\ell) \neq (0,0)\}, \tag{3}$$

$$\Lambda_2 = \{k\ell \,|0 \leqslant k \leqslant p_1, 0 \leqslant \ell \leqslant p_2 \,, (k\ell) \neq (0,0)\} \tag{4}$$

$\varepsilon(i,j) \sim \mathcal{N}(0,1)$ is an iid two dimensional random process, and $\{\alpha_0, \alpha_{01}, \cdots, \alpha_{q_1 q_2}, \beta_{01}, \cdots, \beta_{p_1 p_2}\}$ are the parameters of model. It is clear that $h(i,j)$ is the conditional variance of $x(i,j)$, therefore $x(i,j)$ is conditionally distributed as

$$x(i,j)\,|\psi(i,j) \sim \mathcal{N}(0, h(i,j)) \tag{5}$$

where

$$\psi(i,j) = \{\{x(i-k, j-\ell)\}_{k,\ell \in \Lambda_1}, \{h(i-k, j-\ell)\}_{k,\ell \in \Lambda_2}\}.$$

To compute the parameters of model, maximum likelihood estimation (MLE) can be used [15,17].

2.3 2D-GARCH Model for Contourlet Coefficients

Here, we check the suitability of 2D-GARCH model for contourlet coefficients. Since 2D-GARCH is a heteroscedastic model, first of all, we should study the heteroscedasticity of the contourlet coefficients and their match with the specific type of dependency provided by 2D-GARCH model. In this way, we use the Lagrange-multiplier test proposed in [18] which examines the existence of two dimensional heteroscedasticity and two-dimensional GARCH effect. The results of this hypothesis test are reported by (1) a Boolean decision variable "H" that 1 indicates acceptance of the alternative hypothesis that GARCH effects exist, (2) "pValue" that is the significance level at which this test rejects the null hypothesis, and (3) "GARCHstat" that indicates GARCH test statistic. Table 1 represents the results of applying the LM test [18] to the contourlet transform of Peppers image (eight directional subbands in the finest scale). This table confirms the existence of two dimensional heteroscedasticity in contourlet coefficients. It should be mentioned that similar results have been obtained for other test images but because of limited space, only the results of peppers image have been reported.

Moreover, the compatibility between the histograms of contourlet coefficients and the 2D-GARCH model is examined. The histograms of two contourlet subbands of *Peppers* image and the histograms of the best fitted 2D-GARCH model are presented in Fig. 1. To compare the efficiency of 2D-GARCH model with other statistical models, the best fitted Gaussian and Generalized Gaussian (GG) distributions have been plotted in Fig. 1, too. This figure confirms that 2D-GARCH model provides a better compatibility to the data.

3 Proposed Watermarking Scheme

As mentioned before, watermarking schemes consist of two main steps: watermark insertion and watermark detection. In the following, we describe these steps for the proposed watermarking method.

Table 1. Results of two dimensional [18] hypothesis tests for the presence of 2D-GARCH effect in the eight contourlet subbands of the finest scale for the *Peppers* image.

	H	pValue	GARCHstat
Subband 1	1	0	1.0166e+004
Subband 2	1	0	2.1819e+004
Subband 3	1	0	7.1432e+003
Subband 4	1	0	6.8305e+003
Subband 5	1	0	1.2155e+004
Subband 6	1	0	6.9759e+003
Subband 7	1	0	2.5357e+004
Subband 8	1	0	8.1015e+003

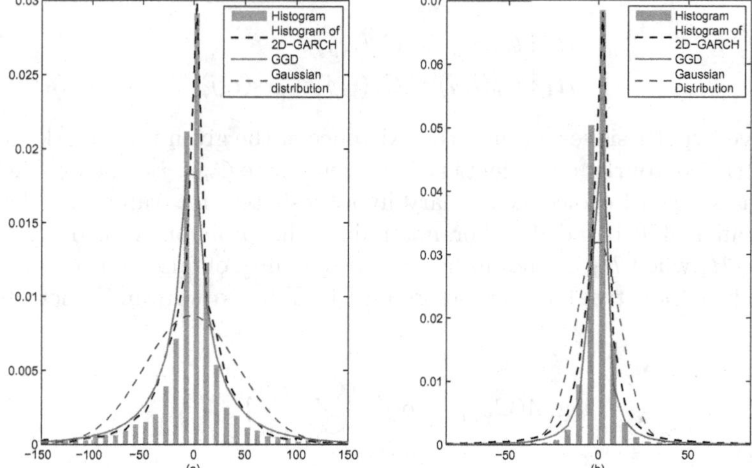

Fig. 1. Histograms of the contourlet coefficients for *Peppers* image: (a) directional subband in the first pyramidal level (b) directional subband in the third pyramidal level

3.1 Watermark Insertion

To embed the watermark into an original image, first, the contourlet transform is applied to the image. Then, to increase the imperceptibility and robustness, the subband with the highest energy is selected to embed the watermark [4]. Additive spread spectrum approach is used to insert the watermark into the selected subband. Let $\mathbf{C_x} = \{C_x(i,j)|i = 1,\ldots,M, j = 1,\ldots,N\}$ denotes the selected subband of the original image. Bold types denote two dimensional vectors. The watermark is embedded into the selected subband using the following formula:

$$C_y(i,j) = C_x(i,j) + \alpha s(i,j), \tag{6}$$

where $\mathbf{C_y} = \{C_y(i,j)|i = 1,\ldots,M, j = 1,\ldots,N\}$ and $\mathbf{s} = \{s(i,j)|i = 1,\ldots,M, j = 1,\ldots,N\}$ denote the watermarked contourlet subband and the watermark sequence, respectively. $\mathbf{s}$ is a bipolar watermark sequence taking the values -1 and 1 with the equal probability. To obtain $\mathbf{s}$, we use a pseudorandom sequence (PRS) generator with an initial state depends on the value of a secret key. γ is the embedding power factor that adjusts the watermark to document ratio (WDR) [19]. After embedding the watermark into the selected contourlet subband, inverse contourlet transform is applied to obtain the watermarked image.

3.2 Watermark Detection

To detect the watermark from a given image, contourlet transform is applied to it and the subband with the highest energy is selected. Then, watermark detector should examine the existence of a known watermark in the related contourlet subband [4,19]. Due to (6), we can formulate the additive watermark detection in the contourlet domain (the most energetic subband) as the binary hypothesis test:

$$\mathcal{H}_0 : C_y(i,j) = C_x(i,j) \tag{7}$$

$$\mathcal{H}_1 : C_y(i,j) = C_x(i,j) + \alpha s(i,j) \tag{8}$$

The above hypothesis test verifies the existence of the given watermark sequence $s(i,j)$ in the contourlet coefficients of the given image $C_y(i,j)$. The log-likelihood ratio test is a good choice as a binary hypothesis test for watermark detection because an LLRT based detector maximizes the probability of detection P_D (deciding $\mathcal{H}_1$ when $\mathcal{H}_1$ is true) for a given probability of false-alarm P_{FA} (deciding $\mathcal{H}_1$ when $\mathcal{H}_0$ is true). So, we design an LLRT based watermark detector and we have:

$$\log\{\Lambda(\mathbf{C_y})\} = \log\{\frac{p(\mathbf{C_y}|\mathcal{H}_1)}{p(\mathbf{C_y}|\mathcal{H}_0)}\} \underset{0}{\overset{1}{\gtrless}} T \tag{9}$$

where $p(\mathbf{C_y}|\mathcal{H}_0)$ and $p(\mathbf{C_y}|\mathcal{H}_1)$ denote the probability density functions (pdfs) of $\mathbf{C_y}$ under the conditions $\mathcal{H}_0$ and $\mathcal{H}_1$, respectively. T is the threshold that controls P_D and P_{FA}. To obtain an optimal watermark detector using (9), an accurate statistical model of the contourlet coefficients is crucial. In Sect. 2.3, the compatibility between 2D-GARCH and contourlet coefficients has been studied and confirmed. So, we design an LLRT based watermark detector using 2D-GARCH model. Suppose the contourlet coefficients of the original image $\mathbf{C_x}$ follow 2D-GARCH (p_1, p_2, q_1, q_2) with the parameters $\{\alpha_0, \alpha_{01}, \cdots, \alpha_{q_1 q_2}, \beta_{01}, \cdots, \beta_{p_1 p_2}\}$. Using (5), (7), and (8), the log-likelihood ratio given in (9) can be written as:

$$\log\{\Lambda(\mathbf{C_y})\} = \log \frac{\prod\limits_{ij\in\Phi} p(C_y(i,j) - \alpha s(i,j)\,|\psi(i,j))}{\prod\limits_{ij\in\Phi} p(C_y(i,j)\,|\psi(i,j))} \tag{10}$$

$$= \log \frac{\prod\limits_{ij\in\Phi} \frac{1}{\sqrt{2\pi h(i,j)}} \exp(\frac{-(C_y(i,j)-\alpha s(i,j))^2}{2h(i,j)})}{\prod\limits_{ij\in\Phi} \frac{1}{\sqrt{2\pi h(i,j)}} \exp(\frac{-C_y(i,j)^2}{2h(i,j)})} \tag{11}$$

where $\psi(i,j)$ and Φ are as defined in Sect. 2.2. Substituting $h(i,j)$ from (2) in (11) and applying some simplifications, the log-likelihood ratio can be formulated as:

$$\begin{aligned}\log\{\Lambda(\mathbf{C_y})\} = &\sum_{ij\in\Phi} \log \sqrt{\frac{\sum_{k\ell\in\Omega_1}\alpha_{k\ell}C_y(i-k,j-\ell)^2+\sum_{k\ell\in\Omega_2}\beta_{k\ell}h(i-k,j-\ell)}{\sum_{k\ell\in\Omega_1}\alpha_{k\ell}(C_y(i-k,j-\ell)-w(i-k,j-\ell))^2+\sum_{k\ell\in\Omega_2}\beta_{k\ell}h(i-k,j-\ell)}}\\ &+\sum_{ij\in\Phi}\Big[\frac{C_y(i,j)^2}{2\alpha_0+\sum_{k\ell\in\Omega_1}\alpha_{k\ell}C_y(i-k,j-\ell)^2+\sum_{k\ell\in\Omega_2}\beta_{k\ell}h(i-k,j-\ell)}\\ &+\frac{-(C_y(i,j)-w_{ij})^2}{2\alpha_0+\sum_{k\ell\in\Omega_1}\alpha_{k\ell}(C_y(i-k,j-\ell)-w(i-k,j-\ell))^2+\sum_{k\ell\in\Omega_2}\beta_{k\ell}h(i-k,j-\ell)}\Big]\end{aligned} \quad (12)$$

4 Experimental Results

In this section, we report the results of simulations. In the proposed method, for multiscale decomposition, 9-7 biorthogonal filters with two levels of pyramidal decomposition are used and in the multi directional decomposition stage, PKVA ladder filters are employed. We carried out extensive simulations on a large number of images but according to the space limitations, we report the results of two 512 × 512 grayscale representative images, namely, *Peppers* and *Living room*. To examine the statistical significance of our results, we used 50 natural images and report the averaged results. Also, we compare the performance of proposed watermark detector (CT-GARCH) with two other detectors: (1) contourlet domain generalized Gaussian based detector (CT-GG). It should be mentioned that using generalized Gaussian distribution for the contourlet coefficients has been proposed in some papers such as [13]. (2) wavelet domain 2D-GARCH based detector (WT-GARCH) [3]. For 2D-GARCH based detectors, 2D-GARCH(1, 1, 1, 1) is used.

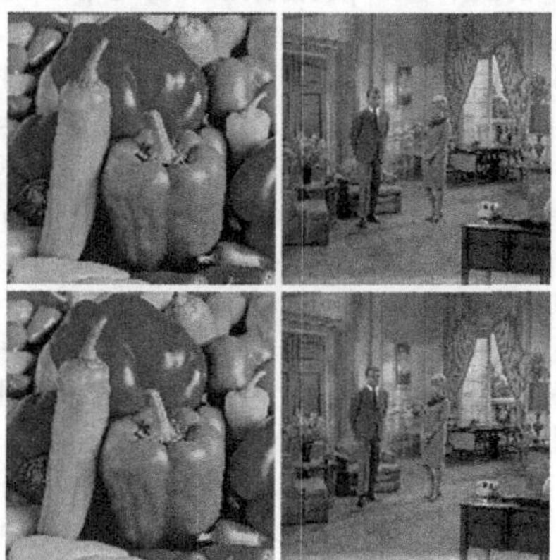

Fig. 2. Test images: "Peppers", and "Living Room" up to down: Original images, and watermarked images using the proposed scheme with WDR = −50 dB

Figure 2 shows two test images and the watermarked version of them with $WDR = -50\,dB$. This figure confirms the imperceptibility of the proposed watermarking scheme. To evaluate the performance of watermark detectors, receiver operating characteristic (ROC) curves that are plots of probabilities of detection P_D in terms of probabilities of false alarm P_{FA} are used.

First, we study the performance of watermark detectors with out any attacks. Figure 3 presents the ROCs of the contourlet domain 2D-GARCH based detector (CT-GARCH), contourlet domain generalized Gaussian based detector (CT-GG), and wavelet domain 2D-GARCH based detector (WT-GARCH) for two test images. It can be inderstood from this figure that the proposed detector provides the highest probability of detection for any chosen value of the false alarm.

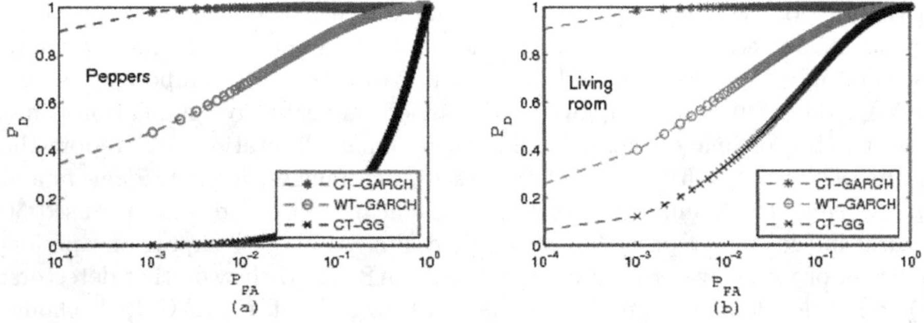

Fig. 3. ROC of the three statistical detectors. Test images are (a) Peppers, and (b) Living Room. (WDR = −50 dB)

In the following, we study the performance of proposed watermark detector (CT-GARCH) and also CT-GG and WT-GARCH detectors under some standard attacks, i.e., JPEG compression, median and Gaussian filtering, scaling, and a combinational attack (combination of the Gaussian filtering (window size = 5 × 5) and additive white Gaussian noise (AWGN) attacks (std = 10)). To quantitatively study the detection performance, we use the area under the ROC curve (AUROC) [2,3,7]. Table 2 reports the AUROC results of the CT-GARCH, CT-GG, and WT-GARCH detectors under these types of attack for WDR = −50 dB. Scaling attack has been applied with scaling factor 0.75. The size of window for median and Gaussian filtering is 5×5. It is clear from Table 2 that the CT-GARCH detector outperforms the other detectors under all types of attacks.

Finally, to study the statistical significance of the reported results, we perform the robustness tests on 50 natural images. Table 3 reports the averaged AUROC results over 50 images that confirms the higher performance of the proposed method (CT-GARCH) compared with CT-GG and WT-GARCH.

Table 2. AUROC results under JPEG compression, Scaling, Median filtering, Gaussian filtering and combination of Gaussian filtering and AWGN attacks (WDR = $-50\,dB$)

JPEG compression attack (QF = 60)			
Image	CT-GG detector	WT-GARCH detector	CT-GARCH detector
Peppers	0.7887	0.8512	0.9814
Living room	0.7527	0.9294	0.7527
Scaling attack (scaling factor = 0.75)			
Peppers	0.7877	0.7516	0.9920
Living room	0.7578	0.8335	0.9986
Median filtering attack (5 × 5)			
Peppers	0.8237	0.8037	0.9971
Living room	0.9994	0.8820	1.0000
Gaussian filtering attack (5 × 5)			
Peppers	0.9439	0.8565	1.0000
Living room	0.9269	0.8808	1.0000
Gaussian Filtering (5 × 5) + AWGN attacks			
Peppers	0.8864	0.7773	0.9649
Living room	0.8299	0.7268	0.9936

Table 3. Average AUROC results for 50 natural images under different types of attacks (WDR = $-50\,dB$)

Attack type	CT-GG detector	WT-GARCH detector	CT-GARCH detector
Compression (QF = 60)	0.8463	0.8454	0.9320
Scaling (SF = 0.75)	0.7863	0.7982	0.9654
Median filtering 5 × 5	0.9324	0.8864	0.9828
Gaussian filtering 5 × 5	0.9288	0.8705	0.9951
Gaussian filtering + AWGN	0.7536	0.7585	0.9708

5 Conclusion

A novel contourlet domain watermark detector has been proposed in this paper. Watermark is additively embedded into the contourlet subband with the highest energy. Watermark detection is formulated as a binary hypothesis test and the optimal detector can be achieved using LRT. Contourlet domain LRT is based on statistical modeling of contourlet coefficients. Greater the precision of the distribution of contourlet coefficients used in the detector results in the higher the reliability of detection of the watermark. For this purpose, statistical properties of contourlet coefficients have been studied and it has been demonstrated that contourlet coefficients are heteroscedastic. Since none of the previously proposed statistical models for contourlet coefficients can capture the heteroscadsticity, 2D-GARCH model has been proposed for contourlet coefficients and its compatibility with contourlet coefficients has been confirmed. Consequently, LRT based contourlet domain watermark detector based on using 2D-GARCH model has been designed. Simulation results demonstrate the high efficiency of proposed method.

References

1. Cox, I.J., Miller, M.L., Bloom, J.A., Fridrich, J., Kolker, T.: Digital Watermarking and Steganography, 2nd edn. Morgan Kaufmann Publishers, Burlington (2008)
2. Rabizadeh, M., Amirmazlaghani, M., Ahmadian-Attari, M.: A New detector for contourlet domain multiplicative image watermarking using Bessel K form distribution. J. Vis. Commun. Image Represent. **40**, 324–334 (2016)
3. Amirmazlaghani, M.: Additive watermark detection in wavelet domain using 2D-GARCH model. Inf. Sci. **370**, 1–17 (2016)
4. Sadreazami, H.R., Ahmad, M.O., Swamy, M.N.S.: A study of multiplicative watermark detection in the contourlet domain using alpha-stable distributions. IEEE Trans. Image Process. **23**(10), 4348–4360 (2014)
5. Amirmazlaghani, M., Rezghi, M., Amindavar, H.: A novel robust scaling image watermarking scheme based on Gaussian Mixture Model. Expert Syst. Appl. **42**, 1960–1971 (2015)
6. Sadreazami, H., Ahmad, M.O., Swamy, M.N.S.: Multiplicative water- mark decoder in contourlet domain using the normal inverse Gaussian distribution. IEEE Trans. Multi. **18**(2), 196–207 (2016)
7. Bian, Y., Liang, S.: Locally optimal detection of image watermarks in the wavelet domain using Bessel K Form distribution. IEEE Trans. Image Process. **22**(6), 2372–2384 (2013)
8. Okman, O.E., Akar, G.B.: Quantization index modulation-based image watermarking using digital holography. J. Opt. Soc. Am. **24**(1), 243–252 (2007)
9. Cheng, Q., Huang, T.S.: Robust optimum detection of transform domain multiplicative watermarks. IEEE Trans. Sig. Process. **51**(4), 906–924 (2003)
10. Do, M.N., Vetterli, M.: The contourlet transform: an efficient directional multiresolution image representation. IEEE Trans. Image Process. **14**(12), 2091–2106 (2005)
11. Song, H., Yu, S., Yang, X., Song, L., Wang, C.: Contourlet-based image adaptive watermarking. Sig. Process.: Image Commun. **23**(3), 162–178 (2008)
12. Jayalakshmi, M., Merchant, S.N., Desai, U.B.: Blind watermarking in contourlet domain with improved detection. In: Proceedings of the Intelligent Information Hiding and Multimedia Signal Processing (IIH-MSP), Pasadena, CA, USA (2006)
13. Qu, H., Peng, Y.: Contourlet coefficient modeling with generalized Gaussian distribution and application. In: Proceedings of the International Conference Audio, Language and Image Processing (ICALIP), July 2008
14. Akhaee, M.A., Sahraeian, S.M.E., Marvasti, F.: Contourlet-based image watermarking using optimum detector in a noisy environment. IEEE Trans. Image Process. **19**(4), 967–980 (2010)
15. Amirmazlaghani, M., Amindavar, H., Moghaddamjoo, A.R.: Speckle suppression in SAR images using the 2-D GARCH model. IEEE Trans. Image Process. **18**(2), 250–259 (2009)
16. Bollerslev, T.: Generalized autoregressive conditional heteroscedasticity. J. Econom. **31**, 307–327 (1986)

17. Amirmazlaghani, M., Amindavar, H.: Statistical modeling and denoising Wigner-Ville distribution. Digital Sig. Process. **23**(2), 506–513 (2013)
18. Amirmazlaghani, M., Amindavar, H.: A novel sparse method for despeckling SAR images. IEEE Trans. Geosci. Remote Sens. **50**(12), 5024–5032 (2012)
19. Mahbubur Rahman, S.M., Omair Ahmad, M., Swamy, M.N.S.: A new statistical detector for dwt-based additive image watermarking using the GaussHermite expansion. IEEE Trans. Image Process. **18**(8), 1782–1796 (2009)

H-264/RTSP Multicast Stream Integrity

Giuseppe Cattaneo[1(✉)], Andrea Bruno[1], and Fabio Petagna[2]

[1] Dipartimento di Informatica, Università degli Studi di Salerno, Fisciano 84084, SA, Italy
{cattaneo,andbruno}@unisa.it

[2] eTuitus, Spinoff dell'Università degli Studi di Salerno, Fisciano 84084, SA, Italy
fabio.petagna@etuitus.it

Abstract. In this paper we discuss about stream integrity problem and specifically we propose a method to protect a live video stream against forgery. In particular we propose to use cryptographic techniques to provide integrity and source identity embedding in the source stream a hash chain built on the top of each frame as produced by the source.

We first present a real life scenario in which stream integrity could be considered an enabler, then we describe our method to perform real time integrity check.

In last section an implementation is sketched to show main implementation parameter.

Keywords: Live video stream · Video integrity · Stream integrity · Digital forensics · H.264 · RTSP

1 Introduction

Live video streaming is a useful way to reduce the distances and enable participation to virtual meeting saving transfer time and therefore earth pollution.

A typical scenario that can benefit of such tool is the remote testimony in a trial. For example in case of testimony of a witness included in a witness relocation program for his own safety, the jury have to hear the testimony, but is risky and expensive to move the testimony. In this case a system to permit remote testimony could be valuable. Unfortunately actual video conference systems suffer of many vulnerability that made these tool unreliable and therefore unusable in legal context. In fact standard video-conference use does not allow endpoint authentication, end-to-end encryption and integrity check (non repudiation).

Moreover the scenario just described in many cases requires multicast communication to make the source video stream available to several destination. Again multicast protocols do not allow standard protocol for privacy like Transport Layer Security (TLS) and therefore are subject to man in the middle attack. As for the author knowledge is literature is not defined a secure multicast protocol.

S. Battiato et al. (Eds.): ICIAP 2017, Part II, LNCS 10485, pp. 558–568, 2017.
https://doi.org/10.1007/978-3-319-68548-9_51

In this paper we present a general method to secure a multicast live video stream. It has been implemented for demo purposes for H.264 stream. Our method takes a standard H.264 stream as input and produce a secure stream that holds the following properties:

P.1 *Integrity check:* The live video stream cannot be modified without been noticed by the client. A stream can be altered using one or more of the following operations: (a) frame content modification, (b) frame injection or deletion.
P.2 *Source authentication:* The source identity can be certified and verified by all the destination endpoints.
P.3 *Real time:* The verification procedure must be available directly on the received stream without introducing significant delay. This is opposite to verification procedure that are performed on the closed file where the stream has been saved.
P.4 *Multicast:* The method must apply also in one-to-many communication protocols. This is opposite to end-to-end security protocols.

The first two properties holds also after that the stream has been stored in a file. As side effect of first properties in a saved stream the frame order is preserved.

In digital forensics area many solution have been proposed for preserving the video acquired as digital evidence [17], for source camera identification [5,10–12] and for video forgery detection [6,9,15], but on the contrary in literature there are few examples of methods to secure real-time streaming [8,16,20,22].

After the method has been designed we implemented a prototype in a real life context to verify the impact produced by our solution. We adopted the H.264 container over Real-time Streaming Protocol (RTSP) because of it wide spread. The RTSP uses the Secure Sockets Layer (SSL) to provide the first three properties, but unfortunately this can be used only on Point-to-Point communications.

In short we propose to add a nonce negotiation during `DESCRIBE` phase of RTSP to Session Description Protocol (SDP) file in addition to the other connection parameter. This nonce will be used to start a hash chain that will allow to check the frame sequence integrity. In addition, periodically, the camera will sign the current element of the hash chain in order to provide property Sect. 1.

We also observed that because RTSP over UDP is a lossy protocol we have proposed a recovery mechanism that can be applied when the number of lost packet is under a fixed threshold.

Finally as further specification we required that the stream encoded by our method should be decoded and viewed with a standard RTSP\H.264 player unaware of such protocol. In other word the protocol should be transparent to the source stream. In the previous scenario this may be the case of the journalist or the audience that want to assist to the testimony but do not interested in checking the stream integrity.

Structure of the Paper: In Sect. 2 we analyse the current state of the art for the problem of stream integrity, and in Sect. 3 we give an overview of the H.264

and RTSP. In Sect. 4 we describe detail our solution and in Sect. 5 we introduce some implementation details to show how the method can be embedded in a real life protocol and the issue we had to face. Finally in Sect. 6 we show our conclusions and some propositions for future works.

2 State of the Art

The topic of video integrity in the field of digital forensics has been extensively discussed in scientific literature, and it mainly concerns the identification of forgery produced by manipulating a video file. In this paper we want focus on the possibility to check the integrity of a live stream in real-time. This specification rules out all the traditional tools that works on saved files.

When the authorities seize a video to be used as evidence in a trial, they have to proof that it has not been tampered in any way, and possibly they have also to identify its source. To identify the source of the video, is possible to use techniques of *Source Camera Identification* like the largely used Pixel Non Uniformity (PNU) proposed by *Fridrich et al.* [11] for digital images and successively been extended to digital video also as proposed by the same authors in [10]. Here the authors show a technique to determine if two videos have been taken by the same camera, in particular they use a normalized cross-correlation on the input noise from the input videos.

Hsu et al. [13] proposed a refinement of this technique by defining the concept of *temporal sensor pattern noise* of a video, with respect to a reference camera, and the correlation is performed using a Gaussian Mixture Model (GMM).

Byram et al. [6] presents a technique to detect duplicate and modified copies of a specific video based only on the image sensor characteristics. The authors in [14] present a forgery detection system specifically designed for surveillance videos [3,7]. They analyze the peculiar characteristics of these videos and then they propose to transform the SPN for each video by applying a *Minimum Average Correlation Energy* (MACE) filter to identify both RGB and infrared video. The manipulation are detected by estimating the scaling factor and calculating the correlation coefficient.

Cattaneo et al. [9] analyzed the possibility to use the PNU to detect the insertion of *alien frames* in a video, when the injected frames have been recorded with a different camera.

Ramaswamy and Rao proposed [16] a method for digital authentication of video sequence compressed using H.264/Advanced Video Codec (AVC) by using traditional digital signature. This method was intended for signing video after it has been saved in a file to guarantee that cannot be applied any further modification to it.

All the systems analysed so far works off-line on a video after it has been saved in a file. Therefore cannot be applied to our scenario.

Among the on-line systems, privacy has been enhanced by [20] where the authors proposed an approach based on the state-of-art symmetric cypher to realize a H.264/AVC and SVC compliant encrypted stream. Even if compliant with standard the stream could not be read by a standard player due to encryption.

3 Operating Environment

In the next section we will give in short an overview of the H.264/AVC/SVC Sect. 3.1 and RTSP/RTP Sect. 3.2 useful for the rest of our work.

Next we introduce the properties of cryptographic function Sect. 3.3 used in our method Sect. 4.

3.1 H.264/AVC/SVC

H.264/AVC was proposed as a standard by ITU-T since 2003 [21] and recently moved to version 11 of the standard [1].

Basically H.264/AVC is similar to previous video coding standards.Those are the basic steps for coding:

1. The picture is subdivided into macroblocks of 16×16 blocks of luma samples
2. Inter or intra prediction (B-frames or P-frames)
3. Transformation and quantization
4. Entropy coding
5. Network Abstraction Layer (NAL) unit assembly

In 2007 [2] was introduced with the the annex G the Scalable Video Codec (SVC), the standardization of an high quality video bitstream that also contains one or more subset bitstreams.

H.264/AVC has a clear structure which distinguishes a coding layer (Video Coding Layer (VCL) and non-VCL) and a NAL. The first one is responsible for the coded representation of the pictures in the stream, while the second formats this representation and provides additional header information.

The NAL units can contain different unit types, and if a player is unable to understand a (NUT) it can skip it. The NAL type from 24 to 31 are unspecified and are available for used defined types.

3.2 RTSP/RTP in Short

RTSP is a network control protocol for real-time streaming frequently used with Real-time Transport Protocol (RTP) as transport protocol which generally runs over User Datagram Protocol (UDP). RTSP defines a set of control sequences useful in controlling multimedia playback.

For the need of this work we show only the control sequences will be necessary in the following Sects. 4 and 5, more details can be found in RFC 2326 [19].

`DESCRIBE` This control sequence is used to ask the server the presentation description typically in SDP format.

`PLAY` This control sequence is used to start all the media streams.

`TEARDOWN` This control sequence is used to terminate all the media streams.

3.3 Cryptographic Functions

Cryptographic Hash Function. A cryptographic hash function is a special class of hash function that has certain properties which make it suitable for use in cryptography. The ideal cryptographic hash function as presented in [18] has five main properties:

- it is deterministic so the same message always results in the same hash
- it is light to compute the hash value for any given message
- it is infeasible to generate a message from its hash value except by trying all possible messages
- any change to a message, even one single bit, must produce a new completely different hash value without any apparent correlation with the old hash value
- it is infeasible to find two different messages with the same hash value.

Digital Signature. A digital signature is a mathematical scheme for demonstrating the authenticity of digital messages or documents. To be valid, digital signatures require the following properties as presented in [18]:

- Authenticity: a valid signature implies that the signer deliberately signed the associated message
- Unforgeability: only the signer can give a valid signature for the associated message
- Non-re-usability: the signature of a document can not be used on another document
- Non-repudiation: the signer can not deny having signed a document that has valid signature
- Integrity: ensure the contents have not been modified.

4 Our Method

In this section we describe in detail our solution to check the integrity of live multicast video stream.

In particular our method can be split in three procedures: first of all the setup procedure Sect. 4.1, in which all the participants negotiate the parameters used during the rest of the process; the sender procedure Sect. 4.2 in which the sender builds an hash chain with video frame digests, and on a time base digitally signs some of these hash values; the receiver procedure Sect. 4.3 in which on the clients the hash chain and the signature carried on by the stream are verified.

At the end we show also the recovery procedure Sect. 4.4 that allows the client to restart the method when some RTSP packets are lost.

4.1 Setup Procedure

This procedure uses the standard SDP like in a traditional RTSP session. Before starting the streaming session with the Secure Remote Camera (SRC) C the procedure share with all the clients a *nonce* value F_0 and its public key Pk_C. The *nonce* value is an arbitrary number that may only be used once which identifies the session and will be used as the starting value F_0 of the hash chain in the next procedure Sect. 4.2. The public key Pk_C is included in a standard X.509 certificate released by an official certification authority to uniquely identify the source. The corresponding private key Sk_C is stored on the server in a secure device.

4.2 Sender Procedure

This procedure is executed on the server-side and, for seek of clarity: can be divided in two sub procedure, (1) *Frame order integrity* and (2) *Source authentication.*

Frame Order Integrity. According to property (P.1) in the Sect. 1, in order to enable stream integrity checking, we implement a hash chain, i.e. we concatenate each frame of the stream with its successor. In facts the hash value for the I-frame F_i is computed, as shown in (1), adding the hash value of the previous I-frame F_{i-1} to the content of the I-frame F_i (HE_{i-1}). The first element of the chain is the hash of the element F_0 generated by setup procedure Sect. 4.1.

$$HE_i = \begin{cases} Hash(F_0) & \text{if } i = 0 \\ Hash(HE_{i-1} + F_i) & \text{if } i > 0 \end{cases} \tag{1}$$

If a malicious user would delete one or more I-frames the chain is broken and the hash value cannot be recomputed (as the hash function cannot be inverted). Analogously if one or more I-frames are injected in the stream the I-frame produced after the injection will report an invalid hash value in the chain.

Therefore the hash chain creates a verifiable and ordered cryptographic link between the current and the already issued I-frames.

Each value HE_i is sent to the RTSP in the NAL with one of the unused NUT along with the content of the corresponding frame F_i.

Source Authentication. Another property of the secure stream is the *Source Authentication.* In every moment the receivers must be able to verify the identity of the sender.

To achieve this goal we use an asymmetric digital signature. Given n, a parameter statically defined on the SRC or adaptively calculated depending the camera CPU load and the network load, every n I-frames in the stream the digital signature DS_{HE_i} of the value HE_i is calculated for the current I-frame F_i using the private key Sk_S. The resulting value is added to the RTSP stream in the NAL with one of the unused NUT in the first available packet. In order to avoid any delay to the input string the value of the digital signature is sent asynchronously with respect to the corresponding frame F_i.

$$DS_{HE_i} = Sign(HE_i, Sk_S) \tag{2}$$

H.264 clients have been designed to skip the entire NAL unit when an unspecified NUT is received. That is why we used this way to send extra data added by procedure Sect. 4.2. As consequence a client that is not aware of our method will be anyway able to process the stream and decode the video.

4.3 Receiver Procedure

This procedure is executed on client-side and is in charge of performing on-line integrity check on each received I-frame. Moreover the procedure must verify the sender identity to detect man-in-middle attacks as requested. This is necessary for multicast protocols where SSL cannot be used. The goal of this procedure is to perform integrity check and to alert the user if any of the properties defined in Sect. 1 is violated by the received stream.

The client calculates its own hash chain analogously to the sender procedure Sect. 4.2 starting from the value F_0 received with SDP. HE_i^r is the hash chain element for the I-frame F_i^r received by the client. Therefore for each received frame the procedure compares HE_i^r with the received value HE_i if $HE_i^r \neq HE_i$ the frame is marked as tampered.

When the client receives a packet with the NUT carrying a signature DS_{HE_i}:

1. It retrieves the its own hash chain element HE_i^r corresponding to frame F_i.
2. It checks if HE_i^r is equals to the value HE_i used for digital signature.
3. It verifies the signature of the received hash chain element HE_i using the public key Pk_C

$$Verify(DS_{HE_i}, HE_i, Pk_C) \tag{3}$$

4. If the signature verification fails all the frames starting from the last signed I-frame that has been successfully verified up to the next signed I-frame that will pass the step (3) will be marked as untrusted or better potentially tampered.

If an attacker tampered an I-frame in stream. We can have two possible situations:

(a) The tampered frame is a signed I-frame.
(b) The tampered frame is not a signed I-frame.

In the first case, by the properties of the digital signature, the attacker is not able to produce a valid signature for the hash chain element HE_i^t for frame F_i^t because he do not know Sk_S, so the verification fails on step (3).

In the second case $F_i \neq F_i^r \Rightarrow HE_i \neq HE_i^r \Rightarrow HE_{i+1} \neq HE_{i+1}^r \Rightarrow \cdots \Rightarrow HE_j \neq HE_j^r$ by the properties of the hash chain, where j is the next signed I-frame. When the client will receive the signed frame F_j the verify phase will fail on step (2) and also the frame F_i^r will be marked as tampered by step (4).

Analogously is possible to demonstrate that if the attacker tampered a sequence of frame he is not able to forge the verification token. In-fact if we can conduct this problem to the following two cases:

1. The tampered sequence includes one or more signed I-frames, and this case is analogous to the case (a) in which is tampered a single signed I-frame and the attacker is not able to reproduce the digital signature.
2. The tampered sequence does not include signed I-frames. In this case, similarly to the case (b), the procedure Sect. 4.3 fails on step (2).

4.4 Restart on Lost Packet

We assumed to use RTSP over UDP that is a connectionless protocol. Therefore some packets may be lost and our method must hand such event.

For our method, the loss of a packet is equivalent to the deletion of a packet, that causes the failure of the procedure 4.3 by the client.

If the lost packets belong to P-frames or B-frames, nothing happens to the stream because those frames are not considered in our method.

Otherwise the lost packets belong to I-frames or signed I-frames. In such a case all the frames until the next frame with a valid signature DS_{HE_i} will be marked as untrusted and the procedure restarts the hash chain with the signed value of HE_i.

5 Implementation Details

In this section we will show some implementation details of the proposed method. We also will give some indication about the the cryptographic function we have chosen to use according to the recommendations of the National Institute of Standards and Technology (NIST) [4].

5.1 Setup Procedure

In Sect. 4.1 we described the setup procedure for our method. Due to the flexibility of SDP, this procedure has been straight embedded in the `DESCRIBE` command of RTSP. The *nonce* is produced an integer random number generator with a 1024 bit precision. Finally the strength of the entire method strictly depends on the robustness of the asymmetric signature algorithm and corresponding key pair used for. Following the recommendation of NIST [4] we adopted for the *RSA* algorithm with a 3072 bit key pair due to its wide availability. More efficient algorithms such as *ECDSA* could be used in order to save space and computational time. In this case the suggested key size by [4] is 192 bit.

5.2 Sender and Receiver Procedure

As far as the hash function concern we adopted the $SHA256$ function which as recommended by NIST [4]. Stronger hash functions like $SHA512$ (with a length of 512 bit), despite of their light computational cost, could produce a deeper impact on network usage as the result of this function is added to each I-frame.

It is important to highlight that the signature it is always a time consuming process. For this reason we designed the method to send signed I-frames is performed only at regular intervals of n instead of signing each frames. The value of n can be adapted to the resource available.

On the other hand the hash function computed for each I-frame, is light and can be even efficiently implemented in a dedicated hardware.

Obviously the implementation and the issues of the receiver procedure are specular to the sender procedure.

5.3 Implementation

In order to verify the feasibility of our method we engineered a real size case study.The server runs on a Raspberry Pi 3 model B with a 1.2 GHz 64-bit quad-core ARMv8 CPU equipped with a webcam module version 2 supported by Video for Linux (V4L) through the kernel module bcm2835-v4l2. The operating system installed on the server was Rasbian Version 8.0 (jessie). The server side application was based upon the framework live555 (http://www.live555.com/).

We performed several runs with different camera configurations, changing resolution and frame-rate. Despite of the low CPU performance, during all the tests the CPU usage was always under 50%, confirming that our method can be implemented even on small embedded CPU.

On the client side we used a standard linux virtual machine with Ubuntu Linux 14.04 with 1 GB RAM and a 2 Core Intel i7 CPU at 2.6 GHz. On the client side we used mplayer to reproduce the stream without checking the integrity and a custom client application to check the stream integrity before passing the frames to a standard player. Figure 1 shows the client output with the video on the right side and the execution trace on the left hand displaying the results of the hash chain verification.

Fig. 1. Client: Example execution with hash verification and mplayer output

Currently we are working on a new version of the server implementation on an AXIS/ACAP platform.

6 Conclusions and Future Works

In Sect. 4 we showed a method that links each I-frame in a stream building an hash chain and while some I-frames are digitally signed. This permits a SRC to send a stream to a set of clients via multicast UDP protocol, preventing an attacher to tamper the stream by adding, modifying or deleting one ore more frames. This is an enhancement with respect to the standard SSL connection properties, because it works on multicast connections over a connectionless weaker protocol like UDP.

In addition the control information added to the original stream are sent only using an unspecified NUT, preserving therefore the compatibility with H.264. This means that the secure stream can be processed by a standard H.264 client. Even if the method has been conceived for live streams, these streams can also be stored in files and verified off-line with the same procedure Sect. 4.3.

The proposed method is robust to the loss of packets and a recovery procedure has been presented in Sect. 4.4 can be executed when packet are lost.

We implemented a prototype of a SRC using a Raspberry PI with a camera module in order to test this method in a real word scenario.

Actually we are planning to add a symmetric encryption algorithm to the stream to preserve privacy limiting the list of authorized client. In-fact only the client that participated to the setup phase, negotiating the session key will be able to decrypt the stream.

Further Information

Part of the work presented in this paper is protected by an Italian patent registered with number 102016000007162 on January 26th 2016 owned by eTuitus.

References

1. I.T.S.G.: Itu-t h.264 (v11): Advanced video coding for generic audiovisual services. ITU-T Series H: Audiovisual And Multimedia Systems October 2016. http://handle.itu.int/11.1002/1000/12904
2. I.T.S.G.: Itu-t H.264: Advanced video coding for generic audiovisual services. ITU-T Series H: Audiovisual And Multimedia Systems November 2007. http://handle.itu.int/11.1002/1000/9226
3. Albano, P., Bruno, A., Carpentieri, B., Castiglione, A., Castiglione, A., Palmieri, F., Pizzolante, R., You, I.: A Secure Distributed Video Surveillance System Based on Portable Devices. In: Quirchmayr, G., Basl, J., You, I., Xu, L., Weippl, E. (eds.) CD-ARES 2012. LNCS, vol. 7465, pp. 403–415. Springer, Heidelberg (2012). doi:10.1007/978-3-642-32498-7_30
4. Barker, E.: Recommendation for key management part 1: general (revision 4). NIST Spec. Publ. **800**(57), 1–160 (2016)
5. Bayram, S., Sencar, H.T., Memon, N., Avcibas, I.: Source camera identification based on CFA interpolation. In: IEEE International Conference on Image Processing, ICIP 2005, vol. 3, pp. III-69. IEEE (2005)

6. Bayram, S., Sencar, H.T., Memon, N.: Video copy detection based on source device characteristics: a complementary approach to content-based methods. In: Proceedings of the 1st ACM International Conference on Multimedia Information Retrieval, pp. 435–442. ACM (2008)
7. Bruno, A., et al.: Secure and distributed video surveillance via portable devices. J. Ambient Intell. Humanized Comput. **5**(2), 205–213 (2014). doi:10.1007/s12652-013-0181-z
8. Castiglione, A., Cepparulo, M., De Santis, A., Palmieri, F.: Towards a lawfully secure and privacy preserving video surveillance system. In: Buccafurri, F., Semeraro, G. (eds.) EC-Web 2010. LNBIP, vol. 61, pp. 73–84. Springer, Heidelberg (2010). doi:10.1007/978-3-642-15208-5_7
9. Cattaneo, G., Roscigno, G., Bruno, A.: Using PNU-based techniques to detect alien frames in videos. In: Blanc-Talon, J., Distante, C., Philips, W., Popescu, D., Scheunders, P. (eds.) ACIVS 2016. LNCS, vol. 10016, pp. 735–746. Springer, Cham (2016). doi:10.1007/978-3-319-48680-2_64
10. Chen, M., Fridrich, J., Goljan, M., Lukáš, J.: Source digital camcorder identification using sensor photo response non-uniformity. In: Electronic Imaging 2007, p. 65051G. International Society for Optics and Photonics (2007)
11. Fridrich, J., Lukáš, J., Goljan, M.: Digital camera identification from sensor noise. IEEE Trans. Inf. Secur. Forensics **1**(2), 205–214 (2006)
12. Goljan, M., Fridrich, J., Filler, T.: Large scale test of sensor fingerprint camera identification. In: IS&T/SPIE, Electronic Imaging, Security and Forensics of Multimedia Contents XI, vol. 7254, pp. 1–12. International Society for Optics and Photonics (2009)
13. Hsu, C.C., Hung, T.Y., Lin, C.W., Hsu, C.T.: Video forgery detection using correlation of noise residue. In: IEEE 10th Workshop on Multimedia Signal Processing, pp. 170–174. IEEE (2008)
14. Hyun, D.K., Lee, M.J., Ryu, S.J., Lee, H.Y., Lee, H.K.: Forgery detection for surveillance video. The Era of Interactive Media, pp. 25–36. Springer, New York (2013). doi:10.1007/978-1-4614-3501-3_3
15. Liao, S.Y., Huang, T.Q.: Video copy-move forgery detection and localization based on Tamura texture features. In: 2013 6th International Congress on Image and Signal Processing (CISP), vol. 2, pp. 864–868. IEEE (2013)
16. Ramaswamy, N., Rao, K.: Video authentication for H. 264/AVC using digital signature standard and secure hash algorithm. In: NOSSDAV 2006. Newport, Rhode Island, USA (2006)
17. Redfield, C.M., Date, H.: Gringotts: securing data for digital evidence. In: Security and Privacy Workshops (SPW), pp. 10–17. IEEE (2014)
18. Schneier, B.: Applied Cryptography: Protocols, Algorithms, and Source Code in C. Wiley, Hoboken (2007)
19. Schulzrinne, H., Rao, A., Lanphier, R.: Real time streaming protocol (RTSP). Internet Engineering Task Force. Technical report, RFC 2326 (1998)
20. Stütz, T., Uhl, A.: Format-compliant encryption of H. 264/AVC and SVC. In: Tenth IEEE International Symposium on Multimedia, ISM 2008, pp. 446-451. IEEE (2008)
21. Team, J.V.: Advanced video coding for generic audiovisual services. ITU-T Rec. H **264**, 14496–14510 (2003)
22. Yeung, S., Lui, J.C., Yau, D.K.: Secure real-time streaming protocol (RTSP) for hierarchical proxy caching. Int. J. Netw. Secur. **7**(3), 1–20 (2007)

PRNU-Based Forgery Localization in a Blind Scenario

Davide Cozzolino, Francesco Marra, Giovanni Poggi, Carlo Sansone(✉), and Luisa Verdoliva

DIETI - University of Naples Federico II, Via Claudio 21, 80125 Naples, Italy
{davide.cozzolino,francesco.marra,giovanni.poggi, carlo.sansone,luisa.verdoliva}@unina.it

Abstract. The Photo Response Non-Uniformity (PRNU) noise can be regarded as a camera fingerprint and used, accordingly, for source identification, device attribution and forgery localization. To accomplish these tasks, the camera PRNU is typically assumed to be known in advance or reliably estimated. However, there is a growing interest for methods that can work in a real-word scenario, where these hypotheses do not hold anymore. In this paper we analyze a PRNU-based framework for forgery localization in a blind scenario. The framework comprises four main steps: PRNU-based blind image clustering, parameter estimation, device attribution, and forgery localization. Each of these steps impacts on the final outcome of the analysis. The aim of this paper is to assess the overall performance of the proposed framework and how it depends on the individual steps.

1 Introduction

The wide diffusion of powerful image editing tools has made image manipulation very easy. This impacts on many fields of life, and is especially dangerous in the forensic field, where images may be used as crucial evidence in court. Therefore, in the last decade, digital image forensics has grown tremendously, and new methodologies have been developed to track an image source and determine its integrity. In particular, the interest has focused on passive techniques, which detect traces of manipulations from the analysis of the image itself, with no need of collaboration on the part of the user. Some of these techniques rely on intrinsic camera properties, like sensor defects or lens aberration, while other rely on statistical features introduced both in-camera (e.g., demosaicking) and out-camera (e.g., JPEG compression) processing.

Some of the most successful camera-based methods rely on the Photo Response Non-Uniformity (PRNU), a sort of camera fingerprint contained in every image taken by a specific device. Its use was first proposed in [17], both for source identification and forgery localization. In this work we focus on PRNU-based methods for forgery detection and localization.

To extract the PRNU pattern, the high-level image content is removed through some sort of high-pass filtering, obtaining the so-called image residual.

S. Battiato et al. (Eds.): ICIAP 2017, Part II, LNCS 10485, pp. 569–579, 2017.
https://doi.org/10.1007/978-3-319-68548-9_52

Even in the residual, however, the PRNU pattern represents a weak signal overwhelmed by intense noise, both of random origin, and deriving from imperfect image content removal. This makes all PRNU-based analyses quite challenging, to begin with the very same camera PRNU estimation. Several approaches have been proposed in the literature to reduce the influence of the scene content on PRNU estimation. In [15] some forms of enhancement are considered, while in [7] the use of a nonlocal denoising filter has been shown to reduce the scene content in the residual image. A systematic analysis of post-processing methods aimed at improving PRNU estimation has been recently presented in [1]. With reference to the forgery localization task, several improvements have been proposed with respect to the basic method of [17]. In [5] a predictor is estimated which locally adapts the statistical decision test by taking into account image features, such as texture, flatness and intensity, thus reducing the probability of false alarms. In [9], instead, the problem is recast in terms of Bayesian estimation, using a Markov random field (MRF) prior to model the strong spatial dependencies and take decisions jointly rather than individually for each pixel. In [6,8] the problem of small forgery detection is addressed, resorting to image segmentation and guided filtering to improve the decision statistics. Further improvements have been recently proposed by considering the use of discriminative random fields [4] or by introducing multiscale analysis [14].

All these methods rely on the assumption that a large number of images are available, which are known to come from the camera of interest. However, such an hypothesis is not reasonable in a real-world scenario. Therefore, in this paper we propose and analyze a framework for image forgery localization in a blind scenario [10]. We only assume to have a certain number of images, whose origin, however, is unknown. Then we estimate one or more PRNU's by means of a blind source clustering algorithm and use them to establish the integrity of the image under test.

In the following Section we describe the PRNU-based framework for blind forgery localization, while in Sect. 3 present experimental results with reference to various clustering approaches [2,3,18], in order to assess the overall performance of the proposed framework and how it depends on the individual steps. Finally, in Sect. 4 we draw some conclusions.

2 Camera-Based Forgery Localization Framework

In both camera identification and forgery localization tasks, the PRNU of the camera of interest is given in advance, or is accurately estimated from a large number of images coming from the camera. However, in many forensic scenarios, and especially in investigation, no information is available on the origin of the images under analysis, neither the probe nor the dataset. Often, however, it is reasonable to assume that the images in the dataset come from just a few different devices. With this assumption, we can pursue PRNU-based forgery localization in a blind scenario, following the framework shown in Fig. 1 and already outlined in [10].

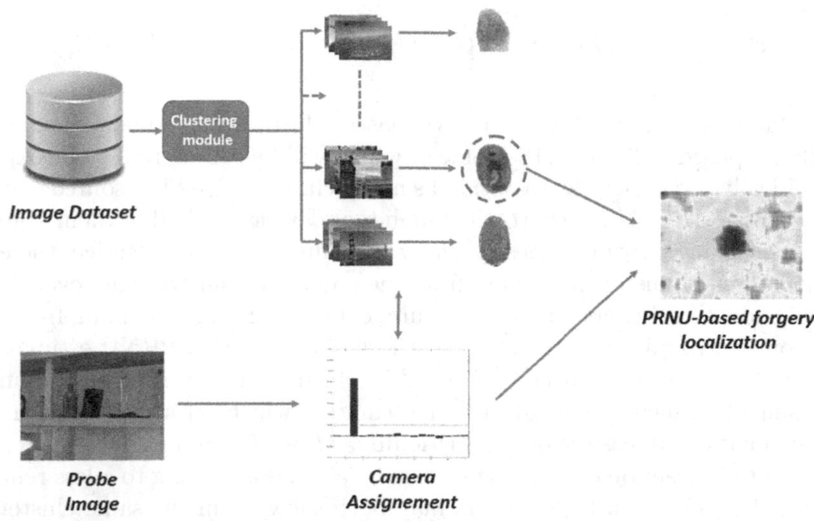

Fig. 1. A framework for PRNU-based forgery localization in a blind scenario.

The considered framework consists of four steps:

1. Residual-based image clustering
2. Cluster PRNU estimation
3. Camera assignment
4. Forgery localization.

The first two steps allow us to group together images coming from the same camera and to estimate their PRNU. Then, in step 3, the test image is associated with one of the clusters (or possibly none) by a PRNU-based correlation test. Finally, the tampered area of the test image is localized by detecting the absence of the selected PRNU. These steps are described in more detail in the following.

2.1 Residual-Based Image Clustering

To perform PRNU-based forgery localization one needs the true PRNU of the camera. Otherwise, it can be estimated by averaging a large number of images taken by the camera of interest. To this end, the first step of the proposed framework aims at grouping together all images of the dataset coming from the same camera. Since these share the same PRNU, they will exhibit a larger correlation than images coming from different cameras. However, before computing correlations, the high-level content of the images, which represents an interference in this context, is removed by high-pass filtering, obtaining the so-called noise residuals.

Let $\mathcal{R} = R_1, R_2, \ldots, R_N$ be the set of all noise residuals in the dataset. We want to partition this set in M distinct clusters, where the number of clusters is not know a priori. Therefore, the output of this step is a partition, P, of the dataset, namely:

$$P = \{C_1, C_2, ..., C_M\} \qquad C_i \cap C_j = \emptyset \; \forall i \neq j, \qquad \bigcup_{i=1}^{M} C_i = \mathcal{R} \tag{1}$$

In the literature, a number of PRNU-based clustering methods have been recently proposed [2,3,10,16,18], some of which will be considered in the experiments. Ideally, we would like to obtain as many clusters as are the source devices in the dataset, $M = M_t$, with M_t the number of devices, and all of them "pure", namely consisting only of images taken by the same device. In practice, the estimated number of clusters may differ from the number of cameras and, even when they coincide, the clusters may not be pure, comprising images coming from different sources. In all cases, the effect is a loss of accuracy in PRNU estimation. When under-partitioning occurs, $M < M_t$, clusters are necessarily "impure", comprising also images coming from other cameras which act as additional noise in the estimation. In case of over-partitioning, $M > M_t$, even pure clusters may comprise only a fraction of all images taken by a camera, leading to a less reliable estimate. The aforementioned effects may both show up in the same clustering experiment. Of course, all deviations from perfect clustering tend to cause a performance loss.

2.2 Camera Fingerprint Estimation

In the second step, each cluster is treated as "pure", and used to estimate both the PRNU and the predictor needed in the localization phase [5].

Given N_m images in the m-th cluster, one can perform a maximum-likelihood (ML) estimate of the PRNU as [5]

$$\widehat{K}_m = \sum_{i=1}^{N_m} \left[\frac{I_i}{\sum_{i=1}^{N_m} I_i^2} \right] R_i \tag{2}$$

In alternative, one can use the simpler sample average

$$\widehat{K}_m = \frac{1}{N_m} \sum_{i=1}^{N_m} R_i \tag{3}$$

which ensures very close performance to the ML case, provided N_m is large enough. On the other hand, when the cluster is too small, both estimates become quite unreliable because the noise residuals, R_i, have a very small signal component overwhelmed by noise. Whatever the estimator, some suitable steps follow to remove non-unique artifacts originated by other camera processes.

Some clustering methods tend to generate a large number of small clusters, even singletons, besides a few large ones. It makes sense to discard such small clusters, due to the ensuing unreliable estimates. Therefore, we introduce a parameter, $N_{\min}$, left to the analyst to set, such that all clusters with $N_m < N_{\min}$ are automatically discarded, avoiding their involvement in the forgery localization process.

Besides the PRNU itself, the localization algorithm proposed in [9] needs a predictor, which establishes the expected value of the correlation for a pristine image. Therefore, for each cluster, we must also estimate the predictor parameters, say Θ_m. To this end, the cluster must be further divided in two subsets, $C_m = C'_m \cup C''_m$. The first one, C'_m, is used to compute an *internal* PRNU, to which images of the second set, C''_m, are correlated. The parameters of the predictor, Θ_m, are then designed to minimize the error between the predicted and observed values of the correlation. Clearly, this further partition of the cluster further stresses the need for it to be large enough. To reduce this problem, we split clusters exactly in half for this task. Note, however, that the final estimate of the cluster PRNU can be carried out from the whole set. Indeed, the test image is completely alien to the cluster, and hence there is no reason to penalize the estimation of the PRNU.

In conclusion, the output of this step is the set of estimated PRNUs and predictor parameters, $\{\widehat{K}_m, \Theta_m, m = 1, \ldots, M\}$.

2.3 Camera Assignment

In this step we try to establish whether the probe image, I_p, is compatible with one of the estimated PRNU's, and which one. This decision is based on the normalized correlation[1]

$$\rho_m = \text{corr}(R_p, I_p \times \widehat{K}_m) \tag{4}$$

between the high-pass image residual, R_p, and each of the scaled fingerprints.

The probe image is assumed to come from the camera with the most correlated PRNU

$$\widehat{K}_{\max} = \arg\max_m \text{corr}(R_p, I_p \times \widehat{K}_m) \tag{5}$$

which is therefore selected to perform forgery localization. However, it is also possible that none of the cameras under analysis originated the probe image, in which case all correlations should be small. To formalize this problem, let us consider the two hypotheses

H_0 : the probe image is alien to the dataset
H_1 : the probe image comes from one of the dataset cameras

To design a statistical test we should know the distribution of ρ under both hypotheses. This is not possible in our blind scenario, therefore we resort to a Neyman-Pearson test, selecting a decision threshold, t, which guarantees a suitably small false alarm probability P_{FA}. Following [13], we assume the normalized correlations to have a Gaussian distribution under H_0

$$\rho \sim N(0, 1/HW) \tag{6}$$

[1] Here, and throughout this work, we assume the images to be perfectly aligned. Otherwise, one can replace normalized correlation with Peak-to-Correlation Energy (PCE) ratio [12], which works correctly also in the presence of image cropping.

where H and W are the image dimensions. Therefore

$$\begin{aligned} P_{FA} &= \Pr(\rho_{\max} > t|H_0) = 1 - (1 - \Pr(\rho_m \le t|H_0))^M \\ &= 1 - (1 - Q(t\sqrt{HW}))^M \simeq MQ(t\sqrt{HW}) \end{aligned} \tag{7}$$

with the latter approximation holding for small M and $Q(t\sqrt{HW}) \ll 1$. By inverting the above relation the desired threshold is obtained.

2.4 PRNU-Based Forgery Localization

In the last step of the framework, a PRNU-based forgery localization technique is applied. Several such methods have been proposed in the last few years, and they all share the same basic idea. When the image is tampered with, for example through the splicing of some alien material, its PRNU is locally removed. Therefore, a sliding-window correlation test is performed, and when the local correlation index falls below a given threshold, a forgery is declared. Since the correlation may also depend on the image content, the threshold must be adapted locally by using the predictor with parameters $\Theta_{\max}$ estimated in step 2.

The output of this localization step is a binary decision mask that highlights the pixels that are considered as tampered. Given such a mask, and the corresponding ground truth mask, one can compute a number of performance indicators. However, it is worth pointing out that the output mask should be always analyzed by a human interpreter. In fact, real-life image forgeries are performed with a purpose, and they possess a semantics that is not easily captured by algorithms. The localization mask should be therefore regarded as a diagnostic tool to support the expert decision.

3 Experimental Results

In this section we evaluate the performance of the proposed PRNU-based framework for blind forgery localization. Experiments are carried out on six cameras: Canon EOS-10D, Canon EOS-450D, Canon Ixus 95IS, Nikon D200, Nikon Coolpix S5100, Sony DSC S780. For each camera we use 50 images as training set to perform the PRNU-based clustering and to estimate the cluster PRNUs. Performance is assessed on 50 more images per camera, different from those of the training set. All images have the same size of 768×1024 pixels, and are cropped from the same region of the full-size images. To study forgery localization, we generate forged versions of the test images by pasting on them, at the center, a square region of 128×128 or 256×256 sampled randomly from another image. In addition, we repeat the experiments using JPEG compressed images with a quality factor of 90. All the noise residuals are extracted by using the BM3D denoising filter [11], and removing non-unique artifacts caused by demosaicing and lens distortions as proposed in [5].

Localization results are given in terms of ROC curves, giving pixel-wise probability of detection, P_D, and probability of false alarm, P_{FA}, as a function of the decision threshold. As a synthetic measure, the area under the ROC curve (AUC) is also computed. Before considering localization, however, we study the performance of previous steps, to understand their impact on the accuracy.

3.1 Image Clustering and PRNU Estimation

We implemented three clustering algorithms, based on Normalized Cuts (NCut) [2], on pairwise nearest neighbor (PCE-PNN) clustering [3,10], and on correlation clustering [18], denoted as Marra2017. Note that NCut requires a threshold parameter to be estimated on a training set, so we consider here an *oracle* version, selecting *a posteriori* the best parameter. For PCE-PNN we used the threshold used by the authors in the original paper. Other PRNU-based clustering methods [3,16] are not considered here because they have been shown in [2,18] to provide a generally worse performance.

Table 1. Performance of clustering algorithms.

	NCut-oracle			PCE-PNN			Marra2017		
Set	ARI	TPR	FPR	ARI	TPR	FPR	ARI	TPR	FPR
Original	0.872	84.31	1.21	0.839	75.74	0.00	0.960	94.79	0.26
JPEG (QF = 90)	0.647	61.07	2.77	0.819	79.02	1.50	0.921	93.58	1.33

Table 1 shows results of clustering algorithms on both original and JPEG compressed images in terms of adjusted rand index (ARI), true positive rate (TPR) and false positive rate (FPR). Marra2017 provides clearly the best results, even better than the oracle version of NCut, with ARI always very close to 1 (perfect clustering).

In Fig. 2 we show a graphical representation of the results. For uncompressed images (left) Marra2017 provides near-perfect results, with just a few extra clusters for the Sony camera, removed because too small ($N_m < N_{\min}$). In this condition, almost all available images can be used to estimate the PRNU's. The other methods show a higher fragmentation, but clusters are large and pure enough to provide good estimations. Using JPEG compressed images, performance impairs for all methods, but only slightly for Marra2017. On the contrary PCE-PNN and NCut-oracle suffer more on this dataset, especially for the Nikon D200 images, that will not allow a good PRNU estimate.

3.2 Image to Cluster Assignment

After clustering the images and estimating the cluster fingerprints, the probe image is correlated with all PRNU's. If the maximum correlation exceeds the decision threshold, t, forgery localization is performed. Together with the 600 test images coming from the selected cameras, we use 600 (negative) images

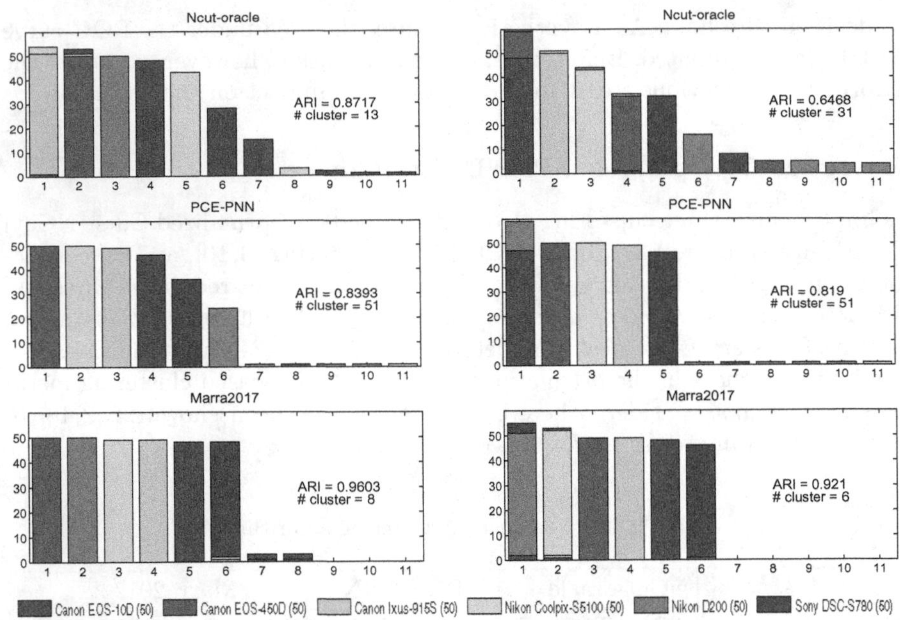

Fig. 2. Clustering results on original images (left) and JPEG compressed images (right) for NCut-oracle, PNN-PCE and Marra2017. Colors refer to the devices (see legend) while bar height indicate number of images in a cluster. (Color figure online)

taken from other sources, and cropped to the same size. Table 2 shows the detection performance for a threshold, t, set so as to obtain a theoretical false alarm probability $P_{FA} = 10^{-3}$. In detail, the FPR is the fraction of negative images that pass the test, while the TPR is the fraction of positive images (taken by one of the cameras in the dataset) recognized as such. The FPR is always very small, compatible with the theoretical level. The TPR is also quite large, but almost 6% of the positives are rejected, a fraction that grows above 10% with JPEG compressed images (almost 20% for PCE-PNN). Considering that Marra2017 provides near-perfect clustering, these errors must be attributed to the intrinsic problems of PRNU estimation. After correct detection, we could still

Table 2. Detection performance on original and JPEG compressed images.

	Original		JPEG (Qf = 90)	
Set	TPR	FPR	TPR	FPR
NCut-oracle	94.3%	0%	89.2%	0%
PCE-PNN	94.0%	0.3%	81.0%	0.7%
Marra2017	93.9%	0%	89.6%	1.5%

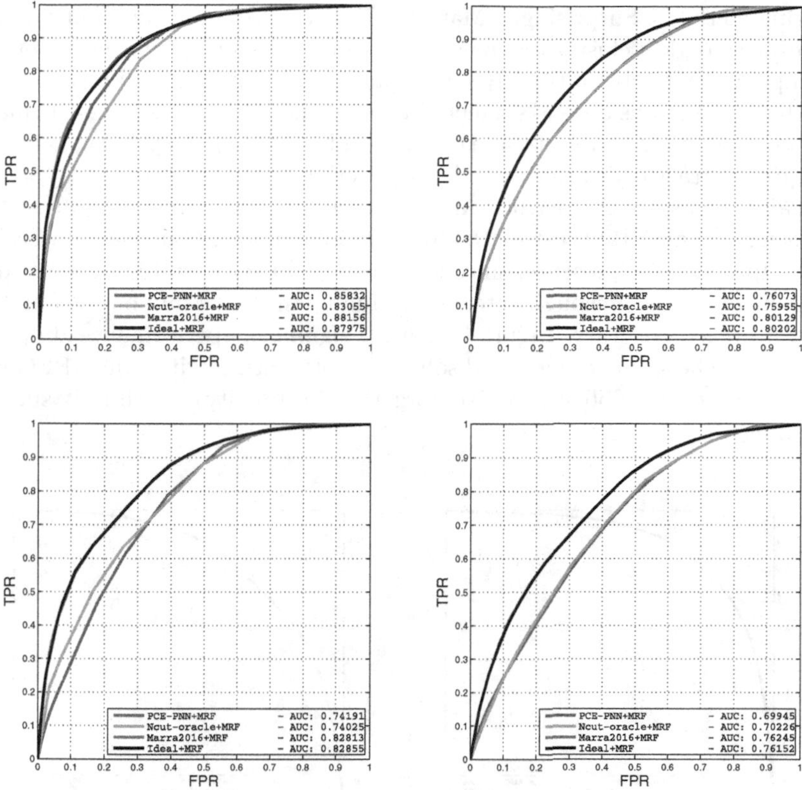

Fig. 3. Forgery localization results on original (top) and JPEG compressed images (down) with forgeries of 256 × 256 (left), and 128 × 128 pixels (right).

have a wrong assignment, that is, the probe image could be associated with a wrong camera/PRNU. However, our experiments show this event to be extremely unlikely, with probabilities lower than 0.1% in all cases and not reported in detail for the sake of brevity.

3.3 Forgery Localization

We conclude this analysis by studying forgery localization performance. Localization is carried out by the algorithm proposed in [9], based on a MRF prior and on the predictor of [5]. Together with the ideal case where the PRNU's are estimated from all available images, the case of real-world imperfect clustering is also considered, with all methods discussed before.

Figure 3 shows the ROC curves for original (top) and JPEG compressed images (down) with the two different forgery sizes. With large forgeries on uncompressed images results are very good. The AUC's are close to 0.9 with both ideal and Marra2017 clustering, and only slightly smaller with the other clustering methods. Surprisingly, Marra2017 provides even a small improvement with respect to ideal clustering, maybe because the discarded images are outliers that impact negatively on the PRNU estimation. As expected, all results impair somewhat when considering smaller forgeries and JPEG compressed images. However, the performance obtained with blind clustering keep being very close (equal for Marra2017) to those of ideal clustering.

Finally, we assess the performance when we renounce clustering altogether, computing a single PRNU estimated by averaging all images in the dataset. This "naive" approach makes sense, since the estimated PRNU will bear traces of all camera fingerprints, although attenuated due to the large number of unrelated images averaged together. Figure 4 shows a significant performance drop with respect to the best clustering-based solution, both with original and JPEG compressed images (only 256 × 256 pixel forgeries, for brevity) which fully supports our findings.

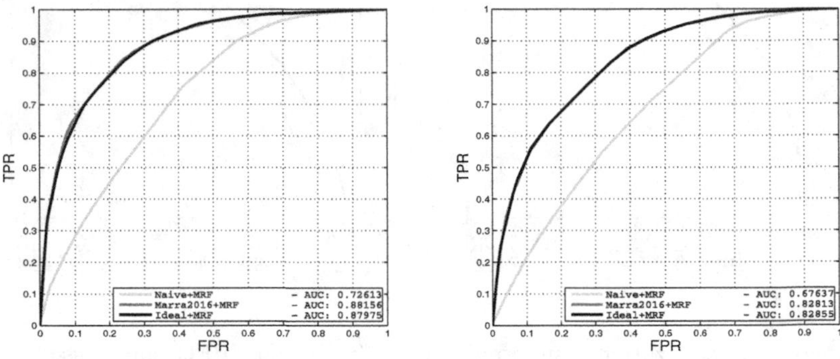

Fig. 4. Results for clustering-based and "naive" solutions on original (left) and JPEG compressed images (right) with 256 × 256 pixel forgeries.

4 Conclusion

In this paper we analyze a Camera-based framework for forgery localization in a blind scenario using a controlled dataset. The framework is composed of different steps, each of which is a possible source of error. The aim of our experiment is to show the performance of each single step due to the errors of the previous steps. As we see, for the original images the performance of all clustering algorithm are high enough to create cluster with a low FPR that assure to estimate pure PRNUs (all the images coming from the same camera device). This allow the forgery detector to perform as well as in the ideal case. In the JPEG compressed dataset, we note a performance drop when the clustering become less accurate and fragmented. The comparison with the *naive* solution say us that the effort in having a good clustering algorithm and pure PRNUs estimation is not pointless.

References

1. Al-Ani, M., Khelifi, F.: On the SPN estimation in image forensics: a systematic empirical evaluation. IEEE Trans. Inf. Forensics Secur. **12**(5), 1067–1081 (2017)
2. Amerini, I., Caldelli, R., Crescenzi, P., Mastio, A.D., Marino, A.: Blind image clustering based on the normalized cuts criterion for camera identification. Sig. Process. Image Commun. **29**(8), 831–843 (2014)
3. Bloy, G.J.: Blind camera fingerprinting and image clustering. IEEE Trans. Pattern Anal. Mach. Intell. **30**(3), 532–534 (2008)
4. Chakraborty, S., Kirchner, M.: PRNU-based forgery detection with discriminative random fields. In: International Symposium on Electronic Imaging: Media Watermarking, Security, and Forensics, February 2017
5. Chen, M., Fridrich, J., Goljan, M., Lukás, J.: Determining image origin and integrity using sensor noise. IEEE Trans. Inf. Forensics Secur. **3**(1), 74–90 (2008)
6. Chierchia, G., Cozzolino, D., Poggi, G., Sansone, C., Verdoliva, L.: Guided filtering for PRNU-based localization of small-size image forgeries. In: IEEE International Conference on Acoustics, Speech and Signal Processing, pp. 6231–6235. May 2014
7. Chierchia, G., Parrilli, S., Poggi, G., Sansone, C., Verdoliva, L.: On the influence of denoising in PRNU based forgery detection. In: 2nd ACM workshop on Multimedia in Forensics, Security and Intelligence, pp. 117–122 (2010)
8. Chierchia, G., Parrilli, S., Poggi, G., Sansone, C., Verdoliva, L.: PRNU-based detection of small size image forgeries. In: International Conference on Digital Signal Processing, pp. 1–6. July 2011
9. Chierchia, G., Poggi, G., Sansone, C., Verdoliva, L.: A Bayesian-MRF approach for PRNU-based image forgery detection. IEEE Trans. Inf. Forensics Secur. **9**(4), 554–567 (2014)
10. Cozzolino, D., Gragnaniello, D., Verdoliva, L.: Image forgery localization through the fusion of camera-based, feature-based and pixel-based techniques. In: IEEE Conference on Image Processing, pp. 5237–5241. October 2014
11. Dabov, K., Foi, A., Katkovnik, V., Egiazarian, K.: Image denoising by sparse 3-D transform-domain collaborative filtering. IEEE Trans. Image Process. **16**(8), 2080–2095 (2007)
12. Goljan, M.: Digital camera identification from images – estimating false acceptance probability. In: Kim, H.-J., Katzenbeisser, S., Ho, A.T.S. (eds.) IWDW 2008. LNCS, vol. 5450, pp. 454–468. Springer, Heidelberg (2009). doi:10.1007/978-3-642-04438-0_38
13. Goljan, M., Fridrich, J., Filler, T.: Large scale test of sensor fingerprint camera identification. In: Proceedings SPIE, vol. 7254, pp. 72540I–72540I-12 (2009)
14. Korus, P., Huang, J.: Multi-scale analysis strategies in PRNU-based tampering localization. IEEE Trans. Inf. Forensics Secur. **12**(4), 809–824 (2017)
15. Li, C.T.: Source camera identification using enhanced sensor pattern noise. IEEE Trans. Inf. Forensics Secur. **5**(2), 280–287 (2010)
16. Li, C.T.: Unsupervised classification of digital images using enhanced sensor pattern noise. In: IEEE Int. Symp. on Circuits and Systems, pp. 3429–3432 (2010)
17. Lukàš, J., Fridrich, J., Goljan, M.: Digital camera identification from sensor pattern noise. IEEE Trans. Inf. Forensics Secur. **1**(2), 205–214 (2006)
18. Marra, F., Poggi, G., Sansone, C., Verdoliva, L.: Blind PRNU-based image clustering for source identification. IEEE Trans. Inf. Forensics Secur. **12**(9), 2197–2211 (2017)

Recognizing Context for Privacy Preserving of First Person Vision Image Sequences

Sebastiano Battiato, Giovanni Maria Farinella(✉), Christian Napoli, Gabriele Nicotra, and Salvatore Riccobene

Department of Mathematics and Computer Science, University of Catania, Viale A. Doria 6, 95125 Catania, Italy
{battiato,gfarinella,napoli,riccobene}@dmi.unict.it

Abstract. The constant increasing evolution of life-logging wearable devices, as well as the fast grow of their market, has introduced relevant changes in the acquisition, storage and automatic understanding of images and videos. Along with the novel users' opportunities, this technology is introducing a large amount of privacy-related concerns, mainly regarding the unaware or unwilling contexts subject that could get recorded by a life-logging device. In this work, we devise an approach to help life-logging wearable devices enforcing restrictions for context-related users' privacy preservation. The proposed approach joins different technological innovations, from computer vision techniques to bluetooth beacon technology and network security.

1 Introduction

Since late 1980 s the world of wearable devices has encountered a tremendous evolution while components miniaturization enabled us to freely interact with a wide range of mobile systems and implement them in many aspects of everyday life [1]. Such devices have begun to overwhelmingly interact with personal information domain. Moreover an increasing amount of such devices is equipped with built-in cameras and has introduced relevant changes in the acquisition, storage and automatic understanding of images and videos. Therefore, while the novel availability of images and videos has encouraged personal creativity, it has also raised a certain amount of privacy concerns, mainly regarding the unaware or unwilling contexts subject, that could get caught on such multimedia contents. Moreover relevant legal implication should be taken into account [2], especially regarding the large amount of data continuously produced by the so called life-logging devices [3–6]. Nowadays, these devices allow their users to continuously record and share online many different kinds of data, as videos, audio, pictures, personal data, as well as collective information or individual activities. On the other hand, while traditional devices as cameras or audio recording devices were only used sporadically and deliberately, modern life-logging devices can record and share their data continuously, therefore tampering with bystanders' expectations about privacy and discretion [7]. For these reasons privacy and discretion aspects have gained great importance; as a matter of facts the typical user of

S. Battiato et al. (Eds.): ICIAP 2017, Part II, LNCS 10485, pp. 580–590, 2017.
https://doi.org/10.1007/978-3-319-68548-9_53

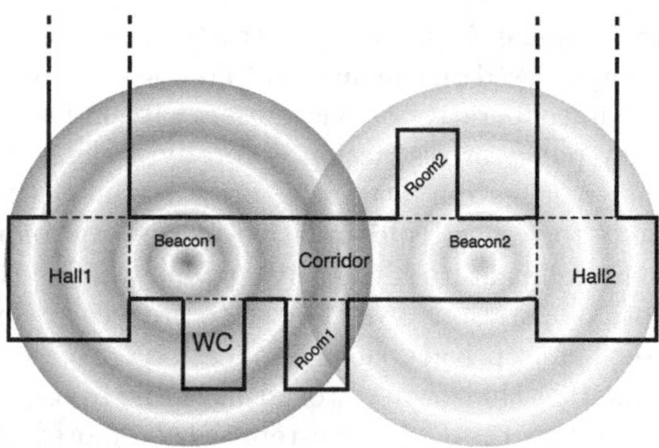

Fig. 1. Contexts recorded into video recorded at the Department of Mathematics and Computer Science of Catania's University Campus.

such life-logging devices may prefer to enforce privacy through location based control of image collection, in order to avoid later burdensome review of all collected media. Finally, automatic face recognition software performances are now almost as good as human abilities [8], therefore on one hand they offer a useful service, on the other hand they can put at even greater risk personal privacy e.g. taking into account the treat represented by malwares which could seize private multimedial contents surreptitiously [9,10].

2 Proposed System

In this work, we present an overall architecture for context related privacy preservation. The system has been designed to work in places affected by an high level of similarity among different contexts. Therefore the presented approach enforces privacy constraints by applying computer vision methods as well as low energy bluetooth technology for context recognition.

We tested the proposed method in nontrivial use cases, therefore, we decided to use a low-end commercial wearable to record portions of our University campus facilities with an high degree of similarity (e.g. offices). Specifically the raw video data were collected wearing the Recon Jet TM smart glasses and recording while walking trough several rooms, lounges and hallways (see Fig. 1).

2.1 Scenario and Communication Protocol

In order to grant users' privacy and enforce all the required security measures, the proposed system has been provided with an ad hoc communication protocol (Fig. 2). The protocol is enforced with the following steps:

1. Environment identification;
2. Generation of session encryption key for row file transmission;
3. Cloud Service for Policies handling;
4. Handled file retrieving.

In our protocol we assume the presence of trustworthy users and untampered device. This restriction is based on the fact that no-one can prevent the recording of image or sound by uncooperative or nasty user, with hidden camera. In these cases, defining privacy policy or restriction is totally useless.

Instead we want to focus our attention on a scenario where a user with wearable device wants to respect rules relative to the environment where it is, obtaining from the environment itself the privacy policy defined by others. In this sense, all the encryption operations are finalized to prevent any image acquisition by unauthorized user before privacy rules application.

Finally, we assume that the "owner" (or at least the bystanders) of a specific location, have uploaded to the Cloud System a set of preferences or rules in order to determine whether or not enforce any privacy-related restriction when the context of interest is recognized.

In the following formalism we will define three agents: a generic wearable device W^n that grabs the environment images, a generic beacon B^m that identifies the particular portion of the environment, and a Cloud Service C that handles the recorded images.

The first phase (*environment identification*) involves both the wearable device and the nearest beacon. The beacon broadcasts continuously its identity, providing its $\mathtt{ID}_{\mathtt{b}}^{m}$ and a cloud-related public key $\mathtt{K}_{\mathtt{Pub}}^{C}$, so that

$$B^m \rightarrow W^n : \mathtt{ID}_{\mathtt{b}}^{m}, \mathtt{K}_{\mathtt{Pub}}^{C}$$

The Cloud Service couple of public and private keys $\mathtt{K}_{\mathtt{Pub}}^{C}\mathtt{K}_{\mathtt{Priv}}^{C}$ is generated by an independent Certification Authority; this step provides the properties of authentication and confidentiality for the Cloud Service. The public key can be obviously retrieved also in other ways.

After the detection of the beacon's presence, the wearable device generates a session key $\mathtt{K}_{\mathtt{S}}^{n,i}$ which is used to encrypt the video recorded, in the follow called $V^{n,i}$, and a timestamp $T^{n,i}$, which is used to identify univocally the video.

In the follow, the encrypted information is represented with the common bracket formalism $\{\cdot\}_k$, where k is the encryption key.

The encrypted video is stored locally until network connection availability or a user interaction. When the connection is available or the device owner decides, the stored cyphered data is upload to the Cloud Service:

$$W^n \rightarrow C : \mathtt{ID}_{\mathtt{b}}^{m}, \{V^{n,i}\}_{\mathtt{K}_{\mathtt{S}}^{n,i}}, \left\{T^{n,i}, \mathtt{K}_{\mathtt{S}}^{n,i}, \mathtt{K}_{\mathtt{R}}^{n,i}, [\mathtt{ID}_{\mathtt{b}}^{l}]_{l\neq m}\right\}_{\mathtt{K}_{\mathtt{Pub}}^{C}}$$

In this transmission the wearable device sends:

- the $\mathtt{ID}_{\mathtt{b}}^{m}$ in clear text;
- the recorded video $\{\tilde{V}^{n,i}\}$ encrypted with the Cloud public key;
- a tuple with a transmission timestamp $T^{n,i}$, the session encryption key, a response encryption key $\mathtt{K}_{\mathtt{R}}^{n,i}$, the list of beacon listened by the device, except $\mathtt{ID}_{\mathtt{b}}^{m}$.

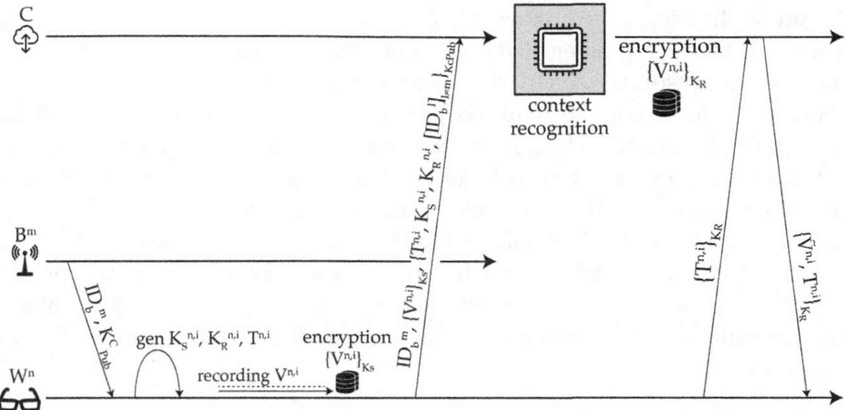

Fig. 2. Network protocol

The Cloud Service, that owns its $\mathtt{K}^{C}_{\mathtt{Priv}}$ key, is the only one able to decrypt the last part of the received communication. It retrieves the session key $\mathtt{K}_{\mathtt{S}}^{n,i}$, so it cat decode the video. By means of the list of beacon ID, it can retrieve the privacy policies previously defined, applying them to the video.

After this described communication, and the related message decoding, the Cloud resident application is able to recognize the context of each image through computer vision algorithms, and apply the required privacy enforcement rules. Only after this process, and the blurring of privacy concerned images, the resulting video can be transmitted back to the wearable device's owner.

Before its transmission from the cloud service to the users client, the processed video is re-encrypted using the response key $\mathtt{K}_{\mathtt{R}}^{n,i}$, to avoid unauthorized accesses.

The wearable device finally requests to the Cloud Service repository the transmission of the handled video:

$$W^n \rightarrow C : \{T^{n,i}\}_{\mathtt{K}_{\mathtt{R}}^{n,i}}$$

$$C \rightarrow W^n : \left\{T^{n,i}, \tilde{V}^{n,i}\right\}_{\mathtt{K}_{\mathtt{R}}^{n,i}}$$

Note that the response key $\mathtt{K}_{\mathtt{R}}^{n,i}$ and the timestamp $T^{n,i}$ can be provided by the wearable device to every authorized user, to realized independently steps 3 and 4. For this reason $\mathtt{K}_{\mathtt{R}}^{n,i}$ must be different from $\mathtt{K}_{\mathtt{S}}^{n,i}$.

3 Visual Context Recognition

In Sect. 2.1 we described the designed communication protocol between the wearable device and a cloud-resident application. In this section we will describe how the cloud application proceeds with the required context recognition. This latter goal is achieved by using dedicated computer vision algorithms as well as machine learning solutions. Once the context is identified, then, the cloud is responsible for applying the required privacy preserving policies. Such policies will be applied by blurring the images regarding contexts for which the users have required a privacy enforcement rule. In the following we will compare two different implementations for the proposed approach. The first uses Bag of Words for feature extraction and k-Nearest Neighbors algorithm for context recognition (see Sect. 3.1). The second approach uses AlexNet for feature extraction and Support Vector Machine for context recognition (see Sect. 3.2). Finally these two implementation are compared on the base of their results and performances (see Sect. 4).

3.1 Bag-of-Words and k-Nearest Neighbors Algorithm

The Bag-of-Words (BoW) [11] method was born for information retrieval in text document analysis. For image processing purposes it is possible to apply the same model by creating a vocabulary of *visual words* constructed as a catalog of visual features. BoW model relies a distance based features clustering. The features are extracted from local regions after keypoint detection. It is possible to apply the BoW model for image classification by the following steps:

1. extract local regions from Points of Interests;
2. compute and extract local descriptors on these local regions;
3. compute a visual vocabulary through the clustering of the local descriptor;
4. represent an image as distribution of its visual word with respect to the computed visual vocabulary.

In this work, the BoW model has been used with Dense-SURF as features. The algorithm has been instructed to use an 8 by 8 pixel grid. The visual vocabulary obtained with k-means clustering is constituted by 1024 visual words.

In our solution we created different classifiers, one for each beacon, in order to assist context recognition. We used the following set-ups. We split the dataset in three parts and we used one or two parts for training and only one for testing.

The k-Nearest Neighbors algorithm [12] (k-NN) is the algorithm we used for classification when BoW is employed as representation. This algorithm is based on the prediction of the class of an image, considering k training data neighbors. In our study we used 1-NN algorithm implementation.

3.2 AlexNet and SVM Algorithm

AlexNet [13] is a convolution neural network (CNN) for objects recognition. AlexNet is composed by 650000 neurones triggered by 60 millions input parameters. The AlexNet model has been trained on a subset of ImageNet dataset composed by 1.2 million images of 1000 categories. We used AlexNet as alternative of BoW for image representation purpose.

Fig. 3. Example of images into dataset

We have coupled AlexNet representation with an SVM [12] classifier. We used SVM in multiclass procedure. This algorithm is based on the construction of detach clusters. In this study we have six detach cluster one for each class.

4 Experimental Settings and Results

The experiments were designed to test the proposed system employing beacon by comparing it's efficacy while used to improve two well known classification methods based on Bag-of-Words [11] and AlexNet [13]. The classic classification methods, therefore, will be taken as reference baseline for the result presented in the next sections. The Bag-of-Words model has been used jointly with the k-Nearest Neighbors [12] algorithm for classification. Similarly, AlexNet has been used and then feed a SVM [12] algorithm for classification purpose. Our dataset is composed by six classes each of them related to a different context (see Fig. 1).

4.1 Dataset

The dataset is composed of video frames (Fig. 3). The frames have been collected from a set of recorded videos. Such videos have been captured with a ReconJet by an operator walking through one wing of the building (Fig. 1). We performed many simulations regarding each one of the presented methods. In order to collect sufficient data and minimize statistical interferences we used several configurations for both the training and the testing set. The dataset has been split in three equally sized partitions (`T1`, `T2`, and `T3`), moreover three mixed partitions have been created: `T12` combining `T1` and `T2`, `T13` combining `T1` and `T3`, and `T23` combining `T2` and `T3`. Each of these partitions have been used independently for training in each simulation while paired with a complementary test set partition (see Table 1).

Table 1. Correlation between beacon, images and classes. N_B is the number of images for the classifier training step in the based method and our solution. N_C is the number of images per class. The parameter K is 1 for single part of the dataset (e.g. T1) and 2 for combined parts (e.g. T12).

Beacon #	N_B	Class	N_C	Policy
-	$K * 4170$	Hall1	$K * 695$	Yes
		Corridor		Yes
		WC		No
		Room1		No
		Room2		Yes
		Hall2		No
Beacon #	N_B	Class	N_C	Policy
1	$K * 2780$	Hall1	$K * 695$	Yes
		Corridor		Yes
		WC		No
		Room1		No
2	$K * 2085$	Corridor	$K * 695$	Yes
		Room2		Yes
		Hall2		No

4.2 BoW Model

In order to compare our beacon based solution with the foremost standards, initially we used a BoW model and k-NN algorithm to obtain a reference baseline. Table 2 shows the results of such a model for each possible combination of training and testing sets (see Table 2). In this phase the classifier has been trained for context recognition among all possible classes with no restriction. Therefore the BoW/k-NN model has been applied to the context of each frame with respect to six different contexts (see Fig. 1). While each test has produced consistent results, on the other hand, it should be noticed that, when we used T3 as test set, the accuracy of this classifier produces less accurate results. Similarly, also when T3 is used for training, the resulting classifier obtains a very low accuracy. Effectively the T3 dataset is affected by a relevant noise (e.g. blurred or overexposed frames, too dark or too bright scenes, etc.). On the other hand we also noticed that using T3 combined with another set among T1 or T2 for training highly increases the classification capabilities of the classifier. We suspect that T3 contributes to train the classifier for context recognition even with noisy data.

4.3 Beacon-Enhanced BoW Model

Table 2 also shows the performances obtained by an improved version of the BoW model which makes use of beacon-driven context recognition (see Sect. 2). In this

Table 2. Performance of the different proposed setups based on BoW/kNN and AlexNet/SVM representation: α^{BoW} is the baseline accuracy of the standard BoW model, α^{AN} is the baseline accuracy of the standard AN model, α_*^{BoW} is the accuracy of our improved BoW-based model with beacon driven context classification, α_*^{AN} is the accuracy of our improved AN-based model.

Training set	Test set	α^{BoW}	α^{AN}	Beacon	α_*^{BoW}	α_*^{AN}
T1	T2	78.56%	74.75%	1	83.17%	84.14%
				2	93.33%	92.04%
T1	T3	67.41%	67.99%	1	80.68%	79.10%
				2	84.56%	82.97%
T2	T1	80.79%	75.97%	1	79.96%	84.10%
				2	96.45%	95.78%
T2	T3	69.40%	69.86%	1	78.13%	86.73%
				2	81.58%	81.97%
T3	T1	73.43%	70.29%	1	82.73%	80.72%
				2	83.31%	87.43%
T3	T2	68.80%	71.70%	1	82.09%	85.42%
				2	77.79%	84.70%
T12	T3	74.68%	75.01%	1	86.33%	84.03%
				2	85.13%	83.88%
T13	T2	83.29%	80.10%	1	88.13%	87.05%
				2	92.71%	89.74%
T23	T1	83.55%	78.03%	1	85.43%	85.58%
				2	95.20%	94.20%

phase we used two classifiers, one for each beacon involved in our experiments. The first classifier has been used to detect the classes associated with context related to the first beacon: *Hall1*, *WC*, *Room1* and *Corridor*. The second classifier has been used to recognize the remaining classes related to the second beacon: *Corridor*, *Room2* and *Hall2*. The data provided to the classifiers where similar to the data used for the standard BoW modes (Sect. 4.2). Moreover, for this second experiment, the device also stored a tag for each frame with a ID list of the beacons in range at recording time. This setup permitted us to obtain an higher system's accuracy (compare columns 3 and 4 of Table 2 w.r.t. the columns 6 and 7). Finally, as in the previous experiment, also in this second scenario the T3 showed the same noise-related issues.

4.4 AlexNet Model

In order to prove the efficacy of the proposed beacon-driven context recognition with respect to the standard image recognition based models, we tested and

Table 3. Improvements with respect to BoW model: α^{BoW} is the baseline accuracy of the standard BoW model, α_1^{BoW} and α_2^{BoW} are the accuracies related respectively to the classes belonging to the first or second beacon in our improved BoW model, $\alpha_{1,2}^{BoW}$ is the average accuracy of our improved BoW model

Training set	α^{BoW}	α_1^{BoW}	α_2^{BoW}	$\alpha_{1,2}^{BoW}$	Improvement
T1	72.98%	81.92%	88.94%	84.92%	~12%
T2	75.09%	79.04%	89.01%	83.31%	~12%
T3	71.11%	82.41%	80.55%	81.61%	~10%
T12	74.68%	86.33%	85.13%	85.82%	~11%
T13	83.29%	88.13%	92.71%	90.09%	~7%
T23	83.55%	85.43%	95.20%	89.61%	~6%

Table 4. Improvements with respect to the AlexNet (AN) model: α^{AN} is the baseline accuracy of the standard AN model, α_1^{AN} and α_2^{AN} are the accuracies related respectively to the classes belonging to the first or second beacon in our AN-based model, $\alpha_{1,2}^{AN}$ is the average accuracy of our AN-based model

Training set	α^{AN}	α_1^{AN}	α_2^{AN}	$\alpha_{1,2}^{AN}$	Improvement
T1	71.37%	81.98%	88.15%	84.62%	~13%
T2	72.91%	85.66%	89.86%	87.46%	~15%
T3	70.99%	82.73%	86.26%	84.24%	~13%
T12	75.01%	84.03%	83.88%	83.96%	~9%
T13	80.10%	87.05%	89.74%	88.20%	~8%
T23	78.03%	85.58%	94.20%	89.27%	~11%

compared a hybrid approach. In this setup we preprocessed the video frames by using AlexNet [13] obtaining for each frame a feature vector. Then we used such a feature vector as input for an SVM classification algorithm. As done previously (see Sects. 4.2 and 4.3), also for this hybrid method we compare the results of an unconstrained test, that we used as comparison baseline, with our improved beacon-driven approach. Similarly to the previous experiments, also this time the noisy dataset T3 affected the classification accuracy of our implemented models. Moreover, despite AlexNet architecture should be robust with respect to such kind of noise, in our experiment we noticed that a strongly noise video recording could tamper it. On the other hand, if T3 is used in conjunction with a low-noise dataset, it seems to improve the accuracy of the classifier (see Sect. 4.5).

4.5 Discussion

The results of the experiments are reported in Tables 2, 3, and 4. In Table 2 we report the performance of the standard Bag-of-Words (α^{BoW}) and AlexNet (α^{AN}) approaches, as well as the performances of our improved models (α_*^{BoW}

and α_*^{AN}). These two latter also make use of beacon driven context classification to improve their accuracy. The same results are reported in columns third and fourth of Table 2. In Table 3 the performance of the implemented BoW-based models are analyzed with respect to the different set of classes (whether if related to beacon 1 or beacon 2): α^{BoW} is the baseline accuracy of the standard BoW model, α_1^{BoW} and α_2^{BoW} are the accuracies related respectively to the classes belonging to the first or second beacon in our improved BoW model, $\alpha_{1,2}^{BoW}$ is the average accuracy of our improved BoW-based model. Table 4 shows the improvement introduced by our modifications to the AlexNet model: α^{AN} is the baseline accuracy of the standard AlexNet model, α_1^{AN} and α_2^{AN} are the accuracies related respectively to the classes belonging to the first or second beacon in our improved AN model, $\alpha_{1,2}^{AN}$ is the average accuracy of our improved AN model (see Tables 3 and 4). Finally, Fig. 4 shows an overview of the implemented methods and the related improvements introduced with the proposed beacon-driven recognition techniques.

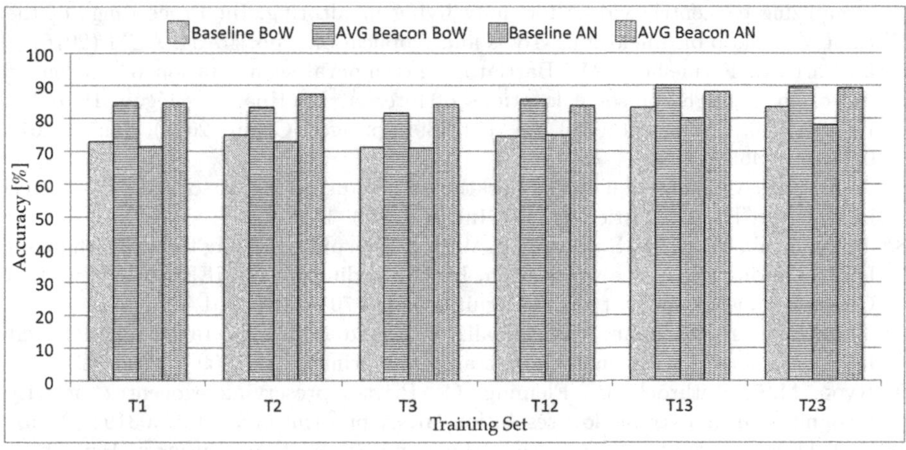

Fig. 4. Comparison of Bag-of-Words and AlexNet representation with and without the exploitation of beacon

5 Conclusions

In this work, we presented a hybrid approach to help life-logging wearable devices enforcing restrictions for context-related users' privacy preservation. The introduction of bluetooth beacon technology have been proven useful to improve the context recognition accuracy of some known image classification solutions based on Bag-of-Words and AlexNet representation. The results showed that the proposed solution is both robust to noise affected datasets as well as efficient for environments that presents an high degree of similarity between different contexts. Moreover, the developed system is highly customizable to enforce the privacy choices of the context owners or bystanders. Finally, the cloud oriented support make it suitable for a wide range of different devices and applications.

References

1. Mann, S.: Wearable computing: a first step toward personal imaging. Computer **30**(2), 25–32 (1997)
2. Cheng, W.C., Golubchik, L., Kay, D.G.: Total recall: are privacy changes inevitable? In: Proceedings of the 1st ACM Workshop on Continuous Archival and Retrieval of Personal Experiences, pp. 86–92. ACM (2004)
3. Allen, A.L.: Dredging up the past: lifelogging, memory, and surveillance. Univ. Chicago Law Rev. **75**(1), 47–74 (2008)
4. Chen, Y., Jones, G.J.: Augmenting human memory using personal lifelogs. In: Proceedings of the 1st Augmented Human International Conference, p. 24. ACM (2010)
5. Ortis, A., Farinella, G.M., D'Amico, V., Addesso, L., Torrisi, G., Battiato, S.: Organizing egocentric videos for daily living monitoring. In: Proceedings of the First Workshop on Lifelogging Tools and Applications, pp. 45–54. ACM (2016)
6. Furnari, A., Farinella, G.M., Battiato, S.: Temporal segmentation of egocentric videos to highlight personal locations of interest. In: Hua, G., Jégou, H. (eds.) ECCV 2016. LNCS, vol. 9913, pp. 474–489. Springer, Cham (2016). doi:10.1007/978-3-319-46604-0_34
7. Teraoka, T.: Organization and exploration of heterogeneous personal data collected in daily life. Hum.-Centric Comput. Inf. Sci. **2**(1), 1 (2012)
8. Taigman, Y., Yang, M., Ranzato, M., Wolf, L.: Deepface: closing the gap to human-level performance in face verification. In: Proceedings of the IEEE Conference on Computer Vision and Pattern Recognition, pp. 1701–1708 (2014)
9. Templeman, R., Rahman, Z., Crandall, D., Kapadia, A.: Placeraider: virtual theft in physical spaces with smartphones. arXiv preprint arXiv:1209.5982 (2012)
10. Ryoo, M.S., Rothrock, B., Fleming, C.: Privacy-preserving egocentric activity recognition from extreme low resolution. arXiv preprint arXiv:1604.03196 (2016)
11. Szeliski, R.: Computer Vision: Algorithms and Applications. Springer, Heidelberg (2011). doi:10.1007/978-1-84882-935-0
12. Bishop, C.M.: Pattern Recognition and Machine Learning. Springer, Heidelberg (2006)
13. Krizhevsky, A., Sutskever, I., Hinton, G.E.: Imagenet classification with deep convolutional neural networks, pp. 1097–1105 (2012)

GRAPHJ: A Forensics Tool for Handwriting Analysis

Luca Guarnera[1], Giovanni Maria Farinella[1], Antonino Furnari[1(✉)], Angelo Salici[2], Claudio Ciampini[2], Vito Matranga[2], and Sebastiano Battiato[1]

[1] Image Processing Laboratory (IPLAB), Department of Mathematics and Computer Science, University of Catania, Catania, Italy
guarneraluca92@gmail.com, {gfarinella,furnari,battiato}@dmi.unict.it
[2] Raggruppamento Carabinieri Investigazioni Scientifiche RIS di Messina, S.S.114 Km 6,400, 98128 Messina, Italy
{angelo.salici,claudio.ciampini,vito.matranga}@carabinieri.it
http://iplab.dmi.unict.it/, http://www.carabinieri.it/

Abstract. Handwriting analysis is a standard forensics practice to assess the identity of a person from written documents. Forensic document examiners consider different features related to the motion and pressure of the hand, as well as the shape of the different characters and the spatial relationship among them. While examiners rely on standard protocols, documents are generally processed manually. This requires a significant amount of time and may lead to a subjective analysis which is difficult to replicate. Automated forensics tools to perform handwriting analysis from scanned documents are desirable to help examiners extract information in a more objective and replicable way. To this aim, in this paper we present GRAPHJ, a forensics tool for handwriting analysis. The tool has been designed to implement the forensics protocol employed by the "Reparto Investigazioni Scientifiche" (RIS) of Carabinieri. GRAPHJ allows the examiner to (1) automatically detect text lines as well as the different words within the document; (2) search for a specific character and detect its occurrences in the handwritten text; (3) measure different quantities related to the detected elements (e.g., character height and width) and (4) generate a report containing measurements, statistics and all parameters used during the analysis. The generation of the report helps to improve the repeatability of the whole process. We also present a set of experiments to assess the compliance of GRAPHJ with respect to conventional handwriting analysis methods. Given a set of handwritten documents, the experiments compare measurements and statistics produced by GRAPHJ to those obtained by an expert forensics examiner performing classic manual analysis.

1 Introduction

Forensic handwriting examination is the analytical process of detecting regularities and singularities of a handwritten text to assess the identity of the writer [1]. The analysis focuses on recognizing the fundamental shapes of the stroke, as well

S. Battiato et al. (Eds.): ICIAP 2017, Part II, LNCS 10485, pp. 591–601, 2017.
https://doi.org/10.1007/978-3-319-68548-9_54

as the relative positions and sizes of letters and words. For handwriting analysis methods it is important to adopt a quantitative approach in order to limit any subjective evaluation due to the personal experience of the examiner.

With this goal in mind, many forensics experts make use of a graphometric-based approach which takes into account the different quantifiable features of handwritten text. Many authors have considered the fundamental principles and techniques of document examination [2–5]. However, handwriting analysis should be robust to variations in the writing process due to several intrinsic and extrinsic causes, such as different writing speeds, dissimulation, tiredness and available space. In this regard, it is well known that the study of character heights is valuable to identify the range of variability of the writer [6]. For instance, Morris [7] underlined the importance of the analysis of dimensional parameters and the comparison of absolute and relative quantities. This kind of analysis, considering speed, slope and style, can be used to identify an attempt of forgery. Hayes [1] shown that dimensions reflect the range of finger and hand movements which are characteristic of individual expression, e.g., some people produce an extremely small writing while others a taller one. Kelly and Lindblom [8] analyzed the ratio of lowercase to uppercase letters, showing that this value can be useful to identify the writer.

In this work, we present GRAPHJ, a useful tool for handwriting analysis. The proposed tool implements several algorithms to perform the analysis of handwritten documents which are currently considered in the protocol used by RIS - Carabinieri in Italy. For instance, GRAPHJ allows to detect text lines and words in the document and to search for all occurrences of a specific character. GRAPHJ is also designed to simplify and improve the documentation of the analysis process, by generating a report containing statistics and measurements. Our approach is similar to the one of Fabiańska et al. [9], but our tool allows for automated detection of elements, in order to minimize the amount of required manual intervention. Aimed at assisting the examiner in analyzing digitalized handwritten documents, GRAPHJ can be considered a multimedia forensics tool [10]. It should be noted that our approach is different from handwriting recognition [11], since we are not interested in digitalizing the analyzed text. To validate GRAPHJ, we performed experiments comparing the produced measurements and statistics to those obtained with classic manual analysis performed by a forensics expert. A video demo of GRAPHJ is reported at our web page http://iplab.dmi.unict.it/graphj/.

2 GRAPHJ

We developed GRAPHJ as a plugin for ImageJ [12], which is a standard framework to perform many specific image processing tasks. The developed plugin allows to automate the standard procedures needed to analyze handwritten documents. The implemented algorithms allow to perform three main tasks: (1) automated detection of elements (text lines and words); (2) automated search of

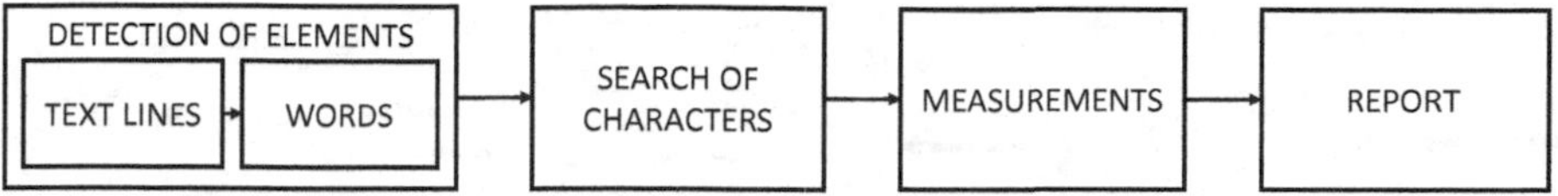

Fig. 1. Diagram of a typical workflow in GRAPHJ.

instances of a given character; (3) automated measurement of quantities (e.g., distance between words and characters, height and width of characters). Furthermore, the examiner can manually intervene to adjust automated detections and measure other quantities such as absolute and relative heights. A typical workflow employed to perform handwriting analysis in GRAPHJ is shown in Fig. 1. Developed algorithms are detailed in the following sections.

2.1 Automated Search of Text Lines

A text line can in general be divided into three areas considering the analysis protocol: a *lower area*, a *median area*, and a *higher area*, as it is illustrated in Fig. 2. Automated search of text lines is performed in two steps. First, all median areas are detected in the document. Second, a lower and higher areas are identified for each detected median area.

The algorithm to search for median areas is illustrated in Fig. 3 and discussed in the following:

1. as a first step, the image is binarized (the resulting binary image is denoted by B) setting to 0 (black color) all pixels whose value exceeds a given threshold T and setting to 1 (white color) all other pixels;
2. a per-row histogram (H_r) is created by counting the number of zero pixels contained in each pixel row of the binary image. The histogram will contain a number of bins equal to the number of rows contained in the original image (i.e., the image height);
3. To detect the central lines of median areas, the algorithm considers all peaks of the histogram which values are above a user-specified threshold s_1. Threshold s_1 is introduced to reduce the influence of noise in the search of median areas;
4. the algorithm hence finds starting and ending rows of each median area. Since histogram values are expected to decay gradually around the peak, this is done by searching for the nearest lower and higher rows which value is over 1/4 of the value of the histogram at the given peak.

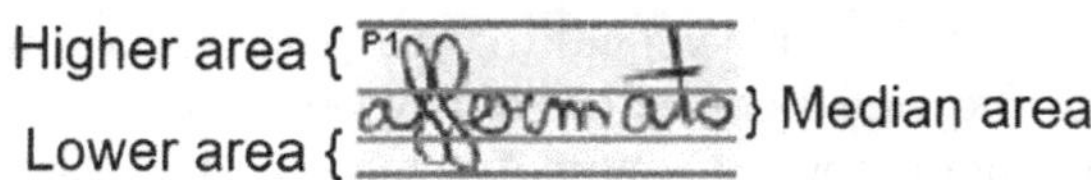

Fig. 2. Higher, median and lower area of a text line.

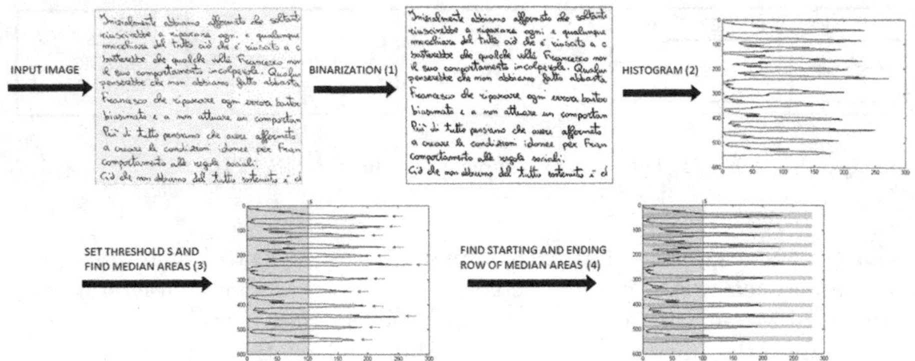

Fig. 3. Automated search of text lines.

Algorithm 1. Median area detection

Input: Binary Image B, Threshold s_1
Output: Histogram H_r, Indices ind
$H_r = \text{histRows}(B)$
$i = 0$
for all $[indMax, valMax] \in \text{findPeaks}(H_r)$ **do**
    if $valMax > s_1$ **then**
        $val = valMax/4$
        for $j = indMax;\ j \geq 0\ \&\&\ H_r[j] > val;\ j = j - 1$ **do**
        end for
        for $k = indMax;\ k < length(A)\ \&\&\ H_r[k] > val;\ k = k + 1$ **do**
        end for
        $ind[i] = j$
        $ind[i + 1] = k$
        $i = i + 2$
    end if
end for

The complete procedure is reported in Algorithm 1, where: histRows(B) computes the per-row histogram of zero pixels as discussed above; findPeaks(H_r) finds the peaks (i.e., local maxima) of histogram H_r and returns both positions ($indMax$) and values ($valMax$). The algorithm returns a list of starting and ending row indexes of all detected median areas (ind).

Once median areas are detected, the algorithm detects row indexes of higher and lower areas. This is done by looking for the nearest higher and lower empty rows. Such rows are easy to detect since they do not contain any black pixel, and hence histogram H_r has value equal to zero at those locations. If no empty rows can be found, the indexes of higher and lower areas are set to correspond to the starting and ending rows of the related median area. Algorithm 2 reports the procedure used to locate indexes of higher and lower areas. The algorithm returns a list of tuples of four values: starting row index of the higher area,

Algorithm 2. Higher and lower areas detection

Input: Indices *ind*, Histogram H_r
Output: List *areas*
Set *areas* to an empty list
for $i = 0;\ i < length(ind);\ i = i + 2$ **do**
  for $j = ind[i];\ j > 0\ \&\&\ H_r[j] \neq 0;\ j = j - 1$ **do**
  end for
  for $k = ind[i+1];\ k < length(H_r) - 1\ \&\&\ H_r[k] \neq 0;\ k = k + 1$ **do**
  end for
  if $i > 0\ \&\&\ (H_r[j] \neq 0\ ||\ j \leq ind[i-1])$ **then**
    $\mathrm{j} = \arg\min_h(H_r[h = ind[i-1]], H_r[h = ind[i]])$
  end if
  if $i < length(ind) - 2\ \&\&\ (H_r[k] \neq 0\ ||\ k \geq ind[i+2])$ **then**
    $\mathrm{k} = \arg\min_h(H_r[h = ind[i+1]], H_r[h = ind[i+2]])$
  end if
  Append tuple $(j, ind[i], ind[i+1], k)$ to list *areas*
end for

starting row index of median area, ending row index of median area, ending row index of lower area.

2.2 Automated Detection of Words

Automated detection of words is performed starting from text lines detected in the binarized image B. The process works in two steps:

1. word boundaries are detected;
2. higher and lower areas are refined for each word.

The first step of the algorithm is illustrated in Fig. 4 and discussed in the following. Let L be a crop of a given text line obtained from the binary image B. A column histogram H_c counting the number of black pixels contained in each column of L is computed. Note that computation of H_c is similar to computation of H_r. To find word boundaries, the algorithm searches for bins in H_c which contain zero values. Such bins represent columns of L not containing any black pixel. If the detected gap is larger than a given threshold s_2, then the starting and ending column indexes of a new word are stored in a list. The algorithm eventually returns a list of tuples of starting and ending indexes (i_s, i_e). The procedure is reported in Algorithm 3.

Once word boundaries are obtained, median, higher and lower areas are detected for each word. This step is performed because words on same text line may have different size and orientation. The procedure works on image crops w of words detected using Algorithm 3. To determine the orientation of a given word, its crop w is rotated by different angles α sampled from interval $[-N, N]$ at step k. For each rotated crop, a row histogram H_w is computed using function *histRows* and its maximum m is computed. The correct orientation is obtained

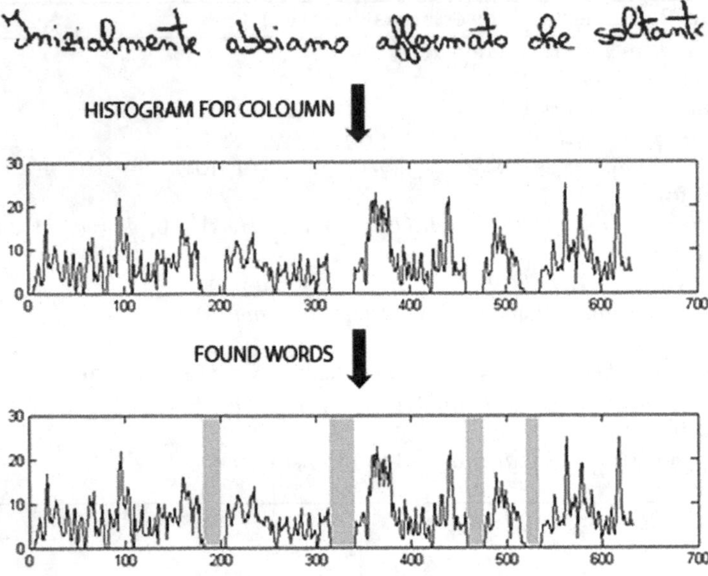

Fig. 4. Automated detection of words.

Algorithm 3. Words detection

Input: Crop L, Threshold s_2
Output: List *words*
Set *words* to an empty list
$H_c = \text{histCols}(L)$
$start = 0$
$i_1 = 0$
$i_2 = 0$
while $i_1 < length(H_c)$ && $i_2 < length(H_c)$ **do**
  for ; $H_c[i_1] \neq 0$; $i_1 = i_1 + 1$ **do**
  end for
  for $i_2 = i_1 + 1$; $H_c[i_2] == 0$; $i_2 = i_2 + 1$ **do**
  end for
  if $i_2 - i_1 \geq s_2$ **then**
    Append tuple $(start, i_1)$ to list *words*
    $start = i_2$
  end if
end while

---

by selecting angle α leading to the highest value for m. This arises from the observation that, if the word is aligned horizontally, histogram H_w will be strongly peaked. Once the correct orientation has been determined, median, higher and lower areas are detected using Algorithm 1 and Algorithm 2. The whole procedure to detect lower, median and higher areas for each word is reported in Algorithm 4, where function $rotate(w, \alpha)$ rotates word w by α degrees.

Algorithm 4. Median, higher and lower areas detection of words

```
Input: Array of image crops words, Angle range width N, Angle step k
Output: List of orientations and line indexes word_areas
for all w ∈ words do
    maxValue = 0
    β = 0
    for α = −N; α <= N; α = α + k do
        w_r = rotate(w,α)
        H_w = histRows(w_r)
        if max(H_w) > maxValue then
            maxValue = max(H_w)
            β = α
        end if
    end for
    w_r = rotate(w,β)
    Determine i_1, i_2, i_3, i_4, indexes of median, higher and lower areas of w_r using
  Algorithm 1 and Algorithm 2
    Append tuple (β, i_1, i_2, i_3, i_4) to list word_areas
end for
```

2.3 Automated Search of Characters

This algorithm allows to search for all occurrences of a specific character in the document. To this end, the system allows the examiner to select a bounding box around the desired character to define a template T. The algorithm hence performs a sliding window search over the whole document to locate possible occurrences of characters. The size of the search window W is selected to be equal to the one of the template. To gain robustness to small rotations, additional candidates are generated by rotating the content of each search window by 10° and −10°. Each window is assigned a score S_W using the procedure outlined in Algorithm 5. Search windows with scores larger than a threshold set by the operator are retained as correctly detected character instances.

The scoring function reported in Algorithm 5 counts the number of black pixels contained in template T which are present in window W.

2.4 Measures

GRAPHJ allows to measure some quantities about words and characters in an automated way. The considered quantities are currently used in the standard protocol. In particular, the algorithm implements two functions:

- automatic computation of the biaxial proportion and relative average;
- automatic computation of the side expansion and relative average.

Automatic Computation of the Biaxial Proportion and Relative Average. Biaxial proportions are the width and the height of the oval characters (see

Algorithm 5. Scoring Function for Automated Search of Characters

```
Input: Template T, Window W, Image height im_height
Output: Score sc
completed=False
sc = 0
while not completed do
    x_T, y_T = find coordinates of next black pixel in template T
    x_W, y_W = find coordinates of next black pixel in template W
    y_T^prev = 0
    y_W^prev = 0
    while y_W < im_height && y_T < im_height do
        if ||y_T^prev - y_T| - |y_W^prev - y_W|| < |y_T^prev - y_T|/2 then
            sc = sc + 1
        end if
        y_T^prev = y_T
        y_W^prev = y_W
        y_T = y_T + 1
        y_W = y_W + 1
    end while
    if no more black pixels to analyze in T or W then
        completed=True
    end if
end while
```

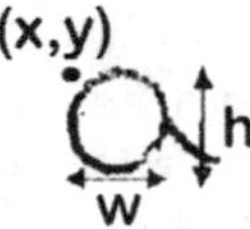

Fig. 5. Coordinate (x,y) and width, height (in pixel)

Fig. 5). To convert such measures from pixels to millimeters, we use the dedicated ImageJ functions. For each character, GRAPHJ computes the average $\rho_i = \frac{w_i}{h_i}$, where w_i and h_i are width and height of the ith characters respectively.

Automatic Computation of the Side Expansion and Relative Average. The side expansion represent the distance between the characters of the word and the distance between words. Distances between words are easily computed using starting and ending word indexes computed using Algorithm 3. Distance between characters is computed in a similar way. To remove the influence of lower and upper termination of characters, those are removed. Figure 6 illustrates the computation of side expansions.

Fig. 6. (a) Computation of space between words (b) Computation of space between characters. Upper and bottom parts of words are removed to facilitate character segmentation.

For each computed distance between characters (denoted as $\mathcal{D}_j^{(C)}$) and words (denoted as $\mathcal{D}_j^{(W)}$), GRAPHJ calculates the following ratios:

$$R_j^{(C)} = \frac{\mathcal{D}_j^{(C)}}{\overline{w}} \qquad R_j^{(W)} = \frac{\mathcal{D}_j^{(W)}}{\overline{w}}$$

where $\overline{w} = \sum_i w_i$ is the average width of oval characters.

3 Experimental Analysis

GRAPHJ was tested on 10 different writing samples. Samples have been written voluntarily by 10 different right-handed subjects. All documents have been written in cursive writing and using similar ink and paper. Every subject was asked to write the same long paragraph of text which was dictated to him. The text included all letters of the Italian alphabet, as well as sentences with different length and complexity.

To compare GRAPHJ performance with standard analysis methods, each sample has been manually analyzed by a forensics expert of RIS. In particular, the examiner measured the heights of two groups of 40 different letters analyzed in a sequential way with a degree of precision of 0.1 mm. In the first group, it is analyzed the height (U) of letters with an upper elongate stroke on the right or on the left side (i.e. “l”, “t”, “d”, “f”, “t”, ...). In the second group, it is analyzed the height fo the body in the median zone (M) of letters without elongate stroke (i.e. “a”, “c”, “o”, “m”, ...).

Table 1 shows the mean μ and standard deviation σ for the two groups of letters. The table compares measurements performed by the forensics expert to those obtained using GRAPHJ on the 10 documents of the dataset. Table 2 reports the mean absolute percentage error related to the measurements obtained on the two groups of letters considering the 10 documents in the dataset. Results show compliance of GRAPHJ analysis to measurements obtained by experts using standard manual techniques. The report generated bt GRAPHJ guarantees repeatability of the process.

A video demo of GRAPHJ is reported at our web page http://iplab.dmi.unict.it/graphj/.

Table 1. Mean and standard deviation for the two groups of letters analyzed by forensics experts using classic manual protocols and using GRAPHJ. All measures are expressed in millimeters.

Sample	Analysis	μ_M	σ_M	μ_U	σ_U
1	Manual	1,99	$\pm 0,59$	4,23	$\pm 0,58$
1	GRAPHJ	1,91	$\pm 0,40$	4,51	$\pm 0,71$
2	Manual	2,00	$\pm 0,40$	5,26	$\pm 0,50$
2	GRAPHJ	1,81	$\pm 0,39$	5,18	$\pm 0,60$
3	Manual	2,15	$\pm 0,55$	4,87	$\pm 0,76$
3	GRAPHJ	2,13	$\pm 0,39$	4,82	$\pm 0,81$
4	Manual	1,81	$\pm 0,53$	6,66	$\pm 0,94$
4	GRAPHJ	1,84	$\pm 0,45$	6,48	$\pm 0,84$
5	Manual	2,03	$\pm 0,32$	4,65	$\pm 0,86$
5	GRAPHJ	2,04	$\pm 0,41$	4,32	$\pm 0,78$
6	Manual	2,14	$\pm 0,47$	4,96	$\pm 0,89$
6	GRAPHJ	2,05	$\pm 0,37$	5,22	$\pm 1,08$
7	Manual	1,70	$\pm 0,55$	4,84	$\pm 0,90$
7	GRAPHJ	1,58	$\pm 0,34$	4,37	$\pm 0,59$
8	Manual	2,47	$\pm 0,84$	6,17	$\pm 1,38$
8	GRAPHJ	2,24	$\pm 0,55$	5,66	$\pm 1,03$
9	Manual	2,09	$\pm 0,59$	5,85	$\pm 0,82$
9	GRAPHJ	1,97	$\pm 0,36$	5,87	$\pm 0,79$
10	Manual	2,38	$\pm 0,43$	7,16	$\pm 0,69$
10	GRAPHJ	2,19	$\pm 0,37$	7,21	$\pm 0,51$

Table 2. Mean absolute percentage error for the two analyzed groups of letters. Results are reported for the 10 documents in the dataset in the two considered areas M (median) and with respect to the height (U).

Sample	**1**	**2**	**3**	**4**	**5**	**6**	**7**	**8**	**9**	**10**
Err%(M)	4,0	9,5	0,9	1,7	0,5	4,2	7,1	9,3	5,7	8,0
Err%(U)	6,6	1,5	1,0	2,7	7,1	5,2	9,7	8,3	0,3	0,7

4 Conclusion

We have presented GRAPHJ, an automated tool to aid the analysis of handwritten documents by forensics experts. The tool has been implemented as a plugin for ImageJ and allows to automate many operations such as detection of elements (e.g., text lines, words and characters) and measurement of quantities (e.g., character height and width). Experiments show that analyses carried out using GRAPHJ are compliant to those obtained by forensics experts using standard manual techniques.

References

1. Allen, M.J.: Forensic handwriting examination: a definitive guide, Reed Hayes, Reed Hayes Publications, Honolulu, HI, 254 p. ($49.95), p. 104 (2008). ISBN: 0-9778415-0-2
2. Evett, I.W., Totty, R.N.: A study of the variation in the dimensions of genuine signatures. J. Forensic Sci. Soc. **25**(3), 207–215 (1985)
3. Huber, R.A., Headrick, A.M.: Handwriting Identification: Facts and Fundamentals. CRC Press, Boca Raton (1999)
4. Abbey, S.E.: Natural variation and relative height proportions. Int. J. Forensic Doc. Examiners **5**, 108–116 (1999)
5. Maciaszek, J.: Natural variation in measurable features of initials. Probl. Forensic Sci. **85**, 25–39 (2011)
6. Koppenhaver, K.M.: Forensic Document Examination: Principles and Practice. Springer Science & Business Media, Heidelberg (2007)
7. Morris, R.: Forensic Handwriting Identification: Fundamental Concepts and Principles. Academic press, Cambridge (2000)
8. Seaman Kelly, J., Lindblom, B.S.: Scientific Examination of Questioned Documents. CRC Press, Boca Raton (2006)
9. Fabiańska, E., Kukicki, M., Zador, G., Dziedzic, T., Bułka, D.: Graphlog-computer system supporting handwriting analysis. Probl. Forensic Sci. **68**, 394–408 (2006)
10. Battiato, S., Giudice, O., Paratore, A.: Multimedia forensics: discovering the history of multimedia contents. In: Proceedings of the 17th International Conference on Computer Systems and Technologies 2016, pp. 5–16. ACM (2016)
11. Plamondon, R., Srihari, S.N.: Online and off-line handwriting recognition: a comprehensive survey. IEEE Trans. Pattern Anal. Mach. Intell. **22**(1), 63–84 (2000)
12. Abràmoff, M.D., Magalhães, P.J., Ram, S.J.: Image processing with ImageJ. Biophotonics Int. **11**(7), 36–42 (2004)

Identity Documents Classification as an Image Classification Problem

Ronan Sicre[1(✉)], Ahmad Montaser Awal[2], and Teddy Furon[1]

[1] Irisa/Inria, Rennes, France
{ronan.sicre,teddy.furon}@inria.fr
[2] AriadNext, Cesson-Sévigné, France
montaser.awal@ariadnext.com

Abstract. This paper studies the classification of identification documents, which is a critical issue in various security contexts. We address this challenge as an application of image classification, a problematic that received a large attention from the scientific community. Several methods are evaluated and we report results allowing a better understanding of the specificity of identification documents. We are especially interested in deep learning approaches, showing good transfer capabilities and high performances.

Keywords: Image forensic · Image classification · Document recognition

1 Introduction

Identity fraud is a major issue in today's societies with serious consequences. The threats vary from small frauds up to organized crimes and terrorist actions. The work presented in this paper is part of a research project IDFRAud[1] proposing a platform for identity documents verification. The first step classifies the query document according to its type and country of origin to prepare the verification of specific security checks, fake detection, document archiving etc. These later processes are out of scope of this paper.

For any supervised classification problem, the first task is to collect annotated data. In our application, these are specimens from various types of documents as well as emitting countries. Obtaining such data in large quantities is not always possible and we should therefore take into account this limitation. Some classes have more samples than others, giving unbalanced datasets. Moreover, query images vary from high quality scans to poor quality mobile phone photos with complex background, various orientations, occlusions, or flares, see Fig. 1.

There are two main approaches in the document classification literature. Methods based on the *layout* are mainly used when documents are composed of

[1] This work is achieved in the context of the IDFRAud project ANR-14-CE28-0012, co-financed by the french DGA: http://idfraud.fr/.

S. Battiato et al. (Eds.): ICIAP 2017, Part II, LNCS 10485, pp. 602–613, 2017.
https://doi.org/10.1007/978-3-319-68548-9_55

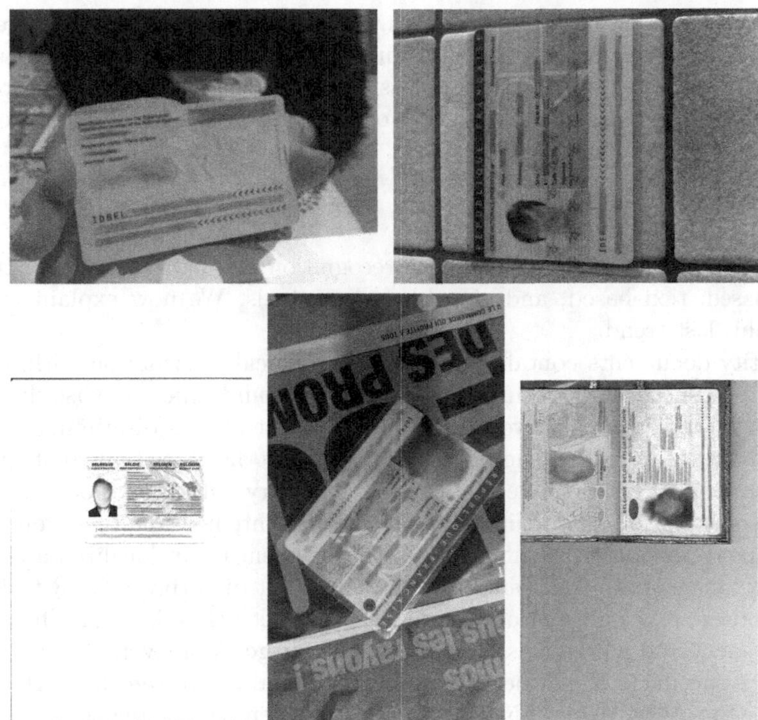

Fig. 1. Sample images from the databases

text blocks, figures, tables, etc. This is the case for journal articles, publications, books, or invoices. Documents are described by their spatial layout, *i.e.* the structure of text blocks, figures, and tables. Such descriptions are finally used to perform classification [2,16] or to compute similarities [8,24]. The second type of approaches is based on *text*. These methods build a description of the text content (extracted with an OCR in the case of scanned documents), such as bag of words or Word2Vec, which is given as input to classifiers [34]. More recently Recurrent Neural Networks (RNN) have been applied to classify documents [17].

Section 2 explains why we have discarded these two classical approaches to propose an alternative based on the visual content of the identity documents. Image recognition has a large spectrum of tasks with applications in search engines, interest object detection, or image categorization/classification, which has been extensively studied over the last decades. The availability of large and/or complex datasets as well as regular international challenges has spurred a large variety of image classification methods. We propose to apply these approaches to deal with identity document classification.

This choice is not obvious as there are few graphical elements in identity documents. Moreover, the portrait photo of the owner is uninformative for classification. However, the recent work on Convolutional Neural Networks (CNN) showed great generic visual descriptions, which are transferable to a large variety of image recognition tasks, such as fine-grain image classification. Thus, our paper studies a wide range of image classification method as well as the transfer capabilities of CNN to the specific task of identity documents classification.

2 Previous Works

The introduction presented document recognition through three main trends: layout-based, text-based, and visual-based methods. We now explain why we choose this last trend.

Identity documents contain textual and graphical information with a given layout. From such well structured documents, one could expect to base their classification on the layout. However, the layout is not always discriminant. Some classes share very similar structure: this is especially the case of different versions of passport or ID card emitted by the same country. Other methods are based on text transcription. Unfortunately, such methods are not adapted to our application due to the following difficulties: The document is not localized a priori in the query image and background information might disturb the OCR tasks, see Fig. 1. Indeed, text information is difficult to extract before knowing the type of the document and where it is localized in the image. Moreover, a large part of the text is specific to the owner of the document and not to the class. Therefore, we prefer to rely on the graphical content of the identity document and we turn towards image classification techniques in search for robustness and diversity.

Image classification has received a large attention from the scientific community, *e.g.* see the abundant literature related to the Pascal VOC [9] and ImageNet [7] challenges. A large part of the modern approaches follow the bag-of-word (BOW) approach [6], represented by a 3 step pipeline: (1) extraction of local image features, (2) encoding of local image descriptors and pooling of these encoded descriptors into a global image representation, (3) training and classification of global image descriptors for the purpose of object recognition. Local feature points, such as SIFT [21], are widely used as local features due to their description capabilities. Regarding the second step, image encoding, BOW were originally used to encode the feature point's distribution in a global image representation [12,16]. Fisher vectors and VLAD later showed improvement over the BOW [14,23]. Pooling has also witnessed many improvement: for instance, spatial and feature space pooling techniques have been widely investigated [18,32]. Finally, regarding the last step of the pipeline, discriminative classifiers such as linear Support Vector Machines (SVM) are widely accepted as the reference in terms of classification performance [4].

Recently, the deep CNN approaches have been successfully applied to large-scale image classification datasets, such as ImageNet [7,15], obtaining state-of-the-art results significantly above Fisher vectors or bag-of-words schemes. These networks have a much deeper structure than standard representations, including several convolutional layers followed by fully connected layers, resulting in a very large number of parameters that have to be learned from training data.

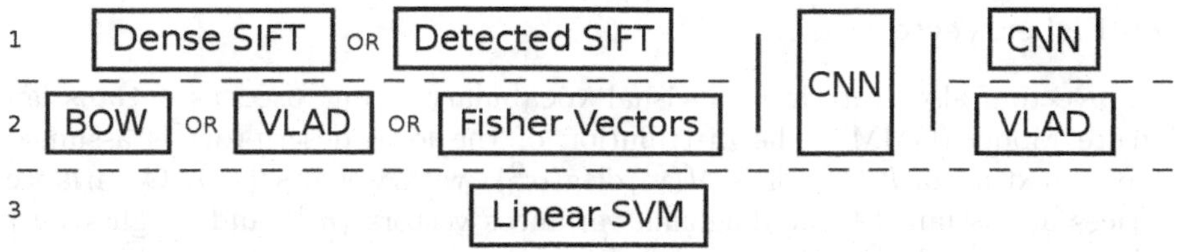

Fig. 2. Classification pipelines composed of (1) feature extraction on the first row, (2) feature encoding on the second row, and (3) classification on the final row.

By learning these networks parameters on large image datasets, a structured representation can be extracted at an intermediate to a high-level [22,35]. Furthermore, Deep CNN representation have been recently combined with VLAD [1,11] or Fisher vectors [5,19] encodings.

It is worth mentioning that other approaches have been proposed in Computer Vision with the aim to build mid-level description [29] or to learn a set of discriminative parts to model classes [10,25,28]. They are highly effective in similar fine-grain classification scenarios but are extremely costly.

3 A Plurality of Methods

To perform image classification, we first follow the BOW-based pipeline. SIFT keypoints are extracted in either a dense fashion or by interest point detection. Dense extraction tends to offer better performance in classification, while interest points are rotation invariant [26]. Then, these features are encoded with BOW, VLAD or Fisher vectors and are used to classify images with SVM.

Secondly, we study CNN based features, where intermediate transferable representations are computed from pre-trained networks. Descriptors are computed using various networks, layers, orientations, and scales. Finally, a VLAD aggregation of activation maps across orientations and scales is proposed. These image descriptors are similarly given as input to SVM to perform classification, see Fig. 2.

3.1 Bag-of-Words

Assume that the local description output vectors in $\mathbb{R}^d$. The Bag of visual Words aims at encoding local image descriptors based on a partition of the feature space $\mathbb{R}^d$ into regions. This partition is usually achieved by using the k-means algorithm on a training set of feature points. It yields a set $\mathcal{V}$, so called *visual vocabulary*, of k centroids $\{\mathbf{v}_i\}_{i=1}^k$, named *visual words*. The regions are the Voronoi cells of the centroids. This process is achieved offline and once for all.

The local descriptors of an image $\{\mathbf{x}_t\}_{t=1}^T$ are quantized onto the visual vocabulary $\mathcal{V}$:

$$\mathsf{NN}(\mathbf{x}_t) = \arg\min_{1 \leq i \leq k} \|\mathbf{x}_t - \mathbf{v}_i\|. \tag{1}$$

The histogram of frequencies of these mappings becomes the global image description whose size is k.

3.2 Fisher Vectors

Fisher vectors also start from a visual vocabulary $\mathcal{V}$ but used as a Gaussian Mixture Model (GMM). The distribution of the local descriptors is assumed to be a mixture of k Gaussian $\mathcal{N}(\mathbf{v}_i, \mathsf{diag}(\boldsymbol{\sigma}_i^2))$ with weights $\{\omega_i\}$. Covariance matrices are assumed to be diagonal, variances vectors $\{\boldsymbol{\sigma}_i^2\}$ and weights $\{\omega_i\}$ are learned from the training set as well.

Fisher vectors considers the log-likelihood of the local descriptors of the image $\{\mathbf{x}_t\}_{t=1}^T$ w.r.t. to this GMM. They are composed of two gradient calculations of this quantity per Gaussian distribution: The gradient G_μ^X w.r.t. $\mathbf{v}_i$ and the gradient G_σ^X w.r.t. to the variance vector $\boldsymbol{\sigma}_i^2$:

$$G_{\mu,i}^X = \frac{1}{T\sqrt{\omega_i}} \sum_{t=1}^{T} \gamma_t(i) \mathsf{diag}(\boldsymbol{\sigma}_i)^{-1}(\mathbf{x}_t - \mathbf{v}_i), \tag{2}$$

$$G_{\sigma,i}^X = \frac{1}{T\sqrt{2\omega_i}} \sum_{t=1}^{T} \gamma_t(i)[\mathsf{diag}(\boldsymbol{\sigma}_i^2)^{-1}(\mathbf{x}_t - \mathbf{v}_i)^2 - \mathbf{1}_d], \tag{3}$$

where $\gamma_t(i)$ represents the soft assignment term, *i.e.* the probability that descriptor $\mathbf{x}_t$ derives from the i-th Gaussian distribution [23], and $\mathbf{a}^2$ denotes the vector whose components are the square of the components of $\mathbf{a}$. The concatenation of these gradients results in a global descriptor of $2kd$ components.

3.3 VLAD

VLAD is similar to Fisher vectors [14] aggregating only the difference between the local descriptors and hard-assigned cluster from the visual vocabulary:

$$\mathbf{d}_i = \sum_{\mathbf{x}_t:\mathsf{NN}(\mathbf{x}_t)=i} \mathbf{x}_t - \mathbf{v}_i. \tag{4}$$

The global descriptor $(\mathbf{d}_1^\top, \ldots, \mathbf{d}_k^\top)^\top$ has a size of dk. A power law, l_2 normalization, and/or PCA reduction are usually performed on Fisher and VLAD [23].

3.4 Convolutional Neural Networks

Deep Convolutional Neural Network [15] are composed of convolutional layers followed by fully connected ones with normalization and/or pooling performed in between layers. There is a large variety of network architectures [30,31], but a usual choice is 5 convolutional layers followed by 3 fully connected layers. The layers parameters are learned from training data.

The works [22,33] showed that extracting intermediate layer produces mid-level generic representations, which can be used for various recognition tasks and a wide range of data [27]. In our case, we use a fast network and a very deep network, both trained on ImageNet ILSVRC data. The fast network from [3] is similar to [15], while the deep network stacks more convolutional layers (19 layers in total) with smaller convolutional filters [30].

Following previous works [22,28,33], image representations are computed by either taking the output of the fully connected intermediate layers or by performing pooling on the output of the last convolutional layer [33].

Unfortunately rotation invariance can not be obtained with such networks. Thus, we enrich our datasets using flipped and rotated versions of each image to artificially enforce such invariance.

Recent works showed that fully connected layers can be kernelized to obtain a fully convolutional networks [20]. Such transformation allows input of various size, which is shown to be beneficial in [13] classification.

After showing the benefits of using several scales and orientations, we propose to aggregate multi-scale information using VLAD across a fixed set of scales and orientations. Specifically, each activation of the feature map is considered as a local descriptors, which are aggregated with equal weights. Unlike the similar NetVLAD [1], our method allows the aggregation over several scales and orientations. To our knowledge, such use of VLAD aggregation over scales and orientation of the activations of various layers has not yet been proposed.

4 Experiments

4.1 Datasets

There is no publicly available dataset of identity documents as they hold sensitive and personal information. Three private datasets are provided by our industrial partner. Images are collected using a variety of sources (scan, mobile photos) and no constraint is imposed. Thus, the documents have any dimension, any orientation, and might be surrounded by complex backgrounds. Figure 1 shows examples of such images.

Preliminary experiments is held on a dataset of 9 classes of French documents (FRA), namely *identity card (front), identity card (back), passport (old), passport (new), residence card (old front), residence card (old back), residence card (new front), residence card (new back), driving licence.* A total of 527 samples are divided into train and test, ranging from 26 to 136 images per class. Then, a larger dataset (Extended-FRA or E-FRA) of the same types of documents with a total of 2399 images (86 to 586 per class) is used. The last dataset consists of 446 samples (8 to 110 per class) of 10 Belgian identity documents (BEL), namely *identity card 1 (front), identity card 1 (back), identity card 2 (front), identity card 2 (back), residence card (old front), residence card (old back), residence card (new front), residence card (new back), passport (new), passport (old).*

4.2 Results

An extensive evaluation is carried out on the image datasets. Three measures are calculated: mean average precision (mAP), overall mean accuracy, and averaged accuracy per class.

Table 1. Evaluation of BOW, VLAD, and Fisher in terms of mAP for detected and dense features, on the FRA dataset.

Encoding	Dim.	Detected SIFT	Dense SIFT
BOW 1k	1k	80.7	79.2
BOW 10k	10k	87.0	85.9
VLAD 16	1k	78.5	81.7
VLAD 64	4k	86.6	90.7
VLAD 256	16k	90.1	91.0
Fisher 16	2k	88.9	88.3
Fisher 64	8k	92.8	93.1
Fisher 256	32k	92.7	92.8

Table 2. Performance of several CNN-based features, on the FRA dataset.

Net. layer	Dim.	mAP	Mean acc.	Acc./class
fast fc7	4k	91.1	85.4	85.6
fast fc6	4k	91.7	81.3	81.6
fast c5 - Avg	256	93.2	89.0	90.5
fast c5 - Max	256	92.9	85.7	88.0
vd19 fc7	4k	87.0	81.9	83.2
vd19 fc6	4k	89.4	85.4	86.0
vd19 c5 - Avg	512	89.6	85.4	86.0
vd19 c5 - Max	512	88.3	83.6	82.3

First, SIFT-based methods are evaluated on the FRA dataset, see Table 1. This comprises BOW, VLAD, and Fisher Vector encodings with several visual vocabulary sizes, and from detected or dense SIFT local descriptors. We note that SIFT descriptors are square-rooted and PCA is applied to obtain 64-dimensional vectors. We observe that Fisher Vector performs better than VLAD, which performs better than BOW. This is expected: the more refined the encoding, the longer the global descriptor, and the better the performances. Even when comparing similar global descriptor dimension, Fisher Vector offers the best performance. Note that Fisher Vector does not improve over 64 Gaussians. Secondly, dense local description overall outperforms detected feature except for the case of BOW encoding. These results agrees with general observations made in computer vision for classification tasks [26].

Then, we evaluate CNN-based descriptors on the same FRA dataset, see Table 2. Two architecture are compared: the 'fast' network [3] and the deep 'vd19' network [30]. Descriptors are obtained by extracting the output of the two first fully connected layers (*fc6* and *fc7*), as well as the last convolutional layer (*c5*). Average and max pooling of *c5* are evaluated as well. Surprisingly, the fast network outperforms vd19. Average pooling is also shown to outperform max pooling for convolutional layer and is preferred in the following experiments. Overall *c5* outperforms *fc6*, which outperforms *fc7*. In fact, lower layers (*c5*) encodes lower level and more generic information, which is less sensible to network training data.

Table 3. Orientation invariance of CNN features, on the FRA dataset.

Net. layer	mAP	Mean acc.	Acc./class
fast fc7	92.5	90.5	91.3
fast fc6	92.3	88.1	86.6
fast c5 - Avg	94.1	90.2	90.7
vd19 fc7	89.9	84.8	85.0
vd19 fc6	90.8	87.2	86.5
vd19 c5 - Avg	91.2	88.7	88.8

Since the CNN feature do not have any rotation invariance mechanism, we propose to enrich the training data collection by adding rotated and flipped images (ending up in 8 distinct descriptors per image), see Table 3. Such process offers a constant improvement for every descriptors.

Further experiments are achieved on the larger E-FRA dataset, see Table 4. Unlike for FRA dataset alone, we observe that *fc6* outperforms *c5*. Unsurprisingly, the more training data the better the performance reaching up to 99% mAP and more than 96% accuracy, when training on E-FRA. More experiment is performed on the BEL dataset, see Table 5. We divide the dataset into three folds, then learn on two third and test on the last one. Scores obtained on all permutations and finally averaged. As for E-FRA, the sixth fully connected layer offers the best performance. Also performances on the BEL dataset are much lower because some classes (residence card (old/front), residence card (old/back), residence card (new/back)) have very few (5 to 12) training samples.

The very recent work of [13] highlighted how input dimension of the image can have a large impact on performance. Therefore, we experiment various input sizes, ($s1 = 224 \times 224, s2 = 544 \times 544, s3 = 864 \times 864$), see Table 6. Concerning convolutional layers, the feature maps are averaged pooled as earlier. However, since fully connected layers require a fixed input feature maps dimension, we kernelize the layers so they are applied at every location of the larger feature map output by the last convolutional layer and finally perform max pooling. We observe a stable gain for every layer, *c5* and *s3* offering the best performance on the FRA dataset. We further note that higher dimensionality (1184×1184, 1504×1504) offers worst results in our experiments.

Since larger scales and multiple orientations encapsulate more precise information, we decide to aggregate the activations of several scales ($s1 = 224 \times 224, s2 = 544 \times 544, s3 = 864 \times 864$, and $s4 = 1184 \times 1184$) and 8 orientations

Table 4. Performance using various combination of the FRA and E-FRA datasets with orientation invariance. Tr Te and E represents Training set of FRA, Testing set of FRA, and E-FRA.

Train/test net. layer	mAP	Mean acc.	Acc./class
Tr/E fast fc7	83.5	81.2	76.7
Tr/E fast fc6	85.6	83.5	78.6
Tr/E fast c5 - Avg	87.8	85.4	83.8
Te/E fast fc7	83.3	83.7	77.8
Te/E fast fc6	84.7	85.8	81.3
Te/E fast c5 - Avg	83.6	86.1	82.1
TrTe/E fast fc7	89.5	86.8	83.9
TrTe/E fast fc6	91.4	89.7	87.6
TrTe/E fast c5 - Avg	90.0	88.5	86.8
TrE/Te fast fc7	97.9	93.6	95.4
TrE/Te fast fc6	99.0	96.3	96.7
TrE/Te fast c5 - Avg	96.6	94.8	95.4
TeE/Tr fast fc7	99.4	96.5	96.4
TeE/Tr fast fc6	99.6	98.0	98.2
TeE/Tr fast c5 - Avg	98.0	94.0	94.4

Table 5. Results obtained on the BEL dataset using 3 folds.

Net. layer	mAP	Mean acc.	Acc./class
fast fc7	71.7	78.6	64.9
fast fc6	73.8	79.0	66.3
fast c5 - Avg	70.9	77.9	60.0

Table 6. Varying scales CNN features with orientation invariance, on the FRA dataset.

Net. layer scale	mAP	Mean acc.	Acc./class
vd19 fc7 s2	94.9	93.9	90.8
vd19 fc7 s3	96.2	95.4	93.1
vd19 fc6 s2	95.1	94.8	91.9
vd19 fc6 s3	95.8	95.4	92.9
vd19 c5 - Avg s2	96.3	95.4	93.1
vd19 c5 - Avg s3	96.3	96.3	94.3

together in a VLAD descriptor. Each activation is centered, PCA reduced to 128 dimensions, and *l2*-normalized. Once concatenated in the VLAD, the final vector is power normalized. Table 7 shows the final performance on the FRA dataset and we observe a stable improvement for all layers reaching a very high performance, around 99% mAP and 98% mean accuracy.

Our application requires fast processing of the scanned documents. We report the computation time of SIFT and CNN features extraction in Table 8. Execution times hold for a single threaded i7 core 2.6 GHz. Note that image dimensions remained unchanged for SIFT features, while images are resized to 224×224 for CNN features using the fast network. CNN features are much faster than SIFT, and keypoints detection is quite slow especially for high-resolution images.

Table 7. VLAD aggregation over scales and orientations, on the FRA dataset.

Net. layer	mAP	Mean acc.	Acc./class
vd19 fc7	98.8	97.2	93.2
vd19 fc6	99.2	97.9	94.8
vd19 c5 - Avg	99.5	98.8	95.6

Table 8. Computation time for detected SIFT, dense SIFT, and CNN features extracted from (224×224) dimensional features on FRA train/test sets.

Features	Detected SIFT	Dense SIFT	CNN
Average per image	54 s/43 s	5.1 s/4.9 s	0.2 s/0.2 s
Total	180 m/240 m	17 m/27 m	40 s/53 s

To conclude, CNN generate highly effective compact description, largely outperforming earlier SIFT-based encoding schemes from the classification performance and run-time point of view. Secondly, our evaluation provides insight regarding the amount and balance of data required to reach very high performance. Finally, the proposed VLAD aggregation across scales an orientations shows superior performance.

5 Conclusion

This paper addressed the problem of identification documents classification as an image classification task. Several image classification methods are evaluated. We show that CNN features extracted from pre-trained networks can be successfully transferred to produce image descriptors which are fast to compute, compact, and highly performing.

References

1. Arandjelović, R., Gronat, P., Torii, A., Pajdla, T., Sivic, J.: NetVLAD: CNN architecture for weakly supervised place recognition. In: CVPR (2016)
2. Bagdanov, A., Worring, M.: Fine-grained document genre classification using first order random graphs. In: ICDAR, pp. 79–83 (2001)
3. Chatfield, K., Simonyan, K., Vedaldi, A., Zisserman, A.: Return of the devil in the details: delving deep into convolutional nets. In: BMVC (2014)
4. Chen, S., He, Y., Sun, J., Naoi, S.: Structured document classification by matching local salient features. In: ICPR, pp. 653–656. IEEE (2012)
5. Cimpoi, M., Maji, S., Vedaldi, A.: Deep filter banks for texture recognition and segmentation. In: Proceedings of the IEEE CVPR, pp. 3828–3836 (2015)
6. Csurka, G., Dance, C.R., Fan, L., Willamowski, J., Bray, C.: Visual categorization with bags of keypoints. In: International Workshop on Statistical Learning in Computer Vision (2004)
7. Deng, J., Dong, W., Socher, R., Li, L.J., Li, K., Fei-Fei, L.: Imagenet: a large-scale hierarchical image database. In: CVPR. IEEE (2009)
8. Eglin, V., Bres, S.: Document page similarity based on layout visual saliency: application to query by example and document classification. In: ICDAR (2003)
9. Everingham, M., Van Gool, L., Williams, C.K.I., Winn, J., Zisserman, A.: The Pascal visual object classes (VOC) challenge. Int. J. Comput. Vis. **88**(2), 303–338 (2010). Springer
10. Felzenszwalb, P.F., Girshick, R.B., McAllester, D., Ramanan, D.: Object detection with discriminatively trained part-based models. Trans. PAMI **32**(9), 1627–1645 (2010)
11. Gong, Y., Wang, L., Guo, R., Lazebnik, S.: Multi-scale orderless pooling of deep convolutional activation features. In: Fleet, D., Pajdla, T., Schiele, B., Tuytelaars, T. (eds.) ECCV 2014. LNCS, vol. 8695, pp. 392–407. Springer, Cham (2014). doi:10.1007/978-3-319-10584-0_26
12. de las Heras, L.P., Terrades, O.R., Llados, J., Fernandez-Mota, D., Canero, C.: Use case visual bag-of-words techniques for camera based identity document classification. In: ICDAR, pp. 721–725, August 2015
13. Herranz, L., Jiang, S., Li, X.: Scene recognition with CNNs: objects, scales and dataset bias. In: CVPR, pp. 571–579 (2016)
14. Jégou, H., Perronnin, F., Douze, M., Schmid, C., et al.: Aggregating local image descriptors into compact codes. Trans. PAMI **34**, 1704–1716 (2012)
15. Krizhevsky, A., Sutskever, I., Hinton, G.E.: Imagenet classification with deep convolutional neural networks. In: NIPS, pp. 1097–1105 (2012)
16. Kumar, J., Doermann, D.: Unsupervised classification of structurally similar document images. In: ICDAR, pp. 1225–1229. IEEE (2013)
17. Lai, S., Xu, L., Liu, K., Zhao, J.: Recurrent convolutional neural networks for text classification. In: AAAI, vol. 333, pp. 2267–2273 (2015)
18. Lazebnik, S., Schmid, C., Ponce, J.: Beyond bags of features: spatial pyramid matching for recognizing natural scene categories. In: CVPR (2006)
19. Liu, L., Shen, C., Wang, L., van den Hengel, A., Wang, C.: Encoding high dimensional local features by sparse coding based fisher vectors. In: NIPS (2014)
20. Long, J., Shelhamer, E., Darrell, T.: Fully convolutional networks for semantic segmentation. In: CVPR, pp. 3431–3440 (2015)
21. Lowe, D.: Object recognition from local scale-invariant features. In: ICCV (1999)
22. Oquab, M., Bottou, L., Laptev, I., Sivic, J., et al.: Learning and transferring mid-level image representations using convolutional neural networks. In: CVPR (2014)

23. Perronnin, F., Sánchez, J., Mensink, T.: Improving the fisher kernel for large-scale image classification. In: Daniilidis, K., Maragos, P., Paragios, N. (eds.) ECCV 2010. LNCS, vol. 6314, pp. 143–156. Springer, Heidelberg (2010). doi:10.1007/978-3-642-15561-1_11
24. Shin, C., Doermann, D.: Document image retrieval based on layout structural similarity. In: IPCV, pp. 606–612 (2006)
25. Sicre, R., Avrithis, Y., Kijak, E., Jurie, F.: Unsupervised part learning for visual recognition. In: CVPR (2017)
26. Sicre, R., Gevers, T.: Dense sampling of features for image retrieval. In: ICIP, pp. 3057–3061. IEEE (2014)
27. Sicre, R., Jégou, H.: Memory vectors for particular object retrieval with multiple queries. In: ICMR, pp. 479–482. ACM (2015)
28. Sicre, R., Jurie, F.: Discriminative part model for visual recognition. CVIU **141**, 28–37 (2015)
29. Sicre, R., Tasli, H.E., Gevers, T.: Superpixel based angular differences as a mid-level image descriptor. In: ICPR, pp. 3732–3737. IEEE (2014)
30. Simonyan, K., Zisserman, A.: Very deep convolutional networks for large-scale image recognition. In: ICLR (2015)
31. Szegedy, C., Liu, W., Jia, Y., Sermanet, P., Reed, S., Anguelov, D., Erhan, D., Vanhoucke, V., Rabinovich, A.: Going deeper with convolutions. In: CVPR (2015)
32. Tasli, H.E., Sicre, R., Gevers, T., Alatan, A.A.: Geometry-constrained spatial pyramid adaptation for image classification. In: ICIP (2014)
33. Tolias, G., Sicre, R., Jégou, H.: Particular object retrieval with integral max-pooling of CNN activations. In: ICLR (2016)
34. Xing, C., Wang, D., Zhang, X., Liu, C.: Document classification with distributions of word vectors. In: APSIPA
35. Zhou, B., Lapedriza, A., Xiao, J., Torralba, A., Oliva, A.: Learning deep features for scene recognition using places database. In: NIPS (2014)

Using LDP-TOP in Video-Based Spoofing Detection

Quoc-Tin Phan[1(✉)], Duc-Tien Dang-Nguyen[2], Giulia Boato[1], and Francesco G. B. De Natale[1]

[1] University of Trento, Trento, Italy
quoctin.phan@unitn.it
[2] Dublin City University, Dublin, Ireland

Abstract. Face authentication has been shown to be vulnerable against three main kinds of attacks: *print, replay,* and *3D mask*. Among those, video replay attacks appear more challenging to be detected. There exist in the literature many countermeasures to face spoofing attacks, but a sophisticated detector is still needed to deal with particularly high-quality video based attacks. In this work, we perform analysis on the noise residual in frequency domain, and extract discriminative features by using a dynamic texture descriptor to characterize video based spoofing attacks. We propose a promising detector, which produces competitive results on the most challenging dataset of video based spoofing.

Keywords: Face anti-spoofing · Local Derivative Pattern · Video based attacks

1 Introduction

Among many applications of biometric authentication, face authentication has been considered as an efficient and reliable access control mechanism. A face authentication system works in less intrusive manner, which requires little cooperation from users. Thanks to the advances in face detection and recognition, a face authentication system can be flawlessly deployed on low-cost devices.

"Fingerprints cannot lie, but liars can make fingerprints." [7]

Like other biometric modalities, a face authentication system can be bypassed easily even at very low cost. We group such spoofing attacks into three main categories: (i) *Print attacks*: the use of printed photo of an authorized user, (ii) *Replay attacks*: a photo or a video of an authorized user is replayed on a digital screen, (iii) *3D mask attacks*: the authorized user's face is simulated by 3D mask. The vulnerability of face authentication system to spoofing attacks has motivated plenty of proposed countermeasures in the past few years.

One of first attempts to detect print attacks is introduced in [1]. By analyzing the total amount of movements over video frames, print attacks can be

S. Battiato et al. (Eds.): ICIAP 2017, Part II, LNCS 10485, pp. 614–624, 2017.
https://doi.org/10.1007/978-3-319-68548-9_56

effectively detected. This explains the great success, i.e., high performance detection accuracy [13], of motion-based methods. While print attacks leave clear and inevitable evidences, photos or videos replayed on a digital screen are more challenging to be detected [4,15]. Replayed photos are generally in higher quality (e.g., color, contrast) compared with printed photos, and replayed videos can easily fool an authentication system since the face in a high quality video and a real face are almost indistinguishable. By the advance of 3D printing techniques, another kind of face spoofing attack has been introduced in [5], the so-called 3D mask attacks. In [5] the vulnerability of face authentication system to spoofing with 3D masks is shown.

In this work, our concentration is placed on video-based attacks which refer to replaying a video on a digital screen. A number of approaches have been proposed [2,6,8–11,14], and we classify them into two main categories: *spatial domain* and *frequency domain* analysis.

Methods performing analyses on spatial domain take into account directly the pixel values of the suspected image. In [14], discriminative features characterizing spoofing attacks are extracted, such as specular reflection, blurriness, chromatic moment, and color diversity. Multiple classifiers are trained based on the concatenation of all extracted features and final decision is given by taking mutual information from multiple classifiers. Another approach exploiting dynamic texture descriptor has been introduced in [9]. In this work, the authors extend Local Derivative Pattern (LDP) to LDP on Three Orthogonal Planes (LDP-TOP) in order to capture highly detailed information on spatial domain as well as subtle face movements over frames. By analyzing image distortion artifacts such as reflection, color distribution, Moiré patterns (i.e., overlapping grids) and face shape deformation, the authors in [8] have developed a countermeasure to spoofing attacks on mobile phones. Most recently, the analysis on disparities between color texture of genuine faces and fake ones is investigated in [2]. The authors show that the use of YCbCr and HSV color spaces results in generally better detection accuracy.

Despite the success of methods relying on spatial domain analysis, discriminative features extract in spatial domain may become scene-dependent. Roughly speaking, instead of addressing only artifacts of spoofing attacks, spatial domain models learn also redundant information of the scene. Frequency domain analysis focuses on periodic patterns, i.e., Moiré patterns, which present as peaks in the spectrum image. Such periodic patterns are independent to image scene. In [6], the authors extract Moiré patterns from still images using a bandpass filter and make analysis on frequency domain. A large-scale dataset dedicated to video-based attacks is introduced in [11]. This is a challenging dataset containing huge number of videos recorded under different environmental conditions. The authors also provide a baseline method analyzing the noise residual of the video in terms of visual rhythms. Another method dedicated to noise analysis is mentioned in [10]. Instead of extracting only *low-level* features which are basically artifacts on the spectrum video, to reduce the sensitivity between intra- and extra-class variations, the authors propose to extract also mid-level features as *visual codebooks* from low-level features.

In this work, we select to analyze artifacts on frequency domain for some reasons. First, it is challenging to extract discriminative features on the spatial domain due to the contamination of the scene. Despite good detection performance on some benchmarking datasets, methods on the spatial domain tend to get overfitted on specific conditions of capturing. This results in the low generality in real applications where attempted attacks might be performed in various conditions. Moreover, most of methods on the spatial domain rely on the reliability of face detection and tracking algorithms, which are not always successful under poor conditions. We propose to analyze the spectrum video by using a dynamic (spatial-temporal) texture descriptor. Dynamic texture descriptor is able to capture not only highly detailed information on spatial domain but also subtle changes over time. Thanks to the success of Local Derivative Pattern on Three Orthogonal Planes (LDP-TOP) [9], we select LDP-TOP as the descriptor. The main difference to [9] is that we use LDP-TOP to analyze discriminative textures of spectrum videos. Our proposed method outperforms two recent and closely related works on large-scale dataset of video-based attacks in terms of detection accuracy. By analyzing only the noise residual, we can skip face detection and tracking which require more computation and depend heavily on environmental conditions.

The paper is structured as follows: Sect. 2 presents in detail the schema of the proposed methods, and experimental analysis is described in Sect. 3. In Sect. 4, we draw some conclusions.

2 The Proposed Method

2.1 The Recaptured Artifacts

Video-based spoofing attacks can easily bypass the authentication system because the face in a high quality video is nearly indistinguishable with the real face. Nevertheless, when the camera records a digital display, the resulting video presents a number of visible artifacts:

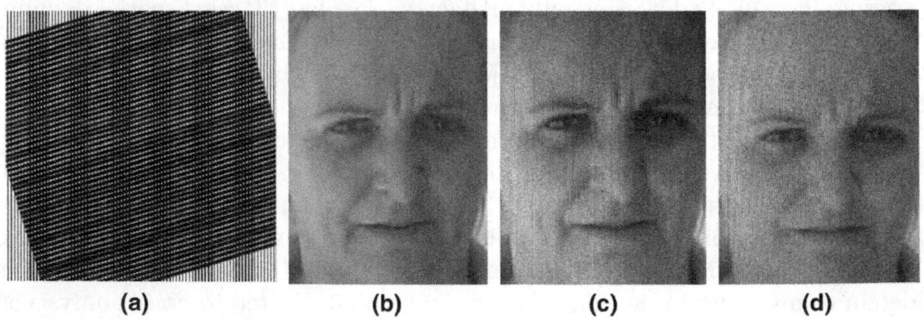

Fig. 1. Examples of Moiré patterns.

- **Moiré pattern**: In recaptured videos, Moiré patterns occur in the form of visible periodic or almost periodic patterns in every video frame. Specifically, the sampling grid of the displaying device is overlaid by the sampling grid of acquisition device resulting the third grid pattern. Misalignment between the two devices also causes different observable forms of Moiré patterns. Shown in Fig. 1(a) is the Moiré pattern generated from two overlaid patterns containing parallel lines. Figure 1(b) is original image shown in Macbook Pro screen, Fig. 1(c) and (d) show images captured from Macbook Pro screen by HTC Desire HD phone and Apple Ipad Air, respectively. Note that the original image is taken from UVAD dataset [11].
- **Flickering effect**: This effect corresponds to horizontal or vertical lines of equal spaces, caused by the desynchronization between the flashing frequencies of displaying and acquisition device. These noticeable lines might move vertically or horizontally over video frames. Shown in Fig. 2(a) is an example of flickering effect observed when capturing the image from a screen display by the Olympus SP 800UZ. The alignment of these effects is highlighted in Fig. 2(b).
- **Other artifacts**: Besides Moiré patterns and flickering effects, a recaptured video is generally blurred compared with the original video. The change in color tone can be also observed according to different acquisition devices.

Fig. 2. Example of flickering effect.

Since aforementioned artifacts are independent to image scene, they should be isolated from the content of the image by using an image filter. Moiré pattern and flickering effect are almost periodic, they are characterized by high-energy peaks in the frequency domain. A practical way of detecting such artifacts is to perform Fourier transform of the suspected image, and apply a matched filter to the transformed image. This naive approach might result in many false detections. On the other hand, blurring artifact implies the increase of low frequency components which are challenging to be detected by simple thresholding.

2.2 Processing Pipeline

The complete processing pipeline of the proposed method is presented in Fig. 3 and includes three main steps:

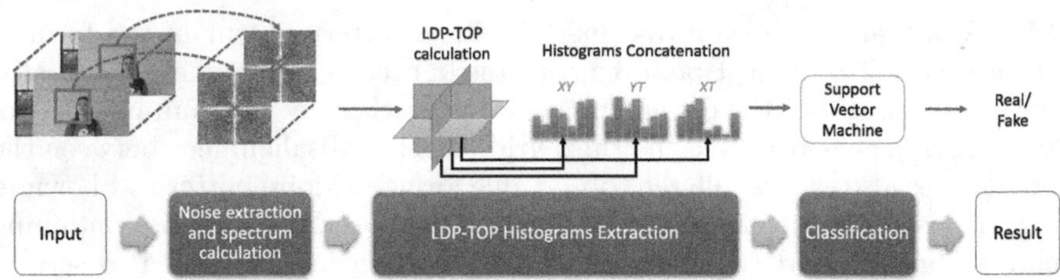

Fig. 3. Schema of the proposed method.

A. Noise Extraction and Spectrum Calculation
Since all artifacts present entirely on every frame, we first define a *region of interest* (RoI) in the spatial domain. Taking into account RoI of size $w \times h$, we can reduce greatly the computation time of our proposed method. Denote the frame t-th of the video V as $V^{(t)}$, we extract its residual by applying a denoising filter $\mathcal{F}$ and subtracting the denoised frame from $V^{(t)}$ to obtain $V_r^{(t)}$.

$$V_r^{(t)} = V^{(t)} - \mathcal{F}(V^{(t)}), \quad 1 \leq t \leq M, \tag{1}$$

where M is the number of frames. The noise residual $V_r^{(t)}$ contains the noise pattern of frame t-th. Let $V_f^{(t)}$ denote the presentation of $V_r^{(t)}$ on the frequency domain. $V_f^{(t)}$ is calculated using 2D Discrete Fourier Transform (DFT).

$$V_f^{(t)} = DFT(V_r^{(t)}) \tag{2}$$

To get the final spectrum video V_s, we collect all Fourier spectrum $|V_f^{(t)}|$, and calculate their logarithmic scale.

$$V_s^{(t)} = \log(|V_f^{(t)}| + 1) \tag{3}$$

B. Histogram Extraction
In this work, we treat the Fourier transformed video (spectrum video) as a three-dimensional texture map, and then apply a sophisticated local descriptor on the spectrum video in order to extract meaningful features. We select our previously proposed Local Derivative Pattern on Three Orthogonal Planes, the so-called LDP-TOP [9], as the descriptor thanks to its success in spoofing detection.

Given the image I, the first-order derivative along each direction $\alpha = \{0°, 45°, 90°, 135°\}$ is denoted as I_α. Let Z_0 be a pixel, and Z_i, $i = 1, \cdots, 8$ be the neighboring pixels around Z_0. The four first-order derivatives at $Z = Z_0$ can be written as:

$$I_{0°}(Z_0) = I(Z_0) - I(Z_4) \qquad I_{45°}(Z_0) = I(Z_0) - I(Z_3)$$
$$I_{90°}(Z_0) = I(Z_0) - I(Z_2) \qquad I_{135°}(Z_0) = I(Z_0) - I(Z_1)$$

Z_1	Z_2	Z_3
Z_8	Z_0	Z_4
Z_7	Z_6	Z_5

Generally, the nth-order directional LDP, $\text{LDP}^n_\alpha(Z_0)$, in direction α at $Z = Z_0$ is defined as:

$$\text{LDP}^n_\alpha(Z_0) = \{f(I^{n-1}_\alpha(Z_0), I^{n-1}_\alpha(Z_1), \cdots, f(I^{n-1}_\alpha(Z_0), I^{n-1}_\alpha(Z_8)\}, \quad (4)$$

where $I^{n-1}_\alpha(Z_0)$ is the $(n-1)^{th}$-order derivative in direction α at $Z = Z_0$, and $f(I^{n-1}_\alpha(Z_0), I^{n-1}_\alpha(Z_i))$ is defined as

$$f(I^{n-1}_\alpha(Z_0), I^{n-1}_\alpha(Z_i)) = \begin{cases} 0, & \text{if } I^{n-1}_\alpha(Z_i) \cdot I^{n-1}_\alpha(Z_0) > 0 \\ 1, & \text{if } I^{n-1}_\alpha(Z_i) \cdot I^{n-1}_\alpha(Z_0) \leq 0 \end{cases}, \quad i = 1, \cdots, 8. \quad (5)$$

Equation (4) encodes $(n-1)^{th}$-order gradient transitions, resulting the nth-order binary pattern on the local region. Binary patterns are represented in 4 histograms, each describing a specific direction. This way, the final histogram contains 4×2^8 bins.

We consider the time window size T_{ws} ($T_{ws} \leq M$) as the number of chronological-order frames. Only the first T_{ws} frames are taken into account for histogram extraction. In Fig. 3, three planes XY, XT, YT are pair-wise orthogonal, where XY corresponds to a frame of the spectrum video. XT, YT refer to horizontal and vertical planes. As a result, we end up three 2D texture maps of size $w \times h$, $w \times T_{ws}$, and $h \times T_{ws}$.

Histogram of LDP-TOP is the concatenation of three LDP histograms from three orthogonal planes. Finally we end up a feature vector of dimension $3 \times 4 \times 2^8$.

C. Classification
We use Support Vector Machine (SVM) [3] to learn and detect video-based attacks. Specifically, we embed the Histogram Intersection Kernel (HIK) as the kernel of SVM. HIK was introduced in [12] to compare color histograms. A HIK between two histogram a and b is simply defined as:

$$K(a, b) = \sum_{i=1}^{n} \min(a_i, b_i), \quad a_i \geq 0, b_i \geq 0. \quad (6)$$

In the next section, we present how parameters are experimentally selected and give some insights on the effectiveness of the proposed method.

3 Experimental Analysis

3.1 Parameter Selection

We validate the effectiveness of the proposed method on the Unicamp Video-Attack Database (UVAD) [11]. This dataset contains valid access and attempted attack videos of 404 different identities. Each video is recorded at high quality, 30 frames per seconds, and 9 s long. The resolution of all videos is fixed to 1366×768, where the face appears approximately in middle of the frame. Six cameras have been used to record real access videos. Each person is recorded by only one camera, but in different scenarios (different backgrounds, lighting conditions and places), generating 808 real access videos in total. For attempted attacks, real access videos are displayed in seven different display screens and recaptured by the same set of cameras used before. Finally, the recapturing process produces 16,268 attempted attack videos.

We evaluate our method using videos from all six cameras: Sony, Kodak, Olympus, Nikon, Canon and Panasonic. The training set contains real access and attempted attack videos from Sony, Kodak and Olympus, resulting in 344 real access and 3528 attempted attack videos. On the other hand, real access and attempted attack videos from Nikon, Canon and Panasonic are used for testing purpose, resulting in 60 real access and 6356 attempted attack videos. This setup is applied in [10]. Specifically, we select 300 (30 real access and 270 attempted attacks) samples of the training set to serve as the development set which is used for decision threshold estimation. Some example video frames are shown in Fig. 4 [1].

Fig. 4. The first row presents example video frames of real access in outdoor (the first two images) and indoor (the last two) condition. The second row presents example video frames of attempted attacks in outdoor (the first two images) and indoor (the last two) condition.

We report statistics mainly in terms of Half Total Error Rate (HTER), and Area Under the Curve (AUC). The decision of *positive* or *negative* is simply made by comparing the output score of the test sample with a decision threshold. This threshold directly causes two kinds of error: False Rejection Rate (FRR) referring to rejecting real faces, and False Acceptance Rate (FAR) referring to accepting spoofed faces. HTER is applied as threshold-dependent performance measurement, and is defined as the average of FRR and FAR. On the other hand, AUC is calculated as the area under the ROC curve, and is invariant to decision threshold.

[1] The set of identities involved in real access videos are disjoint with those involved in attempted attacks. That is why we do not show the attempted attack of the same identity.

In order to select the best configuration in our proposed method, we run experiments on UVAD under various settings. We test the effectiveness of three denoising filters: Median, Gaussian and Wiener, and use second-order and third-order LDP-TOP to extract features from spectrum videos. We consider the RoI of 256×256 pixels which is located in the center of every frame. We set the time window size T_{ws} to 100. The window size of all denoising filters are identically 7×7, and Gaussian standard deviation is set to 2. Figure 5 depicts DET (Detection Error Tradeoff) curves of all configurations.

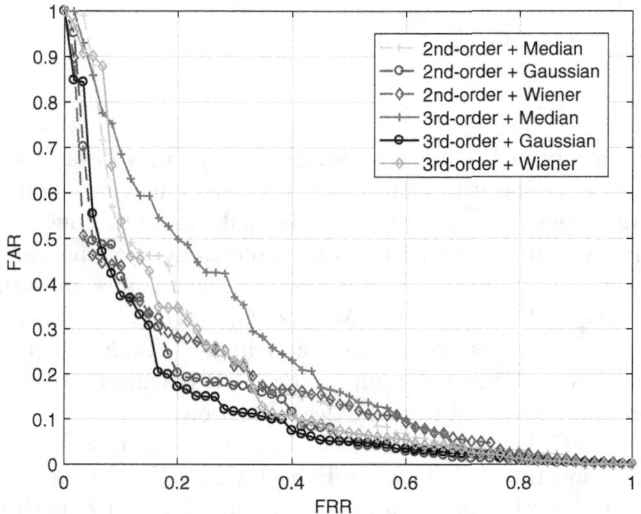

Fig. 5. DET curves under different configurations.

As can be seen in Fig. 5, Gaussian filter allows best error tradeoff in both of second-order and third-order LDP-TOP. More specifically, third-order LDP-TOP can produce better detection performance compared with second-order LDP-TOP. Median filter is particularly powerful in removing outlier pixels (i.e., salt and pepper noise) and preserving edges, but it appears less efficient in our analysis. Numeric results on UVAD are shown in Table 1.

3.2 Comparisons with State-of-the-Art Approaches

In this section, we discuss about performance comparisons on UVAD. Since our method is dedicated to noise analysis, we compare our results mainly with the two recent and closely related works: Visual Rhythm (VR) [11] and Visual Codebooks (VC) [10].

VR based approach captures noise signatures in terms of 2D maps which are basically generated by traversing the spectrum video in horizontal, vertical and zigzag directions. This can be considered as a baseline method in UVAD

Table 1. Results the proposed method on UVAD.

	HTER (%)	AUC (%)
(a) 2nd-order LDP-TOP		
Median	29.39	80.01
Gaussian	25.45	86.19
Wiener	30.15	83.24
(b) 3rd-order LDP-TOP		
Median	33.34	74.93
Gaussian	23.69	87.58
Wiener	24.25	81.70

dataset. The authors in [11] report results in a different dataset configuration where the size of testing set is much smaller than our consideration which is described in the previous section. In order to make more thorough and coherent analysis, we run visual rhythm extraction on our dataset configuration by using the implementation provided by the authors. We select Gaussian as the denoising filter and Gray-Level Co-Occurrence Matrix (GLCM) as the texture descriptor since this is the best reported configuration in [11]. Table 2 depicts detection performance of VR on UVAD. As can be seen, VR reaches its best performance with visual rhythm extracted from vertical direction.

According to VC, low-level features are extracted from small cuboids in the spectrum video, and the authors apply Bag-of-Visual-Word to map onto a more discriminative mid-level representation. We report all statistics in [10] since those were collected in the same dataset configuration.

Table 2. Results of VR on UVAD.

	HTER (%)	AUC (%)
Horizontal	52.01	51.02
Vertical	28.09	73.48
Zigzag	41.28	71.77

Shown in Table 3 are performance comparisons with VR and VC. In our method we choose the best configuration in which the Gaussian denoising filter and third-order LDP-TOP are applied. Compared with VR and VC, our proposed method achieves lower detection error, as shown in Table 3. Moreover, we consider only the RoI of size 256×256, which is much smaller than the entire volume of the video. There exist plenty of methods contributing to spatial domain analysis, however, reproducing all results of them on UVAD is beyond the scope of this paper. We emphasize more on methods on the frequency domain and

Table 3. Results of all methods on UVAD.

	FAR (%)	FRR (%)	HTER (%)
Correlation [1]	81.60	14.56	48.06
LBP [4]	27.41	66.04	46.72
VR [11]	44.52	**11.67**	28.09
VC [10]	44.73	15.00	29.87
Proposed method	**7.38**	40.00	**23.69**

show here only results of baseline methods on the spatial domain [1,4]. It is evident that methods in [1,4] perform poorly in detecting video based spoofing attacks.

3.3 Computational Complexity

Considering the computational complexity of the method, we provide here the analysis of the most impactful steps. Denote N the total number of pixels in RoI, $N = w \times h$. The complexity of the filtering step is $\mathcal{O}(N)$. Fast Fourier Transform requires $\mathcal{O}(N \times \log(N))$, and third-order LDP-TOP requires $\mathcal{O}(N)$ computations. Finally, the computational complexity is bounded to $\mathcal{O}(T_{ws} \times N \times \log(N))$, where T_{ws} is simply the number of considered frames.

Our testing is run on the computer with the following configuration: Intel(R) Xeon(R) CPU E5-2630 v3 2.40GHz; 64 Gb of RAM; Linux Ubuntu 14.04 LTS 64 bit installed. By considering small RoI, it is shown that the proposed approach can be applied for real-time applications.

The Matlab implementation can be obtained via: github.com/quoctin/anti_video_spoofing.

4 Conclusions

We have proposed a novel approach for detecting video based spoofing attacks. A recaptured video basically contains discriminative artifacts such as blurring, Moiré patterns and flickering effects. Those signatures are present in the frequency domain and can be analyzed by using dynamic texture descriptor. Thanks to the superiority of Local Derivative Pattern on Three Orthogonal Planes (LDP-TOP), we achieve promising results compared with related works. Future extension of this work will be devoted to designing sophisticated filters for extracting aforementioned artifacts. We will also find a mechanism to reduce the dimension of feature vectors.

References

1. Anjos, A., Marcel, S.: Counter-measures to photo attacks in face recognition: a public database and a baseline. In: International Joint Conference on Biometrics 2011, October 2011

2. Boulkenafet, Z., Komulainen, J., Hadid, A.: Face spoofing detection using colour texture analysis. IEEE Trans. Inf. Forensics Secur. **11**(8), 1818–1830 (2016)
3. Chang, C.C., Lin, C.J.: LIBSVM: a library for support vector machines. ACM Trans. Intell. Syst. Technol. **2**, 27:1–27:27 (2011)
4. Chingovska, I., Anjos, A., Marcel, S.: On the effectiveness of local binary patterns in face anti-spoofing. In: International Conference of Biometrics Special Interest Group, pp. 1–7, Septemper 2012
5. Erdogmus, N., Marcel, S.: Spoofing face recognition with 3D masks. IEEE Trans. Inf. Forensics Secur. **9**(7), 1084–1097 (2014)
6. Garcia, D.C., de Queiroz, R.L.: Face-spoofing 2D-detection based on moiré pattern analysis. IEEE Trans. Inf. Forensics Secur. **10**(4), 778–786 (2015)
7. Marcel, S., Nixon, M.S., Li, S.Z.: Handbook of Biometric Anti-Spoofing: Trusted Biometrics Under Spoofing Attacks. Springer, New York (2014)
8. Patel, K., Han, H., Jain, A.K.: Secure face unlock: spoof detection on smartphones. IEEE Trans. Inf. Forensics Secur. **11**(10), 2268–2283 (2016)
9. Phan, Q.T., Dang-Nguyen, D.T., Boato, G., De Natale, F.G.B.: Face spoofing detection using LDP-TOP. In: IEEE International Conference on Image Processing, pp. 404–408, Septemper 2016
10. Pinto, A., Pedrini, H., Schwartz, W.R., Rocha, A.: Face spoofing detection through visual codebooks of spectral temporal cubes. IEEE Trans. Image Process. **24**(12), 4726–4740 (2015)
11. Pinto, A., Schwartz, W.R., Pedrini, H., Rezende de Rocha, A.: Using visual rhythms for detecting video-based facial spoof attacks. IEEE Trans. Inf. Forensics Secur. **10**(5), 1025–1038 (2015)
12. Swain, M.J., Ballard, D.H.: Color indexing. Int. J. Comput. Vis. **7**(1), 11–32 (1991)
13. Tirunagari, S., Poh, N., Windridge, D., Iorliam, A., Suki, N., Ho, A.T.S.: Detection of face spoofing using visual dynamics. IEEE Trans. Inf. Forensics Secur. **10**(4), 762–777 (2015)
14. Wen, D., Han, H., Jain, A.K.: Face spoof detection with image distortion analysis. IEEE Trans. Inf. Forensics Secur. **10**(4), 746–761 (2015)
15. Zhang, Z., Yan, J., Liu, S., Lei, Z., Yi, D., Li, S.Z.: A face antispoofing database with diverse attacks. In: 5th IAPR International Conference on Biometrics (ICB), pp. 26–31, March 2012

A Classification Engine for Image Ballistics of Social Data

Oliver Giudice[1(✉)], Antonino Paratore[2], Marco Moltisanti[1], and Sebastiano Battiato[1]

[1] Image Processing Laboratory, University of Catania, Catania, Italy
{giudice,moltisanti,battiato}@dmi.unict.it
[2] iCTLab s.r.l., Spin-off of University of Catania, Catania, Italy
http://www.ictlab.srl

Abstract. Image Forensics has already achieved great results for the source camera identification task on images. Standard approaches for data coming from Social Network Platforms cannot be applied due to different processes involved (e.g., scaling, compression, etc.). In this paper, a classification engine for the reconstruction of the history of an image, is presented. Specifically, machine learning techniques and a-priori knowledge acquired through image analysis, we propose an automatic approach that can understand which Social Network Platform has processed an image and the software application used to perform the image upload. The engine makes use of proper alterations introduced by each platform as features. Results, in terms of global accuracy on a dataset of 2720 images, confirm the effectiveness of the proposed strategy.

Keywords: Social networks · Image forensics · JPEG · Digital ballistic

1 Introduction

Image Forensics traditionally refers to a number of different tasks on digital images aiming at producing evidence on the authenticity and integrity of data (e.g., forgery detection) and on the identification of the acquisition device (camera identification) [1–3]. To solve the forgery detection task, some approaches stand above the others: a group of them looks at the structure of the file (e.g., JPEG blocking artifacts analysis [4,5], hash functions [6], JPEG headers analysis [7], thumbnails [8] and EXIF analysis [9], etc.); others try to identify the device that acquired the image by making use of PRNU patterns [10,11], or focus on statistical analysis of the DCT coefficients [12–14]. Some in-depth studies [15,16] showed that it is possible to coarsely solve the camera identification task, using the DCT coefficients as a feature. Hence it is clear the importance of the JPEG pipeline in retrieving information about the history of an image. Nowadays Social Networks allow their users to upload and share large amounts of images: just on *Facebook* about 1 billion images are shared every day. What happens when a picture is shared on a social platform? How does the upload

S. Battiato et al. (Eds.): ICIAP 2017, Part II, LNCS 10485, pp. 625–636, 2017.
https://doi.org/10.1007/978-3-319-68548-9_57

process affect the JPEG elements of the image? A Social Network is yet but another piece of software that alters images for bandwidth, storage and layout reasons. These kind of alterations, specifically scaling and re-compression, have been proved to make state-of-the-art approaches for camera identification less precise and reliable [17,18]. Recent studies [19–21] have shown that, although the platform heavily modifies an image, this processing leaves a sort of fingerprint on the image itself. All those studies focus on the analysis of too few Social Networks and specific unrealistic scenarios making their works not general enough. In order to improve state of the art and to deeply understand how SNSs process images, a dataset of images from different camera devices was collected, under controlled conditions. We selected ten SNSs through which we processed the collected images by mean of an upload and download process. By doing this, a dataset of images has been obtained, in order to identify any alterations on JPEG elements. The main discovery of our study was that alterations observed are platform dependent (server-side) but also related to the application carrying out the upload (client-side). This evidence can be fundamental for investigation purposes to understand not only the provenience of an image, but also if it has been uploaded from a given device (e.g., Android, iOS). All the observed alterations allowed to build an automatic classifier, based on two K-NN classifiers and a decision tree fitted on the built dataset. Starting from an input image, the proposed approach can predict the SNS that processed the image and the client application through which the image has been uploaded. The remainder of the paper is structured as follows: in Sect. 2, we describe how the dataset has been built, which social platforms have been considered and what kind of upload methods have been used; in Sect. 3, an in-depth analysis on dataset images is reported in order to find alterations that can be coded into a fingerprint for a SNS processing; in Sect. 4, our approach for image ballistics on social image data is presented with the obtained classification results. Finally, conclusions and reasoning about possible future works on the topic are discussed.

2 A Dataset of Social Imagery

The alterations introduced on images by SNS can be thought as a unique fingerprint left by the SNS. The aim of our study is to discover those fingerprints by analyzing the behavior of the most popular SNSs that allow image sharing. Hence, 10 platforms have been selected. First of all, *Facebook* (www.facebook.com) and *Google+* (http://plus.google.com) were taken into account as being the two most popular platforms where users can share their statuses and multimedia content to a network of friends. *Twitter* (http://www.twitter.com) and *Tumblr* (http://www.tumblr.com) were considered as being representative of the microblogging concept. We included also *Flickr* (https://www.flickr.com) and *Instagram* (https://www.instagram.com) as platforms focused on sharing high quality artistic photos with capabilities of image editing and filtering. *Imgur* (http://www.imgur.com) and *Tinypic* (http://www.tinypic.com) were also taken into consideration even if they are not properly SNSs but are very popular platforms for image sharing: users usually link images hosted on them from forums and web sites all over the Internet. Finally *WhatsApp* and *Telegram* were also selected as being the two most popular mobile messaging platforms that, by allowing users to create chat groups, are another big place for image sharing on the Internet. Specifically, the last two services are often involved in forensic investigations. To discover how SNSs process images, we collected a set of photos with the camera devices listed in Table 1. Images were acquired representing three different types of scenes: outdoor scenes with buildings (artificial environment), outdoor scenes without buildings (natural environment) and indoor scenes. When taking a picture, we captured two versions: a High Quality (HQ) photo at the maximum resolution allowed by the device, and a Low Quality (LQ) photo (see also Table 1). Capturing images in this way, a dataset with a good variability in terms of contents and resolutions was obtained. Images collected so far were uploaded to each of the considered platforms with two different methods: with a web browser, and with iOS and Android native apps. No further discrimination is needed for web browsers because we observed that alterations are not browser-dependent. Each download was performed by searching for the image file URL in the HTML code of the page showing the image itself. At the end of this phase 2400 images were properly collected. The second upload method was carried out with iOS and Android native apps of each social platform, except for *Tinypic* and *Imgur* that do not possess an official app in stores. Moreover, the upload has been done by choosing images in two ways: by searching in the gallery for a previously acquired image (images from local gallery) and by acquiring the image with the camera app embedded in the app itself (embedded camera app). After uploading all images as described above, all of them were downloaded through the "URL searching technique" previously described. 320 more images processed through 8 platforms were thus obtained. All uploads were performed with default settings. The overall dataset consists of 2720 images in JPEG format and it is available at the following web address http://iplab.dmi.unict.it/DigitalForensics/social_image_forensics/.

3 Dataset Analysis

The main aim of our work is to find a fingerprint left by SNSs on JPEG structure elements, after an upload/download process, in order to build a classifier for image ballistics. To achieve this goal, all information contained in the JPEG file specification has been analyzed: image filename, image size, meta-data and JPEG compression information. We observed that each upload/download process through the considered SNSs produces different alterations among the above-mentioned elements that could be taken into account as fingerprints of the process itself. Details of these alterations will be described in the following Subsections.

3.1 Image Filename Alterations

The analysis of the filename of an image and the comparison with known patterns during an investigation on storage devices can provide information about the platform from which it could be downloaded and the date when it was uploaded. For this reason, we first evaluated if and how each platform modifies the file name. We observed that all platforms except *Google+* do a rename.

Table 1. Devices used to carry out image collection. For each device the corresponding Low Quality (LQ) and High Quality (HQ) resolutions are reported.

Model	Device type	Low resolution	High resolution
Canon Eos 650D	Dedicated device	720 × 480	5184 × 3456
QUMOX SJ4000	Dedicated device	640 × 480	4032 × 3024
Sony Powershot A2300	Dedicated device	640 × 480	4608 × 3456
Samsung Galaxy Note 3 Neo	Android 4 phone	640 × 480	3264 × 2448
HTC Desire 526g	Android 5 phone	640 × 480	3264 × 2448
Huawei G Play Mini	Android 6 phone	640 × 480	4208 × 3120
iPhone 5	iOS 6 phone	640 × 480	2448 × 3264
iPad mini 2	iOS 8 pad	640 × 480	800 × 600

Table 2. Renaming scheme for an uploaded image with original filename IMG_2641.jpg. The new file name for each platform is reported (Image IDs are marked in bold).

Social	Rename (image ID in bold)	Image lookup	Other information
Facebook	11008414_**746657488782610**_850837898930766639_n.jpg	YES	Upload resolution
Flickr	26742193671_**8a63f10c85**_h.jpg	YES	Download resolution (h = 1600)
Tumblr	tumblr_**o3q9ghRCRh1vnf44lo9**_1280.jpg	YES	Download resolution (1280)
Imgur	04 - **Dw0KXG2**.jpg	YES	
Twitter	**CdqCPQ-WAAAzrHI.jpg**	YES	
WhatsApp	IMG-20160314-WA0038.jpg	NO	Receiving date (2016-03-14)
Tinypic	1zqdirm.jpg	NO	
Instagram	1689555_**169215806798447**_744040439_n.jpg	YES	Upload resolution
Telegram	422114602_5593965449613038107.jpg	NO	

As an example, in Table 2 the new names for an uploaded file with name "IMG_2641.jpg" are reported. The column "image lookup" describes the presence into the new filename of an ID useful to reconstruct an URL that points to the web location where the image file is stored.

Table 3. Alterations on JPEG files. The EXIF column reports how JPEG meta-data are edited: maintained, modified or deleted. The File Size column reports if a resize is applied and the corresponding conditions. The JPEG compression column reports if a new JPEG compression is carried out and the corresponding conditions (if any).

Social	EXIF		File size		JPEG compression	
	Camera data	Other data	Resize	Resize condition	Re-compression	Re-compression condition
Facebook	Delete	Delete	Yes	LQ: $M > 960$ HQ: $M > 2048$	Yes	Always
Google+	Maintain	Maintain/edit	Yes	$M > 2048$	Yes	$M > 2048$
Flickr	Delete	Maintain/edit	Yes	Depends on options	Yes	Depends on options
Tumblr	Maintain	Maintain/edit	Yes	$M > 1280$	Yes	$M > 1280$
Imgur	Delete	Delete	No	Never	Yes	Image size (MB) > 5.45 MB
Twitter	Delete	Delete	Yes	$M > 2048$	Yes	Always
whatsApp	Delete	Delete	Yes	$M > 1600$	Yes	Always
Tinypic	Maintain	Maintain/edit	Yes	$M > 1600$	Yes	$M > 1600$
Instagram	Delete	Delete	Yes	$M > 1080$	Yes	Always
Telegram	Delete	Delete	Yes	$M > 2560$	Yes	Always

3.2 Image Size Alterations

A stronger evidence than file naming is the resize of the uploaded images on each platform. A fine-grained test was performed by using synthetic images derived from our dataset and resized at different scales.

On most platforms, resizing is applied if and only if the input image matches certain conditions. This condition is linked to the length in pixels of the longest side M of the original image, where $M = max(width, height)$. If M is greater than a threshold, a resizing algorithm is applied and the resulting image has its longest size equal to the threshold. In Table 3, such conditions and the corresponding thresholds for each platform are reported. *Tumblr* does not rescale uploaded images, while in *Flickr* the threshold is set by the user. When the images are resized, the longest side will be set to a fixed value that identifies, in some sense, the platform that made the operation (see Table 3).

3.3 Meta-Data Alterations

The best evidence to obtain information, for investigation purposes, are meta-data embedded in JPEG files. These meta-data are technically known as EXIF and can store information like the device that acquired an image, the date and time of acquisition and also the GPS coordinates. For our purposes, we divided EXIF data into two categories: "camera data" which contains all those key-valued that allow to identifying the device that acquired the image and "other data" for every other EXIF information.

In Table 3, the results of the analysis on EXIF data are resumed for each platform. In particular, it is reported if "camera data" and "other data" are deleted, maintained or just edited throughout the processing. Unfortunately, most of the SNSs delete all meta-data, specifically those related to camera data.

3.4 Image JPEG Compression Alterations

The images considered in our dataset are all encoded in JPEG format, both the original versions and the downloaded ones. Thus, an analysis on how the SNS processing affects the JPEG compression has been carried out. We focused on the Discrete Quantization Tables (DQTs) used for JPEG compression (extracted by DJPEG: an open source tool part of libjpeg project [22]).

Considering how platforms affect DQTs, it is possible to divide them into two categories:

- Platforms that always re-compress images (*Facebook*, *Twitter*, *Telegram*, *WhatsApp*, *Instagram*);
- Platforms that re-compress images at a given condition (*Google+*, *Tumblr*, *Tinypic*, *Imgur*).

The compression follows the same rules we described for resizing. In fact, a threshold-based evaluation is performed on the longest image side and, if it is bigger than the threshold, the image is compressed using a DQT that will be different from the original one. This is not true for all the considered platforms; *Flickr* allows the user to choose the threshold (if any), while on *Imgur* the threshold is fixed in terms of size in MegaBytes; specifically, if the input image size is greater than 5.45 MB, than the re-compression is performed, otherwise nothing happens (see also Table 3).

4 Image Ballistics of Social Data

Starting from the results of the analysis reported in previous Sections, regarding the alterations on JPEG elements of processed images, it is possible to assess that such alterations bring pieces of information about the history of the image but they could be insufficient, if considered alone, for investigation purposes. Hence, we encoded all the observed alterations into a set of features to be used as input for an automatic classifier. The following elements are then embedded into proper numerical features:

- The DQTs coefficients divided in 64 coefficients for the Chrominance table and 64 for the Luminance one, which represent the JPEG compression alterations. These coefficients were investigated separately with PCA and we obtained an explained variance of 99% for the first 32 coefficients of the luminance table and the first 8 coefficient of the chrominance one;
- Image size (width and height in pixels), which brings information about size alterations;
- Number and typology of EXIF data (key-value couples), which describes meta-data alterations (both camera and other data);
- Number of markers in JPEG files as defined in [23].

PRNU was not taken into consideration among our features, because, as already mentioned, the heavy processing done on images by SNSs degrades PRNU approaches for camera identification in terms of accuracy [17].

4.1 Implementing Image Ballistics: A Classification Engine

Given a JPEG image I, our objectives are to define:

1. if there is a compatibility between the non-related JPEG elements of I (i.e. filename, EXIF data) and the processing pipeline of SNSs;
2. if there is a compatibility between the JPEG elements of I and the processing pipeline of SNSs;
3. which SNS is compatible with the JPEG elements of the image, with a certain degree of confidence, and what is the uploading source in terms of operating system (OS) and application.

We represent each image I as a 44-dimensional vector

$$\boldsymbol{v} = \{w, h, |E|, m, l_j, c_k\}, \tag{1}$$

where

- $w \times h$ is the size in pixels of I;
- $E = \{key, value\}$ is an associative array containing the EXIF metadata, thus $|E|$ is the number of metadata found in the structure of I;
- m is the number of JPEG markers in I;
- $l_j, j = 0, \ldots, 31$ are the first 32 coefficients of the luminance quantization table;
- $c_k, k = 0, \ldots, 7$ are the first 8 coefficients of the chrominance quantization table.

Moreover, we define $fn\,(I)$ as the filename of the image I.

At the first stage, we consider $fn\,(I)$ and E. If there is a matching between $fn\,(I)$ and the renaming patterns observed in Sect. 3.1, our approach confirms the compatibility between I and the SNS with the matched pattern. Also, E is taken into account, looking for the "Exif.Image.UniqueCameraModel" key. If it is set, then our system returns that value.

Thus, the whole dataset representation is

$$\boldsymbol{V} = \{\boldsymbol{v}_1, \ldots, \boldsymbol{v}_N\}$$

where N is the total number of images. In order to train the SNS and Upload Scenario classifiers, we augment this representation with the corresponding labels. Thus, the final representation for a generic image I_i is

$$\boldsymbol{I}_i = \{\boldsymbol{v}_i, sns_i, uc_i, sm_i\}$$

where sns_i is the SNS, uc_i is the client application and sm_i is the image selection method.

Our classifier performs a two-steps analysis. First, we implement an Anomaly Detector to exclude the images that have not been processed by SNSs, then we run in parallel a K-NN Classifier and a Decision Tree [24] to asses respectively the SNS of origin and the uploading scenario (OS + application).

Given the representations $\boldsymbol{v}_{I_1}$ of an image I_1 and $\boldsymbol{v}_{I_2}$ of an image I_2, we define the cosine distance between $\boldsymbol{v}_{I_1}$ and $\boldsymbol{v}_{I_2}$

$$d(\boldsymbol{v_1}, \boldsymbol{v_2}) = \frac{\boldsymbol{v_1} \cdot \boldsymbol{v_2}}{|\boldsymbol{v_1}||\boldsymbol{v_2}|} \tag{2}$$

as a measure of similarity between I_1 and I_2. Therefore, it is possible to build a distance matrix $\boldsymbol{D}$ of size $N \times N$ where the element d_{ij} is equal to the distance between the images I_i and I_j. We will refer to the $r-$th row of this matrix as $\boldsymbol{D}_r$ and to the $c-$th column as $\boldsymbol{D}^c$. It is important to note that $\forall\ I_i, I_j,\ 0 \leq d(\boldsymbol{v_i}, \boldsymbol{v_j}) \leq 1$, and specifically, the more is the similarity, the more the distance will be closer to 1. Exploiting this property, we define the Anomaly Detector as

$$a\left(\boldsymbol{v}_i, \boldsymbol{D}\right) = \begin{cases} (\boldsymbol{v}_i, i) & if \sum_{j=1}^{K} d_{ij} < T \\ \text{not processed} & otherwise \end{cases} \tag{3}$$

where $T \in [0, K]$ is defined as the Anomaly Threshold. In other words, since the more two images are similar, the more their distance will be closer to 1, we make sure that at least $\lfloor K \rfloor$ samples in our dataset are similar to the query image representation. Then, when $a\left(\boldsymbol{v}_i, \boldsymbol{D}\right) = 0$, the representation is far apart the samples, and we can state that probably the image has not been processed.

The output of a is then used as input by K-NN (4) and Decision Tree algorithms [24].

$$knn\left(\boldsymbol{v}_i, i\right) = sns_j \mid d_{ij} = \min \boldsymbol{D}^i \tag{4}$$

$$dt\left(\boldsymbol{v}_i, i\right) = \left(uc_j, sm_j\right) \tag{5}$$

where uc_j and sm_j are the leaves obtained following the path with $\boldsymbol{v}_i$ as input. Hence, the classification scheme, shown in Fig. 1, can be formalized as follows

$$C(\boldsymbol{v}_i, \boldsymbol{D}) = knn\left(a\left(\boldsymbol{v}_i, \boldsymbol{D}\right)\right) \oplus dt\left(a\left(\boldsymbol{v}_i, \boldsymbol{D}\right)\right) \tag{6}$$

K-NN algorithm looks for the closest sample in the dataset, and assigns the same SNS to the query image. A Decision tree (Eq. 5) builds classification in the form of a tree structure. It breaks down a dataset into smaller and smaller subsets while at the same time an associated decision tree is incrementally developed. The final result is a tree with decision nodes. The algorithm used for building the decision tree is the ID3 [24] which employs a top-down, greedy search through the space of possible branches with no backtracking. ID3 uses Entropy to construct a decision tree by evaluating $\boldsymbol{v} \in \boldsymbol{V}$.

Finally, the output of the K-NN Classifier sns_j is processed through a SNS Consistency Test. Let be $S = \{sns_1, \ldots, sns_n\}$ the set of SNSs that operates a re-compression at the condition $max(w,h) > C_{sns_i}$ where C_{sns_i} is the conditional threshold for the $i-$th SNS and w and h as listed in Table 3.

Given that $sns_j \in S$, if $max(w,h) < C_{sns_j}$ it is an anomaly. The test is then repeated for the next most probable prediction from the SNS Classifier until the corresponding condition is satisfied or the loop stalls on the same

SNS prediction. In this last case, the result of the classification is "not sure"; otherwise, a SNS prediction is reached and outputted (sns_j) with the predicted upload client application (uc_j) and image selection method (sm_o).

Figure 1 shows a schematic representation of the proposed approach.

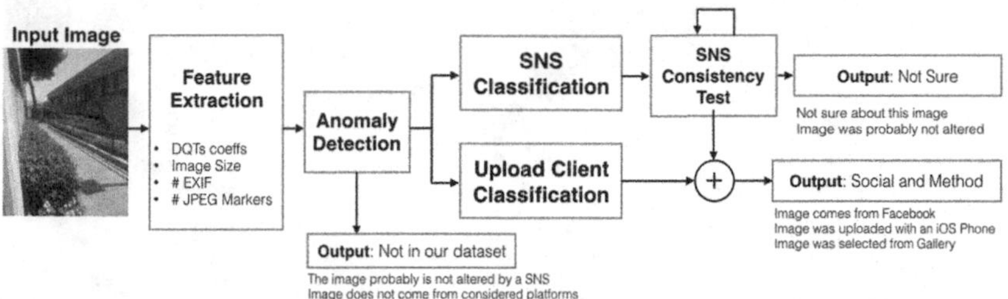

Fig. 1. Classification scheme for Image Ballistics in the era of Social Network Services. The proposed approach encodes JPEG information from an input image into a feature vector that is then processed through machine learning techniques in order to predict the most probable SNS from which the input image was downloaded and the correspondent upload method.

4.2 Classification Results

In this Section, validation results for the proposed approach are reported to demonstrate its goodness. The anomaly detector was validated by taking from our dataset 240 random images that suffered alterations, and 240 images that did not pass through any alteration. The anomaly detector achieved the best error rate, equal to 3.37%, with K = 3 and T = 2.90. The entire approach for image ballistics described in the previous Section was then tested through a 5-fold cross validation test. Best Ks and T were found through grid-search hyper-parameter tuning method. In Fig. 2, confusion matrices reporting the average value through the 5 runs are shown.

The accuracy obtained for the SNS classification task was 96% with best K equal to 3 while the accuracy value for the upload client classification task was 97.69% with an accuracy of 91% for the prediction of image selection method, given iOS or Android native app as prior.

Different approaches with other classifiers (like linear and non linear SVM) or combination of classifiers (like hierarchical or cascade approaches) were also tested, but the overall results were slightly worse. The classification scheme reported in Fig. 1 was the best approach we obtained throughout our tests.

In our experiments, we observed that, as happens for different camera devices of the same model [16], different images, from the same platform, have slightly differences in DQT coefficients. This demonstrated the effectiveness of K-NN over other methods for giving to the approach the resilience against little differences while detecting the most-similar SNS fingerprint. We also built a new test set composed of 20 images randomly downloaded from each considered SNS on which

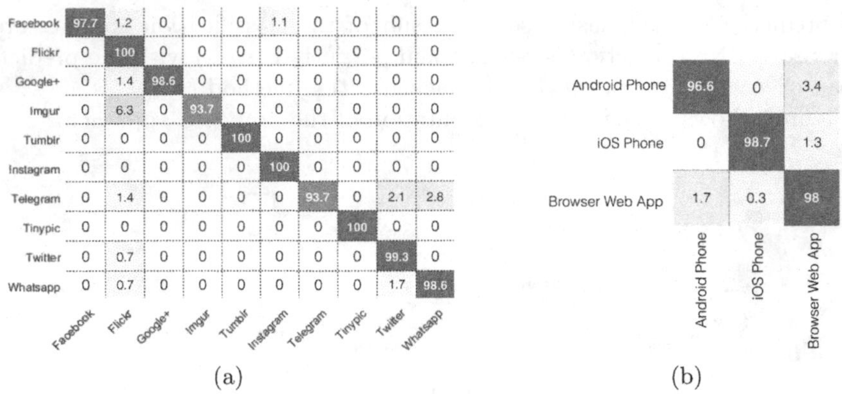

Fig. 2. Confusion Matrices obtained from 5-cross validation on our dataset. The reported values, are the average accuracy values (%) in 5 runs of cross validation test. (a) Confusion Matrix for Social platform Classification, (b) Confusion Matrix for upload method classification.

we achieved an accuracy in SNS prediction of 94% that is quite similar to the validation results.

Another consideration is needed about the SNSs fingerprints described in this work and regarding the fact that all the alterations observed can change according to software development and releases. For these reasons, the proposed approach is justified for being able to readapt through time, just by updating the reference dataset.

5 Conclusions and Future Works

In this work, we presented a dataset for image ballistic and proposed a classification engine to discover if an image has been processed by a Social Network Service and, if the answer is positive, by which SNS among the 10 considered platforms. The proposed approach performed the task of Image Ballistics with good accuracy by predicting the SNS that process an image and the corresponding upload method, with an accuracy respectively of 96% and 97.69%.

We think that this work can open new perspectives on the field of Image Forensics: the approach can be upgraded by considering other formats (e.g., PNG) and new features related to image contents.

References

1. Piva, A.: An overview on image forensics. ISRN Sig. Process. **2013**, 22 (2013)
2. Stamm, M.C., Wu, M., Liu, K.J.R.: Information forensics: an overview of the first decade. IEEE Access **1**, 167–200 (2013)

3. Battiato, S., Giudice, O., Paratore, A.: Multimedia forensics: discovering the history of multimedia contents. In: Proceedings of the 17th International Conference on Computer Systems and Technologies 2016, pp. 5–16 ACM (2016)
4. Bruna, A.R., Messina, G., Battiato, S.: Crop detection through blocking artefacts analysis. In: Maino, G., Foresti, G.L. (eds.) ICIAP 2011. LNCS, vol. 6978, pp. 650–659. Springer, Heidelberg (2011). doi:10.1007/978-3-642-24085-0_66
5. Luo, W., Qu, Z., Huang, J., Qiu, G.: A novel method for detecting cropped and recompressed image block. In: IEEE International Conference on Acoustics, Speech and Signal Processing, vol. 2, pp. II.217–II.220, April 2007
6. Battiato, S., Farinella, G.M., Messina, E., Puglisi, G.: Robust image alignment for tampering detection. IEEE Trans. Inf. Forensics Secur. **7**(4), 1105–1117 (2012)
7. Kee, E., Johnson, M.K., Farid, H.: Digital image authentication from JPEG headers. IEEE Trans. Inf. Forensics Secur. **6**(3), 1066–1075 (2011)
8. Kee, E., Farid, H.: Digital image authentication from thumbnails. In: IS&T/SPIE Electronic Imaging. International Society for Optics and Photonics, p. 75 410E (2010)
9. Gloe, T.: Forensic analysis of ordered data structures on the example of JPEG files. In: 2012 IEEE International Workshop on Information Forensics and Security (WIFS), pp. 139–144, December 2012
10. Chen, Y., Thing, V.L.: A study on the photo response non-uniformity noise pattern based image forensics in real-world applications. In: Proceedings of the International Conference on Image Processing, Computer Vision, and Pattern Recognition (IPCV), p. 1 (2012)
11. Dirik, A.E., Karaküçük, A.: Forensic use of photo response non-uniformity of imaging sensors and a counter method. Opt. Express **22**(1), 470–482 (2014)
12. Redi, J.A., Taktak, W., Dugelay, J.-L.: Digital image forensics: a booklet for beginners. Multimedia Tools Appl. **51**(1), 133–162 (2011)
13. Galvan, F., Puglisi, G., Bruna, A.R., Battiato, S.: First quantization matrix estimation from double compressed JPEG images. IEEE Trans. Inf. Forensics Secur. **9**(8), 1299–1310 (2014)
14. Battiato, S., Messina, G.: Digital forgery estimation into DCT domain: a critical analysis. In: Proceedings of the First ACM Workshop on Multimedia in Forensics, ser. MiFor 2009, pp. 37–42. ACM, New York (2009)
15. Farid, H.: Digital image ballistics from JPEG quantization: a followup study. Department of Computer Science, Dartmouth College, Techncial report TR2008-638 (2008)
16. Kornblum, J.D.: Using JPEG quantization tables to identify imagery processed by software. Digital Invest. **5**, S21–S25 (2008). The Proceedings of the Eighth Annual DFRWS Conference
17. Goljan, M., Chen, M., Comesaña, P., Fridrich, J.: Effect of compression on sensor-fingerprint based camera identification. Electron. Imaging **2016**(8), 1–10 (2016)
18. Rosenfeld, K., Sencar, H.T.: A study of the robustness of PRNU-based camera identification. In: IS&T/SPIE Electronic Imaging. International Society for Optics and Photonics, p. 72 540M (2009)
19. Moltisanti, M., Paratore, A., Battiato, S., Saravo, L.: Image manipulation on facebook for forensics evidence. In: Murino, V., Puppo, E. (eds.) ICIAP 2015. LNCS, vol. 9280, pp. 506–517. Springer, Cham (2015). doi:10.1007/978-3-319-23234-8_47
20. Castiglione, A., Cattaneo, G., Santis, A.D.: A forensic analysis of images on online social networks. In: International Conference on Intelligent Networking and Collaborative Systems, pp. 679–684 (2011)

21. Caldelli, R., Becarelli, R., Amerini, I.: Image origin classification based on social network provenance. IEEE Trans. Inf. Forensics Secur. **12**(6), 1299–1308 (2017)
22. DJPEG - LibJPEG open-source project on GITHUB. https://github.com/LuaDist/libjpeg
23. Miano, J.: Compressed Image File Formats: JPEG, PNG, GIF, XBM, BMP. Addison-Wesley Professional, Boston (1999)
24. Quinlan, J.R.: Induction of decision trees. Mach. Learn. **1**(1), 81–106 (1986)

Join Cryptography and Digital Watermarking for 3D Multiresolution Meshes Security

Ikbel Sayahi(✉), Akram Elkefi, and Chokri Ben Amar

REGIM: REsearch Group on Intelligent Machines,
National School of Engineers (ENIS), University of Sfax,
BP 1173, 3038 Sfax, Tunisia
{phd.ikbe.sayahi,akram.elkefi,chokri.benamar}@ieee.org
http://www.regim.org/members-1/

Abstract. The main objective of this paper is to combine cryptography with digital watermarking to secure 3D multiresolution meshes. The result is a new crypto-watermarking algorithm which contains 3 parts: the first part is called watermark preparation and it aims to encrypt the logo using AES (Advanced Encryption Standard) algorithm, to combine the encrypted logo with a binary sequence obtained by transforming the description of the host mesh using ASCII (American Standard Code for Information Interchange) code and to encode the hole watermark using a convolutional encoder. As for the second part, it is said mesh preparation and it consists in applying a wavelet transform to the mesh in order to generate a wavelet coefficient vector. All these coefficients will be presented in the spherical coordinate system. Finally, the third part of our algorithm intervenes to insert the data in the mesh using the LSB (Least Signifiant Bit) method. Found results prove that we are able to insert a high amount of data without influencing the mesh quality. The application of the most popular attacks does not prevent a correct extraction of data already inserted. Our algorithm is, then, robust against these attacks.

Keywords: Digital watermarking · 3D multiresolution meshes · AES algoithm · LSB method · Robustness

1 Introduction

Solving security problems increased from the day when technology has made available speed computer networks and remote multimedia databases allowing the sharing and the transmission of digital data. This is due to the abundant use of networks and remote databases for the transfer and the share of digital data in general and 3D meshes in particular. Digital watermarking and cryptography are, then, announced as a solutions for security problems which justify the appearance, in one hand, of watermarking approaches such as works published in [3,4,6,7,16]. Despite the variety of techniques used in these approaches, 3D watermarking domain still has different deficiencies.

S. Battiato et al. (Eds.): ICIAP 2017, Part II, LNCS 10485, pp. 637–647, 2017.
https://doi.org/10.1007/978-3-319-68548-9_58

In the other hand, it is remarkable that cryptography is absent in the field of 3D representation despite its abundant use in securing documents shared mainly for images. Hence come our idea to join cryptography with 3D watermarking in order to ensure the security of 3D multiresolution meshes.

In this context, we propose this paper which aim to design a new crypto-watermarking system for 3D multiresolution meshes. The proposed algorithm uses cryptographic tool (AES algorithm) and a watermarking tools (wavelet transform, spherical coordinate system, error correcting codes and LSB method) to insert a maximum of data without influencing the mesh quality and while ensuring robustness against any treatment threatening 3D mesh security.

2 Related Works

Sharing digital data, and specially 3D meshes, between remote users poses a great security problems. To solve these problems, many solutions have been proposed such as digital watermarking, cryptography and steganography. In this paper, we are interested only on digital watermarking and cryptography.

In fact, to secure 3D meshes, several 3D watermarking approaches appeared. The goal is always enhancing insertion rate, invisibility and robustness against attacks through the use of various techniques and methods. To classify these approaches, we use the inserting domain as a criterion. The first category being is the spatial approaches such as the approaches of Hitendra published in [3], Tsai et al. in [16] and Wang et al. in [4]. These approaches modify, during embedding, either the topological or the geometric information. As for the second category, it uses a transformed domain: Frequency domain [6] and multiresolution domain [7]. In this case, frequency and multiresolutions coefficients are targeted during embedding. Despite the considerable improvements brought by algorithms proposed over the last decade, the field of digital watermarking still suffer from deficiencies. This is due, firstly, to the difficulty to find a good compromise between invisibility, high insertion rate and robustness which are contradictory (the increase of the insertion rate causes either a deterioration of the mesh quality or reduces the level of robustness). Secondly, manipulating 3D multiresloution meshes is not an easy task in comparing them with other types of meshes. This is mainly explained by the sensitivity of handling this type of meshes which is due to the existence of the multi-resolution appearance in the 3D meshes.

In addition to that, cryptography has proven its efficiency in protecting digital data specially image. This idea is justified by the abundant use of cryptographic algorithm in the field of image processing to ensure security ([1] and [13] are examples). In spite of the effectiveness and the very encouraging results of applying cryptography to image, it is not used until now in the field of 3D representation. This can be justified by the particular presentation of 3D multiresolution meshes and to the difficulty of manipulating these data type.

As a result, we aim in this paper to combine cryptography and 3D watermarking in order to protect 3D multiresolution meshes.

3 Techniques and Methods

3.1 Encrypting Logo: AES Algorithm

The encryption algorithm AES is the standard system block encryption, that aims to replace the DES which becomes vulnerable. The efficiency, the speed, the reliability and the resistance of this algorithm to various attacks encourage us to use it to encrypt the logo to be embedded in the 3D mesh. This iterative encryption standard used only one key whose size depends on the size of the data to be encrypted. Thus, three types of AES algorithm can be distinguished according to the key length, the block data sizes and the round number (number of iterations executed during encryption) [5]. Specially in this work, we will use the AES with key length equal to 128. In this case, the block data size should be equal to 128 bits with 10 rounds.

As shown in Fig. 1, to encrypt a data block with AES algorithm, an AddRoundKey step should take place in order to apply an XOR logic operation between a subkey and the block of data. Just after, there will be an execution of the rounds. For each of these rounds, the following four operations must be applied [8]:

- SubByte: At this level, each byte of the block is replaced by another value from an Sbox which is a fixed table given in design.
- ShiftRow: In this step, the lines are shifted cyclically with different Offsets.
- MixColumn: In this step, each column is treated as a polynomial, multiplied on $GF(2^8)$ by a matrix.
- AddRoundKey: which is a Simple XOR logic operation between the current data and the subkey of the current round.

The AES algorithm performs a final supplemental routine which consists in applying a SubByte, ShiftRow, and AddRoundKey steps before producing the final encrypted data.

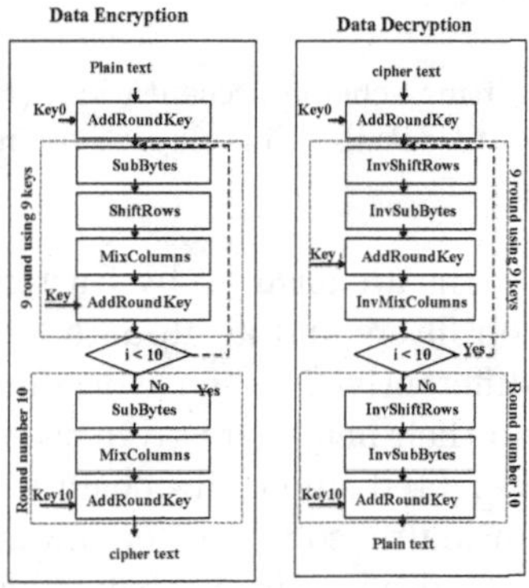

Fig. 1. AES encryption and decryption scheme.

3.2 Wavelet Transform

The main idea of the multiresolution analysis is to decompose a signal (3D mesh in our case) using two basic functions (prediction and update) into a lower resolution mesh and a set of wavelet coefficients needed to reconstruct the original mesh in the synthesis step [11]. All these coefficients are grouped in a wavelet coefficient vector (WCV) (see formula 1).

$$WCV = \begin{pmatrix} D_1 \\ \vdots \\ D_i \end{pmatrix} = \begin{pmatrix} d_1^x & d_1^y & d_1^z \\ \vdots & \vdots & \vdots \\ d_i^x & d_i^y & d_i^z \end{pmatrix} \quad (1)$$

To obtain the WCV, a lifting scheme, which is a wavelet transform of second generation, is used. As shown in Fig. 2, the lifting scheme is divided into three steps:

- The poly-phase transform: it divides the vertices of the mesh M_j into two parts: M_j^0 (vertices colored in green) and M_j^1 (vertices colored in red).
- Prediction: it takes as input M_j^0 (vertices colored in green) and predicts the positions of the vertices of M_j^1 (vertices colored in red). The prediction errors correspond to the wavelet coefficients C_j^k
- Update: it takes as an input the wavelet coefficient C_j and modifies the position of vertices of M_j^0 to obtain the low frequency mesh M_{j-1} .

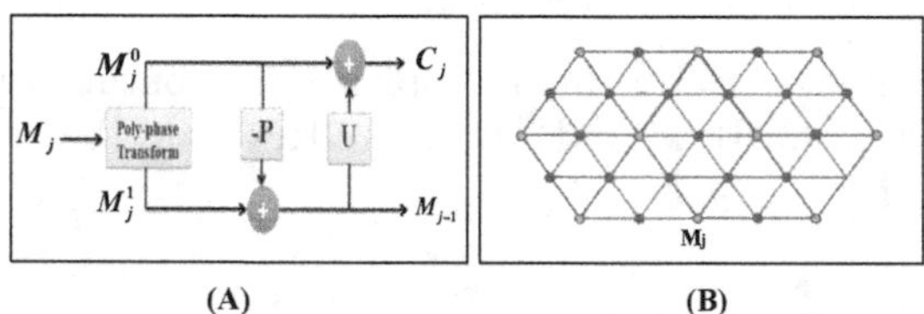

Fig. 2. Lifting scheme: (A): Lifting scheme presentation, (B): Mesh to be decomposed into a coarse version and a set of details (Color figure online)

Prediction and update functions are ensured by Buttery filters. These latter are composed of three prediction filters and an update filter. The prediction filters are the same but oriented differently. The sense depends only of the positions of the point to be predicted and their neighbors given that the predicted coefficients of the treated point are a weighted sum of its neighbors coefficients.

The choice of working in multiresolution domain is due to the fact that the insertion of information, by modifying the wavelet coefficient's vector, is done at different levels of resolution. This significantly eliminates the interaction between inserted data.

3.3 Concolutional Error Correcting Code

Convolutional codes are known for their simplicity, power and efficiency, which justifies their frequent use especially in fixed and mobile communications. These characteristics incite us to use a convolutional error correcting code in our watermarking algorithm to improve robustness criterion.

Convolutional Encoders. A convolutional encoder can work either with a state machine or a trellis presentation [10].

For the first case, let M be the state machine of the encoder. Each state of M is a particular state of the registers which are initially assumed to be zero. Thus, each branch represents the state change of the encoder according to the arrival of a new bit from the watermark to be inserted. These branches are then identified by the value of the input bit that causes the change of state and the codeword produced at the arrival of this bit.

As for the second case, the representation of trellis can be compared to a state machine repeated numerous times. Indeed, each vertex corresponds to a state of the encoder. Each edge is a transition and has as label the corresponding output of the encoder.

Convolutional Decoder. To decode data, in this work, Viterbi algorithm is used. This Choice is argued by the ability of this algorithm to correct randomly generated errors which is the case of the errors resulting from the application of attacks to the host mesh.

To decode a message using Viterbi algorithm, these steps should be followed [10]:

- Step1: From the first floor of trellis diagram, where two branches come to each state, we calculate Hamming distance between the word of the branch and the received codeword of the two paths that arrive at these states. For each state, we keep the way whose distance is the smallest (the survivor) and eliminate the others.
- Step2: Repeat step 1 for each floor of the trellis.
- Step3: When the origin of all survivors passes through a single path, it represents part of the decoded word.

4 Proposed Approach

The originality of this work is to combine cryptography with 3D watermarking to secure multiresolution 3D meshes shared by remote users. This crypto-watermarking algorithm allows copyright protection and indexation.

Our algorithm includes two steps. The first is insertion and it consists in inserting data into 3D meshes. This data is composed of a logo and a text describing the 3D object. As for the second, it is said extraction and it allows excerpting already inserted data from the mesh. Once Extracted, a verification of the authenticity takeS place.

4.1 Embedding

As it is already said, the goal of this step is to embed data in the host mesh without influencing its quality. To this end, three sub-step should take place which are: Watermark preparation, Host mesh preparation and insertion. As shown in Fig. 3, Data to insert is composed of a logo (grey level image) and a mesh description (text).

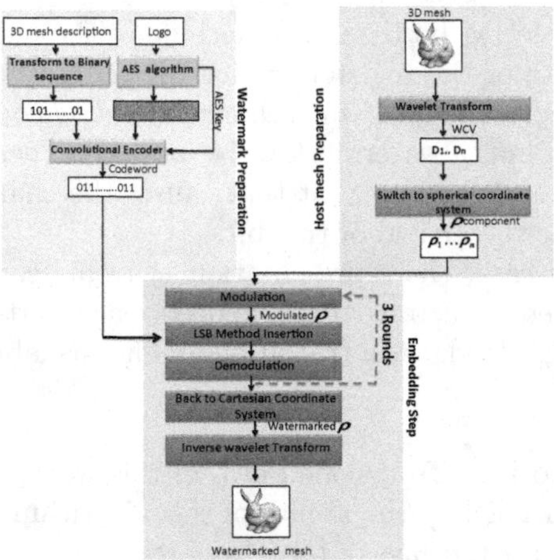

Fig. 3. Our Embedding step scheme.

Watermark Preparation. In this sub-step, The logo will be encrypted using AES algorithm. The result being is a grey level image having the same size which refer to the encrypted logo. This latter, after being transformed to a vector, will be presented in the form of a binary sequence. As for the text describing the host mesh, it will be transformed to a binary sequence using the ASCII code. To start encoding using the convolutional encoder, the watermark is generated by combining the two binary sequences already obtained.

Host Mesh Preparation. To prepare the mesh for watermarking, a wavelet transform takes place to generate the wavelet coefficient vector which is targeted by the insertion step. Each of these coefficients will be transformed to the spherical coordinate system to start embedding.

Embedding. This sub-step includes:

- Applying a modulation to the ρ component, generated by the transformation to the spherical coordinate system, to have a modulated ρ.
- Inserting a bit to the modulated ρ component using LSB method. This method substitute the least significant bit of ρ by the bit to be inserted.

- Applying demodulation to the watermarked ρ component.
  The last three point (modulation, insertion and demodulation) are repeated thrice in order to be able to insert 3 bits in each wavelet coefficient.
- Representing again the watermarked coefficients in the Cartesian coordinate system.
- Applying an inverse wavelet transform to obtain the watermarked mesh.

4.2 Extraction

This step aims to extract correctly data from the watermarked and attacked mesh after a dissemination phase. To this end, we use only the watermarked mesh. Our algorithm is said then Blind.

Thus, the watermarked and attacked mesh undergoes a wavelet transform in order to generate the watermarked wavelet coefficient vector. For each wavelet coefficient, a transformation to the spherical coordinate system takes place to obtain the watermarked ρ component. Having this component, we apply thrice the following steps:

- Modulation.
- Representation of the modulated coefficient in the form of binary sequence.
- Extraction of the Least Significant Bit which present a bit of the watermark.
- Demodulation.

After collecting all the bits that form the watermark, a decoding step using a convolutional decoder occurs to correct possible alterations of the watermark. Finally, a verification of the authenticity takes place after the decryption of the logo.

5 Results

To evaluate our crypto-watermarking algorithm, two points should be considered. The first being is encryption system evaluation. The second is the watermarking system evaluation which includes assessing the insertion rate, the invisibility and the robustness against attacks.

5.1 Encryption System Evaluation

During insertion, the watermark is composed of a logo encrypted using AES 128 algorithm and a text describing the host mesh. Figure 4 shows the original and the encrypted logo.

In order to assess the performance of AES algorithm, we calculate the entropy and the PSNR between the original and encrypted logo. The Entropy value is equal to 7.997 (close to 8) and a PSNR value is 7.590. These values prove the effectiveness of the encryption system used.

Fig. 4. Results of applying AES algorithm on the logo. (a) Original logo. (b) Encrypted logo.

Table 1. Comparison with literature of our invisibility and insertion rate results

Approaches	Insertion rate	MSQE	PSNR
[14]	765	0,007	–
[7]	10650	0.2×10^{-3}	–
[3]	21022	2.7×10^{-5}	–
[15]	199	3.2×10^{-5}	–
[16]	172974	1.2×10^{-5}	–
[11]	250000	$1,2 \times 10^{-6}$	126,35
Our approach	337929	2×10^{-7}	131.3

5.2 Watermarking Algorithm Evaluation

As already mentioned, the evaluation of our watermarking algorithm is done through the following points:

Insertion Rate and Invisibility. In this section, we present our results in term of the amount of data inserted and the influence of our algorithm on the mesh quality. Thus tests show that we are able to insert about 337929 bits and having an MSQE equal to 2×10^{-7} and a PSNR equal to 131.3 db. We can say, then, that our algorithm allows a good compromise between invisibility and insertion rate.

Comparison with recent literature has shown that our algorithm presents a real improvement in term of insertion rate and invisibility. Results presented in Table 1 confirm that recently released values, of MSQE, PSNR and insertion rate, are negligible in comparison with our results. We can conclude that our algorithm provides a significant size of the information to be inserted while keeping the mesh quality.

Robustness Against Attacks. In this section, we present results of the experimentation using the most popular attacks.

- Similarity Transformation Attacks:
  The application of rotation, translation and uniform scaling of the watermarked mesh doesn't prevent the correct extraction of the hole inserted data. Correlation is always equal to 1. Our algorithm is, then, robust against similarity transformation attacks.

Table 2. Correlation values after applying noise addition attack.

Noise level	10^{-1}	10^{-2}	10^{-3}	10^{-4}	10^{-5}	10^{-6}	10^{-7}	10^{-8}
Correlation in [12]	0.02	–	0.6	–	0.8	–	–	–
Correlation in [4]	–	–	–	0.05	0.3	–	–	–
Obtained correlation	0.12	0.2	0.38	0.57	0.9	1	1	1

Table 3. Correlation values with applying smoothing attacks.

dFactor	10^{-3}	10^{-4}	10^{-5}	10^{-6}	10^{-7}	10^{-8}	10^{-9}	10^{-10}
Correlation in [12]	–	–	–	–	–	0.18	0.31	0.43
Correlation in [15]	–	–	–	0.3	0.4	0.5	0.8	1
Obtained correlation	0.09	0.21	0.35	0.42	0.8	0.92	1	1

Table 4. Correlation values with applying coordinate quantization attacks.

Quantization level	2	3	4	5	10	12	13	14
correlation in [15]	–	–	–	–	0.7	0.85	–	–
correlation in [9]	–	–	–	–	0.14	0.628	0.954	1
Obtained correlation	0.01	0. 2	0.27	0.35	0.56	0.8	0.91	1

- Noise Addition Attack:
  Results of this attack are presented in Table 2. Results shows that our crypto-watermarking algorithm resists to the noise addition attack with a level down to 10^{-4}. These results are better then those published in [4,12].
- Smoothing Attack:
  The results of evaluating our approach against smoothing attack are summarized in Table 3. Indeed, we have good correlation values when the deformation factor is down to 10^{-6} which present an improvement compared to correlation values of [12,15].
- Coordinate Quantization:
  Table 4 shows our results in the case of applying Coordinate quantization attack. Our algorithm present better correlation values then those published in [9,15]. In fact, we have good correlation values when the quantization level is between 10 and 14.
- Simplification:
  The results of simplification attack are presented in Table 5. These results show that despite the application of simplification to the watermarked mesh, we were able to recover all inserted information. The correlation values are equal to 1 whatever the number of iterations applied. Our algorithm is robust against this type of attack.
- Compression Attack: To evaluate the robustness of our algorithm against this attack, we present Table 6. Presented results show that our watermarking

Table 5. Correlation values with applying simplification attacks.

iteration Number	1	2	3	4	5	6
[3]	–	–	–	0.79	0.68	0.61
[2]	–	0.6	0.45	0.25	0.1	0.05
[15]	1	0.95	0.92	–	–	–
Obtained correlation	1	1	1	1	1	1

Table 6. Correlation values with applying compression attack.

Bit/vertex	0.1	0.5	1	1.5	2	2.5	3
Correlation in [9]	0.07	0.34	0.4	0.6	0.89	0.9	1
Obtained correlation	0.3	0.42	0.6	0.78	0.9	1	1

algorithm is robust against the compression attack. In fact, we got correlation values close to 1 from a rate greater than 1.5 bit/vertex. These values turned into 1 when the rate reached 2.5 bit/vertex, which justifies a real improvement in the 3D watermarking area having a lack of compression results in recent works.

All results already presented allows studying the ability of our algorithm to extract correctly inserted data despite the application of attacks. Obtained correlation values Seem improved in comparison with some works recently published which prove that our algorithm ensure robustness against attacks.

6 Conclusion

In this paper, we propose a new crypto-watermarking algorithm for 3D multiresolution meshes. The particularity of this work is to combine cryptography with 3D watermarking to secure 3D multiresolution meshes. This choice is supported by the efficiency that cryptography has proven in securing images.

Our approach allows inserting a high quantity of data (logo encrypted with AES algorithm and a text describing the mesh) in multiresolution meshes. These latter undergo a wavelet transform to generate the wavelet coefficient vector. Embedding modifies these coefficients after being presented in the spherical coordinate system. Just after, an inverse wavelet transform is applied and the watermarked mesh is reconstructed.

Experimentation results prove that our blind crypto-watermarking algorithm has kept the mesh quality despite the high insertion rate. Applying the most popular attacks to the watermarked meshes did not prevent the correct extraction of inserted data. As a perspective, we will use other cryptographic tools while targeting huge meshes.

References

1. Norouzi, B., Mirzakuchaki, S., Seyezadeh, S.M., Mosavi, M.R.: A simple, sensitive and secure image encryption algorithm based on hyper-chaotic system with only one round diffusion process. Multimed. Tools Appl. **71**(3), 1469–1497 (2014)
2. Cho, D.J.: Watermarking scheme of mpeg-4 laser object for mobile device. Int. J. Secur. Appl. **9**(1), 305–312 (2015). doi:10.14257/ijsia.2015.9.1.29
3. Garg, H., Khandelwal, K.K., Gupta, M., Agrawal, S.: Uniform selection of vertices for watermark embedding in 3-d polygon mesh using ieee754 floating point representation. In: International Conference on Communication Systems and Network Technologies, pp. 788–792 (2014)
4. Wang, J.T., Chang, Y.C., Yu, C.Y., Yu, S.S.: Hamming code based watermarking scheme for 3d model verification. In: International Symposium on Computer, Consumer and Control, pp. 1095–1098 (2014)
5. Karthigaikumar, P., Rasheed, S.: Simulation of image encryption using aes algorithm. IJCA Spec. Issue Comput. Sci. - New Dimensions Perspect. **4**, 166–172 (2011)
6. Basyoni, L., Saleh, H.I., Abdelhalim, M.B.: Enhanced watermarking scheme for 3D mesh models. In: International Conference on Information Technology, pp. 612–619 (2015)
7. Zaid, A.O., Hachani, M., Puech, W.: Wavelet-based high-capacity watermarking of 3-d irregular meshes. Multimed. Tools Appl. **74**(15), 5897–5915 (2015). doi:10.1007/s11042-014-1896-3
8. Mahajan, P., Sschdeva, A.: A study of encryption algorithms AES, DES and RSA for security. Global J. Comput. Sci. Technol. Netw. Web Secur. **13**(15), 15–22 (2013)
9. Sayahi, I., Elkefi, A., Amar, C.B.: A multiresolution approach for blind watermarking of 3D meshes using spiral scanning method. In: Graña, M., López-Guede, J.M., Etxaniz, O., Herrero, Á., Quintián, H., Corchado, E. (eds.) ICEUTE/SOCO/CISIS -2016. AISC, vol. 527, pp. 526–535. Springer, Cham (2017). doi:10.1007/978-3-319-47364-2_51
10. Sayahi, I., Elkefi, A., Amar, C.B.: Blind watermarking algorithm based on spiral scanning method and error correcting codes. Int. J. Multimed. Tools Appl. **76**(15), 1–24 (2016)
11. Sayahi, I., Elkefi, A., Amar, C.B.: Blind watermarking algorithm for 3D multiresolution meshes based on spiral scanning method. Int. J. Comput. Sci. Inf. Secur. **14**(6), 331–342 (2016)
12. Sayahi, I., Elkefi, A., Koubaa, M., Amar, C.B.: Robust watermarking algorithm for 3d multiresolution meshes. In: International Conference on Computer Vision Theory and Applications, pp. 150–157 (2015)
13. Tariq, S., Ayesha, Q.: Encrypting grayscale images using s8 sboxes chosen by logistic map. Int. J. Comput. Sci. Inf. Secur. **14**(4), 440–444 (2016)
14. Zhou, X., Zhu, Q.: A DCT-based dual watermarking algorithm for three-dimensional mesh models. In: International Conference on Consumer Electronics, Communications and Networks, pp. 1509–1513 (2012)
15. Ying, Y., Pintus, R., Rushmeier, H., Ivrissimtzis, I.: A 3D steganalytic algorithm and steganalysis-resistant watermarking. IEEE Trans. Vis. Comput. Graph. **23**(2), 1002–1013 (2016)
16. Tsai, Y.Y.: An efficient 3D information hiding algorithm based on sampling concepts. Multimed. Tools Appl. **74**(34), 1–17 (2015). doi:10.1007/s11042-015-2707-1

Kinect-Based Gait Analysis for People Recognition Over Time

Elena Gianaria(✉), Marco Grangetto, and Nello Balossino

Computer Science Department, University of Turin,
Corso Svizzera 185, 10149 Turin, Italy
{elena.gianaria,marco.grangetto,nello.balossino}@unito.it

Abstract. Gait has recently attracted a great interest in the biometric field because, contrary to other classical biometric traits such as fingerprint, iris or retina, it allows to capture samples at a distance, through inexpensive and not intrusive technologies that do not need any subject's collaboration. In spite of this advantage, such technique is still not widespread for human identification task, because it is considered not to exhibit the fundamental characteristic of being invariant in the lifetime of each individual. But is this assertion really true? In this paper we investigate if gait can be considered invariant over time for an individual, at least in a time interval of few years, by comparing gait samples of several subjects three years apart. We train a Support Vector Machine with gait samples of 10 subjects, then we employ it for recognizing the same subjects with gait samples collected three years later. In addition, we try to recognize the subjects carrying three different accessories: a shoulder bag, a backpack and a smartphone.

Keywords: Gait recognition · Computer vision · Kinect · Biometrics

1 Introduction

Biometrics is the science that combines the study of human physiology with computer science for measuring and analyzing human characteristics. Each individual has unique traits, both from the physical point of view, such as the color of iris, the retina, the shape of hand and the fingerprints, and under the behavioral point of view, such as the timbre of voice, the handwriting and the walking style [1]. The biometric recognition aims to exploit these features, taken singularly or in combination, for identifying individuals. Nowadays, there are many contexts in which it is necessary to identify a person: for access control to protected systems, in order to ensure access only to authorized persons, for identification of people responsible for fraud, crimes, and so on. Unlike the access control case, in which the subject directly and actively cooperates with the system because he/she want to gain an access, in the case of people identification the collaboration of the subject cannot be guaranteed. Therefore, in situations like an un-controlled surveillance scenario it is important to exploit a human biometric

S. Battiato et al. (Eds.): ICIAP 2017, Part II, LNCS 10485, pp. 648–658, 2017.
https://doi.org/10.1007/978-3-319-68548-9_59

feature which can be captured at a distance and without the subject's collaboration: the walking style, also called gait, is a promising candidate for this task.

The study of gait for recognition purposes has recently gained interest due to it advantages compared to other biometric features [2]. First of all, gait can be capture at a distance, unobtrusively and, as said before, without the subject's cooperation. In addition, the motion capture does not require sophisticated and expensive instrumentations: a simple and inexpensive RGB or RGBD (video and depth) camera is sufficient. As a counterpart, gait is not considered as a valid system for recognizing individuals over time, because the walking style of an individual could change during lifetime, like other behavioral features [3].

In this paper we investigate if gait can really not be considered a distinctive biometric trait during the years, or rather if is possible to recognize a person comparing his gait sequences with the ones observed few years before. In order to acquire gait samples of different individuals we employ the Microsoft Kinect sensor, a device originally designed for gaming that is quickly become a valuable support for scientific research on gesture recognition, gait and motion analysis, and so on [4]. This device is a RGBD camera which provides, in addiction to RGB and depth streams, the *skeleton tracking* feature, that allows to track a human body in real-time. Such "skeleton" is a body model composed by 20 joints, each of which with coordinates in 3D space. By means of the 3D skeleton data, we have designed a recognition method related to the distance between joints and the sway of joints: the peculiarity of our method is that the recognition for sways does not depend on gait trajectories, letting the acquisition of skeleton be independent from the mutual position between Kinect sensor and shooting subject. We have collected two datasets of gait samples three years apart: we employ the old one for training our classification system, and the new one for testing. These datasets have been designed for reproducing realistic unconstrained surveillance scenarios, letting people walk in no preset path, in direction of the camera or away from it, and even carrying basic accessories: in this way we also analyze how these challengingly scenarios can influence the accuracy of gait recognition task.

The rest of this paper is organized as follows: Sect. 2 presents an overview of the related work on gait recognition systems; in Sect. 3 the proposed method for gait features extraction and classification is described, while Sect. 4 presents the datasets of gait sequences we have employed in our study. The experimental analysis is worked out in Sect. 5 and the conclusions are reported in Sect. 6.

2 Related Work

In this paper, we propose an application of gait analysis in the fields of surveillance [5] and forensic science [6], but it has also been widely applied in the health care field, for example for postural control [7], rehabilitation [8], falls detection for elderly people [9], and so on. Regardless of the specific application context, in the literature we can find two ways for collecting data on human body used for analyzing the gait [10]. In the past decades, the most common methods were the appearance-based ones, that employ the body silhouette from 2D images for reconstructing the shape of human being. This kind of methods are quite inexpensive in terms of computational cost, but they are usually not robust to variation of scale and viewpoint of the camera. A notable example of this approach is the work of Wang *et al.* [11], where the background subtraction procedure has been combined with the image segmentation in order to isolate the spatial silhouettes of a walking person. Recently, Hu *et al.* [12] have reduced the gait features space taken from different views by implementing a unitary projection method. Ortells *et al.* [13] have extracted reliable gait measures from corrupted silhouettes in gait sequences by applying a weighted averaging method.

The second kind of methods are based on the model, using some technique for reconstructing a 3D model of the body: they require more computational power than the appearance-based techniques, with the advantage of being independent on viewpoint and scale. Some examples of model-based methods are those of Bouchrika and Nixon [14] which have proposed a methodology based on elliptic Fourier descriptors for extracting body joints and build motion templates for gait description. Argyropoulos and Araujo [15] have used a channel coding approach for constructing a model of gait employed in their gait recognition system. Jung *et al.* [16] have developed a system to enable face acquisition and recognition in surveillance environments, by analyzing the 3-D gait trajectory.

Thanks to the spread of RGBD cameras, there was a significant increase of model-based approaches for gait studies. To mention a few, Ahmed *et al.* [17] have defined two types of features, i.e. joint relative distance and joint relative angle, which are robust against pose and view variations. Dikovski *et al.* [18] have investigate the role of different types of features and body parts in the gait recognition process. Andersson and Araujo [19] have extracted anthropometric attributes, gait kinematic and spatio-temporal parameters.

We have already explored the gait recognition task through Kinect sensor in other works, in particular in [20], where we discriminate between pair of subjects with similar anthropometric features, and in [21], where we face people recognition in a controlled surveillance scenario. In the present paper we collect new gait data on some subjects we had already studied in these works, and we verify if they are still recognized by the old datasets collected few years ago. To the best of our knowledge, the present paper is the first attempt of recognizing individuals by their walking style using gait data from the past.

3 Proposed Method

In our analysis we employ the Microsoft Kinect sensor as input device for capturing the gait sequences. In this section we describe how we collect, clean and manage the skeleton data provided from this sensor for performing gait recognition task. Kinect provides both RGB and depth streams and, thanks to the skeletal tracking capability, it allows to follow in real-time the body movement represented by a human skeleton map. In detail, the body model (shown in Fig. 1) is constructed in each frame of the video stream, and it is made up of 20 joins, labeled $J_0, \ldots, J_{19}$. Each joint is a 3D point in the Kinect coordinates system (x, y, z), centered on the sensor, where x, y and z represent the horizontal, vertical and depth direction respectively. Thus, we define $J_k^i = (J_{k,x}^i, J_{k,y}^i, J_{k,z}^i)$ the coordinates of the k-th joints at time i and a gait sequence as an interrupted sequence of skeleton maps, from when the user is detected by the sensor, until it comes out of the field of view. Moreover, the sensor provides an estimate of the floor plane where the user is walking, given by the equation $ax + by + cz + d = 0$, where $\mathbf{n} = (a, b, c)$ is the normal vector and d is the height of the camera center with respect to the floor.

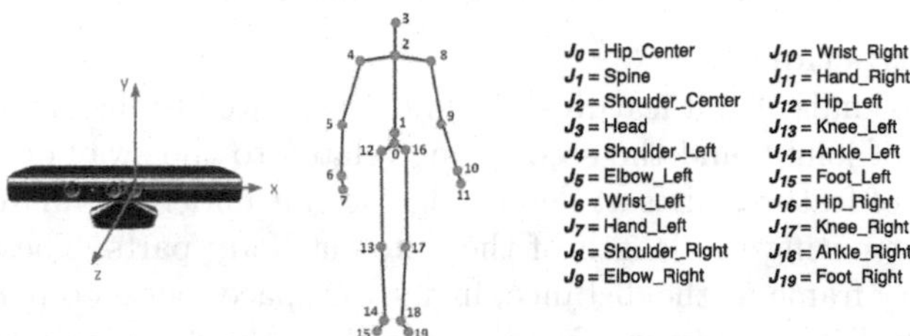

Fig. 1. The Kinect skeleton body model.

Collection of Skeleton Maps

We collect the skeleton map only if all of its joints are fully tracked: the sensor provides in automatic this kind on information. For each sample, we also take track of the direction of the walk, distinguishing if it is toward or away from the camera. Kinect does not provide any automatic mechanism for recognizing frontal and rear poses, then we have checked this condition by simply monitoring the position of the center of mass (joint J_0), and in particular its depth coordinate z: if this value decreases over time, the subject is walking toward the camera (we remind that the origin of the coordinates system is the sensor), otherwise he/she is walking in the other direction. In this way we can properly collect data on left/right arms and legs when a change of direction is detected. Obviously this mechanism fails if the subject walks backwards, but within the scope of this paper we assume that this case is not applicable. For comparing gait samples from different viewpoints and with no linear walking paths, we also apply a translation and a rotation to each joint: our objective is to obtain, for each collected walk, a sample in linear direction and totally in front of the sensor, that is a sample where the axis z coincides with the walking direction. To this end, for each frame we need to detect the angle of walking direction with respect to axis z, and to rotate all the joints of such angle. We can define the direction of walk as the tangent line to the trajectory of J_0, in the point $(J_{0,x}^i, J_{0,z}^i)$, then the i-th walking direction angle θ_i is estimated by approximating the derivative with the incremental ratio:

$$\theta_i = \tan^{-1}\left(\frac{J_{0,x}^{i+1} - J_{0,x}^i}{J_{0,z}^{i+1} - J_{0,z}^i}\right). \tag{1}$$

For changing the local coordinates system we first translate the axes origin to J_0 and then we rotate the axes according to the walking direction and the floor normal $\mathbf{n}$. The transformation matrix is defined as

$$\mathbf{T}_i = \begin{bmatrix} \cos\theta_i & 0 & \sin\theta_i \\ a & b & c \\ -\sin\theta_i & 0 & \cos\theta_i \end{bmatrix} \cdot \begin{bmatrix} 1 & 0 & 0 & -J^i_{0,x} \\ 0 & 1 & 0 & -J^i_{0,y} \\ 0 & 0 & 1 & -J^i_{0,z} \end{bmatrix} \cdot \tag{2}$$

This process is repeated for each frame, obtaining a gait sequence in a new coordinate system (X, Y, Z), where axis Y is the normal vector of the Kinect floor plane and Z coincides with the walking direction.

Features Extraction

We extract two kinds of gait features, the first one related to the distance between pair or group of joints, and the second one related to the sway of single joints. The first kind of features aims at describing the gait under the anatomical point of view, by computing the length of the different body parts. These length are computed every frame as the distance, in the 3D space, between pairs or groups of joints. Each distance features is represented by the temporal average of the corresponding collected measures, in order to obtain a single value for each kind of distance in a gait sequence. We collect 27 distance features, summarized in Tables 1 and 2.

Table 1. Classical anthropometric measures.

Description	Joints	
Torso	$J^i_3, J^i_2, J^i_1, J^i_0$	
Height	$J^i_3, J^i_2, J^i_1, J^i_0, J^i_{12}, J^i_{13}, J^i_{14}, J^i_{15}$	
	Left joints	*Right joints*
Arm	$J^i_4, J^i_5, J^i_6, J^i_7$	$J^i_8, J^i_9, J^i_{10}, J^i_{11}$
Leg	$J^i_{12}, J^i_{13}, J^i_{14}, J^i_{15}$	$J^i_{16}, J^i_{17}, J^i_{18}, J^i_{19}$

The second kind of features aims at characterize the gait style, by computing the sway of body joint, along the lateral (X axis) and vertical (Y axis) directions. For each walking sample, composed by few strides, we consider the time series $J^i_{k,X}$ and $J^i_{k,Y}$ of each joint: for each one of these two directions, we compute the temporal average and the median absolute deviation. As a result, for each one of the 19 joints (the center of mass is excluded, because it coincides with the system origin after the change of coordinates) we obtain 4 sway features (temporal average and median absolute deviation for both lateral and vertical directions), for a total of 76 sway features. Considering both distance and sway features we collect 103 gait parameters. We decided to choose these particular features because in our previous investigation [21] we have noticed that the classical gait parameters (such as stride length and walking speed) are poorly estimated by Kinect sensor, due mainly to the limited depth range, which allows

Table 2. Distances between adjacent (a) and not-adjacent (b) joints ($L = left, R = right, C = center$).

Description	Joints	
C shoulder-spine-hip	J_2^i, J_1^i, J_0^i	
L-C-R shoulder	J_4^i, J_2^i, J_8^i	
L-C-R hip	$J_{12}^i, J_0^i, J_{16}^i$	
Description	**Left joints**	**Right joints**
Shoulder-elbow	J_4^i, J_5^i	J_8^i, J_9^i
Elbow-wrist	J_5^i, J_6^i	J_9^i, J_{10}^i
Shoulder-elbow-wrist	J_4^i, J_5^i, J_6^i	J_8^i, J_9^i, J_{10}^i
Hip-knee	J_{12}^i, J_{13}^i	J_{16}^i, J_{17}^i
Knee-ankle	J_{13}^i, J_{14}^i	J_{17}^i, J_{18}^i
Hip-knee-ankle	$J_{12}^i, J_{13}^i, J_{14}^i$	$J_{16}^i, J_{17}^i, J_{18}^i$

(a)

Description	Joints
L-R elbow	J_5^i, J_9^i
L-R wrist	J_6^i, J_{10}^i
L-R hand	J_7^i, J_{11}^i
L-R knee	J_{13}^i, J_{17}^i
L-R ankle	J_{14}^i, J_{18}^i
L-R foot	J_{15}^i, J_{19}^i

(b)

to perform only 3 or 4 strides in a single gait sequence. On the contrary, features like distance between elbows and knees or head oscillation appears to be more effective for recognition task.

We use these features for building a features matrix containing the features of all the gait sequences collected for each subject: this matrix will be used for the classification task. We construct this matrix as follows. First of all, we start with a matrix where each row contains the 103 features collected in a gait sequence. Then we normalize the features by column, in order to obtain values in the range [0,1]. Finally we reduce the number of features by means of *Principal Component Analysis* (PCA): the number of components is chosen by selecting the minimum number of components having sum of their explained variances greater than a fixed threshold. The classification task is carried out through the *Support Vector Machine* (SVM), a consolidate machine learning technique for performing supervised classification. Considering the purpose of the paper, the traing phase is accomplished usign the samples in our old dataset, and the testing samples have been collected recently.

4 Datasets

In this section we introduce the datasets used in our experimentation. In [20] we have collected a dataset of 20 subjects (called *KinectUNITO'13*): each subject has been recorded while walking along a straight corridor, in front or rear to the camera. For each subject we have collected 20 gait samples, 10 for each of the two directions (approaching to or moving away from the camera). In this particular configuration each gait sequence follows approximately a straight path. For our study we also collect a brand new dataset, slightly different from the previous one. Due to the transformation matrix in Eq. 2, we can compare gait sequences with different walking paths, not only straight. For this reason, the new dataset (called *KinectUNITO'16*) was acquired in a big lecture hall, letting the subjects to follow a curvilinear path. In addition, for reproducing a more realistic uncontrolled surveillance scenario we also collect gain sequences where the subjects carry some accessory: a shoulder bag, a backpack and a smartphone, for a total of 4 different scenarios (with any accessory or with one of these accessories). For each of these 4 scenarios we collect 4 walking sequences, 2 from the frontal view and 2 from the rear one, for a total of 16 samples per subject. This dataset contains the gait sequences of 10 subjects, 8 males and 2 females aged from 30 to 50, all of them involved in the old trial. Such scenarios have been chosen for a particular reason: in fact, the Kinect sensor is set up only for recognizing people standing in front of the camera, any other configuration represents a challenge for the skeleton tracking capability. The major inaccuracies in skeleton acquisition takes place when the arms are partially occluded, as in rear poses; when something covers the arm, like a shoulder strap; or when the arm is bent for keeping something. We have added scenarios concerning all these conditions to the dataset for analyzing in detail how much serious is the performance loss with such obstacles.

The gait sequences for both datasets have been collected using the Kinect for Windows v1.

5 Experimental Results

In this section we present the experiments we have worked out for analyzing the performance of the proposed method in terms of person identification accuracy. We apply the methods described in Sect. 3 on the samples acquired from the two datasets presented in Sect. 4: in particular, for the classification task we employ C-SVM, where parameter C has been computed through 10-fold cross-validation procedure, with both linear (LIN) and radial basis functions (RBF) as kernel functions.The performance has been evaluated as portion of samples classified correctly among all samples. We use the samples in *KinectUNITO'13* for training the SVM, while the samples in *KinectUNITO'16* are used in the testing phase, accordingly to the paper purpose, that is to investigate if gait can be a distinctive biometric trait during the years: to do so we try to recognize an individual by comparing his current gait sequences with the ones observed few years before. As reported in Sect. 3, the Kinect sensor has some difficulty in tracking the skeleton in rear samples. For this reason, in a first attempt we employ only the front samples of the two datasets. Later we will also add the rear samples, and we will compare the performances for analyzing how much the rear poses may have an impact on the recognition task.

We start by describing the first classification experiment. Considering only front samples, we train the SVM with 10 samples per subject. For the testing phase we have four different sub-cases, that we label as follow: walking (a) without any objects, (b) with a bag, (c) with a backpack, (d) with a smartphone. For each subject we have two front samples per case, for a total of 8 gait sequences.

As described in Sect. 3, we apply the PCA in order to reduce the number of features: in this case we fix this threshold to 90%, with a reduction of features from 103 to 20. As kernel function for SVM, we select the linear one, because we observe better performances for this dataset. The results of the classification task are shown in Fig. 2, where the x-axis represents the subjects and the y-axis the recognition accuracy in terms of correctly classified samples: each color of the block represent the number of correctly classified samples in each sub-category. We remind that each sub-category contains 2 samples, for a total of 8 samples per subject. We can notice that 7 subject over 10 have been recognized in more the 50% of cases: three of them in more the 75% of cases and one subject is correctly classified in all 8 cases. The worst results concern the subjects 2 and 7, which are recognized only in one case. As expected, the classification task is more effective if the individual carries anything: in this case, all the samples of 7 subjects have been correctly classified, and just 1 subject has been never classified in the correct way. Also the backpack is not a great obstacle: 6 subjects have been always well classified, and just 1 subject has been always misclassified. The bag decreases more the accuracy (4 subjects always classified correctly, but 5 subjects always misclassified), but the worst performance is obtained when an individual is walking when talking on the phone: in this scenario only 3 subjects have been classified correctly, while all the other ones have been always misclassified.

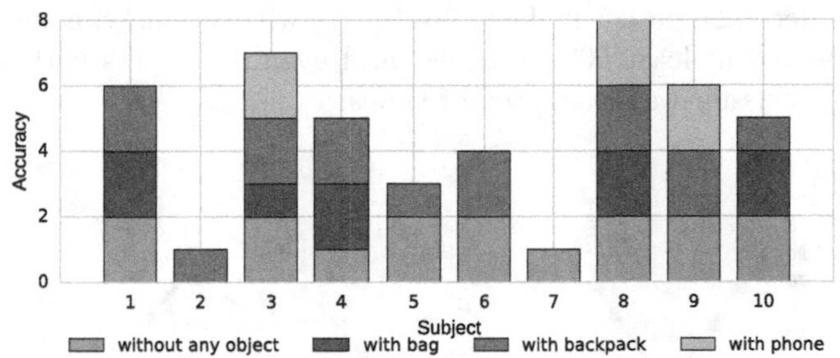

Fig. 2. Classification accuracy considering only front samples.

In the second classification experiment we also consider the rear samples of the two datasets: the training set is now composed by 20 samples per subject, and the testing set contains 16 samples per subject, 4 for each sub-case listed above. In this experiment we fix threshold for PCA to 80%, because we have noticed that higher values produce over-fitting of data: the features reduce from 103 to 14. As kernel function we select the RBF, because we have observed that in this case it produces an accuracy higher than the linear kernel. The accuracy of this second classification task is shown in Fig. 3. As in the first experiment, we notice that 7 subject over 10 have been correctly classified in more the 50% of cases. We can also observe a slight improvement on classification for subject 2, and a marked improvement for subject 7. On the other hand, for some subject

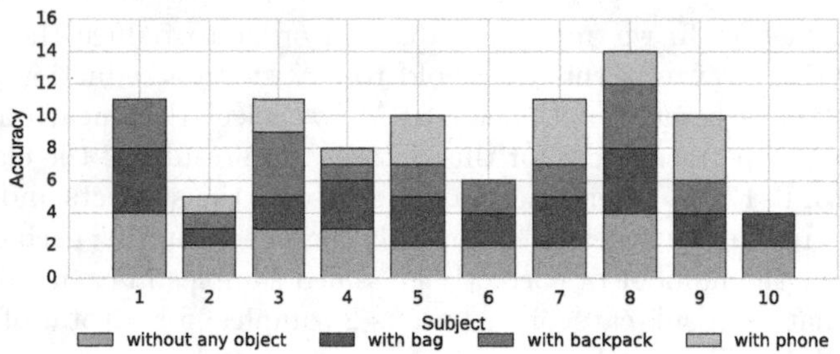

Fig. 3. Classification accuracy considering both front and rear samples.

(e.g. the number 10) the recognition rate decreases when we consider also the rear samples: such worsening is due to the limits of Kinect sensor, that is less efficient in acquiring rear poses, with a negative impact on the training phase. If we analyze each sub-case, we can notice that the classification accuracy for an individual carrying anything is still high, at least 50% for each subject. For the sub-case of backpack, the accuracy is become slightly worse respect to the previous experiment, with 3 subjects classified correctly in less than 50% of cases: during the skeleton tracking we have noticed that the straps of backpack sometimes occlude an arm. On the other hand, the accuracy of classification if an individual is carrying a bag or is making a phone call improves when we also consider rear poses: in fact 5 subjects with a smartphone have been correctly classified at least 50% of times, and even 9 subjects with a bag have been correctly classified at least 50% of times.

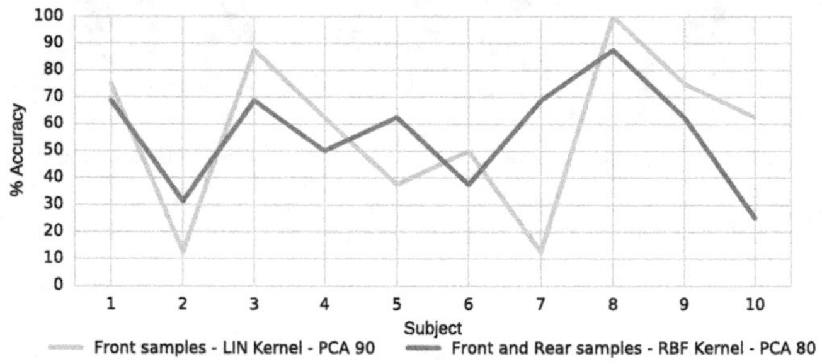

Fig. 4. Comparison between classification with only front samples (pink line) and with both front and rear samples (light blue line). (Color figure online)

Finally, in order to understand if the rear poses acquired from Kinect sensor can be considered for gait analysis, in Fig. 4 we compare the accuracy rate of the two experiments, without considering the distinction in sub-cases: in such figure each subject is reported in the x-axis, while the y-axis shows the percentage of samples correctly classified in the two experiments (the pink line for the first one, and the light blue line for the second one). In 7 subjects over 10 the use of only front samples gives a better accuracy, then apparently the rear poses decrease the performances of classification task. But if we look the graphs more carefully, we notice that the line of front poses presents more peaks, both negative and positive, respect to the line with also rear poses, that results more stable, with only a negative peak of accuracy lower than 30%. Obviously the accuracy in second case should be lightly due to the employment of more samples for training the SVM, but we have to consider that the rear samples are less accurate, because of the limits of the Kinect sensor in acquiring the figure in rear pose.

6 Conclusion

In this paper we have proposed a gait recognition system that collects a rich set of static and dynamic gait features, by computing the distance between joints and the sway of joints respectively, from a set of gait samples acquired through the Kinect sensor: the distinctive trait of our system is that it makes the gait samples invariant with respect to the acquisition viewpoint. The recognition systems is based on the SVM supervised classification, preceded by a phase of feature space reduction through PCA. The experimental analysis has been performed using two datasets acquired three years away from each other: both datasets include samples in front and rear poses, but the second one also includes gait samples in which people carry some object. The objective of study is double: we want to understand if gait can be considered an invariant biometric trait over years in our lifetime, and we also want to analyze how the use of unusual gait samples, i.e. rear poses or people carrying objects, can modify the accuracy of gait recognition. Results show that gait allows to recognize a person even after years, or at least after few years, making it compliant to forensic and security applications: just think of those situations where the perpetrator of crime has been observed through surveillance cameras years before trial. Moreover we have also observed that the presence of accessories makes the recognition task worse in case of only frontal acquisitions, but the addition of rear poses during the learning task can make the recognition task more stable, at the cost of lowering a little its accuracy. Future works will be devoted to the extension of experiments to a larger dataset of subjects, and the exploitation of the proposed gait recognition approach in security and forensic applications, as a support to investigation.

References

1. Ashbourn, J.: Biometrics - Advanced Identity Verification: The Complete Guide. Springer, Heidelberg (2002)
2. Boulgouris, N.V., Hatzinakos, D., Plataniotis, K.N.: Gait recognition: a challenging signal processing technology for biometric identification. IEEE Signal Process. Mag. **22**(6), 78–90 (2005)
3. Jain, A.K., Ross, A., Prabhakar, S.: An introduction to biometric recognition. IEEE Trans. circ. Syst. Video Technol. **14**(1), 4–20 (2004)

4. Zhang, Z.: Microsoft kinect sensor and its effect. IEEE MultiMedi. **19**(2), 4–10 (2012)
5. Ran, Y., Zheng, Q., Chellappa, R., Strat, T.M.: Applications of a simple characterization of human gait in surveillance. IEEE Trans. Syst. Man. Cybern. Part B (Cybernetics) **40**(4), 1009–1020 (2010)
6. Bouchrika, I., Goffredo, M., Carter, J., Nixon, M.: On using gait in forensic biometrics. J. Forensic Sci. **56**(4), 882–889 (2011)
7. Massion, J.: Postural control system. Curr. Opin. Neurobiol. **4**(6), 877–887 (1994)
8. Zhou, H., Hu, H.: Human motion tracking for rehabilitationa survey. Biomed. Signal Process. Control **3**(1), 1–18 (2008)
9. Maki, B.E.: Gait changes in older adults: predictors of falls or indicators of fear? J. Am. Geriatr. Soc. **45**(3), 313–320 (1997)
10. Wang, J., She, M., Nahavandi, S., Kouzani, A.: A review of vision-based gait recognition methods for human identification. In: Digital Image Computing: Techniques and Applications (DICTA), pp. 320–327. 2010 International Conference on IEEE (2010)
11. Wang, L., Tan, T., Ning, H., Hu, W.: Silhouette analysis-based gait recognition for human identification. Pattern Anal. Mach. Intell. IEEE Transactions **25**(12), 1505–1518 (2003)
12. Hu, M., Wang, Y., Zhang, Z., Little, J.J., Huang, D.: View-invariant discriminative projection for multi-view gait-based human identification. IEEE Trans. Inf. Forensics Secur. **8**(12), 2034–2045 (2013)
13. Ortells, J., Mollineda, R.A., Mederos, B., Martín-Félez, R.: Gait recognition from corrupted silhouettes: a robust statistical approach. Mach. Vis. Appl. 1–19 (2016)
14. Bouchrika, I., Nixon, M.S.: Model-based feature extraction for gait analysis and recognition. In: Gagalowicz, A., Philips, W. (eds.) MIRAGE 2007. LNCS, vol. 4418, pp. 150–160. Springer, Heidelberg (2007). doi:10.1007/978-3-540-71457-6_14
15. Argyropoulos, S., Tzovaras, D., Ioannidis, D., Strintzis, M.G.: A channel coding approach for human authentication from gait sequences. IEEE Trans. Inf. Forensics Secur. **4**(3), 428–440 (2009)
16. Jung, S.-U., Nixon, M.S.: On using gait to enhance frontal face extraction. IEEE Trans. Inf. Forensics Secur. **7**(6), 1802–1811 (2012)
17. Ahmed, F., Paul, P.P., Gavrilova, M.L.: Dtw-based kernel and rank-level fusion for 3d gait recognition using kinect. Vis. Comput. 1–10 (2015)
18. Dikovski, B., Madjarov, G., Gjorgjevikj, D.: Evaluation of different feature sets for gait recognition using skeletal data from kinect. In: Information and Communication Technology, Electronics and Microelectronics (MIPRO) 2014 37th International Convention on, pp. 1304–1308. IEEE (2014)
19. Andersson, V., Araujo, R.: Person identification using anthropometric and gait data from kinect sensor. In: Proceedings of the Twenty-Ninth Association for the Advancement of Artificial Intelligence Conference, AAAI (2015)
20. Gianaria, E., Balossino, N., Grangetto, M., Lucenteforte, M.: Gait characterization using dynamic skeleton acquisition. In: Multimedia Signal Processing (MMSP) 2013 IEEE 15th International Workshop on, pp. 440–445 September (2013)
21. Gianaria, E., Grangetto, M., Lucenteforte, M., Balossino, N.: Human classification using gait features. In: Biometric Authentication, pp. 16–27 Springer (2014)

Imaging for Cultural Heritage and Archaeology

ARCA (Automatic Recognition of Color for Archaeology): A Desktop Application for Munsell Estimation

Filippo L.M. Milotta[1,2(✉)], Filippo Stanco[1], and Davide Tanasi[2]

[1] Image Processing Laboratory (IPLab),
Department of Mathematics and Computer Science,
University of Catania, Catania, Italy
{milotta,fstanco}@dmi.unict.it

[2] Center for Virtualization and Applied Spatial Technologies (CVAST),
Department of History, University of South Florida, Tampa, FL, USA
milotta@mail.usf.edu, dtanasi@usf.edu

Abstract. Archaeologists are used to employing the Munsell Soil Charts on cultural heritage sites to identify colors of soils and retrieved artifacts. The standard practice of Munsell estimation exploits the Soil Charts by visual means. This procedure is error prone, time consuming and very subjective. To obtain an accurate estimation the process should be repeated multiple times and possibly by other users, since colors might not be perceived uniformly by different people. Hence, a method for objective and automatic Munsell estimation would be a valuable asset to the field of archaeology. In this work we present ARCA: Automatic Recognition of Color for Archaeology, a desktop application for Munsell estimation. The following pipeline for Munsell estimation aimed towards archaeologists has been proposed: image acquisition of specimens, manual sampling of the image in the ARCA desktop application, automatic Munsell estimation of the sampled points and creation of a sampling report. A dataset, called ARCA108, consisting of 22,848 samples has been gathered, in an unconstrained environment, and evaluated with respect to the Munsell Soil Charts. Experimental results are reported to define the best configuration that should be used in the acquisition phase. Color tolerance values of the proposed framework are also reported.

Keywords: Color standardization · Munsell · Color space conversion · Digital archaelogy · Color specification

1 Introduction

At the beginning of the 20th century, Munsell [1] established a system for specifying colors more precisely and showing the relationships among them. The Munsell color order system is based on the color-perception attributes of hue, value and chroma. Munsell defined numerical scales with visually uniform steps

S. Battiato et al. (Eds.): ICIAP 2017, Part II, LNCS 10485, pp. 661–671, 2017.
https://doi.org/10.1007/978-3-319-68548-9_60

for each of these attributes. Hue is that attribute of a color by which we distinguish blue from red, yellow from green, and so on. Hues are naturally ordered in this scale: red (R), yellow-red (YR), yellow (Y), green-yellow (GY), green (G), blue-green (BG), blue (B), purple-blue (PB), purple and red-purple (RP). Black, white and the grays between them are called neutral colors (N). Value indicates the lightness of a color in a scale of value ranges from 0 (pure black) to 10 (pure white). Chroma is the degree of departure of a color from the neutral color of the same value. The scale starts from 0, for neutral colors, but there is no arbitrary end to the scale, as new pigments gradually become available. However, limits for representable chroma values have been defined by the so called MacAdam limits [2]. Specifying color by the Munsell system is a practice limited to opaque objects, such as soils or painted surfaces. This practice provides a simple visual method as an alternative to the more complex and precise method based on the CIE system and on spectrophotometry. For this reason, the Munsell system is adopted in contexts in which the recording or identification of colors of specimens (i.e., flowers, minerals, soils) is required [3]. The Munsell charts are appropriate for almost all jobs requiring color specification by visual means, as stated by specific neurobiological researches that demonstrated how that system has successfully standardized color in order to match the reflectance spectra of Munsells color chips with the sensitivity of the cells in the lateral geniculate nucleus (LGN cells), responsible for color specification [4]. In archaeology Munsell charts are widely used as the standard for color specification of organic materials, colored glass, soil profiles, rock materials, textiles, metals, colored glasses, paintings and principally pottery. Archaeologists are used to employing Munsell Soil Charts directly on a cultural heritage or excavation site to identify the colors of the soils and of the artifacts retrieved. Indeed, it is very useful in the examination, classification and genesis analysis of soils [5–7]. For which regards the interpretation of pottery the precise color specification of such parts like treated surfaces, clay body, core, and outer layers like painting and slip, it is fundamental for defining its stylistic and technical features. Color specification might be exploited to bind the artifacts to a specific culture, society or civilization or even to a certain period of time [8].

As previously mentioned, the standard practice of Munsell estimation exploiting the Soil Charts is by visual means. The two adjacent constant-hue charts or chips between which the hue of the specimen lies have to be chosen. Then, by moving the masks from chip to chip to find the most similar one to the specimen, one can estimate its value, chroma and hue [3]. As can be seen, this procedure is error prone, time consuming and very subjective. In order to obtain a more accurate estimation, the process described above should be repeated more than once and possibly also by other users, since colors might not be perceived uniformly by different people [9]. Hence, an objective and automatic Munsell estimation method would be a valuable improvement to the field of archaeology.

Digital cameras have been used before to acquire pictures of soil specimens in a laboratory with controlled lighting conditions. Then, the Munsell notation has been exploited to estimate the mineral and organic composition of

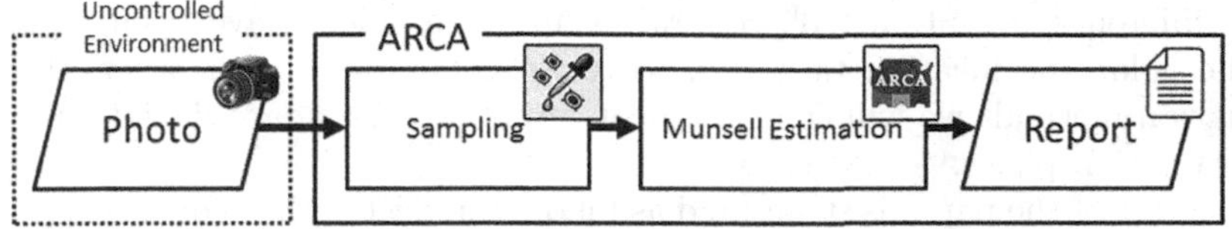

Fig. 1. Pipeline of the proposed ARCA application.

the specimens [10–15]. However, all of these works still require a strictly controlled environment for the digital acquisition of the images. Suggested controls for a perfect estimation are related to artificial and natural lightning conditions, specimen and camera positions, angle of view, setting of the working plane and background with proper opaque and black materials to avoid light reflection [3]. Prepare a perfectly controlled environment is difficult, time-consuming and potentially expensive. With the spread of smartphones with ever more sensors onboard, particularly high resolution cameras, new methods exploiting the Munsell system have been developed. In [16] a mobile phone application for Munsell estimation under strictly controlled illumination conditions is presented. In [17] a similar setting is discussed, but focused on a Complementary Metal Oxide Semiconductor (CMOS) sensor assembled on a smartphone. Also in these cases, a controlled environment is required. In the past we have performed several experiments about this topic in a such environment [12,14,15,18,19]. However, to the best of our knowledge, a method using an uncontrolled setting for image acquisition is still missing. In this work, we present ARCA: Automatic Recognition of Color for Archaeology, a desktop application for Munsell estimation. ARCA is the core of the pipeline of our proposed method, consisting in the image acquisition of specimens, manual sampling of the image in a user-friendly way for archaeologists, Munsell estimation of the sampled points and creation of a sampling report (Fig. 1). We focused on the need of archaeologists to have a practical and tested application that might help them in the color specification task during an excavation. Through the proposed pipeline, archaeologists do not need expensive tools (i.e., spectophotometer, Munsell Soil Charts, color checker) or a laboratory with a controlled environment for the acquisition in order to perform color estimation. They just need to take a picture of the specimen, and moreover, no strict constraints need to be applied in advance. Then, from the ARCA application, they are able to select multiple samples at once and the system will estimate the Munsell notation for them in an objective and deterministic way.

A dataset of 108 images, called for this reason ARCA108, consisting of a total of 22,848 samples, have been gathered in order to evaluate in an uncontrolled environment what the best configuration in which the image acquisition should be done. This dataset represents a new valuable asset for color specification research purposes. The Munsell system is usually exploited to establish and evaluate the color and gloss tolerance of specimens [20,21]. We compared all the samples with Munsell reference values exploiting the CIEDE2000 (ΔE_{00})

color difference definition [20]. Several accuracy problem have been reported for the color specification task [22], so to be comparable with other Munsell estimation methods we will consider mean values and standard deviations from the evaluation phase.

The rest of the paper is structured as follows: in Sect. 2 the acquisition phase, validation phase and ARCA desktop application will be described. The experimental results are given in Sect. 3 and then final remarks and considerations conclude the paper in Sect. 4.

2 Material and Methods

Two main phases can be distinguished in our experiments: acquisition and validation. In the former, we wanted to simulate the most common situation of Munsell field-estimation as possible, while in the latter the main aim was to validate the proposed system, in order to prove its reliability in the Munsell estimation process. In the following subsections, the acquisition phase, ARCA desktop application and validation phase are detailed. The order in which they are presented is coherent with the proposed pipeline: acquire, sample and estimate.

2.1 Acquisition Phase

No strict constraints have been added in the acquisition phase, in order to allow an easy replicability of the process shown in this work. Two kinds of devices have been employed in our experiments: a professional reflex and a common smartphone. The reflex model was a Canon EOS 1200D (mounting an EFS 18–55 mm zoom lens model) with a resolution of 18 megapixels, while the smartphone model was a Nexus 5X with a main camera resolution of 12.2 megapixels. The subjects of the taken pictures were the following Munsell Soil Color Charts (Year 2000 Revised Washable Edition): *GLEY1*, *GLEY2*, *10R*, *2.5YR*, *5YR*, *7.5YR*, *10YR*, *2.5Y*, *5Y*. A Gretag-Macbeth color checker has been also employed, in order to evaluate the gains of have reference colors during photos acquisition.

Our acquisition was set in Tampa, Florida (US), in GPS coords 28°03'47.9"N 82° 24'40.9"W, on March 8, that was an almost sunny day, with some cloud cover (Fig. 2(a)). It was performed from 10:30 am to 12:30 pm and with an unguided approach (Fig. 2(b)), so without any fixed positions or angles of view for the camera or subjects. We acquired the 9 charts of the Munsell Soil Color Charts, with the following possible settings:

- 2 kinds of devices: professional DSLR (Digital Single Reflex Camera) and common smartphone;
- 3 automatic white balancing algorithms (executed by the devices in the image capture phase): automatic, sunny (corresponding to standard illuminant D65: $\sim 6,500 K^{\circ}$) and cloudy (corresponding to standard illuminant D75: $\sim 7,500 K^{\circ}$);
- 1 fluorescence presetting: direct sunlight;

- 1 ISO setting: 400 ISO;
- 1 focus setting: autofocus;
- 2 kind of subject: the chart itself and the chart with a Gretag-Macbeth color checker nearby.

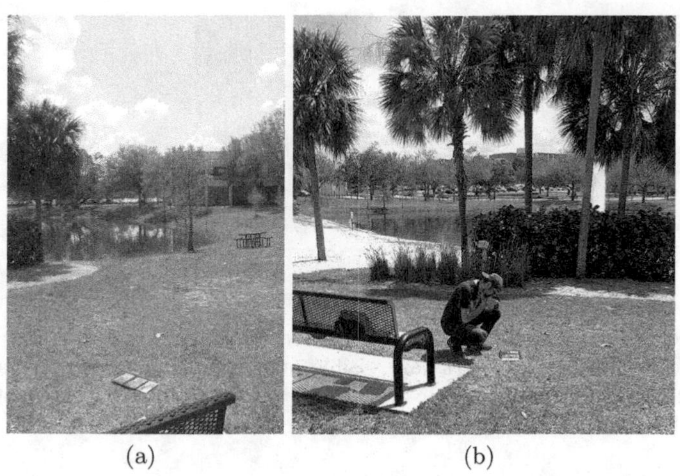

(a) (b)

Fig. 2. (a) The day in which acquisition phase has been performed was a sunny day with minor cloud cover. (b) Photos have been taken with an unguided approach.

In this way, we obtained a total of 12 configurations for each Munsell chart, gaining a total of 108 images. The resolution of the images is 5184 × 3456 pixels and 3840 × 2160 pixels for pictures taken by a DSLR camera and smartphone, respectively. All the images were saved in the standard JPG format, with a lossless setting for the quality (the highest possible).

The gathered dataset has been publicly released with the name ARCA108 and it is freely available at http://iplab.dmi.unict.it/ARCA108/.

2.2 ARCA Desktop Application

The current version of the ARCA desktop application has been developed in Matlab Graphical User Interface Design Environment (GUIDE). From the GUI the user is able to perform several actions: open an image, zoom in/out to focus on a detail of an image, sample the image (through pick-point or draw-a-region-by-freehand), remove the current sample, estimate the Munsell notation of the sample and save the report. The ARCA application GUI is shown in Fig. 3. Multiple samples can be selected through the pick-point tool; their Munsell notation estimation will be done at once when launched. After every estimation, indexed markers are added on the image, so the user can track all the samples with their own Munsell estimation. Samples are also highlighted with a red border (this is particularly useful when draw-a-region tool is used). Munsell conversion and ΔE_{00} computations for a validation phase are performed exploiting the publicly

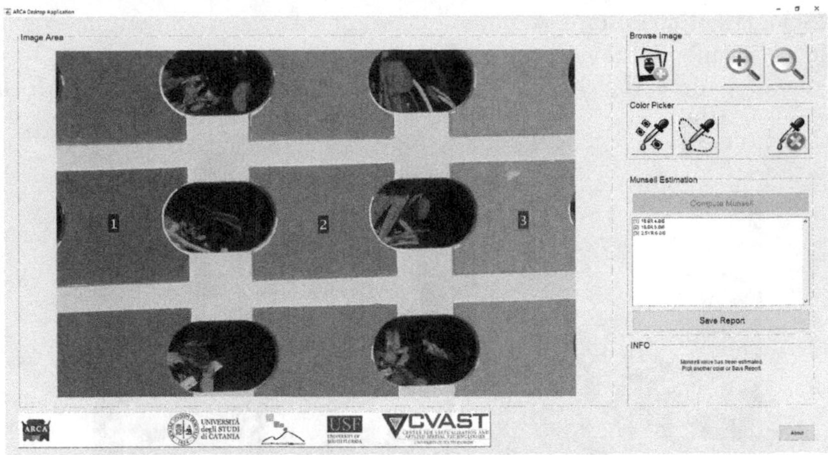

Fig. 3. Screenshot of the ARCA desktop application GUI. Three samples have been taken on the current image; marks are visible on the image so the user can visually track the estimated Munsell values. Now the user can keep sampling the image (adding new Munsell estimations) or save a report of the current estimation.

available Matlab toolbox by Centore [23,24], that has been proved to be comparable with other not open-source conversion methods [25–28]. Finally, when the report of the estimation is going to be created, the user must provide a name for the report and a directory will be created with that name. The report is made up of three elements: the starting image with the indexed markers on it, a Matlab file and a textual report containing the list of Munsell estimations.

2.3 Validation Phase

We evaluated the system comparing the expected Munsell value of each chip in the Munsell charts with its observed one. We performed the sampling from the charts importing the images in our ARCA desktop application and manually picking points which were visually near to the centroid of each chip. We considered a patch of 49×49 pixels around the picked centroid, for a total of 2,401 pixels per chip. As done in [16], the Munsell charts labeled as *GLEY1* and *GLEY2* have not been evaluated, since they contain neutral colors very similar to one another's and with very low chroma values. We sampled 238 chips (for each one of the 12 configurations), and for each sampled chip we computed mean, median and mode of the extracted patch. So, by also taking into account the RGB value in the centroid, we obtained 4 RGB values for each sampled chip. Using the Munsell toolbox by Centore [24] the sampled RGB values have been converted to the Munsell color space. We have also considered a discretized version of the converted RGB values, computed by rounding the converted values to the closest Munsell reference values in the Munsell charts. In this way, we obtained a total of 22,848 Munsell observed values to be compared with the 238 expected ones.

3 Results

In the experimental setting, 12 possible configurations were defined (Sect. 2.1). We repeat that a "configuration" is one of the possible combination of the following settings: Device:[Reflex/Smartphone] + WhiteBalancing:[Auto/Sunny/Cloudy] + Subject:[Solo_Chart/With_Macbeth]. Moreover, for each sampled chip, 4 order statistics were investigated: mean, median, mode and centroid value have been exploited in the Munsell computation (Sect. 2.3). Since Munsell references are a discrete set of values, it is also possible to apply a discretization to the continuous Munsell values obtained after the conversion, so the order statistics to be taken into account become 8. Hence, several questions can be raised, and will be answered in the following subsections:

1. What is the best configuration, among the 12 defined?
2. What is the best order statistic, among the 8 defined?
3. How much is worthwhile the application of the discretization?
4. Is the error in the Munsell notation estimation *acceptable*?

3.1 Best Configuration

For each one of the 12 possible configurations, 7 Munsell charts were acquired. The average value of the ΔE_{00} between the Munsell reference chips and the 8 order statistics from every chip in the acquired charts has been computed. Results are shown in Fig. 4(a). From this chart it is possible to assess that the best configuration is [Reflex, Auto White Balancing, Solo Chart]. Instead, among the configurations that exploit the smartphone as device, the best configuration is [Smartphone, Sunny White Balancing, With Macbeth]. It is interesting, and almost surprising, to notice how the use of a color checker together with a reflex professional camera increases the ΔE_{00} distance, while together with a general purpose smartphone it has a positive influence decreasing the distance. Hence, in our best configuration none expensive color checker is needed.

3.2 Best Order Statistic

The average value of the ΔE_{00} between the Munsell reference chips and the 8 order statistics from every chip in the whole dataset has been computed. Results are shown in Fig. 4(b). The values of the order statistics, respectively with and without quantization, is almost similar, besides for 2.5Y and 5Y Munsell charts where it is almost the same in both the cases. The mean slightly outperforms the other order statistics. Additional evidence coming from this chart is that quantization decrease the ΔE_{00} distance in almost the totality of the cases. This state directly brings to the successive question.

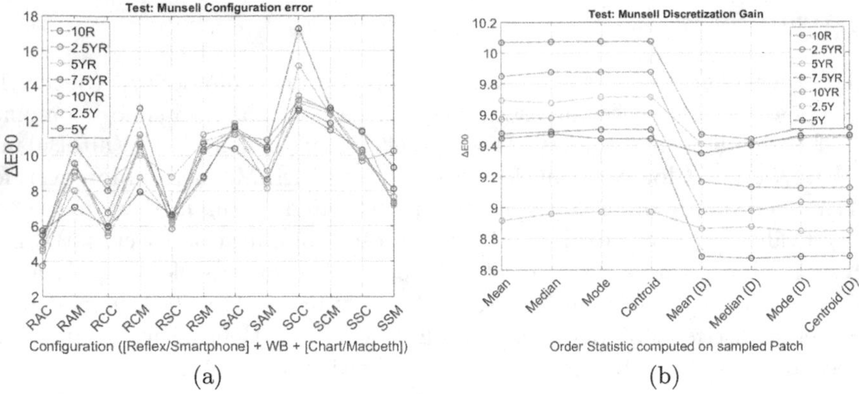

Fig. 4. Validation plots. (a) Investigation of the best configuration among the 12 tested. (b) Investigation of the best order statistic to be used on the patch during the Munsell estimation. Note how the discretization decreases the ΔE_{00} in almost the totality of the cases, as expected.

3.3 Discretization

Munsell Soil Charts contain a discrete set of reference values, but conversion from RGB to Munsell System generates values in a continuous system. Archaeologists are used to employ only the discrete values, not the continuous ones, so a discretization is needed. Since we consider all the values near to a reference one as the same, the discretization to the closest Munsell reference value will matter in the ΔE_{00} computation. We counted how many times discretization actually decreased or increased the initial ΔE_{00}. In the 59.42% of the cases a positive gain has been obtained. Moreover, the negative gain is usually obtained with low chroma values, which are the most ambiguous to be classified. From the result shown in Fig. 4(b) and this other cue it is possible to assess that it is worthwhile to apply discretization, as expected.

3.4 Color Tolerance

The issue related to the amount of acceptable error on a Munsell estimation is known as color tolerance. The tolerance ranges change with respect to the application for which the estimation is made for. The standard definition is that same colors should have $\Delta E_{00} = 1$ [20]. In the industrial field two colors can be considered the same (imperceptible differences) only if the ΔE_{00} is lesser than 2. However, this strong criteria are usually relaxed introducing "tolerable" ranges: until 3–4 CIELAB units can be considered the same colors, until 5–6 CIELAB units the colors are hardly distinguishable, higher than 6 CIELAB units classification performance starts to decrease [16]. Moreover, the colors printed in the Munsell reference Soil Charts are usually affected by an intrinsic error from ~1 to ~4 CIELAB units, where higher error is found in elder Charts [22].

Related works employing smartphones during acquisition phase in a controlled environment have reported an error in the estimation of 3.75 ± 1.8 CIELAB units [16,17]. In Table 1 the mean and standard deviation values of ΔE_{00} computed during the validation phase have been reported. As previewed by Fig. 4(a), the best configuration is [Reflex, Auto White Balancing, Solo Chart], which has 4.95 ± 2.89 CIELAB units of error. Performances drastically drop with other configurations. The best configuration for smartphone, that is [Smartphone, Sunny White Balancing, With Macbeth], has 8.20 ± 2.71 CIELAB units of error. To summarize, taking into account all the previous considerations about intrinsic error of Munsell Charts and the unconstrained experimental setting, the error obtained with our best configuration (employing the reflex) seems reasonable.

Table 1. Mean and standard deviation of Munsell estimation for each one of the 12 defined configurations.

	Configurations ([Reflex/Smartphone] + WB + [Chart/Macbeth])											
	RAC	RAM	RCC	RCM	RSC	RSM	SAC	SAM	SCC	SCM	SSC	SSM
Mean	*4.950*	8.992	6.626	10.338	6.691	10.169	11.381	9.436	13.786	12.264	10.352	8.198
St. D.	2.887	2.667	2.807	3.191	*2.562*	2.914	2.888	3.266	3.220	2.756	2.998	2.719

4 Conclusions

In this work, ARCA: Automatic Recognition of Color for Archaeology, a desktop application for Munsell estimation, has been presented. We focused on the need of archaeologists to have a practical and tested application that might help them in the color specification task during an excavation. The following pipeline for Munsell notation estimation, aimed at archaeologists, has been proposed: image acquisition of specimens, manual sampling of the image in the ARCA desktop application, automatic Munsell estimation of the sampled points and creation of a sampling report. Differently from our previous works [12,14,15], we performed the whole experiments in an uncontrolled environment. A dataset of 22,848 samples has been gathered under the uncontrolled environment assumption and evaluated with respect to the Munsell reference Soil Charts. This dataset has been called ARCA108 and it represents a new valuable asset for color specification research purposes. We defined 8 possible order statics for characterize the samples and 12 possible configurations during the acquisition phase . Experimental results shown that the defined order statistics reach very similar results, and that discretization of the converted Munsell notation decreases the error of ∼1 CIELAB unit. The best configuration among the tested ones is [Reflex, Auto White Balancing, Solo Chart], with 4.95 ± 2.89 CIELAB units of error. Compared to other related works, taking into account intrinsic error of Munsell reference Soil Charts and the uncontrolled experimental setting, this result is encouraging and reasonable. We proved that ARCA can represent for archaeologists a valid tool for color specification. ARCA allows archaeologists to select multiple samples and estimate the corresponding Munsell notation at once, in a fast, objective and deterministic way, avoiding the error-prone and time-consuming procedure of Munsell Estimation by visual means and without any expensive tool like spectophotometer, Munsell Soil Charts or Gretag-Macbeth color checker. For future works, we are planning to improve the ARCA application (i.e., image processing algorithms for noise reduction, deployment of a mobile version), to expand the validation phase acquiring other Munsell Soil Charts from Tropical Soils edition and, most of all, to conduct a color specification test-case on archaeological soils and pottery.

Acknowledgments. The method described in this work was filed with the United States Patent and Trademark Office on April 19, 2017, as "Automatic Digital Method for Classification of Colors in Munsell Color System", and assigned Serial No. 62/487,178.

References

1. Munsell, A.H.: Atlas of the Munsell Color System. Howland and Company, Wadsworth (1915)
2. MacAdam, D.L.: The theory of the maximum visual efficiency of colored materials. JOSA **25**(8), 249–252 (1935)
3. ASTM. Standard practice for specifying color by the Munsell system. ASTM International D 1535-14 (2014)
4. Conway, B.R., Livingstone, M.S.: A different point of hue. Proc. Nat. Acad. Sci. USA **102**(31), 10761–10762 (2005)
5. Soil Survey Staff USA. Soil Taxonomy: A basic system of soil classification for making and interpreting soil surveys. US Government Printing Office (1975)
6. Michéli, E., Schad, P., Spaargaren, O.: World Reference Base for Soil Resources 2006: A Framework for International Classification, Correlation and Communication. Food and agriculture organization of the United nations (FAO), Rome (2006)
7. Sánchez-Marañón, M., Soriano, M., Melgosa, M., Delgado, G., Delgado, R.: Quantifying the effects of aggregation, particle size and components on the colour of mediterranean soils. Eur. J. Soil Sci. **55**(3), 551–565 (2004)
8. Gerharz, R.R., Lantermann, R., Spennemann, D.R.: Munsell color charts: a necessity for archaeologists. Aust. J. Hist. Archaeol. **6**, 88–95 (1988)
9. Goodwin, C.: Practices of color classification. Mind Cult. Act. **7**(1–2), 19–36 (2000)
10. Aydemir, S., Keskin, S., Drees, L.R.: Quantification of soil features using digital image processing (DIP) techniques. Geoderma **119**(1), 1–8 (2004)
11. Viscarra Rossel, R.A., Fouad, Y., Walter, C.: Using a digital camera to measure soil organic carbon and iron contents. Biosyst. Eng. **100**(2), 149–159 (2008)
12. Stanco, F., Tanasi, D., Bruna, A., Maugeri, V.: Automatic color detection of archaeological pottery with munsell system. In: Maino, G., Foresti, G.L. (eds.) ICIAP 2011. LNCS, vol. 6978, pp. 337–346. Springer, Heidelberg (2011). doi:10.1007/978-3-642-24085-0_35
13. O'Donnell, T.K., Goyne, K.W., Miles, R.J., Baffaut, C., Anderson, S.H., Sudduth, K.A.: Determination of representative elementary areas for soil redoximorphic features identified by digital image processing. Geoderma **161**(3), 138–146 (2011)

14. Stanco, F., Tanasi, D., Gueli, A.M., Stella, G.: Computer graphics solutions for dealing with colors in archaeology. In: Conference on Colour in Graphics, Imaging, and Vision, vol. 2012, pp. 97–101. Society for Imaging Science and Technology (2012)
15. Stanco, F., Gueli, A.M.: Computer graphics solutions for pottery colors specification. In: IS and T/SPIE Electronic Imaging, pp. 86600S–86600S. International Society for Optics and Photonics (2013)
16. Gómez-Robledo, L., López-Ruiz, N., Melgosa, M., Palma, A.J., Capitán-Vallvey, L.F., Sánchez-Marañón, M.: Using the mobile phone as munsell soil-colour sensor: an experiment under controlled illumination conditions. Comput. Electron. Agric. **99**, 200–208 (2013)
17. Han, P., Dong, D., Zhao, X., Jiao, L., Lang, Y.: A smartphone-based soil color sensor: for soil type classification. Comput. Electron. Agric. **123**, 232–241 (2016)
18. Stanco, F., Tenze, L., Ramponi, G., De Polo, A.: Virtual restoration of fragmented glass plate photographs. In: Proceedings of the 12th IEEE Mediterranean Electrotechnical Conference 2004, MELECON 2004, vol. 1, pp. 243–246. IEEE (2004)
19. Bruni, V., Crawford, A., Vitulano, D., Stanco, F.: Visibility based detection and removal of semi-transparent blotches on archived documents. VISAPP **1**, 64–71 (2006)
20. ASTM. Standard practice for calculation of color tolerances and color differences from instrumentally measured color coordinates. ASTM International D 2244-05 (2005)
21. ASTM. Standard practice for establishing color and gloss tolerances. ASTM International D 3134-97 (1997)
22. Sánchez-Marañón, M., Huertas, R., Melgosa, M.: Colour variation in standard soil-colour charts. Soil Res. **43**(7), 827–837 (2005)
23. Centore, P.: An open-source inversion algorithm for the munsell renotation. Color Res. Appl. **37**(6), 455–464 (2012)
24. The Munsell and Kubelka-Munk toolbox by Paul Centore. http://www.munsellcolourscienceforpainters.com/MunsellAndKubelkaMunkToolbox/MunsellAndKubelkaMunkToolbox.html. Last updated on February 2017, last visited on April 2017
25. Rheinboldt, W.C., Menard, J.P.: Mechanized conversion of colorimetric data to munsell renotations. JOSA **50**(8), 802–807 (1960)
26. Simon, F.T., Frost, J.A.: A new method for the conversion of cie colorimetric data to munsell notations. Color Res. Appl. **12**(5), 256–260 (1987)
27. WallkillColor Munsell Conversion Software. http://www.wallkillcolor.com/. Accessed Apr 2017
28. BabelColor PatchTool Software. http://www.babelcolor.com/products.htm#PRODUCTS_PT. Accessed Apr 2017

Two-Stage Recognition for Oracle Bone Inscriptions

Lin Meng(✉)

Department of Electronic and Computer Engineering, Ritsumeikan University, Kusatsu, Shiga 525-8577, Japan
menglin@fc.ritsumei.ac.jp

Abstract. This paper designs a two-stage recognition system for recognizing oracle bone inscriptions (OBIs) that were inscribed on cattle bone or turtle shells with sharp objects about 3000 years ago. First, the system reduces noise and extracts the skeleton and line feature (line points), then two-stage recognition is proposed for recognition. The first stage is line feature recognition, which calculates the distance between the line points of the target OBI image and those of the templates for ranking the five shortest distance templates as candidates in Hough space. The second stage involves a checkpoint recognition method that sets several checkpoints in an image and checks the checkpoint hit rate of the candidates and target image for checking their similarity again. The experimental results show that almost 90% of the inscriptions on the third-most similar templates were recognized. And the proposed method can recognize well, even if the noise exists, the original inscriptions images are tilted and broken, and some noise exists in the thinning results.

Keywords: OBI recognition · Two-stage recognition · Checkpoint

1 Introduction

Oracle bone inscriptions (OBIs) are the oldest kind of characters and were inscribed on cattle bone or turtle shells about 3000 years ago. OBIs were discovered in 1899; few papers describe them, and the aging process has made the inscriptions less legible. Hence, understanding the inscriptions is important for researching world history, character evaluations, etc.

The most common OBI storage method involves rubbing, where the OBI surface is reproduced by placing a piece of paper over the subject and then rubbing the paper with rolled ink. Figure 1(a) shows an example of oracle bone rubbing with the lower part showing an enlarged view. Figure 1(b) shows two kinds of OBIs (King and Day) which are cut from the rubbing literature [2] artificially. We found that the character sizes are uneven, some smaller and bigger noises exist, some characters are broken, and the inclinations of the characters are non-uniform. Hence, OBI recognition is very difficult, and common OCR (optical character recognition) for OBI recognition is not easy. Our proposed

S. Battiato et al. (Eds.): ICIAP 2017, Part II, LNCS 10485, pp. 672–682, 2017.
https://doi.org/10.1007/978-3-319-68548-9_61

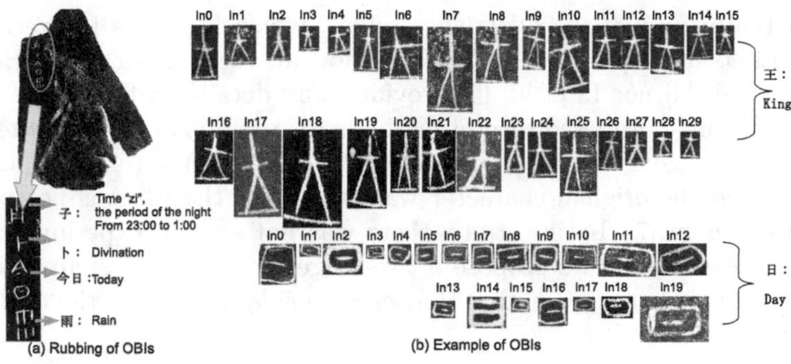

Fig. 1. Example of OBIs.

method is intended to help archaeologists understand OBIs. That is why we list the third-most similar templates, not the most similar ones.

We have designed a two-stage recognition system that consists of line features and checkpoint recognition for recognizing OBIs. The system recognizes inscriptions from an inscription template database that contains images of normalized inscriptions similar to a dictionary. The template database is created by expert belonging to the College of Letters of Ritsumeikan University. More than 2000 normalized OBI templates are stored in the database [1].

In preprocessing, we use Gaussian filtering and labeling for reducing noise, affine transformation and Thinning [12] for extracting the skeleton of the OBIs, Hough transform [14] and clustering for extracting the line feature(line points).

The first stage of recognition calculates the shortest distance between the extracted line points of the original and the template OBI images in Hough space and lists the five shortest distance templates as the candidate templates. In the second stage of recognition, we propose a checkpoint recognition method that sets some checkpoints in an image and checks the checkpoint hit rate of the candidates and target image for checking their similarity again.

The contributions of this paper are:

1. A good OBI recognition rate.
2. A two-stage recognition method using line feature and checkpoint recognition.

2 Related Work

As technologies evolve, various researchers have attempted to recognize OBIs by image processing. However, few English papers have reported on OBIs. We do know that the recognition rate needs to be improved. Li et al. [3] presented a recognition method that treats OBIs as a non-directed graph for recording the features of end-points, three-cross-points, five-cross-points, blocks, net-holes, etc. Li et al. [4] presented another recognition method using non-directed graph too. However, due to the age of OBIs, some of the holes and cross-points that occur are not actually part of the OBIs themselves, which increases the recognition difficulty. Li et al. [5] proposed a DNA method for recognizing OBIs. However, neither Li et al. [4] nor Li et al. [5] provided any details on their experiments. We have previously proposed several methods for recognizing OBIs by template matching and by using Hough transform [6–8]. However, the template matching was weak when the original character was tilted and the tilt was not properly processed either in [7]. In one study, there was not enough experimental work [8]. In the present work, we propose a two-stage recognition system to be used from noise reduction to recognition that can consider the tilt of the character.

3 Recognition Flow

Figure 2 shows the OBI recognition flow, which consists of preprocessing and recognition processing.

In preprocessing, the noise reduction is performed on the original image, and the skeleton of the noise-reduced image is extracted by affine transformation and Thinning. Then, the feature is extracted by Hough transform and clustering. The template images are applied by Thinning, Hough transform, and clustering for extracting the skeleton of the templates. Figure 3 shows the results of line extraction of the template and original image "Rain", which consists of four line points, as shown in (b) and (d).

Recognition processing consists of two stages. In the first stage, the distances between the line points of the original image and the results of the template images are calculated, and the five shortest distances for getting the candidate

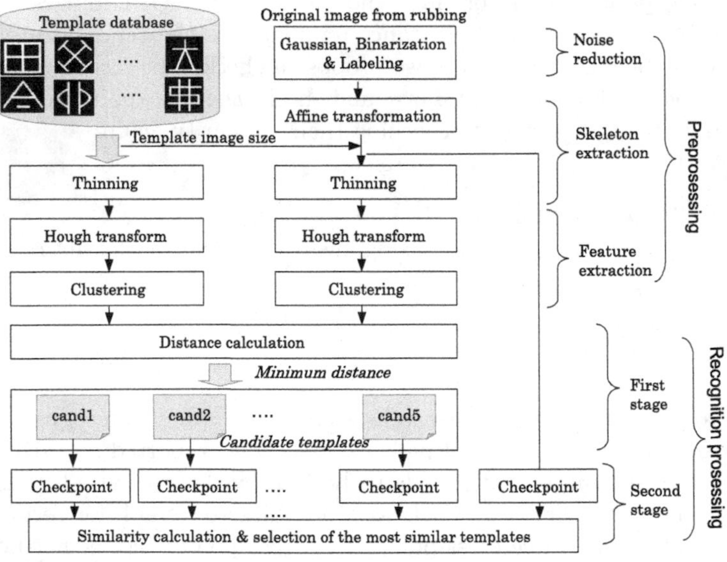

Fig. 2. OBI recognition Flow.

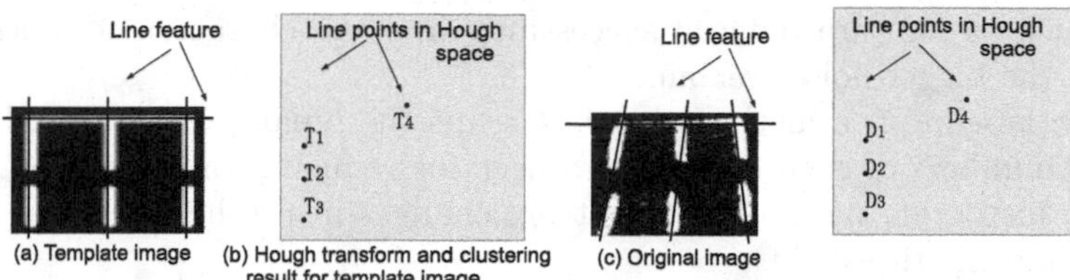

Fig. 3. Line feature extraction results.

templates are searched for. Sometimes the clustered line point numbers are different from the correct numbers, causing the correct result to not be in the top three ranks. For improving the recognition accuracy, in the second stage, we propose a checkpoint method for measuring the similarity between the five candidates and the original image. The checkpoint method sets some checkpoints and checks how many checkpoints match in the character for renewing the ranking.

4 Preprocessing

Preprocessing consits of noise reduction, skeleton extraction and line feature extraction.

4.1 Noise Reduction Processing

Due to aging, big and small artefacts exist on OBI rubbings. Hence noise reduction processing is a difficult and important processing in the recognition. Figure 4(a) is the original image, which is a rubbing image cut from literature [2]. As shown, both smaller noises such as fog and some bigger noises exist in the image.

We use Gaussian filtering bulrring and binarization for reducing the smaller noises. Gaussian filtering is shown in Fig. 4(b). By analysing, we fond the histogram of the Gaussian filtering results divided into two peaks, hence the Otsu method [10] was used to decide the threshold for binarization and reduce the smaller noise. Figure 4(c) shows the results of binarization, where the smaller

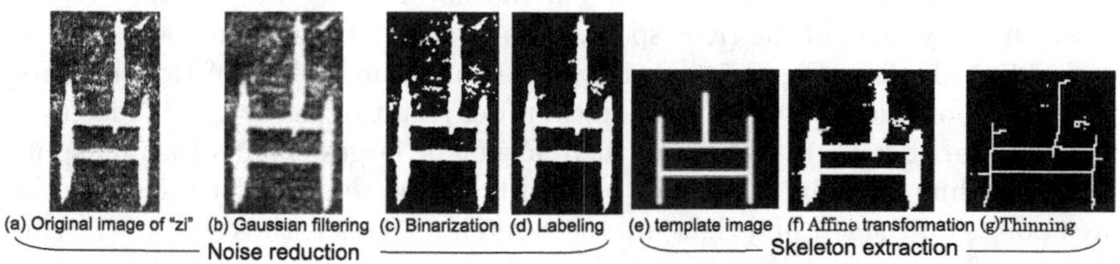

Fig. 4. Results of noise reduction and skeleton extraction of the symbol "zi".

noises (such as fog) are reduced successfully and the character becomes clear. However, the bigger noises remain.

Lastly, labeling is a method that last scans the binarized image and counts the pixel numbers of each connected object for reducing the bigger noise [11]. We use a histogram method to detect big changes in the histogram of objects for detecting the threshold.

If an object's pixel number is less than a threshold the object is treated for noise reduction. Figure 4(d) shows the result of labeling. As shown, the bigger noises are reduced successfully and the character becomes more clear.

4.2 Skeleton Extraction

Skeleton extraction includes affine transformation and Thinning. For recognition, the size of original image and the templates need to be the same. The affine transformation is a map that transforms points and vectors in the original image space into points and vectors in the database image space. Equation (1) shows the changing method, where (x_i, y_i) is the pixel axis in the original image, (x_c, y_c) is the character center axis of the template image, θ is the angle to which the original image will be changed as a result of the rubbing, and M is the size of the extension from the original image size and to the template image size. Figure 4(f) shows the affine transformation results, namely, the correct changing of the labeling result space of Fig. 4(d) into the template space of Fig. 4(e).

After the affine transformation, we extract the skeleton from the original image using Hilditch's algorithm. The method considers each of the eight neighborhoods (p2, p3,...,p9) of the target pixel as one pixel (p1) and decides whether to peel it off or keep it a skeleton. Figure 4(g) shows the thinning results.

$$\begin{pmatrix} x_i^* \\ y_i^* \end{pmatrix} = \begin{pmatrix} \cos\theta - \sin\theta \\ \sin\theta \quad \cos\theta \end{pmatrix} \begin{pmatrix} M \times x_j \\ M \times y_j \end{pmatrix} + \begin{pmatrix} x_c \\ y_c \end{pmatrix} \tag{1}$$

4.3 Line Feature Extraction

This subsection shows line feature (line points) extraction by Hough transform and Clustering. Hough transform is widely used for extracting lines from images, by transforming the (x, y) space to the (r, θ) space using Eq. 2.

Every point in the (x, y) space will be transformed into a curve in the (r, θ) space by changing the θ from 0 to 2π. The method records the time the curve is passed in every pixel of the (r, θ) space. Finally, the recording time will be used for deciding the line. Figure 5(e, f) shows the three dimensions of Hough transform results on the (r, θ) space for the thinning result of the original (Fig. 5(a)) and template (Fig. 5(c)). We found there are eight largest points that make up the feature line point in Fig. 5(e, f), especially in the template of Fig. 5(f). The three points on the left and right are the same line. This is the case of (θ) being 0° and 360°. Therefore, the line points are five.

If we catch the eight largest points correctly and transform the (r, θ) space into (x, y) space again, it will be possible to generate the results of the Hough

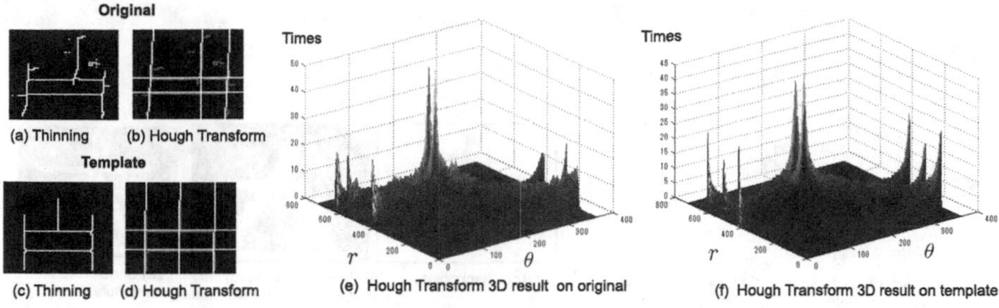

Fig. 5. Hough transform.

transform in Fig. 5(b, d). The times at which we find the largest points of the template and original are the same in Fig. 5(e, f). Hence, deciding on the largest points and calculating the distance between the original image and the temp image is helpful for recognizing the OBIs.

Below are definitions pertaining to the line feature point decision. Algorithm 1 shows the line points decision. $C_{(r,\theta)}$: all points of the (r, θ) space; $LC_{(r,\theta)}$: the largest point in (r, θ) space that still does not be checked; $LPs_{(r,\theta)}$: a set of larger points that is decided as the line feature point. It keeps the area of every point by radius; $SDis(LP, LC)$: the distances between $LC_{(r,\theta)}$ and all of the $LPs_{(r,\theta)}$; $MinDis_{(r,\theta)}$: the smallest distance of $SDis(LP, LC)LP_{(r,\theta)}$; $SLP_{(r,\theta)}$: a point of $LP_{(r,\theta)}$ that generates the $MinDis$ with LC.

$$r = x\cos\theta + y\sin\theta \tag{2}$$

Algorithm 1. Line point decision

$C \Leftarrow SortC$
while $C \neq NULL$ **do**
  Search LC ; Generate $SDis$ by using LC and LPs, Search $MinDis$
  if $MinDis$ is lower than the radius of SLP **then**
    do nothing
  else {$MinDis$ is lower than (the radius of SLP+30)}
    record (the radius of $SLP \Leftarrow MinDis$) into LPs
  else
    input the LC into LP and keep the axis
  end if
end while

5 Recognition

After the preprocessing, the line points are decided in the (r, θ) Hough space, which are shown in Fig. 6(a, b). The system uses two-stage recognition. The first

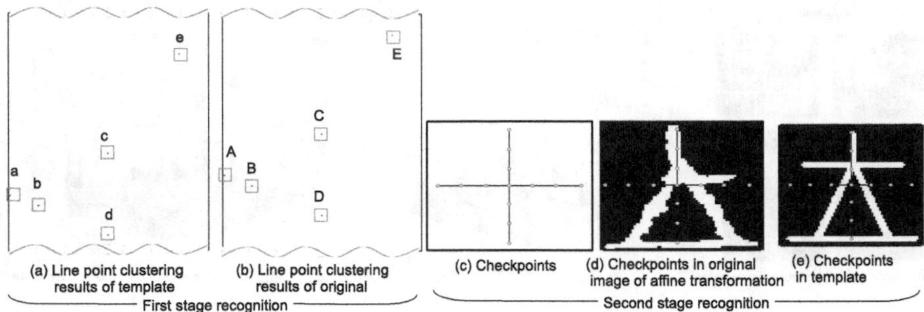

Fig. 6. Two-stage recognition.

stage calculates the shortest distance of the line points for the templates and the original image and lists the five shortest distance templates as candidates. The second stage uses checkpoints for searching the three most similar templates from the five candidates again for the recognition.

5.1 First-Stage Recognition Using Distance Calculation

By calculating the distance between every clustering result (line points in Hough space of two graphs) of the template and the original image (as in Fig. 6(a, b)), the line points in two graphs can obtain a bipartite graph. In the distance calculation, all patterns (line points) of the distances are calculated in the bipartite graph, and the shortest distances are selected as the distance. The shortest distance of Fig. 6(a, b) is Eq. 3.

$$Dis = \overline{aA} + \overline{bB} + \overline{cC} + \overline{dD} + \overline{eE} \tag{3}$$

However, the line points of the template and the original are often different. Hence, the shortest distances are normalized by line feature point number by Eq. 4. The smaller point number is defined as $DisLines$, which we used for calculating the distance. The larger point number is defined as $MatchLines$. When the line points of the template and the original are the same, then $DisLines = MatchLines$.

$$NorDis = \frac{\frac{SumDis}{DisLines} \times MatchLines}{DisLines} = \frac{SumDis \times MatchLines}{DisLines^2} \tag{4}$$

5.2 Second Stage Recognition Using Checkpoints

The checkpoint method uses several checkpoints for recorrecting the recognition rank of the first stage.

Figure 6(c) shows the checkpoints that are described in the image frame. The recognition collects the checkpoint hit rate of the candidates and the target image for checking the similarity of the candidates and target image again. Figure 6(d) shows an example of checkpoints in the original image of affine transformation result and the template. We found the checkpoint hit rate of the target image and its template are similar; hence, we use the checkpoint hit rate for renewing the similarity rank and improving the recognition accuracy again.

6 Experimental Results

We used 30 templates as the dictionary and 29 kinds of original OBIs about 576 characters to measure the performance of our proposed method. Figure 7 shows the template images with the character number shown below the images. Part of an original OBI image is shown in Fig. 1. The original OBI images are are obtained randomly, scaned and cut from the rubbing literature [2] artificially in this paper. We are challenge to capture the original OBI images from the scaned rubbing automatically by using gabor filter now [15].

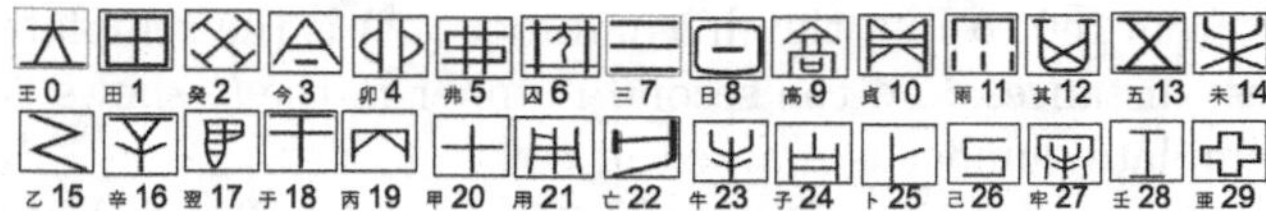

Fig. 7. Template image.

6.1 Recognition Results

Figure 8 shows the recognition rate of our method. We extract the shortest distance rate (rank1), second-shortest distance rate (rank2), and the third-shortest

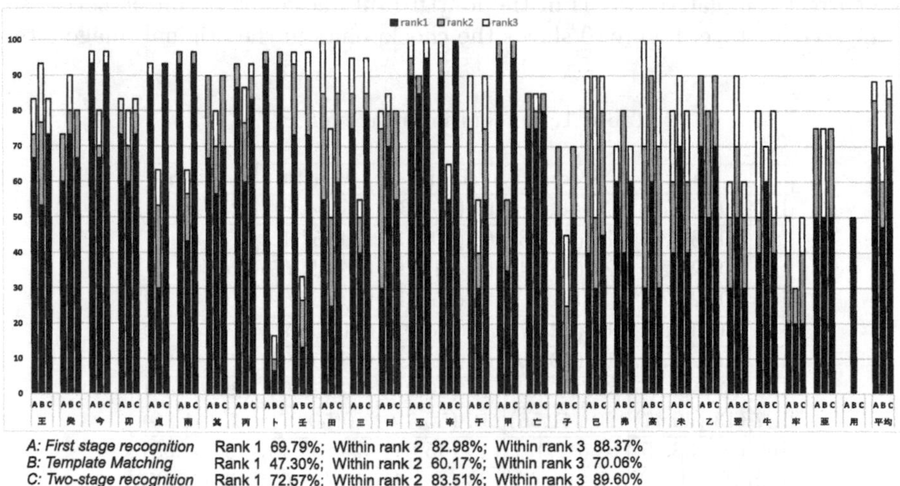

Fig. 8. Recognition Rate.

distance rate (rank3). Rank 1 means how many rate OBIs are recognized which has the shortest distance with its template. Rank 2 and Rank 3 have the similar meaning with Rank1 The horizontal axis of Fig. 7 shows the character of the original OBIs. Every OBI results has A which shows the recognition result of the template matching, B which shows the result of only the first-stage recognition and C which shows the result of the two-stage recognition.

The average column (most right) shows that more than 70% of ROIs are recognized as the shortest distance, and 10% are recognized as the second-shortest distance time in the two-stage recognition system. The proposal has a good recognition rate compared to template matching and has a good performance in the first rank compared to when using only first-stage recognition.

6.2 Recognition Analysis

We list the recognition results of the five shortest distances and the template number for the first-stage recognition in Table 1. Every column shows the input original image from In0 to In9, the first number (t*) is the template number of the five shortest distances, and the second number in the () shows the distance. The correct template number is marked in red.

Figure 9 lists the recognition results of 王 for input 0 to 3 and the templates. In these figures, from (1) to (5) shows binarization, labeling, affine transformation, Thinning and inverse Hough transform results, respectively (the inverse Hough transform was not used in recognition; they are included to make it easier to understand the proposal). These results are listed on the clustering result image; the squares point out the line feature points. Figure 9(a) shows that parts of the OBIs are broken, and in (b), there is noise still exist. However, the cluster extracts the line feature correctly and the inscriptions are recognized.

In Table 1, In0 to In6 and In9 are recognized in the first rank in the first stage. The correct template of In7 is in the fourth rank; however, it can be corrected in the second stage. Figure 10 shows the checkpoints in the original image after

Table 1. Recognition results of 王

	First	Second	Third	Fourth	Fifth
In0	t0 (31.76)	t3 (40.36)	t4 (49.03)	t6 (54.60)	t23 (56.35)
In1	t0 (34.34)	t16 (46.33)	t3 (50.60)	t20 (51.15)	t13 (54.76)
In2	t0 (21.22)	t20 (35.64)	t3 (41.43)	t16 (41.68)	t24 (50.91)
In3	t0 (28.30)	t3 (50.71)	t9 (55.83)	t29 (57.69)	t16 (59.00)
In4	t0 (9.18)	t13 (52.27)	t13 (54.91)	t21 (56.26)	t16 (57.02)
In5	t0 (24.71)	t13 (43.06)	t3 (55.46)	t16 (58.91)	t29 (63.61)
In6	t0 (16.57)	t3 (41.63)	t13 (50.83)	t16 (52.04)	t29 (57.90)
In7	t13 (40.67)	t20 (41.56)	t16 (50.79)	t0(50.97)	t3 (51.86)
In8	t13 (38.15)	t0 (42.42)	t16 (45.07)	t28 (54.40)	t3 (60.63)
In9	t0 (11.86)	t3 (40.97)	t28 (45.81)	t13 (51.62)	t22 (56.05)

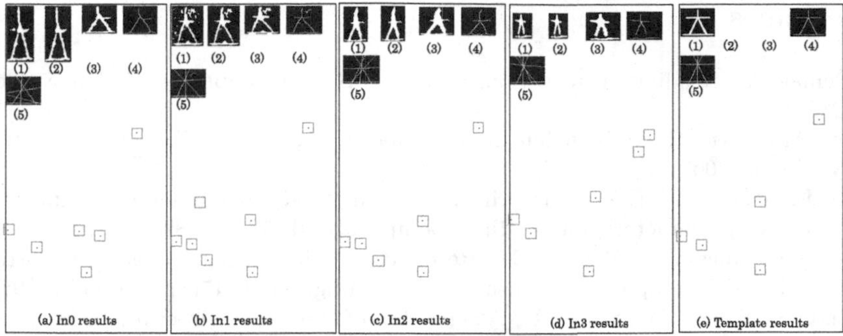

Fig. 9. Recognition analysis of 王 in first stage.

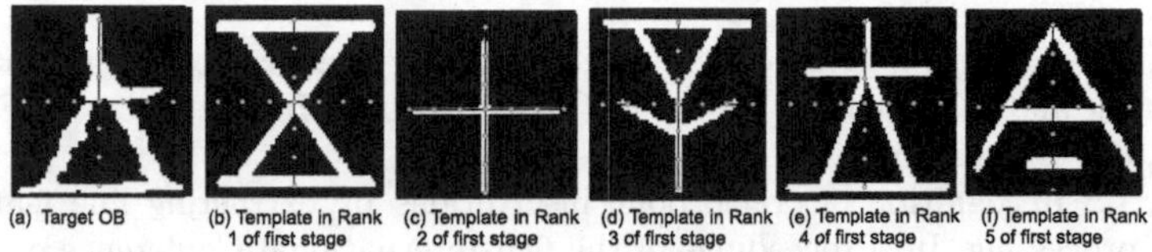

Fig. 10. Recognition analysis of 王 in second stage.

affine transformation and the five shortest distance templates. By checking the checkpoint hit rates, the proposed method corrects the target template, which moves it into first rank. Hence, the proposal achieved a good recognition rate.

7 Conclusion

We propose a two-stage recognition system that consists of line points and checkpoint recognition for recognizing OBIs. Firstly, the system reduces the noise of OBIs using Gaussian filtering and labeling and extracts the skeleton using affine transformation and Thinning. Then, the line points of the inscriptions are extracted by Hough transform and clustering. The system calculates the shortest distance of the line feature points between the original image and lists the five shortest distance templates as the candidate templates. Finally, the checkpoint is applied for reranking the similarity of the five candidates. The experimental results show that in total almost 90% of inscriptions were recognized on the third-most similar templates. And the proposed method can recognize inscriptions well, even if noise exists, the original inscription images were tilted and broken, and the whisker exists in the thinning results. Analyzing the reason for failure and increasing the recognition rate are major future work. Increasing the dataset of templates and original OBIs will be an important future works too.

Acknowledgments. This work was supported by a Grant-in-Aid for Scientists (15K00163) and Grant-in-Aid for Young Scientists (26870713) from JSPS.

References

1. Ochiai, A.: Reading History from Oracular Bone Inscriptions. Chikuma Shobo (2008)
2. Pu, M.Z., Xie, H.Y.: Shanghai Bo Wu Guan Cang Jia Gu Wen Zi. Shanghai Bo Wu Guan (2009)
3. Li, F., Woo, P.Y.: the coding principle and method for automatic recognition of Jia Gu Wen characters. Int. J. Hum.-Comput. Stud. **53**(2), 289–299 (2000)
4. Li, Q.S., Yang, Y.X., Wang, A.M.: Recognition of inscriptions on bones or tortoise shells based on graph isomorphism. Comput. Eng. Appl. **47**(8), 112–114 (2008)
5. Li, Q.S., Yang, Y.X.: Sticker DNA algorithm of oracle-bone inscriptions retrieving. Comput. Eng. Appl. **44**(28), 140–142 (2008)
6. Meng, L., Fujikawa, Y., Ochiai, A., Izumi, T., Yamazaki, K.: Recognition of oracular bone inscriptions using template matching. Int. J. Comput. Theory Eng. **8**(1), 53–57 (2016)
7. Meng, L., Izumi, T., Oyanagi, S.: Recognition of oracular bone inscriptions by clustering and matching on the hough space. J. Inst. Image Electron. Eng. Jpn **44**(4), 627–636 (2015)
8. Meng, L.: Recognition of oracle bone inscriptions by extracting line features on image processing. In: Proceedings of the 6th International Conference on Pattern Recognition Applications and Methods (ICPRAM 2017), pp. 606–611 (2017)
9. Ochiai, A.: Oracle Bone Inscriptions Database (2014). http://koukotsu.sakura.ne.jp/top.html
10. Sezgin, M., Sankur, B.: Survey over image thresholding techniques and quantitative performance evaluation. J. Electron. Imag. **13**(1), 146–165 (2015)
11. He, L.F., Chao, Y.Y., Suzuki, K.: A run-based two-scan labeling algorithm. IEEE Trans. Image Process. **17**(5), 749–756 (2008)
12. Schneider, P.K., Eberly, D.H.: Thinning Methodologies: A Comprehensive Survey. Morgan Kaufmann Publishers, Burlington (2003)
13. Lam, L., L, S.W., and Suen, C. Y. : Geometric tools for computer graphics. IEEE Trans. Pattern Mach. Intell. 14(9), 869–885 (1992)
14. Ballard, D.H.: Generalizing the hough transform to detect arbitrary shapes. Pattern Recogn. **13**(2), 111–122 (1981)
15. Ken, T.: Filters for Common Resampling Tasks, Graphics Gems I, pp. 147–165. Academic Press (1990)

Imaging Solutions for Improving the Quality of Life

Real Time Indoor 3D Pipeline for an Advanced Sensory Substitution Device

Anca Morar(✉), Florica Moldoveanu, Lucian Petrescu, and Alin Moldoveanu

University Politehnica of Bucharest, Bucharest, Romania
{anca.morar, florica.moldoveanu, alin.moldoveanu}@cs.pub.ro, petrescu.lucian.24@gmail.com

Abstract. In this paper, we present the indoor 3D pipeline of an assistive system for visually impaired people, whose goal is to scan the environment, extract information of interest and send it to the user through haptics and sounds. The particularities of indoor scenes, containing man-made objects, with many planar faces, led us to the idea of developing the 3D object recognition algorithms around a planar segmentation, based on normal vectors. The 3D pipeline starts with acquiring depth frames from a range camera and synchronized IMU data from an inertial sensor. The pre-processing stage computes normal vectors in the 3D points of the scanned environment and filters them to reduce the noise from the input data. The next stages are planar segmentation and object labeling, which divides the scene into ground, ceiling, walls and generic objects. The whole 3D pipeline works in real-time on a consumer laptop at approximately 15 fps. We describe each step of the pipeline, with the focus on the labeling stage, and present experimental results and ideas for further improvements.

Keywords: Pre-processing · Planar · Segmentation · Labeling · Depth map · Indoor

1 Introduction

The research in the sensory substitution domain has experienced a spectacular growth in recent years. In particular, the plethora of available consumer 3D cameras has led to an increased interest for the development of new computer vision based assistive systems. In this context, we propose fast algorithms on the CPU and GPU for the basic labeling of indoor scenes, with the purpose of helping visually impaired people navigate in unknown environments. There are two main technologies that estimate the distance from camera to surrounding objects: structured light and time-of-flight (ToF). Both technologies have a series of limitations, such as sensor error which increases with the distance to the objects and the impossibility of estimating the depth for reflective surfaces or for highly illuminated scenes. Another category of 3D cameras is the one based on stereovision. However, in indoor scenes, where there is a lack of texture (for example, in case of walls or big objects that have the same color), the structured light or ToF cameras are more reliable. In the first stage of the project we

S. Battiato et al. (Eds.): ICIAP 2017, Part II, LNCS 10485, pp. 685–695, 2017.
https://doi.org/10.1007/978-3-319-68548-9_62

evaluated a series of acquisition devices, based on their capabilities and their form factor and we decided to choose the Structure Sensor camera, due to its reduced dimensions, light weight and its acceptable performance, which is similar to that of Kinect version 1.

The system acquires depth images from the Structure Sensor and information about camera rotation from an inertial sensor (LPMS). The information obtained from the two devices is synchronized based on the timestamps of the frames. The pre-processing stage computes normal vectors in the 3D points of the point cloud. The point cloud is obtained by transforming positions (*x*, *y*, *depth*) from the depth map in the camera space, based on the camera's intrinsic parameters. A very important step in the pipeline is the segmentation of the scene into planar surfaces. The output of the segmentation is afterwards merged into objects that are labeled based on their geometric properties. The steps of the pipeline are detailed in Sect. 3.

In the following section, we provide a short survey of indoor labeling methods. Next, the algorithms of our pipeline are described in detail. In the last sections, we present experimental results and draw the conclusions.

2 State of the Art

Most assistive systems for visually impaired people acquire images from a video camera and process them by employing geometric based heuristics or through machine learning.

Stoll et al. [1] perform a simple conversion of depth images into sounds: they down-sample the depth map by averaging the values of the pixels from a neighborhood. Each cell in the new depth map is encoded into a sound which is loud if the object is close and soft if the object is far away. Blessenohl et al. [2] employ geometric heuristics based on local normal vectors for ground and wall detection. Taylor et al. [3] perform edge detection, planar segmentation, and then identify ground and walls with simple heuristics. Liu et al. [4] propose an obstacle detection method for RGB-D images which identifies the ground with a multi-scale voxel plane segmentation. Next, the obstacles are detected through region-growing and by using context information.

Lai et al. [5] use sliding window detectors on the pixels in a RGB-D image and combine the results with geometric properties for object indoor labeling. Wang et al. [6] tackle the problem of insufficient 3D training data for machine learning detection algorithms and propose a Support Vector Machine (SVM) based scheme to transfer 2D existing image labels to point clouds. Kim et al. [7] propose the use of Conditional Random Fields (CRF) over voxels obtained from a 3D point cloud. Anand et al. [8] extract geometric features from processed 3D point clouds and use a maximum-margin learning method [9] for contextually guided semantic labeling and object search. Lai et al. [10] introduce the HLP3D classifiers and use Markov Random Fields (MRF) to detect both small objects and large furniture. Wang et al. [11] perform a graph-based planar segmentation and train cascade decision trees to detect ground, walls and tables. The See ColOr system [12] uses a detecting and tracking hybrid algorithm that learns the properties of natural objects through a training phase. Huang et al. [13] build patches from a 3D point cloud and use geometric features such as coplanarity and color coherence for the grouping of patches into objects. Deng et al. [14] propose the incorporating of global co-occurrence constraints, relative height relationship constraints and local support relationship constraints into a CRF for semantic segmentation of RGBD images. Gupta et al. [15] perform contour detection, bottom-up grouping, object detection and scene classification by training an additive kernel SVM. Yang et al. [16] use both geometric modeling and machine learning for indoor scene understanding. Usually, artificial intelligence-based systems have a better accuracy, but require a lot of manually annotated training data and introduce a heavy computational load.

3 3D Processing Algorithms

The proposed workflow is described in Fig. 1 and detailed in this section.

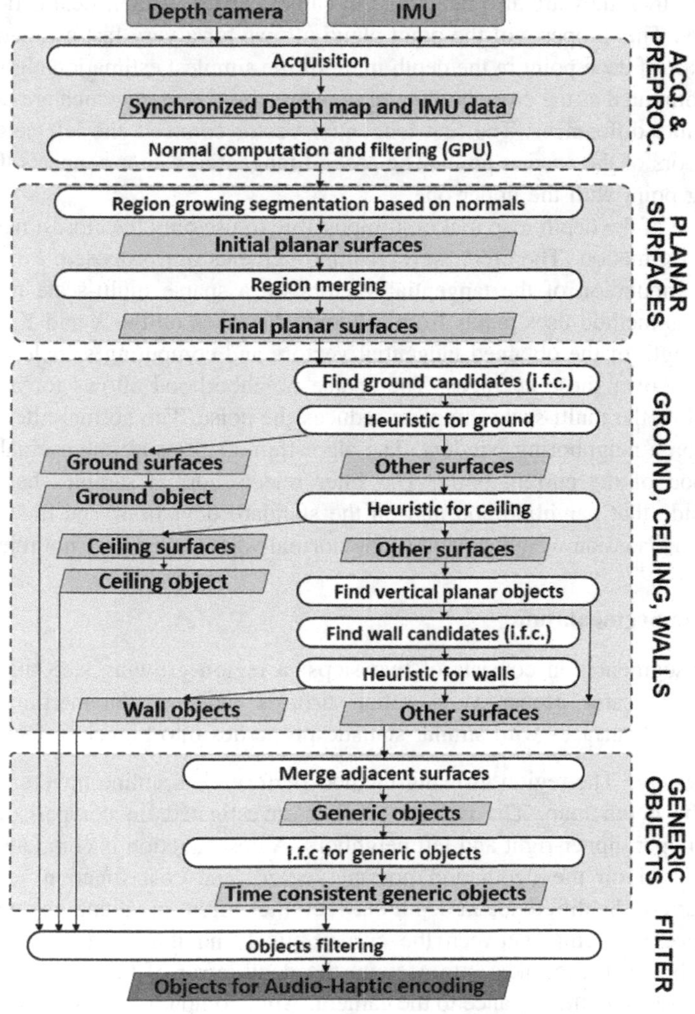

Fig. 1. Workflow of our 3D processing

3.1 Pre-processing

As previously mentioned, the input from structured light cameras is noisy and contains areas with un-sampled pixels. Therefore, the filtering of the input data is of utmost importance. The problem of the un-sampled pixels can be solved with an inpainting algorithm [17]. Several inpainting methods have been implemented and tested, but were not included in the final pipeline, because the obtained improvements were too small in comparison with the computational load.

The normal estimation is the basis for the planar segmentation. The point cloud obtained from a depth camera is organized, in the sense that there is a one to one correspondence between the pixels in the depth map and the points in the point cloud. The camera intrinsic parameters allow for the computation of a 3D point in camera space, from the pixel position in image space. In this paper, we use the terms 'pixel' and 'point' interchangeably, due to this correspondence. If two points are neighbors in image space, then they are also neighbors in camera space, if their depths do not vary considerably. This property of the point cloud allows for a very fast estimation of the normal vector of each point in the depth map. In the simplest estimation algorithm, the normal is computed as the cross product of two tangential vectors which are determined based on finite differences. The first tangential vector connects the left neighbor and right neighbors of the current point. The second tangential vector connects the bottom neighboring point with the upper one.

The noise of the depth map makes it impossible to use only the closest neighbors in the normal estimation. Therefore, several improvements are proposed. First, we perform the computation of the tangential vectors on a sparse multi-scale neighboring window. This method uses pixels from different distances on the X and Y axes, averaging the length of the obtained tangential vectors and computing sample normals as cross product over these averages. The sparse neighborhood allows for a fast computation, while the multi-scale sampling reduces the noise. The normal filtering is also performed on a neighboring window. The algorithm analyses all the normals inside a neighborhood of the current point. The filter rejects normal samples based on the average inside that neighborhood and on the standard deviation. The final normal is obtained as a Gaussian weighted sum of the normal vectors that were not rejected [18].

3.2 Planar Segmentation

The planar segmentation consists of two steps: a region growing step that connects pixels with the same properties in initial surfaces and a region merging step that combines initial surfaces with similar statistic properties into bigger surfaces.

Region Growing. The region growing method performs a scanline traversal of all the pixels in the depth map. The current pixel is investigated, in comparison with its upper-left, upper, upper-right and left neighbors. A cost function is computed for each neighbor, based on the depth and normal vector. The cost function represents a weighted sum of the difference in depth between the current pixel and its neighbor and the difference in normal between the current pixel and the surface containing the neighbor. This cost also takes into account the depth error of the acquisition device, which increases with the distance to the camera. After computing the similarity costs of the pixel with its neighbors, we determine the neighbor with the minimum associated cost. If this minimum cost is lower than a given threshold, the current pixel is added to the region containing the correspondent neighbor. During the region growing step we also determine the adjacency of the surfaces. Two surfaces are considered adjacent if at least two pixels, one from the first surface and one from the second surface, are neighbors, and their depth values do not differ significantly.

Region Merging. The region growing step is very sensitive to noise, leading thus to an over-segmentation. The region merging step solves this problem to some extent. The region growing outputs 3D planar surfaces. These surfaces can be further merged based on their statistic properties. We compute the average normal and position of the surface, as the averages of all the points in the surface. The average normal vectors represent a good metric for the comparison of two surfaces. Another good metric for comparing two adjacent surfaces is the distance from the position of the camera, i.e., (0, 0, 0) in camera space, to the theoretic plane that contains the current surface. The equation of the plane containing a surface can be computed based on the average normal $\overline{N}(x_N, y_N, z_N)$ and the average position $P(x_p, y_p, z_p)$ of the points inside that surface.

The similarity cost between two surfaces, S_1 and S_2, is computed as a weighted sum of the difference in normal, C_{normal}, and the difference in distance from the camera to the theoretic plane containing each surface, C_{dist}. If this similarity cost is lower than a given threshold, then the surfaces are merged into a bigger surface. The region growing step is an iterative process, while this region merging is recursive, following a BFS traversal. A more detailed description of the segmentation algorithm, together with some experimental results, is given by Morar et al. [18].

3.3 Labeling

The input of the labeling stage consists of a vector of planar surfaces, an adjacency map for these surfaces, and a segmentation map, i.e. an array with the resolution of the depth image which contains for each pixel the ID of the surface it belongs to. In this step, we merge surfaces into objects and add them to one of the following categories: ground, ceiling, walls and generic objects. Next, we filter the objects that will be sent to sound and haptic encoding, based on user options.

Ground. In world space coordinates, the ground stands on a horizontal plane which is positioned at the lowest y coordinate, in a system where the Y axis is oriented upwards. However, in camera space, the situation is a bit different, due to the orientation of the camera. The inertial sensor allows the computation of the normal vectors in world space. The normal of a horizontal surface in world coordinates is close to (0, 1, 0). If the dot product between the average world normal of a surface and the (0, 1, 0) vector is bigger than a given threshold, then that surface is considered horizontal.

For a stable ground detection, we employ inter-frame consistency and identify the most suitable ground candidates based on the ground surfaces detected in the previous frames. We compute a similarity cost between each horizontal surface and the ground surfaces obtained in the previous frame. This cost depends on the difference in normal, the intersection of the surfaces and the difference in distances from the camera to the planes containing the surfaces. The intersection of the surfaces is computed based on the number of pixel pairs (P_1, P_2), both having the same x and y coordinates in image space, one belonging to the surface from the current frame and the other to the surface from the previous frame. This intersection cost is normalized with the number of pixels contained by each surface, as in [18]. If the similarity cost for a horizontal surface and one of the ground surfaces from the previous frame is smaller than a threshold, then the current horizontal surface is marked as a candidate for ground.

The next step is the detection of the first ground region. We consider the ground to belong to the horizontal plane with the biggest distance to the camera, d_{max}. The distance d for a region R is computed in camera space, based on its average position, (x_R, y_R, z_R), and the reference vertical normal for the current frame, $\bar{N_v}$ (x_{Nv}, y_{Nv}, z_{Nv}):

$$d = -x_{Nv} \cdot x_R - z_{Nv} \cdot z_R - z_{Nv} \cdot z_R. \tag{1}$$

$\bar{N_v}$ is the vertical normal in world space, i.e., (0, 1, 0), multiplied by the camera orientation matrix obtained from the IMU. d is a signed distance, positive for horizontal regions below the (*XOZ*) plane and negative for the ones above (*XOZ*).

Figure 2 illustrates this heuristic: R_3 is detected as the first ground region, because it has the biggest distance, d_3, from the camera to its theoretic plane, P_{R3}.

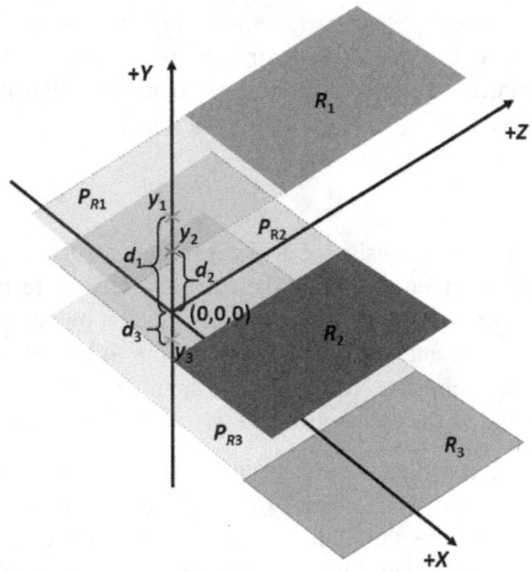

Fig. 2. Example of horizontal regions (R_1, R_2, R_3) which enter the ground heuristic. Region R_3 is the best ground candidate (it has the lowest y coordinate, y_3, in world space, i.e., the biggest negative distance from the camera to its theoretic plane, P_{R3})

We then compute the difference between the d_{max} and each of the distances from the camera to the planes corresponding to the other horizontal surfaces. If the difference is lower than a given threshold $T_{Gstrict}$, then the current horizontal surface is also marked as belonging to the ground. The surfaces that were marked as ground candidates in the inter-frame consistency step have a high probability of belonging to the ground. Therefore, the conditions for these surfaces are a little bit more relaxed, with the use of the threshold $T_{Grelaxed}$ instead of $T_{Gstrict}$. All the ground surfaces are merged into a single ground object and take the id of the ground object from the previous frame.

Ceiling. The heuristic for ceiling is very simple. All the surfaces with the average world normal close to $(0, -1, 0)$ and a distance to the ground higher than a threshold are considered to belong to the ceiling object.

Walls. The walls are usually big objects perpendicular to the ground. However, the region growing and merging algorithms could still lead to over-segmentation, especially when the walls are not perpendicular to the view direction. Therefore, in the first step of the wall detection we merge adjacent surfaces which are perpendicular to the ground into bigger vertical objects. The surfaces are merged only if their average normal vectors are similar and if the difference in distances from the camera to their corresponding theoretic planes is lower than a threshold. Since the normal is very sensitive to noise, even after a lot of filtering, the distance from the camera to the theoretic plane of a surface is also sensitive. Therefore, when comparing two adjacent surfaces that have similar orientation (similar average normal vectors), we don't use their own normal vectors to compute the distance, but we obtain a combined normal, weighted based on the size of each surface. If N_1 is the normal of the first surface, which contains $|S_1|$ pixels, and N_2 is the normal of the second surface, containing $|S_2|$ pixels, then the average normal is computed as follows:

$$N = \frac{N_1 \cdot |S_1| + N_2 \cdot |S_2|}{\|N_1 \cdot |S_1| + N_2 \cdot |S_2|\|}. \tag{2}$$

The next step in the wall detection is identifying the most probable wall candidates based on interframe consistency. A similarity cost is computed for each vertical object from the current frame with all the wall objects from the previous frame. If the minimum cost is lower than a given threshold, then the current vertical object takes the ID of the wall object from the previous frame corresponding to this minimum cost and is added to the most probable wall candidates. The heuristic for walls is also very simple. If the size of a vertical object or if its height is big enough, then that object is considered a wall. We compute the width and height in camera space of each vertical object, based on the bounding box in image space. We determine the bottom-left and upper-right corners of the bounding box and compute the 3D positions of these points, $C_1(x_{c1}, y_{c1}, z_{avg})$ and $C_2(x_{c2}, y_{c2}, z_{avg})$, taking into account the average depth of the object, z_{avg}. The difference on the Y axis, $|y_{c1}-y_{c2}|$, represents the height of the object in camera space. The width is computed differently. For example, a wall that is not perpendicular to the camera direction could have a very small width in image space, even if in reality it would have a big width, unfolded on the Z axis in camera space. We therefore compute the width of the object based on the difference of C_1 and C_2 on the X axis, but also on the difference between the maximum and minimum depth values of the object:

$$width_{cs} = \sqrt{(x_{c1} - x_{c2})^2 + (z_{\max} - z_{\min})^2}. \quad (3)$$

In order to avoid cases when the bounding box is too large compared to the object, for example because of some outliers, we weight these dimensions with the ratio between the actual size of the object, i.e., its number of contained pixels, and the area of the image space bounding box. Similar to the ground heuristic, vertical objects that were labeled as wall candidates in the inter-frame consistency phase have a high probability of belonging to the wall class. Therefore, their conditions for size and height are more relaxed than for the other vertical objects.

Generic Objects. All remaining surfaces are merged into generic objects, based on their adjacency. Here, the only condition for uniting two adjacent surfaces is their depth intervals $[z_{min}, z_{max}]$ not to be disjoint. Next, we apply an inter-frame consistency step for the generic objects in order to propagate the IDs of the objects from the previous frame.

Figure 3 presents the output of the pre-processing, segmentation and labeling.

Fig. 3. Output of the steps in the pipeline: the normal map (bottom-left), the output of the planar segmentation, where each region is colored based on its ID (bottom-right), the output of the labeling, where each object is colored based on its ID (upper-right) and the output of the labeling, where each object is colored based on its type: red for ground, yellow for ceiling, green for wall and blue for generic object (upper-left). (Color figure online)

Filtering of Objects. The labeling algorithm could lead to a lot of generic objects and walls, that must be encoded into sounds and haptics. If the obtained processed scene is too complex, the user might get disoriented by the multitude of audio or haptic signals. Therefore, in a configuration step, the user can choose the maximum number of objects to be encoded and decide how these objects should be selected. In the filtering phase, we compute the importance of each generic object as a weighted sum of its size, average depth, and deviation from the view direction. The weights are established by the user. Thus, the biggest objects, or the closest ones, or the objects closest to the direction where the user is looking, could be chosen for encoding.

4 Experimental Results

We evaluated the output of the system in rudimentary scenarios, indoor rooms with only boxes, hallways, and then we increased the complexity by adding other generic objects, such as static and even dynamic human beings. The number of objects and their position was visually inspected and compared to the real scene. We also achieved a qualitative evaluation on an indoor video of 230 frames, each with the resolution of 640 × 480. We performed a manual annotation one every ten frames from the video, obtaining 23 annotated frames. The results of the manual annotation were compared pixelwise with the results of the 3D pipeline. The configuration step, that filters the most important objects, was not included in the 3D pipeline. Table 1 presents the results of this evaluation. For each labeled pixel in the frames obtained with the automatic processing, we assessed whether its label type is similar with the one obtained from the manual annotation. For each object type, we measured the percentage this type was correctly detected, and the false positives, i.e., the percentage this type was wrongly labeled as another type. For example, the ceilings were correctly labeled in 98.24% of cases, and wrongly identified as walls for 0.26% pixels and as generic objects for 1.84% pixels.

Table 1. Performance of object recognition

Object type	Percentage of recognition (%)			
	Ground	Ceiling	Wall	Generic
Ground	99.43	0	0.06	0.5
Ceiling	0	98.24	1.09	0.66
Wall	0	0.26	99.27	0.31
Generic	5.87	1.84	9.43	82.84

In this evaluation, we counted only the true positives and the false positives. The false negatives were not determined because the 3D pipeline eliminates very small regions in the segmentation phase and very small objects in the labeling step. Therefore, the automatic processed frames contain less labeled pixels than the manual annotated ones. These elimination steps were introduced due to the fact that very small objects are not considered of interest for the encoding phase.

As can be observed from the table, the results are promising for true positives, especially for ground (99.43% accuracy), ceiling (98.24% accuracy) and walls (99.27% accuracy). However, the walls were wrongly identified as generic objects in 9.43% of the cases. The noisy input leads to small regions that cannot be included in the big vertical objects in the labeling step, and therefore, are not included into wall objects. This problem can be however reduced in the object filtering phase: parts of wall that are far away from the user or very small can be excluded from the output.

During the evaluation, other small issues were observed. The radial error of the acquisition device introduces some errors in the computation of the point cloud and of the normal vectors at the left and right borders of the image. These distortions affect the ground plane sometimes, and lead to some small parts of the ground not being included into the ground object, but considered as generic objects. These occasional errors could be reduced by adding a confidence parameter to each generic object: if a generic object has not been seen in at least a couple of consequent frames, then it is not reported at output. The last issue is the fact that objects and walls do not always hold the same unique ID throughout a whole video, but change it from time to time. This could be caused by the sudden movement of the camera, the drift in the IMU sensor, or the strict conditions for inter-frame consistency. A way of solving this problem, which was not introduced until now because of its memory requirements and time complexity, is global reconstruction. This small issue has not been addressed yet because the sound and haptic encoding do not necessarily require time-consistent objects.

The whole 3D pipeline runs at approximately 15 fps on an Intel Core i7-4720HQ Processor with a GTX 970M GPU, on a Windows 10 OS. The application was written in C++ using OpenGL. The pre-processing step is implemented in parallel on the GPU (with compute shaders), while the segmentation and labeling run on the CPU.

5 Conclusions

In this paper, we described a succession of innovative 3D video processing algorithms with the final goal of extracting objects of interest when navigating in an indoor environment.

In the future, we plan to combine information from the IMU with visual odometry, in order to obtain a robust camera motion estimation, to add a fast global reconstruction, and to try another segmentation, on the GPU, in order to reduce the CPU load. We also plan to expand the annotated data base and implement a benchmarking application for a more thorough evaluation of the pipeline's accuracy and for comparisons with future similar methods. This initial pipeline is intended for navigation. However, in the future we plan to integrate an algorithm that recognizes objects [19], in order to obtain a semantic labeling that would allow the blind users not only to navigate, but also to perceive the environment.

Acknowledgement. This work has received funding from the European Union's Horizon 2020 research and innovation program under grant agreement 643636 (Sound of Vision research project – www.soundofvision.net).

References

1. Stoll, C., Germain, R.P.: Navigating from a depth image converted into sound. Appl. Bionics Biomech. (2015)
2. Blessenohl, S., Morrison, C.,Criminisi, A., Shotton, J.: Improving indoor mobility of the visually impaired with depth-based spatial sound. In: IEEE International Conference on Computer Vision Workshop (2015)
3. Taylor, C.J., Cowley, A.: Parsing indoor scenes using RGB-D imagery. In: Robotics: Science and Systems (2012)
4. Liu, H., Wang, J., Wang, X., Qian, Y.: iSee: obstacle detection and feedback system for the blind. In: UbiComp/ISWC 2015 Adjunct, pp. 197–200 (2015)
5. Lai, K., Bo, L., Ren, X., Fox, D.: Detection-based object labeling in 3D scenes. In: International Conference on Robotics and Automation (2012)
6. Wang, Y., Ji, R., Chang, S.F.: Label propagation from imagenet to 3D point clouds. In: IEEE Conference on Computer Vision and Pattern Recognition (2013)
7. Kim, B.S., Kohli, P., Savarese, S.: 3D scene understanding by voxel-CRF. In: IEEE International Conference on Computer Vision (ICCV) (2013)
8. Anand, A., Koppula, H.S.: Contextually guided semantic labeling and search for three-dimensional point clouds. Int. J. Robot. Res. **32**(1), 19–34 (2013)
9. Taskar, B., Chatalbashev, V., Koller, D.: Learning associative markov networks. In: International Conference on Machine Learning (2004)
10. Lai, K., Bo, L., Fox, D.: Unsupervised feature learning for 3D scene labeling. In: IEEE International Conference on Robotics and Automation (2014)
11. Wang, Z., Liu, H., Wang, X., Qian, Y.: Segment and label indoor scene based on RGB-D for the visually Impaired. In: Gurrin, C., Hopfgartner, F., Hurst, W., Johansen, H., Lee, H., O'Connor, N. (eds.) MMM 2014. LNCS, vol. 8325, pp. 449–460. Springer, Cham (2014). doi:10.1007/978-3-319-04114-8_38
12. Gomez, J.D.R., Bologna, G.: See ColOr: an extended sensory substitution device for the visually impaired. J. Assistive Technol. (2014)
13. Huang, H., Jiang, H., Brenner, C., Mayer, H.: Object-level segmentation of RGBD data. ISPRS Ann. Photogrammetry Remote Sens. Spat. Inf. Sci. **II**(3), 73–78 (2014)
14. Deng, Z., Todorovic, S., Latecki, L.J.: Semantic segmentation of RGBD images with mutex constraints. In: IEEE International Conference on Computer Vision (2015)
15. Gupta, S., Arbelaez, P., Girshick, R., Malik, J.: Indoor scene understanding with RGB-D images: bottom-up segmentation, object detection and semantic segmentation. Int. J. Comput. Vis. **112**(2), 133–149 (2015)
16. Yang, S., Maturana, D., Scherer, S.: Real-time 3D scene layout from a single image using convolutional neural networks. In: IEEE International Conference on Robotics and Automation (2016)
17. Petrescu, L., Morar, A., Moldoveanu, F., Moldoveanu, A.: Kinect depth inpainting in real time. In: International Conference on Telecommunications and Signal Processing (2016)
18. Morar, A., Moldoveanu, F., Petrescu, L., Balan, O., Moldoveanu, A.: Time-consistent segmentation of indoor depth video frames. In: Accepted for publication at the International Conference on Telecommunications and Signal Processing (2017)
19. Petroniu, M., Trascau, M., Mocanu, I.: Object recognition with kinect sensor. In: 20th International Conference on Control Systems and Science, Bucharest (2015)

Contactless Physiological Data Analysis for User Quality of Life Improving by Using a Humanoid Social Robot

Roxana Agrigoroaie(✉) and Adriana Tapus

Autonomous Systems and Robotics Laboratory, U2IS ENSTA-ParisTech, Université Paris-Saclay, Palaiseau, France
{roxana.agrigoroaie,adriana.tapus}@ensta-paristech.fr

Abstract. Robots have increasingly been used for improving the quality of life of people with special needs, such as elderly, people suffering from Mild Cognitive Impairment (MCI), and people with different physical and cognitive abilities. In this paper, we propose a method for extracting and analyzing physiological data by using contactless sensors (i.e., RGB and thermal cameras). The physiological parameters that can be analyzed are: the respiration rate, blinking rate, and the temperature variation across different regions on the face (i.e., the forehead, the nose, the perinasal region, the left and right periorbital regions, and the entire face). These parameters, together with the action units (AUs) can be indicators of the current internal state of an user. We analyze data from three different scenarios and report the results obtained.

Keywords: Physiological data · Human-robot interaction · Quality of life

1 Introduction

In the past years, many research works concentrated on developing assistive technology for people with special needs (e.g., children and adults suffering of Autism Syndrome Disorder (ASD), elderly, people with different physical abilities, people suffering from Mild Cognitive Impairment (MCI)) [1–4]. One such project is EU Horizon2020 ENRICHME project[1]. The purpose of this project is to develop an assistive robot for elderly with MCI. A robot should be aware of the internal emotional and physiological state of the human users so as to better serve them and in the same time to improve their quality of life.

Determining the physiological internal state of a person is a research topic that received a lot of attention in the last decade [5,6]. Most of the devices used to gather physiological data are invasive. To counteract this, other methods have been developed, which use contactless sensors (e.g., RGB-D and thermal cameras). These cameras can be used to extract different physiological parameters that provide valuable information for determining the internal state of a person.

[1] www.enrichme.eu

S. Battiato et al. (Eds.): ICIAP 2017, Part II, LNCS 10485, pp. 696–706, 2017.
https://doi.org/10.1007/978-3-319-68548-9_63

The overall goal of the robot is to understand the context and the internal state of the user so as to adapt its behavior for a more natural human-robot interaction (HRI). The novelty of our work consists in combining different algorithms for extracting physiological parameters in a non-invasive way with the final goal of adapting the behavior of a humanoid robot for improving the quality of life of users.

Our paper is structured as follows: Sect. 2 presents a description of the physiological parameters and why they are important in the context of assistive robotics based on the literature; Sect. 3 briefly describes the sensors used for data acquisition; Sect. 4 describes how the data was recorded and analyzed; Sect. 5 shows the results, while Sect. 6 presents a short discussion based on the obtained results; finally, Sect. 7 concludes the paper and offers a perspective on future work.

2 Physiological Parameters

For a better understanding of human behavior, we must look at the underlying physiological activity. Some of the physiological data that we can look at are: the electrocardiogram (ECG), the electroencephalogram (EEG), the pupillary dilation, the electrodermal activity (EDA), the respiration rate (RR), the heart rate, the blinking rate (BR), etc. Moreover, there are some other non-verbal and para-verbal indicators of the current internal state of a person (e.g., facial expressions, prosody). In this paper, we look at the blinking rate, the respiration rate, facial expressions, and the temperature variation on the face during an interaction.

In this context, three types of blinking are identified in the literature [7]: spontaneous (without external stimuli and internal effort), reflex (it occurs in response to an external stimulus), and voluntary (similar to the reflex blink but with a larger amplitude). The type of blinking that is of interest for assistive applications is the spontaneous blinking. The authors of [8] found an average resting BR of 17 blinks/min, with a higher BR during a conversation (mean BR of 26) and lower BR during reading (mean BR of 4.5). In [9], the authors showed that the BR relates to both the task that an individual has to perform and to the difficulty of that task (e.g., during a mental arithmetic task, the more difficult the task, the higher the BR). Blinking has also been used in a deception detection approach [10].

While performing a certain activity, either physical or cognitive, it is important to make sure that the individual performing the task is not too stressed. The temperature variation on different regions of interest on the face could indicate the presence of different emotions (e.g., stress, fear, anxiety, joy, pain) [11]. In [12], the authors have used a thermal camera to reliably detect stress while interacting with a humanoid robot. The authors in [13] have shown some of the limitations and the problems that can arise when using thermal imaging for determining the temperature variation across different regions on the face.

One aspect in which the quality of life of an individual can be increased is by promoting more physical and cognitive exercises. An important parameter to monitor during these activities is the RR. The number of breaths per minute (BPM) is how RR is measured; where a breath is made up of two phases: inspiration and expiration. The average resting RR for adults lies between 12–18 BPM [14]. The RR can be measured either by using contact sensors (e.g., respiration belt [15], electrocardiogram [16]), or by using a thermal camera [17,18].

Emotional responses can be detected on the face as well. The easiest and most natural way of communicating our emotions is by using facial expresssions. Ekman and Friesen [19] have developed the Facial Action Coding System (FACS) for describing these facial expressions. The coding system defines atomic facial muscle actions called action units (AUs). There are 44 AUs and 30 of them are related to the contractions of specific muscles (12 on the upper face and 18 on the lower face). AUs can occur singly or in combination. The combinations of different AUs define different emotions [20] (e.g., happiness AU6 - cheek raiser + AU12 - lip corner puller).

3 Sensors

For the non-intrusive and contactless acquisition of data necessary for analyzing the physiological features presented in Sect. 2, we have used two sensors:

- An ASUS Xtion Live Pro RGB-D camera
- An Optris PI 640 USB-powered thermal camera

The RGB-D camera provides a 640 × 480 RGB image at up to 30 Hz. We used the configuration with a frame rate of 25 Hz. The 45 × 56 × 90 mm thermal camera has an optical resolution of 640 × 480 pixels and a spectral range from 7.5 to 13 µm. It is capable of measuring temperatures ranging from −20 °C to 900 °C at a frame rate of 32 Hz.

4 Methodology: Data Extraction and Analysis

The algorithms that we developed for the extraction and analysis of the physiological parameters, and the AUs are described. For the extraction of all parameters previously mentioned, the face of an individual is required. Therefore we developed a Robot Operating System (ROS) [21] package that can detect and track faces in a video feed. The face detection algorithm uses Dlib [22], an open source library with applications in image processing, machine learning, etc. The same library can also be used to determine the location of 68 feature points of interest on a face. We used these feature points to define regions of interest (ROIs) on the face.

4.1 Blink

For the detection of blinking we used the RGB data. Our ROS package detected the face and the feature points around the eyes (see Fig. 1a). The only ROI in this case is made up of the eyes. In order to determine if a person blinks or not, we looked at the distance between the eyelids.

As can be seen in Fig. 1b, each eyelid is characterized by two feature points. First, we computed the distance between the upper and lower eyelids for both points (e.g., in Fig. 1b the distance between the feature points 37 and 41). When a person is very close to the camera the face region is very large, but as the distance between the camera and the person increases, the face region decreases. Therefore, the distance between the eyelids can be very small. To counteract this, we squared the sum of the two distances for each eye. As there are multiple eye shapes and sizes, we had a period of 30 s in which we recorded the distances for both eyes, at the end of which we computed the mean eyelid distance for each eye. Only after this mean is available, we can detect if a person is blinking or not. We consider that a person has its eyes closed if the current distance is smaller than half the mean distance for that eye. When the person reopens the eyes we consider that he/she blinked. Knowing that a blink lasts for a period of 100–300 ms, our module is capable of making the distinction between a blink and keeping the eyes closed.

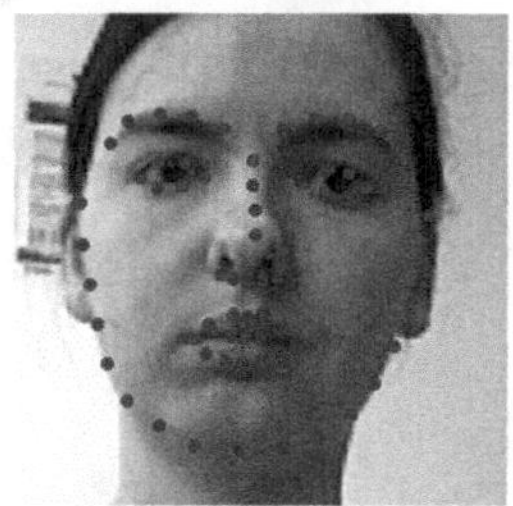

(a) Feature points

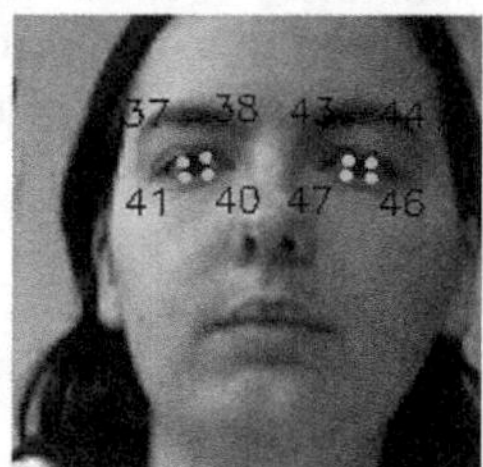

(b) Eyes feature points

Fig. 1. Facial feature points

We developed two methods to detect BR. For both methods, we are using a time period of one minute. In the first one, we simply count the number of blinks that we detected. For the second method, we saved all eyelids distances (see Fig. 3a for input signal) in a file. On the saved values we applied signal processing algorithms to detect the blinks. The steps that we applied are:

- We applied a low pass Butterworth filter with a sampling frequency of 25 Hz (the frame rate of our RGB camera), a cutoff frequency of 1.75 (see (1)) with the purpose of filtering out small variations in eyelid distance.
- We emphasized the moments with the largest change in distance by applying a differences and approximate derivative.
- Finally, all peaks were detected and counted (see Fig. 3b); in case the number of blinks for the left and right eye are different, the minimum of the two is considered to be the BR (a difference can appear in case a person is winking).

To compute the optimal cutoff frequency we applied (1) (see Eq. 9 in [23]), where f_s represents the sampling frequency, and f_c, the cutoff frequency.

$$f_c = 0.071 * f_s - 0.00003 * f_s^2 \tag{1}$$

4.2 Temperature Variation

The face detection algorithm was trained with RGB images, as a result the detector does not always work when using a thermal image. Therefore, we developed a program, which enables us to manually select the face region in the first frame and it is then tracked for the entire duration of the interaction. Once the face region was determined, the 68 feature points can be localized on the face and the ROIs can be defined (see also Fig. 2):

- the entire face: a region that covers the entire face.
- the forehead: a region with a width equal to the distance between the middle of the two eyebrows, and a height of 50 pixels.
- the left, and right periorbital region: a region with a size of 15 × 15 pixels around the inner corner of each eye.
- the perinasal region: a region defined between the corners of the nose and the distance between the tip of the nose and the upper lip.
- the nose: a region defined between the corners of the nose and the distance between the tip of the nose and the root of the nose.

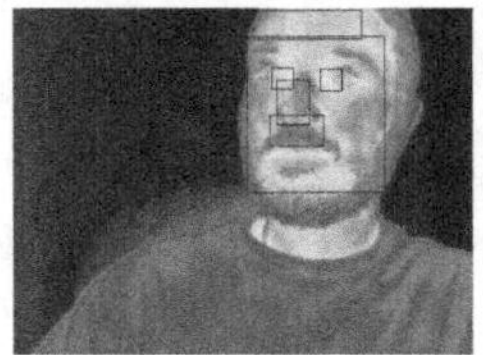

Fig. 2. ROIs in the thermal data. Temperatures range from 20 °C in dark purple to 40 °C in light yellow (Color figure online)

From each of these regions the average temperature was extracted. Our previous work [13] shows that there are 3 ROIs that need more consideration when performing the analysis. These regions are: the two periorbital regions and the entire face. One problem that was encountered was the presence of glasses. As the glasses do not transfer the heat from the face, their temperature is lower than the rest of the face. Moreover, in cases when the person turns its head there might be situations when the ROIs include parts of the background. All these values have to be discarded so that they do not influence the temperature variation. We previously found that all temperatures below 30 °C are associated with either the background or the glasses.

Once these temperatures were removed, we applied a low pass Butterworth filter (f_s = 32 Hz, f_c = 2.24 Hz) in order to eliminate temperature variations, which are associated to the movement of the person.

4.3 Respiration

The RR was determined using the temperature variation in the perinasal region defined in Sect. 4.2. As previously mentioned, a respiration is composed of two phases: inspiration and expiration. The temperature variation between these two phases is of interest when computing the RR. The following procedure (based on [18]) was implemented.

First, the mean temperature from the ROI was stored in a circular buffer of 30 s. Once the buffer was full, the RR could be estimated. As the resting RR lies between 12–18 BPM (0.2 Hz and 0.3 Hz), a Butterworth bandpass filter was applied in order to eliminate all other frequencies. Before applying a Fast Fourier Transform (FFT) on the signal a Hann window function was performed. Using the maximum frame rate of the camera, i.e., 32 Hz, and a duration of 30 s, a maximum resolution of 2 BPM (0.033 Hz) can be obtained. In order to improve this resolution, a quadratic interpolation was applied on the maximum magnitude and its neighbors of the frequency spectrum, which resulted after applying the FFT. The RR corresponds to the index of the maximum magnitude after applying the interpolation.

4.4 AUs

The detection of AUs can provide valuable input on the emotional state of a person. The module for AU detection was previously developed in our laboratory. The detector was trained using the CMU-Pittsburgh AU-Coded Face Expression Image Database [24]. The database consists of 2105 image sequences from 182 adult subjects of varying ethnicity, which perform multiple tokens of most primary FACS action units. For the training, a support vector machine was used and the OpenCV Viola Jones face detection algorithm. Our detector is capable of detecting the following AUs: AU1 (inner brow raiser), AU2 (outer brow raiser), AU4 (brow lowerer), AU5 (upper lid raiser), AU6 (cheek raiser), AU7 (lid tighten), AU12 (lip corner puller), AU15 (lip corner depressor), AU20 (lip stretcher), AU23 (lip tighten), and AU25 (lips part).

Given a video frame, the detector provides the prediction confidence for each of the previously mentioned AUs. By applying an empirically found threshold we select only the relevant AUs in that frame. Once this is accomplished we can know the emotional state of a person. Given the detected AUs, we are able to detect the following emotions [20]: surprise (AU1 + AU2 + AU5 + AU26 - jaw drop), fear (AU1 + AU2 + AU4 + AU5 + AU20 + AU25), happiness (AU6 + AU12), and sadness (AU6 + AU15).

5 Results

The algorithms previously described have been tested on 3 different data sets. The first data set was recorded during a demo that we performed with two participants at a care facility (Lace Housing, UK), part of the ENRICHME project.

The two participants (P1, P2) are older people who interacted with the Kompaï robot through a tablet mounted on the torso of the robot. As the two cameras (RGB-D, and thermal) were mounted on the head of the robot, the participants did not look directly at them.

The second and third datasets were recorded during two experiments that we performed in our laboratory for detecting deception. For the second dataset, the participants (P3 - P5) interacted with the same Kompaï robot, in an interview-like scenario. Therefore, all participants looked directly at the two cameras which were mounted below the head of the robot. The age of the participants varied between 20 and 40 years old. The last dataset used was recorded while the Meka M1 robot gave the instructions for the participants (P6 - P9) in a pen and paper task. The cameras were positioned on the table in front of the participants, while the robot was positioned a little on the participants' right side. Due to this positioning none of the participants looked directly at the cameras. All participants showed increased head movement during the interactions. Some of them had facial hair, while others wore glasses. We encouraged them to leave the glasses on, as we did not want to induce further stress.

Table 1. Blinking rate and respiration rate results

Participant	Recording duration [seconds]	BR manual annotation [blinks]	BR1 [blinks]	BR2 [blinks]	RR [BPM]
P1	120	16	29	16	16.87
P2	120	24	53	37	17.72
P3	147	95	108	78	11.78
P4	145	141	160	30	19.02
P5	134	29	53	74	27.58
P6	125	35	43	21	13.58
P7	110	23	101	73	12.24
P8	125	17	17	72	18.73
P9	105	12	19	52	22

5.1 Blinking

For testing our blinking algorithms we determined the mean distance for each eye and saved all the distances in a file. The processing was performed offline. We manually annotated the data (column BR manual annotation in Table 1) and compared it with the first blinking detection algorithm (column BR1), as well as with the second blinking detection algorithm (column BR2). As it can be seen, the first algorithm did not perform as well as the second in the case of older participants (P1, and P2). Neither of the two algorithms was able to detect all blinks, which can be explained by the movement of the participants or by the fact that most of them had the tendency to look downwards. However, for the first algorithm, the mean difference between the ground truth and the detected blinks (18.55) is half of that corresponding to the second algorithm (38.33).

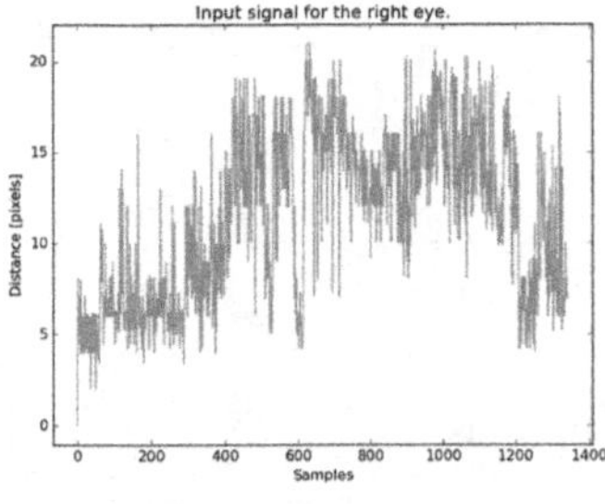

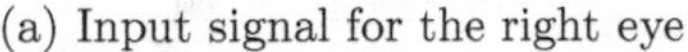

(a) Input signal for the right eye

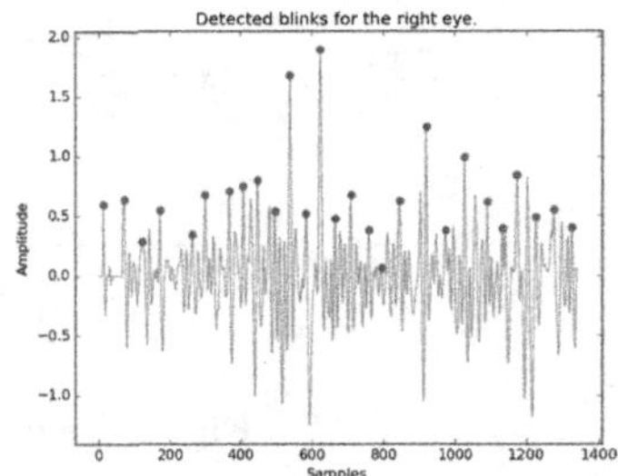

(b) Blinks detected for the right eye

Fig. 3. Blinking results

5.2 Thermal Data

Thermal data variation in different ROIs on the face is a good indicator of the internal state of a person. Given an input signal of the mean temperature in a region (Fig. 4a), by applying a low-pass Butterworth filter, a smoother version of the signal can be obtained. Using a linear regression the general trend of the signal can be obtained. Temperature increase in the periorbital region could be an indicator of anxiety [11].

5.3 Respiration

The results given by the respiration module are shown in Table 1. As no physical sensors were used, these results cannot be compared with the real RR of the participants. Figure 5a represents the input signal composed of the mean temperature values in the perinasal region of one of the participants. The output of the algorithm is displayed in Fig. 5b. The index corresponding to the maximum value after applying the interpolation gives the respiration rate. This point is represented in the figure, with the text label "Maximum".

5.4 AUs

As the database with which the AU detector was trained was composed of adults, AUs are not properly detected in case of older people. A person can be detected as being sad (due to the presence of AU4 and AU15), when actually the person is neutral. This problem appears due to the presence of wrinkles.

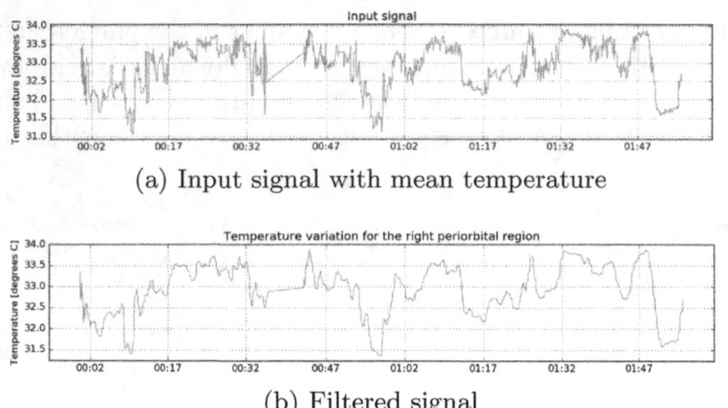

(a) Input signal with mean temperature

(b) Filtered signal

Fig. 4. Temperature variation analysis

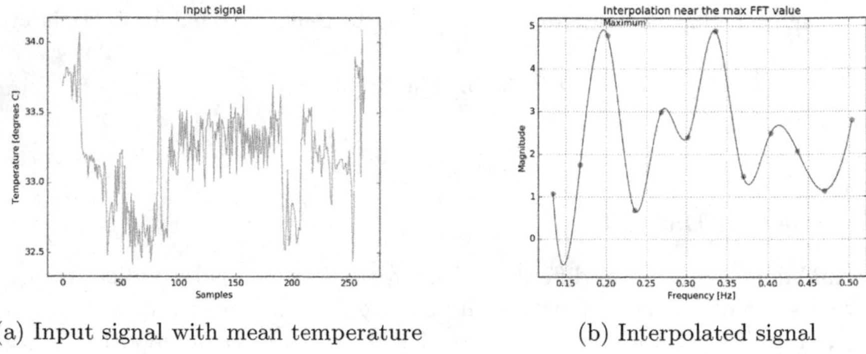

(a) Input signal with mean temperature

(b) Interpolated signal

Fig. 5. RR computing steps

6 Discussion

In a HRI scenario, in order to ensure a natural interaction the robot needs to know the internal emotional state of the person it interacts with. The algorithms presented in this paper enable a robot to estimate that state. All the modules were developed using ROS and have been tested on different robots in our laboratory (i.e., Meka M1 and the Kompaï robots). The results of the blinking detection algorithm show that in a real scenario, where a person is not forced to look directly at the camera, this can be very challenging. The person might move its head, the distance between the camera and the person can change, or the person could be looking downwards. An adaptive algorithm could provide a better performance than the algorithm we have used. Moreover, a mobile pan-tilt platform could be used to enable the cameras mounted on the robot to follow the gaze of a person.

In order to ensure ourselves that the results we get are in accordance with the real physiological parameters we are planning to perform a series of experiments in which to compare the values given by our physiological parameters analysis modules with the values given by physical sensors. However, our camera sensors are not considered medical devices as per the definition given by the EEC Council Directive 93/42[2].

The temperature variation module, in its current state, does not work in a real-time interaction. The analysis can be performed only offline. One way to improve this can be to perform the analysis every 20–30 s. This could enable the robot to know if there are important changes in the facial temperature of the person it interacts with.

7 Conclusion and Future Work

In conclusion, we have presented a series of algorithms that have been applied to extract and analyze physiological parameters (i.e., BR, RR, temperature variation across different ROIs on the face, and AUs). These algorithms have been used to detect the internal state of participants in experiments carried out in our laboratory. Some of our future work includes to perform a mapping between the RGB-D and thermal cameras, for face detection and feature points detection. Moreover, we plan to train a new AU detector based on the facial features provided by the Dlib library. This could enable us to detect multiple AUs.

Acknowledgments. This work was funded and done in the context of the EU Horizon2020 ENRICHME project, Grant Agreement No: 643691.

References

1. Chevalier, P., et al.: Impact of sensory preferences of individuals with autism on the recognition of emotions expressed by two robots, an avatar and a human. Auton. Robots Spec. Issue Assistive Rehabil. Rob. **41**, 613–635 (2016)
2. Tapus, A., et al.: Exploratory study: children's with autism awareness of being imitated by Nao robot. In: Interaction Studies (2012)
3. Tapus, A., et al.: Improving the quality of life of people with dementia through the use of socially assistive robots. In: Proceedings of the 1st International Symposium on Quality of Life Technology Intelligent Systems for Better Living (2009)
4. Mataric, M., Tapus, A., Winstein, C., Eriksson, J.: Socially assistive robotics for stroke and mild TBI rehabilitation. In: Advanced Technologies in Rehabilitation (2009)
5. Hernandez, J., et al.: Bioglass: physiological parameter estimation using a head-mounted wearable device. In: Proceedings of Wireless Mobile Communication and Healthcare (2014)
6. Hernandez, J., et al.: Bioinsights: extracting personal data from 'still' wearable motion sensors. In: 12th IEEE International Conference on Wearable and Implantable Body Sensor Networks (BSN) (2015)
7. Hart, W.M.: The eyelids. In: Adler's Physiology of the Eye. Mosby (1992)
8. Bentivoglio, A.R., et al.: Analysis of blink rate patterns in normal subjects. In: Movement Disorders (1997)

[2] http://eur-lex.europa.eu.

9. Tanaka, Y., Yamaoka, K.: Blink activity and task difficulty. In: Perceptual and Motor Skills (1993)
10. Abouelenien, M., et al.: Analyzing thermal and visual clues of deception for a non-contact deception detection approach. In: Proceedings of the 9th International Conference on Pervasive Technologies Related to Assistive Environments (2016)
11. Ioannou, S., Gallese, V., Merla, A.: Thermal infrared imaging in psychophysiology: potentialities and limits. Phychophysiology **51**(10), 951–963 (2014)
12. Sorostinean, M., Ferland, F., Tapus, A.: Reliable stress measurement using face temperature variation with a thermal camera in human-robot interaction. In: 15th IEEE-RAS International Conference on Humanoid Robots, Humanoids (2015)
13. Agrigoroaie, R., Ferland, F., Tapus, A.: The ENRICHME project: lessons learnt from a first interaction with the elderly. In: Agah, A., Cabibihan, J.-J., Howard, A.M., Salichs, M.A., He, H. (eds.) ICSR 2016. LNCS (LNAI), vol. 9979, pp. 735–745. Springer, Cham (2016). doi:10.1007/978-3-319-47437-3_72
14. Fei, J., Pavlidis, I.: Virtual thermistor. In: 29th Annual International Conference of the IEEE Engineering in Medicine and Biology Society - EMBS (2007)
15. Merritt, C.R., et al.: Textile-based capacitive sensors for respiration monitoring. IEEE Sens. J. **9**, 71–78 (2009)
16. Park, S.-B., et al.: An improved algorithm for respiration signal extraction from electrocardiogram measured by conductive textile electrodes using instantaneous frequency estimation. Med. Biol. Eng. Comput. **46**, 147–158 (2008)
17. Barbosa-Pereira, C., et al.: Remote monitoring of breathing dynamics using infrared thermography. In: 37th Annual International Conference of the IEEE Engineering in Medicine and Biology Society - EMBS (2015)
18. Chauvin, R., et al.: Contact-free respiration rate monitoring using a pan-tilt thermal camera for stationary bike telerehabilitation sessions. IEEE Syst. J. **10**, 1046–1055 (2014)
19. Ekman, P., Friesen, W.V.: Facial Action Coding System. Consulting Phychologists Press, Palo Alto (1978)
20. Ekman, P., et al.: Facial action coding system - investigator's guide. In: Research Nexus, Salt Lake City - USA, p. 174 (2002)
21. Quigley, M., et al.: ROS: an open-source robot operating system. In: ICRA Workshop on Open Source Software (2009)
22. King, D.: Dlib-ml: a machine learning toolkit. J. Mach. Learn. Res. **10**, 1755–1758 (2009)
23. Yu, B., et al.: Estimate of the optimum cutoff frequency for the Butterworth low-pass digital filter. J. Appl. Biomech. **15**, 318–329 (1999)
24. Kanade, T., et al.: Comprehensive database for facial expression analysis. In: Proceedings of the International Conference on Automatic Face and Gesture Recognition, pp. 46–53 (2000)

Exploiting Social Images to Understand Tourist Behaviour

G. Gallo[1,2], G. Signorello[1,3], G.M. Farinella[1,2], and A. Torrisi[3(✉)]

[1] CUTGANA, University of Catania, Catania, Italy
[2] Department of Mathematics and Computer Science (DMI), University of Catania, Catania, Italy
[3] Department of Agriculture, Food and Environment (Di3A), University of Catania, Catania, Italy
alessandro_torrisi@hotmail.com

Abstract. In this paper we propose to exploit the georeferenced images publicly available on social media platforms as a source of information to get insights of the behavior of tourists. At first, the metadata of the georeferenced images are explored through Visual Analytics tools to identify trends, patterns and relationships among the information acquired by social media. Then data mining techniques are used to generate a traveling model of the area of interest. Finally, we consider sites that are likely to be jointly visited and analyze how the tourist flow goes from one to another in these cases. To confirm the effectiveness of the proposed analysis we have tested the proposed methods in a case study.

Keywords: Social media · Tourism management

1 Introduction

Tourism plays an important role in the economic development of many geographical areas. It requires adequate analysis of tourists behaviors to encourage innovation and promotion of the products along to market trends and needs. Researchers have studied the touristic flows with interviews and opinion polls at entrances of sites of interest, such as museums and national parks. This data gathering technique is expensive and is limited in terms of spatial and temporal coverage and it does not ensure long-term prospects for the activation of policies and marketing strategies.

The direct interviews of tourists require a staff of people to conduct the interviews and the related analysis. Hence, it can be performed only for a limited number of days to limit costs. Since stratified information about gender, nationality, ages, etc. of visiting persons is of special relevance, the size of the significative samples for a direct interview would be prohibitively high.

Tickets at the entrance of the attractions are another way to count the number of visitors. However, a huge number of natural and monumental areas are just open spaces and there is no possibility to count people through ticketing.

S. Battiato et al. (Eds.): ICIAP 2017, Part II, LNCS 10485, pp. 707–717, 2017.
https://doi.org/10.1007/978-3-319-68548-9_64

This is typically the case of many highly recognized public areas inscribed on the World Heritage List [1].

The aforementioned considerations suggest the research for new effective methods to achieve a comprehensive understanding of tourists' travel behavior. To this aim, a solution comes from the huge amount of georeferenced information voluntarily posted by tourists on social media during their trips. Georeferencing refers to the association of GPS coordinates to a given digital format data. Looking at the digital images posted in social media, the geotagging tells where any image has been taken. Geotagging also allows the images to be organized and geographically arranged on a map system.

There are, by now, several very popular web photo sharing services (Flickr [2], Instagram [3], 500Pixels [4], etc.). Among these we choose to work with the Flickr portal because of the rich information for each picture posted on this platform and for the ready availability of suitable Application Program Interface (API) for data extraction. The most relevant information we look at is of course the geographical coordinates record, but information about the data when each picture has been taken and posted together with information about the user (age, nationality, gender, etc.) are also collected. These information can then be used to build statistics and recommendation systems for tourist attractions and services. The relevance of such systems in the e-tourism domain is discussed in [5]. The authors of [6] present a method to capture travel information from geotagged photos on Flickr. First, they conduct a cluster analysis in order to identify the major areas of interest visited by inbound tourists in Hong Kong. Then, a reconstruction of tourist movements is obtained according to the time information of the pictures extracted on a daily basis. To enrich this information with a model of tourists flow from a location to another, a Markov model is employed. Different travel routes followed by tourists have been discovered by the authors with this approach. The authors in [7] propose a recommendation system by combining topic models and Markov chains. Their final tourist behavior model provide a set of personalized travel routes that match the user's current location, user's interest, and user's spare time. Another travel route recommendation system is proposed in [8] where the authors describe an approach to build structured models of human travel as a function of spatially-varying latent properties of locations and travel distances. Similarly to what is proposed in [6], the authors make the assumption that human travel can be treated as a Markov process. By discretizing the desirability of locations and the distance between them, it is possible to analyze affinities between locations. In addition, individuals are grouped into clusters with distinct travel models. The use of latent models is hence exploited to make predictions in new situations. Another approach in this context is the one described in Torrisi et al. [9]. The authors propose to use Flickr data to build a Visual Analytics tool in order to analyze the distribution of tourists with respect to gender, nationality (e.g., domestic vs foreign), also considering the date of the visit (e.g., distribution of pictures for month/year).

In this paper we introduce a simple model to learn tourist behavior by exploiting Flickr georeferenced data. Flickr data of selected areas are collected and organized to perform visual analysis. We process the information of the georeferenced images to infer the tourist flow through a graph representation whose nodes are related to parts of the area to be monitored. All the images are associated with a specific node of the graph taking into account the GPS coordinates. Two cells, i.e., two nodes in the graph, are connected by a weighted edge: the weight is proportional to the number of single Flickr users that have posted photos taken in both cells. Each connection on the graph is hence equivalent to a likely route covered by tourists during a trip. This procedure allows us to infer information about the major routes followed by tourists during their visits. Statistical association rules that model the tourist traffic among neighboring sites are obtained using the Apriori data mining algorithm [10]. The most followed paths are then selected for further analysis. The proposed method is useful to the tourist managers to conduct planning activities regarding the transportation system, tour operators and positioning of the accommodation and info points.

2 Proposed Method

In this section we detail the method used to extract insights about tourist behavior from social images.

2.1 Data Extraction and Exploration

Social media platforms generate a huge amount of data every day. An automated procedure to have a constant access to this data for tourism management purposes is hence very helpful. In addition to the browsing mode, Flickr allows users to download public data through its web based API [11]. To use the API, a structured query is submitted to the service and a response is returned in XML, XML-RPC, JSON or PHP format. Flickr offers clear documentation and an intuitive API Explorer that allows to try API methods through the browser. Tutorials and examples of how to use the API with different programming languages are also available.

The first thing to do in order to extract data from a social platform is the selection of an area of interest. Flickr API's allow two ways to retrieve geotagged pictures. The first way retrieves pictures located inside a circle specified by latitude, longitude of its center and a radius. The other way is to retrieve the pictures located inside a bounding box specified with the coordinates of the bottom-left and top-right corners. There is however a problem: Flickr server does not provide more than about 4000 distinct records for each query. This required us to adopt a simple recursive subdivision strategy using the bounding box approach. In particular, the initial box that includes the entire area of interest is splitted quadtree-likewise until we have a reasonable confidence that all the pictures in the area have been taken. This first gathering of records is refined and integrated using other Flickr API's methods. In particular, using *flickr.photos.getInfo* and *flickr.people.getInfo*, we complete the records of each picture with more information about time, location and about gender and provenience of the photographer.

We adopt Visual Analytics methods on the data obtained insofar in order to generate insights for further analysis. Latitude-longitude pairs allow to obtain a representation of spatial distribution of photographers inside the area of interest. By combining this chart with the nationality of photographers it is also possible to have proportion of photographers for each region or roughly distinguish them between domestic or foreign. However, one of the drawback of this type of representation is that we do not have a density estimation of the tourists presents at a given point. In other words, photos taken in the same point are superimposed in a trivial 2D visualization. To overcome this problem, we consider a probability density function estimation to investigate the properties of the data points. In particular, the Parzen-Rosenblatt window method is used [12]. Intuitively, this approach counts the number of samples belonging to a specified square region $R_{(lon,lat)}$ with size $h \times h$ surrounding a geolocalized point of interest with GPS coordinates (lot, lan). In our study, we consider a Gaussian kernel centered on geolocalized point of interest to compute the Parzen-Rosenblatt probability density function. Hence, given the whole set of images' coordinates, the probability density function in a location $x = (lon, lat)$ is estimated as:

$$p(x) = \frac{1}{n}\sum_{i=1}^{n} \frac{1}{2\pi h^2} \exp\left(-\frac{||x - x_i||^2}{2h^2}\right) \tag{1}$$

The probability density obtained in this way can be superimposed to a map to give information about the most popular locations travelled by tourists. We also obtain other visual charts useful to analyze the distribution of photographers with respect to the gender or the date of the visit.

2.2 Data Mining Techniques to Analyze Travel Preferences

The previous described analytics tools are useful to understand the interest of tourists on a given area. The creation of summary and statistics of this data can help tourism managers to have an idea about the tourist population. However, this kind of analysis do not provide details about the most followed paths chosen by tourists during their trips. To this aim we take into account the location where pictures have been acquired as well as time information of collected images. In this way we are able to reconstruct the tourist trip considering the sites they have visited and the duration of their trip (e.g., daily or weekly excursion).

For this kind of analysis the density distribution described in the previous sub-section is not very helpful. What is needed here is some kind of "discretization" of the locations inside the area of interest. We hence build a list of "sites" that in this context are simply clusters of the locations where the pictures have been taken. These clusters are obtained with a naive, but effective, method: the area of interest is partitioned into square cells of uniform dimension. The pictures belonging to a "site" are those whose coordinates fall within one of these cells. This approach opens an issue: how to select the scale ("granularity") of the cells of the discretizing grid? The choice of the proper scale is indeed application-dependant: in large areas (e.g., natural parks, wildlife areas, etc.) it is reasonable to choose quite large cells (cell diameters in the order of thousands of meters), on the other hand in smaller areas (urban city centers) the cell diameter should not exceed the hundreds of meters.

We keep track of the trips taking into account the temporal sequence of images of each photographer (Fig. 1). For all the trips that have been carried out on the area of interest in a specific time interval by the same photographer, we achieve a weighted diagram of the routes followed by the photographers. The proposed method is formally modeled by a directed graph. We define a graph $G = (N, E)$ composed by N different nodes and E edges. An oriented edge from i to j indicates tourist flow from the site i to the site j. Two main parameters are involved to model the grid graph:

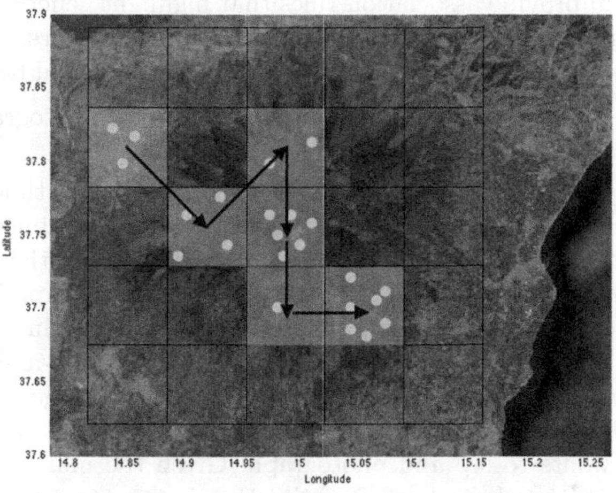

Fig. 1. Reconstruction of tourists trips. Yellow points represent the images taken by a photographer inside the area of interest. Each image is associated to a cell of the grid taking into account its GPS coordinates. The trip of the photographer is then reconstructed using the temporal sequence of the shot images. (Color figure online)

- M: it represents the granularity of the grid. The higher the value of M, the greater the number of nodes in the grid graph (while the area size of the sites of interest decreases). In the example in Fig. 1, M is equal to 5.
- Δt: it indicates the duration of the trip. In this way we can choose to model trips that have a daily or weekly basis.

The graph is represented through a weighted adjacency matrix. Each entry of the matrix indicates the amount of traffic flow between pairs of sites. To understand data an analyst may look to the graph to have a comprehensive view of all the paths followed by tourists inside the selected area. A limit in this kind of visualization is that it erases the "sequentiality" of each photographer visit to the cells. In other words, given a pair of sites (i, j) it cannot be directly estimated from the adjaceny matrix what are the preferred routes. Some tourists could stop at a site, others might continue for one or more sites. Understanding how much a particular path (composed by 2 or more nodes) is taken by travelers is a useful information to tourism managers to conduct planning and tourism management of the territory. Because of this we need to extract association rules between sets of nodes that have a certain probability of being jointly visited during a trip. To learn association rules among tourist sites we employ the Apriori algorithm [10]. The Apriori algorithm attempts to infer frequent subsets of sites which are common to the different tourists with a minimum support. To this aim, Apriori uses a bottom-up approach, where frequent subsets of tourist sites are extended one site at a time and groups of candidates are tested against the data. This iterative process continues until no further successful extensions of the sites subsets are found. The frequent tourist sites item sets computed by the Apriori algorithm provide association rules that highlight general trends related to which sites are visited jointly with a specific confidence score. In our settings, for a given rule $s^i_{(lon,lat)} \rightarrow s^j_{(lon,lat)}$, confidence is proportional to the likelihood that the site $s^j_{(lon,lat)}$ is visited during the same trip of a photographer who has visited the site $s^i_{(lon,lat)}$.

The last step of our analysis takes into account one at each time the strongest association rules generated by the Apriori algorithm. The aim is to get more details on the tourist flow generated between nodes (sites) of the obtained association rules. In other words, it would be useful to have a comprehensive knowledge of the routes covered by tourists to move from one site of interest to another one along these "most travelled paths". They could follow the main path, others may choose auxiliary routes. For example, a tourist could get away from the main route and choose a customized way as it presents particular naturalistic details that he wants to see and photograph. Given the main path, which are the possible paths that branch off from it? How many tourists are affected by these detour paths? Which part of the route is more appropriate to locate an accommodation facility for tourists? These are just some examples of possible questions that require the need to know travel information along specific paths. The easiest and quickest way to get this type of information involves the use of buffers in ArcGIS [13]. From the geometric point of view, the buffer is a polygon whose perimeter identifies a territorial area which is located at a distance with respect to the path of interest, between a minimum and a maximum value. Our analysis includes the creation of buffers of increasing size from the main path related to association rules to locate different bands of territory and see the evolving of the tourist flow in these areas.

3 Case Study

The case study considered here is the area around Mount Etna, an active volcano on the east coast of Sicily. The summital area of the volcano is a protected natural area of approximately 59.000 ha. The overall natural environment surrounding the highest active volcano in Europe is very attractive for visitors from all the world but because of its large extension and of its openess there is no way to directly know the tourists behavior.

The region of interest that is included in the bounding box for data extraction includes the entire area of the volcano and the twenty small towns on its proximities. Flickr allows to add a search filter based on the date on which the photos have been taken. Another parameter that could be used in the search is "accuracy": it is a measure of how accurate are the GPS coordinates of the image with respect to the point where the image was taken. We have conducted different experiments with different levels of accuracy. It can take any value between 1 (World level) and 16 (Street level). Filtering images using the accuracy parameter has the only effect of reducing, with higher levels, the number of retrieved images. It does not affect the quality of the visual content in the images as we have observed in few experiments. Since for the present application to have large datasets is relevant we choose the minimum accuracy level. The final dataset is composed by 30,692 images taken by 2932 different photographers. Figure 2

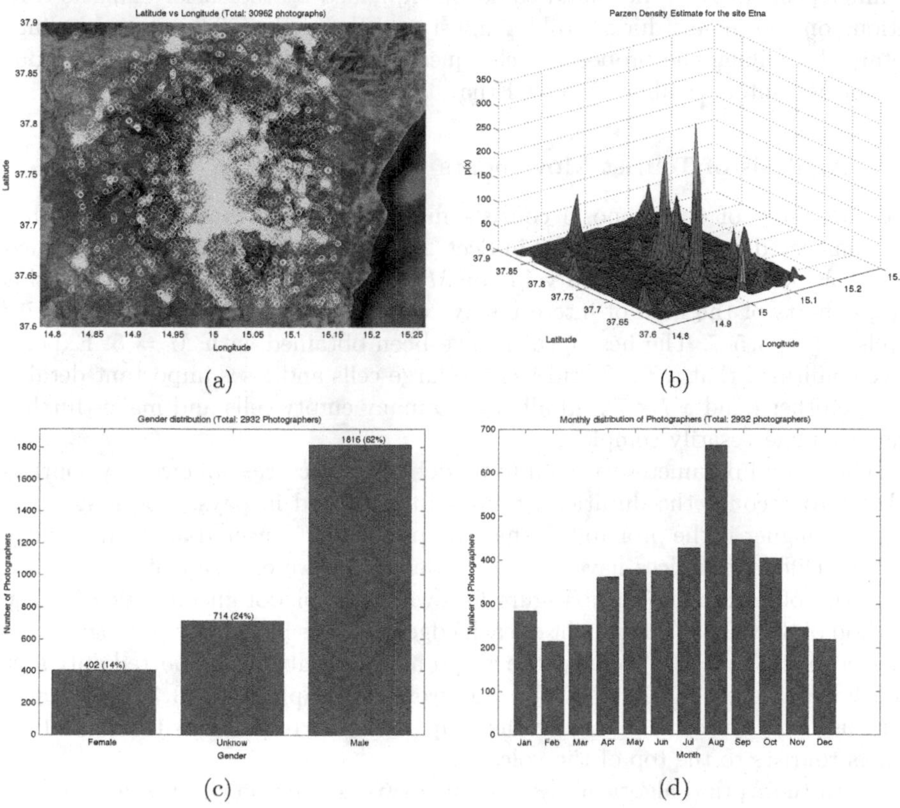

Fig. 2. Some examples of Visual Analytics charts of the collected dataset. (a) Distribution of photographs inside the area of interest and (b) its Parzen-Rosenblatt probability density estimation. (c) Gender and (d) monthly distribution of photographers.

shows some insights of the collected data. In particular, Fig. 2(a) reports the distribution of pictures inside the area of interest. With this information it is simple to infer that the most visited sites are located in the eastern and the southern sides of the volcano. More details among the tourist concentration in these areas are obtained through the application of the Parzen-Rosenblatt probability density function (Fig. 2(b)). The two peaks we get through this approach correspond respectively to two sites on the volcano. The highest one is related to "Rifugio Sapienza", the starting point of many different guided tours. It also hosts the entrance of the cableway facilities. The second peak corresponds to the central area surrounding the main active crater of the volcano. This area is of particular interest for tourists since mount Etna is an active volcano and tourists go as closer to the crater during the eruptions as safety permits. Some details about photographers are shown in Fig. 2(c) and (d). The histograms report the gender distribution and the seasonal distribution. Both gender and seasonal distributions agree with the general distribution of Flickr users. Indeed, there is no evidence that Mount Etna gets more visits from men than from women. The summer peak on the other hand could be explained by the easier climatic conditions on the mount during July, August and September. The above findings obtained applying the proposed techniques confirm what was the guess about the tourists interests about Mount Etna.

3.1 Analysis of Tourist Movements

The next step of our proposal consists in analyzing the tourist traffic within the considered area. As described in Sect. 2.2, the area of interest is subdivided into a $M \times M$ cells grid. The value of M should be chosen carefully. To this aim, experts of the area of interest may consulted. We tried three granularity levels: $M = 3, 5, 7$. The best results have been obtained with $M = 5$. Experts have confirmed that a 3×3 grid has too large cells and loses important details. On the other hand a 7×7 grid allows too many empty cells and makes further analysis unnecessarily complex.

The second parameter we need to model the trajectories followed by tourists takes into account the duration of the trip measured in days. The longer the trip, the higher is the probability that completely unconnected areas are jointly visited. Different choices have been considered and we experimentally selected only trips of maximum 7 days. Figure 3 shows the graph obtained considering the method of Sect. 2.2. The weight of each edge indicates how many photographers have covered the route. To make the graph more readable, we reported only arcs that have a weight ≥ 10. As can be seen from the graph in Fig. 3, tourist traffic is focused on the central part of the map which correspond to the path that brings tourists to the top of the volcano.

With the Apriori algorithm we obtained two association rules which confirm that the edge between the red nodes in Fig. 3 is the most travelled one. In particular the two symmetric rules that characterize the path between the two red nodes have $Support = 8\%, Confidence = 22\%$ in one case and $Support = 8\%, Confidence = 25\%$ in the other case.

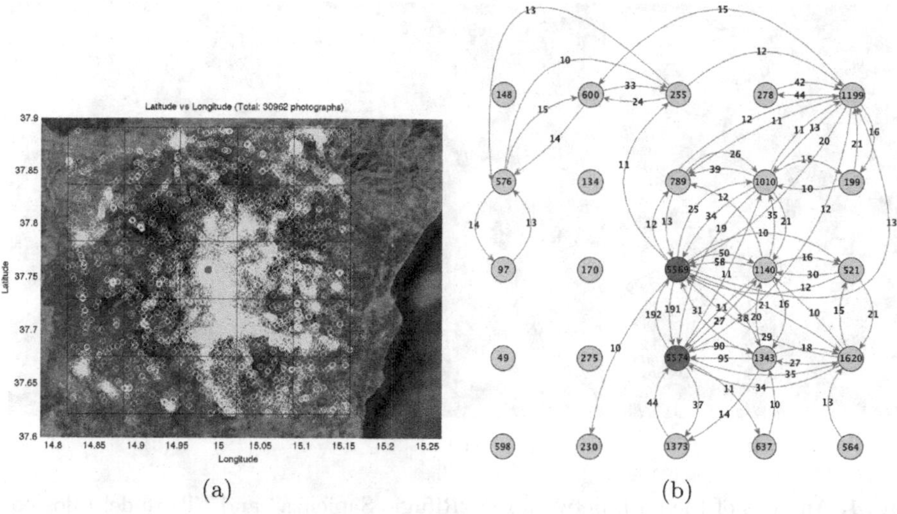

Fig. 3. Travel model of photographers inside the mount Etna area. (a) The area of interest is subdivided in 25 different sites of interest. (b) Weighted graph of the paths followed by tourists during their trips. Red points represent the endpoints of a path that is the most covered by tourists in both directions. (Color figure online)

In the final step of our study we zoom on the edge between the red nodes. The idea is to analyze which paths are chosen during the crossing between these two sites. The Etna paths are developed on recent and historical lava flows, in wooded areas or in areas without tree vegetation. These often present a variety of slopes resulting from the changing morphology of the volcano. The paths that we focus on are those that connect "Rifugio Sapienza" (node (4, 3) in the graph) with "torre del filosofo" (node (3, 3) in the graph). More specifically, "Rifugio Sapienza" at about 2000 m above the sea level is an area easily reached with cars or buses and local guides offer many trekking opportunities toward the higher areas starting from there. "Torre del filosofo" (english: Philosopher tower) at about 2900 m above the sea level is the mytical place where Heraclitus killed himself jumping into the lava flow. More realistically is a hill from where the main crater at 3000 m above the sea level can be easily and safety observed. To investigate tourist behaviors in this area we have identified what is the main route taken by the guides using ArcGIS. We have considered four different buffers around the main path (Fig. 4(a)). We have considered a maximum radius of 200 m from the center of the path, with incremental steps of 50 m. Since the GPS coordinates of acquisition devices may differ by several meters from the actual value, the first buffer covers a distance of 50 m around the main path. Figure 4 shows the details about the areas belonging to each buffer. The box in Fig. 4(a) and (b) contains 7259 images. Most of the pictures are located outside the path of interest and the four buffers. This means that tourists move largely away from the main road to explore the surroundings. The photos within the

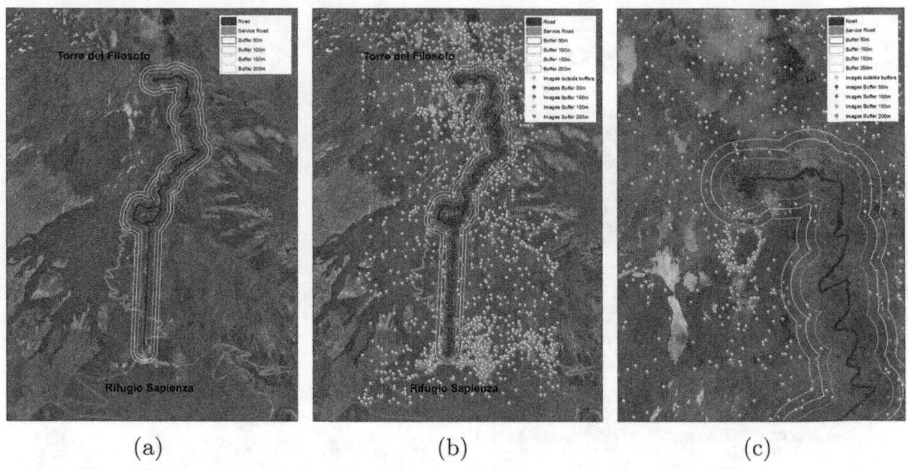

Fig. 4. Analysis of the path between the "Rifugio Sapienza" and "Torre del Filosofo" sites. (a) The dark blue line refers to the main path to be investigated. Four different buffers are considered with an incremental radius from the center of the main route. (b) Images taken inside the area of interest are shown on the map with the colour of the corresponding buffer. (c) Zoom of the final section of the route. (Color figure online)

four buffers are only 2333. The first and the second buffers contain both the 32% of the whole set of images and then this percentage decreases to 19% in the third buffer and to 17% in the last one.

The data taken into account in the case study reveal, through the suggested tool, some interesting insights. Our analysis confirms that the tourist population on Mount Etna is mainly distributed on the eastern and southern versants of the volcano. Although Etna area has been interested by a continuous tourism development, its versants have different degrees of anthropization. The west side of the mountain is strongly linked to traditional agricultural activities; it represents an area of territory less developed from the tourism point of view although it is the most suitable to practice a tourism regarding naturalistic aspects. The southern and eastern sides, on the contrary, offer greater receptive hospitality structures. The presence of nature trails and ski facilities further contributes to making this area the most popular choice among tourists. The seasonal distribution of visits helps to understand when the highest presence of tourists is realized. Our analysis confirms the seasonal nature of the Etna territory. It highlights that the winter tourism (from November to March) is practiced almost exclusively by winter sports enthusiasts. They usually take less photographs than a tourist who prefers a hiking tourism feasible mostly from April to October. The second part of our methodology produces a travel model of the tourism movements within the area of interest.

With the buffer analysis we have noticed there is considerable amount of pictures taken outside the main path of interest. This indicates that many tourists customize their way to go to the north part of the volcano.

4 Conclusion

In this paper we have proposed a framework composed by different methods which is useful to analyze the tourist flow present in an area of interest and extracts useful clues that helps the tourism managers to conduct a reliable activity for tourism planning. In particular, we based our analysis on the data that are today available on social media platforms. We have built a tool that analyzes the spatial distribution of the tourists. Using data mining techniques we have also proposed a method to infer tourist flow within the area of interest. The method has been applied to a specific case study, taking into account the area around the mount Etna, one of sites included in the World Heritage List. The finding supports and confirms the experts' knowledge about this region.

References

1. Unesco, World Heritage List. http://whc.unesco.org/en/list/. Accessed 19 June 2017
2. Yahoo!, Flickr. http://www.flickr.com/. Accessed 19 June 2017
3. Instagram. https://www.instagram.com/. Accessed 19 June 2017
4. Five hundred px, 500px. https://500px.com/. Accessed 19 June 2017
5. Rabanser, U., Ricci, F.: Recommender systems: do they have a viable business model in e-tourism? In: Frew, A.J. (ed.) Information and Communication Technologies in Tourism 2005, pp. 160–171. Springer, Heidelberg (2005). doi:10.1007/3-211-27283-6_15
6. Huy, Q.V., Gang, L., Rob, L., Ben, H.Y.: Exploring the travel behaviors of inbound tourists to Hong Kong using geotagged photos. Tour. Manag. **46**, 222–232 (2015)
7. Kurashima, T., Iwata, T., Irie, G., Fujimura, K.: Travel route recommendation using geotagged photos. Knowl. Inf. Syst. **37**(1), 37–60 (2012)
8. Guerzhoy, M., Hertzmann, A.: Learning latent factor models of human travel. In: NIPS Wokshop on Social Network and Social Media Analysis: Methods, Models and Applications (2012)
9. Torrisi, A., Signorello, G., Gallo, G., De Salvo, M., Farinella, G.M.: Mining social images to analyze routing preferences in tourist areas. In: Workshop on Visualisation in Environmental Sciences (EnvirVis). The Eurographics Association (2015)
10. Agrawal, R., Srikant, R.: Fast algorithms for mining association rules in large databases. In: Proceedings of the International Conference on Very Large Data Bases, pp. 487–499 (1994)
11. Yahoo!, Flickr API. https://www.flickr.com/services/api/. Accessed 19 June 2017
12. Bishop, C.M.: Pattern Recognition and Machine Learning. Springer, New York (2006)
13. Esri, ArcGIS. https://www.arcgis.com/. Accessed 19 June 2017

Showing Different Images to Observers by Using Difference in Retinal Impulse Response

Daiki Ikeba[1], Fumihiko Sakaue[1], Jun Sato[1](✉), and Roberto Cipolla[2]

[1] Department of Computer Science, Nagoya Institute of Technology, Nagoya 466-8555, Japan
ikeba@cv.nitech.ac.jp, {sakaue,junsato}@nitech.ac.jp

[2] Department of Engineering, University of Cambridge, Cambridge CB2 1PZ, UK
cipolla@eng.cam.ac.uk

http://www.cv.nitech.ac.jp, http://www.eng.cam.ac.uk

Abstract. In this paper, we propose a novel method for displaying different images to individual observers by using the difference of temporal response characteristics of these observers. The temporal response characteristics of human retina have individuality, and thus each observer observes slightly different image, even if the same image is presented to these observers. In this paper, we describe the variety of temporal response characteristics of observers by using the linear combination of impulse response of observers in time domain, and derive an observation model of observers in time domain. By using the observation model, we propose a method for representing arbitrary different images to individual observers. Experimental results from a special camera which reproduces the impulse response of human retina show that our proposed method can represent different and arbitrary images to multiple observers.

1 Introduction

The multiplex image representation enables us to show multiple different images to individual observers [1–4]. They are often used for displaying 3D images to human observers, in which left and right eyes of the observer observe different images with a disparity. Many applications can be realized in future by using the multiplex image representation. If the multiplex image is presented by a TV in a living room, multiple observers can watch different TV programs simultaneously by using a single TV set as shown in Fig. 1(a). If the display on a vehicle can present multiplex images, the passengers can enjoy amusement programs on the display, while the driver of the vehicle is observing a navigation image on the same display as shown in Fig. 1(b).

Unfortunately, the existing multiplex image representation methods require special equipments, such as stereo glasses which separate multiplex images into single images. The parallax barriers [5,6] and lenticular lenses [7] are also used for separating multiplex images according to the viewpoint of observers. However,

S. Battiato et al. (Eds.): ICIAP 2017, Part II, LNCS 10485, pp. 718–729, 2017.
https://doi.org/10.1007/978-3-319-68548-9_65

the positions of the viewpoints are fixed in these methods, and thus observers cannot observe objective images at arbitrary viewpoints. For moving observers, Perlin et al. [8] proposed a method, which controls light rays of the display according to the viewpoint of the observer, so that the observer can see appropriate images at any viewpoint. More recently, complex light field displays were developed and used for showing dense light fields toward multiple observers simultaneously [9–14]. These methods are very powerful when we want to display the same 3D information to all the observers. However, if we want to display different images to individual observers regardless of their viewpoints, these existing methods fail, since all these methods were based on the geometric difference of viewpoints.

(a) (b)

Fig. 1. Multiplex image representation. (a) Multiple observers can enjoy different TV programs simultaneously by using a single television. (b) Passengers can enjoy amusement programs on a vehicle display, while a driver is observing a navigation image on the same display.

Recently, Nonoyama et al. [3] proposed a new method for displaying multiplex images without using the geometric properties of light fields. In their method, they used not the characteristics of equipments, but the characteristics of human observers for displaying different images to individual observers. In particular, they used the difference of spectral sensitivity of human vision for displaying different images to individual observers. Their method does not require special glasses nor the fixation of viewpoints. However unfortunately, it cannot be used when the spectral sensitivities of observers are similar to each other. In addition, their method requires special display equipments which can control the spectral distribution of output light at each image pixel.

In order to overcome these problems, we in this paper propose a novel method for displaying multiplex images by using the temporal characteristics of human vision. In our method, we use the difference in the reaction time of human retina. Suppose a light ray is received by a photoreceptor on a human retina. Then, the response of the photoreceptor in time domain is not identical in all human observers. That is, there are individuality on reaction time of retinas. In this paper, we show that by using the difference of the temporal characteristics of human retinas, we can present different images to individual human observers simultaneously by using a single display. The proposed method does not require special equipments unlike the existing methods, and thus it can be used in various applications.

2 Observation Model

2.1 Temporal Impulse Response of Human Retina

We first consider the temporal response characteristics of human retinas. Suppose a photoreceptor on a human retina receives an impulse light as shown in Fig. 2(a). Then, the photoreceptor responds to the light and generates an output signal. It is known that the response of the photoreceptor is not an impulse at a moment, but a continuous signal which remains a certain period of time as shown in Fig. 2(b) [15]. This temporal response caused by an impulse light is called an impulse response of retina. As shown in Fig. 2(b), the impulse response of the human retina becomes large gradually, and after that it becomes small gradually. It is reported that the impulse response of retina depends on observers, and thus, the response curves are different from each other [15].

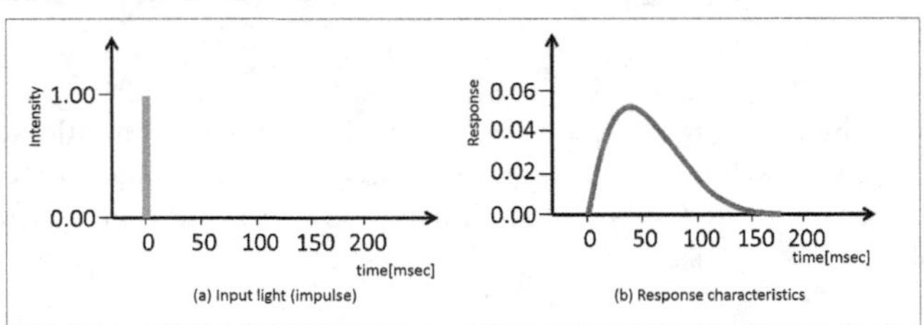

Fig. 2. Temporal response characteristics of human retina when the retina receives an impulse light signal. (a) Shows an input impulse light, and (b) shows the temporal response of human retina, i.e. impulse response.

Since the impulse response is continuous, we sample it with a fixed sampling period, and represent it by an n-dimensional vector $\mathbf{m}$ as follows:

$$\mathbf{m} = [m_1, \cdots, m_n]^\top \tag{1}$$

2.2 Observation Model from Impulse Response

We next consider the response of the human retina when the retina receives sequential light signals.

Let us consider the case where sequential light signals shown in Fig. 3(a) are received by a human retina. The light signals are sampled and represented by a discrete vector $\mathbf{x} = [x_1, \cdots x_n]^\top$. Since the impulse response of the retina has a

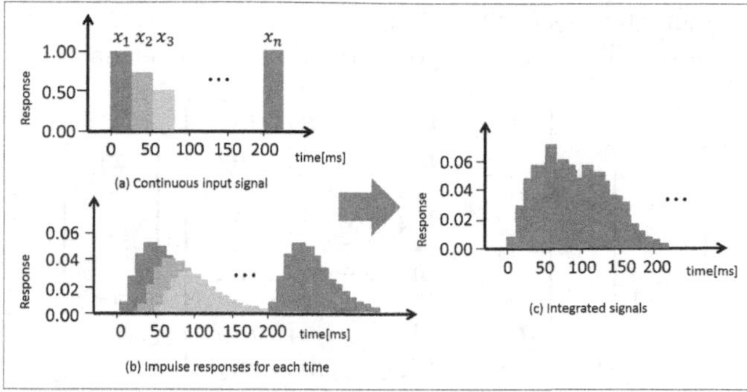

Fig. 3. Temporal response of human retina when the retina receives sequential light signals. (a) Shows sequential input light signals, and (b) shows impulse responses for each input signal. (c) Shows the integral of impulse response for the sequential input signals.

certain duration, the response of the retina at time t depends not only on the input light signal at time t but also on the past input light signals as shown in Fig. 3(b). As a result, the response of the retina at time t can be described by the sum of the impulse response multiplied by the sequential input signals as follows:

$$y_t = \sum_{i=1}^{n} x_{t-i} m_{1+i} \tag{2}$$

Therefore, the sequential response y_t of the retina can be represented by a set of linear equations on the impulse response m_t and the input light signals x_t as follows:

$$\begin{bmatrix} y_1 \\ y_2 \\ \vdots \\ y_{n-1} \\ y_n \\ y_{n+1} \\ \vdots \\ y_{2n-1} \end{bmatrix} = \begin{bmatrix} m_1 & & & & \\ m_2 & m_1 & & & \\ \vdots & m_2 & & & \\ m_{n-1} & \vdots & & m_1 & \\ m_n & m_{n-1} & \cdots & m_2 & m_1 \\ & m_n & & \vdots & m_2 \\ & & & m_{n-1} & \vdots \\ & & & m_n & m_{n-1} \\ & & & & m_n \end{bmatrix} \begin{bmatrix} x_1 \\ x_2 \\ \vdots \\ x_{n-1} \\ x_n \end{bmatrix} \tag{3}$$

2.3 Observation Model for Periodic Signals

We next consider a specific case where the input light signal is periodic. In this case, not only the input light signal but also the impulse response becomes periodic because of their linearity.

Let us consider the case where a periodic signal $[x_1, \cdots, x_n]^\top$ is emitted from the light source. Then, the response of the human retina can be described as follows:

$$\begin{bmatrix} y_1 \\ y_2 \\ \vdots \\ y_{n-1} \\ y_n \end{bmatrix} = \begin{bmatrix} m_1 & m_n & & m_3 & m_2 \\ m_2 & m_1 & & m_4 & m_3 \\ m_3 & m_2 & & \vdots & \vdots \\ \vdots & m_3 & \cdots & m_n & m_{n-1} \\ m_{n-1} & \vdots & & m_1 & m_n \\ m_n & m_{n-1} & & m_2 & m_1 \end{bmatrix} \begin{bmatrix} x_1 \\ x_2 \\ \vdots \\ x_{n-1} \\ x_n \end{bmatrix} \tag{4}$$

Then, the equation can be rewritten as follows:

$$\mathbf{Y} = \mathbf{M}\mathbf{X} \tag{5}$$

where $\mathbf{Y} = [y_1 \cdots y_n]^\top$, $\mathbf{X} = [x_1, \cdots x_n]^\top$ and $\mathbf{M}$ is a matrix which consists of the impulse response m_t. In this paper, we call $\mathbf{Y}$ as observed response, $\mathbf{M}$ as observation matrix and $\mathbf{X}$ as input signal. In the following sections, we describe a method for representing multiplex images by using the observation model shown in Eq.(5).

3 Multiplex Image Representation from Impulse Response

We next consider a method for embedding multiple images into a single set of sequential images by using the observation model derived in the previous section. As described in the previous section, each person has individual impulse response of retina, and thus, observed images are different from each other. Thus, by using the property of retina, we can present arbitrary different images to individual observers.

Suppose there are two people, whose observation matrices are $\mathbf{M}_A$ and $\mathbf{M}_B$ respectively. Then, when we display a signal $\mathbf{X}$ to these two people, their observed signals $\mathbf{Y}_A$ and $\mathbf{Y}_B$ can be described as follows:

$$\mathbf{Y}_A = \mathbf{M}_A\mathbf{X} \tag{6}$$

$$\mathbf{Y}_B = \mathbf{M}_B\mathbf{X} \tag{7}$$

Equations (6) and (7) show that the observed image signal $\mathbf{Y}_A$ and $\mathbf{Y}_B$ are different from each other, even if the input signal $\mathbf{X}$ is the same. In other words, we can represent different images by using the difference in the impulse response of retina.

Suppose we want to show image signal $\mathbf{Y}_A$ and $\mathbf{Y}_B$ to observer A and B respectively. Then, a display signal $\mathbf{X}$ for showing these objective signals to observers can be estimated by minimizing the following cost function E.

$$E = \left\| \begin{bmatrix} \mathbf{Y}_A \\ \mathbf{Y}_B \end{bmatrix} - \begin{bmatrix} \mathbf{M}_A \\ \mathbf{M}_B \end{bmatrix} \mathbf{X} \right\|^2 \tag{8}$$

The cost function E can be minimized by the least mean square method.

The above method can be extended to a method for more than 2 observers in a straightforward manner. Suppose there are N observers, whose observation matrices are $\mathbf{M}_i(i = 1, \cdots, N)$ and their objective signals are $\mathbf{Y}_i(i = 1, \cdots, N)$ respectively. Then, a display signal $\mathbf{X}$ for showing these objective signals to observers can be estimated by minimizing the following cost function E:

$$E = \left\| \begin{bmatrix} \mathbf{Y}_1 \\ \mathbf{Y}_2 \\ \vdots \\ \mathbf{Y}_N \end{bmatrix} - \begin{bmatrix} \mathbf{M}_1 \\ \mathbf{M}_2 \\ \vdots \\ \mathbf{M}_N \end{bmatrix} \mathbf{X} \right\|^2 \tag{9}$$

Note that, the image signal $\mathbf{X}$ is presented by using display devices, such as projectors, and thus, the range of image intensity is limited by the display devices. In general, they cannot display negative intensity, and their maximum intensity is also limited. Thus, the display signals have the following limitation:

$$0 \leq x_i \leq I_{max} \tag{10}$$

where, I_{max} is the maximum intensity of the display device. Thus, the cost function E must be minimized with this limitation for estimating the display signal $\mathbf{X}$. The minimization can be achieved by using the conditional least mean square method.

4 Contrast Adjustment of Objective Images

4.1 Contrast Adjustment for a Whole Image

By using the method described in the previous section, we can achieve multiplex image projection. However, because of the limitation of intensity in display devices, it is difficult to present original objective images. Therefore, we next consider a method for modifying objective images so that the display devices can present objective images properly under the existence of the limitation of display intensity.

In general, the display devices can present objective images properly, if the contrast of the objective images is small, since the intensity range of display images becomes small in this case. However, the contrast of the observed images also becomes small in this case, and thus the visibility of presented images degrades. Thus, we in this section consider the optimization of contrast of observed images under the limitation of intensity of display images.

Suppose we have two observers, and their objective images are $\mathbf{Y}_A^i$ and $\mathbf{Y}_B^i$ $(i = 1, \cdots, K)$ respectively, where K denotes the number of pixels in an image. Then, the contrast of these objective images can be modified linearly as follows:

$$\begin{bmatrix} \mathbf{Y}'^i_A \\ \mathbf{Y}'^i_B \end{bmatrix} = \begin{bmatrix} \alpha \mathbf{Y}_A^i + m\mathbf{1} \\ \beta \mathbf{Y}_B^i + n\mathbf{1} \end{bmatrix} \tag{11}$$

α and β are scale coefficients, and m and n are modified zero levels of these images, where $\mathbf{1} = [1, \cdots, 1]^\top$. Note, α, β, m and n are common in images.

By using the modification of objective image contrast, the image observation shown in Eqs.(6) and (7) can be rewritten as follows:

$$\begin{bmatrix} \alpha \mathbf{Y}_A^i + m\mathbf{1} \\ \beta \mathbf{Y}_B^i + n\mathbf{1} \end{bmatrix} = \begin{bmatrix} \mathbf{M}_A \\ \mathbf{M}_B \end{bmatrix} \mathbf{X}^i \tag{12}$$

where, $\mathbf{X}^i$ denotes an input signal of the i-th image pixel. By minimizing the difference between the left term and the right term in Eq.(12), we can estimate not only the sequential display images $\mathbf{X}^i$ $(i = 1, \cdots, K)$ but also the contrast parameters α, β, m and n. However, the contrast of objective image becomes extremely small, if there are no conditions for the contrast parameters. Thus, we consider a penalty on contrast degradation, and define a new cost function E_2 for estimating the display image as follows:

$$E_2 = \sum_{i=1}^{K} \left\| \begin{bmatrix} \alpha \mathbf{Y}_A^i + m\mathbf{1} \\ \beta \mathbf{Y}_B^i + n\mathbf{1} \end{bmatrix} - \begin{bmatrix} \mathbf{M}_A \\ \mathbf{M}_B \end{bmatrix} \mathbf{X}^i \right\|^2 - w\alpha - w\beta \tag{13}$$

where, w denotes a weight for the penalty of contrast degradation. By estimating the display images $\mathbf{X}^i$ $(i = 1, \cdots, K)$ and the contrast parameters α, β, m and n, which minimize the cost function E_2, we can derive appropriate display images adjusting their contrast.

4.2 Pixel-Wise Contrast Adjustment

In the previous section, we described a method for adjusting contrast of display images, in which the contrast parameters α, β, m and n are identical for all the pixels in display images. For adjusting the contrast more efficiently and for obtaining larger contrast in the display images, we next derive a method for adjusting the contrast of each image pixel independently. By adjusting the contrast pixel by pixel, we may be able to obtain larger contrast in the display images.

Suppose we have two observers, and their objective sequential intensities at i-th image pixel are $\mathbf{Y}_A^i$ and $\mathbf{Y}_B^i$ respectively. Then, if we modify the contrast of each image pixel separately, the observation model shown in Eq.(12) can be rewritten as follows:

$$\begin{bmatrix} \alpha^i \mathbf{Y}_A^i + m^i\mathbf{1} \\ \beta^i \mathbf{Y}_B^i + n^i\mathbf{1} \end{bmatrix} = \begin{bmatrix} \mathbf{M}_A \\ \mathbf{M}_B \end{bmatrix} \mathbf{X}^i \tag{14}$$

where, α^i, β^i, m^i and n^i are the contrast parameters of i-th image pixel. By minimizing the difference between the right term and the left term, we can estimate the pixel-wise contrast parameters and the display images simultaneously. However, the contrast parameters vary drastically pixel by pixel, if there are no constraints in contrast change. Thus, we consider smoothness constraints on the spatial variation of contrast parameters, and define a cost function E_3 for display image estimation as follows:

$$E_3 = \sum_{i=1}^{K} \left\| \begin{bmatrix} \alpha^i \mathbf{Y}_A^i + m^i \mathbf{1} \\ \beta^i \mathbf{Y}_B^i + n^i \mathbf{1} \end{bmatrix} - \begin{bmatrix} \mathbf{M}_A \\ \mathbf{M}_B \end{bmatrix} \mathbf{X}^i \right\|^2 + \sum_{i=1}^{K} (w_1(\mathcal{L}(\alpha_i) + \mathcal{L}(\beta_i)) + w_2(\mathcal{L}(m_i) + \mathcal{L}(n_i)) - w_3(\alpha_i + \beta_i)) \quad (15)$$

where $\mathcal{L}$ denotes the absolute Laplacian of a contrast parameter, which represents the spatial variation of contrast parameters, and the last term in Eq.(15) represents a penalty on contrast degradation as before. w_1, w_2 and w_3 denote weights for each constraint. The philosophical background of this cost function is that as far as the local structure of intensity is preserved, gradual changes in global structure of intensity in an image do not affect visual impression of observers so much. Thus, by estimating the display images $\mathbf{X}^i$ $(i = 1, \cdots, K)$ and the contrast parameters α_i, β_i, m_i and n_i $(i = 1, \cdots, K)$ which minimize E_3, we can estimate appropriate display images and contrast parameters simultaneously. By using the pixel-wise contrast adjustment, we can avoid the contrast degradation of observed images caused by the limitation of display intensity.

5 Experimental Results

5.1 Multiplex Image Representation

We next show experimental results from the proposed multiplex image representation method. In this experiment, we present two different images to two different observers, observer A and observer B, by using the proposed method. For evaluating the observed images objectively, we used a virtual observer made of a camera, whose impulse response can be changed arbitrary, and used it for simulating two different human observers. These two observers have different impulse responses as shown in Fig. 5. These impulse responses were synthesized based on the real impulse response of human retina [16]. In human retina, there are three types of photoreceptor, i.e. L-cone, M-cone and S-cone, for sensing object colors. Thus, in this experiment, we set the impulse response of these three photoreceptors separately as shown in Fig. 5.

In order to observe a sequential input light with these impulse responses, we used a variable exposure time camera developed by Uda et al. [17]. This camera consists of an LCoS device and an image sensor, and can control the exposure parameters in subframe. In this experiment, the amount of exposure was controlled in subframe according to the impulse responses shown in Fig. 5. Then, two different observers were simulated by using two different set of impulse responses. The sequential display images were computed by using the proposed method, and projected from a projector to a screen. The projected sequential images were observed by the variable exposure time camera as shown in Fig. 4.

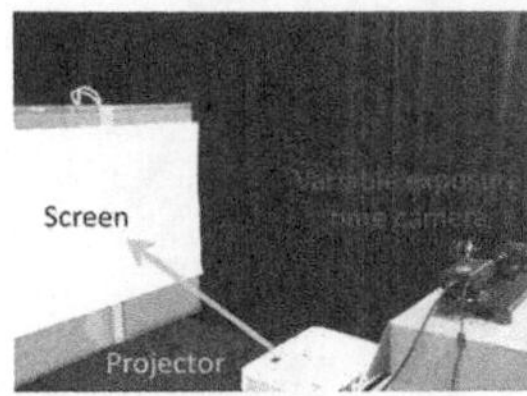

Fig. 4. Experimental setup

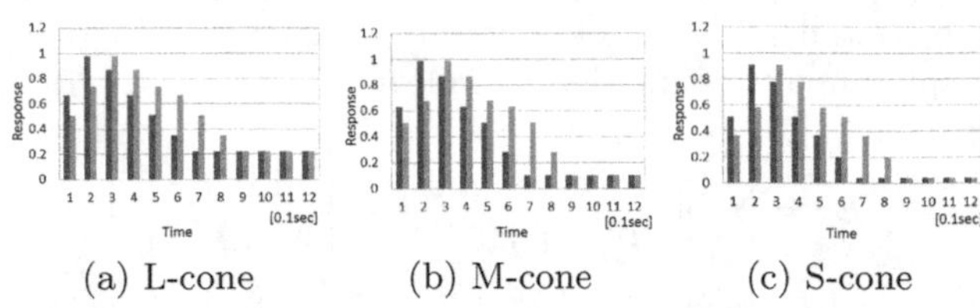

(a) L-cone (b) M-cone (c) S-cone

Fig. 5. Impulse responses of two different observers used in our experiments. The blue one shows the impulse response of observer A, and the red one shows the impulse response of observer B. (Color figure online)

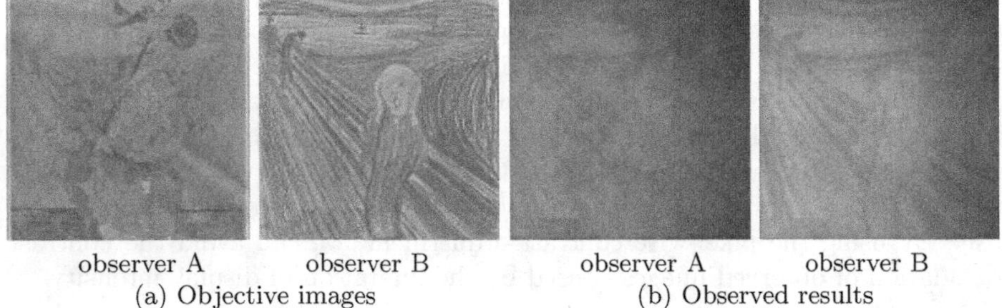

observer A observer B observer A observer B
(a) Objective images (b) Observed results

Fig. 6. Objective images and observed result. (a) Shows objective images for observer A and observer B, and (b) shows images observed by these two observers.

Figure 6(a) shows objective images for these two observers. The set of sequential display images was derived from these objective images by using the proposed method. Figure 7(a), (b), (c) and (d) show four example images in the derived sequential display images. These sequential display images were projected from the projector to the screen, and observed by the variable exposure time camera. The images observed by the variable exposure time camera with two different impulse responses are shown in Fig. 6(b). By comparing the images in Fig. 6(b) with the objective images in Fig. 6(a), we find that two observers can see different objective images properly, even if the projected image is identical.

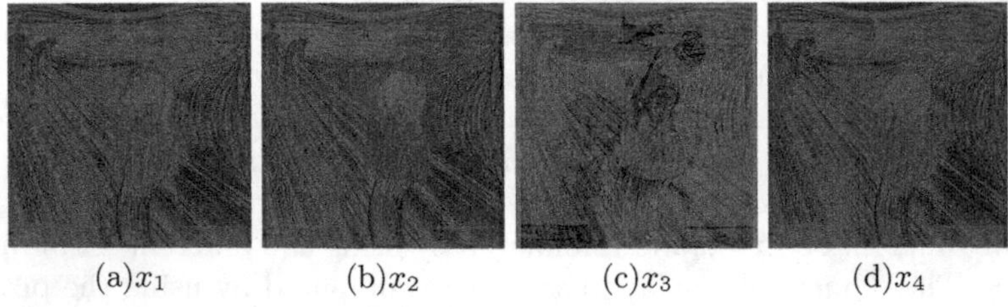

(a)x_1 (b)x_2 (c)x_3 (d)x_4

Fig. 7. Examples of sequential display images at four different instants.

(a) Objective images (b) Observed results

Fig. 8. Another result from the proposed multiplex image representation method. (a) Shows objective images for observer A and observer B, and (b) shows images observed by these two observers.

Figure 8 shows other results of multiplex image representation in the proposed method. As shown in these results, the proposed method enables us to present different images to individual observers without wearing special glasses and without fixing viewpoints.

5.2 Comparison

We next compare the pixel-wise contrast adjustment in the proposed method with the whole image contrast adjustment. The display images with the pixel-wise contrast adjustment were derived by minimizing E_3 in Eq.(15), while the display images with the whole image contrast adjustment were derived by minimizing E_2 in Eq.(13). The observed images from these two sets of display images are shown in Fig. 9. As shown in Fig. 9 the observed images from the pixel-wise contrast adjustment have much larger contrast than the observed images from the whole image contrast adjustment. This is because the pixel-wise contrast adjustment can modify image contrast more flexibly preserving the local intensity structure of original images.

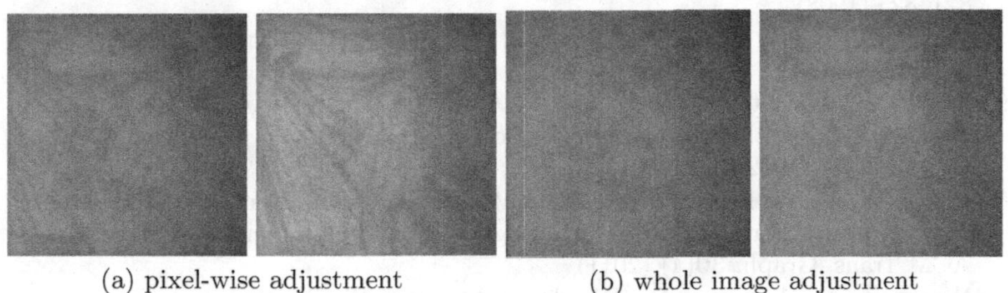

(a) pixel-wise adjustment (b) whole image adjustment

Fig. 9. Comparison of the pixel-wise contrast adjustment and the whole image contrast adjustment. (a) shows observed images from the pixel-wise contrast adjustment, and (b) shows observed images from the whole image contrast adjustment.

6 Conclusion

In this paper, we proposed a novel multiplex image representation method by using the difference of temporal impulse response of human retinas. We first derived a temporal observation model of human retina based on its impulse response. Then, we proposed a method for presenting different images to individual observers simultaneously by displaying a set of sequential images to these observers. For presenting multiplex images with enough image contrast, we proposed a method for adjusting pixel-wise image contrast. Experimental results with a variable exposure time camera show that the proposed method can present different images to individual observers in real environment. The proposed method requires the impulse response of observers, and thus we will develop a method for measuring the impulse response of human observers in the future work.

References

1. Hamada, T., Nagano, K., Scritters, T.: A multiplexed image system for a public screen. In: Proceedings of Virtual Reality International Conference, pp. 321–323 (2010)
2. Kakehi, Y., Iida, M., Naemura, T., Shirai, Y., Matsushita, M., Ohguro, A.: Lumisight table: an interactive view-dependent tabletop display. Comput. Graph. Appl. **25**, 48–53 (2005)
3. Nonoyama, M., Sakaue, F., Sato, J.: Multiplex image projection using multi-band projectors. In: Proceedings of IEEE Workshop on Color and Photometry in Computer Vision (2013)
4. Muramatsu, K., Sakaue, F., Sato, J.: Estimating spectral sensitivity of human observer for multiplex image projection. In: Proceedings of International Conference on Computer Vision Theory and Applications, pp. 183–191 (2016)
5. Berthier, A.: Images stereoscopiques de grand format. Cosmos **34**, 205–210, 227–233 (1896)
6. Ives, F.: A novel stereogram. J. Franklin Inst. **153**, 51–52 (1902)
7. Lippmann, G.: Epreuves reversibles donnant la sensation du relief. J. Phys. **7**(4), 821–825 (1908)
8. Perlin, K., Paxia, S., Kollin, J.S.: An autostereoscopic display. In: ACM SIGGRAPH (2000)
9. Masiaa, B., Wetzsteinb, G., Didykc, P., Gutierrez, D.: A survey on computational displays: pushing the boundaries of optics, computation, and perception. Comput. Graph. **37**(8), 1012–1038 (2013)
10. Wetzstein, G., Lanman, D., Heidrich, W., Raskar, R.: Layered 3D: tomographic image synthesis for attenuation-based light field and high dynamic range displays. ACM Trans. Graph. **30**(4) (2011)
11. Wetzstein, G., Lanman, D., Hirsch, M., Rasker, R.: Tensor displays: compressive light field synthesis using multilayer displays with directional backlighting. In: ACM Transactions on Graphics (2012)
12. Hirsch, M., Wetzstein, G., Rasker, R.: A compressive light field projection system. In: ACM Transactions on Graphics (2014)

13. Lanman, D., Luebke, D.: Near-eye light field displays. In: ACM SIGGRAPH, Asia (2013)
14. Huang, F., Chen, K., Wetzstein, G.: The light field stereoscope immersive computer graphics via factored near-eye light field displays with focus cues. In: ACM SIGGRAPH (2015)
15. Cao, D., Zele, A., Pokorny, J.: Linking impulse response functions to reaction time. Vis. Res. **47**, 1060–1077 (2007)
16. Uchikawa, K.: Comparison of temporal integration of chromatic response different hues. Vision **5**, 1–9 (1993)
17. Uda, S., Sakaue, F., Sato, J.: Variable exposure time imaging for obtaining HDR images. In: Proceedings of International Conference on Computer Vision Theory and Applications, pp. 118–124 (2016)

A Framework for Activity Recognition Through Deep Learning and Abnormality Detection in Daily Activities

Irina Mocanu[1(✉)], Bogdan Cramariuc[2], Oana Balan[1], and Alin Moldoveanu[1]

[1] Computer Science Department, University Politehnica of Bucharest, Spaliul Independentei 313, 060042 Bucharest, Romania
{irina.mocanu,oana.balan,alin.moldoveanu}@cs.pub.ro

[2] IT Center for Science and Technology, Av. Radu Beller 25, 011702 Bucharest, Romania
bogdan.cramariuc@citst.ro

Abstract. Activity recognition plays a key role in providing activity assistance and care for users in intelligent homes. This paper presents a two layer of convolutional neural networks to perform human action recognition using images provided by multiple cameras. We consider one PTZ camera and multiple Kinects in order to offer continuity over the users movement. The drawbacks of using only one type of sensor is minimized. For example, field of view provided by Kinect sensor is not wide enough to cover the entire room. Also, the PTZ camera is not able to detect and track a person in case of different situations, such as the person is sitting or it is under the camera. Also the system will identify abnormalities that can appear in sequences of performed daily activities. The system is tested in Ambient Intelligence Laboratory (AmI-Lab) at the University Politehnica of Bucharest.

Keywords: Activity recognition · Convolutional neural network · RGB images · Optical flow · Abnormal activities

1 Introduction

The percentage of elderly in todays societies keeps growing. As a consequence, we are faced with the problem of supporting older adults to compensate for their loss of cognitive autonomy in order to continue living independently in their homes as opposed to being forced to live in a care facility. Smart environments have been developed and proven to be able to support end-users and society in general in this era of demographic change [1].

Dementia is frequently invoked for aging population. One symptom of dementia consists in repetition of simple actions such as wandering. Researchers are trying to detect symptoms of dementia at early stages.

This can be detected using an activity recognition system integrated with abnormality detection in daily activities. Activity recognition of older persons

S. Battiato et al. (Eds.): ICIAP 2017, Part II, LNCS 10485, pp. 730–740, 2017.
https://doi.org/10.1007/978-3-319-68548-9_66

daily life should be done indirectly in a comfortable manner using non-intrusive sensors capable to track the user all day long.

The objective of this paper is the development of an non-intrusive technology in support of independent living of older adults in their home. The main focus is on monitoring their daily living activities and recognition of unusual situations. User tracking is performed using multiple cameras: one PTZ camera and multiple Kinect sensor. Also, abnormality in daily activities is detected by analysing the sequence of activities and the duration of each performed activity using a learned user behaviour.

We consider only one person in the supervised room. Generally, users prefer noninvasive sensors. Visual and audio data gathered from cameras, infrared sensors or microphones are more intrusive and can pose privacy concerns, but the acquired performance can be higher. In this paper, we don't treat privacy implications, we perform activity recognition for a single user.

The rest of the paper is organized as follows: Sect. 2 describes some existing work. System description is given in Sect. 3. Section 4 presents evaluation of the proposed system. Conclusions and future work are given in Sect. 5.

2 Related Works

The main methods of daily activity recognition are based on images analysis acquired from one or more cameras. It requires processing phase for extracting a set of features that will be used with a machine learning algorithm for performing activity recognition.

The main issues are (i) selecting image processing methods in order to acquire good results regardless of the image quality (resolution, noise, the influence of lighting, occlusion between objects), (ii) selecting the machine learning algorithms (iii) optimization of the response time in order to obtain a real-time method.

Frequently used methods are based on Hidden Markov Models (HMM) [2], [3] and Bayesian networks [4] - they can have characterized additional properties. They are not suitable for modeling complex activities that have large state and observation spaces. Parallel HMM are proposed to recognize group activities by factorizing the state space into several temporal processes. Methods based on motion (optical flow) [5] allow analysis of actions in time, modeling long time activities with variable structure of their sub-components. A number of descriptors such as: Histogram of Oriented Gradients (HOG), HOG3D, extended SURF, Harris3D or cuboids [6,7] have been proposed for image description.

Methods from deep learning show that could revolutionize the results obtained in machine learning [8], by eliminating of the featured extracted from the images. Convolutional neural networks (CNN) is a main method from "deep learning" [9] - represent a class of supervised learning algorithms that can learn a hierarchy of features by building a set of high-level features from a lot of low features [10]. Paper [11] uses CNN for human posture recognition. In [12] CNN are used for study different features to be applied for different sport actions. Also researches are at the beginning. One major problem with this approach consist in choosing the architecture of the CNN. In [13] is presented a method based on a hierarchically CNN that recognize human actions, showing high accuracy and favorable execution time for real time execution. Another approach is based on a combination of motion and appearance features of the human body, based on different methods of temporal aggregation using CNN [14], showing the method is more robust in errors estimation due to a higher accuracy for human posture recognition.

A combination of CNN and hand-crafted features is used in paper [15] that combines HOG and CNN obtaining results that outperform CNNs.

Approaches that use multiple CNNs provide better accuracy for activity recognition compared to the use of a single CNN. In paper [16] is described an architecture using two different recognition streams: spatial and temporal, that are combined by late fusion. The spatial stream performs action recognition from still video frames, and the temporal stream is trained to recognise action from motion in the form of dense optical flow. Decoupling the spatial and temporal networks is exploiting the availability of large existing amounts of annotated image data using pre-training spatial network with the ImageNet dataset [17]. Evaluation of the method described in [16] is performed on large action recognition databases (UCF-101 [18] and HMDB-51 [19]) and results compare better with other similar existing methods (both in accuracy and time).

This paper describes a system for daily activity recognition using two layers of CNNs having as input images provided by multiple cameras that are used for user tracking. The recognition of daily activities is integrated with a method for abnormalities detection in daily activities, especially regarding the wandering symptoms that can appear in case of dementia illness.

3 System Description

The proposed system will track and supervise one user in a smart room. In order to eliminate "black spots" for user detection, this system will consider multiple cameras. Thus, the user will be tracked by multiple cameras, a PTZ camera and a set of Kinect sensors, as given in [20]. RGB images captured from these cameras will be analysed in order to perform daily activity recognition. Based on the promising existing results from [16], we use a two layer CNNs for activity recognition: one spatial CNN and other optical CNN. At the end their results are fusioned using a SVM. There are situations for which the user behaviour must be analysed in order to detect some possible abnormalities. After the activity recognition step, sequences of recognised activities will be analysed in order to detect some changes in user behaviour (for example, wandering behaviour). Thus, we consider a sequence of activities - each activity is being composed of a set of events. Each event will be analysed if it is relevant or not for the current activity (based on the previously observed behaviour of the user). A similarity measure between activities is composed based on the duration of the events and on the number of occurrences in the training event sequences.

The system is designed and tested in the AmI-Lab at the University Politehnica of Bucharest. This laboratory is a rectangular room of 8.5 m × 4.5 m that is equipped with regular office furniture. The laboratory is equipped with various tracking sensors, installed in a non-intrusive manner, as described in [20].

The sensors relevant to this project are Microsoft Kinect sensors and a Samsung PanTilt-Zoom surveillance camera. The schema of the laboratory and the positions of the sensors are presented in the Fig. 1:

- 9 Microsoft Kinect sensors are installed in the AmI-Lab. These sensors contain an RGB camera and a depth sensing circuit. They are able to deliver up to 30 fps of RGB and depth information. The range field according to the official specifications is 0.7 to 6 m. There are areas of the room that are outside of the Kinect sensors view (i.e. room corners).
- One pan-tilt-zoom Samsung H.264 Network PTZ Dome Camera. The purpose of this device is the tracking of a person throughout the environment. The PTZ camera allows the free movement of the optical sensor: a 360° angle on the horizontal plane, an 180° angle on the vertical plane. The pan angle generally covers the horizontal orientation of the camera, while the tilt controls the vertical orientation. It is also capable of performing optical zoom (up to 12X). Based on its specifications the streaming rate is 30 fps.

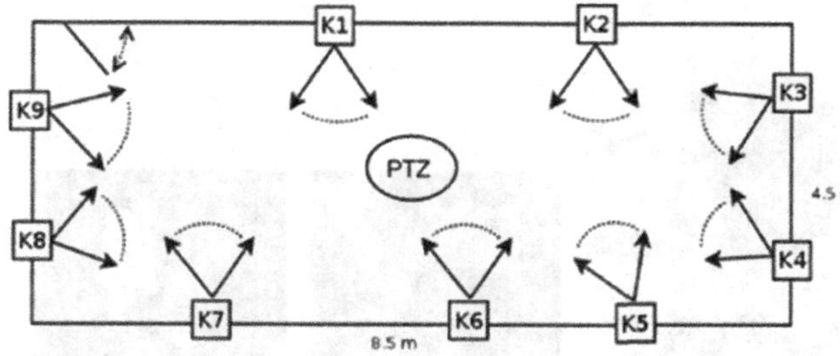

Fig. 1. AmI-Lab sensor configuration, as given in [20]

The set of Kinect sensors can continuously track the user inside the room. Also, there are some blind spots the corner of the room, where the Kinect sensors cannot offer information. The PTZ camera has the ability to track the user more accurately than the Kinect sensors. Thus, we combine the PTZ camera with the set of Kinect sensors in order to perform human activity recognition.

3.1 Human Tracking

Supervising a person into a smart room needs to track the user into the room at every moment. This implies to detect and track the user in different poses (standing, sitting, lying), also when located in dead angles into the room (there might be portions of the room which are not covered by the tracking system) or in uncomfortable angles in which the detection has to be performed (a person might be sitting right under the camera or might be keeping his hands crossed in front of the Kinect sensor). All this information is obtained by analyzing images with cluttered background (usually, a room is filled with objects, which sometimes increases the difficulty of detecting a person). In order to avoid such situations, we propose a solution for user tracking by collecting images from multiple and different types of sensors: multiple Kinect sensors and a PTZ camera, as it is presented in [21]. The information gathered from these two types of sensors are combined in order to obtain continuity over the users movement.

For tracking the user, first he needs to be detected and after that user tracking is performed. Integral Channel Features method is used for person detection [22]. 10 channels, as proposed in [21] are used for the person detection: three channels for the color components in the CIELUV color space (L, U, V channels), 6 channels for gradient orientation bins and one channel for gradient magnitude. We used random feature pools of size 5000. A simple boosted classifier and a cascade of 5 boosted classifiers are used for image classifier.

Optical flow computed using the Farneback method is used for human tracking. The flow image contains a large blob representing the moving person. The flow image is converted to grayscale and a threshold is applied (Fig. 2). The largest blob is extracted from the previous image. This blob represents the moving person with a high probability.

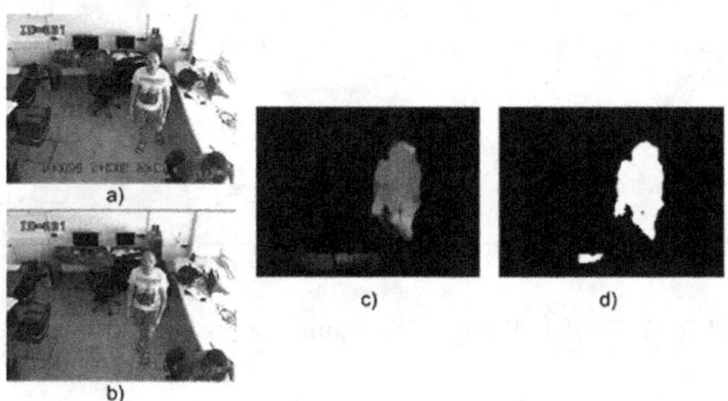

Fig. 2. Steps in selecting the region of a moving person using optical flow. (a), (b) original successive frames; (c) - optical flow; (d) - blob extraction

Tracking the user with both the PTZ and the Kinect sensors is made as follows: once a Kinect detected a person, it will transmit the location of the person to the PTZ camera. The PTZ will perform the necessary camera movements in order to center the subject. After that, person detection is performed on the acquired images.

3.2 Activity Recognition

Convolutional neural networks represent the earliest successfully established deep architecture. A convolutional neural network classifier comprises a convolutional neural network for feature extraction and a classifier in the last step for classification. Each unit from a CNN layer receives inputs from a set of units located in a small neighbourhood from the previous one. Hidden layers are organized with several planes (called feature map) within which all the units share the same weights. There are two kinds of hidden layers in CNN: convolution layer and subsampling layer. Feature maps can extract different types of features from a previous layer. Thus, convolution layers are used as features extractors. A subsampling layer follows a convolution layer. Each feature map in subsampling layers performs subsampling on the feature map from the previous one, reducing the resolution of the feature map. The number of units from the output layer is equal to the number of predefined activities classes.

Based on the good results obtained for activity recognition described in [16], we use two layers of CNNs: a spatial CNN and an optical CNN, as in Fig. 3.

The Spatial CNN processes individual video frames, performing action recognition from still images. Based on existing results, spatial CNNs are essentially for image classification, as suggested in the proposed architecture used for large-scale image recognition methods [17].

The optical CNN receives as input a set of optical flow images between several consecutive frames (obtained as described in Sect. 3.1). This input will describe the motion between video frames, which makes the recognition easier, as the network does not need to estimate motion implicitly.

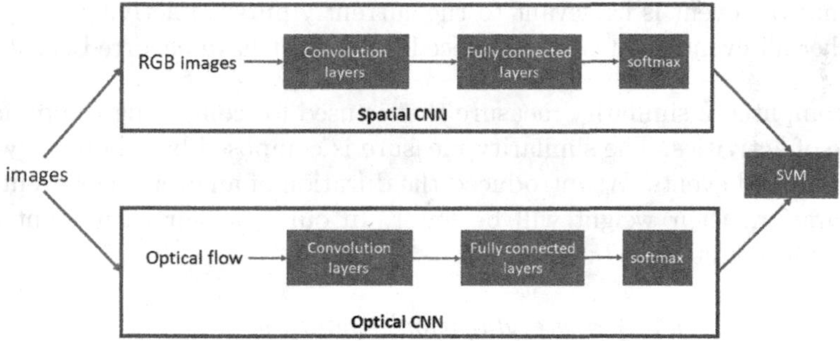

Fig. 3. Two architecture layer for activity recognition

For spatial and temporal CNN architecture, we use the model described in [17]. The CNN architecture corresponds to the CNN-M-2048 architecture from [23] combined with the architecture proposed in [17]. The weights of the hidden layers use the rectification activation function; maxpooling is performed over 33 spatial windows with stride 2 and the settings for local response normalisation are used as in [17].

Fusion for the final classification is performed with a multi-class linear Support Vector Machine (SVM) [24]. The SVM fusion is composed of a multi-class linear Support Vector Machine trained with the output of the CNNs from the validation dataset. For testing, the SVM predicts the class with the largest score.

3.3 Observing Abnormal Activities

Once the daily activities are identified and the behaviour of the user is learned it is necessary to identify abnormal situations during the daily life of the user. We propose abnormalities detection in the activities performed by people with early stages of dementia. In order to detect anomalies in daily activities we consider the algorithm described in [25].

To analyze and derive useful information about the structure of the performed activity, we consider a sequence of activities. Each activity is composed of a set of events. These events are entered in a matrix $\boldsymbol{m}$, which contains one row and one column for each of the possible events that may occur in the activity. If an event $\boldsymbol{e_i}$, occurs in a sequence, then ***m[i,i]*** is incremented by 1.

Based on the matrix $\boldsymbol{m}$, the following aspects can be given:

1. if ***m[i,j]*** $=$ $\boldsymbol{n}$ ($\boldsymbol{n}$ is the number of given sequences of activities) then event e_i will occur in every sequence
2. if ***m[i,i]*** $\geq$ ***m[i,j]*** $=$ ***m[j,j]*** and ***m[j,i]*** $=$ ***0*** then e_j will appear after the e_i, otherwise ordering of events e_i and e_j are irrelevant in order.

To decide if an activity performed by the user has any abnormality, whenever a new event occurs, the following need to be checked.

1. whether the event is irrelevant to the currently pursued activity
2. whether all events that need to precede the event have occurred earlier.

We computed a similarity measure that is used for comparing the performed sequence of activities. The similarity measure is composed based on the weights of the contained events. We introduced the duration of an event. For events with lower duration, their weight will be lower. In our case, for each event e_i, we associated a weight $\boldsymbol{w(e_i)}$ as in Eq. 1:

$$w(e_i) = w(e_i\_duration) * w(e_i\_number) \tag{1}$$

where:

- $\boldsymbol{w(e_i\_duration)}$ is the weight associated with event e_i based on its duration relative to the total duration of the actions (from the sequence)
- $\boldsymbol{w(e_i\_number)}$ is the weight associated with event e_i based on its number of occurrences in the training event sequences (it is equal with $n_{e_i}/\boldsymbol{n}$, where n_{e_i} is the number of occurrences of e_i in the training event sequences and n the number of training event sequences.

The similarity measure $\boldsymbol{sm}$, for a sequence $e = \{e_1, e_2, \cdots, e_n\}$ is defined as in Eq. 2:

$$sm = \sum_{k=1}^{n} (w(e_k) * n_{e_k}) \tag{2}$$

The similarity measure is computed based on the idea the more number of times of an event appears in the actions set, the more similar the activity is to the one existing in the training set. If the value of $\boldsymbol{sm}$ is greater than a threshold, the sequence will be considered similar with the activities usually performed by the user.

4 Evaluation Results

The CNN for activity recognition was evaluated on the MSR Daily Activity dataset 3D [26]. This database was captured using a Kinect device. The dataset consists of 320 scenes each having the depth map, the skeleton position and the RGB video. There are 16 actions in the dataset: walk, sit down, stand up, sit still, use laptop, read book, drink, write on paper, eat, toss paper, play guitar, use vacuum cleaner, play game, call phone, cheer up, lie down on sofa. Each action is performed by 10 subjects for a number of two times: once in standing position and once in sitting position. We extend the database with a set of images with images captured from the AmI-Lab, named AmI Database. In this laboratory there is a cluttered background - many objects can be distinguished in the background, including wooden panels, which can be easily mistaken for patches of human skin, due to the similarity of colors.

For both CNNs we used a CNN architecture for both object detection and recognition [17]. The parameters used for the training phase are based on the similarly CNN architecture from [27] which uses stochastic gradient with momentum (0.8). For each iteration, the network forwards a batch of 128 samples. All convolutions use rectified linear activation functions. Weights are initialization using automatically determined based on the number of input neurons. To reduce the chances of overfitting, we apply dropout on the fully-connected layers with a probability of 76%. The learning rate is set to 103 and it is decreased by a factor of 50 at every epoch, the training is stopped after (40 epochs).

We train the multi-class Support Vector Machine using the implementation from the scikit-learn toolbox [28]. We used a linear kernel and default scikit-learn regularization parameter C = 1 with the square of the hinge loss as loss function.

Activity recognition was tested first onthe MSR Daily Activity Dataset. In this case, we use only part of these activities. We are interested only in three general types of such activities: walking, standing and sitting. Sit down activities are composed of: sit still, read book and eat. All other activities from the MSR Daily Activity dataset are ignored.

We divided the dataset into training, validation and test sets. The training set contains 7 subjects. The validation set is used in order to obtain the configuration with the highest accuracy. The confusion matrix obtained for these three types of activities is given in Table 1. Based on the results given in Table 1, the recognition rate is good. Also, there are some misclassified activities. This happens between activities walking and standing. These situations appear in case of a lower speed of the user when these two activities can't be distinguished.

In order to see if our model is suitable for real-time classification applications, we made experiments which measure the classification time. The timed obtained for classification of about 0.0017 s per frame which, from our point of view, is low enough for real time classification.

Table 1. Confusion matrix for activity recognition tested over the extended MSR database

MSR data set	Walking	Sitting	Standing
Walking	**0.97**		0.03
Sitting		**1**	
Standing	0.08		**0.92**

As can be seen from Table 1, the proposed solution that combines the two convolutional networks together with the SVM improves the existing results compared with the recognised activities from the analysed database, as given in [26].

For testing the algorithm of finding abnormalities in daily activities we consider different activities composed of sequence of events. In our case an event will be an activity recognised by the two layer of CNNs: walking, standing and sitting. We create a training set with images acquired in Ami-Lab from by tracking the user with the set of cameras existing there: the PTZ and the set of Kinects. We recorded 20 h of activities composed of walking, sitting and standing. The recorded activities are splitted in sequences of 30 min based on which we computed the matrix $\boldsymbol{m}$ and the value of n_{e_i}. For testing the algorithm, we collect another 5 h of activities, also splitted in sequences of 30 min. For testing database, users perform activities in completely different order than the existing ones in the training database. Also, the activities are executed with different duration, especially shorter walking activities were interleaved between standing and sitting ones. The effectiveness of the detected abnormality in a sequence of daily activities depends on the training event sequences. The training event sequences should reflect all the possibilities for sequences of activities that can be performed without any abnormality.

5 Conclusion and Future Work

Activity recognition and abnormality detection in daily activities is an important field of research with real applicability in medical applications. This paper describes a system that supervises a user using multiple cameras, recognises performed activities and detects abnormalities regarding his daily behaviour. Activity recognition is performed through deep learning using a two layer of CNNs, one spatial CNN and other optical CNN, at the end their results are fusioned using a SVM. Abnormality in daily activities is detected by analysing the sequence of activities and the duration of each performed activity using a learned user behaviour. As future work, we will extend the set of activities that are used for both recognition and abnormality detection. Also, for a better detection of abnormalities in daily activities, different attributes will be associated with daily activities (for example: number of performed steps or the users trajectory through the room).

Acknowledgments. This work has been funded by University Politehnica of Bucharest, through the Excellence Research Grants Program, UPB GEX. Identifier: UPB-EXCELENTA-2016 Optimizarea Activitatilor Zilnice Folosind Deep Learning Implementata pe Sisteme Reconfigurabile/Daily Activities Using Deep Learning Implemented on FPGA, Contract number 8/26.09.2016 (code 341).

References

1. Popescu, D., Dobrescu, R., Maciuca, A., et al.: Smart sensor network for continuous monitoring at home of elderly population with chronic diseases. In: 20th Telecommunications Forum, pp. 603–606 (2012)
2. Chiperi, M., Mocanu, I., Trascau, M.: Human tracking using multiple views. In: 40th International Conference on Telecommunications and Signal Processing. IEEE Press (2017)
3. Zhu, C., Sheng, W.: Human daily activity recognition in robotassisted living using multisensor fusion. In: IEEE International Conference on Robotics and Automation, pp. 2154–2159 (2009)
4. Vinh, L., Lee, S., Le, H., Ngo, H., Kim, H., Han, M., Lee, Y.-K.: Semimarkov conditional random fields for accelerometer-based activity recognition. Appl. Intell. **35**, 226–241 (2011)
5. Lara, O.D., Labrador, M.A.: A mobile platform for real time human activity recognition. In: Proceedings of IEEE Conference on Consumer Communications and Networks, pp. 667–671 (2012)
6. Tang, K., Fei-Fei, L., Koller, D.: Learning latent temporal structure for complex event detection. In: IEEE Conference on Computer Vision and Pattern Recognition, pp. 1250–1257 (2012)
7. Wang, H., Schmid, C.: Action recognition with improved trajectories. In: IEEE International Conference on Computer Vision, pp. 3551–3558 (2013)
8. Jiang, Z., Lin, Z., Davis, L.S.: A unified tree-based framework for joint action localization, recognition and segmentation. Comput. Vis. Image Underst. **117**, 1345–1355 (2013)
9. Can Deep Learning Revolutionize Mobile Sensing? http://niclane.org/pubs/lane_hotmobile15.pdf
10. Krizhevsky, A., Sutskever, I., Hinton, G.E.: ImageNet classification with deep convolutional neural networks. In: Proceedings of the 25th International Conference on Neural Information Processing Systems, pp. 1097–1105 (2012)

11. Ji, S., Xu, W., Yang, M., et al.: 3D convolutional neural networks for human action recognition. IEEE Trans. Pattern Anal. Mach. Intell. **35**, 221–231 (2013)
12. Toshev, A., Szegedy, C.: DeepPose: human pose estimation via deep neural networks. In: IEEE Conference on Computer Vision and Pattern Recognition, pp. 1653–1660 (2014)
13. Sun, L., Jia, K., Chan, T., et al.: DL-SFA: deeply-learned slow feature analysis for action recognition. In: IEEE Conference on Computer Vision and Pattern Recognition, pp. 2625–2632 (2014)
14. Jin, C.-B., Li, S., Do, T.D., Kim, H.: Real-time human action recognition using CNN over temporal images for static video surveillance cameras. In: Ho, Y.-S., Sang, J., Ro, Y.M., Kim, J., Wu, F. (eds.) PCM 2015. LNCS, vol. 9315, pp. 330–339. Springer, Cham (2015). doi:10.1007/978-3-319-24078-7_33
15. P-CNN: Pose-Based CNN Features for Action Recognition. https://hal.inria.fr/hal-01187690/file/PCNN_cheronICCV15.pdf
16. Zhang, T., Zeng, Y., Xu, B.: HCNN: a neural network model for combining local and global features towards human-like classification. Int. J. Pattern Recogn. Artif. Intell. **30**, 1–19 (2016)
17. Two-Stream Convolutional Networks for Action Recognition in Video. https://papers.nips.cc/paper/5353-two-stream-convolutional-networks-for-action-recognition-in-videos.pdf
18. Large scale visual recognition challenge. http://www.image-net.org/challenges/LSVRC/2010/
19. UCF101: A dataset of 101 human actions classes from videos in the wild. http://crcv.ucf.edu/data/UCF101.php
20. Kuehne, H., Jhuang, H., Garrote, E., Poggio, T., Serre, T.: HMDB: a large video database for human motion recognition. In: IEEE International Conference on Computer Vision, pp. 2556–2563 (2011)
21. Ismail, A.A., Florea, A.M.: Multimodal indoor tracking of a single elder in an AAL environment. In: van Berlo, A., Hallenborg, K., Rodríguez, J., Tapia, D., Novais, P. (eds.) mbient Intelligence - Software and Applications. AISC, vol. 219, pp. 137–145. Springer, Heidelberg (2013). doi:10.1007/978-3-319-00566-9_18
22. Integral Channel Features. http://pages.ucsd.edu/ztu/publication/dollarBMVC09ChnFtrs_0.pdf
23. Return of the devil in the details: delving deep into convolutional nets. http://www.bmva.org/bmvc/2014/files/paper054.pdf
24. Crammer, K., Singer, Y.: On the algorithmic implementation of multiclass kernel-based vector machines. J. Mach. Learn. Res. **2**, 265–292 (2001)
25. Karamath, H.A., Amalarethinam, D.I.G.: Detecting abnormality in activities performed by people with dementia in a smart environment. Int. J. Comput. Sci. Inf. Technol. **5**, 2453–2457 (2014)
26. Wang, J., Liu, Z., Wu, Y., Yuan, J.: Mining actionlet ensemble for action recognition with depth cameras. In: IEEE Conference on Computer Vision and Pattern Recognition, pp. 1290–1297 (2012)
27. Liu, W., Jia, Y., Sermanet, P., Reed, S., Anguelov, D., Erhan, D., Vanhoucke, V., Rabinovich, A.: Going deeper with convolutions. In: Proceedings of the Computer Vision and Pattern Recognition (2015)
28. Scikit-learn. http://scikit-learn.org/

Combining Color Fractal with LBP Information for Flood Segmentation in UAV-Based Images

Loretta Ichim and Dan Popescu(✉)

Faculty of Automatic Control and Computers, University Politehnica of Bucharest, Bucharest, Romania
{loretta.ichim, dan.popescu}@upb.ro

Abstract. The paper presents a method for patch classification to the end of flooded areas segmentation from aerial images. As patch descriptors color fractal dimension and color local binary patterns were proposed, so both color and texture information is combined. The remote images were taken by the aid of an Unmanned Aircraft System (MUROS) implemented by an authors' team. The algorithm of remote image segmentation has two phases: the learning and the segmentation phase. The class representative consists of a set of intervals created in the learning phase. The classification is made by a voting criterion which takes into consideration the weights calculating from both descriptors. The results were obtained on 100 images with high resolution from the orthophotoplan created with the images taken in a real mission. The accuracy of segmentation was better than in the separate approaches (single fractal or local binary patterns descriptors).

Keywords: Aerial images · Color fractal dimension · Color LBP · Flooded areas · Image segmentation · Texture analysis

1 Introduction

Floods produce important and frequent economic damages in many economic domains and, especially, in agriculture (rural zones). For flood detection and evaluation on small areas, Unmanned Aircraft System (UAS) is an excellent solution [1–4]. Comparing with other solutions like the satellites, manned aircrafts and helicopter solutions, this has the following advantages: high resolution on the ground, low cost and possibility of operating on cloudy weather. Therefore, we proposed the UAS remote imaging tasks to detect the flooded zones and to evaluate the damages in agriculture. The system is able to perform the following functions: (a) the mission planning and its transmission on board unmanned aerial platform, (b) pre-flight system configuration and verification, (c) UAV launching, (d) video data acquisition from cameras installed on board, (e) real-time transmission of video data to GCS (Ground Control Station) at distance via internet, (f) storage and displaying video data to GCS, (g) aircraft command and (h) the control in several modes (manual, semiautomatic and automatic), (i) control data transmission from GCS to UAV and (j) mission review and on board transmission

S. Battiato et al. (Eds.): ICIAP 2017, Part II, LNCS 10485, pp. 741–752, 2017.
https://doi.org/10.1007/978-3-319-68548-9_67

during the flight. By programming an optimal flight path and using a high resolution commercial camera, a great accuracy can be obtained.

Aerial image interpretation is often based on its separation in regions of interest having a specific meaning for the application. This process is named image segmentation and it is achieved as a result of the pixel classification or, sometimes, of the patch classification [4]. The images from UAV (Unmanned Aerial Vehicle) are characterized by textural aspects on different scales (water, buildings, forest, crops, soil, etc.) and therefore, image segmentation by textural properties can be used. There are many works that address the aerial texture segmentation, based on different methods, but relatively few are those that deal with chromatic textures. Original Local Binary Pattern (LBP), fractal dimension, and cooccurence matrices do not encode color information which is essential for flood detection. The color indication can improve the classification tasks [5–8]. The most authors consider the texture feature on separate color channels and interpret them separately. Color textural information can be successfully applied to a range of applications including the segmentation of natural images, medical imaging and product inspection. This approach produced far superior results when encompasses both color and texture information in an adaptive manner. Relative recently, inter-correlated information from color channels was considered [5–8]. Thus, in [5, 6] the authors proposed algorithms to evaluate the Haralik's features [9] from a special co-occurrence matrix which take into account the occurrence of the pixel values from two different color channels. For example, the authors in [8] use so-called color co-occurrence matrix to extract the second order characteristics for road segmentation. Fractal dimension is a method widely used to capture intensity and texture information. In the past, the computation of fractal dimension of images has been defined for only binary and gray scale images. Recently, a color version has been proposed by Ivanovici and Richard [7] that capture information from the three different color channels. Color fractal dimension was applied with good results in medical applications, like: skin images of psoriasis lesions, prostate cancer [10] or melanoma.

The algorithm for color fractal dimension [7] was considered as an extension of the differential box counting algorithm for gray level [11]. Based on the fact that DBC is an important descriptor of local texture [12] we used color fractal dimension (CFD) to characterize local color texture in remote sensing. On the other hand another important local descriptor of texture is the LBP histogram [13]. The authors in [14] extended the classical method of LBP for gray scale and proposed a new methodology of LBP histogram evaluation for pairs of complementary color information.

In this paper we proposed a new method, based on combining spatial and color information provided by fractal and LBP fusion for the flood detection and segmentation on small areas like farmland. The images are acquired by an UAS, implemented by the authors in MUROS project [15], and processed at distance via GSM/internet. The contributions refer to fractal dimension evaluation for color texture in a 5D-space (fractal color dimension - FCD) and to flood segmentation based on a voting scheme which uses both FCD and color LBP). We extend LBP to the entire color space representation of images.

2 Proposed Method

For flooded areas detection we proposed a method that uses the decomposition of the orthophotoplan, created from images taken by UAV, into non-overlapping patches. Each patch is investigated upon a classification score that combines texture information from color fractal dimension and color LBP.

2.1 Color Fractal Dimension

Color fractal dimension (CFD) can be considered as an extension of the Gray Level Fractal Dimension (GFD) used for characterization of gray level fractals and also gray level textures. For example, an algorithm to evaluate GFD is the Differential Box-Counting (DBC) algorithm [11] which derives from the classical box counting algorithm. Some works [4, 12] prove that GFD has a good efficiency in classification and segmentation of monochromatic textures. For GFD calculation, a gray level relief is created. The base of the relief is the 2D (matrix) spatial position of pixels and the height is the gray level. This relief is covered with 3D boxes of sizes $r \times r \times s$ where r is the division factor and s (1) is the box height,

$$s = \frac{rI_{\max}}{n}, \tag{1}$$

where $I_{\max}$ represents the maximum value of the gray level value and $n \times n$ is the matrix dimension. Let be p (i, j), the maximum value of intensity (2) on the square S_r (i, j) of dimension $r \times r$ centered in (i, j):

$$p(i,j) = \max\{I(u,v)/(u,v) \in S_r(i,j)\} \tag{2}$$

Similarly, we considered the minimum value (3) of intensity q (i, j) on the square S_r (i, j) of dimension $r \times r$ centered in (i, j):

$$q(i,j) = \min\{I(u,v)/(u,v) \in S_r(i,j)\} \tag{3}$$

For each r, the differences (4) are summed over the entire image (5):

$$dif_r(i,j) = p(i,j) - q(i,j) \tag{4}$$

$$N_r = \sum_i \sum_j dif_r(i,j) \tag{5}$$

DBC is calculated as (6):

$$DBC = -\lim_{r \to 0} \frac{\log N_r}{\log r} \tag{6}$$

Next, the DBC algorithm behaves the same as the classical binary box counting algorithm (log-log representation of N_r and r; DBC is the slope of the regression line). Usually, n and r are powers of 2 and log has the base 2.

Let us consider the *RGB* color space. A pixel of the image [*I*] in this color space can be considered as a vector (7) of 5 dimension, *P* (*i, j*):

$$P(i,j) = [i,j,R(i,j),G(i,j),B(i,j)], \tag{7}$$

where (*i, j*) represents the pixel position in the matrix representation, *R* (*i, j*), *G* (*i, j*) and *B* (*i, j*) respectively the color components on red, green and blue color channels. The color level relief is covered with hyper-boxes of sizes $r \times r \times s_R \times s_G \times s_B$ where r is the division factor and s_K (8) is the box height on the color channel *K* (*K* = *R, G, B*),

$$s_K = \frac{rI_{K\max}}{n}, \tag{8}$$

In Eq. (8) $I_{K\max}$ represents the maximum value of the intensity level on channel K and $n \times n$ is the matrix dimension. Similarly with (2) and (3) we considered the maximum (9) and the minimum (10) values on all color channels, on the square S_r (*i, j*):

$$p_{RGB}(i,j) = \max\{I_R(u,v), I_G(u,v), I_B(u,v)/(u,v) \in S_r(i,j)\} \tag{9}$$

$$q_{RGB}(i,j) = \min\{\, I_R(u,v), I_G(u,v), I_B(u,v)/(u,v) \in S_r(i,j)\} \tag{10}$$

The difference (11) is used to calculate *CBC* (12):

$$dif_{RGB,r}(i,j) = p_{RGB}(i,j) - q_{RGB}(i,j) \tag{11}$$

$$CBC = -\lim_{r \to 0} \frac{\log \sum_i \sum_j dif_{RGB,r}(i,j)}{\log r} \tag{12}$$

In case of local texture classification by fractal approach, the CBC is approximated as the slope of the regression line in log – log representation. So, CFD is considered as local feature to classify the patches in flood – noted as water (*W*) – and non flood (*nW*) – for example, vegetation (*V*) and soil (*S*).

2.2 Color Local Binary Patterns

It is well known that LBP histogram is an efficient descriptor of textures [13]. Like the fractal dimension, LBP histogram can be extended to the color representation of images. So, the color LBP histogram is considered as local descriptor and it is used for patch classification. The color histogram is obtained by concatenating the LBP histograms on R, G and B color channels [14]. If the gray level histogram is represented in 10 points, then the color LBP histogram (CLBP) contains 30 points.

2.3 Image Segmentation

For image segmentation both color local descriptors CLBP and CFD are taken into account by a joint voting scheme. The aerial images are first integrated in an orthophotoplan and then decomposed in non-overlapping boxes (patches) of dimension 128 × 128 pixels. For each patch CLBP and CFD are calculated.

There are two phases in the classification process: the learning phase and the testing phase. In the learning phase, for each class C, from the characteristics (feature values) calculated on a learning set of patches, an interval from minimum and maximum values [C_{min}, C_{Max}] is created. In CFD case, this is a simple interval as representative for the class C, while in the LBP case it is a set of 30 intervals (each for a position in the concatenated histogram) which create a channel inside the cumulative histogram representation (Fig. 1). The upper segments in the histogram from Fig. 1 represent the intervals [min, max] obtained from the learning patch values. In the cumulative histogram, the grouping interval is of 25 values.

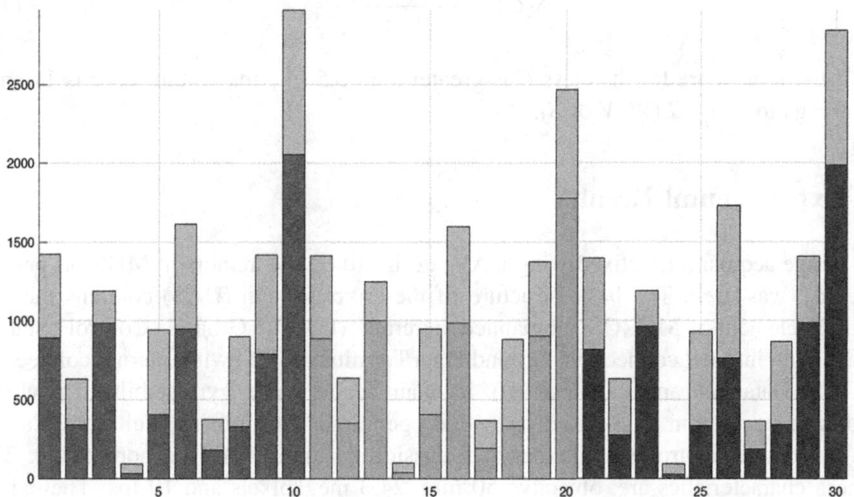

Fig. 1. Representative of the class *C* in the color LBP case (cumulative histogram) (Color figure online)

Based on these intervals, a voting scheme for patches classification as belonging to class *C* is considered. First, for CLBP criterion, a score of matches in the representative channel for this class, $S_{CLBP}(C)$, is established for a testing patch *P*. Obviously, $S_{CLBP}(C) \in [0, 30]$, or in the normalized representation (13),

$$S_{CLBP}(C) \in [0, 1] \tag{13}$$

Value of 1 represents a total matching of the patch CLBC histogram in the representative of the class *C* (30 matches).

For CFD criterion we consider a maximum weight of 1 at the middle of the interval [min, max] and the weights of 0.5 at the both ends of the interval: min and max. So we can consider linear weights for each interval $[\min, \frac{\min + \max}{2}]$ (14) and $[\frac{\min + \max}{2}, \max]$ (15):

$$S_{CFD} = 0.5 + \frac{x - \min}{\max - \min}, \quad \textit{if } x \leq \frac{\max + \min}{2} \tag{14}$$

$$S_{CFD} = 0.5 + \frac{\max - x}{\max - \min}, \quad \textit{if } x \geq \frac{\max + \min}{2} \tag{15}$$

In Eqs. (14) and (15) x represents the CFD value for the tested patch. It can be seen that $S_{CFD}\max = 1$, if $x = \frac{\max + \min}{2}$. The total score for the class C is considered as the average of the two partial scores (16):

$$S(C) = \frac{S_{CLBP} + S_{CFD}}{2} \tag{16}$$

If the total score for the class C is greater than 0.5 (the maximum score is 1), then P belongs to class C (W, V or S).

3 Experimental Results

For image acquisition, a fixed-wing UAV, designed by the authors in MUROS project [15, 16], was used. The basic structure of the entire system (UAS) contains the following elements: MUROS unmanned aircraft (UAV), Ground Control Station (GCS) with internet connection, Ground Data Terminal (GDT) with internet connection to GCS, Data link and Launcher (L). Its main features are: gyro-stabilized payload, automatic navigation, GIS-based, extended operational range using multiple GCs and GDTs, remote control via Internet and mission planning software application. The camera characteristics are: objective 50 mm, 24.3 megapixels and 10 fps. The experimental model for UAV MUROS and the payload with camera for image acquisition are presented in Fig. 2.

a)

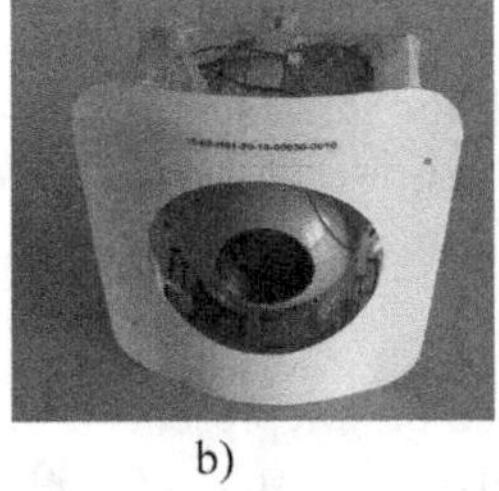

b)

Fig. 2. Support for image acquisition: (a) UAV MUROS, (b) Payload with camera

In order to segment and evaluate the flood damage, the images taken from UAV are concatenated in on orthophotoplan and then decomposed in patches of dimension 128 × 128 pixels. In the learning phase, from the learning images (Fig. 3), a set of 20 patches -10 patches with flood (class *W*) and 10 patches with other different regions of interest like vegetation (*V*) and soil (*S*), representing non flood (class *nW*) - like in Fig. 4, is considered for defining the class *W* in the segmentation process.

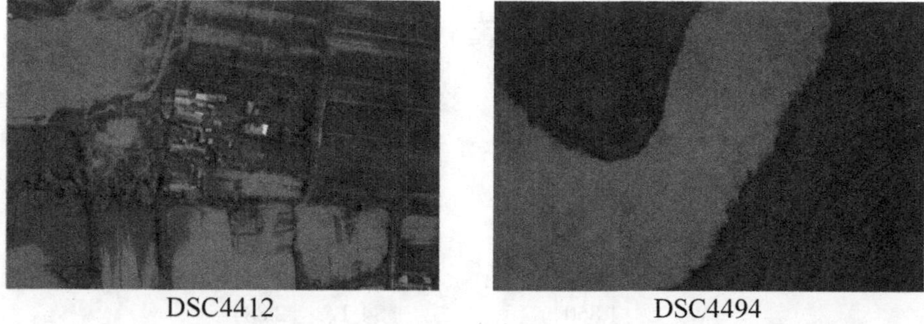

Fig. 3. Examples of images for learning

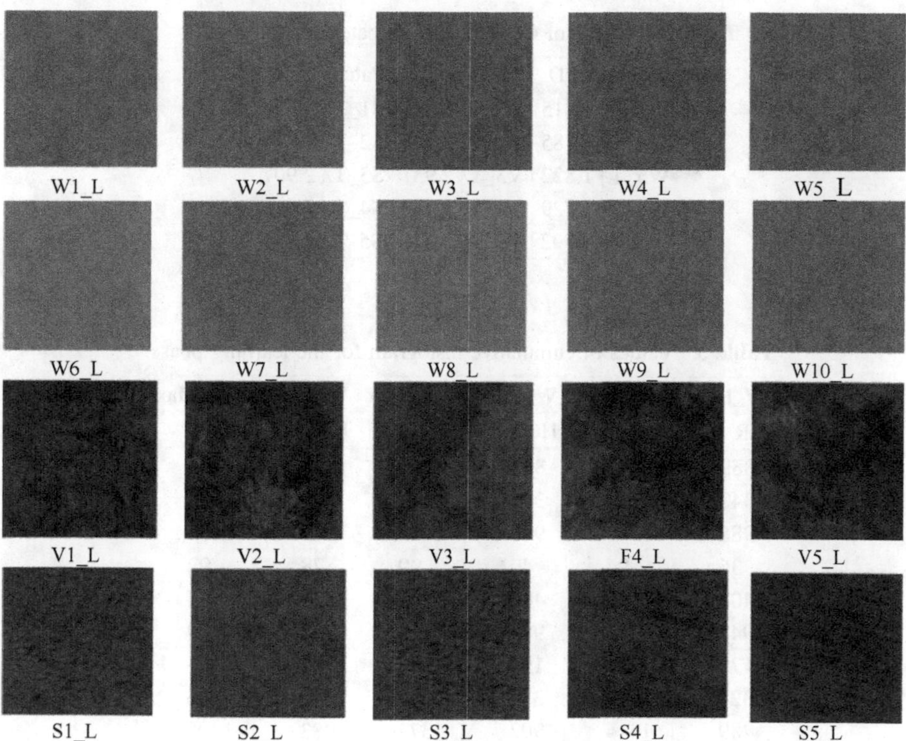

Fig. 4. Examples of patches for defining the classes

CFD and CLBP are calculated for each learning patches and the corresponding intervals [min, max] for the W class are presented in Table 1 (CFD) and, respectively, in Table 3 (CLBP).

Table 1. Values of CFD for training.

Patch water	CFD W	MIN MAX	Patch nonwater	CFD
W1_L	1.713	Wmin = 1.713	V1_L	2.980
W2_L	1.754	Wmax = 1.859	V2_L	3.011
W3_L	1.762		V3_L	2.944
W4_L	1.773		V4_L	2.953
W5_L	1.785		V5_L	2.983
W6_L	1.707		S1_L	2.926
W7_L	1.821		S2_L	2.941
W8_L	1.842		S3_L	2.897
W9_L	1.850		S4_L	2.884
W10_L	1.859		S5_L	2.916

Table 2. Values of CFD for testing patches from Fig. 6.

Patch	CFD	Patch	CFD	Patch	CFD
W1_T	1.815	V1_T	3.031	S1_T	1.850
W2_T	1.785	V2_T	2.944	S2_T	2.897
W3_T	1.822	V3_T	2.931	S3_T	2.907
W4_T	1.729	V4_T	2.984	S4_T	2.833
W5_T	1.792	V5_T	3.011	S5_T	2.825

Table 3. Values of cumulative histogram for the learning phase.

W_Lmin	W_LMax	W_Lmin	W_LMax	W_Lmin	W_LMax
HR		HG		HB	
889	1419	884	1410	814	1395
342	630	349	635	276	627
985	1187	998	1246	962	1186
36	94	40	99	28	92
407	939	404	945	335	931
948	1614	986	1595	763	1729
179	375	192	365	186	361
328	899	333	879	340	866
789	1414	902	1372	983	1405
7813	11063	7910	11066	7929	11340

Table 4. Score values for segmentation.

Patch	R score	G score	B score	S_{CLBP}	Norm. SCLBP	S_{CFD}	S(W)	Classification result
W Class								
W1_T	9	10	10	29	0.97	0.80	0.89	W
W2_T	5	5	4	**14**	0.47	0.99	0.73	W
W3_T	8	7	8	23	0.77	0.75	0.76	W
W4_T	8	7	7	22	0.73	0.61	0.67	W
W5_T	9	9	6	24	0.80	0.96	0.88	W
V Class								
V1_T	1	1	1	3	0.10	−7.50	−3.70	nW
V2_T	2	2	1	5	0.17	−6.93	−3.38	nW
V3_T	0	0	1	1	0.03	−6.84	−3.41	nW
V4_T	1	1	0	2	0.07	−7.20	−3.57	nW
V5_T	1	1	1	3	0.10	−7.39	−3.65	nW
S Class								
S1_T	1	1	1	3	0.10	**0.56**	0.33	nW
S2_T	1	2	1	4	0.13	−6.61	−3.24	nW
S3_T	2	2	2	6	0.20	−6.68	−3.24	nW
S4_T	1	1	1	3	0.10	−6.17	−3.04	nW
S5_T	1	2	2	5	0.17	−6.11	−2.97	nW

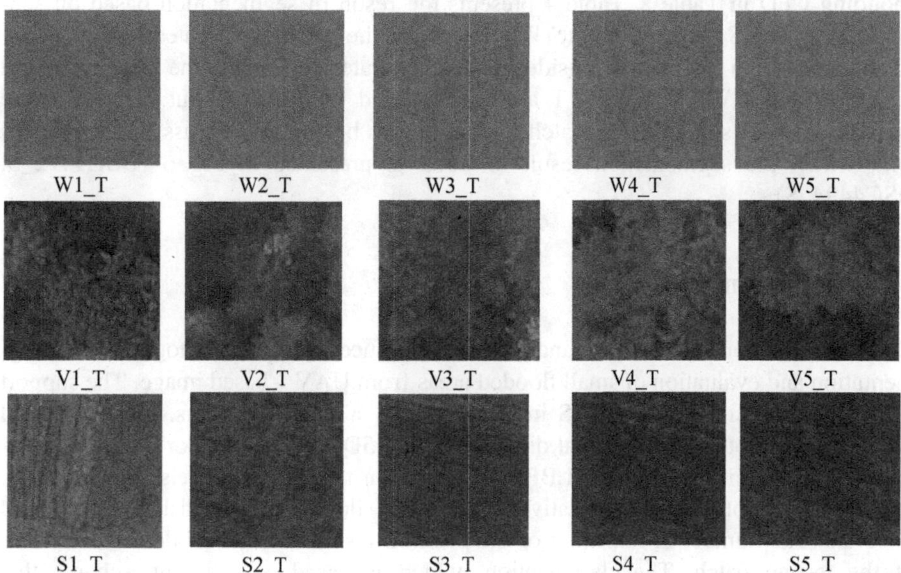

Fig. 5. Examples of patches for test

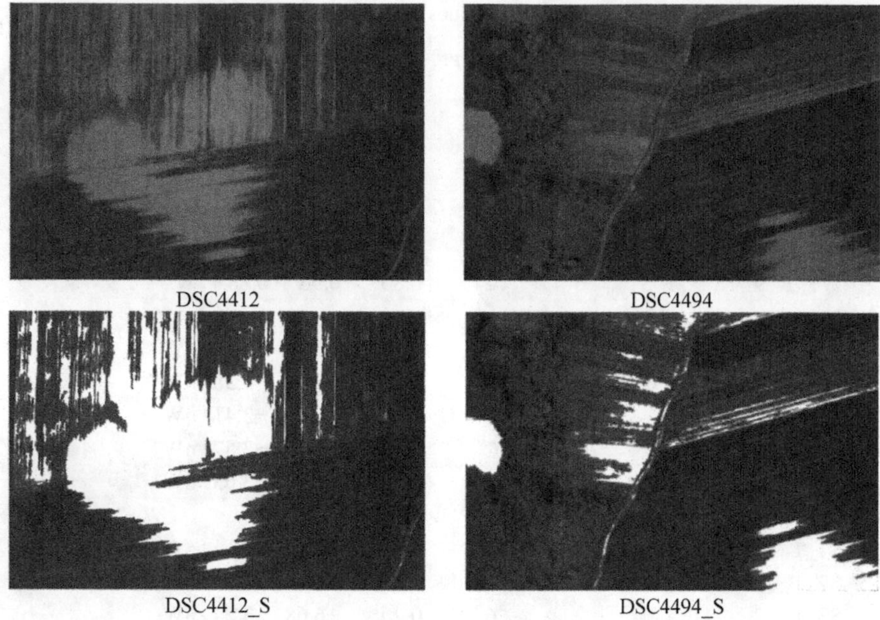

Fig. 6. Segmented flooded areas.

It can be observed that, in the learning phase the classes *W* and *nW* are well separated. For the testing phase, a set of 100 patches in W class and 100 patches in *nW* class was used. A set of 15 patches, as example is presented in Fig. 6 and the corresponding CFD in Table 2. Table 4 presents the result of segmentation based on separate (S_{CFD} and S_{CFD}) and total ($S(W)$) scores for the patches presented in Fig. 5.

It can be seen that if it is considered only separate scores, with the same threshold 0.5, the patches W2_T and S1_T are misclassified (gray color), but the total score correctly classifies them. If the patches identified as belong to the class *W*, are marked with white, the segmentation result of flood is presented in Fig. 6 (DSC4412_S, DSC4494_S).

4 Conclusions

Color information and texture analysis can be successfully used together for segmentation and evaluation of small flooded areas from UAV – based image. The support for image acquisition was a UAS implemented by a team of authors. We introduced two local descriptors, color fractal dimension, in a 5D space, and color binary patterns histogram by concatenating the LBP histograms on the color channels (R, G and B). For each descriptor, a representative of the class flood was established as a set of intervals. The number of matching or the position inside of intervals defines the score of the testing patch. The classification, which is based on a voting scheme that combines both scores, one based on CFD and other on CLBP, gives better results than the separate criteria.

Acknowledgements. The work has been funded by Romanian National Authority for Scientific Research and Innovation, UEFISCDI, project SIMUL, number BG49/2016 and Data4Water H2020, TWINN 2015 Project.

References

1. Feng, Q., Liu, J., Gong, J.: Urban flood mapping based on unmanned aerial vehicle remote sensing and random forest classifier-a case of Yuyao. China. Water **7**, 1437–1455 (2015). doi:10.3390/w7041437
2. Tamminga, A.D., Eaton, B.C., Hugenholtz, C.H.: UAS-based remote sensing of fluvial change following an extreme flood event. Earth Surf. Proc. Land. **40**, 1464–1476 (2015). doi:10.1002/esp.3728
3. Ahmad, A., Tahar, K.N., Udin, W.S., Hashim, K.A., Darwin, N., Hafis, M., Room, M., Hamid, N.F.A., Azhar, N.A.M., Azmi, S.M.: Digital aerial imagery of unmanned aerial vehicle for various applications. In: IEEE International Conference on Control System, Computing and Engineering (ICCSCE 2013), 535–540 (2013)
4. Popescu, D., Ichim, L.: Image recognition in UAV application based on texture Analysis. In: Battiato, S., Blanc-Talon, J., Gallo, G., Philips, W., Popescu, D., Scheunders, P. (eds.) ACIVS 2015. LNCS, vol. 9386, pp. 693–704. Springer, Cham (2015). doi:10.1007/978-3-319-25903-1_60
5. Khelifi, R., Adel, M., Bourennane, S.: Multispectral texture characterization: application to computer aided diagnosis on prostatic tissue images. EURASIP J. Adv. Sig. Proc. **118**, 1–13 (2012). doi:10.1186/1687-6180-2012-118
6. Losson, O., Porebski, A., Vandenbroucke, N., Macaire, L.: Color texture analysis using CFA chromatic co-occurrence matrices. Comput. Vis. Image Underst. **117**, 747–763 (2013)
7. Ivanovici, M., Richard, N.: Fractal dimension of color fractal images. IEEE Trans. Image Process. **20**, 227–235 (2011)
8. Popescu, D., Ichim, L., Gornea, D., Stoican, F.: Complex image processing using correlated color information. In: Blanc-Talon, J., Distante, C., Philips, W., Popescu, D., Scheunders, P. (eds.) ACIVS 2016. LNCS, vol. 10016, pp. 723–734. Springer, Cham (2016). doi:10.1007/978-3-319-48680-2_63
9. Haralick, R., Shanmugam, K., Dinstein, I.: Textural features for image classification. IEEE Trans. Syst. Man. Cybern., 610–621 (1973)
10. Yu, E., Monaco, J.P., Tomaszewski, J., Shih, N., Feldman, M., Madabhushi, A.: Detection of prostate cancer on histopathology using color fractals and probabilistic pairwise Markov models. In: Annual International Conference of the IEEE Engineering in Medicine and Biology Society, pp. 3427–3430 (2011)
11. Sarker, N., Chaudhuri, B.B.: An efficient differential box-counting approach to compute fractal dimension of image. IEEE Trans. Syst. Man Cybern. **24**, 115–120 (1994)
12. Chaudhuri, B.B., Sarker, N.: Texture segmentation using fractal dimension. IEEE Trans. Pattern Anal. Mach. Intell. **17**, 72–77 (1995)
13. Ojala, T., Pietikäinen, M., Harwood, D.: A comparative study of texture measures with classification based on feature distributions. Pattern Recogn. **29**, 51–59 (1996)

14. Porebski, A., Vandenbroucke, N., Hamad, D.: LBP histogram selection for supervised color texture classification. In: IEEE International Conference on Image Processing, Melbourne, VIC, pp. 3239–3243 (2013)
15. MUROS - Teamnet International. http://www.teamnet.ro/grupul-teamnet/cercetare-si-dezvoltare/muros/
16. Popescu, D., Ichim, L., Stoican, F.: Unmanned aerial vehicle systems for remote estimation of flooded areas based on complex image processing. Sensors **17**(3), 1–24 (2017). doi:10.3390/s17030446

Interconnected Neural Networks Based on Voting Scheme and Local Detectors for Retinal Image Analysis and Diagnosis

Traian Caramihale[1], Dan Popescu[1(✉)], and Loretta Ichim[1,2]

[1] Faculty of Automatic Control and Computers, Politehnica University of Bucharest, Bucharest, Romania
traian90@gmail.com,
{dan.popescu,loretta.ichim}@upb.ro
[2] Stefan S. Nicolau Institute of Virology, Bucharest, Romania

Abstract. This paper presents a method for accurate detection and localization of important regions of interest (ROIs) for retinal disease diagnosis: optic disc, macula, blood vessels, exudates and hemorrhages. To this end, the image processing is made on patches of different dimensions, obtained by combining sliding box and non-overlapping box algorithms. Two interconnected neural networks are considered: one for feature selection and class representative establishment (learning phase) and one for the classification phase: ROIs detection, localization, segmentation and evaluation. The last one is based on a voting scheme of local detectors with different weights, taking into account the results from associated confusion matrices. The information is fused from the selected features (textural type and fractal type, which are computed for different color channels). We tested 200 images from different public databases (40 for the learning phase and 160 for the classification phase). The experimental results indicate a good accuracy for all analyzed regions of retinal images.

Keywords: Image segmentation · Neural networks · Local image processing · Retinal disease diagnosis · Voting scheme

1 Introduction

Different eye diseases can cause partial or even total loss of vision, therefore affecting all daily activities. Analysis of retinal images can help detecting the presence of ophthalmological diseases. The localization and analysis of eye components, like optic disc (OD), macula (MA) or blood vessels (BV) represent important steps in diabetic retinopathy detection. Moreover, identifying exudates (EX), hemorrhages (HE) or blood vessels occlusions can determine the diseases' stage of evolution, as they are all caused by diabetic retinopathy.

An important parameter regarding blood vessels is represented by the occlusions identification. One of the causes of the occlusions is diabetic retinopathy. The occlusions can lead to the appearance of hemorrhages and are an indicator of the diseases' evolution, being noticed in early stages.

S. Battiato et al. (Eds.): ICIAP 2017, Part II, LNCS 10485, pp. 753–764, 2017.
https://doi.org/10.1007/978-3-319-68548-9_68

Computer-aided diagnosis can benefit from automatizing the detection process of the eye components, also having the advantages of computational power, speed and continuous improvement. An automated system can help ophthalmologist to see more patients and to deliver better analysis results. Students can also benefit using the system as a training tool.

Different approaches for regions of interest identification in retinal images have been published [1–11].

Maninis et al. [1] developed a framework (DRIU) for retinal images analysis which performs optic disc recognition and blood vessel segmentation. The method uses deep convolutional neural networks with 2 sets of layers. The results obtained on 4 image datasets are similar to human-equivalent diagnosis. Melinscak et al. [2] develop a GPU implementation of deep max-pooling convolutional neural networks for blood vessels segmentation. The classification is done at the pixel level. Multiple layers are alternated and an activation function determines the type of each pixel. The accuracy of the algorithm is 0.9466 on the DRIVE database.

In [3], a CNN learning model for OD localization is presented. Cascade classifiers are used to improve the algorithms' speed. A fast OD detection was achieved on 8 datasets, with high sensitivity. Similarly, in [4, 5] various neural networks classifiers are validated in the detection process of exudates and lesions. The multi-layer perceptron performs the best under different constraints. Other classifiers include support vector machine and radial basis functions.

Van Grinsven et al. [6] train CNNs to detect hemorrhages. Their goal is to decrease training time. They select only the most representative samples during each step of the training phase, therefore reducing the redundant information.

Sopharak et al. [7] perform exudates detection using a fuzzy k-means clustering technique. They use different image features as parameters, obtained after a preprocessing phase. They obtain an accuracy of 99% on a private dataset.

A machine learning based method is developed in [8] for the detection of age-related macular degeneration. Support vector machine and neural networks are trained with histogram based feature descriptors obtained after applying Gabor filters on choroidal images. Different stages of age-related macular degenerations are identified on images from 21 patients.

Previously, we developed several algorithms [9–11] for the detection and analysis of different eye component in retinal images. We used a variety of techniques for selecting the regions of interest based on textural and fractal features: voting schemas, data clustering, image segmentation and box dividing algorithms, among others.

In this paper we propose a new method to detect the regions of interest in retinal images. We use two interconnected neural networks, the first for features computation and selection (learning phase) and the second for regions of interest detection and analysis (classification phase), based on the results from the first network. This implementation is suitable for parallelization, in order to improve execution speed by adding more computational power. The classification phase uses of a voting scheme based on the weights of local detectors, which are determined by associated confusion matrices. Thus the patches containing ROIs are selected. We also perform segmentation on each ROI and evaluate specific parameters which help determining the presence and severity of the diseases.

2 Methodology and Algorithm

2.1 Learning Phase

During the learning phase, we seek to select representative features for each class of ROI. These features will be used in the automated ROI detection. All the functions in this phase will be performed by the nodes of an artificial neural network (ANN).

The ANN imitates the behavior of human neural networks, having multiple interconnected nodes distributed in different layers, with each node performing a processing function. The connections between the nodes can be weighted, giving more importance to some input information. We will use a multi-layer ANNs both in the learning and in the classification phase.

The input of the neural network is an eye-fundus image (IM). The input layer decomposes the image into the R, G and B components, which will be further analyzed. The features we are investigating are easier to compute and have a better visual representation on grayscale images.

As most of the regions of interest are much smaller than the size of retinal images, we will apply a box dividing algorithm to generate patches at the first layer (Layer I) of the ANN. The box dividing algorithm adapts the size of the patches ($s \times s$) and the *displacement* (pixel distance, both vertical and horizontal, between two neighboring patches) taking into account the image size and the searched region of interest. Considering our previous researches [9–11], analyzing eye-fundus images from various datasets (MESSIDOR [12], STARE [13], DRIVE [14] and HRF [15]), with varying image sizes, and experimenting with different values for the size and *displacement*, we decided to have 3 cases for the box dividing algorithm:

- The OD and MA are considered in the first case, as they are single representatives of their class in each image. They are required to be entirely contained in one patch. The sliding box algorithm is used, allowing the patches to overlap. In this way, the ROI will also be split across multiple neighboring patches, but it will surely be fully covered by one single patch. The formulas for the patch size and displacement for OD (and similar for MA) have been experimentally determined and can be found in (1–2):

$$s_{OD} = \left\lceil \frac{image_{HEIGHT} + image_{WIDTH}}{2} \times 4\% \right\rceil \tag{1}$$

$$displacement_{OD} = \left\lceil \frac{s_{OD}}{2} \right\rceil \tag{2}$$

- The second case is represented by EX and HE. These ROIs are dependent of the diseases' presence and have different sizes and shapes. They are not required to be fully contained in a single patch, as their location and covered surface is more important for evaluation of the diseases' evolution. Therefore, the patches are not allowed to overlap (the displacement equals 0 pixels), to prevent redundancy of the analyzed information. The formulas for the size and displacement for EX (and similar for HE) can be found in (3–4):

$$size_{EX} = \left\lceil \frac{s_{OD}}{2} \right\rceil \tag{3}$$

$$disp_{EX} = 0 \tag{4}$$

– The BVs represent the third case. The vessels have a variety of shapes and widths in the same eye fundus image. Using only a single sized patch may not be sufficient. The main focuses regarding the blood vessels are their segmentation and the detections of occlusions. We will consider 3 patch sizes for the detection of segments of BV, starting with the values in (2) divided by 2, and further dividing the patch size by 2 two more times (as experimentally determined).

Using the selected values, we obtained good results on images from all of the 4 specified datasets [12–15]. Each of the patches is labeled with an identifier taking into account the row and column of the patch in the obtained grid of patches.

All the patches are passed to the second layer (Layer II), where the mean gray level co-occurrence matrix for each patch is generated. The GLCM shows the occurrence probability of a pair of pixel intensities in an image, with a given degree of similarity and under spatial constraints. The standard GLCM is computed with a given displacement between the pixels and a fixed direction. For the scope of this paper, we chose a displacement of 5 pixels and 8 complementary directions, as shown in [10]. The GLCM we use is an average of the 8 obtained GLCMs.

Haralick et al. [16] use the GLCM to define important statistical features which include contrast, energy, correlation, variance, homogeneity and entropy among others. In addition to the before mentioned features we also compute mean-intensity, a simple first order statistical feature and 3 fractal features [16]: fractal dimension, mass fractal dimension and lacunarity. All the formulas for the features can be found in [9–11, 16, 17]. The features are computed in the nodes of the third layer (Layer III). There is a node for every feature (11) computed on each of the 3 colors.

At the fourth layer (Layer IV) we have a node for each pair G_k (feature, color component), with k =1, 33. At this point, we manually label all the patches that contain ROIs (OD, MA, EX, HE, BV) with the help of an ophthalmologist, considering the remaining patches as background (BK). On each node, the values for the respective group for all the patches are stored and sorted descending. Then we tend to group the data from the ordered tables into clusters, using a K-means clustering algorithm, as described in [9]. The scope is to identify clusters of feature values that characterize each of the required ROI. We focus on identifying clusters only in the extreme positions of the sorted tables (top or bottom), because using the relative position of middle clusters is not suitable for automated detection.

We use the obtained clusters to select unique signatures of features that describe each class of ROI. For each pair G_k we will compute the probabilities P_{Gk} with which a previously labeled patch R_i is assigned to each class of regions of interest C_j conditioned by the clustered feature values V_{Gk} of that respective pair, as in (5):

$$P_{Gk}(R_i, C_j) = P(R_i \in C_j | V_{Gk}(R_i)) \quad (5)$$

The algorithm generates the confusion matrices based on the computed probabilities in the fifth layer (Layer V). During the learning phase, the confusion matrix for a pair G_k is computed as the average value of the confusion matrices obtained for each analyzed image (40 images in the learning phase) for the respective pair. In Table 1, the confusion matrix for (contrast, green) can be seen:

Table 1. Average confusion matrix for (contrast, green).

ROI/C_j	C_{OD}	C_{MA}	C_{HE}	C_{EX}	C_{BV}	C_{BK}
OD	0.83	0	0	0.08	0.05	0.04
MA	0.08	0.21	0.11	0.07	0.24	0.29
HE	0.08	0.12	0.27	0.07	0.21	0.25
EX	0.1	0.2	0	0.75	0.05	0.07
BV	0.05	0.04	0.03	0.02	0.71	0.15
BK	0.08	0.12	0.12	0.09	0.14	0.45

The values of the cells on the main diagonal show the probability of correct membership to a class. For example, 21% of the patches containing the macula can be found in one of the clusters obtained from the sorted tables of contrast values computed on the green component. The rest of the patches values are placed in other clusters.

The background included all the other patches that have not been labeled as a ROI.

For a pair to be considered as representative for a class, the probability for the respective class should be at least 80% for OD and MA and at least 70% for the other ROIs. These threshold values have been experimentally determined.

As it can be seen in Table 1, the contrast for the green component is a representative feature for both the OD, EX (the first 2 top clusters) and BV (the bottom cluster). The OD patches tend to be organized in the top clusters. Once we remove the patches that contain OD, the new top clusters will mostly (75%) include exudates patches. Therefore, we impose that the OD detection is done prior to the exudates detection in the classification phase, so that we can remove the detected patches.

Considering the proposed implementation which was described in the current section, the following ANN model for the learning phase is used (Fig. 1).

2.2 Optic Disc and Macula Segmentation

For a single region of interest in a retinal image, like the optic disc or macula, we selected only those features that had a probability of correct detection for the OD and MA of above 80%. These features are mean intensity, contrast, mass fractal dimension (top clusters, green) and lacunarity (top cluster, red), for the OD. As for the MA, the selected features are energy (top cluster, green), mass fractal dimension (bottom cluster, green) and lacunarity (bottom clusters, green and blue).

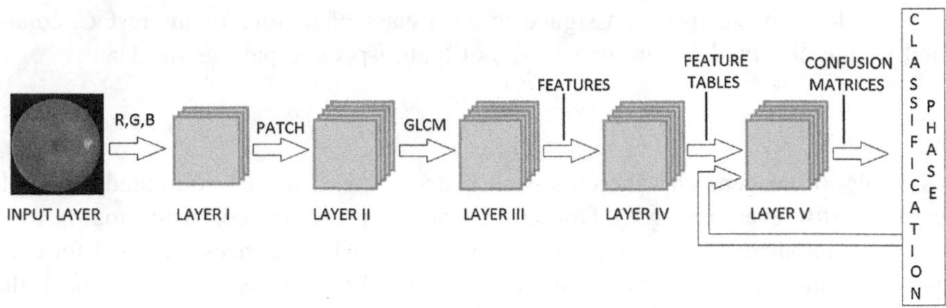

Fig. 1. ANN used in the learning phase

The ANN used in the learning phase (Fig. 1) will be modified to compute only the selected parameters for this phase. The sorted and clustered tables and the confusion matrices are received as an input by a new 3-layer neural network, which performs the classification and evaluation of ROIs (Fig. 2).

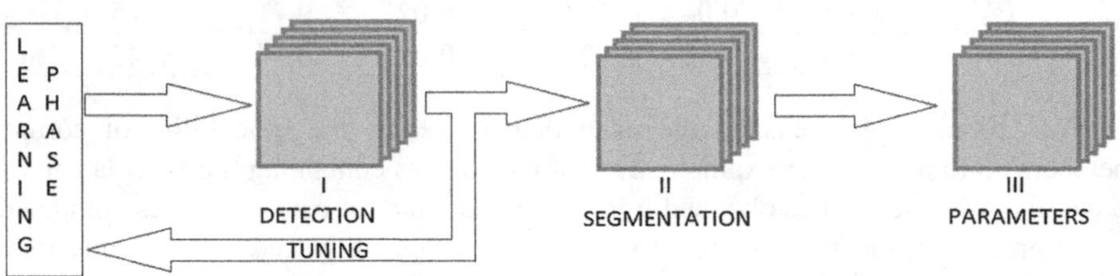

Fig. 2. ANN used in the classification phase

The first layer (DETECTION - I) will identify the patch that best contains the ROI (OD or MA). A voting schema based on local detectors is used. For each feature, the detector will rank the patches according to the cluster direction (top or bottom) and will weight those ranks using the values for the searched ROI in the associated confusion matrices. The score for a patch are computed as a mean of the values given by the 4 local detectors for each ROI. The patch with the best score will be selected as the one whose contains the best representation of the OD or MA, respectively. Based on experimental threshold values and patches neighbors' selection, we determine additional patches that contain parts of the searched ROI, as well as false positives, in order to continuously tune the values in the confusion matrices. The second layer (SEGMENTATION - II) performs the segmentation of the OD and MA. We will apply the algorithm described in the learning phase on the selected patch from the previous layer, with a new size of 3×3 pixels and no overlapping. The same ranking process as in the current classification phase will be applied, but this time the list of scores is split only in two clusters. All the patches in the top cluster are considered to have ROI background. The segmentation of the ROI is performed by putting together the selected patches (from the top cluster). For the optic disc segmentation, we also take into consideration that some patches containing blood vessels will be rejected. Having performed the segmentation of both the OD and the MA, we will compute the disc-to-macula (fovea) distance to disc-diameter ration (DM/DD) in one of the nodes in the third layer (PARAMETERS - III). A value bellow 4 of this parameter (increased optic disc cup size) marks the possible presence of glaucoma.

2.3 Hemorrhages and Exudates Classification

The threshold for the probability of correct detection in this case is 70%. These selected signatures of features are mean intensity, contrast, mass fractal dimension and energy (top clusters, green) for the EX [10] and mean intensity, mass fractal dimension (top cluster, red), lacunarity (bottom cluster, red) and mass fractal dimension (top cluster, green) for HE [9].

A similar neural network as in Sect. 2.2 will also be used in this case, but changes will be brought to each layer. In the first layer (DETECTION - I), the patches that have been previously identified as OD or MA are removed from the sorted tables. The local detectors will rank the patches for the selected signatures of features. The difference in this case is that all the patches with 3 or more associated detectors will be considered candidates. The final vote will take into account 3 threshold values:

- If the best rank is lower than a first threshold value, it is determined that the image does not contain the ROI.
- All patches with a rank bellow a second threshold (different values for 3 or 4 detectors) will be removed.
- If the distance between two consecutive ranks is higher than a third threshold, the algorithm stops.

The second layer (SEGMENTATION - II) is similar to the one described in the OD/MA classification phase, but it will process multiple patches selected as ROI. After the segmentation is done, we compute the number of hard exudates, hemorrhages and the percentage of coverage in two of the nodes in the third layer (PARAMETERS - III). Additionally, in another node we compute the macula to exudates distance. The parameters are important in determining the presence of macular edema.

2.4 Blood Vessels Segmentation

A correct detection rate of 70% is also used in this case to select the features from the confusion matrices: mean intensity, mass fractal dimension, energy (top cluster, red) and contrast (bottom cluster, green).

As specified before, the detection of the blood vessels is more complicated because of the complex structure of the system. Therefore, the algorithm (as described in Sects. 2.2 and 2.3) is executed with three different patch sizes, as it has been experimentally observed that such a division can cover the whole blood vessel system. The neural network described in Fig. 2 will not make use of the PARAMETERS layer.

As in the EX/MA classification phase, in the first layer (DETECTION - I), we consider patches that have at least 3 associated feature detectors. Out of the candidate patches we select only those for which the detectors compute a rank that passes the last two threshold validations from Sect. 2.3.

After the detection is performed for all the 3 patch sizes, the selected patches are segmented in the SEGMENTATION layer. The segmented blood vessel system is reconstructed from all segmentations. Having already performed the detection of the optic disc, we can determine the location of the major blood vessels.

3 Experimental Results

The proposed method has been validated on 200 images from 4 different retinal images datasets (MESSIDOR, DRIVE, STARE and HRF), as seen in Fig. 3. Although the images present differences in size, crop, brightness, color and clarity, the experimentally determined sizes (formulas 1–4) and signatures of features are applicable across all the analyzed databases.

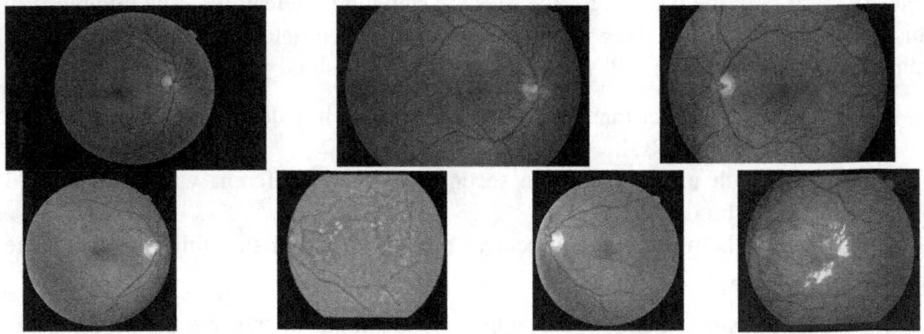

Fig. 3. Selected images (Color figure online)

The obtained results for the optic disc and macula classification and segmentation can be seen in Fig. 4. As it can be noted from the figure, each patch is uniquely labeled in an image, so we can also select the neighboring overlapping patches that contain parts of the ROI. This approach is suitable when we need to remove the patches from the feature tables for further analysis (as it described for contrast on green in the learning phase in Sect. 2.1). When performing the segmentation of the OD, the blood vessels might be considered as black background. Therefore we adjusted the algorithm to fill any empty spaces within the boundaries of the OD.

The position of the centers of the OD and MA (foveae) are computed as the center of the smallest ellipse that fully fits each ROI. Having the two center points, we compute the distance between the OD and the MA, and the OD disc diameter. The DM: DD parameter's variables are shown in Fig. 5. A value below 4 of this parameter shows the possible presence of glaucoma (left image – no glaucoma, right image - glaucoma).

In Fig. 6, we can see the results of the segmentation for the exudates and hemorrhages. We compute the coverage percentage of the EX and HE in each identified patch as the ratio between the number of white pixels and the total number of pixels in a

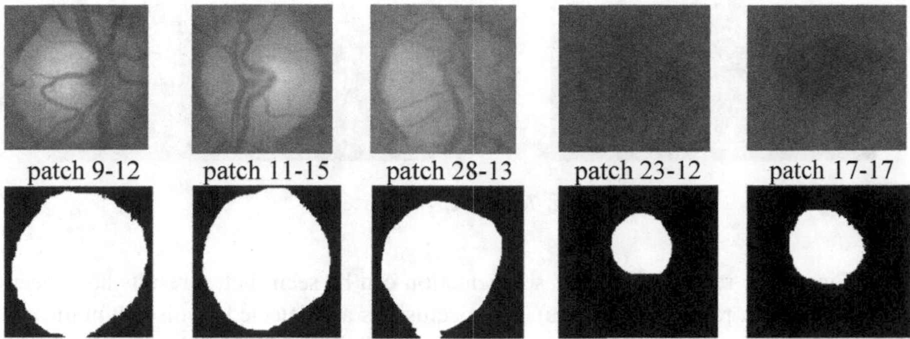

Fig. 4. OD and MA segmentation

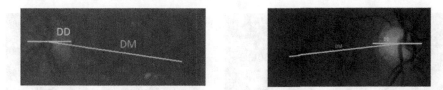

Fig. 5. DM:DD parameter

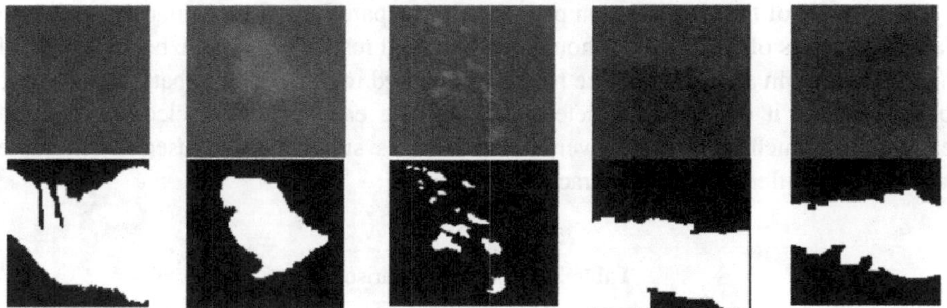

Fig. 6. EX and HE segmentation

patch. These two parameters, along with the coverage percentage in the image, show the evolution stage of the disease. The number of hard exudates is also an indicator used by ophthalmologists to identify diabetic retinopathy, which we provide by identifying shapes based on the segmentation.

The MA to EX distance (Fig. 7) is an indicator of macular edema. This parameter can be determined after the localization of exudates is performed. We select a circular neighborhood surrounding the macula, with the foveae as its center and a diameter equal to the size of the MA patch. Distances in pixels to the exudates in the neighborhood are computed. If the minimum distance is lower than the diameter of the optic papilla, the risk of macular edema is high.

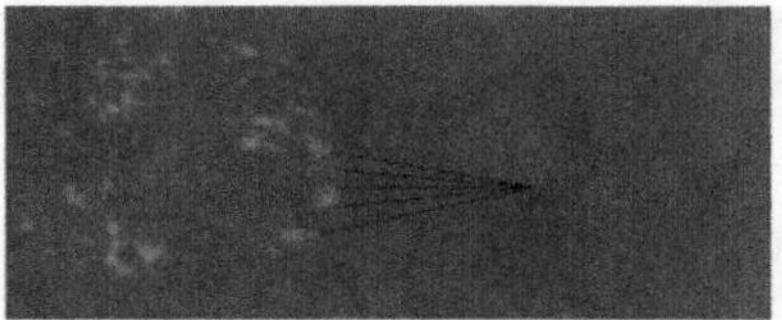

Fig. 7. MA-EX distance

In Fig. 8, the results of the BV segmentation can be seen (better results have been obtained on high resolution images). The occlusions are detected as discontinuities in the blood vessels (black pixels between white pixels, with the blood vessel having a minimum imposed width of 3 pixels).

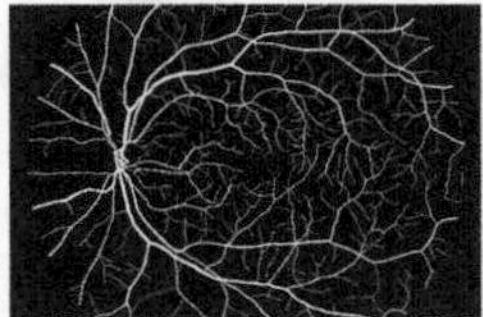
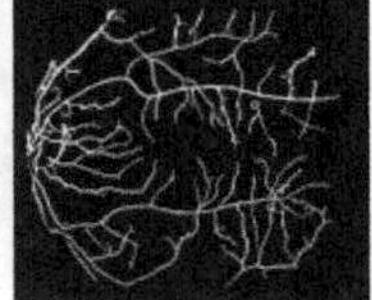
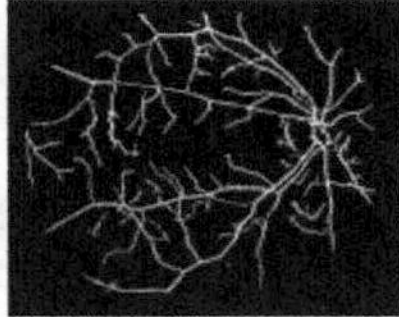
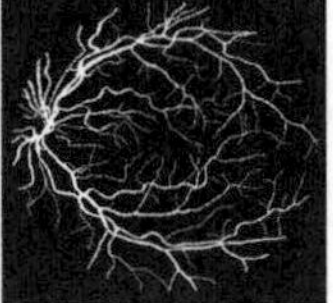

Fig. 8. Segmented BV systems

Our results of the classification process are compared to other similar researches in Table 2 in terms of accuracy. Although the obtained results are not the best for each of the ROI, the main advantage of the method described in this paper is that it not specific for a ROI, but it can localize, detect and evaluate each ROI, by selecting different features and patch sizes. Moreover, the method is suitable to be used on multiple datasets with different image characteristics.

Table 2. Results comparison.

Method	OD	MA	EX	HE	BV
Current	98.8%	97.3%	97.5%	96.71%	95.31%
[2]	–	–	–	–	97.49%
[3]	99.2%	–	–	–	–
[4]	–	–	97.01%	–	–
[6]	–	–	–	97.2%	–
[7]	–	–	99.11%	–	–
[8]	–	94.4%	–	–	–
[9]	98.6%	97.3%	97.2%	96.46%	–
[10]	96.66%	94%	93.68%	–	–
[15]	–	–	–	–	95%

4 Conclusion

In this paper we developed a method based on neural networks for the localization and evaluation of all the regions of interest in retinal images. The main advantage of this method, in comparison with other researches and with our previous work is that it can be used, with minor modifications, for every required ROI. In the learning phase we identified the textural and fractal features used to generate confusion matrices. Local detectors, with different weights, based on the confusion matrices, are used to rank the patches obtained from box dividing algorithms. The regions of interest are segmented and further used for various parameters computation. The results show an improvement from our previous work, being similar with those of other equivalent algorithms, which are suitable for a lower number of ROIs. The speed of the algorithm can be increased by adding additional nodes. Our next focus will be on building a system for monitoring the evolution eye-diseases, using the current algorithm as a step in the monitoring process.

References

1. Maninis, K.K., Pont-Tuset, J., Arbeláez, P., Van Gool, L.: Deep retinal image understanding. In: Ourselin, S., Joskowicz, L., Sabuncu, M., Unal, G., Wells, W. (eds.) MICCAI 2016. LNCS, vol. 9901, pp. 140–148. Springer, Cham (2016). doi:10.1007/978-3-319-46723-8_17
2. Melinscak, M., Prentasic, P., Loncaric, S.: Retinal vessel segmentations using deep neural networks. In: 10th International Conference on Computer Vision Theory and Applications, pp. 577–582 (2015)
3. Alghamdi, H.S., Tang, H.L., Waheeb, S.A., Peto, T.: Automatic optic disk abnormality detection in fundus images: a deep learning approach. In: Ourselin, S., Joskowicz, L., Sabuncu, Mert R., Unal, G., Wells, W. (eds.) MICCAI 2016. LNCS, vol. 9900, pp. 1–8. Springer, Cham (2016). doi:10.1007/978-3-319-46720-7
4. Garcia, M., Sachez, C.I., Lopez, M.I., Lopez, C.I., Abasolo, D., Homero, R.: Neural network based detection of hard exudates in retinal images. Comput. Methods Programs Biomed. **93**, 9–19 (2009)
5. Garcia, M., Valverde, C., Lopez, M.I., Poza, J., Homero, R.: Comparison of logistic regression and neural networks classifiers in the detection of hard exudates in retinal images. In: 35th Annual International Conference of the IEEE EMBS, pp. 5891–5894 (2013)
6. van Grinsven, M.J.J.P., van Ginneken, B., Hoyng, C.B., Theelen, T., Sanchez, C.I.: Fast convolutional neural network training using selective data sampling: application to hemorrhage detection in color fundus images. IEEE Trans. Med. Images **25**, 1273–1284 (2016)
7. Sopharak, A., Uyyanonvara, B., Barman, S.: Automatic exudate detection from non-dilated diabetic retinopathy retinal images using fuzzy c-means clustering. Sensors **9**(3), 2148–2161 (2009)
8. Deng, J., Xie, X., Terry, L., Wood, A., White, N., Margrain, T.H., North, R.V.: Age-related macular degeneration detection and stage classification using choroidal oct images. In: Campilho, A., Karray, F. (eds.) ICIAR 2016. LNCS, vol. 9730, pp. 707–715. Springer, Cham (2016). doi:10.1007/978-3-319-41501-7_79

9. Caramihale, T., Popescu, D., Ichim, L.: Detection of regions of interest in retinal images using artificial neural networks and k-means clustering. In: 22nd International Conference on Applied Electromagnetics and Communications (ICECOM), pp. 1–6 (2016)
10. Popescu, D., Ichim, L., Caramihale, T.: Texture based method for automated detection, localization and evaluation of the exudates in retinal images. In: Arik, S., Huang, T., Lai, W. K., Liu, Q. (eds.) ICONIP 2015. LNCS, vol. 9492, pp. 463–472. Springer, Cham (2015). doi:10.1007/978-3-319-26561-2_55
11. Popescu, D., Ichim, L., Caramihale, T.: Computer-aided localization of the optic disc based on textural features. In: The 9th International Symposium on Advanced Topics in Electrical Engineering (ATEE), pp. 307–312 (2015)
12. Decencière, E., et al.: Feedback on a publicly distributed database: the Messidor database. Image Anal. Stereol. **33**, 231–234 (2014)
13. Hoover, A., Goldbaum, M.: Locating the optic nerve in a retinal image using the fuzzy convergence of the blood vessels. IEEE Trans. Med. Imaging **22**, 951–958 (2003)
14. Stall, J.J., Abramoff, M.D., Niemeijer, M., Viergever, M.A., van Ginneken, B.: Ridge based vessel segmentation in color images of the retina. IEEE Trans. Med. Imaging **23**, 501–509 (2004)
15. Odstrcilik, J., Jan, J., Gazárek, J., Kolář, R.: Improvement of vessel segmentation by matched filtering in colour retinal images. In: Dössel, O., Schlegel, W.C. (eds.) World Congress on Medical Physics and Biomedical Engineering, pp. 327–330. Springer, Heidelberg (2009). doi:10.1007/978-3-642-03891-4_87
16. Haralick, R.M., Shanmugam, K., Dinstein, I.: Textural features for image classification. IEEE Trans. Syst. Man Cybern. **SMC-3**, 610–621 (1973)
17. Sarker, N., Chaudhuri, B.B.: An efficient differential box-countig approach to compute fractal dimension on image. IEEE Trans. Syst. Man Cybern. **24**, 115–120 (1994)

A Unified Color and Contrast Age-Dependent Visual Content Adaptation

M'Hand Kedjar(✉), Greg Ward, Hyunjin Yoo, Afsoon Soudi, Tara Akhavan, and Carlos Vazquez

IRYStec Software Inc., 3 Place Ville Marie, Suite 400, Montreal, Quebec H3B 2E3, Canada
{mhand.kedjar,hyunjin.yoo,afsoon,tara}@irystec.com, gward@lmi.net, carlos.vazquez@etsmtl.ca
http://www.irystec.com

Abstract. We present a unified color and contrast content adaptation (UCA) method designed to modify images while preserving observers' preferences for white color and detail level. Our method is based on newly released models of the color matching functions (CMF) and contrast sensitivity function (CSF) incorporating age-dependent components. Using CIE-2006 physiological model in combination with an extension of Barten's CSF model, our technique adjusts the display white point and local contrast to correspond to the viewer's age-related color and contrast response preference. The results of a subjective evaluation show the effectiveness of the method on a large range of image types and confirmed that combining both color and contrast is the preferred approach for content adaptation to viewers' characteristics.

Keywords: Color matching functions · Metamerism · Contrast sensitivity · Human visual system (HVS)

1 Introduction

Human color and contrast perception differ from person to person, not only in terms of vision deficiency, but also among color-normal observers. However, this variability is not taken into account in current display technologies, and it is assumed that a single average standard observer can represent the entire population.

Historically, the objective of colorimetry was to integrate properties of the human color vision into the measurement of visible light to define visual equivalents of colored stimuli. More recently, its objective has been to provide procedures that enable quantifying color matches and differences [1]. In that sense colorimetry is based on the assumption that everyone's color response can be quantified with the CIE standard observer functions, which predict the average viewer's response to the spectral content of light. However, individual observers may have slightly different response functions, which may cause disagreement

S. Battiato et al. (Eds.): ICIAP 2017, Part II, LNCS 10485, pp. 765–778, 2017.
https://doi.org/10.1007/978-3-319-68548-9_69

about which colors match and which do not. For colors with smoothly varying (broad) spectra, the disagreement is generally small, but for colors mixed using a few narrow-band spectral peaks, differences can be as large as 10 CIELAB units [2].

The human visual system (HVS) works over a remarkably wide range of illumination thanks to two classes of photoreceptors, rods and cones [3]. In daylight, vision relies mainly on three types of cones photoreceptors, and is referred as photopic vision. In low light conditions, only rods are active, and this is referred as scotopic vision. In between photopic and scotopic conditions, both cones and rods operate simultaneously, and this is referred as mesopic vision. Contrast sensitivity is a very important measure, especially under dim illumination, the presence of fog or glare, or when the contrast between objects and their background is reduced. For instance, an activity that requires a good contrast sensitivity for safety is driving at night. Numerous studies showed that older people have a significantly lower contrast sensitivity than young people [4,5]. This suggests that activities that rely on visual performance could be impacted by this decline in sensitivity. One of the many challenges is the reduced visual performance in executing everyday tasks for older people compared to younger ones. The effect of age on human vision has been extensively studied in order to identify and understand the mechanisms of this degradation [6,7]. The contrast sensitivity function (CSF), in particular, has become very popular within the vision research community, and thanks to recent advances in image processing, it becomes possible to model in real time its settings according to the effective age of the observer. In terms of displays, it allows to increase the image contrast following the preference of the observer.

Although the large variability of color and contrast perception among individual observers is a well-established fact, age is not the only factor that contributes to a change in the sensitivity of the HVS. In the same age group, the inter-observer variability of the sensitivity is high [8]. For instance, the response of the optical components of the HVS may be altered dramatically by surgical treatments. Furthermore, CIE-2006 age parameters do not match the real observer properties and cannot thoroughly describe age-induced effects of visual perception [9]. There is not enough correlation between the real age and the age-dependent parameters incorporated in CIE-2006 model [9].

In this study, we aim at developing a unified visual method to correct the effect of age on color perception and contrast sensitivity, with a strong emphasis on the personalization of the viewing experience. Given the large differences between different viewers, we will introduce a method to determine the "effective age" of a specific observer. The rest of the paper is organized as follows. In Sect. 2, we will mention some of the previous studies in contrast and color enhancement, with more emphasis on those that target specific observers. Section 3 will provide a detailed overview of our method. In Sect. 4, we will describe the subjective user study we conducted, and discuss the results. We will finish in Sect. 5 by drawing some conclusion and paths for future studies.

2 Background

Color and contrast enhancement are among the most studied fields in image processing. Their aim is to improve the image visual quality based on a specific metric, a particular application, or targeted to a specific observer's preference. Many algorithms have been proposed by the research community. One of the simplest techniques is histogram equalization, which does not provide good results in preserving the local details and natural look and color of the image [10]. To overcome these limitations, researchers developed other histogram algorithms such as local and adaptive techniques [11]. Another more sophisticated method is the Retinex theory and its variations [12,13], whose objective is to imitate the perception of the human visual system by performing a comparison between the target pixel and the referenced white point.

Some of the techniques that attempt to improve the image content to a particular set of viewers have also been developed. Peli and Peli [14] were among the first to investigate the use of data from spatial frequency content for contrast enhancements targeting visually impaired people. Lawton [15] used the CSF of the target patient to adapt the content by enhancing the contrast to the most important frequencies of that individual. More recently, Choudhury and Medioni [16] presented an approach which aims at improving the visual quality of images for both people with normal vision and patients with low-vision. This method separates the image into illumination and reflectance components and then corrects only the illumination component while trying to achieve color constancy. In [17], an adaptive color and contrast enhancement method for digital images is proposed. The intensity, contrast and color are modified and applied to the three channels in RGB space, based on features of HVS. In [18], researchers proposed a data-dependent gamma correction for applications to elderly vision. This approach converts RGB components by setting the gamma according to each pixel value of the hue, saturation and lightness contrast.

All those methods - either addressed only one aspect of image enhancement - or did not incorporate a personalization aspect to include the observer preference. Our method, on the other hand, is designed to enhance an image color and contrast based on user preference. Moreover, it incorporates a luminance retargeting component that adapts to the ambient light and the target display peak luminance. Finally, it is designed to be fully compatible with the next generation of wide color gamut displays.

3 Age-Based Content Adaptation Method

Two parts of our age-based visual content adaptation method consist of a color modification based on the observer white preference and a contrast enhancement based on the observer's preferred level of detail. The first stage of our method is a white balancing technique that allows for observers color variation, combined with a gamut expansion algorithm that maps the content from the sRGB color space to the target display wide gamut. The second stage consists of a retargeting technique that modifies the local contrast based on the observer preferred level of detail. The Fig. 1 presents the detailed components of our algorithm, which will be explained in the next section.

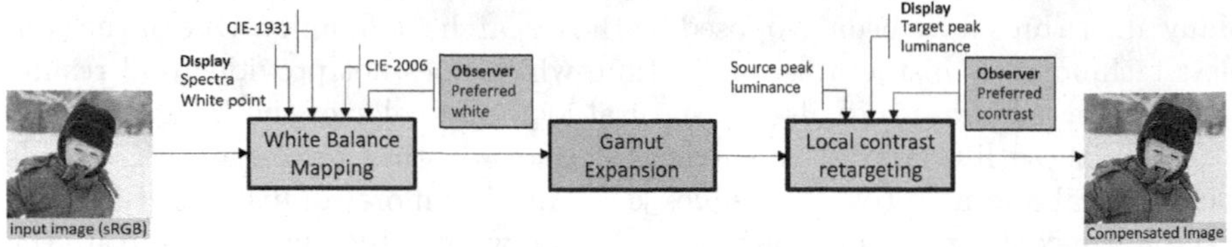

Fig. 1. The diagram of the proposed age-dependent content adaptation algorithm

3.1 White Balance

We make use of the CIE-2006 model of age-based observer color-matching functions (CMF), which establishes a method for computing LMS cone-responses to spectral stimuli [19]. We use this model to discover the range of expected variation rather than predict responses from age alone.

The most important step in this first part is to adjust the display white point to correspond to the viewer's age-related color response and preference. Our two inputs are CIE-2006 observer age and black body temperature. From these parameters and detailed measurements of the OLED RGB spectra and default white balance, we compute the white balance multipliers (r, g, b) using the procedure presented in Fig. 2.

Display Calibration. The first step of the white balance method is to normalize the OLED primary spectra $r(\lambda)$, $g(\lambda)$ and $b(\lambda)$ so they sum to the current display white point $x_w = 0.3127$, $y_w = 0.3290$, $z_w = 1 - x_w - y_w$. This is achieved by multiplying by the CIE 1931 standard observer curves $\bar{x}(\lambda)$, $\bar{y}(\lambda)$ and $\bar{z}(\lambda)$, and solving for the RGB scaling $\boldsymbol{J} = (J_r, J_g, J_b)^T$ that produces the measured xy-chromaticity.

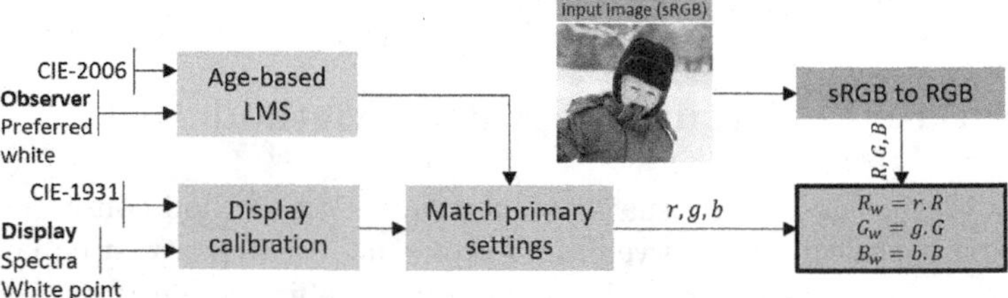

Fig. 2. The diagram of the adaptive white balance method

$$M_{\bar{x}\bar{y}\bar{z}\to SPD_{r,g,b}}\boldsymbol{J} = \begin{pmatrix} \bar{x}(\lambda)r(\lambda) & \bar{x}(\lambda)g(\lambda) & \bar{x}(\lambda)b(\lambda) \\ \bar{y}(\lambda)r(\lambda) & \bar{y}(\lambda)g(\lambda) & \bar{y}(\lambda)b(\lambda) \\ \bar{z}(\lambda)r(\lambda) & \bar{z}(\lambda)g(\lambda) & \bar{z}(\lambda)b(\lambda) \end{pmatrix} \quad \boldsymbol{J} = y_w^{-1}\left(x_w, y_w, z_w\right)^T \tag{1}$$

where $M_{\bar{x}\bar{y}\bar{z}\to SPD_{r,g,b}}$ is the 3×3 matrix from emission to response spectra. The normalized spectra are then given by: $(r,g,b)_n(\lambda) = (r,g,b)(\lambda)J_{r,g,b}$.

Age-Based LMS. From the LMS cone responses $\overline{LMS}_{age}(\lambda) = (\bar{l}_{age}(\lambda), \bar{m}_{age}(\lambda), \bar{s}_{age}(\lambda))$ for the given age, based on the CIE-2006 physiological model, and the black body spectrum for the specified target color temperature $M_e(\lambda, T)$ ([20], p. 83), we compute the age-based LMS cone responses

$$\mathbf{w}_{\overline{LMS}} = \frac{\sum_\lambda \left[\overline{LMS}_{age}(\lambda)M_e(\lambda,T)\right]}{\sum_\lambda \overline{LMS}_{age}(\lambda)} \tag{2}$$

Matching Primary Settings. Next, we compute the 3×3 matrix corresponding to the LMS cone responses to the OLED RGB primary spectra, and solve the linear system to determine the RGB factors $\mathbf{x} = (x_r, x_g, x_b)$ that achieve the desired black body color match.

$$M_{LMS\to SPD_n(\lambda)_{r,g,b}}\mathbf{x} = \begin{pmatrix} \bar{l}(\lambda)r_n(\lambda) & \bar{l}(\lambda)g_n(\lambda) & \bar{l}(\lambda)b_n(\lambda) \\ \bar{m}(\lambda)r_n(\lambda) & \bar{m}(\lambda)g_n(\lambda) & \bar{m}(\lambda)b_n(\lambda) \\ \bar{s}(\lambda)r_n(\lambda) & \bar{s}(\lambda)g_n(\lambda) & \bar{s}(\lambda)b_n(\lambda) \end{pmatrix}\mathbf{x} = \mathbf{w}_{\overline{LMS}} \tag{3}$$

Finally, divide these white balance factors by the maximum of the three, such that the maximum factor is 1. With $m = max(x_r, x_g, x_b)$, we get the linear factors: $r = \frac{x_r}{m}$, $g = \frac{x_g}{m}$ and $b = \frac{x_b}{m}$.

These are the linear factors we will apply to each RGB pixel to map an image to the desired white point. Note that there are two degrees of freedom in the input, age and color temperature, and two degrees of freedom in the output, since one of the RGB factors is always 1.0.

The left graph in Fig. 3 shows the color matching functions (CMF) for some ages using the CIE-2006 age model. The right graph displays the difference in

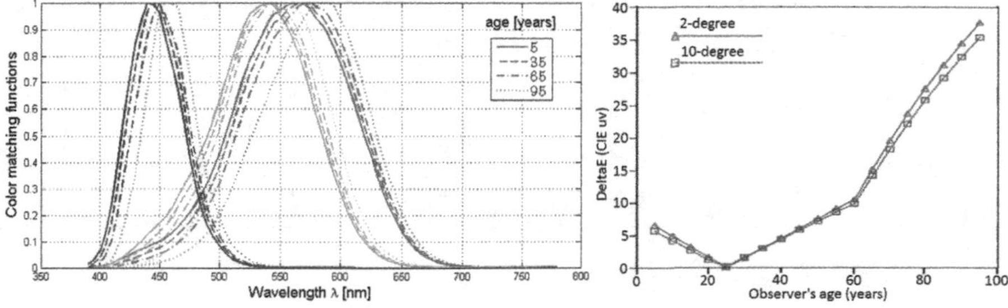

Fig. 3. Left: LMS cone responses using CIE-2006 age model. Right: Difference in D65 white appearance relative to a 25 year-old reference subject on a Samsung AMOLED display (Galaxy Tab S 10.5) for 2° and 10° patches.

D65 white appearance relative to a 25 years-old reference subject on a Samsung AMOLED display (Galaxy Tab S 10.5) for 2° and 10° patches. These results illustrate the large variability among individual observers.

The result of this stage is an image compensated for the viewer white balance preference.

The next main goal of this first part is to map the image to the target display wide gamut. We use a hybrid color mapping (HCM) [21] that is designed to preserve a selected region in chromaticity space while exploiting the larger gamut of the intended target display. This method preserves the earth and flesh tones while expanding the most saturated colors to the destination gamut.

3.2 Contrast Sensitivity Function

Human visual system ability to perceive and identify objects in the environment varies as a function of the object size, distance, contrast and orientation [7]. The contrast sensitivity function (CSF) extends and enriches the limited information given by the measures of acuity by evaluating the visual efficiency of an individual for the perception and identification of objects over a wide range of sizes, distances and orientations [7]. The CSF is usually derived by measuring the minimum contrast needed to detect sinusoidal grating patterns with different spatial frequencies expressed in cpd (cycles/degree) [7]. Since low-contrast thresholds are associated with high levels of visual sensitivity, the reciprocal of the threshold is calculated and plotted as a function of the spatial frequency.

Numerous studies report a consistent pattern of change related to age in the CSF gathered under well-lit conditions (photopic vision) [4,6]. For targets of spatial frequency between 4 and 18 cpd, the decline of the contrast sensitivity is about 0.3 log units on average across the second half of adult life [4,5,22]. The left graph in the Fig. 4 shows an example of the CSF as a function of spatial frequency (cpd), while the right graph shows the sensitivity loss between an observer of a given age a, and our reference observer of 25 years. The two graphs are computed with the help of Eqs. 7 and 8 (see Sect. 3.)

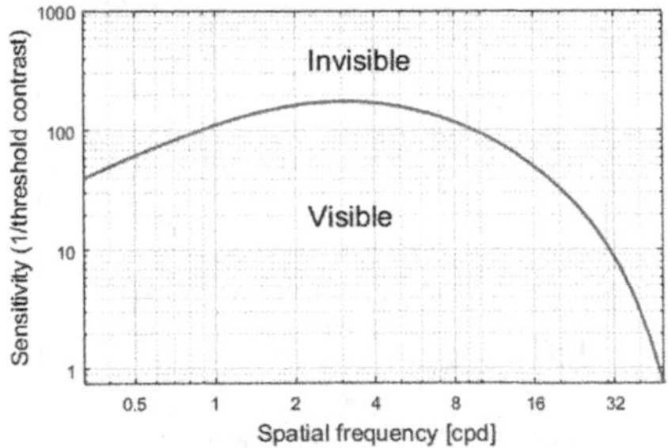

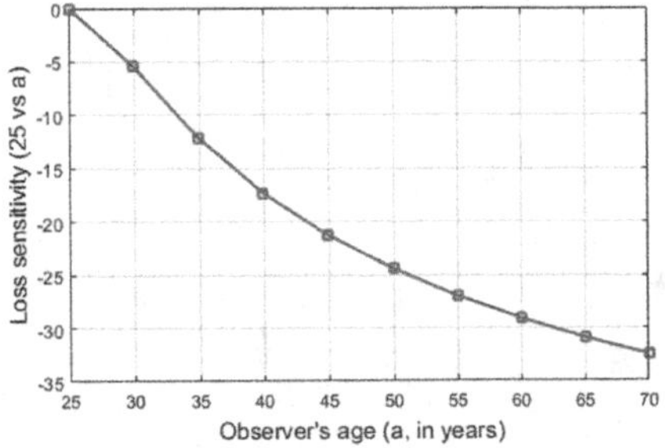

Fig. 4. Left: Illustration of the contrast sensitivity function and the threshold contrast. Right: sensitivity loss for an observer compared to our reference observer of 25 years. Spatial frequency is 10 cpd and the adaptation luminance at 100 cd/m^2

Local Contrast Retargeting. In this section, we will define the most important parts related to the local contrast retargeting. For a more complete treatment of the topic, please refer to the excellent studies in [23, 24].

Measures of contrast usually used by the vision community is the Michelson Contrast M, and the logarithmic contrast G when calculating image contrast in a multi-scale representation [24].

$$M = \frac{L_{max} - L_{min}}{L_{max} + L_{min}}, \quad G = \frac{1}{2} log_{10}\left(\frac{L_{max}}{L_{min}}\right) \tag{4}$$

where L_{max} and L_{min} are the maximum and the minimum luminance values of a sine wave.

The local contrast modification will follow the methodology developed by Wanat and Mantiuk [24], where the authors proposed a luminance retargeting technique that modifies the perceived colors and contrast of an image to match the appearance under different luminance levels. The authors formulated the condition for matching contrast as the matching of their supra-thresholds contrast, i.e. that the difference between the physical contrast M and the detection threshold M_t must be constant across different luminance levels:

$$M - M_t = \tilde{M} - \tilde{M_t} \tag{5}$$

where M and $\tilde{M}$ are the Michelson Contrast observed at different luminances.

The detection threshold M_t is estimated by the CSF function

$$M_t = \frac{\Delta L}{L} = \frac{1}{CSF_{hdrvdp}(\rho, L_a)} = \frac{1}{S \cdot CSF(\rho, L_a)} \tag{6}$$

where ρ is the spatial frequency in cycles per degrees, L_a is the adaptation luminance in cd/m^2, and S is the absolute sensitivity of the HVS, necessary to adjust the CSF for a particular experiment. To compute the CSF from Eq. 6, we make use of the formula by Mantiuk et al. [23], and we assume it is equal to the CSF of our reference age of 25 years.

To integrate the age factor in Eq. 6, we employ a new model of the contrast sensitivity function [25], which is an extension of the one proposed by Barten [26] with age dependencies a.

$$CSF_B(\rho, L_a, a) = \frac{1}{m_t(\rho, a)} = \frac{M_{opt}(\rho, a)}{2k(a)} \sqrt{\frac{(XYT)(\rho)}{\Phi_{ph}(a) + \Phi_0(a)/M_{lat}^2(\rho, a)}} \tag{7}$$

where m_t is the modulation threshold, $M_{opt}(\rho)$ the optical MTF (modulation transfer function) that describes the behavior of the input signal passing through the optical elements of the eye, k the signal to noise ratio, X, Y, and T are the spatial and temporal integration area of the eye, Φ_{ph} is the photon noise that describes the statistical fluctuations in the number of incident photons absorbed by the photoreceptors. Φ_0 is the neural noise, and $M_{lat}(\rho)$ is the lateral inhibition term [26]. The block diagram of the used model is shown in Fig. 5, where the

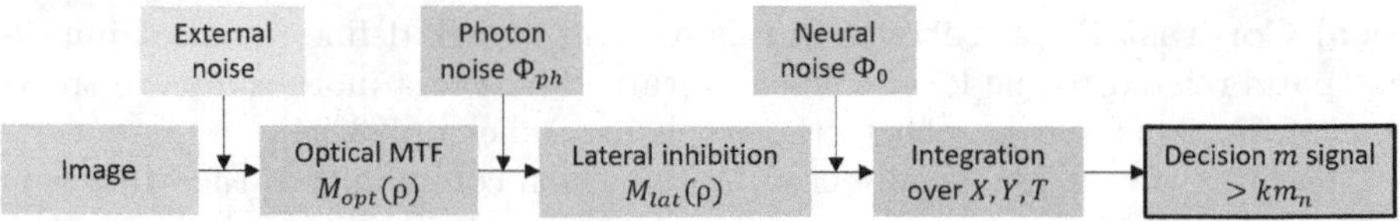

Fig. 5. Block diagram of the processing of information and noise according to Barten's CSF model [26] (Color figure online)

components that are age dependent are depicted in green, and m_n denotes the average modulation of the internal noise. We use the model from Eq. 7 to predict the sensitivity loss between our reference observer (25 years), and an observer at a given age a.

$$CSF_{loss}(\rho, L_a, a) = CSF_B(\rho, L_a, 25) - CSF_B(\rho, L_a, a) \tag{8}$$

The CSF loss function will depend on both the spatial frequency f and the adaptation luminance L_a. Thus, the CSF from the Eq. 6 will be modified depending on the observer's level of detail preference which corresponds to the observer's effective age.

$$M_t = \frac{1}{CSF_{hdrvdp}(\rho, L_a, a)} = \frac{1}{CSF_{hdrvdp}(\rho, L_a) - CSF_{loss}(\rho, L_a, a)} \tag{9}$$

Figure 6 shows CSF data measurements for two age groups (24y, 73y), and for 3 luminance levels (0.107, 3.38, 107 cd/m^2) as a function of the spatial frequency (in cpd) [5]. We notice the sensitivity loss for the elderly group compared to the young. The loss occurs at all luminance levels, and is more pronounced for higher spatial frequencies.

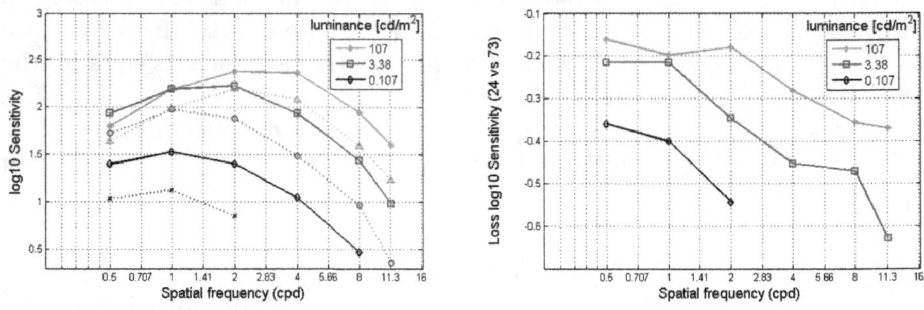

Fig. 6. Left: Data measurements from the paper [5] for two age groups, 24y (continuous line) and 73y (dotted line), and 3 luminance levels. Right: Sensitivity loss (young vs old).

To perform the local contrast retargeting, the image is decomposed into low-pass and high-pass bands using the Laplacian pyramid, and computed for the log luminance values. The localized broadband contrast is defined by [24]:

$$c(x, y) = \sqrt{(g_\sigma * [l(x, y) - (g_\sigma * l)(x, y)]^2)(x, y)} \tag{10}$$

where $*$ denotes the convolution and g_σ is the Gaussian kernel and σ is the standard deviation of g_σ.

The spatial frequency ρ can be calculated from σ (given in pixels) and the angular display resolution of the display R_{ppd} (in pixels per visual degree) as:

$$\rho = 2^{-(k+1)} R_{ppd} \tag{11}$$

where $k = 1, ..., N$ is the level of the pyramid, and $k = 1$ represents the finest level.

Contrast retargeting is performed as a local enhancement of the Laplacian pyramid [24]:

$$\tilde{P}_k(x, y) = P_k(x, y).m_k(x, y) \tag{12}$$

P_k refers to the source image pyramid level, and m is the contrast modification defined by:

$$m_k(x, y) = \frac{c_k(x, y) - G(M_t) + G(\tilde{M}_t)}{c_k(x, y)} \tag{13}$$

The reconstruction of the enhanced image in terms of the contrast is done by summing all processed levels of the pyramids $\tilde{P}_k(x, y)$ with the addition of the base band (which was not modified by the local contrast step).

Figure 7 shows the result of our algorithm when applied to a test image. We easily notice the skin color and white are more natural in the processed image, as well as the details which are more pronounced. The compensated image looks natural in both color and contrast, and the content is personalized to the specific observer's preferences.

Fig. 7. Proposed unified content adaptation - Left: unprocessed, Right: compensated for both color and contrast. (Effective age for color: 57, contrast: 54). (Color figure online)

4 Experimental Validation

We subjectively evaluated the performance of our unified age-dependent content adaptation model using the pairwise comparison approach introduced in [27]. The experiment was run in the same room, which has a relatively constant illuminance of 50 Lux. Based on internal measurements, this corresponds to a target luminance of 32 cd/m^2 for the visualization tablet. This value was kept constant during the experiment. The visualization tablet is the Samsung AMOLED Galaxy Tab S 10.5, with a resolution of 2560×1600, and was at 50 cm distance from the observer, about 90° relative to the observer's eyes.

To determine the observer's effective age, the observer was presented with two videos that show the changing of an image's color and details separately. In the first video, they were offered a 1-D axis control to find their preferred white point settings, which corresponds to the age-related color dimension. In the second video, their preferred level of detail setting, which corresponds to the age-related contrast dimension.

We used 10 images processed by 3 different color models, our unified color and contrast content adaptation -UCA, only color adaptation -CCA, and original image - SDS (same drive signal). For comparative studies, previous research [28] indicated that a size between 8 and 25 subjects is sufficient to provide statistically significant results. We asked 30 naive observers to compare the presented result. Our observers were asked to pick their preferred image of the pair. For each observer, 30 total pairs of images were displayed using the Samsung AMOLED Galaxy Tab S, 10 pairs for CCA:UCA, 10 pairs for CCA:SDS, and 10 pairs for SDS:UCA. Displaying the image pairs was randomized, and a black image is shown for 2 s between each consecutive image pairs. The observers were instructed to select one of the displayed images as their preferred image based on the overall feeling of color and details level. Our observers consist of 7 females and 23 males from the age of 23 to 60. On average, the whole experiment took about 10 min to complete.

Figure 8 shows processing results for color and contrast (UCA) with original images (SDS) and only color content adaptation (CCA). A few of the images include both indoor and outdoor scenes with a lot of details, and portraits where people's faces are close to the camera. Our method keeps the white color, skin and earth tones natural while enhancing the local contrast for better visualization.

We used the pairwise comparison method with just noticeable difference (JND) evaluation in our experiment. This method has been recently used for subjective evaluation in the literature [24,27,29]. We used the Bayesian method of Silverstein and Farrel [30], which maximizes the probability that the pairwise comparison results accounts for the experiment under the assumptions of equal variances and uncorrelated distributions. During an optimization procedure, a quality value for each image is calculated to maximize the probability, modeled by a binomial distribution. Since we have 3 conditions for comparison (UCA, CCA, SDS), this Bayesian approach is suitable. It is more common when we compare a large number of conditions and it is known for being robust to unanimous answers.

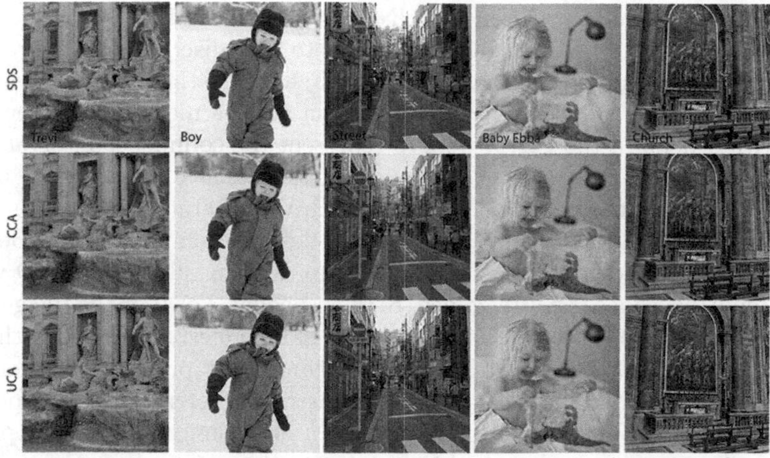

Fig. 8. Content adaptation examples with original images. UCA - our unified content adaptation, SDS - original image, and CCA - only color content adaptation. (Effective age for color: 38, contrast: 40) (Color figure online)

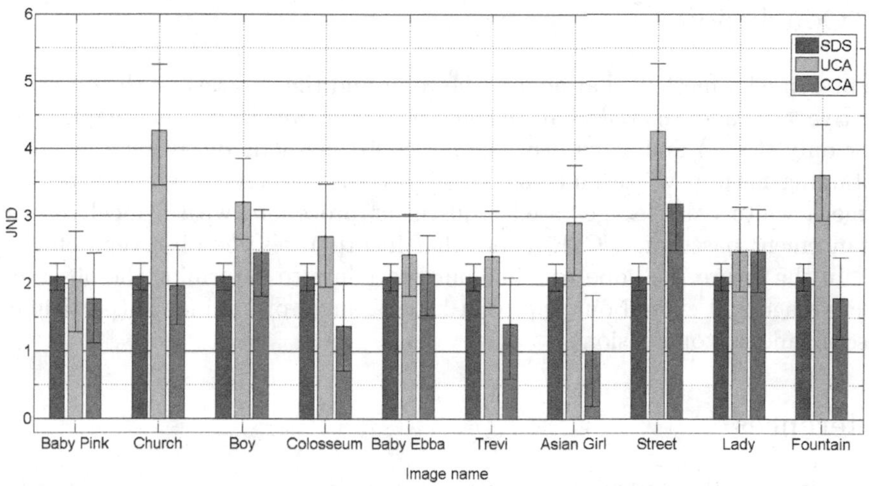

Fig. 9. Subjective evaluation results of pairwise comparison representing as JND values for each 10 images including error bars which denote 95% confidence intervals calculated by bootstrapping. UCA: our proposed unified age-based color and contrast visual content adaptation, SDS: original image, and CCA: only color content adaptation. (Color figure online)

Figure 9 shows the results of the subjective evaluation calculating the JND value using the definition in [27]. For discriminating between choices, only relative differences can be used. Thus, the absolute JND values are not very relevant taken alone. A method that has a high JND value is desirable over methods with smaller JND values, where 1 JND corresponds to 75% discrimination threshold. The Fig. 9 represents the JND value for each scene, rather than the average value, because JND is a relative value that can also be meaningful when compared with others. In the Fig. 9, we also represent the confidence intervals with 95% probability for each JND. To calculate the confidence intervals, we used a numerical method, known as bootstrapping [31]. We generated 500 random sampling, then computed 2.5^{th} and 97.5^{th} percentiles for each JND point. Because both JND and confidence intervals for JND are relative values, the JND values for the mode we choose as a reference, SDS, are equal. For 9 of the images, our proposed unified age content adaptation is the most preferred method with JND differences of 0.31−2.2 between it and the second most preferred method. We also notice when only color is compensated, observers preferred the unprocessed in 6 of the images. The reason of this result in our opinion, is that people are more familiar in seeing unprocessed images in a variety of displays. For instance, in the images where there is water (Trevi), and sky (Colosseum), the colors in the processed images are very different from the SDS mode, which makes it difficult for people to decide without having a reference.

5 Conclusion

We presented a method that adapts the color and the contrast to the observer's effective age, and showed that this method was preferred over adapting the color only (CCA) and same drive signal (SDS) methods on average in images containing natural white and high detail levels. The key to our method is to merge a white balance age-based adjustment procedure with a local contrast enhancement based on a CSF model that incorporates the observer's effective age. In the future, we hope to refine our procedure to determine the observer's effective age and extend our approach for all ranges of luminance (particularly mesopic and scotopic vision).

References

1. Brainard, D.H., Stockman, A.: Colorimetry. In: Handbook of Optics, Volume III - Vision and Vision Optics, 3rd edn. Michael Bass (2010)
2. Fairchild, M.D., Wyble, D.R.: Mean observer metamerism and the selection of display primaries. In: Proceedings of 15th IS&T/SID Color Imaging Conference, pp. 151–156, November 2007
3. Barbur, J.L., Stockman, A.: Photopic, mesopic, and scotopic vision and changes in visual performance. In: Dartt, D., Besharse, J., Dana, R. (eds.) Encyclopedia of the Eye , pp. 323–331. Academic Press, Oxford (2010). https://doi.org/10.1016/B978-0-12-374203-2.00233-5
4. Owsley, C., Sekuler, R., Siemsen, D.: Contrast sensitivity throughout adulthood. Vis. Res. **23**(7), 689–699 (1983)

5. Sloane, M.J., Owsley, C., Alvarez, S.L.: Aging, senile miosis and spatial contrast sensitivity at low luminance. Vis. Res. **28**(11), 1235–1246 (1988)
6. Schieber, F.K.D., Kline, T.J.B., Fozard, J.L.: The relationship between contrast sensitivity and the visual problems of older drivers. Technical paper 920613, pp. 1–7 (1992)
7. Schieber, F.: Vision and aging. In: Birren, J.E., Warner Schaie, K., Abeles, R.P., Salthouse, T.A. (eds.) Handbook of the Psychology of Aging (AP 2006) 6th edn., pp. 129–154 (2006)
8. Mantiuk, R.K., Ramponi, G.: Human vision model including age dependencies. In: 2015 23rd EUSIPCO, Nice, pp. 1616–1620 (2015)
9. Sarkar, A., Blondé, L., Le Callet, P., Autrusseau, F., Stauder, J., Morvan, P.: Modern displays: why we see different colors, and what it means? In: EUVIP, p. 16. IEEE (2010)
10. Gonzalez, R.C., Woods, R.E.: Digital Image Processing, 3rd edn., pp. 122–143. P. Hall (2008)
11. Kim, Y.T.: Contrast enhancement using brightness preserving bi histogram equalisation. IEEE Trans. Consum. Electron. **43**(1), 1–8 (1997)
12. Land, E., McCann, J.: Lightness and retinex theory. J. Opt. Soc. Am. **61**, 1–11 (1971)
13. Marini, D., Rizzi, A.: A computational approach to color adaptation effects. IVC **18**(13), 1005–1014 (2000)
14. Peli, E., Peli, T.: Image enhancement for the visually impaired. Opt. Eng. **23**, 47–51 (1984)
15. Lawton, T.B.: Image enhancement filters significantly improve reading performance for low vision observers. Ophthalmic Physiol. Opt. **12**, 193–200 (1992)
16. Choudhury, A., Medioni, G.: Color contrast enhancement for visually impaired people. In: 2010 IEEE Computer Society Conference on CVPR - Workshops, San Francisco, CA, pp. 33–40 (2010)
17. Wang, Y., Luo, Y.: Adaptive color contrast enhancement for digital images. Opt. Eng. **50** (2011). http://dx.doi.org/10.1117/1.3655500
18. Ueda, C., Azetsu, T., Suetake, N., Uchino, E.: Gamma correction-based image enhancement for elderly vision. In: 15th ISCIT, Nara, pp. 141–144 (2015)
19. Stockman, A., Sharpe, L.: Physiologically-based color matching functions. In: Proceedings of ISCC/CIE Expert Symposium 2006, CIE Publication x030:2006, pp. 13–20 (2006)
20. Hunt, R.W.G., Pointer, M.R.: Measuring Colour, 4th edn. Wiley Publishing, Hoboken (2011). ISBN 978-1-119-97537-3
21. Ward, G., Yoo, H., Soudi, A., Akhavan, T.: Exploiting wide-gamut displays. In: CIC24. IRYStec, Inc., USA (2016)
22. Elliott, D.B.: Contrast sensitivity decline with aging: a neural or optical phenomenon? Ophthalmic Physiol. Opt. **7**, 415–419 (1987)
23. Mantiuk, R.K., Joong Kim, K., Rempel, A.G., Heidrich, W.: HDR-VDP-2: a calibrated visual metric for visibility and quality predictions in all luminance conditions. ACM TOG **30**(4), 40 (2011)
24. Wanat, R., Mantiuk, R.K.: Simulating and compensating changes in appearance between day and night vision. ACM Trans. Graph. **33**(4), 1–12 (2014)
25. Joulan, K., Brémond, R., Hautiére, N.: Towards an analytical age-dependent model of contrast sensitivity functions for an aging society. Sci. World J. **2015**, 1–11 (2015)
26. Barten, P.G.J.: Contrast Sensitivity of the Human Eye and its Effects on Image Quality. SPIE, Bellingham (1999)

27. Eilertsen, G., Wanat, R., Mantiuk, R.K., Unger, J.: Evaluation of tone mapping operators for HDR-video. Comput. Graph. Forum **32**(7), 275–284 (2013). Wiley Online Library
28. Macefield, R.: How to specify the participant group size for usability studies: a practitioner's guide. J. Usability Stud. **5**(1), 34–45 (2009). http://dl.acm.org/citation.cfm?id=2835429
29. Rezagholizadeh, M., Akhavan, T., Soudi, A., Kaufmann, H., Clark, J.J.: A retargeting approach for mesopic vision: simulation and compensation. In: JIST (2015)
30. Silverstein, D.A., Farrell, J.E.: Efficient method for paired comparison. JEI **10**(2), 394–398 (2001)
31. Varian, H.: Bootstrap tutorial. Math. J. **9**(4), 768–775 (2005)

Deep Appearance Features for Abnormal Behavior Detection in Video

Sorina Smeureanu[1,2], Radu Tudor Ionescu[1,2(✉)], Marius Popescu[1,2], and Bogdan Alexe[1,2]

[1] University of Bucharest, 14 Academiei Street, Bucharest, Romania
raducu.ionescu@gmail.com, popescunmarius@gmail.com
[2] SecurifAI, 24 Mircea Voda, Bucharest, Romania

Abstract. We propose a novel framework for abnormal event detection in video that is based on deep features extracted with pre-trained convolutional neural networks (CNN). The CNN features are fed into a one-class Support Vector Machines (SVM) classifier in order to learn a model of normality from training data. We compare our approach with several state-of-the-art methods on two benchmark data sets, namely the Avenue data set and the UMN data set. The empirical results indicate that our abnormal event detection framework can reach state-of-the-art results, while running in real-time at 20 frames per second.

Keywords: Abnormal event detection · Deep features · Convolutional neural networks · One-class SVM

1 Introduction

Abnormal event detection in video is a challenging task in computer vision, as the definition of what an abnormal event looks like depends very much on the context. For instance, a car driving by on the street is regarded as a normal event, but if the car enters a pedestrian area, this is regarded as an abnormal event. A person jogging on the beach (normal event) versus running outside from a bank (abnormal event) is another example. Although what is considered abnormal depends on the context, we can generally agree that abnormal behaviour should be represented by unexpected events that occur less often than familiar (normal) events. As it is generally impossible to find a sufficiently representative set of anomalies, the use of traditional supervised learning methods is usually ruled out. Hence, most abnormal event detection approaches [1,4,11,13–16,25,27] learn a model of familiarity from a given training video and label events as abnormal if they deviate from the model. We approach abnormal behavior detection in a similar manner, and propose to build a model of normality by using a one-class Support Vector Machines (SVM) [20] classifier. The outliers detected by our approach will be labeled as abnormal events at test time. Although it seems straightforward to apply one-class SVM, related works have adopted different approaches, for example dictionary learning [4,5,7,14,17] or locality sensitive

S. Battiato et al. (Eds.): ICIAP 2017, Part II, LNCS 10485, pp. 779–789, 2017.
https://doi.org/10.1007/978-3-319-68548-9_70

hashing filters [26]. Nevertheless, we show in this paper that we can achieve state-of-the-art results by using one-class SVM. Before training our normality model, we extract deep features by using convolutional neural networks (CNN) pre-trained on the ILSVRC benchmark [18]. Deep learning models reach impressive performance levels on object recognition from images [3,9,21,23]. Although the features learned by CNN models are not particularly designed for computer vision tasks outside the original purpose, the knowledge embedded in the CNN features is quite general and it can easily be transferred to various tasks, for example to the task of predicting the difficulty of an image [10]. To the best of our knowledge, we are the first to transfer pre-trained CNN features to the task of abnormal behavior detection in video.

We perform abnormal event detection experiments on the Avenue [14] and the UMN [16] data sets in order to compare our approach with several state-of-the-art methods [5,6,14,16,19,22,26]. The empirical results on the Avenue data set indicate that our model is able to surpass the state-of-the-art methods [6,14] for this data set. As for the UMN data set, we are able to reach state-of-the-art performance on two of the three video scenes. Although we show that we can obtain better performance by employing deeper models [9,21,23] for feature extraction, we choose to use the VGG-f model [3] which allows us to process the video in real-time at 20 frames per second on a standard CPU.

We organize the paper as follows. We present related work on abnormal event detection in Sect. 2. We describe our learning framework in Sect. 3. We present the abnormal event detection experiments in Sect. 4. Finally, we draw our conclusions in Sect. 5.

2 Related Work

Abnormal event detection is usually formalized as an outlier detection task [1,4,5,7,11,13–17,22,25–27], in which the general approach is to learn a model of normality from training data and consider the detected outliers as abnormal events. Some abnormal event detection approaches [4,5,7,14,17] are based on learning a dictionary of normal events, and label the events not represented by the dictionary as abnormal. Other approaches have employed deep features [25] or locality sensitive hashing filters [26] to achieve better results.

Interestingly, there have been some approaches that employ unsupervised steps for abnormal event detection [7,17,22,25]. The approach presented in [7] is to build a model of familiar events from training data and incrementally update the model in an unsupervised manner as new patterns are observed in the test data. In similar fashion, Sun et al. [22] train a Growing Neural Gas model starting from training videos and continue the training process as they analyze the test videos for anomaly detection. Ren et al. [17] use an unsupervised approach, spectral clustering, to build a dictionary of atoms, each representing one type of normal behavior. Their approach requires training videos of normal events to construct the dictionary. Xu et al. [25] use Stacked Denoising Auto-Encoders to learn deep feature representations in a unsupervised way. However, they still employ multiple one-class SVM models to predict the anomaly scores. The approach proposed in [6] is to detect changes on a sequence of data from the video to see which frames are distinguishable from all the previous frames. As the authors want to build an approach independent of temporal ordering, they create shuffles of the data by permuting the frames before running each instance of the change detection. As we employ pre-trained CNN features, our feature extraction step is also unsupervised with respect to the approached task.

3 Method

3.1 Feature Extraction

In many computer vision tasks, higher level features, such as the ones learned with convolutional neural networks (CNN) [12] are the most effective. To build our appearance features, we consider a pre-trained CNN architecture able to process the frames as fast as possible, namely VGG-f [3]. Considering that we want our detection framework to work in real-time on a standard desktop computer, not equipped with expensive GPU, the VGG-f [3] is an excellent choice as it can process about 20 frames per second on CPU. We hereby note that better anomaly detection performance can be achieved by employing deeper CNN architectures, such as VGG-verydeep [21], GoogLeNet [23] or ResNet [9].

The VGG-f model is trained on the ILSVRC benchmark [18]. We use the pre-trained CNN model to extract deep features as follows. Given the input video, we resize the frames to 224×224 pixels. We then subtract the mean imagine from each frame and provide it as input to the VGG-f model. We remove the fully-connected layers (identified as *fc6*, *fc7* and *softmax*) and consider the activation maps of the last convolutional layer (*conv5*) as appearance features. While the fully-connected layers are adapted for object recognition, the last convolutional layer contains valuable appearance and pose information which is more useful for our anomaly detection task. Ideally, we would like to have at least slightly different representations for a person walking versus a person running. Interestingly, Feichtenhofer et al. [8] have also found that the *conv5* features are more suitable for action recognition in video, a task closely related to ours.

Finally, we reshape each activation map into an 169 dimensional vector and concatenate the vectors corresponding to the 256 filters of the *conv5* layer into a single feature vector of 43264 ($13 \times 13 \times 256$) components. The resulted feature vectors are normalized using the L_2-norm. It is important to note that unlike other approaches [4,25], we apply the same steps in order to extract features from video, irrespective of the data set.

3.2 Learning Model

We use the one-class SVM approach of Schölkopf et al. [20] to detect abnormal events in video. The training data in our case is composed of a few videos representing only normal events. We consider each video frame as an individual and independent sample, disregarding the temporal relations between video frames. Let $\mathcal{X} = \{x_1, x_2, \ldots, x_n \mid x_i \in \mathbb{R}^m\}$ denote the set of training frames. In this formulation, our one-class SVM model will learn to separate a small region capturing most of the normal frames from the rest of feature space, by maximizing the distance from the separating hyperplane to the origin. This results in a binary classification function g which captures regions in the input space where the probability density of normal events lives:

$$g(x) = sign\left(\sum_{i=1}^{n} \alpha_i k(x, x_i) - \rho\right), \tag{1}$$

where x is a test frame that needs to be classified either as normal or abnormal, $x_i \in \mathcal{X}$ is a training frame, k is a kernel function, α_i are the weights assigned to the support vectors x_i, and ρ is the distance from the hyperplane to the origin. If we desire a score reflecting the abnormality level of a frame, we can simply remove the (sign) transfer function from Eq. (1). The coefficients α_i are found as the solution of the dual problem:

$$\min_{\alpha} \frac{1}{2} \sum_{i=1}^{n} \sum_{j=1}^{n} \alpha_i \alpha_j k(x_i, x_j) \text{ subject to } 0 \leq \alpha_i \leq \frac{1}{\nu n}, \sum_{i=1}^{n} \alpha_i = 1, \tag{2}$$

where $\nu \in [0, 1]$ is a regularization parameter that controls the percentage of outliers to be excluded by the learned model. As noted by Schölkopf et al. [20], the offset ρ can be recovered by exploiting that for any α_i that is not at the lower or upper bound, the corresponding sample x_i satisfies:

$$\rho = \sum_{j=1}^{n} \alpha_j k(x_j, x_i). \tag{3}$$

Since we already represent the frames in a high dimensional space ($m = 43264$) by extracting CNN features, we no longer have to embed the samples into a higher dimensional space. Hence, we decide to use the linear kernel function in our one-class SVM model, which corresponds to the feature map $\phi(x) = x$:

$$k(x, z) = \langle x, z \rangle. \tag{4}$$

4 Experiments

4.1 Data Sets

We show abnormal event detection results on two benchmark data sets.

Avenue. We first consider the Avenue data set [14], which contains 16 training and 21 test videos. In total, there are 15328 frames in the training set and 15324 frames in the test set. Each frame is 640×360 pixels. Locations of anomalies are annotated in ground truth pixel-level masks for each frame in the testing videos.

UMN. The UMN Unusual Crowd Activity data set [16] consists of three different crowded scenes, each with 1453, 4144, and 2144 frames, respectively. The resolution of each frame is 320×240 pixels. In the normal settings people walk around in the scene, and the abnormal behavior is defined as people running in all directions. As in [5], we use the first 400 frames in each scene for training.

4.2 Evaluation

We employ ROC curves and the corresponding *area under the curve* (AUC) as the evaluation metric, computed with respect to ground truth frame-level annotations, and, when available (only for the Avenue data set), pixel-level annotations. We define the frame-level and pixel-level AUC as in previous works [5,6,14,15]. At the frame-level, a frame is considered a correct detection if it contains at least one abnormal pixel. At the pixel-level, the corresponding frame is considered as being correctly detected if more than 40% of truly anomalous pixels are detected. We use the same approach as [6,14] to compute the pixel-level AUC. The approach consists of resizing each frame to 160×120 pixels, and uniformly partitioning each frame to a set of non-overlapping 10×10 patches. Corresponding patches in 5 consecutive frames are stacked together to form a spatio-temporal cube, each with resolution $10 \times 10 \times 5$. We remove the cubes with less than 5 non-zero values as [6,14]. The frame-level scores produced by our framework are assigned to the remaining spatio-temporal cubes. The results are smoothed with the same filter useb by [6,14] in order to obtain our final pixel-level detections.

Although many works [5,7,14,15,25,26] include the Equal Error Rate (EER) as evaluation metric, we agree with [6] that metrics such as the EER can be misleading in a realistic anomaly detection setting, in which abnormal events are expected to be very rare. Thus, we do not use the EER in our evaluation.

4.3 Implementation Details

We extract deep appearance features from the training and the test video sequences. We consider the pre-trained VGG-f [3] and VGG-verydeep [21] models provided in MatConvNet [24]. To learn a model of normality, we employ the one-class SVM implementation from LibSVM [2]. In all the experiments, we set the regularization parameter of one-class SVM to 0.2, which means that the model will have to single out 80% of the training frames as normal (the other 20% are outliers). Setting such a high value for the regularization parameter ensures that our model will not overfit the training data.

In Table 1, we present preliminary results on the Avenue data set to provide empirical evidence in favor of the CNN features that we choose for the subsequent experiments. The results indicate that better performance can be obtained with the *conv5* features rather than the *fc6* or *fc7* features. For the speed evaluation, we measure the time required to extract features and to predict the anomaly scores on a computer with Intel Core i7 2.3 GHz processor and 8 GB of RAM using a single core. We present the number of frames per

Table 1. Abnormal event detection results in terms of frame-level and pixel-level AUC on the Avenue data set. We show results with different CNN architectures and features from different layers. The number of frames per second (FPS) is computed by running the models on a computer with Intel Core i7 2.3 GHz processor and 8 GB of RAM using a single core.

Method	Frame AUC	Pixel AUC	Time (FPS)
VGG-verydeep fc7 + one-class SVM	84.3%	93.3%	1.84
VGG-verydeep fc6 + one-class SVM	84.5%	93.4%	1.84
VGG-verydeep conv5 + one-class SVM	85.3%	93.9%	1.85
VGG-f fc7 + one-class SVM	82.7%	92.5%	19.4
VGG-f fc6 + one-class SVM	82.8%	92.7%	19.5
VGG-f conv5 + one-class SVM	84.6%	93.5%	19.7

second (FPS) in Table 1. Although, we are able to report better results with the VGG-verydeep [21] architecture, we choose the shallower VGG-f [3] architecture for the rest of the experiments, as its processing time is about 10 times shorter. Using a single core, our final model is able to process the test videos in real-time at nearly 20 FPS.

4.4 Results on the Avenue Data Set

We first compare our abnormal behavior detection framework based on deep features with two state-of-the-art approaches [6,14]. The frame-level and pixel-level AUC metrics computed on the Avenue data set are presented in Table 2. Compared to the method of Del Giorno et al. [6], our framework yields an improvement of 6.3%, in terms of frame-level AUC, and an improvement of 2.5%, in terms of pixel-level AUC. We also obtain better results than Lu et al. [14], as our framework gains 3.7% in terms of frame-level AUC and 0.6% in terms of pixel-level AUC. Overall, our method is able to surpass the performance of both state-of-the-art methods.

Figure 1 illustrates the frame-level anomaly scores, for test video 4 in the Avenue data set, produced by our framework based on VGG-f features and one-class SVM. According to the ground-truth anomaly labels, there are two

Table 2. Abnormal event detection results in terms of frame-level and pixel-level AUC on the Avenue data set. Our framework is compared with two state-of-the-art approaches [6,14].

Method	Frame AUC	Pixel AUC
Lu et al. [14]	80.9%	92.9%
Del Giorno et al. [6]	78.3%	91.0%
VGG-f conv5 + one-class SVM	84.6%	93.5%

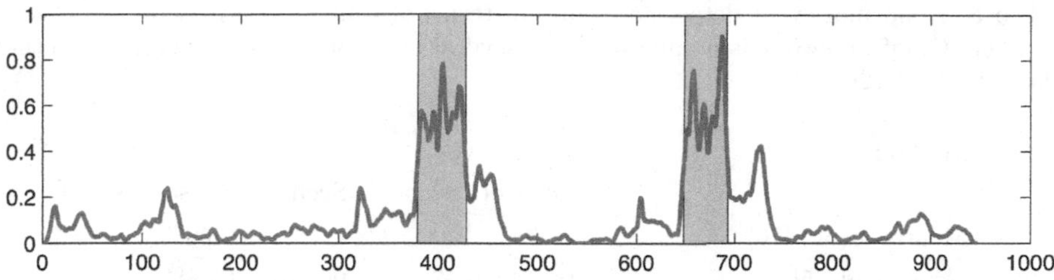

Fig. 1. Frame-level anomaly detection scores (between 0 and 1) provided by our framework for test video 4 in the Avenue data set. The video has 947 frames. Ground-truth abnormal events are represented in cyan, and our scores are illustrated in red. Best viewed in color.

Fig. 2. True positive (top row) versus false positive (bottom row) detections of our framework based on VGG-f features and one-class SVM. Examples are selected from the Avenue data set. Best viewed in color.

abnormal events in this video. In Fig. 1, we notice that our scores correlate well to the ground-truth labels, and we can easily identify both abnormal events by setting a threshold of around 0.4, without including any false positive detections.

We show some examples of true positive and false positive detections in Fig. 2. The true positive abnormal events are *a person throwing an object* and *a person running*, while false positive detections are *two persons walking synchronously* and *a person carrying a backpack.*

4.5 Results on the UMN Data Set

On the UMN data set, we compare our framework with several supervised methods [5,16,19,22,26]. In Table 3, we present the frame-level AUC score for each individual scene, as well as the average score for all the three scenes. On the first scene and the last scene, the performance of our one-class SVM framework

Table 3. Abnormal event detection results in terms of frame-level AUC on the UMN data set. Our framework is compared with several state-of-the-art supervised methods [5,16,19,22,26].

Method	Frame AUC			
	Scene 1	Scene 2	Scene 3	All scenes
Mehran et al. [16]	-	-	-	96.0%
Cong et al. [5]	99.5%	97.5%	96.4%	97.8%
Saligrama and Chen [19]	-	-	-	98.5%
Zhang et al. [26]	99.2%	98.3%	98.7%	98.7%
Sun et al. [22]	99.8%	99.3%	99.9%	99.7%
VGG-f conv5 + one-class SVM	98.8%	93.6%	98.9%	97.1%

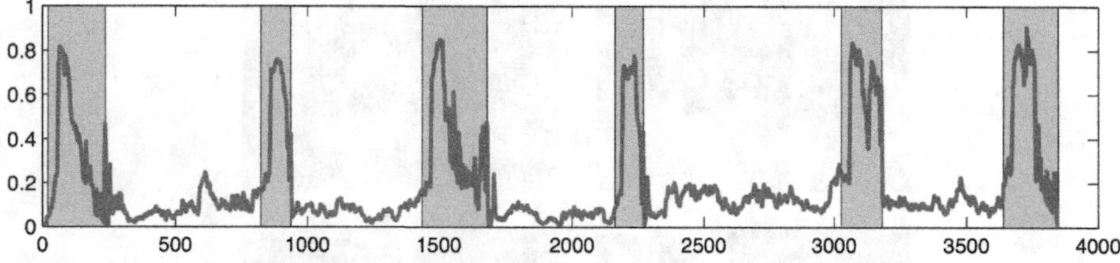

Fig. 3. Frame-level anomaly detection scores (between 0 and 1) provided by our framework for the second scene in the UMN data set. The test sequence has 3854 frames. Ground-truth abnormal events are represented in cyan, and our scores are illustrated in red. Best viewed in color.

based on deep features is on par with the state-of-the-art approaches. For the third scene, we are able to surpass the performance reported by [5,26]. However, we obtain a much lower performance for the second scene, perhaps due the significant illumination changes in this scene. Our approach yields an overall frame-level AUC of 97.1%, which represents an improvement of 1.1% over the approach of Mehran et al. [16]. However, the best approach [22] on the UMN data set is nearly 2.6% better than our approach.

As illustrated in Fig. 3, our approach is able to correctly identify the abnormal events in the second scene without any false positives, by applying a threshold of around 0.4. However, our approach does not detect the abnormal events right from the beginning. We believe that the changes in illumination when people enter the room have a negative impact on our approach. These observations are also applicable when we analyze the false positive detections presented in Fig. 4. Indeed, the example in the bottom right corner of Fig. 4 illustrates that our method triggers a false detection when a significant amount of light enters the room as the door opens. The true positive examples in Fig. 4 represent *people running around in all directions.*

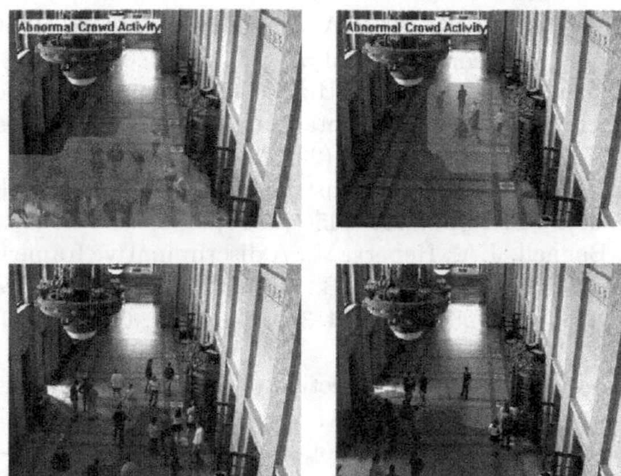

Fig. 4. True positive (top row) versus false positive (bottom row) detections of our framework based on VGG-f features and one-class SVM. Examples are selected from the second scene of the UMN data set. Best viewed in color.

5 Conclusion and Future Work

In this work, we have proposed a novel framework for abnormal event detection in video that is based on extracting deep features from pre-trained CNN models, and on using one-class SVM to learn a model of normality. We have conducted abnormal event detection experiments on two data sets in order to compare our approach with several state-of-the-art approaches [5,6,14,16,19,22,26]. The empirical results indicate that our approach gives better performance than some of these approaches [6,14,16], while processing the video online at 20 FPS.

Although our model can reach very good results, it completely disregards motion information and the temporal structure in video. In future work, we aim to improve our performance by including motion features into our framework. One possible approach would be to employ convolutional two-stream networks [8] to extract both motion and appearance features. We also aim to evaluate our framework on other data sets.

Acknowledgments. We thank reviewers for their helpful comments. This research is supported by University of Bucharest, Faculty of Mathematics and Computer Science, through the 2017 Mobility Fund, and by SecurifAI through Project P/38/185 funded under the Competitiveness Operational Programme POC-A1-A1.1.1-C-2015.

References

1. Antic, B., Ommer, B.: Video parsing for abnormality detection. In: Proceedings of ICCV, pp. 2415–2422 (2011)
2. Chang, C.C., Lin, C.J.: LibSVM: a library for support vector machines. ACM Trans. Intell. Syst. Technol. **2**, 27:1–27:27 (2011). Software available at. http://www.csie.ntu.edu.tw/cjlin/libsvm

3. Chatfield, K., Simonyan, K., Vedaldi, A., Zisserman, A.: Return of the devil in the details: delving deep into convolutional nets. In: Proceedings of BMVC (2014)
4. Cheng, K.W., Chen, Y.T., Fang, W.H.: Video anomaly detection and localization using hierarchical feature representation and Gaussian process regression. In: Proceedings of CVPR, pp. 2909–2917 (2015)
5. Cong, Y., Yuan, J., Liu, J.: Sparse reconstruction cost for abnormal event detection. In: Proceedings of CVPR, pp. 3449–3456 (2011)
6. Del Giorno, A., Bagnell, J.A., Hebert, M.: A discriminative framework for anomaly detection in large videos. In: Leibe, B., Matas, J., Sebe, N., Welling, M. (eds.) ECCV 2016. LNCS, vol. 9909, pp. 334–349. Springer, Cham (2016). doi:10.1007/978-3-319-46454-1_21
7. Dutta, J.K., Banerjee, B.: Online detection of abnormal events using incremental coding length. In: Proceedings of AAAI, pp. 3755–3761 (2015)
8. Feichtenhofer, C., Pinz, A., Zisserman, A.: Convolutional two-stream network fusion for video action recognition. In: Proceedings of CVPR, pp. 1933–1941 (2016)
9. He, K., Zhang, X., Ren, S., Sun, J.: Deep residual learning for image recognition. In: Proceedings of CVPR, pp. 770–778, June 2016
10. Ionescu, R.T., Alexe, B., Leordeanu, M., Popescu, M., Papadopoulos, D., Ferrari, V.: How hard can it be? Estimating the difficulty of visual search in an image. In: Proceedings of CVPR, pp. 2157–2166, June 2016
11. Kim, J., Grauman, K.: Observe locally, infer globally: a space-time MRF for detecting abnormal activities with incremental updates. In: Proceedings of CVPR, pp. 2921–2928 (2009)
12. Krizhevsky, A., Sutskever, I., Hinton, G.E.: ImageNet classification with deep convolutional neural networks. In: Proceedings of NIPS, pp. 1106–1114 (2012)
13. Li, W., Mahadevan, V., Vasconcelos, N.: Anomaly detection and localization in crowded scenes. IEEE Trans. Pattern Anal. Mach. Intell. **36**(1), 18–32 (2014)
14. Lu, C., Shi, J., Jia, J.: Abnormal event detection at 150 FPS in MATLAB. In: Proceedings of ICCV, pp. 2720–2727 (2013)
15. Mahadevan, V., Li, W.X., Bhalodia, V., Vasconcelos, N.: Anomaly detection in crowded scenes. In: Proceedings of CVPR, pp. 1975–1981 (2010)
16. Mehran, R., Oyama, A., Shah, M.: Abnormal crowd behavior detection using social force model. In: Proceedings of CVPR, pp. 935–942 (2009)
17. Ren, H., Liu, W., Olsen, S.I., Escalera, S., Moeslund, T.B.: Unsupervised behavior-specific dictionary learning for abnormal event detection. In: Proceedings of BMVC, pp. 28.1–28.13 (2015)
18. Russakovsky, O., Deng, J., Su, H., Krause, J., Satheesh, S., Ma, S., Huang, Z., Karpathy, A., Khosla, A., Bernstein, M., Berg, A.C., Fei-Fei, L.: ImageNet large scale visual recognition challenge. Int. J. Comput. Vis. **115**(3), 211–252 (2015)
19. Saligrama, V., Chen, Z.: Video anomaly detection based on local statistical aggregates. In: Proceedings of CVPR, pp. 2112–2119 (2012)
20. Schölkopf, B., Platt, J.C., Shawe-Taylor, J.C., Smola, A.J., Williamson, R.C.: Estimating the support of a high-dimensional distribution. Neural Comput. **13**(7), 1443–1471 (2001)
21. Simonyan, K., Zisserman, A.: Very deep convolutional networks for large-scale image recognition. In: Proceedings of ICLR (2014)
22. Sun, Q., Liu, H., Harada, T.: Online growing neural gas for anomaly detection in changing surveillance scenes. Pattern Recogn. **64**(C), 187–201 (2017)
23. Szegedy, C., Liu, W., Jia, Y., Sermanet, P., Reed, S., Anguelov, D., Erhan, D., Vanhoucke, V., Rabinovich, A.: Going deeper with convolutions. In: Proceedings of CVPR, pp. 1–9 (2015)

24. Vedaldi, A., Lenc, K.: MatConvNet - convolutional neural networks for MATLAB. In: Proceeding of ACMMM (2015)
25. Xu, D., Ricci, E., Yan, Y., Song, J., Sebe, N.: Learning deep representations of appearance and motion for anomalous event detection. In: Proceedings of BMVC, pp. 8.1–8.12 (2015)
26. Zhang, Y., Lu, H., Zhang, L., Ruan, X., Sakai, S.: Video anomaly detection based on locality sensitive hashing filters. Pattern Recogn. **59**, 302–311 (2016)
27. Zhao, B., Fei-Fei, L., Xing, E.P.: Online detection of unusual events in videos via dynamic sparse coding. In: Proceedings of CVPR, pp. 3313–3320 (2011)

Author Index

Printed in the United States
By Bookmasters